AF553816

MICROBIAL BIOTECHNOLOGY

Editors:

J.P. Tewari

Department of Agricultural Food
and Nutritional Science
Faculty of Agriculture and Forestry
University of Alberta
Alberta, Edmonton
Canada T6G 2P5

T.N. Lakhanpal

Department of Biosciences
Himachal Pradesh University
Shimla - 171 005
H.P.

Jagjit Singh

Environmental Building Solutions Ltd.
30 Kirby Road, Dustable
Bedfordshire
London LU6 3JH
U.K.

Rajni Gupta

Department of Botany
University of Delhi
Delhi - 110 007

B. P. Chamola

Applied Mycology Laboratory
Department of Botany
University of Delhi
Delhi - 110 007

MICROBIAL BIOTECHNOLOGY

Edited by:

J.P. TEWARI
T.N. LAKHANPAL
JAGJIT SINGH
RAJNI GUPTA
B.P. CHAMOLA

A.P.H. PUBLISHING CORPORATION
4435-36/7, ANSARI ROAD, DARYA GANJ
NEW DELHI-110 002

Published by
S.B. Nangia
A P H Publishing Corporation
4435-36/7, Ansari Road, Daryaganj
New Delhi 110002
Ph.: 23274050
E-mail : aphbooks@gmail.com

2024

Rs. 2995/-

Printed at
Balaji Offset
Navin Shahdara, Delhi 110032

Preface

This book entitled "Advances in Microbial Biotechnology" is being published to honour an outstanding scientist/mycologist and human being Professor Krishna Gopal Mukerji for his endless endeavour in mycology, plant pathology, microbial ecology and microbial biotechnology.

This book contains a wide ranging articles on microbial biotechnology, microbial ecology and mycology which emphasises the need for research in this direction and use of microbes in biotechnology. We have tried to synthesize all the informations which will be useful for researchers and students alike.

This volume contains 32 articles contributed by 62 authors and covers various aspects of microbial biotechnology. Chapter 1 describes a genus named after Professor K.G. Mukerji. Chapter 2-10 deals with various biochemical and molecular aspects of different microbes. Chapter 11 gives a very exhaustive data on aflatoxins and their effects on human health. Chapter 12 discusses a very recently advanced technology on protoplast fusion in certain fungi. Chapter 13 explains the microbial health hazards inside the buildings and palaces. Chapter 14 and 15 deals with mycoherbicides. Chapter 16 and 17 deals with dry rot fungi and their control. Chapter 18 and 19 deals with biocontrol. Chapter 20 deals with rhizosphere biology and chapter 21 deals with plant surfaces biology. Chapter 22 deals with management of soil borne diseases. Chapter 23-28 deals with various aspects of mycorrhizal biology. Chapter 29 deals with nematophagous fungi. Chapter 30 describes head smut of *Paspalum* and its management. Chapter 31 gives a monographic account of the genus *Morchella*. The last chapter gives a very interesting account on taxonomy, ecology and development of very rearly known fungi the Laboulbeniomycetes.

We are grateful to all authors for their contributions to this book. We would like to thank Mr. Sanjeev Hasija of M/s New Apcon for his excellent computer typesetting of this book in stipulated time. We are also thankful to all members associated with Applied Mycology Laboratory (# 15), Department of Botany, University of Delhi, who encouraged and helped us in various ways in completing this volume.

Many technical hands have helped us in the completion of this uphill task. We are grateful to Professor N.N. Bhandari and Professor K.M.M. Dakshini, Department of Botany, University of Delhi for their active cooperation and valuable suggestions during preparetion of this volume. We are also grateful to Professor K.R. Shivanna, Head, Department of Botany for releasing the book.

It is possible that in a work of this nature, some mistakes might have crept in text inadvertently and for these we own undiluted responsibility.

The editors are also thankful to Shri S.B. Nangia, the publisher for his enthusiastic cooperation in publishing this book in stipulated time using the latest printing technology.

We take great pleasure in presenting this Festschrift Volume to Professor K.G. Mukerji as a token of affection and regards. We wish him many more happy and fruitful years of research and a peaceful and prosperous life with his family members.

J.P. Tewari
T.N. Lakhanpal
Jagjit Singh
Rajni Gupta
B.P. Chamola

May 4, 1999

Prof. K.G. Mukerji and his Contributions

B.P. Chamola, J.P. Tewari and I. Tewari

Professor Dr. Krishna Gopal Mukerji was born on 4th May, 1934 at Lucknow as fourth child of Mr. Davendra Nath Mukerji and Mrs. Leela Mukerji. He received his early education in Lucknow. He obtained his B.Sc. and M.Sc. degree from Lucknow University in 1953 and 1955 respectively. He started his teaching career as Lecturer in Botany at Lucknow University in 1955, from where he also obtained his Ph.D. Degree in Botany (Soil Microbiology) for his thesis entitled "Microfungi of Usar Soil of India" in 1962. Professor Mukerji was appointed Lecturer in Botany at Banaras Hindu University (BHU) in 1962, where he worked for two years. He joined Department of Botany, University of Delhi, Delhi - 110 007 as Lecturer in 1964. Professor Mukerji was appointed as Reader and Professor in the years 1969 and 1982 respectively. Professor Mukerji shouldered the responsibility of the Head, Department of Botany, University of Delhi during the years 1985 to 1988.

In 1970 Professor Mukerji married Abha Mukerji, who stood by him all through his career and has been virtually the backbone for his academic and professional achievements. Their son Amit pursued the career of his own choice and he is doing Ph.D. in Plant Molecular Biology from University of Alberta, Canada after doing his B.Sc. (Hons.), M.Sc. and M.Phil. in Botany from University of Delhi.

Professor Mukerji is actively engaged in research since September, 1957 on different aspects of Microbial Ecology with particular reference to fertility of soil and increase in productivity of plants. The publications of Professor Mukerji shows the significant contribution made by him in different areas of Mycology, Plant Pathology, Microbial Ecology and Microbial Biotechnology. During his research he investigated/discovered/reported several species and genera of fungi and mycorrhizal fungi from Indian soils.

Professor Mukerji has keen interest in culture collection and received advanced training in Taxonomy of Fungi at the Commonwealth Mycological Institute (CMI), Kew, Surrey, England (from Jan.-April, 1968), where he was also offered a permanent post of Mycologist in 1968. He also worked at Institute of Seed Pathology at Copenhagen, Denmark (from Jan.-August, 1974); Institute of Meeresforschung, Bremerhaven, Am Handelshafen, Bremen, West Germany (December, 1973) and Centrralblureau voor Schimmelcultures, Baarn, Netherlands (September, 1974).

Professor Mukerji, visited various Institutes, Culture Collections and Laboratories in Australia, Austria, Canada, Denmark, England, France, Germany, Hong Kong, Italy, Japan, Malaysia, Netherland, New Zealand, Philippines, Sweden, Switzerland, Thailand and USA as visiting Professor/Scientist. During his visit he was visiting Professor at the Department of Botany, University of Basel, Switzerland (August, 1985); Institute of Biotechnology and Microbiology, Glasgow University, Stratethclyde; Commonwealth Mycological Institute, Ferry Lane, Kew, Surrey; Berkbeck College, London; Biotechnology Department, Portsmouth Polytechnic, Portsmouth, U.K. (September, 1985; British Council Senior Visitor); Department of Soil Science and Geology, University of Laval, Quebec, Canada (August, 1986); Biotechnology Research Centre, Dhaka University, Dhaka, Bangladesh (February, 1987;

UNESCO Expert); Department of Plant Sciences, Faculty of Agriculture and Forestry, University of Alberta, Edmonton, Alberta, Canada (April, 1987 and September, 1991); Department of Botany, University of Florida, Gainesville, Florida, USA (May, 1987); Department of Soil Science and Microbiology, University of Torino, Torino, Italy (July, 1987); Department of Biological Sciences, Faculty of Education, Shizuoka University, Shizuoka, Japan (September, 1990) and Department of Biology, the Chinese University of Hong Kong, Shatin, N.T. Hong Kong (May, 1993).

Professor Mukerji has about 40 years of research experience on Taxonomy and Ecology of fungi from soil, rhizosphere, root including mycorrhiza, plant surface, dung and seeds. About 50 and more than 100 students have been already awarded their Ph.D. and M.Phil. degrees respectively of Delhi University on various aspects of mycology, plant pathology, microbial ecology and microbial biotechnology under supervision of Professor Mukerji.

Professor Mukerji is a widely travelled botanist and has attended several International, National Conferences and Symposia and presented several key note adresses and invited papers of which the important ones are :

International : First International Mycological Congress (8-15 September, 1971) Exeter, UK; First National Congress of Ecology (8-14 September, 1974) Hague, Netherlands; First International Symposium on Microbial Ecology (21-26 August, 1977) Dunedin, New Zealand; Third International Symposium on the Microbiology of Leaf Surfaces (1-15 September, 1981) University of Aberdeen, Aberdeen, Scotland, U.K.; Second International Symposium on Microbial Ecology (7-12 September, 1981) University of Warwick, Warwick, U.K.; Sixth North American Conference on Mycorrhiza (24-29 June, 1984) Oregon, USA; Fourth International Symposium on the Microbiology of the Phyllosphere (2-6 September, 1985) Wageningen, Netherlands; International Symposium on Root in Forest Soils : Biology and Symbiosis (4-8 August, 1986) Victoria, Canada; International Workshop on "Managed Forest Land" (August 8-11, 1986) Parksville BC, Canada; Biotechnology Symposium (February 2-5, 1987), Dhaka University, Bangla Desh, Dhaka; 7th North American Conference on Mycorrhiza (May 3-8, 1987) Gainesville, Florida, USA; XIV International Botanical Congress (July 24 August 2, 1987) West Berlin, Fed. Rep. Germany; International conference on Integrated Management and Rural Development of Arid and Semi Zones (September 17-22, 1987) Yangling, Shaanxi, China; First Asian Conference on Pharmaceutical Education, Research and Drug Industry (May 26-29, 1988) National Unviersity, Singapore, Symposium on Plant Roots and their Environment (August 21-28, 1988), Agriculture University, Uppsala, Sweden; Regional Symposium on Recent Development in Tree Plantations of Humid/Subhumid Tropics of Asia (June 5-9, 1989) Selangor, Malaysia; Regional Seminar on Microbial Research (December 1-5, 1989) Royal Nepal Academy of Science and Technology, Kathmandu, Nepal; Fourth Mycorrhiza Seminar of Japan - Japan Mycorrhizal Research Society (September 16 to 22, 1990) Osaka, Japan; Asian Conference on Mycorrhizae (March 12-17, 1991) Chiang Mai, Thailand; Third International Society of Root Research Symposium on "Root Ecology and its Practical Application" (September 2-7, 1991), Vienna, Austria; International Symposium on "Rehabilitation of Tropical Rainforest Ecosystems : Research and Development Priorities" (September 2-7, 1992); Kuching, Sarawak, Malaysia; International Symposium on "MPTS for Rural Livelihood" (May 3-6, 1993), Holiday Inn, Manila, Phillippines; Third International Conference on Biodeterioration of Cultural Property (July 4-7, 1995) Bangkok, Thailand and Asian Conference on Lichenology (Dec. 27, 1996 to 3rd Jan., 1997) Kathmandu, Nepal.

National : National Symposium on Fungal Ecology in Relation to Human Welfare (November 28-30, 1987) Osmania University, Hyderabad; Biotechnology in Forestation, (January 14-16, 1988) TERI, Delhi; First Asian Conference on Mycorrhiza, (January 27-30, 1988) Madras; National Symposium (April 12-14, 1990), Osmania University, Hyderabad.

Following conference/seminars were organised under chairman and co-chairmanship of Professor Mukerji.

1. Mycorrhiza Round Table (March 13-15, 1987), J.N.U., New Delhi-Chairman.
2. First Asian Conference on Mycorrhiza (January 27-30, 1988), Madras, India–Co-chairman.
3. International Conference on Research in Plant Sciences and its Relevance to Future (March 7-11, 1988), Delhi–Chairman.

Professor Mukerji is a member of numerous societies and associations and also associated with these societies as :

Treasurer, Mycological Society of India, 1973-76; Member, Editorial Board, Kavaka, The Transactions of the Mycological Society of India, 1979-81; Vice-President, Mycological Society of India, 1980-81; Vice-President, Association of Tropical Microbial Ecology, 1980-84; President, Society for Advancement of Botany, 1987; President, Society for Environment Scientists, 1988-91; Member Editorial Board, Phytologia, Jour. Indian Bot. Soc. and Jour. Phytol. Research (since 1988); Chief Editor, Jour. Rec. Adv. Appl. Science (since 1988); Member, Advisory Committee, Indian J. Microbial. Ecol. (1991 - to date); Editor, Frontiers in Applied Microbiology (1985-1994); Member Advisory Board (Board of Editors) CRC Critical Reviews on Biological Control of Plant Pests, Diseases and Weeds, USA (since 1990); Councillor, International Society of Root Research (1988-91, Sweeden; 91-94, U.S.A.); Associate, Nitrogen Fixing Tree Association, U.S.A. (NFTA) (since 1990); Vice President, Council on Biodeterioration of Cultural Property (1992-1998).

Professor Mukerji has already authored or co-authored more than 400 research papers on various aspects of Mycology, Plant Pathology, Microbial Ecology and Microbial Biotechnology. He has co-authored, edited and co-edited more than 35 books on various aspects of Microbial Ecology and Microbial Biotechnology.

Professor Mukerji will complete 65 years of age on 4, May 1999. This does not mean that he has retired on this date, instead he has started a new phase of his career. A lot of credit for Professor Mukerji's success goes to his family. First and foremost of all credit goes to his mother and father, the late Mrs. Leela Mukerji and late Mr. D.N. Mukerji, his wife Mrs. Abha Mukerji and their son, through their unswering support and affection has also moulded Professor Mukerji's career.

It is a privilege to know Professor Mukerji as distinguished mycologist and microbial ecologist, a person of repute for his research contributions all over the world.

Publications of K.G. Mukerji

BOOKS

i) Mukerji, K.G. and Juneja, R.C. 1974. *Fungi of India*, Emkay Publications India Ltd., Delhi.

ii) Lakhanpal, T.N. and Mukerji, K.G. 1981. *Taxonomy of the Indian Myxomycetes*, J. Cramer, Germany.

iii) Mukeri, K.G.; Agnihotri, V.P. and Singh, R.P. (eds.) 1984. *Progress in Microbial Ecology*, Print House (India) Ltd., Lucknow, India.

iv) Mukerji, K.G.; Pathak, N.C. and Singh, V.P. (eds.) 1985. *Frontiers in Applied Microbiology*, Vol. I. Print House (India) Ltd., Lucknow, India.

v) Mukerji, K.G.; Singh, V.P. and Garg, K.L. (eds.) 1986. *Frontiers in Applied Microbiology*, Vol. II. Print House (India) Ltd., Lucknow, India.

vi) Mukerji, K.G. and Khosla, J. 1986. *Index to Indian Plant Diseases*. Tata Mc Graw Hill Publication, New Delhi, India.

vii) Mukerji, K.G. and Garg, K.L. (eds.) 1988. *Biocontrol of Plant Diseases*, Vol. I. CRC Press, USA.

viii) Mukerji, K.G. and Garg, K.L. (eds.) 1988. *Biocontrol of Plant Diseases*, Vol. II. CRC Press, USA.

ix) Verma, A.K.; Oka, A.K.; Mukerji, K.G.; Tilak, K.V.B.R. and Raj. J. (eds.) 1988. *Mycorrhiza Round Table*. Manuscript Report, 201 e. IDRC, Canada.

x) Mukerji, K.G.; Singh, V.P. and Garg, K.L. (eds.) 1989. *Frontiers in Applied Microbiology*, Vol. III Rastogi and Company, Meerut, India.

xi) Mukerji, K.G.; Singh, V.P. and Garg, K.L. (eds.) 1989. *Frontiers in Applied Microbiology*, Vol. IV Rastogi and Company, Meerut, India.

xii) Arora, D.K.; Rai, B.;Mukerji, K.G. and Knudson, G.R. (eds.) 1991. *Handbook of Applied Mycology*, Vol. I. Soil and Plants. Marcel and Dekker Inc., USA.

xiii) Arora, D.K.; Ajello, L. and Mukerji, K.G. (eds.) 1991. *Handbook of Applied Mycology*, Vol. II. Human, Animals and Insects. Marcel and Dekker Inc., USA.

xiv) Arora, D.K.; Mukerji, K.G. and Marth, E.H. (eds.) 1991. *Handbook of Applied Mycology*, Vol. III. Foods and Feeds. Marcel and Dekker Inc., USA.

xv) Arora, D.K.; Elander, R.P. and Mukerji, K.G. (eds.) 1992. *Handbook of Applied Mycology*, Vol. IV. Fungal Biotechnology. Marcel and Dekker Inc., USA.

xvi) Mukerji, K.G.; Tewari, J.P.; Arora, D.K. and Saxena, G. (eds.) 1992. *Recent Developments in Biocontrol of Plant Diseases*, Aditya Books Pvt. Ltd., India.

xvii) Mukerji, K.G.; Bhatnagar, A.K.; Tripathi, S.C.; Bansal, M. and Saxena, M. (eds.) 1992. *Currect Concepts in Seed Biology.* Naya Prokash, Calcutta India.

xviii) Mukerji, K.G.; Srivastava, A.K.; Singh, K.P. and Garg, K.L. (eds.) 1992. *Advances in Medical Mycology.* Aditya Books Pvt. Ltd. India.

xix) Bhandari, N.N. and Mukerji, K.G. 1993. *The Haustorium.* John Wiley and Sons Ltd., U.K.

xx) Mukerji, K.G. and Singh, V.P. (eds.) 1994. *Concepts in Applied Microbiology and Biotechnology.* Aditya Books Pvt. Ltd., India.

xxi) Garg, K.L.; Garg, N. and Mukerji, K.G. (eds.) 1994. *Recent Advances in Biodeterioration and Biodegradation*, Vol. I. Naya Prokash, Calcutta, India.

xxii) Garg, K.L.; Garg, N. and Mukerji, K.G. (eds.) 1994. *Recent Advances in Biodeterioration and Biodegradation*, Vol. II. Naya Prokash, Calcutta, India.

xxiii) Mukerji, K.G.; Mathur, B.; Chamola, B.P. and Chitralekha, P. (eds.) 1996. *Advances in Botany.* APH Publishing Corporation, New Delhi - 110 002.

xxiv) Upadhyay, R.K.; Mukerji, K.G. and Rajak, R.L. (eds.) 1996. *IPM System in Agriculture*, Vol. I. Principles and Perspectives, Aditya Books Pvt. Ltd., New Delhi - 110 002.

xxv) Upadhyay, R.K.; Mukerji, K.G. and Rajak, R.L. (eds.) 1996. *IPM System in Agriculture*, Vol. II. Biocontrol in Emerging Biotechnology, Aditya Books Pvt. Ltd., New Delhi - 110 002.

xxvi) Mukerji, K.G. (ed.) 1996. *Concepts in Mycorrhizal Research.* Kluwer Academic Publishers, The Netherlands.

xxvii) Upadhyay, R.K. and Mukerji, K.G. (eds.) 1996. *Toxins in Plant Disease Development and Emerging Biotechnology.* Science Publisher, Inc. USA.

xxviii) Upadhyay, R.K.; Mukerji, K.G. and Rajak, R.L. (eds.) 1997. *IPM System in Agriculture*, Vol. III. Cereals, Aditya Books Pvt. Ltd., New Delhi - 110 002.

xxix) Upadhyay, R.K.; Mukerji, K.G. and Rajak, R.L. (eds.) 1997. *IPM System in Agriculture*, Vol. IV. Pulses, Aditya Books Pvt. Ltd., New Delhi - 110 002.

xxx) Upadhyay, R.K.; Mukerji, K.G., Chamola, B.P. and Dubey, O.P. (eds.) 1998. *Integrated Pest and Disease Management.* APH Publishing Corporation, New Delhi - 110 002.

xxxi) Mukerji, K.G.; Chamola, B.P., Upreti D.K. and Upadhyay, R.K. (eds.) 1999. *Biology of Lichens.* Aravali Books International, New Delhi - 110 020.

xxxii) Mukerji, K.G.; Chamola, B.P. and Upadhyay, R.K. (eds.) 1999. *Biotechnological Approaches in Biocontrol of Plant Pathogens.* Kluwer Academic/Plenum Press, New York, USA.

xxxiii) Upadhyay, R.K.; Mukerji, K.G. and Rajak, R.L. (eds.) 1999. *IPM System in Agriculture*, Vol. V. Oil Seeds, Aditya Books Pvt. Ltd., New Delhi, India.

xxxiv) Mukerji, K.G.; Chamola, B.P. and Singh, J. (eds.) 1999. *Mycorrhizal Biology.* Kluwer Academic/Plenum Press, New York, USA.

(xxxv)Upadhyay, R.K.; Mukerji K.G. and Dubey, O.P. (eds.), 1999. *IPM System in Agriculture*, Vol. VI. Cash Crops. Aditya Books Pvt. Ltd., New Delhi, India.

(xxxvi) Upadhyay, R.K.; Mukerji, K.G. and Dubey, O.P. (eds.). *1999. IPM System in Agriculture*, Vol. VII. Key Pathogens. Aditya Books Pvt. Ltd., New Delhi, India (In Press).

(xxxvii) Upadhyay, R.K.; Mukerji, K.G. and Dubey, O.P. (eds.). *1999. IPM System in Agriculture*, Vol. VIII. Key Pests and Diseases. Aditya Books Pvt. Ltd., New Delhi, India (In Press).

RESEARCH PAPERS

1. Rai, J.N. & Mukerji, K.G. 1961. New Records of microfungi from usar soils of India. Curr. Sci., 30: 345.
2. Rai, J.N.; Mukerji, K.G. & Tewari, J.P. 1961. Two new records in soil fungi. Curr. Sci. 30: 231-232.
3. Rai, J.N.; Tewari, J.P. and Mukerji, K.G. 1961. A new *Helicostylum* from Indian soils. Can. J. Bot. 39: 1282-1284.
4. Rai, J.N. & Mukerji, K.G. 1962a. A new species of *Chaetomium* from Indian soils. Can. J. Bot. 40: 857-860.
5. Rai, J.N. & Mukerji, K.G. 1962b. *Sporotrichum carthusioviride*. Rai and Mukerji, a new species from Indian soils. Mycopath. Mycol. appl. 18: 122-126.
6. Rai, J.N.; Mukerji, K.G. & Tewari, J.P. 1963. *Tripterospora tetraspora* sp. nov., a new cleistothecial Ascomycetes. Can. J. Bot. 41: 327-329.
7. Rai, J.N.; Tewari, J.P. & Mukerji, K.G. 1964a. *Achaetomium*, a new genus of Ascomycetes. Can. J. Bot., 42 : 693-697.
8. Rai, J.N.; Tewari, J.P. & Mukerji, K.G. 1964b. A new *Aspergillus* from Indian soils. *A. striatus* sp., nov. Can. J. Bot. 42: 1521-1524.
9. Rai, J.N.; Tewari, J.P. & Mukerji, K.G. 1964c. Cultural and taxonomic studies on two rare species of *Aspergillus - A. paradoxus* and an interesting strain of *A. variecolor* from Indian soils. Mycopath. Mycol. Appl. 26: 369-376.
10. Mukerji, K.G. 1966a. Studies on the effect of hydrogen ion concentration on the growth and sporulation of certain soil fungi. Mycopath. Mycol. Appl. 28: 312-316.
11. Mukerji, K.G. 1966b. Production of cleistothecia in a sterile strain of *Thielavia setosa* Dade. Mycopath. Mycol. Appl. 28: 317-320.
12. Mukerji, K.G. 1966c. Ecological studies on the microorganic population of Usar soils. Mycopath. Mycol. Appl. 29: 339-349.
13. Mukerji, K.G. 1967. *Aspergillus niger* growing on coconut oil. J. Gen. Appl. Microbiol. 13: 407-408.
14. Agarwal, M.K.; Mukerji, K.G. & Shivpuri, D.N. 1968. Studies on the allergenic fungal spores in Delhi atmosphere. Aspects of Allergy and Appl. Immun. 1:91-97.
15. Dhawan, S. & Mukerji, K.G. 1968. The development of ascospore in *Rhytidihysterium rufulum* (Speg.) Petrak, Preslia 40: 415-416.

16. Mukerji, K.G. 1968a. Fungi of Delhi. II. Some interesting records. Proc. Natl. Inst. Sci. India 34B : 71-81.

17. Mukerji, K.G. 1968b. The position of the genus *Achaetomium* in Pyrenomycetes. Proc. Natl. Inst. Sci. India 34B: 288-292.

18. Mukerji, K.G. 1968c. Fungi of Dehli. IV. A new species of *Piptocephalis.* Mycologia 60: 326-336.

19. Mukerji, K.G. 1968d. A Descriptions of pathogenic bacteria and fungi, Nos. 181-190, Commonwealth Mycological Institute, Kew, England.

20. Mukerji, K.G. 1968e. Observations on the mutual relationships among soil microorganisms. J. Gen. Appl. Microbiol, 14:243-250.

21. Mukerji, K.G. & Dhawan, S. 1968. Fungi of Dehli. X. A new species of *Eutryblidiella* from India. Nova Hedwigia 16 : 433-437.

22. Mukerji, K.G. & Ranga Rao, V. 1968a. Gentamycin as antibiotic in dilution plates for the isolation of soil fungi. Plant and Soil 29: 331-332

23. Mukerji, K.G. & Ranga Rao, V. 1968b. *Volutella lini* sp. nov. Trans. Brit. Mycol. Soc. 51: 337-339.

24. Ranga Rao, V. & Mukerji, K.G. 1968. Ascus development in *Chaetomium bostrychodes* Zoph. Preslia 40: 304-305.

25. Agarwal, M.K.; Shivpuri, D.N. & Mukerji, K.G. 1969a. Studies on the allergenic fungal spores of the air of Delhi, India, metropolitan area: Botanical aspects (aeromycology). J. Allergy 44: 198-204.

26. Agarwal, M.K.; Shivpuri, D.N. & Mukerji, K.G. 1969b. Species specific antigens in fungi. Aspects of Allergy and Appl. Immun. 2 : 69-80.

27. Kapoor, S. & Mukerji, K.G. 1969. Fungi of Delhi. VIII. Two unrecorded members of the Hypocreales. J. Indian Bot. Soc. 48: 255-257.

28. Magan, G.; Vishwanathan, L.; Venkitasubramanian, T.A. & Mukerji, K.G. 1969. Studies on Aflatoxin production by Indian strains of *Aspergillus flavus.* Link ex. Fries. J. Gen. Microbiol. 59:119-129.

29. Mukerji, K.G. 1969a. Fungi of Delhi I. A new species of the genus *Sepedonium* Link. Mycopath. Mycol. Appl. 37: 304-312.

30. Mukerji, K.G. 1969b. Fungi of Delhi. VI. Two members of Mucorales. Ceska Mycol. 23: 65-67.

31. Mukerji, K.G. 1969c. On the nomenclature of the powdery mildew of *Dalbergia sisoo* Roxb. Mycologia. 61: 181-184.

32. Mukerji, K.G. 1969d. White rot of *Zephyranthes* species. Sci. and Cult. 36: 163-165.

33. Mukerji, K.G.; Agarwal, M.K. & Saxena, A.S. 1969. Where from the allergenic fungal spores are coming? Aspects of Allergy and Appl. Immun. 2: 181-189.

34. Mukerji, K.G.; Bedi, K.; Tewari, J.P. & Tewari, I.K. 1969. Studies on the Indian speceis of *Xylaria* Hill ex Grey. and *Poronia* Willd. ex. Fr. Phytomorphology 19: 219-224.

35. Mukerji, K.G. & Kapoor, S. 1969a. Fungi of Delhi. V. Some interesting Loculoascomycetes. Ceska Mycol. 28: 256-261.

36. Mukerji, K.G. & Kapoor, S. 1969b. Fungi of Delhi. VII Some interesting members of the Sphaeriales. J. Indian Bot. Soc. 48: 228-231.

37. Mukerji, K.G. & Kapoor, S. 1969b. Fungi of Delhi. XIII. Two Interesting records from soils. J. Gen. Appl. Mycrobiol. 15: 261-265.

38. Mukerji, K.G. & Tewari, J.P. 1969. White rot of onion in Lucknow. PANS London 15B: 235-236

39. Rai, J.N.; Tewari, J.P. & Mukerji, K.G. 1969. Mycoflora of mangrove mud. Mycopath. Mycol. Appl. 38: 17-31.

40. Ranga Rao, V. & Mukerji, K.G. 1969a. Fungi of Delhi. IX. Additions to our knowledge of soil fungi. J. Indian. Bot. Soc. 48: 258-261.

41. Ranga Rao, V. & Mukerji, K.G. 1969b. Cytology of the ascus development in the genus *Chaetomium* Fuckel. Can. J. Bot. 47: 869-871.

42. Saxena, A.S.; Mukerji, K.G. & Agarwal, M.K. 1969. Spread of disease due to fungi. Aspects of Allergy and Appl. Immun. 2: 175-180.

43. Dakshini, K.M.M.; Tandon, R.K. & Mukerji, K.G. 1970. A new species of *Phyllachora*. Mycologia 62: 296-300.

44. Mukerji, K.G. 1970. Fungi of Delhi III. Some interesting ascomycetes. Mycopath. Mycol. Appl. 62: 301-306.

45. Ranga Rao, V. & Mukerji, K.G. 1970. Cytology of ascus in *Ascotricha guamensis*. Mycologia 62: 301-306.

46. Saxena, A.S. & Mukerji, K.G. 1970a. *Kernia bifurcotricha* sp. nov. Trans. Brit. Mycol. Soc. 54: 16-148.

47. Saxena, A.S. & Mukerji, K.G. 1970b. Fungi of Delhi. XIV. Imperfect state of *Kernia geniculotricha* Seth. Acta Bot. Neerlandica 19: 49-52.

48. Saxena, A.S. & Mukerji, K.G. 1970c. *Didymostilbe ellisii* sp. nov. Trans. Br. Mycol. Soc. 55: 504-504.

49. Saxena, A.S. & Mukerji, K.G. 1970d. Fungi of Delhi. XV. *Lophotrichus indicus* sp. nov. Acta Bot. Neerlandica 19: 722-726.

50. Gopal, S.; Magan, K.K.; Venkitasubramanian. T.A. & Mukerji, K.G. 1971. Physiology of *Trigonella* infected with *Peronospora trifoliorum*. Biol. Plantarum 13: 396-401.

51. Ranga Rao, V. & Mukerji, K.G. 1971a. Cytology of the ascus in *Achaetomium globosum* and *A. leuteum*. J. Gen. and Appl. Microbiol. 17: 311-319.

52. Ranga Rao, V. & Mukerji, K.G. 1971b. Fungi in the root zone of four cultivars of wheat. Ann. Inst. Pasteur 121: 533-544.

53. Ranga Rao, V. & Mukerji, K.G. 1971c. Cytology of the ascus in *Achaetomium strumarium*. Bot. Gaz. 132: 79-183.

54. Ranga Rao, V. & Mukerji, K.G. 1971d. Studies on charcoal rot disease of *Abelmoschus esculentus.* II. Fungal flora of the root zone of healthy and infected plants. Ann. Inst. Pasteur 122: 81-90.

55. Ranga Rao, V. & Mukerji, K.G. 1971e. Nuclear behaviour during the development of ascus in *Chaetomium bostrychodes.* Trans. Mycol. Soc. Japan 12: 91-98.

56. Sharma, K.R. & Mukerji, K.G. 1971. Prevalence of *Candida albicans* on *Gossypium* leaves. J.I.B.S. 51: 291-297.

57. Gupta, N.C.; Nanda, P.; Ranga Rao, V. & Mukerji, K.G. 1972. Studies on charcoal rot disease of *Abelmoschus esculentus.* IV. Pycnidiospore germination. Trans. Br. Mycol. Soc. Japan, 13: 275-283.

58. Mukerji, K.G. & Bhandari, N.N. 1972. Fungi of Delhi XI. *Sporormia cainia* sp.n. from India. Trans. Mycol. Soc., Japan 12: 99-103.

59. Padma, R. & Mukerji, K.G. 1972. Fungi in the root region of *Rauwolfia serpentina* and *Rauwolfia canesens.* Indian Phytopathology 25: 104-107.

60. Ranga Rao, V.; Jayakar, M.; Sharma, K.R. & Mukerji, K.G. 1972. Effect of foliar spray of Morphactin on fungi in the root zone of *Capsicum annum.* Plant and Soil 37: 179-182.

61. Ranga Rao, V. & Mukerji, K.G. 1972. Studies on charcoal rot disease of *Abelmoschus esculentus.* I. Soil Host-parasite relationships. Trans. Mycol. Soc. Japan, 13: 265-274.

62. Ranga Rao, V.; Subba Rao, N.S. & Mukerji, K.G. 1972. Inhibition of Nodulation in gram (*Cicer arietenum*) by mophactin. Indian J. Microbiol. 12: 264-266.

63. Saxena, A.S. & Mukerji, K.G. 1972. Fungi of Delhi. XVII. A new speceis of *Chaetomium.* Mycologia 64: 1326-1330.

64. Sharma, K.R. & Mukerji, K.G. 1972. Succession of fungi on cotton leaves. Ann. Inst. Pasteur 122: 425-455.

65. Behera, N.; Rikhy, M.; Sharma, K.R. & Mukerji, K.G. 1973. Fungi of Delhi. XXXI. Some interesting records. Proc. India Nat. Sci. Acad. 39: 710-718.

66. Lakhanpal, T.N. & Mukerji, K.G. 1973. Morphology of some Indian species of Xylariaceae and Clavicipitaceae. Ceska Mycol. 27: 169-173.

67. Mukerji, K.G. 1973a. The nomenclatural status and synonymy of *Acremoniella serpentina* and *Sporormia cainia.* Trans. Mycol. Soc. Japan. 14: 175-176.

68. Mukerji, K.G. 1973b. Additions to Plant Diseases of Delhi. Angewandte Botanik. 47: 205-213.

69. Ranga Rao, V.; Ganapathy, P.S. & Mukerji, K.G. 1973. Studies on charcoal rot disease of *Abelmoschus esculentus.* III. Pyenidical formation of the isolate as influenced by light, plant tissue and tissue extracts. Phytopath. Z. 76: 123-127.

70. Ranga Rao, V.; Mukerji, K.G. & Subba Rao, N.S. 1973. Effect of Chlorflurenol, a Morphactin on root nodulation in *Pisum sativum.* Z. Pflanzephysiol. 69: 84-86.

71. Ranga Rao, V.; Subba Rao, N.S. & Mukerji, K.G. 1973a. Inhibition of *Rhizobium in vitro* by non-nodulating legume roots and root extracts. Plant and Soil 39: 449-452.

72. Ranga Rao, V.; Subba Rao, N.S. & Mukerji, K.G. 1973b. *In vitro* effects of some growth regulators on *Rhizobium*. J. Gen. Appl. Microbiol. 19: 55-58.

73. Rikhy, M.; Malhotra, G. & Mukerji, K.G. 1973. Fungi of Delhi XXIV. *Bahusandhika compacta* sp. nov. and *Endosporostilbe minuta* sp. nov. Rev. de Mycol. 38: 91-94.

74. Rikhy, M. & Mukerji, K.G. 1973. Fungi of Delhi, XXVI, Three new ascomycetes. Kavaka 1: 91-94.

75. Saxena, A.S. & Mukerji, K.G. 1973a. Fungi of Delhi XIX. *Bahupaathra minuta* sp. nov. Mycopath. Mycol. Appl. 49: 201-204.

76. Saxena, A.S. & Mukerji, K.G. 1973b. Fungi of Delhi. XVI. Further additions to Indian species of *Chaetomium*. Ceska Mycol. 27: 162-164.

77. Saxena, A.S. & Mukerji, K.G. 1973c. Fungi of Delhi. XVII. Three unrecorded members of Ascomycetes. Ceska Mycol. 27: 162-164.

78. Sharma, K.R.; Behera, N. & Mukerji, K.G. 1973. *Podoxyphium indicum* sp. nov. and *Polyschema indica* sp. nov. Norw. J. Bot. 20: 27-29.

79. Sharma, K.R. & Mukerji, K.G. 1973a. *Phoma exigua* Desm. A heterothallic Deuteromycete, Mycologia 65: 709-712.

80. Sharma, K.R. & Mukerji, K.G. 1973b. Isolation of Myxomycetes from soils. Current Sci. 42(6): 213-215.

81. Sharma, K.R. & Mukerji, K.G. 1973c. Microbial colonisation of aerial parts of plants - a review. Acta Phytopath. 8: 425-461.

82. Behera, N. & Mukerji, K.G. 1974. Fungi of Delhi. XXV. *Chlamydoabsidia dasgupti* sp. nov. and *Polyschema indica*. Sharma, Behera and Mukerji. Norw. J. Bot. 21: 1-3.

83. Mukerji, K.G. 1974. Phylloplane-pathogens versus non pathogens. In "Current Trends in Plant Pathology" (eds. S.P. Raychaudhuri and J.P. Verma), pp. 20-23.

84. Mukerji, K.G. & Saxena, A.S. 1974. Notes on *Achaetomium, Anixiella, Boothiella, Chaetomium, Lophotrichus, Pseudeurotium, Pycnidiophora* and the classification of the Chaetomiales. Beiheft. Nova Hedwegia 47: 373-404.

85. Mukerji, K.G. & Singh, N. 1974. Fungi of Delhi. XXI. *Chaetomium dehlianum* sp. nov. Friesia 10: 265-269.

86. Ranga Rao, V. & Mukerji, K.G. 1974. On the morphogenesis of *Schizophyllum commune* f. *radiatum*. Beihft. Nova Hedwegia 17: 405-412.

87. Rikhy, M. & Mukerji, K.G. 1974. Studies on Indian Endogonaceae. I. Four new records. Trans. Mycol. Soc. Japan 15: 47-52.

88. Sharma, K.R.; Behera, N. & Mukerji, K.G. 1974. A comparison of three techniques for the assessment of Phylloplane Microbes. Trans. Mycol. Soc. Japan 15: 223-233.

89. Sharma, K.R. & Mukerji, K.G. 1974a. *Candida albicans* - A natural inhabitant of the Phyllosphere. Jap. J. Ecol. 24: 60-63.

90. Sharma, K.R. & Mukerji, K.G. 1974b. Incidence of pathogenic fungi on leaves. Indian Phytopath. 27: 558-566.

91. Lakhanpal, T.N. & Mukerji, K.G. 1975. Taxonomic studies on Indian Myxomycetes. VII. The genus *Licea.* Kavaka 3: 101-105.

92. Mukerji, K.G. 1975. Descriptions of Pathogenic bacteria and fungi. Set. 46 Nos. 453, 455, 456, 458-460. Commonwealth Mycological Institute, Kew, Surrey, England.

93. Rikhy, M.; Malhotra, G. & Mukerji, K.G. 1975. Fungi of Delhi. XXVIII. New Additions to Indian species of *Chaetomium*. Proc. INSA 42B: 29-33.

94. Juneja, R.C.; Nayyar, V.L. & Mukerji, K.G. 1976. Further additions to Plant Diseases of Delhi. Angew. Botanik 50: 43-48.

95. Lakhanpal, T.N. & Mukerji, K.G. 1976a. Taxonomic studies on Indian Myxomycetes. IV. Some new records of Liceales. Proc. INSA 42B: 34-40.

96. Lakhanpal, T.N. & Mukerji, K.G. 1976b. Taxonomic studies of Indian Myxomycetes. VI. The *Metatrichia* R. Ing. Proc. INSA 42B: 34-40.

97. Lakhanpal, T.N. & Mukerji, K.G. 1976c. Taxonomic studies on Indian Myxomycetes. I. The order Echinosteliales in India. Norw. J. Bot. 23: 107-111.

98. Lakhanpal, T.N. & Mukerji, K.G. 1976d. Taxonomic studies on Indian Myxomycetes. V. Some new reocrds of Trichiales. Proc. INSA 42B: 125-129,.

99. Lakhanpal, T.N. & Mukerji, K.G. 1976e. Experimental studies on Indian myxomycetes. I. Sporangial development in *Licca scyphoides: Clastoderma debaryanum* and *Macbrideola cornea.* Trans. Mycol. Soc. Japan. 17: 106-120.

100. Lakhanpal, T.N. & Mukerji, K.G. 1976f. Experimental studies on Indian Myxomycetes. II. Cultural studies on some species of *Didymium.* Trans. Mycol. Soc. Japan 17: 121-125.

101. Lakhanpal, T.N. & Mukerji, K.G. 1976g. Taxonomic studies on India Myxomycetes. IX. The genus *Lycogala.* Kavaka 4: 55-58.

102. Malhotra, G. & Mukerji, K.G. 1976. Fungi of Delhi XXIX. Three new species of *Chaetomium* from decaying wood. Rev. de. Mycol. 40: 1979-1984.

103. Rajeshwari, R.; Kaur, D.; Mukerji, K.G. & Bhandari, N.N. 1976. Histochemical changes in stem galls of *Ipomoca pestegrides* infected with *Albugo impomoee-panduratae* (Swine) Wing. Ind. J. Bot. 1: 75-81.

104. Sharma, K.R. & Mukerji, K.G. 1976a. Effect of carbon sources on the physiology or reproduction in *Phoma exigua.* Incompatibility News letter 7: 48-55.

105. Sharma, K.R. & Mukerji, K.G. 1976b. Microbial ecology of *Sesamum orientale* L. and *Gossypium hirsutum* L. In : Microbiology of Aerial Plant Surface, pp. 375-390. (Eds. C.H. Dickinson and T.F. Prece). Academic Press, London.

106. Lakhanpal, T.N. & Mukerji, K.G. 1976a. Taxonomic studies on Indian Myxomycetes. III. Two new Myxomycetes from India. Acta Bot. Indica 5: 58-61.

107. Lakhanpal, T.N. & Mukerji, K.G. 1977b. Taxonomic studies on Indian Myxomycetes. XII. Two new speceis of *Comatrichia.* Trans. Mycol. Soc. Japan 18: 125-133.

108. Lakhanpal, T.N. & Mukerji, K.G. 1977c. Taxonomic studies on Indian Myxomycetes. VIII. Some new records of Stemonitales. Ind. Phytopath. 30: 28-31.

109. Mukerji, K.G. 1977. Phycomycetes. Chapter X. In: Thallophytes (in Hindi) (ed. P. Chandola). Central Hindi Directorate, Govt. of India, Delhi, pp. 74-91.

110. Bhatnagar, R.K. & Mukerji, K.G. 1978. Amino acid composition of two species of *Aspergillus* producing different types of aflatoxins. Indian Phytopath. 31: 374-375.

111. Lakhanpal, T.N. & Mukerji, K.G. 1978a. Taxonomic Studies on Indian Myxomycetes. X. Some new records of Physaraceae. J.I.B.S. 57: 86-92.

112. Lakhanpal, T.N. & Mukerji, K.G. 1978b. Taxonomic Studies on Indian Myxomycetes. XIII. Three new speceis of *Lamproderma*. Kavaka 6: 9-14.

113. Lakhanpal, T.N. & Mukerji, K.G. 1978c. Taxonomic Studies on Indian Myxomycetes. XI. Some new records of Didymeaceae. J.I.B.S. 57: 174-180.

114. Lakhanpal, T.N. & Mukerji, K.G. 1978d. Taxonomic Studies on Indian Myxomycetes. XV. Some new species of *Didymium*. Acta Bot. Indica 6 (Suppl.) : 16-21.

115. Malhotra, G. & Mukerji, K.G. 1978. Fungi of Delhi XXVIII. Four Pyrenomycetes from bark. Trans Mycol. Soc. Japan 19(3): 283-288.

116. Mohan, M. & Mukerji, K.G. 1978a. Abnormal conidium development in *Alternaria tenuissima*. Indian Phytopath. 31: 247.

117. Mohan, M. & Mukerji, K.G. 1978b. Some biologically active extra cellular products of blue-green algae. Physco. 18: 73-82.

118. Mohan, M. & Mukerji. K.G. 1978c. Seed microflora and some important diseases of cauliflower and cabbage. Seeds and Farms 4: 19-21.

119. Mohan, M.; Mukerji, K.G. & Rani, K. 1978a. Diseases of Bajra. Seeds and Farms 4: 35-38.

120. Mohan, M.; Mukerji, K.G. & Rani, K. 1978b. Diseases of *Sorghum*. Seeds and Farms 4: 23-30.

121. Mukerji, K.G. 1978. Taxonomy of Chaetomiales in relation to its morphology and cytology. In : Taxonomy of Fungi (ed. C.V. Subramanian) Porc. Int. Symp. Taxonomy of Fungi 1973, pp. 1: 258-262.

122. Mukerji, K.G. & Mohan, M. 1978. Harmful Seed-borne fungi. School Science 16: 258-262.

123. Nath, R. & Mukerji, K.G. 1978a. Direct observation of leaf surface propagules. School Science 17: 15-17.

124. Nath, R. & Mukerji, K.G. 1978b. Direct observation of leaf surface propagules. Angawandte Botanik 52: 301-304.

125. Rani, Kavita; Mohan, M. & Mukerji, K.G. 1978. Studies on seed-borne fungi on *Sorghum* Seeds. Seed Research 6: 38-42.

126. Singh, N. & Mukerji, K.G. 1978. Studies on Indian Coprophilous Fungi III. The genus *Arachnomyces*. Indian J. Mycol. Res. 16: 283-289.

127. Vön Arx. J.A.; Mukerji, K.G. & Singh, N. 1978a. *Leucosphaera,* new genus of the Psedoeuretiaceae. Persoonia 10: 141-143.

128. Vöx Arx, J.A.; Mukerji, K.G. & Singh, N. 1978b. A new coprophilous ascomycete from India. Personia 10: 144-146.

129. Behera, N. & Mukerji, K.G. 1979. Vertical distribution of fungi in two different soils of Delhi. Curr. Sci. 48: 954-955.

130. Bhandari, N.N.; Rajeshwari, R.; Kaur, D. & Mukerji, K.G. 1979. Histological and histochemical changes in leaves of *Achyranthes aspera* L. infected with *Albugo bliti* (Bei Bern.) Kutze. In : Physiology of Parasitism. (eds. G.P. Agarwal and K.S. Bilgrami), pp. 441-450.

131. Bhatnagar, R.K. & Mukerji, K.G. 1979. Toxins : Mycotoxins. School Sci. 18: 25-30.

132. Bhatnagar, R.K.; Mukerji, K.G.; Magon, K.K. & Venkitasubramanian, T.A. 1979. Effect of Lithium on Aflatoxin biosynthesis by certain strains of *Aspergillus flavus* Group. European. J. Appl. Microbial. Biotechnol. 7: 99-102.

133. Lakhanpal, T.N. & Mukerji, K.G. 1979a. Taxonomic studies on Indian Myxomycetes from India. Kavaka 7: 59-62.

134. Lakhanpal, T.N. & Mukerji, K.G. 1979b. Taxonomic studies on Indian Myxomycetes. XVIII. Two new species of *Didymium.* Rev. de. Mycol. 43: 236-241.

135. Lele, A. & Mukerji, K.G. 1979. Metabolic changes caused by virus infection in Cucumber and French bean. Indian Phytopath. 32: 236-241.

136. Malhotra, G. & Mukerji, K.G. 1979. Fungi of Delhi. XXX. A note on the genus *Sordariella.* India Phytopath. 32: 325-327.

137. Mathur, M. & Mukerji, K.G. 1979a. Diseases of Sunhemp. Seeds and Farms 5:19-21.

138. Mathur, M. & Mukerji, K.G. 1979b. Seed borne fungi. II. Three new records and a new species of *Curvularia.* Proc. INSA, 45A : 147-149.

139. Mukerji, K.G.; Mohan, M. & Ibrahim, G. 1979. Seed-borne fungi. Some New records Acta Bot. Ind. 7: 87-89.

140. Nannenga-Bremekamp, N.E.; Mukerji, K.G. & Singh, N. 1979. On two coprophillious myxomycetes from India. Proc. K. Ned. Akad. Wet. Ser. C. 82: 217-221.

141. Shahda, W.T.A. & Mukerji, K.G. 1979. Diseases of *Clarkia* and Dianthus. Seeds and Farms 5: 29-30.

142. Sharma, K.R. & Mukerji, K.G. 1979. The ecology of microfungi on leaves. In: Recent Advances in the Biology of Microorganisms (eds. K.S. Bilgrami and K.M. Vyas) pp. 69-80.

143. Bhattacharjee, M. & Mukerji, K.G. 1980. Studies on Indian Endogonaceae. III. Further records. Acta. Bot. Indica 8: 99-102.

144. Singh, N. & Mukerji, K.G. 1979. Studies on Indian Coprophilious Fungi. I. Some rare records. J.I.B.S. 58: 163-167.

145. Bhattacharjee, M.; Mukerji, K.G. & Mishra, S. 1980. Studies on Indian Endogonaceae. II. The genus *Glomus.* Sydowia 33: 14-17.

146. Mukerji, K.G. & Khanna, M. 1980. Fungi of Delhi. XII. *Chaetomium lawrence-amesii.* Acta Mycologia 16: 255-256.

147. Mukerji, K.G.; Mohan, M. & Bawa, J. 1980. Fungi of Delhi. XXXII, A Homothalic strain of *Syncephalastrum recemosum* Coh. ex Schroeter. Proc. Indian Natn. Sci. Acad. 46: 387-388.

148. Nath, R.; Mathur, M. & Mukerji, K.G. 1980. Morphological changes in sunhemp virus-infected and morphactin treated plants of *Crotalaria juncea* L. Acta agron. Acad. Sci. Hung. 29(1-2) : 104-107.

149. Rajeevalochana, M. & Mukerji, K.G. 1980. A dwarf strain of *Phycomyces blakesleanus* from dear dung. Phycomyces 4: 83-84.

150. Shahda, W.T. & Mukerji, K.G. 1980. Formation of two new less pathogenic variants of *Alternaria brassicicola* by the use of a new dithiocarbamate fungicide. Incompatibility News Letter 12: 71-75.

151. Bhattacharjee, M.; Mukerji, K.G.; Tewari, J.P. & Skoropad, W.P. 1981. Structure and hyperparasitism of a new species of *Gigaspora.* Trans. Br. Mycol. Soc. 77: 184-188.

152. Kaur, D.; Mukerji, K.G. & Bhandari, N.N. 1981. Histopathology and histochemistry of Diseased Plants. A review, In: Advancing Frontiers of Mycology and Plant Pathology. (ed. K.S. Bilgrami), 11: 183-200. Today and Tomorrow, Delih.

153. Mathur, M. & Mukerji, K.G. 1981a. Antagonistic behaviour of *Cladosporium spongisum* against *Phyllactinia dalbergiae* on *Dalbergia sisoo.* Angewandte Botanik 55: 75-77.

154. Mathur, M. & Mukerji, K.G. 1981b. Two new leaf spot diseases of *Crotalaria juncea.* Angeandte Botanik 55: 78-81.

155. Lakhanpal, T.N. & Mukerji, K.G. 1981. Taxonomic studies on Indian Myxomycetes. II. The development of cortical scales in *Lycogala* species and their bearing on the delimination of species. Trans. Mycol. Soc. Japan 22: 81-87.

156. Mukerji, K.G. & Gupta R. 1981. Ecology of two Deuteromycetous yeasts on the leaf surface of *Corchorus olitorius* L. plants. In : Advances in Biotechnology. Pergamon Press, Canada pp. 521-527.

157. Von Arx, J.A. & Mukerji, K.G. 1981. *Faurelina indica* sp. nov. Sydowia 34: 39-41.

158. Bakshi, I.S. & Mukerji, K.G. 1982. Biological Clock. School Science 23: 1-4.

159. Bhatnagar, R.K.; Ahmad, S.; Kohli, K.K.; Mukerji, K.G. & Venkitasubramanian, T.A. 1982. Induction of Polysubstrate monomygenase and aflatoxin production by phenobarbitone in *Aspergillus parasiticus,*NRRL 3240. Biochemi. Biophys. Res. 104 : 1287-1293.

160. Bhatnagar, R.K.; Ahmad, S.; Mukerji, K.G. & Venkatasubramanian, T.A. 1982. Role of blastospores in protecting *Aspergillus parasiticus* NRRL 3240 from high levels of Aflatoxins. Appl. Environ. Microbiol. 44: 579-582.

161. Bhattacharjee, M. & Mukerji, K.G. 1982. The SEM structure of *Sclerocystis coremioides*. Nova Hedwigia 36: 101-104.

162. Gupta, R.; Mohindra, P. & Mukerji, K.G. 1982. Some further additions to plant diseases of Delhi. Nova Hedwigia 36: 104-112.

163. Gupta, R. & Mukerji, K.G. 1982a. Nigeran production in some *Aspergillus* and *Penicillium* species. Folia Microbiol. 27: 38-42.

164. Gupta, R. & Mukerji, K.G. 1982b. Microbial communities on the leaf surface of four varieties of *Corchorus olitorius* L. Angewandte Botanik 56: 109-120.

165. Mukerji, K.G. & Bhattacharjee, M. 1982. Vesicular Arbuscular Mycorrhiza- A review. In : Advances in Mycology and Plant Pathology (eds. S.B. Chattopadhyayaand and N. Samaj pati), Oxford and IBH Publishing Co., Delhi pp. 111-130.

166. Mukerji, K.G.; Bhattacharjee, M. & Mohan, M. 1982. Ecology of the Indian Endogonaceae. Angewandte Botanik 56: 121-132.

167. Mukerji, K.G. & Subba Rao, N.S. 1982. Plant Surface Microflora and Plant Nutrition. In : Advances in Agricultural Microbiology (ed. N.S. Subba Rao) Oxford and IBH Publishing Co., New Delhi pp. 111-138.

168. Mohindra, P. & Mukerji, K.G. 1982. Fungal Ecology of Termite Mounds. Rev. Ecol. Biol. Soil. 19: 351-362.

169. Saraswat, R.P. & Mukerji, K.G. 1982a. Effect of morphactin on shoot morphogenesis in *Pinus caribae* Mor. J. Assam Sci. Soc. 25: 73-75.

170. Tewari, J.P.; Skoropad, W.P.; Mukerji, K.G. & Mishra, S. 1982. Surface morphology of *Glomus macrocarpum* chlamydospores. Trans. Br. Mycol. Soc. 79: 364-366.

171. Bakshi, I.S. & Mukerji, K.G. 1983. Mitochondria, structure and function. School Science 21: 18-19.

172. Gupta, R. & Mukerji, K.G. 1983. Fungi of Delhi. XXXIV. *Zygorhynchus japonicus,* a new record from India and its hyerparasite. Acta Mycol. 19: 193-195.

173. Gupta, R., Rajeevalochana, M. & Mukerji, K.G. 1983. Dimorphism in some *Chaetomium* species. Bibliotheca Mycol. 91: 463-465.

174. Malini, R.;Mukerji, K.G. & Venkitasubramanian, T.A. 1983. Microbial interactions and Aflatoxin production by *Aspergillus falvus.* Bibliotheca Mycol. 91: 285-290.

175. Malini, R.; Venkitasubramanian, T.A. & Mukerji, K.G. 1983a. Effect of some antibiotics on aflatoxin produciton by *Aspergillus flavus.* Proc. Symp. Mycotoxin in food and feed, Bhagalpur pp. 283-288.

176. Malini, R.; Venkitasubramanian, T.A. & Mukerji, K.G. 1983b. Microbial interactions and aflatoxin produciton. Proc. Symp. Mycotoxin in food and feed, Bhagalpur pp. 297-300.

177. Mukerji, K.G. 1983. Biocontrol of Plant Diseases. In : Recent Advances in Plant Pathology (eds. A. Hussain, K. Singh, V.P. Singh and V.P. Agnihotri), Print House (India) Ltd., Lucknow pp. 1-20.

178. Mukerji, K.G.; Bhattacharjee, M. & Tewari, J.P. 1983. New species of vesicular-arbuscular mycorrhizal fungi. Trans. Br. Mycol. Soc. 81: 641-643.

179. Mukerji, K.G. & Kochar, B. 1983. Vesicular-arbuscular mycorrhiza in rape seed plant. Proc. 6th Intern. Rape seed Conf. Paris. Vol. II : 945-950.

180. Mukerji, K.G.; Verma, V. & Gupta, R. 1983. Biology of sexual reproduction in some ascomycetes, Phytomorphology 33: 107-122.

181. Nath, R. & Mukerji, K.G. 1983. Microbial ecology of the morphactin treated leaves of *Crotalaria juncea* L. Acta Acad. Sci. Hung. 32: 138-139.

182. Rajeevalochana, M. & Mukerji, K.G. 1983. Microbial ecology of the morphactin treated leaves of *Crotalaria juncea* L. Acta Acad. Sci. Hung. 32: 138-139.

183. Saraswati, R.P. & Mukerji, K.G. 1983. Leaf blight of *Achyranthes aspera* L. Indian Phytopath. 36: 573-574.

184. Zaidi, R. & Mukerji, K.G. 1983. Incidence of vesicular-arbuscular mycorrhiza (VAM) in diseased and healthy plants. Indian J. Plant Pathol. 1: 24-26.

185. Behera, N. & Mukerji, K.G. 1984a. Studies on soil microfungi in relation to edaphic factors. Acta Bot. Indica 12: 153-156.

186. Behera, N. & Mukerji, K.G. 1984b. Seasonal fluctuations of Aspergilli in Delhi soils. Indian J. Mycol. Path. 14: 283-285.

187. Bhatnagar, R.K.; Malini, R.; Ahmad, S.; Venkitasubramanian, T.A. & Mukerji, K.G. 1984. Ecology of Aflatoxin producing *Aspergillus* species. In : Progress in Microbial Ecology (eds. K.G. Mukerji, V.P. Agnihotri and R.P. Singh), Print House (India) Lucknow pp. 273-291.

188. Kaur, D.; Bhandari, N.N. & Mukerji, K.G. 1984. A histochemical study of cytoplasmic changes during wall layer formation in the oospore of *Albugo candida,* Phytopath. Z. 103: 117-130.

189. Malini, R.; Mukerji, K.G. & Venkitasubramanian, T.A. 1984. Effect of Aluminium and Nickel on Aflatoxin production by *Aspergillus flavus.* Folia Microbiol. 29: 104-107.

190. Mohindra, P. & Mukerji, K.G. 1984. Cestum Mosaic Virus : Disease symptoms and metabolic changes due to infection, Jour. Assam Sci. Soc. 26: 38-42.

191. Mukerji, K.G.; Sabharwal, A.; Kochar, B. and Ardey, J. 1984. Vesicular-arbuscular mycorrhiza : Concepts and Advances. In: Progress in Microbial Ecology (eds. K.G. Mukerji, V.P. Agnihotri and R.P. Singh) Print House (India) Lucknow pp. 273-291.

192. Nannenga-Bremekamp, N.E.; Mukerji, K.G. & Pasricha, R. 1984. Note on Indian Myxomycetes - Three new species and comments on others. Proc. K. Nederl. Akad Wet. C. 87: 471-482.

193. Shahda, V.T. & Mukerji, K.G. 1984. Variation to less pathogenic types as cuased by a new dithiocarbamate fungicide in *Alternaria brassicola.* Acta Agro Scient. Hungary 33: 448-452.

194. Singh, V.P.; Garg, K.L. & Mukerji, K.G. 1984. Architecture of ribosomes in Prokaryotes and Eukaryotes. School Science 22: 36-37.

195. Ardey, J. & Mukerji, K.G. 1985. Intensity of VAM infection in compositae. Proc. 6th North, Am. Conf. Mycorrhiza p. 408.

196. Bakshi, I.S., Mukerji, K.G. & Venkitasubramanian, T.A. 1985. The nitrate reductae of fungi. In: Frontiers Appl. Microbiol (eds. K.G. Mukerji, N.C. Pathak and V.P. Singh). 1: 129-136.

197. Behera, N. & Mukerji, K.G. 1985a. Seasonal variation and distribution of microfungi in forest soils of Delhi. Folia Geobotanica et Phytotaxonomica 20: 291-311.

198. Behera, N. & Mukerji, K.G. 1985b. Ecology of micromycetes in forest soils of Delhi. Acta Mycol. 21: 101-108.

199. Garg, A.P.; Gandotra, S.; Mukerji, K.G. & Pugh, G.J.F. 1985. Ecology of Keratinophillic fungi. Proc. Indian Acad. Sci. 94: 149-163.

200. Kochar, B. & Mukerji, K.G. 1985. Development of vesicular arbuscular mycorrhiza in field grown barley. Proc. 6th North Am. Conf. Mycorrhiza, p. 409.

201. Mathur, A.; Mathur, S.K. & Mukerji, K.G. 1985. Phylloplane fungi of *Crotolaria juncea* during growth and decomposition of leaves. Indian Phytopath. 38: 683-685.

202. Mukerji, K.G. & Ardey, J. 1985. Studies on Vesicular arbuscular mycorrhiza. I. *Kalanchoe spicata* and *Eclipta alba*. Acta Bot. Indica 13: 101-103.

203. Mukerji, K.G.; Singh, V.P. & Garg, K.L. 1985. Biology of reproduction in Zygomycetes. Jour. Pl. Sci. 1: 31-47.

204. Pasricha, R. & Mukerji, K.G. 1985. Studies on growth and sporulation of *Didymium muscorum* Lakhanpal and Mukerji. Acta. Bot. Indica 13: 224-227.

205. Sabharwal, A. & Mukerji, K.G. 1985. Occurrence and characteristics of VAM in ecologically different sites. Proc. 6th North Am. Con. Mycorrhiza 316.

206. Ardey, J. & Mukerji, K.G. 1986. Fungi associated with the roots of herbaceous plants. Current Science 55: 934-935.

207. Bhatnagar, R.K.; Ahmad, S.; Mukerji, K.G. & Subramanian, T.A.V. 1986a. Pyridine nucleotides and tedox state regulation of aflatoxin biosynthesis in *Aspergillus parasiticus* NRRL 3240. Jour. Appl. Bacteriol. 60: 135-141.

208. Bhatnagar, R.K.; Ahmad, S.; Mukerji, K.G. & Subramanian, T.A.V. 1986b. Nitrogen metabolism in *Aspergillus parasiticus* NRRL 3240 and *A. flavus* NRRL 3537 in relation to aflatoxin production. Jour. Appl. Bacteriol. 60: 203-211.

209. Mukerji, K.G. & Kapoor, A. 1986. Occurrence and importance of vesicular arbuscular mycorrhizal fungi in semi arid regions of India. For. Ecol. Management 16: 117-126.

210. Mukerji, K.G. & Singh, V.P. 1986. Unusual oospore in *Peronospora trifoliorum*. Phytomorphology 36: 383.

211. Rani, R. & Mukerji, K.G. 1986. Influence of VA mycorrhizal inoculation on growth of *Zea mays* and *Avena sativa*. Rec. Advan. Appl. Sci. 1: 100-102.

212. Saxena, G. & Mukerji, K.G. 1986. Formation of Sclerotia in a non-sclerotial soil strain of *Botrytis cinerea*. Pl. Cell Incomp. News Letter 18: 28-32.

213. Bhatnagar, R.K.; Malini, R.; Mukerji, K.G. & Subramanian, T.A.V. 1987. Secondary metabolism in *Aspergillus* with special reference to aflatoxin production by *Aspergillus parasiticus.* Frontiers in Appl. Microbiol. 2: 127-150.

214. Jagpal, R. & Mukerji, K.G. 1987. Large scale cropping on Indian arid lands. Proc. Int. Symp. Dry Land Farming, Yangling Saanxi, China (September 17-22, 1987) Vol. 1: 56-77.

215. Mittal, Neelima; Saxena, G.; Jagpal, R. & Mukerji, K.G. 1987. Interaction between nematophagous fungi and vesicular arbuscular mycorrhizal fungi. Pl. Cell Incomp. News Letter 19: 54-59.

216. Mukerji, K.G. & Jagpal, R. 1987. VAM in reforestation in Indian Waste Lands. Proc. 4th North Amer. Conf. Mycorrhizeae, May 3-8, 1987, Gainesville, Florida pp. 161-163.

217. Mukerji, K.G. & Kapoor. A. 1987. Vesicular-arbuscular mycorrhiza-Patterns of modern research. Frontiers in Applied Microbiol. 2 : 189: 200.

218. Nigam, N.; Rai, B. & Mukerji, K.G. 1987. Studies on Indian coprophilous fungi. IV. Two interesting members of Mucorales. Nova Hedwigia 44: 523-526.

219. Pasricha, R. & Mukerji, K.G. 1987. Mycorrhizal status of some ornamental plants at their reproductive and vegetative state. Acta Bot. Indica 15: 311-313.

220. Rani, Rekha; Jagpal, R. & Mukerji, K.G. 1987. Mycorrhizal status of some ornamental plants at their reproductive and vegetative stage. Acta. Bot. Indica 15: 311-313.

221. Saxena, G.; Dayal, R. & Mukerji, K.G. 1987a. Nematophagous fungi. Frontiers in App. Microbiol. 2 : 257-286.

222. Saxena, G.; Dayal, R. & Mukerji, K.G. 1987b. Interaction of nematodes with nematophagous fungi: Induction of trap formation, attraction and detection of attractants. FEMS Microbiol Ecology 45: 319-327.

223. Singh, K.; Verma, A.K. & Mukerji, K.G. 1987. Vesicular arbuscular mycorrhizal fungi in diseased and healthy plants of *Vicia faba.* Acta Bot. Indica 15: 305-310.

224. Thapa, B.; Mukerji, K.G.; Chaturvedi, C. & Kothari, S.K. 1987. Effect of various herbicides on vesicular-arbuscular infection in *Mentha arvensis* L. var. Mas-1. Indian Perfumer 31(4): 347-350.

225. Bali, M.; Nigam, N. & Mukerji, K.G. 1988. Interaction of vesicular-arbuscular mycorrhiza with rhizospher and rhizoplane fungi of *Gossypium hirsutum* and *Corchorus olitorium.* In: Mycorrhiza Round Table. IDRC Canada, Mans. Rep. 201e: 347-355.

226. Ganju, V.; Nair, L.N. & Mukerji, K.G. 1988. Cultivation fo *Pleurotus sajorcaju* (Fr.) Singer on different substrates. Jour. Rec. Adv. Appl. Sci. 3: 478-492.

227. Gupta, N.; Saxena, G.; Singh, V.P. & Mukerji, K.G. 1988. Factors affecting distribution of nematophagus fungi in Delhi. Acta Botanica Indica 16: 46-50.

228. Jagpal, R. & Mukerji, K.G. 1988a. VAM fungi - A tool for reforestation. J. Phytol. Res. 1: 31-35.

229. Jagpal, R. & Mukerji, K.G. 1988b. Morphology and anatomy of endomycorhizal fungi. In: Mycorrhiza Round Table. IDRC Canada, Manuscript Report 201e: 190-210.

230. Jagpal, R.; Rani, R.; Bali, M. & Mukerji, K.G. 1988. Morphology and anatomy of endomycorrhizal fungi. In : Mycorrhiza Round Table. IDRC Canada, Manuscript Report 201e: 190-210.

231. Kapoor, A. & Mukerji, K.G. 1988. Influence of VAM fungi and P levels of soybean growth in fumigated microplots. In : Biofertilizers-Potentials and Problems (eds. S.P. Sen and P. Palit). Naya Prokash, Calcutta pp. 259-263.

232. Mittal, N.; Saxena, G. & Mukerji, K.G. 1988. Few predators from Delhi soils. Jour. Rec. Adv. Appl. Sci. 3: 383-388.

233. Nigam, N. & Mukerji, K.G. 1988. Biological control-concepts and practices. In : Biocontrol of Plant Disease (eds. K.G. Mukerji and K.L. Garg). Vol. 1. CRC Press, Inc., Florida USA pp. 1-12.

234. Nigam, N.; Rani, B. & Mukerji, K.G. 1988a. Effect of non-volatile metabolites of some dominant phylloplane fungi on the sporangial germination of *Peronospora arborescens* and phylloplane fungi of opium poppy. Jour. Rec. Adv. Appl. Sci. 3: 433-443.

235. Nigam, N., Rai, B. & Mukerji, K.G. 1988b. Effect of acidic precipitation on soil fungal population. Acta Botanica Indica, 16: 42-45.

236. Nigam, N.; Rai, B. & Mukerji, K.G. 1988c. Effect of leaf injury and capsule incision on associated microflora of opium poppy. Acta Botanica Indica 16: 195-536.

237. Panigrahi, A.; Choudhury, R. & Mukerji, K.G. 1988a. Effect of air pollution on some growth parameters of *Melilotus indica* (L) Al. Acta Botanica Indica 16: 1-7.

238. Panigrahi, A.; Choudhury, R. & Mukerji, K.G. 1988b. Effect of acidic precipitation on soil fungal population. Acta Botanica Indica 16: 42-45.

239. Rani, R. & Mukerji, K.G. 1988. Indian vesicular-arbuscular mycorrhizal fungi. In : Mycorrhiza Round Table (eds. A.K. Verma. A.K. Oka, K.G. Mukerji, K.V.B.R. Tilak and Janak Raj) IDRC, Canada, Manuscript Report 201e: 166-180.

240. Saxena, G.; Daya, R. & Mukerji, K.G. 1988. Studies on colony interaction between nematophagous and some selected fungi. Allonia 28: 185-191.

241. Saxena, G. & Mukerji, K.G. 1988. Biological control of nematodes. In: Biocontrol of Plant Diseases (eds. K.G. Mukerji and K.L. Garg) Vol. 1. CRC Press, Inc., Florida, USA pp. 113-127.

242. Saxena, M.; Mukerji, K.G. & Raj. H.G. 1988. Positive correlation exists between glutathione S-transferase activity and aflatoxin formation in *Aspergillus flavus*. Biochem. J. 254: 567-570.

243. Singh, K.; Verma, A.K. & Mukerji, K.G. 1988. Study on healthy and diseased hosts of fodder legume *Vicia faba* in relation to VAM. In: Mycorrhiza Round Table. IDRC Canada. Mans. Rep. 201e: 535-549.

244. Bali, M. & Mukerji, K.G. 1989a. Effect of VAM fungi on *Fusarium* wilt of cotton and jute. Proc. 1st Asian Conf. Mycorrh. Madras 1988, pp. 233-234.

245. Bali, M. & Mukerji, K.G. 1989b. Growth responses in mycorrhizal Jute plants. Proc. Nat. Symp. Fungal Ecol. Hyderabad 1987, pp. 35-38.

246. Bakshi, I.S.; Mukerji, K.G. & Venkitasubramanian, T.A. 1989a. An interrelationship between nitrate reductase and glutamine synthetase in the fungus *Aspergillus parasiticus*. Joar. Rec. Adv. Appl. Sci. 4: 574-580.

247. Bakshi, I.S.; Mukerji, K.G. & Venkitasubramanian, T.A. 1989b. Control of nitrate reductase in the fungus *Aspergillus flavus*. Jour. Rec. Adv. Appl. Sci. 4: 659-665.

248. Biswas, G.; Raj, H.G. & Mukerji, K.G. 1989. Glutathione levels and Y glutamyl transpectidase activities afltoxigenic strains of *Aspergillus flavus*. J. Toxicol: Toxin Reviews 8(1and2): 329-338.

249. Choudhury, R.K.; Panigrahi, A. & Mukerji, K.G. 1989. Effect of some atmospheric pollutans on microorganisms. In: Frontiers in Appl. Microbiol. Vol. III. pp. 85-116.

250. Jagpal, R. & Mukerji, K.G. 1989a. Distribution of VAM fungi in some Bihar soils. Jour. Rec. Adv. Appl. Sci. 4: 609-612.

251. Jagpal, R. & Mukerji, K.G. 1989b. VAM in bamboo. Proc. Nat. Symp. Fungal Ecol. Hyderabad 1987 pp. 47-50.

252. Jagpal, R.; Panigrahi, A. & Mukerji, K.G. 1989. Effect of automobile pollution on VAM spore population of soil. Proc. Ist Asian Conf. Mycorrh. Madras 1988, pp. 46-47.

253. Jagpal, R.; Sharma, P.D. & Mukerji, K.G. 1989. Distribution of VAM in relation to pollution. Proc. Ist Asian Conf. Mycorrh. Madras 1988, pp. 48-49.

254. Kapoor, A.; Singh, V.P. & Mukerji, K.G. 1989. Studies on the phosphatases on mycorrhizal and non-mycorrhizal *Trigonella* roots. Proc. Ist Asian Conf. Mycorrh. Madras 1988, pp. 125-127.

255. Mittal, N.; Saxena, G. & Mukerji, K.G. 1989a. Interaction between *Arthrobotrys musiformis* and some soil fungi. Incompatibility News Letter, 21: 49-58.

256. Mittal, N.; Saxena, G. & Mukerji, K.G. 1989b. Ecology of nematophagous fungi : Distribution in Dehli. J. Phytol. Res. 2: 31-37.

257. Mohan, M.; Panigarhi, A. & Mukerji, K.G. 1989. Detached-leaf culture technique for estabilishing pathogenecity of seed-borne pathogens of cauliflower. Biome 4: 49-57.

258. Monga, P.K.; Lakhanpal, T.N. & Mukerji, K.G. 1988 (89). Leaf decomposition in *Crotolaria juncea, Corchorus olitorius* and *Arachis hypogea* in relation to soil fertility. Ind. J. Mycol. Pl. Pathol. 18: 169-170.

259. Mukerji, K.G. & Bali, N. 1989. Status of endomycorrhizal research in India. Proc. 1st Asian Conf. Mycorrhiza Madras 1988, pp. 6-10.

260. Panigarhi, A. & Mukerji, K.G. 1989. Yield loss assessment caused by *Peronospora arborescens* in opium poppy. Indian Phytopath. 42: 110-115.

261. Panigrahi, A. & Mukerji, K.G. 1989. The effect of air pollutants on microorganisms. Proc. Nat. Symp. Fungal Ecol. Hyderabad 1987. pp. 62-65.

262. Rani, R. & Mukerji, K.G. 1989a. Distribution of VA mycorrhizal fungi in some Indian soils. Proc. Nat. Symp. Fungal Ecology, Hyderabad 1987. pp. 62-65.

263. Rani, R. & Mukerji, K.G. 1989b. Taxonomy of vesicular arbuscular mycorrhizal fungi. Proc. 1st Asian Conf. Mycorrh. Madras 1988, pp. 103-105.

264. Saxena, G. & Mukerji, K.G. 1989a. Few namatophagous fungi from Varanasi. Proc. Nat. Symp. Fungal Ecol. Hyderabad 1987 pp. 39-46.

265. Saxena, G. & Mukerji, K.G. 1989b. Ecology of nematophagous fungi. Jour. Rec. Adv. Appl. Sci. 4: 624-632.

266. Saxena, G.; Dayal, R. & Mukerji, K.G. 1989. Nutritional studies on nematode trapping fungi. Folia Microibol. 34(1): 42-49.

267. Saxena, M.; Mukerji, K.G.; Mallic, J.; Allameh, A. & Raj. H.G. 1989. Metabolism of Glutathione in Microorganisms. Frontiers in Appl. Microbiol. Vol. III : 117-134.

268. Saxena, M.; Allameh, A.; Mukerji, K.G. & Raj. H.G. 1989. Studies on glutathione transferases of *Aspergillus flavus* group in relation to aflatoxin production. J. Toxicon. Toxin Review 8: 319-328.

269. Bali, M. & Mukerji, K.G. 1989. Influence of VAM on the growth of two cultivars of *Capsicum annum*. Proc. 1st Asian Conf. Mycorrh. Madras 1988. pp. 262-264.

270. Bali, M. & Mukerji, K.G. 1990. Interaction between soil microflora and VAM fungi in relation to growth of cotton. Proc. 8th NACOM Jackson, Myeming, U.S.A. pp. 14.

271. Bansal, M.; Jagpal, R. & Mukerji, K.G. 1990. VAM in fine root dynamic. Proc. 8th NACOM Jackson, Wyoming, U.S.A. pp. 15.

272. Jagpal, R.; Bansal, M. & Muerji, K.G. 1990. Growth depression in mycorrhizal *Trifolium* plants Proc. 8th NACOM, Jackson, Myoming, U.S.A. pp. 151.

273. Kapoor, A. & Mukerji, K.G. 1990. A strategy for selection and application of VAM fungi. Curr. Trends Mycol. Res. pp. 139-140.

274. Kaur, S.; Bakshi, I.S. & Mukerji, K.G. 1990. Experimental studies on myxomycete *Didymium nigripes* (Link). Fr. Jour. Rec. Adv. Appl. Sci. 5: 91-92.

275. Mukerji, K.G. & Kapoor, A. 1990. Aeromycoflora of Berhampur University Campus, Orissa. J. Indian Bot. Soc. 69: 417-420.

276. Mukerji, K.G. & Kapoor, A. 1990. Taxonomy of VAM fungi with special reference to Indian taxa. Pers. Mycol. Res. 11: 7-16.

277. Rani, R. & Mukerji, K.G. 1990. The distribution of vesicular arbuscular mycorrhizal fungi in India. Acta Microbiologica 37: 3-7.

278. Sakhuja, S.; Bali, M.; Nigam, N.; Bansal, M.; Ranga Rao, V. & Mukerji, K.G. 1990. Fungi of Delhi. XXXVII. Three interesting members of Mucorales. Jour. Res. Adv. App. Sci. 5: 83-90.

279. Suvercha & Mukerji, K.G. 1990a. Microbial ecology of rhizosphere and phylloplane. In : Economic Plants and Microbes (ed. R.P. Purkayastha) Naya Prokash, Calcutta pp. 155-162.

280. Suvercha & Mukerji, K.G. 1990b. Phosphatase in mycorrhizospheres and non-mycorrhizospheres of wheat polyploids. Proc. 8th NACOM. Jackson, Wyoming, pp. 273.

281. Bansal, M. & Mukerji, K.G. 1991. Endomycorrhiza in young *Leucaena* forest. Jour. Rec. Adv. Appl. Sci. 6: 35-42.

282. Bansal, M.; Jagpal, R. & Mukerji, K.G. 1991. Viability of VAM fungal spores. Jour. Rec. Adv. Appl. Sci. 6: 65-68.

283. Bali, M. & Mukerji, K.G. 1991. Interaction between VA mycorrhizal fungi and root microflora of jute. In : Plant Roots and Their Environment (eds. Mc Michael and Hans Persson) Elseviar, Amsterdom. pp. 396-401.

284. Jagpal, R. & Mukerji, K.G. 1991a. Reforestation in waste land using vesicular arbuscular mycorrhizae. In : Recent Development in Tree Plantations of Humid/Subhumid Tropics of Asia, University Pertanian Malaysia, Selangor, Malaysia, pp. 488-494.

285. Jagpal, R. & Mukerji, K.G. 1991b. VAM fungi in reforestration. In : Plant Roots and their Environment (eds. Mc Michael and Hans Persson) Elsevier. Amsterdom. pp. 309-313.

286. Kapoor, A. & Mukerji, K.G. 1991. The distribution of vesicular arbuscular mycorrhizae in relation to plants in Indian Soils. Jour. Rec. Adv. Appl. Sci. 6: 1-28.

287. Mittal, N.; Sharma, M.; Saxena, G. & Mukerji, K.G. 1991. Effect of VA mycorrhiza on gall formation in tomato roots. Incompatibility Newsletter. 23: 39-43.

288. Mukerji, K.G. 1991. Role of VA mycorrhiza in afforestation in the tropics. Mycorrhiza News 3: 4.

289. Mukerji, K.G.; Jagpal, R.; Bali, M. & Rani, R. 1991. The importance of mycorrhiza for roots. In : Plants Roots and Their Environment (eds. Mc Michael and Hans Persson) Elsevier, Amsterdom. pp. 290-308.

290. Nigam, N. & Mukerji, K.G. 1991. A new mycovirus. In : Recent Advances in Plant Biology (eds. C.P. Malik and Y.P. Abrol), Narendra Publishing House, Delhi pp. 235-338.

291. Panigrahi, A. & Mukerji, K.G. 1991. Microbial ecology of the rhizosphere microflora of *Crotolaria retusa* L. in relation to flyash. In : Plant Roots and Their Environment, (eds. Mc Michael and Hans Persson) Elsevier, Amsterdom pp. 314-328.

292. Rani, R. & Mukerji, K.G. 1991. Effect of VA mycorrhiza inoculated on roots of *Zea mays.* In : Plant Roots and Their Environment, (eds. Mc Michael and Hans Persson) Elsevier, Asmterdom pp. 494-497.

293. Saxena, G. & Mukerji, K.G. 1991. Distribution of nematophagous fungi in Varanasi, India. Nova Hedwegia 52 (3-4): 487-495.

294. Saxena, G.; Mittal, N.; Mukerji, K.G. & Arora, D.K. 1991. Nematophagous fungi in the biological control of nematodes. In : Hand Bood of Applied Mycology (eds. D.K. Arora, L. Ajello and K.G. Mukerji) Marcel Dekker Inc. USA, Vol. II : 707-735.

295. Saxena, M.; Allameh, A.; Mukerji, K.G. & Raj, H.G. 1991. Epoxidation of Aflatoxin BI by *Aspergillus falvus* microsomes *in vitro:* interaction with DNA and formation of Aflatoxin-Glutathione conjugate. Chem. Biol. Interactions 78: 13-22.

296. Suvercha; Mukerji, K.G. & Arora, D.K. 1991. Ectomycorrhiza. In : Hand Book of Applied Mycology (eds. D.K. Arora, B. Rai, K.G. Mukerji and G.R. Knudson) Marcel Dekker Inc. USA, Vol. I. 187-217.

297. Suvercha & Mukerji, K.G. 1991. Mycorrhizal colonisation in roots of tetraploid and hexaploid wheat species. In : Plant Roots and their Environment (eds. Mc Michael and Hans Persson) Elsevier, Amsterdom, pp. 198-505.

298. Bansal, M.; Saxena, M.; Mukerji, K.G.; Allameh, A. & Raj. H.G. 1992. Atlatoxin in Seeds. In : Current Concepts in Seed Biology (eds. K.G. Mukerji, A.K. Bhatnagar, S.C. Tripathi, M. Bansal and M. Saxena) Naya Prokash, Calcutta, pp. 157-176.

299. Bansal, M. & Mukerji, K.G. 1992. Effect of VAM and phosphorus fertilizers on *Leucaena* root productivity. In : Root Ecology and its Practical Application (eds. L. Kutschera, E. Hibi, E. Lichtenegger, H. Person and M. Sabotik) Vienna pp. 543-546.

300. Kapoor, A. & Mukerji, K.G. 1992. Distribution of vesicular arbuscular mycorrhiza in root zone of host plants insome Indian soils. In : Root Ecology and its Practical Application (eds. L. Kutschera, E. Hibi, E. Lichtenegger, H. Person and M. Sabotik) Vienna pp. 605-609.

301. Kaushik, A.; Dixon, R.K. & Mukerji, K.G. 1992. Vesicular arbuscular mycorrhizal relationsships of *Prosopis juliflora* and *Zizyphus jujuba.* Phytomorphology 133-147.

302. Mago, P.; Agenes, C.A. & Mukerji, K.G. 1992. VA Mycorrhizal status of some Indian Bryophytes. Phytomorphology 42: 231-239.

303. Mittal, N.; Saxena, G. & Mukerji, K.G. 1992. Biological control of *Meloidogyne incognita* by *Paecilomyces lilacinus.* In : Root Ecology and its Practical Application (eds. L. Kutschera, E. Hibi, E. Lichtenegger, H. Person and M. Sabotik) Vienna pp. 546-550.

304. Mittal, N.; Srivastava, A.K. & Mukerji, K.G. 1992. Medical mycology in Indian context: Comments conclusion and suggestion. In : Advances in Medical Mycology (eds. K.G. Mukerji, A.K. Srivastava, K.P. Singh and K.L. Garg). Aditya Books Pvt. Ltd., New Delhi. pp. 211-227.

305. Mukerji, K.G. & Dixon, R.K. 1992. Mycorrhizae in reforestration. In : Proc. Int. Symp. Rehabilitation of Prop. Rainforest Ecosystem, Malaysia pp. 66-82.

306. Mukerji, K.G.; Sharma, M.; Kuashik, A. & Raut, P. 1992. Taxonomy of VAM fùngi. In : Mycorrhizae : An Asian Overview. TERI, New Delhi pp. 12-22.

307. Nigam, N. & Mukerji, K.G. 1992a. Biocontrol of powdery mildews. In : Recent Developments in Biocontrol of Plant Diseases (eds. K.G. Mukerji. J.P. Tewari. D.K. Arora and G. Saxena) Aditya Books Pvt. Ltd., New Delhi. pp. 17-22.

308. Nigam, N. & Mukerji, K.G. 1992b. Root region microflora of opium poppy. In : Root Ecology and its Practical Application (eds. L. Kutschera, E. Hibi, E. Lichtenegger, H. Person and M. Sabotik) Vienna pp. 363-366.

309. Raj, H.K.; Saxena, M.; Allamai, A. & Mukerji, K.G. 1992. Metabolism of foreign compounds by fungi. In : Hand Book of Applied Mycology (eds. D.K. Arora, R.P. Elander and K.G. Mukerji) Marcel Dekker Inc USA Vol. IV: 881-904.

310. Sakhuja, S. & Mukerji, K.G. 1992. Effect of leaf litter decomposition of *Leucaena leucocephala* on fertility of soil. In : Roots Ecology and its Practical Application (eds. Kutschera, E. Hiibi, E. Lichtenegger, H. Persson and M. Sabotik). Veren fur Wnzzehforschun, Vienna pp. 447-449.

311. Sharma, J.; Khanna, R.N.; Sakhuja, S. & Mukerji, K.G. 1992. Formation of two new variants of *Collectotrichum falcatum* by the use of a new quinonoid fungicide. Plant Cell Incompatibility Newsletter, 24: 52-53.

312. Sharma, M. & Mukerji, K.G. 1992. Mycorrhiza - Tool for Biocontrol. In : Recent Developments in Biocontrol of Plant Diseases (eds. K.G. Mukerji, J.P. Tiwari, D.K. Arora, and G. Saxena) Aditya Books Pvt. Ltd., New Delhi pp. 52-80.

313. Singh, K. & Mukerji, K.G. 1992. Recent trends in classification of seed borne fungi. In : Vistas in Seed Biology (eds. T. Singh and R.C. Trivedi). Print Well Publishers, Jaipur pp. 1-8.

314. Singh, V.P.; Smith, J.E.; Harran, G.; Saxena, R.K. & Mukerji, K.G. 1992. Biological aspects of Aflatoxin detoxification. Indian J. Microbiol. 32: 357-369.

315. Bansal, M. & Mukerji, K.G. 1993. Dead fine roots - a neglected biofertiliser. In : Plant Nutrition - from Genetic Engineering to Field Practice (ed. N.J. Barrow) Kluwer Academic Publisher pp. 547-550.

316. Bakshi, I.S. & Mukerji, K.G. 1993. Role of fungi in acid production. Prop. Appl. Microbiol. IV: 101-114.

317. Biswas, G.; Raj, H.G.; Allameh, A.; Saxena, N.; Srivastava, N. & Mukerji. K.G. 1993. Comparative lometoc studies on Aflatoxin B, binding of pulmonary and hepatic DNA of rat and hamster receiving the carcinogen intratracheally. Ferat. Carcino. Muta. 13: 259-268.

318. Garg, K.L. & Mukerji, K.G. 1993. Deterioration of wooden materials by micro-organisms. In : Recent Advances in Biodeterioration and Biodegradation. Vol. I (eds. K.L. Garg N. Garg, and K.G. Mukerji) Naya Prokash, Calcutta pp. 429-479.

319. Jagpal, R. & Mukerji, K.G. 1993. Role of mycorrhiza in growth of plants and its technique of isolation and inoculation. In : Nursery Technology for Agroforestry : Applications of Arid and Semi Arid Regions (eds. S. Purand and P.K. Khosla). Oxford and IBH Publishing Co. Pvt. Ltd., New Delhi pp. 263-271.

320. Kaur, S.; Pathak, A. & Mukerji, K.G. 1993. Studies on Indian Laboulbeniomycetes. 1. Three unrecorded species of the genus *Laboulbenia* Mont. et Robin. Crypt. Botany 3: 153-156.

321. Lakhanpal, T.N.; Sharma, R. & Mukerji, K.G. 1993. Myxomycetes from the moist chambers. Front. Appl. Microbiol. IV. 193-204.

322. Mago, P. & Mukerji, K.G. 1993. VAM as an effective biofertilizer for mints. Botany 2000-Asia News Letter 2:4.

323. Mago, P.; Pasricha, P. & Mukerji, K.G. 1993. Survey of vesicular-arbuscular mycorrhizal fungi in scale leaves of modified stems. Phytomorphology 43: 81-85.

324. Mittal, N.; Saxena, G. & Mukerji, K.G. 1993. Compatibility of nematophagous fungi with nematodes. Incompatibility Newsletter 25: 31-34.

325. Neeraj; Mukerji, K.G.; Sharma, B.C. & Varma, A.K. 1993. A new species of *Gigaspora* from desert of Rajasthan, India. World J. Microbiol. Biotech. 9: 291-294.

326. Sharma, K.; Khanna, R.N.; Sakhuja, S. & Mukerji, K.G. 1993. Two new variants of *Collectotrichum falcatum* formed due to the use a new flavonoid fungicide. Incompatibility News Letters, 25: 35-37.

327. Saxena, R.K.; Sinha, U.; Singh, V.P. & Mukerji, K.G. 1993. Fungal biotechnology. Front. Apply. Microbiol. IV: 91-100.

328. Bansal, M. & Mukerji, K.G. 1994. Efficacy of root litter as a biofertiliser. Biol. Fertil Soils 18(3): 228-230.

329. Bansal, M. & Mukerji, K.G. 1994. Positive correlation between VAM enduced changes in root exudation and mycorrhizosphere mycoflora. Mycorrhiza 5: 39-44.

330. Bansal, M., Srivastava, D. & Mukerji, K.G. 1994. Root litter biodegradation. In : Recent Advances in Biodeterioration and Biodegradation. Vol. II (eds. K.L. Gard, N. Garg and K.G. Mukerji) Naya Prakash, Calcutta pp. 363-376.

331. Bansal, M. & Mukerji, K.G. 1994. Positive correlation between VAM induced charges in root exudation and mycorrhizosphere mycoflora. Mycorrhiza 5(1): 49-44.

332. Choudhury, R.K., Panigrahi, A. & Mukerji, K.G. 1994. Pollution and Microbial equillibrium. In: Frontiers in Microbiol Technology (ed. P.S. Bisen) P.S. Publishers and Distributors, Delhi. pp. 38-44.

333. Garg, M.; Garg, K.L. & Mukerji, K.G. 1994. Microbial degradation of environmental pollutants. In : Recent Advances in Biodeterioration and Biodegradation. Vol. I. (eds. K.L. Garg, N. Garg and K.G. Mukerji) Naya Prokash, Calcutta pp. 327-362.

334. Kaur, S. & Mukerji, K.G. 1994. Effect of light on growth and development of plasmodium in *Didymium signipes*. J. Phytol. Res. 7(2): 121125.

335. Kumar, R.N.; Gupta, R.; Mukerji, K.G. & Upadhyay, R.K. 1994. Eradication of a noxious weed, *Euphorbia geniculata* by root pathogen. Plant Protection Bulletin 46: 37-38.

336. Pasricha, R., Kumar, R.N. & Mukerji, K.G. 1994. Effect of water stress on growth and sporulation of coprophilous fungi. Nova Hedwigia 59: 157-162.

337. Pasricha, R., Kumar, R.N. & K.G. Mukerji, 1994. Nutrition requirements of *Pilobolus crystallation*. J. Indian. Bot. Soc. 73: 365-366.

338. Gupta, Rajni & Mukeri, K.G. 1995. Studies on weeds of Delhi. III. Comp. Newsletter 27: 16.

339. Gupta, Rajni & Mukerji, K.G. 1995. Weeds of Delhi - Cyperacae and Euphorbiacece. J. Phytol. Rep. 8: 137-140.

340. Kaur, S. & Mukerji, K.G. 1995a. Studies on Indian laboulbeniales II. Three unrecorded species. Nova Hewdwigia 61:

341. Kaur, S. & Mukerji, K.G. 1995b. Studies in Indian Laboulbeniales IV. Three species of laboulbenia. Mycoscience 36: 311-314.

342. Kumar, R.N.; Passicha, R.; Singh, N. & Mukerji, K.G. 1995. Taxo-ecological studies of Coprophilous fungi: a review. In : Advances in Ecology and Environmental Sciences (eds. P.C. Mishra, N. Behera, B.K. Senapati and B.C. Guru) Ashish Publishing House, New Delhi, pp. 35-54.

343. Mittal, N., Saxena, G. & Mukerji, K.G. 1995. Integrated control of root-knot diesease in three crop plants using chitin and *Paecilomyces lilacinus.* Crop Protection 14(8): 647-651.

344. Mukerji, K.G.; Kumar, R.N. & Singh, N. 1995. Studies on Indian Coprophillous fungi. IV. Species of genera *Apisordaria* and *Cereophora.* Phytomorphology 45 (1,2) : 87-105.

345. Mukerji, K.G.; Garg, K.L. & Mishra, A.K. 1995. Fungi in deterioration of museum objects. Proc. 3rd International Conference on Biodeterioration of Cultural Property, Bangkok, July 4-7, 1995, pp. 307-318.

346. Sarwar N. & Mukerji K.G., 1995. Mycorrhizal technology in forestation. In Mycorrhizae: Biofertilizers for the Future (eds. A. Adholeya and S. Singh) TERI, New Delhi. pp. 342-344.

347. Sharma, M. & Mukerji, K.G. 1995. Interaction between vesicular-arbuscular mycorrhizal fungi and free living nitrogen fixers and its effect on growth of jute: In : Mycorrhizae: Biofertilizers for the Future (eds. A. Adholeya and S. Singh) TERI, New Delhi. pp. 342-344.

348. Srivastava, D. & Mukerji, K.G. 1995. Field response of mycorrhizal and non-mycorrhizal *Medicago sativa* var. local in the F_1 generation. Mycorrhiza 5(3) : 219-221.

349. Allameh, A.; Saxena, M.; Bansal, B.; Mukerji, K.G. & Raj, H.G. 1996. Aflatoxins: Metabolites and biological effects. In : Concepts in Applied Microbiology and Biotechnology (eds. K.G. Mukerji, V.P. Singh and S. Dwivedi) Aditya Books Pvt. Ltd., New Delhi pp. 229-246.

350. Annapurna, K.; Tilak, KVBR and Mukerji K.G. 1996 Arbuscular mycorrhizal Symbosis recognition and specificity. In : Concepts in Mycorrhizal Research (ed. K.G. Mukerji) Kluwer Academic Publ. Netherland pp. 77-90.

351. Bansal, M. & Mukerji, K.G. 1996. Root exudates in rhizosphere biology. In: Concepts in Applied Microbiology and Biotechnology (eds. K.G. Mukerji, V.P. Singh and S Dwivedi) Aditya Books Pvt. Ltd., New Delhi. pp. 97-119.

352. Gupta, R.; Mukerji, K.G.; Kuhad, R.C. & Saxena, R.K. 1996. Plant surface mycoflora–its role in decomposition and soil fertility. In: Concepts in Applied Microbiology and Biotechnology (eds. K.G. Muerji, V.P. Singh and S. Dwivedi) Aditya Books Pvt. Ltd., New Delhi pp. 120-137.

353. Gupta, R. & Muerji, K.G. 1996. Weeds of Delhi II. Certain members of gramineae. J. Phytol. Res: 9(1) : 43-48.

354. Gupta, R. & Muerji, K.G. 1996. Vesicular-arbiscular mycorrhizae–plants of compositae: occurence, identity and significance. Comp. Newsletter 29 : 40-46.

355. Gupta, R. & Mukerji, K.G. 1996. Studies on the vesicular arbuscular mycorrhizal association in the root of common Delhi grasses. Phytomorphology 46 (2) : 117-121.

356. Kumar, R.N.; Mukerji, K.G. & Upadhyay, R.K. 1996. Role of plant generation in integrated pest management. In: IPM System in Agriculture. Vol. I. Principles and Perspectives (eds. R.K. Updhyay, K.G. Mukerji and R.L. Rajak) Aditya Books Pvt. Ltd. New Delhi pp. 414-429.

357. Kumar, R.N. & Mukerji, K.G. 1996. Integrated disease management future prospective. In: Advances in Botany (ed. K.G. Mukerji, Binny Mathur, B.P. Chamola and P. Chitralekha). APH Publ. Corpn. New Delhi pp. 335-347.

358. Kapoor, R.; Pathak, A.; Mago, P. & Mukerji, K.G. 1996. VAM in two crucifers, Cruciferae New Letter, INRA France 18: 116-117.

359. Kaur, S. & Muerji, K.G. 1996. Studies on Indian Laboulerals III. Three unrecorded diseases genera. Mycoscience 37: 105-108.

360. Kaur, S. & Mukerji, K.G. 1996. Physiology and cultivation of myxomycetes. In: Current Research in Plant Sciences (eds. T.A. Sarma, S.S. Saini, M.L. Trivedi & M. Sharma). Bisen Singh Mahendra Pal Singh, Dehra Dun Vol II : 81-95.

361. Kaushik, A. & Mukerji, K.G. 1996. Mycorrhiza in control of diseases. In : Disease Scenario in Crop Plants (eds. V.P. Agnihotri, Om Prakash and A.K. Mishra). Int. Books of Periodicals Supply Service, New Delhi pp. 217-231.

362. Kumar, R.N.; Mukerji, K.G. & Upadhyay, R.K. 1996. Role of plant quarantine in integrated pest management. In: IPM System in Agriculture. Vol. I. (eds. R.K. Upadhyay, K.G. Mukerji and R.L. Rajak). Aditya Books, Pvt. Ltd., New Delhi, India pp. 413-429.

363. Kumar, S. & Mukerji, K.G. 1996. Microbial degradation of pesticides : Environmental and ecological significance. In : Advances in Botany (eds. K.G. Mukerji, Binny Mathur, B.P. Chamola and P. Chitralekha) APH Publishing Corpn., New Delhi pp. 349-585.

364. Kumar, S.; Muerji, K.G. and Lal, R. 1996. Molecular aspects of pesticide degradation by microorganisms. CRC Critical Reviews in Microbiology 22(1): 1-26.

365. Lakhanpal, T.N.; Shama, R. & Mukerji, K.G. 1996. Status of myxomycetes systematics in India. In: Concepts in Applied Microbiology and Biotechnology (eds. K.G. Mukerji, V.P. Singh and S. Dwivedi). Aditya Books Pvt. Ltd., New Delhi, pp. 1-15.

366. Mukerji, K.G. 1996. Taxonomy of endomycorrhizal fungi. In: Advances in Botany (eds. K.G. Mukerji, B. Mathur, B.P. Chamola and P.. Chitralekha) APH Publ. Corpn. New Delhi pp. 213-221.

367. Mukerji, K.G.; Chamola, B.P. & Sharma, M. 1996. Mycorrhiza in control of plant pathogens. In : Management of Threatening Plant Diseases of National Importance. Malhotra Publishing House, New Delhi pp. 1-18.

368. Mukerji, K.G.; Chamola, B.P.; Kaushik, A.; Sarwar, N. & Dixon, R.K. 1996. Vesicular arbuscular mycorrhiza: Potential biofertilizer for nursery raised multipurpose tree species in tropical soils. An. For. 4 : 12-20.

369. Mukerji, K.G. & Kaur, S. 1996. *Arcyria rufosa* :A new Indian myxomycete. Mycotaxon 59 : 479-481.

370. Mukerji, K.G. & Sharma, M. 1996. Mycorrhiza in forestation. In: Fungi in Biotechnology (ed. A. Prakash) pp. 20-27.

371. Mukhopadhyay, B.; Kumar, R.N. & Mukerji, K.G. 1996. Chemical environment of the leaf surface in relation to clestothecial development of powdery mildew. Acta Botanica Indica 23 : 261-262.

372. Sarwar, N. & K.G. Mukerji 1996. New vistas in plant biotechnology. In: Perspectives in Biological Sciences, Raipur University, Raipur pp. 1-21.

373. Saxena, R.K.; Singh, V.P. & Mukerji, K.G. 1996. Heavy metal immobilisation by fungi: Implication for agriculture. In: Concepts in Applied Microbiology and Biotechnology (eds. K.G. Mukerji, V.P. Singh and S. Dwivedi) Aditya Book Pvt. Ltd., New Delhi pp. 138-144.

374. Singh, V.P. & Mukerji, K.G. 1996. Mycotoxin biotransformations: Biological aspects of detoxification. In: Concepts in Applied Microbiology and Biotechnology (eds. K.G. Mukerji, V.P. Singh and S. Dwivedi) Aditya Books Pvt. Ltd., New Delhi pp. 247-260.

375. Suvercha & Mukerji, K.G. 1996. VAM formation in relation to phosphate acquisition and translocation. In: Concepts in Applied Microbiology and Biotechnology (eds. K.G. Mukerji, V.P. Singh and S. Dwivedi) Aditya Books Pvt. Ltd., New Delhi pp. 145-158.

376. Srivastava, D.; Kapoor, R.; Srivastava, S.K. & Mukerji, K.G. 1996. Vesicular arbuscular mycorrhiza – an overview. In : Concepts in Mycorrhizal Research (ed. K.G. Mukerji). Kluwer Academic Publishers, Netherland pp-1-39.

377. Sharma, O.P.; Upadhaya, R.K.; Jalali, B.L. & Mukerji, K.G. 1996. Decision making tools and their role in successful implementation of integrated pest management. In : IPM System in Agriculture Vol. I. (eds. R.K. Upadhyay, K.G. Mukerji and R.L. Rajak), Aditya Books Pvt. Ltd., New Delhi, India pp. 95-117.

378. Upadhyay, R.K.; Sharma, O.P.; Mukerji, K.G. & Dubey, O.P. 1996. Biological control in integrated pest management needs and opportunities. In : IPM System in Agriculture. Vol. I. (eds. R.K. Upadhyay, K.G. Mukerji and R.L. Rajak), Aditya Books Pvt. Ltd., New Delhi, India pp. 400-412.

379. Upadhyay, R.K.; Mukerji, K.G. & Rajak, R.L. 1996. Integrated Pest Management. In: IPM System in Agriculture. Vol. I. (eds. R.K. Upadhyay, K.G. Mukerji and R.L. Rajak) Aditya Books Pvt. Ltd., New Delhi, India pp. 1-23.

380. Dixon, R.K.; Mukerji, K.G.; Chamola, B.P. & Kaushik, A. 1997. Vesicular arbuscular mycorrhizal symbosis in relation to forestation in arid land. An. For. 5 : 1-9.

381. Gaur, A.; Adhotaya, A. & Mukerji, K.G. 1997.Role of microorganism in crop improvement. Souvenir and Abstracts Nat. Symp. Microbes in Plant Improvement and Environmental Protection, New Delhi pp. 8-14.

382. Kapoor, R. & Mukerji; K.G. 1997. Aromatic oil yield increase in VA mycorrhizal plants. Botany 2000 Asia New Letter 5(4): 5-6

383. Kumar, R.N.; Upadhyay, R.K. & Mukerji, K.G. 1997. Strategies in biological control of plant management. In : IPM System in Agriculture. Vol. II (eds. R.K. Upadhyay, K.G. Mukerji and R.L. Rajak), Aditya Books Pvt. Ltd. New Delhi, India pp. 370-422.

384. Kaur, S.; Pandey, A. and Mukerji, K.G. 1997. Development of hulle cells in *Emericilla nidulans* – A new concept. Phytomorphology 47: 453-457.

385. Mittal, N.; Saxena, G.; Mukerji, K.G. & Upadhyay, R.K. 1997. Biocontrol of nematodes by nematode destroying fungi. In: IPM System in Agriculture. Vol. II (eds. R.K. Upadhyay, K.G. Mukerji and R.L. Rajak) Aditya Books Pvt. Ltd., New Delhi, India pp. 453-502.

386. Mukerji, K.G.; Upadhyay, R.K. & Kaushik, A. 1997. Mycorrhiza and integrated disease management. In: IPM System in Agriculture. Vol. II. (eds. R.K. Upadhyay, K.G. Mukerji and R.L. Rajak) Aditya Books Pvt. Ltd., New Delhi, India pp. 453-502.

387. Nigam, N.; Kumar, R.N.; Mukerji, K.G. & Upadhyay, R.K. 1997. Fungi: A tool for biocontrol. In : IPM System in Agriculture. Vol. II. (eds. R.K. Upadhyay, K.G. Mukerji and R.L. Rajak). Aditya Books Pvt. Ltd., New Delhi, India pp. 503-526.

388. Pathak, A. & Mukerji, K.G. 1997. Studies on Indian Laboulbeneomycetes. VIII. Two new species, Phytomorphology 47 : 333-337.

389. Sumeet & Mukerji, K.G. 1997. Protoplast isolation from *Trichoderma harzianum.* In: Frontiers in Plant Science, (ed. Irfan A. Khan) Hyderabad pp. 567-570.

390. Bansal, M. & Mukerji, K.G. 1998. The role of endomycorrhiza in fine root litter degradation. In : Root Demographics and their Efficiancies in Sustainable Agriculture, Grassland and Forest Ecosystem (ed. J.G. Box Jr.). Kluwer Academic Press, Dordrecht, Netherlands pp. 393-401.

391. Dutt, P.A.; Mukerji, K.G.; Upadhyay, R.K.; Chamola, B.P. & Kumar, R.N. 1998. Biological control of sheath blight and blast of rice. In: Integrated Pest and Disease Management (eds. R.K. Upadhyay, K.G. Mukerji, B.P. Chamola and O.P. Dubey) APH Publishing Corporation, New Delhi, India pp. 83-97.

392. Gaur, A.; Adholaya A. & Mukerji, K.G. 1998. A comparision of AM fungi inoculants using *Capsicum* and pallenthes in marginal soil amended with organic matter. 7 : 307-312.

393. Gupta, R.; Mukerji, K.G. & Upadhyay, R.K. 1998. Mycoherbicides. An overview. In: Integrated Pest and Disease Management (eds. R.K. Upadhyay, K.G. Mukerji, B.P. Chamola and O.P. Dubey), APH Publishing Corporation, New Delhi, India pp. 515-532.

394. Kapoor, R. & Mukerji, K.G. 1998. Microbial interactions in mycorrhizosphere of *Anethum gravellens* L. Phytomorphology 48 : 388-389.

395. Kaur, S.; Mukerji, K.G.; Upadhyay, R.K. & Sharma, O.P. 1998. Pea diseases: Status and management. In: IPM System in Agriculture. Vol. IV. (eds. R.K. Upadhyay, K.G. Mukerji and R.L. Rajak). Aditya Books Pvt. Ltd., New Delhi, India pp. 451-476.

396. Mittal, N.; Kaushik, A.; Suvercha & Mukerji, K.G. 1998, Biofertilizers. In: Problems of Wasteland Development and Role of Microbes, (ed. A.K. Mishra) Amifen Publication, Bhuwneshwar, India pp. 129-160.

397. Mukerji, K.G.; Mandeep & Varma, A. 1998. Mycorrhizosphere microorganism: Screening and Evaluatation. In: Mycorrhiza Manual (ed. A. Varma) pp. 6 : 85-97.

398. Mukerji, K.G. & Mandeep 1998. Mycorrhizal relationships of wetlands and rivers associated plants. In : Ecology of Wetlands and Associated Systems (eds. S.K. Majumdar, G. W. Miller and F.J. Bremmer). The Pennsylvania Academy of Sciences, Pennsylvania, USA pp. 240-257.

399. Mukerji, K.G.; Upadhyay, R.K. & Rawat, R. 1998. Biology of cereal powdery mildews and management option. In: IPM System in Agriculture. Vol. III. (eds. R.K. Upadhyay, K.G. Mukerji and R.L. Rajak). Aditya Books Pvt. Ltd., New Delhi, India pp. 61-74.

400. Naqvi, N.S. & Mukerji, K.G. 1998. Mycorrhization of micropropagated *Leucaena leucocephala* (Lam.) de wit. Symbiosis 24 : 103-114.

401. Rawat, R. & Mukerji, K.G. 1998. Vesicular arbuscular mycorrhizal association in sugarcane. Phytomorphology 48: 309-316.

402. Sarwar, N. & Mukerji, K.G. 1998. Mycorrhiza in forestation in arid, semi-arid and wastelands. In: Ecological Techniques and Approached to Vulnerable Environment (ed. R.B. Singh). Oxford and IBH Publishing Co. Pvt Ltd., New Delhi, India pp. 263-271.

403. Sharma, M.; Mittal, N.; Kumar, R.N. & Mukerji, K.G. 1998. Fungi : Tool for plant disease management. In : Microbes for Health, Wealth and Sustainable Environment (ed. A. Verma), Malhotra Publishing House, New Delhi.

404. Sharma O.P.; Jalali, B.L.; Upadhyay, R.K. & Mukerji, K.G. 1998. *Botrytis* grey mould of chickpea and its management. In : IPM System in Agriculture. Vol. IV. (eds. R.K. Upadhyay, K.G. Mukerji and R.L. Rajak). Aditya Books Pvt. Ltd. New Delhi, India pp. 149-162.

405. Upadhyay, R.K.; Mukerji, K.G.; Chamola, B.P. & Dube, A.P. 1998. Integrated Pest Management: Concept and Prospects. In : Integrated Pest and Disease Management (eds. R.K. Upadhyay, K.G. Mukerji, B.P. Chamola and O.P. Dubey). APH Publishing Corporation, New Delhi, India pp. 1-10.

406. Chamola, B.P.; Giri, B. & Mukerji, K.G. 1999. Vesicular arbuscular mycorrhizal fungi: Biofertilizer for future. In: From Ethnomycology to Fungal Biotechnology : Exploiting Fungi from Natural Resources for Novel Products (eds. J. Singh and K.R. Aneja). Plenum Publishing Corp. UK pp. 225-234.

407. Gupta, R. & Mukerji, K.G. 1999. Mycorrhizal allelopathy in *Trigonella foenum graceum*. In: From Ethnomycology to Fungal Biotechnology (eds. J. Singh and K.R. Aneja). Plenum Publishing Corp. UK pp. 235-246.

408. Gupta, R. & Mukerji, K.G. 1999. How harmful the Mycotoxins are? In : Glimpses in Plant Sciences (eds. K.R. Aneja and M.C. Charaya), pp. 17-33.

409. Gupta, R. & Mukerji, K.G. 1999. Degradation of Agrochemicals in soil by rhizosphere microflora. In: New Trends in Microbial Ecology (eds. B. Rai and M.S. Dhkar) pp. 1-18.

410. Gupta, R. & Mukerji, K.G. 1999. Host parasite specificity of pathogenesis. In : Biotechnological Approaches in Biocontrol of Plant Pathogens (eds. K.G. Mukerji, B.P. Chamola and R.K. Upadhyay). Kluwer Academic/Plenum press, USA pp. 1-29.

411. Joshi, V. & Mukerji, K.G. 1999. Seed borne mycoflora of two under exploited legumes; *Vigna umbellata* and *Psophocarpus tetragonolobus* from north-eastern parts of India. In: From Ethnomycology is Fungal Biotechnology: Exploiting Fungi from Natural Resources for Novel Products (eds. J. Singh and K.R. Aneja), Plenum Publishing Corp. UK pp. 257-268.

412. Kapoor, R. & Mukerji, K.G. 1999. VAM in increasing yield of aromatic plants. In: From Ethnomycology to Fungal Biotechnology: Exploiting Fungi from National Resources for Novel Products (eds. J. Singh and K.R. Aneja). Plenum Publishing Corp. U K pp. 197-204.

413. Kaur, S. & Mukerji, K.G. 1999. Bacteria of biocontrol agents of insects. In: Biotechnological Approaches in Biocontrol or Plant Pathogens (eds. K.G. Mukerji, B.P. Chamola and R.K. Upadhyay). Kluwer Academic/Plenum Publishers, USA, pp. 99-114.

414. Kaur, S. & Mukerji, K.G. 1999. Biological control of bacterial plant diseases. In: Biotechnological Approaches in Biocontrol of Plant Pathogens (eds. K.G. Mukerji, B.P. Chamola and R.K. Upadhyay). Kluwer Academic/Plenum Press, USA pp. 157-176.

415. Mandeep & Mukerji, K.G. 1999. The application of VAM fungi in aforestation. In: From Ethnomycology to Fungal Biotechnology: Exploiting Fungi from Natural Resources for Novel Products (eds. J. Singh and K.R. Aneja). Plenum Publishing Corp. UK pp. 213-224.

416. Mittal, M.; Saxena, G. & Mukerji, K.G. 1999. Biological control of root knot nematode by nematode-destroying fungi. In: From Ethnomycology to Fungal Biotechnology: Exploiting Fungi from Natural for Novel Products (eds. J. Singh and K.R. Aneja). Plenum Publishing Corp. U.K. pp. 143-172.

417. Mukerji, K.G. 1999. Mycorrhiza in control of plant pathogens: Molecular approaches. In: Biotechnological Approaches in Biocontrol of Plant Pathogens (eds. K.G. Mukerji, B.P. Chamola and R. K. Upadhyay). Kluwer Academic/Plenum Press, USA pp. 135-155.

418. Mukerji, K.G. & Bansal, M. 1999. Mycorrhizal root litter as a biofertilizer. In: From Ethnomycology to Fungal Biotechnology: Exploiting Fungi from Natural Resources for Novel Products (eds. J. Singh and K. R. Aneja). Plenum Publishing Corp. UK, pp. 205-212.

419. Mukerji, K.G. & Chamola, B.P. 1999. Lichens: an overview. In: Biology of Lichens (eds. K.G. Mukerji, B.P. Chamola, D.K. Upreti and R.K. Upadhyay). Aravali Books International, New Delhi, India pp. 1-19.

420. Mukerji, K.G.; Upadhyay, R.K.; Saharan, G.S.; Sokhi, S.S. & Khangura, R.K. 1999. Diseases of rape seed-mustard and their integrated management. In: IPM System in Agriculture. Vol. V (eds. R.K. Upadhyay, K.G. Mukerji and R.L. Rajak). Aditya Books Pvt. Ltd, New Delhi, India pp. 91-136.

421. Pandey, A. & Mukerji, K.G. 1999. Storage fungi in edible agricultural commodities. In: From Ethnomycology to Fungal Biotechnology: Exploiting Fungi from Natural Resources for Novel Products (eds. J. Singh and K.R. Aneja): Plenum Publishing Corp. UK pp. 247-256.

422. Patil, M.A.; Mayee, C.D.; Takankhar, V.G; Mukerji, K.G. & Upadhyay, R.K. 1999. Sunflower diseases and their management options. In: IPM System in Agriculture. Vol. V. (eds. R.K. Upadhyay, K.G. Mukerji and R.L. Rajak). Aditya Book Pvt. Ltd., New Delhi, India pp. 281-318.

423. Rawat, R.; Pandey, A.; Saxena, G. & Mukerji, K.G. 1999. Rhizosphere biology of root knot deseased *Abelmoschus esculentus* in relation to its biocontrol. In: From Ethnomycology to Fungal Biotechnology: Exploiting Fungi from Natural Resources for Novel Products (eds. J. Singh and K.R. Aneja). Plenum Publishing Corp., UK pp. 173-184.

424. Sharma, M. & Mukerji, K.G. 1999. VAMycorrhizae in control of fungal pathogens. In: From Ethnomycology to Fungal Biotechnology: Exploiting Fungi from Natural Resources for Novel Product (eds. J. Singh and K.R. Aneja). Plenum Publishing Corp., UK pp. 185-196.

425. Sumeet & Mukerji, K.G. 1999. Effectiveness of lytic enzyme in isolation of protoplasts of *Trichoderma harzianum*. In: From Ethnomycology to Fungal Biotechnology: Exploiting Fungi from Natural Resources for Novel Products (eds. J. Singh and K.R. Aneja). Plenum Publishing Corp., UK pp. 127-136.

426. Sumeet & Mukerji, K.G. 1999. Protoplast fusion in disease control. In: Biotechnological Approaches in Biocontrol of Plant pathogens (eds. K.G. Mukerji, B.P. Chamola and R.K. Upadhyay). Kluewer Academic/Plenum Press, USA pp. 177-198.

427. Tilak, K.V.B.R.; Singh, G. and Mukerji, K.G. 1999. Biocontrol–Plant growth promoting rhizobacteria: mechanism of action. In: Biotechnological Approaches in Biocontrol of Plant Pathogens (eds. K.G. Mukerji, B.P. Chamola and R.K. Upadhyay). Kluwer Academic/Plenum Press, USA pp.115-134.

List of Contributors

1. Agrawal, Neema
International Centre for Genetic Engineering and Biotechnology
Aruna Asaf Ali Marg,
New Delhi - 110 067, INDIA

2. Andersen, Jorgan - Bech
Hussvamp Laboratorie ApS
Bygstubben 7, DK - 2950 Vedback,
DENMARK

3. Aneja, K.R.
Department of Microbiology
Kurukshetra University
Kurukshetra - 136 119
Haryana, INDIA

4. Archana, A.
Department of Microbiology
S.S.N. College, Alipur
University of Delhi
Delhi - 110 036, INDIA

5. Bakshi, Inderjeet Singh
Department of Botany
G.T.B. Khalsa College
University of Delhi
Delhi - 110 019, INDIA

6. Bali, Ashima
Department of Microbiology
University of Delhi South Campus
Benito Juarez Road
New Delhi - 110 021, INDIA

7. Bardhan, Saswati
Department of Plant Pathology
Bidhan Chandra Krishi Viswavidyalaya
Kalyani - 741 235
West Bengal, INDIA

8. Bhadraiah, B.
Mycology and Plant Pathology Laboratory
Department of Botany
Osmania University
Hyderabad - 500 077, INDIA

9. Bhatnagar, R.K.
International Centre for Genetic Engineering and Biotechnology
Aruna Asaf Ali Marg
New Delhi - 110 067, INDIA

10. Bose, Protiti
Laboratory of Bioenergetics
Institute of Microbiology and Biotechnology
Barkatullah University
Bhopal - 462 026
M.P., India

11. Chamola, B.P.
Applied Mycology Laboratory
Department of Botany
University of Delhi
Delhi - 110 017, INDIA

12. Choudhury, Ranginee
Department of Genetics
University of Delhi South Campus
Benito Juarez Road
New Delhi - 110 021, INDIA

13. Dasgupta, B.
Department of Plant Pathology
Bidhan Chandra Krishi Viswavidyalaya
Kalyani - 741 235
West Bengal, INDIA

14. Davidson, W. Sheba
Department of Microbiology
University of Delhi South Campus
Benito Juarez Road
New Delhi - 110 021, INDIA

15. Elborne, Steen Andrew
Hussvamp Laboratoriet ApS
Bygstubben 7, DK-2950 Vedback,
DENMARK

16. Giri, Bhoopander
Applied Mycology Laboratory
Department of Botany
University of Delhi
Delhi - 110 017, INDIA

17. Gupta, Rajni
Department of Botany
University of Delhi
Delhi - 110 007, INDIA

18. Gupta, Rani
Department of Microbiology
University of Delhi South Campus
Benito Juarez Road
New Delhi - 110 021, INDIA

19. Hoagland, Robert E.
USDA - ARS
Southern Weed Science Research Unit
P.O. Box 350
Stoneville, MS 38776,
USA

20. Jain, A.K.
Department of Plant Pathology
Jawaharlal Nehru Agricultural University
College of Agriculture
Rewa, M.P., INDIA

21. Jharia, H.K.
Department of Plant Pathology
Jawaharlal Nehru Agricultural University
Jabalpur - 482 004, INDIA

22. Johri, B.N.
Department of Microbiology
G.B. Pant University of Agriculture and Technology
Pantnagar - 263 145, U.P., INDIA

23. Joshi, Nalini
Department of Genetics
University of Delhi South Campus
Benito Juarez Road
New Delhi - 110 021, INDIA

24. Joshi, Vandana
National Bureau of Plant Genetic Resources
Pusa Campus, New Delhi - 110 012, INDIA

25. Kapoor, Rupam
Applied Mycology Laboratory
Department of Botany
University of Delhi
Delhi - 110 007, INDIA

26. Katiyar, R.
Department of Botany
University of Delhi
Delhi - 110 007, INDIA

27. Katiyar, S.K.
Department of Botany
University of Delhi
Delhi - 110 007, INDIA

28. Kaur, H.
Centre for Biotechnology
Faculty of Science
Hamdard University
New Delhi - 110 062, INDIA

29. Kaur, Surinder
Applied Mycology Laboratory
Department of Botany
University of Delhi
Delhi - 110 007, INDIA

30. Khan, Mohd. Sarosh
Laboratory of Bioenergetics
Institute of Microbiology and Biotechnology
Burkatullah University
Bhopal - 462 026
M.P., INDIA

31. Khanna, Y.P.
International Centre for Genetic Engineering and Biotechnology
Aruna Asaf Ali Marg
New Delhi - 110 067, INDIA

32. Khare, M.N.
Department of Plant Pathology
Jawaharlal Nehru Agricultural University
Jabalpur - 482 004
M.P., INDIA

33. Kumar, Raj
Department of Botany
University College
Kurukshetra University
Kurukshetra - 136 119
Haryana, INDIA

34. Kumar, Sanjeev
Centre for Biotechnology
Faculty of Science
Hamdard University
New Delhi - 110 062, INDIA

35. Lakhanpal, T.N.
Department of Biosciences
Himachal Pradesh University
Summer Hill, Shimla - 171 005
H.P., INDIA

36. Low, Gordon A.
University of Abertay Dundee
Bell Street
Dundee, DD1 JHG
Scotland, UK

37. Manoharachary, C.
Department of Botany
Osmania University
Hyderabad - 500 007
A.P., INDIA

38. Mehrotra, N.K.
Department of Botany
Lucknow University
Lucknow - 226 007
U.P., INDIA

39. Mittal, Neelima
Department of Botany
University of Delhi
Delhi -110 017, INDIA

40. Narula, A.
Centre for Biotechnology
Faculty of Science
Hamdard University
New Delhi - 110 062, INDIA

41. Oughton, Douglas
Oscar Faber
Marlborough House
Upper Marlborough Road
St. Albans, Herts, AL1 3UT
London, UK

42. Palfreyman, John W.
University of Abertay Dundee
Bell Street, Dundee, DD1 JHG
Scotland, UK

43. Pandey, Anjula
National Bureau of Plant Genetic Resources
Pusa Campus, New Delhi - 110 012
INDIA

44. Pathak, Anupama
Department of Botany
A.N.D. College, Govindpuri
University of Delhi
New Delhi - 110 019, INDIA

45. Ramarao, P.
Mycology and Plant Pathology Laboratory
Department of Botany
Osmania University
Hyderabad - 500 007, INDIA

46. Satyanarayana, T.
Department of Microbiology
University of Delhi South Campus
Benito Juarez Road
New Delhi - 110 021, INDIA

47. Saxena, Geeta
Department of Botany
S.S.N. College, Alipur
University of Delhi
Delhi - 110 036, INDIA

48. Saxena, R.K.
Department of Microbiology
University of Delhi South Campus
Benito Juarez Road
New Delhi - 110 021, INDIA

49. Selvapandiyan, A.
International Centre for Genetic Engineering and Biotechnology
Aruna Asaf Ali Marg
New Delhi - 110 067, INDIA

50. Sen, Chitreshwar
Department of Plant Pathology
Bidhan Chandra Krishi Viswavidyalaya
Kalyani - 741 235
West Bengal, INDIA

51. Shad, O.S.
Department of Biosciences
Himachal Pardesh University
Summer Hill, Simla - 171 005
H.P., India

52. Sharma, Neeta
Department of Botany
Lucknow University
Lucknow - 226 007
U.P., INDIA

53. Singh, C.S.
Division of Microbiology
Indian Agricultural Research Institute
New Delhi - 110 012, INDIA

54. Singh, Jagjit
Environmental Building Solutions Ltd.
Bedfordshire, London, LU6 3JH,
UK

55. Singh, Kiran
Laboratory of Bioenergetics
Institute of Microbiology and Biotechnology
Barkatullah University
Bhopal - 462 026, M.P., INDIA

56. Srivastava, P.S.
Centre for Biotechnology
Faculty of Science
Hamdard University
New Delhi - 110 062, INDIA

57. Srivastava, Sheela
Department of Genetics
University of Delhi South Campus
Benito Juarez Road,
New Delhi - 110 021, INDIA

58. Sudha
School of Life Sciences
Jawaharlal Nehru University
New Delhi - 110 067, INDIA

59. Sumeet
Applied Mycology Laboratory
Department of Botany
University of Delhi
Delhi - 110 007, INDIA

60. Varma, A.
School of Life Sciences
Jawaharlal Nehru University
New Delhi - 110 067, INDIA

61. Westwood, Christopher
University of Abertay Dundee
Bell Street, Dundee, DD1 JHG
Scotland, UK

62. White, Nia A.
University of Abertay Dundee
Bell Street, Dundee, DD1 JHG
Scotland, UK

Contents

Preface v

Prof. K.G. Mukerji and his Contributions vii

Publications of K.G. Mukerji xi

List of Contributors xli

1. *Synnmukerjiomyces thermophile* a new Thermophilic Hyphomycete from India *(K.R. Aneja and Raj Kumar)* 1

2. Biology of Delta-Endotoxins Produced By *Bacillus thuringiensis* *(Neema Agarwal, Y.P. Khanna, A. Selvapandiyan and R.K. Bhatnagar)* 7

3. Active Efflux - A Mechanism for Metal Ion Resistance in Bacteria (*Sheela Srivastava, R. Choudhury and N. Joshi*) 25

4. Single - Cell Proteins (*Inderjeet Singh Bakshi*) 43

5. Fungal Allelopathy - An Overview *(Rajni Gupta)* 53

6. Involvement of Cellular Mechanisms in the Development of Drug Resistance in Bacteria (*Kiran Singh, Protiti Bose and Mohd. Sarosh Khan*) 67

7. Microbial Enzymes for Detergent Application *(R.K. Saxena, W. Sheba Davidson and Rani Gupta)* 77

8. Microbial Phytases in Food Biotechnology (*Ashima Bali and T. Satyanarayana*) 93

9. Molecular Probes in Detection and Monitoring of Fluorescent Pseudomonads (*Bhavdish N. Johri*) 111

10. Bacterial Thermostable Cellulase-Free Xylanases in Environment Friendly Paper Pulp Bleaching Technology *(A. Archana and T. Satyanarayana)* 131

11. Aflatoxins: Knowing and Combating a Major Health Hazard (*Chitreshwar Sen, Saswati Bardhan and B. Dasgupta*) 141

12. Protoplast Fusion for Improvement of Fungal Strains (*Sumeet*) 187

13. Health Problems in Buildings - A Holistic Approach (*Jagjit Singh and Douglas Oughton*) 203

14. Plant Pathogens and Microbial Products as Agents for Biological Weed Control (*Robert E. Hoagland*) 213

15. Microbial Herbicides : Factors in Development 257
(*Surinder Kaur*)

16. Microwave Treatment and other New Methods for Eradicating the Dry Rot Fungus (*Serpula lacrymans*) Wulf. : Fr. (Schroet.) 279
(*Jorgen Bech-Andersen and Steen Andrew Elborne*)

17. *Serpula lacrymans*, the Dry Rot Fungus: A Much Misunderstood Organism 287
(*John W. Palfreyman, Gordan A. Low, Christopher Westwood and Nia A. White*)

18. Microbial Technology in the Biocontrol of Post Harvest Diseases 309
(*N.K. Mehrotra and Neeta Sharma*)

19. Microbes in Biocontrol of Heavy Metal Pollution 329
(*S. K. Katiyar and R. Katiyar*)

20. Root Exudation and its Implication on Rhizosphere Mycoflora 351
(*Rupam Kapoor*)

21. Microbial Ecology of Plant Surfaces - Living and Conserved Materials 363
(*Anjula Pandey*)

22. Management of Soil Borne Pathogens through Plants and their Products 381
(*M.N. Khare and H.K. Jharia*)

23. Interactions Between Soil Mycoflora and VA Mycorrhizal Fungi in Relation to Growth of *Vigna umbellata* 391
(*Vandana Joshi*)

24. Arbuscular Mycorrhizal Technology for the Future 409
(*C. Manoharachary*)

25. Vesicular Arbuscular Mycorrhizal Fungi Under Salinity and Drought Conditions 421
(*Bhoopander Giri and B.P. Chamola*)

26. Mineral Nutrition in VA Mycorrhizal Plants 431
(*P. Ramarao, C. Manoharachary and B. Bhadraiah*)

27. Mycorrhiza Under Stress Conditions 445
(*C.S. Singh*)

28. Mycorrhiza Aided Biological Hardening of *in vitro* Raised Plantlets 469
(*Sudha, H. Kaur, A. Narula, Sanjeev Kumar, P.S. Srivastava and A. Varma*)

29. Nematophagous Fungi with Special Reference to Interaction with Nematodes 485
(*Geeta Saxena and Neelima Mittal*)

30. Head Smut of *Paspalum scrobiculatum* L. : Transmission and Management 509
(*A.K. Jain and M.N. Khare*)

31. A Re-appraisal of the Genus *Morchella* in India 517
(*T.N. Lakhanpal and O.S. Shad*)

32. The Laboulbeniomycetes : Developmental Patterns 525
(*Anupama Pathak*)

1 *Synnmukerjiomyces thermophile* a new Thermophilic Hyphomycete from India

K.R. Aneja and Raj Kumar

CONTENTS

ABSTRACT 2

1. INTRODUCTION 2
2. *SYNNMUKERJIOMYCES THERMOPHILE* 2
3. REFERENCES 5

Advances in Microbial Biotechnology
J.P. Tewari, T.N. Lakhanpal, Jagjit Singh, Rajni Gupta & B.P. Chamola (eds.)
APH Publishing Corporation, New Delhi - 110 002, India.

ABSTRACT

A new fungus *Synnmukerjiomyces thermophile* isolated from north-Indian soils is described and illustrated. The fungus produces synnematous conidiomata with annellidic conidiogenesis. Conidia show a characteristic thick, smooth and hyaline outer wall layer produced continuously with the inner layer of the conidiogenous cell.

1. INTRODUCTION

During a survey of thermophilous fungi in north-Indian soils, an interesting synnematous hyphomycete was isolated. Live culture of the fungus was sent to the International Mycological Institute for confirmation of its identification. Study of the fungus and consultation of the monographs (2, 3, 6) led to the conclusion that the fungus is invalidly placed in *Stilbella* Lindau. The fungus is characterised by synnematous conidiomata composed of parallel hyphae bearing cylindrical, percurrently proliferating conidiogenous cells with annellidic conidiogenesis. The conidia are hyaline, ellipsoidal, guttulate, aseptate, thick-walled, produced holoblastically in succession from the conidiogenous cells. As we have failed to find a genus of synnematous fungi to which this fungus could be assigned, we propose it as new.

The morphological characteristics and nature of the synnemata are reported in this paper. Observations were carried out by camera-lucida drawings and light microscopy. A thorough description is provided. Attention is given to the conidiogenesis pattern and cultural characteristics.

2. *SYNNMUKERJIOMYCES THERMOPHILE* (LINDAU) ANEJA & KUMAR gen. et. sp. nov. (Figs. 1-9)

Etym. The genus is named in honour of Prof. K.G. Mukerji, eminent mycologist, Delhi University, Delhi, India in recognition of his contributions to mycology. The specific epithet denote its ability to grow at high temperature.

Coloniae in YpSs agaro discretus ad 45°C. Mycelio plerumque submersis, fusco, hyphae aggregatus, mycelio superficialis septatis, levis 2.0-4.0 μm crassa. Synnematio determinato, conspicuo, distincto, superficialibus, erectus, fuscus vel hepaticus usque ad 450-600 μm longa 7-60 μm crassa ad basi, et 18-120 μm crassa ad convexo apicem. Cellulo conidiogeno hyalinis terminalis, cylindracis vel proliferatus, 15-37 x 5-7 μm. Conidio hyalinis, guttulatae, solitaris, ellipsoides, aseptati parieti crassi, 7-25 x 4-6 μm.

Colonies on Yeast phosphate Soluble starch agar (YpSs) (1) at 45°C are discrete. Mycelium mostly submerged, composed of brownish grey, aggregated hyphae, superficial mycelium septate, smooth, 2.0-4.0 μm wide. Synnemata parallel determinate, conspicuous, distinct, superficial, erect, greyish-brown to pale brown, up to 450-600 μm long, 7-60 μm wide at the base, 18-120 μm wide at the convex apex. Constituent hyphae usually unbranched, smooth, septate, pale brown 2.0-3.0 μm wide, compacted together at the base then splaying out to form

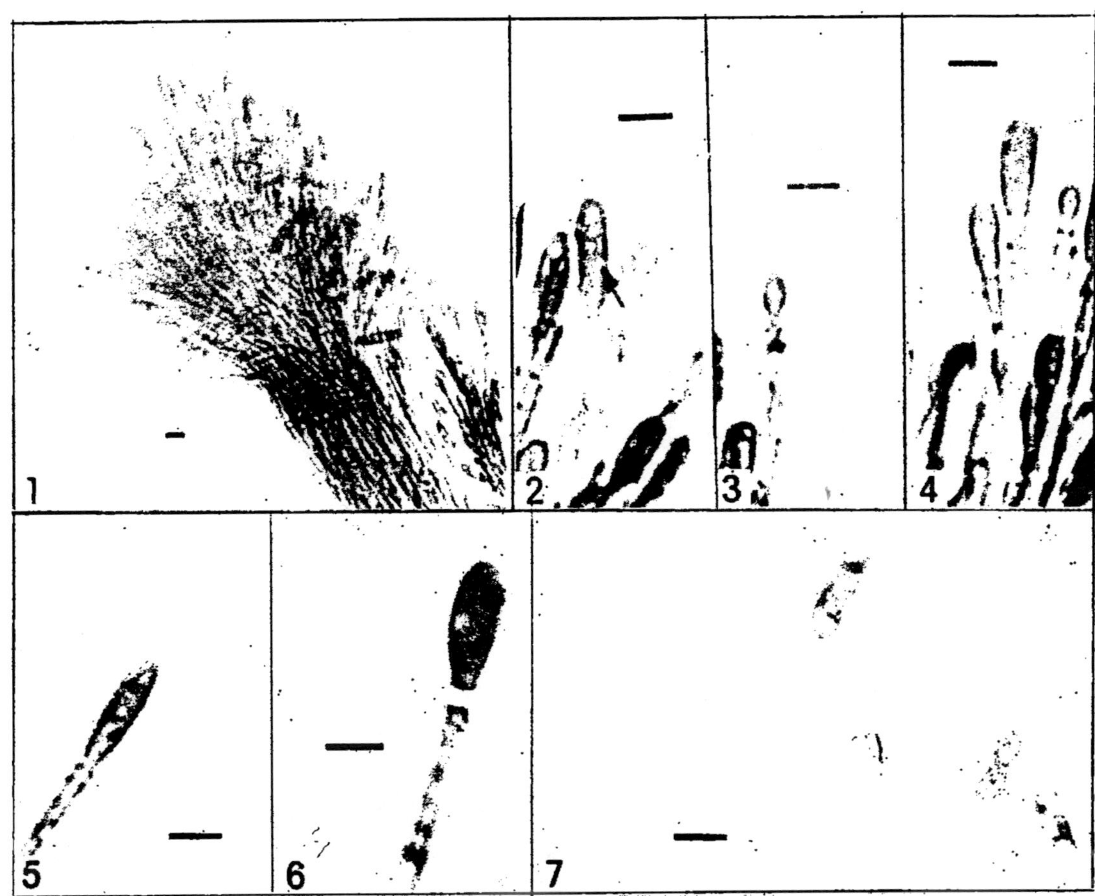

Fig. 1-7. Scale bars 10 µm. *Synnmukerjiomyces thermophile*

Fig. 1. Synnema with conidial head

Fig. 2. Conidiogenous cell producing first conidium holoblastically (arrow)

Fig. 3. Percurrent proliferation in conidiogenous cell with two annellations

Fig. 4. Conidia in different stages of development

Fig. 5. Conidium still attached to the annellate conidiogenous cell

Fig. 6. Schizolytie secession of a conidium.

Fig. 7. Detached conidia

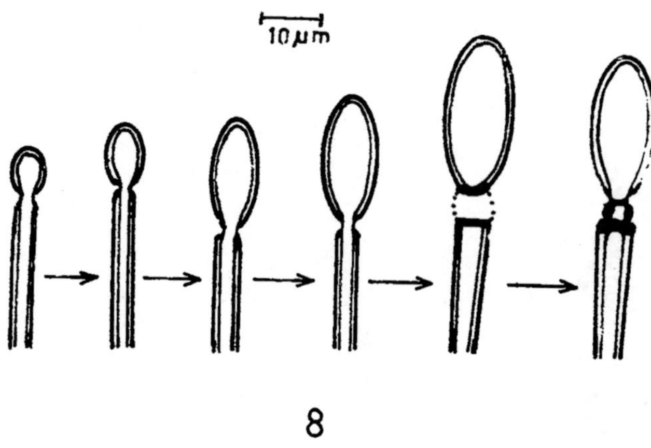

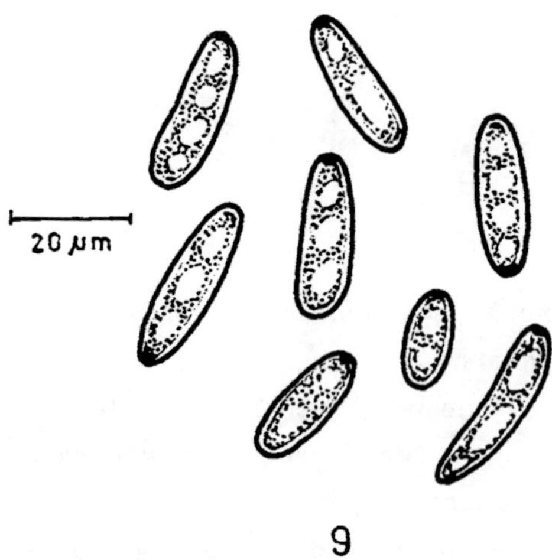

Fig. 8-9 : *Synnmukerjiomyces thermophile.*

Fig 8. Succession of holoblastic conidial development with schizolytic secession, and percurrent growth of conidiophore.

Fig. 9. Thick walled, guttulate conidia.

a wide convex conidial head Conidiogenous cells hyaline, terminal, cylindrical, percurrent proliferation, 15-37 x 5-7 μm. Conidia hyaline, solitary, ellipsoidal, guttulate, aseptate, thick walled, 7-25 x 4-6 μm. Cultures derived from a single conidium reached on average 53 mm diameter, after 120 hrs. on YpSs at 45°C developing a brownish-grey colour. The maximum temperature for growth was at 56°C and minimum 24°C above and below which no growth occurred, thus fitting well into the working definition of a thermophilic fungus given by Cooney and Emerson (1), namely, "one that has a maximum temperature for growth at or above 50°C and a minimum temperature for growth at or above 20°C".

The synnemata of *S. thermophile* are composed of parallel hyphae which flays out in the upper part to form a coremioid apex. In the terminology of Seifert (6), the synnemata are determinate and medium, having a height to breadth ratio of about 10:1. The conidiogenesis appears typically annellidic, the first conidium is produced holoblastically, the outer layer of each successive conidium is produced continuously with the inner layer of the conidiogenous cell. Conidia show a thick hyaline outer layer, which could be interpreted as a succession of holoblastic conidial ontogeny, delimitation by transverse septum, schizolytic secession (4, 5). A thick, smooth, hyaline outer wall layer, participates in the percurrent proliferation of conidiogenous cell in this fungus.

ACKNOWLEDGEMENTS

The authors wish to thank Dr. B.C. Sutton of the International Mycological Institute, Egham, UK for helpful comments on the fungus and University Grants Commission, New Delhi for financial assistance.

3. REFERENCES

1. Cooney, D.G. and Emerson, R. 1964. *Thermophilic Fungi: An Account of their Biology, Activities and Classification*. W.H. Freeman and Company, San Francisco and London.

2. Ellis, M.B. 1971. *Dematiaceous Hyphomycetes*. Commonwealth Mycological Institute, Kew, Surrey, UK.

3. Ellis, M.B. 1976. *More Dematiaceous Hyphomycetes*. Commonwealth Mycological Institute, Kew, Surrey, UK.

4. Minter, D.W., Kirk, P.M. and Sutton B.C. 1982. Holoblastic phialides. *Trans. Brit. Myco Soc.* **79** : 75-93.

5. Minter, D.W., Kirk, P.M. and sutton, B.C. 1983. Thallic phialides. *Trans.Brit. Myco. Soc.* **80**: 39-66.

6. Sefert, K.A. (1985. *A monograph of Stilbella and some allied Hyphomycetes*. Studies in Mycology 27, Centraalbureau voor Schimmelcultures, Baarn, Netherlands.

2 Biology of Delta-Endotoxins Produced by *Bacillus thuringiensis*

Neema Agarwal, Y.P. Khanna, A. Selvapandiyan and R.K. Bhatnagar

CONTENTS

ABSTRACT 8

1. INTRODUCTION 8
2. CLASSIFICATION & INSECTICIDAL TOXNIS 9
3. TOXIN STRUCTURES 9
4. INSECT MIDGUT : TARGET ORGAN FOR *BT* DELTA-ENDOTOXIN 12
5. MODE OF ACTIONS OF *BT* TOXIN 13
6. DEVELOPMENT OF RESISTANCE TO INSECTICIDAL TOXIN 18
7. REFERENCES 19

Advances in Microbial Biotechnology
J.P. Tewari, T.N. Lakhanpal, Jagjit Singh, Rajni Gupta & B.P. Chamola (eds.)
APH Publishing Corporation, New Delhi - 110 002, India.

ABSTRACT

Crops are hosts to several insect pests. These insects inflict heavy damage to crops leading to reduced productivity. Generally these pests are controlled by spraying chemical insecticides. Environmental degradation and harmful consequences of indiscriminate usage of chemical insecticides are being increasingly documented. These concerns have led to the search of environmental friendly and safer pesticides. Amongst these, biopesticides produced by Gram-positive bacterium *Bacillus thruingiensis* and *Bacillus sphaericus* have been the focus of intense research. Polypeptides procuded by this group of bacteria are safe, biodegradable, highly specific and potent. The present review describes some of the basic characteristics of insecticidal toxins produced by these bacteria and also the underlined reason for their high degree of lethality towards susceptible insects.

1. INTRODUCTION

The ecological basis for biological pest control is the increase in pressure on insect populations by insect pathogens. Crop plants are host to several different insects. These insects inflict heavy losses to crop productivity. Conventionally these pests are controlled by employing chemical pesticides. There are several disadvantages associated with their usage. These include effect on non-target organisms, persistence in the environment and potential health hazards to human consumer.

Valuable alternative to chemical pesticides for the control of insect pests in agriculture and forestry are biological pesticides. Biopesticides exhibit high specificity towards the target insects are environment friendly and biodegradable. Bacculovirus and bacteria based pesticides constitute major biopesticides employed in agricultural products. Of these bacteria are the most potential and successful group of organisms for the effective control of insect pests. For the last 35 years members of the genus *Bacillus, B. thuringiensis, B. sphaericus* and *B. popillae* have been used and studied most extensively as bioinsecticide. Of several species of *Bacillus, B. thuringiensis* (*Bt*) is the best known and most widely used insecticide.

Bacillus thuringiensis (*Bt*) was first isolated in 1911 by Berlinger from diseased larvae of the Mediterranean flour moth. It is a Gram-positive bacterium inhabiting a wide range of ecological relevant niches, e.g., soil, plant surfaces, dead insects and stored grain. It produces parasporal crystalline inclusion bodies during the stationary growth phase. These inclusions consist of proteins (protoxins) some of which exhibit specific insecticidal activity hence referred to as insecticidal crystal proteins ((ICPs). They are also called Cry (crystal) proteins or delta-endotoxins. These accumulate in cytoplasm and may account for upto 30% of the dry weight of the sporulated culture. ICPs produced by *Bt* are broadly classified as alpha-, beta- and gamma-exotoxins and delta-endotoxins. Among these polypeptides, the delta-endotoxins are most extensively studied toxin.

These are highly toxic to a wide variety of agriculturally important insect pests. The activity spectrum of *Bt* extends to all major classes of insects such as lepidoptera (caterpillars of butterflies and moths), diptera (larval mosquitoes and blackflies) and coleoptera (larval and adult beetles).

ICP coding genes of *cry* genes are generally located on large (>30MDa), conjugative or mobilizable plasmids usually present in low copy numbers (2-6) (56). Most sub-species of *Bt* contain multiple toxin producing genes. For example, the native plasmids of *Bt* subspecies *aizawai* and *kurstaki* HD1 have five separate ICP genes located at multiple sites of same plasmid or different plasmids (42).

2. CLASSIFICATION OF INSECTICIDAL TOXINS

Earlier, Cry proteins were classified based on their spectrum of insecticidal activity (27). On this basis, two main categories of toxins are recognized: the majority of *Bt* toxins are named 'Cry' and kill only insect cells *in vivo* and *in vitro*, while 'Cyt' toxins are active only against diptera *in vivo* but have a broad spectrum cytolytic activity *in vitro*. The Cry group consists of a large family of related proteins: CryI (L), CryII (D, L), CryIII (C), CryIV (D,L), CryV(L,C) and CryVI(N) (L–Lepidoptera; D–Diptera; C–Coleoptera and N–Nematode) (13).

Subsequently several toxins from different isolates were described and analysed in detail. Their activity spectrum and protein sequences were established and a nomenclature was proposed based exclusively on amino acid identity of toxins (8). To date, over 50 gene sequences have been determined and classified into 15 families.

3. TOXIN STRUCTURE

Delta-endotoxins have modular structure and the domains are structurally independent.

3.1 Primary Structure

Primary structure of *Bt* delta-endotoxin varies with the gene that encodes the protein. CryI and Cry VI toxins are 125-138kDa proteins whereas CryII, CryIII, CryIV and CryV toxins are 67-80kDa proteins. The Cyt toxins which are significantly different both in their structure and biological activities from the Cry toxins are 25-28kDa protein.

All the Cry proteins are thought to have a common evolutionary origin because of the observed high amino acid sequence homology (27). Amino acid sequence comparison of three lepidopteran active CryI toxins, CryIA (a), CryIA(b) and CryIA(c) revealed that the first 282 N-terminal amino acids. In addition, dendrogram constructed between different toxins based on amino acid sequence groups them into families. These are divergent but nevertheless within a group ancestory and origin could be traced. It has been suggested that toxins with newer specificities arose as a result of mutations, gene duplication and gene fusion.

3.2 Tertiary Structure

The delta-endotoxins are packaged into intracellular crystals of 0.6-1 micron which are visible under light microscope(1). The morphological shape of parasporal crystals varies in different strains. The parasporal inclusions produced by *Bt* var. *kurstaki* and var. *berliner* are tetragonal pyramids, having a molecular weight of 130kDa and are toxic to the insects of the order lepidoptera. *Bt* var. *israeliensis* produces irregular parasporal crystals which are toxic to dipteran, *Bt* var. *tenebrionis* and var. *san diego* produce flat rectangular crystals which are active against coleopterans.

The X-ray crystal structure of coleoptera-active delta-endotoxin (CryIIIA) from *Bt* var. *tenebrionis* and lepidopteran specific CryIA(a) have been determined (23, 33) (Fig. 1). The 3-D analysis has allowed tentative prediction of the interaction of insecticidal protein with the receptor located at the midgut of susceptible larvae. The atomic structure revealed that the toxin comprises three distinct domains which are organised from the N-to C-termini, as a seven helix bundle, a three-sheet domain and a beta-sandwich.

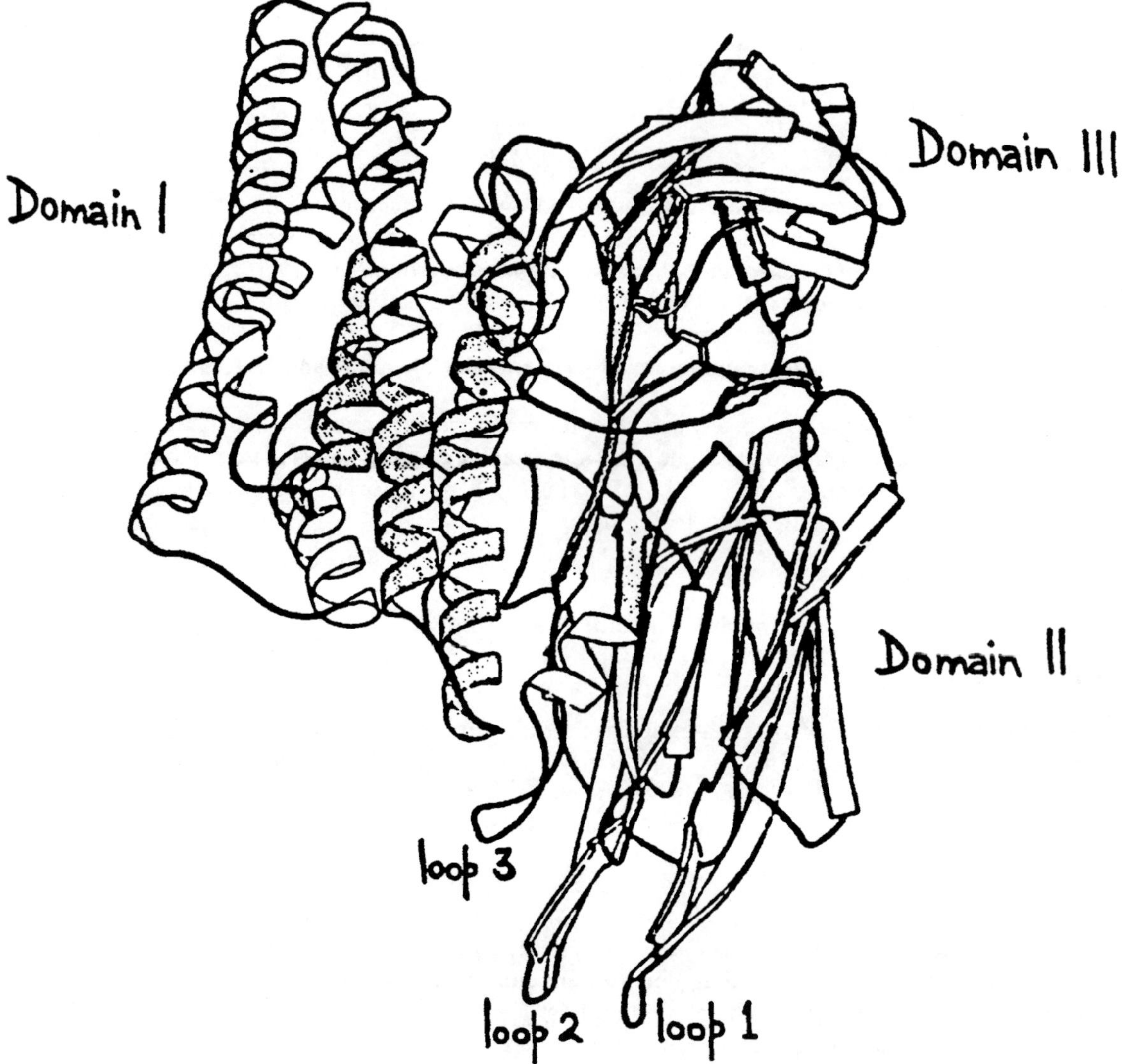

Fig.1. Schematic ribbon diagram of Cry IIIA toxin

3.2.1 Domain I

From the N-terminus there is seven helix bundle in which a central hydrophobic helix designated as alpha-5 is completely surrounded by six outer helices (alpha-1,2,3,4,6, 7). The side of the outer helices facing the central alpha-5 bears hydrophobic residues down their entire length, thus they are amphipathic. Helices are long especially alpha-3 to alpha 7 which contain respectively 8, 7, 6, 9 & 7 complete helical turns, hence would be long enough to span the 30A^0 thick hydrophobic region of the membrane bilayer (33) (Fig. 2).

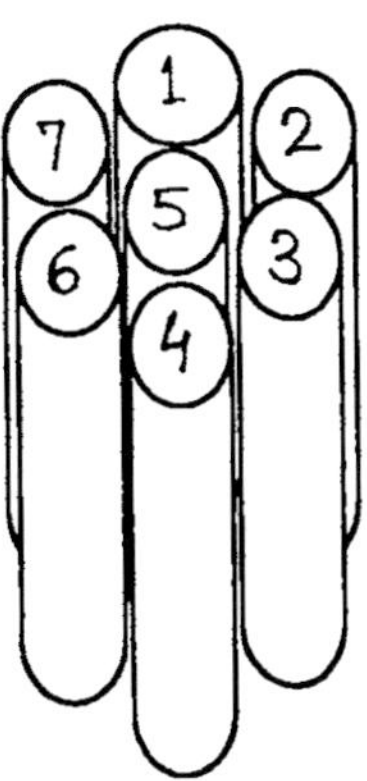

Fig.2. Diagramatic representation of alpha helices of domainI of CryIIIA toxin. Six outer helices surround the central alpha-5 helix which is speculated to form a pore in the membrane

Domain I has been implicated in the channel formation in the membrane in the current delta-endotoxin mode of action model. It has been proposed that after the toxin binds to the receptor, domain I undergoes a conformational change, allowing the hydrophobic surfaces of the helices to face the exterior of the bundle, leading to insertion into the membrane and the formation of the ion channels (31).

Point mutations in the region encoding the central alpha-5 helix of CryIA(c) toxin resulted in the selective loss of activity against different lepidopteran insects without affecting binding to larval midgut vesicles (58). Different characteristics of domain I are necessary to achieve successful integration of the toxin in distinct target membranes. Lepidopteran and coleopteran insects have very different midgut pH conditions (alkali versus acidic respectively) (10, 32) and there may also be differences in the protein and phospholipid compositions of the membranes, suggesting that distinct features of domain I are required for forming channels in different targets.

3.2.2 Domain II

It consists of three antiparallel beta-sheets laid side-by-side and packed around a hydrophobic core. This domain is the least conserved one in the toxin molecule (23). Domain II is considered to be the receptor binding domain thus responsible for determining the specificity of the toxin. Reciprocol hybrid genes between closely related toxins CryIA(a) and CryIA(c) resulted in chimeric toxins with altered specificity (19, 44). Geometrically structure predicts two protruding loops believed to interact with the receptor at the midgut of susceptible larvae. Site directed mutagenesis in these loops affected the binding of the toxin to the brush border membrane vesicles of the susceptible insects (34, 40, 46). The loops 1 and 2 correspond to the regions that displayed the largest structural differences (hypervariable regions) between CryIA(a) and CryIIIA toxins (23) (Fig. 1).

Analysis of specificity binding domain is being pursued by several different strategies including phage display (11). Precise localization of the specificity determining region on the *Bt* toxin would help in constructing broad spectrum toxins by changing the specificity of the pre existing toxins. Such toxins will be immensely useful in integrated pest management program.

3.2.3 Domain III

Domain III is a beta sandwich of two antiparallel beta sheets. The functions of domain III is not critically established yet. However, recent evidences suggest that domain III is responsible in establishing specificity of the toxin. Both beta sheets are important determinants for the proper folding of the toxin (23). It has been proposed that it stabilizes the toxin by protecting against proteolysis and thus may act as chaperonin (33). Several reports provide evidence for the involvement of domain III in receptor binding (2). When the arginine face of one of the five highly conserved regions corresponding to the third domain of the Cry toxin was mutated, the results indicated that these residues in this region are involved in protein structure while others affect the toxin function as an ion channel. It has been suggested that the conserved block 4 of the delta-endotoxin might be involved in a regulatory aspect of ion channel activity, possibly as a voltage sensor or in gating mechanism along with the role in structural stability and integrity of delta-endotoxins (4).

3.3 Identifications of Biologically Active Regions of Cry Toxins

Several studies, mainly the homolog scanning mutagenesis have been performed on the genes encoding Cry proteins to find out the sub-peptide of active polypetide which will retain biological activity. Domain I (54, 55) and helix alpha-5 peptides (17), expressed independently, retain their ability to form cation channels in planar lipid bilayers. Exchanging sequence segments within domain II and domain III between toxins of different specificities produce active hybrids showing altered target specificity (9, 18, 44). These investigations have not yet led to identification of an active peptide, suggesting that conformation of the active polypeptide is critical in establishing biological activity of protein.

4. INSECT MIDGUT: TARGET ORGAN FOR *Bt* DELTA-ENDOTOXIN

The larval lepidopteran midgut is a simple tubular epithelium (Fig. 3). The tissue is composed of two major cell types: a columnar cell with a microvillate apical border, which is believed to play a role in nutrient absorption; and a goblet cell, containing a large vacuolar cavity, permanently linked to the apical surface by a value (7, 10).

In the lepidopteran larvae, K^+ represents the major ionic component of the intestinal contents, intestinal tissue and hemolymph. The ionic homeostasis of the gut and of the hemolymph is accomplished mainly by the activity of electrogenic K^+ pump (20).

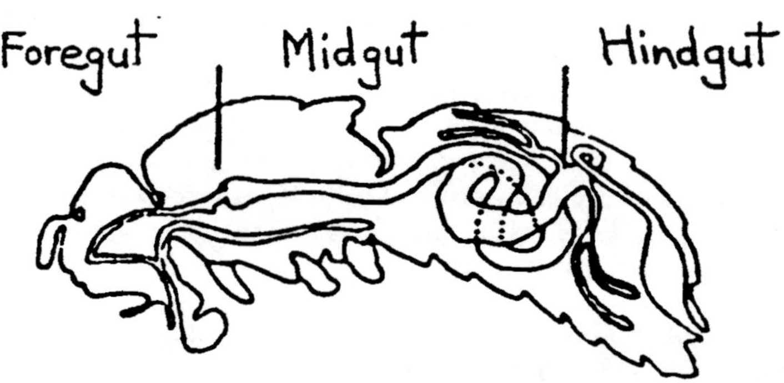

Fig. 3. Generalised diagram of an insect alimentary canal

The K^+ pump is known to be located at the luminal side in the apical membrane of the goblet cell, pumping K^+ from the cytoplasm into the vacuolar cavity and thence to the gut lumen via the valve. The potent electrogenic K^+ transport generates a transepithelial potential difference of over 150mV (lumenal positive to cell). The pump comprises a plasma membrane proton pump (of the V-ATPase family) which pumps protons from cytoplasm to cavity and a H^+/K^+ exchanger, which produces a net K^+ flux in the same direction as the active H^+ flux (57).

A co-transport mechanism exists between K^+ and several neutral basic amino acids (such as valine, isoleucine, serine). The amino acid carriers located on the lumenal side of the midgut columnar cells couple the downhill electrochemical gradient of K^+ to the active intracellular accumulation of amino acid which represent the main fuel for enterocyte metabolism.

5. MODE OF ACTION OF *Bt* TOXIN

5.1 Solubilization

Upon ingestion of the delta-endotoxins by the susceptible insect larvae, the crystalline inclusions are dissolved and acted upon by specific protease present in the midgut lumen to yield the active toxin (Fig. 4). *Bt* parasporal inclusions show differential solubility: CryI toxins are soluble at pH 9.5; CryII toxins require a pH of about 12 for complete solubilization; CryIIIA protoxin solubilizes at pH 7 to 8 CryIVa, CryIVB and CryIVD toxins are soluble at pH 12, CryIA (c) is soluble at pH values less than 3.5 and above 9.5 Under *in vitro* conditions, the crystal proteins can be dissolved using a reducing (dithiothreitol) and denaturing agent (1% SDS). The specificity of a delta-endotoxin seems to be determined by both extrinsic factors such as, insect midgut pH, activator proteases and density & nature of receptors and intrinsic factors such as, toxin structure, functional domain, such as binding domain and cytolytic domain, within the toxin region of the protein.

5.2 Proteolytic Activation

The protoxin released after solubilization are biologically inactive, in the midgut until they are converted into active toxins when acted upon by toxin specific gut proteases.

The larger protoxins, CryI and CryIV (130-140kDa), all require enzymatic processing to become active toxins (27). This action, executed by gut proteases, yields a core 60-70kDa proteinase resistant toxin fragments (6). During activation, C-terminal half (approximately 500 residues) of the larger protoxins is removed and limited cleavage occurs at the N-terminus (5,26). The smaller protoxins, CryII, CryIII, CryIVD and CryV do not undergo protease mediated C-terminal cleavage because these proteins are naturally truncated. They are processed only at the N-terminus where about 50 residues are removed to yield 60-65kDa active toxin (3). The activated toxins show a conserved C-terminus which corresponds to the beta sheet of domain III. Its position precludes further processing from C-terminus. Even a 4 to 8 residue deletion from this site results in loss of activity (3, 24). This is because inner sheet plays a critical part in the structural integrity and stability of the toxins through intraction with the helical bundle.

Upon activation of toxins, residues near the amino and carboxyl termini interact and such interactions have been found to be essential for specificity. It is likely that the tertiary structural interactions between non-contiguous regions of delta-endotoxin are important for the protection of the conserved hydrophobic regions implicated in toxicity (24).

Protoxin activation can be duplicated *in vitro* using a combination of alkaline/reducing buffers and proteases. These *in vitro* activated toxins have been used in conjunction with insect cell lines to investigate the specificity and mechanism of action of a number of toxins.

Ingestion of delta-endotoxin by susceptible larvae

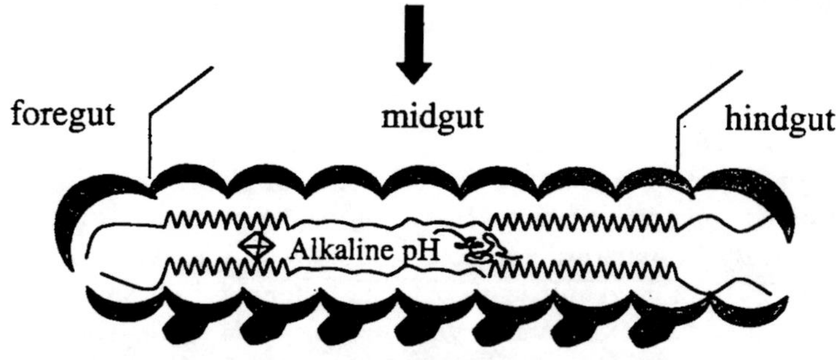

Solubilization in midgut releases protoxin

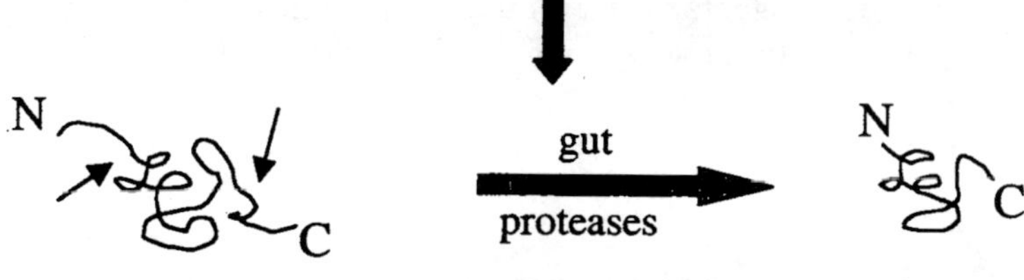

Activation of protoxin

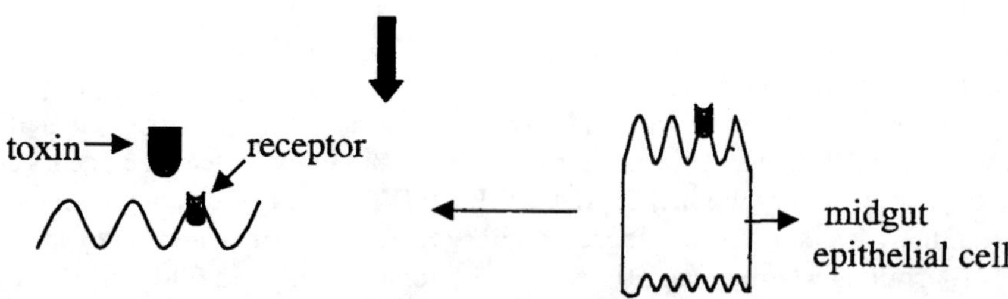

Activated toxin binds specifically to receptor located on brush border membrane of midgut epithelial cell

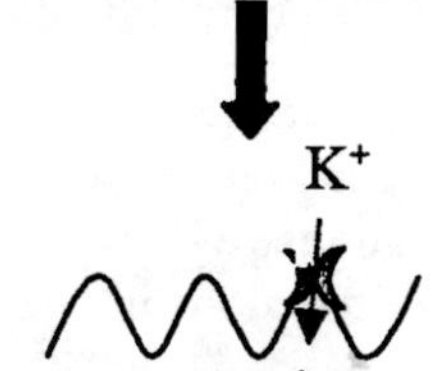

Formation of ion channel

Fig. 4. Flow chart illustrating the mode of action of insecticidal toxin

5.3 Toxin-Membrane Interaction

In order to study the early events in the action of *Bt* toxins, *in vitro* assays have been designed based on the use of brush border membrane vesicles (BBMVs). BBMVs from the insect gut constitute a powerful system in which receptor binding and mode of action of *Bt* toxins can be studied *in vitro*. Two approaches were used to investigate the interactions of *Bt* toxin with BBMV from susceptible and resistant insects.

5.3.1 (a) 125I-labeled toxin-vesicle binding assays

BBMVs are incubated with 125_I-labeled toxin at room temperature in phosphate buffered saline containing 0.1%BSA. After removing the unbound toxin, bound radioactivity is counted in a gamma detector.

5.3.1.1 Binding experiments

(i) *Direct binding*: BBMVs are incubated with a series of concentrations of 125I-labeled Cry toxins at room temperature. After incubation, the samples are centrifuged at 4^{0}C and the pellet is washed with cold Tris-BSA buffer. The bound radioactivity is measured using liquid scintillation counter (38). The binding data is analyzed by using the LIGAND computer program (37). Saturation curves are determined in these experiments; a saturable binding to BBMVs indicates that the toxin has at least one specific binding site. Scathchard plots (bound Vs bound-free ratio) determine the number and type (high/low affinity) of populations of binding sites. Nonspecific binding is determined in presence of 100:1 mixture of unlabeled and radiolabeled toxins. Specific binding is calculated by determining the difference between the total binding value and the nonspecific binding value.

(ii) Competitive binding–

(a) Homlogous: BBMVs are incubated with fixed concentrations of 125I-labeled toxins that previously have been mixed with a series of concentrations of unlabeled homologous competitor toxins in Tris-BSA buffer. The levels of radioactivity bound to BBMVs are determined after centrifugation. The displacement curves (unlabeled competitor vs. %125I-labeled toxin specific binding) show the percentages of specific binding of labeled toxin, where 100% corresponds to the level of binding without competitor and 0% corresponds to the level of binding following the addition of a 300-fold access of unlabeled homologous toxin.

(b) Heterologous: BBMVs are incubated with fixed concentrations of 125I-labeled toxins that previously have been mixed with a series of concentrations of unlabeled heterologous competitor toxins in Tris-BSA buffer. The levels of radioactivity bound to BBMVs are determined after centrifugation. Heterologous competition experiments determine the sharing of the binding site by the two toxins, if any. IC_{50s} (the concentrations of the unlabeled toxin that inhibited 50% specific binding of radiolabeled toxin) are expressed as the molar ratio of unlabeled competitors to labeled toxins. IC_{50} values determine the affinity of the toxin for the binding site (15).

5.3.2 (b) Protein Blot Analysis

This helps in identifying the protein molecule in the BBMVs to which the toxin binds. BBMVs are heated in a cracking buffer containing SDS and DTT. Proteins present in the BBMVs are then separated by running a SDS-PAGE. These proteins are transferred to nitrocellulose membrane and incubated with 3% BSA. The membrane is then incubated with the labeled toxin, the highlighted band corresponds to the protein specific for the toxin (16).

5.4 Receptor For *Bt* Toxin

Several reports suggest specific binding and of the *Bt* toxin to midgut BBMV. Although the purified ICP have the ability to induce ion leakage in synthetic phospholipid vesicles, the toxin concentration needed to exert these effects is at least 1000-times higher than *in vivo* biocide concentration. It has been proposed that the receptor functions as a binding protein to increase the local concentration of the toxin in the vicinity of the membrane (45, 54). A number of putative receptors have also been identified.

While different *Bt* toxins have significantly different insecticidal spectra *in vivo*, in most cases, the activity spectrum correlates with the presence of specific receptor in BBMV preparations from susceptible insects (25, 51, 52). If the binding of receptor to the toxin is changed then the larvae develop resistance to *Bt* insecticidal toxin (50, 53). This supports the role of specific toxin receptors in conferring toxicity of a particular delta-endotoxin to an insect. In some cases the binding of toxin to BBMV could not be corroborated with the *in vivo* toxicity results. These results suggest that although the specific binding of toxin to midgut epithelial receptor is a prerequisite for insecticidal activity, other factors such as the ability of the toxin to integrate into the midgut epithelial cells, appear to contribute to the *in vivo* difference among the structurally related delta-endotoxins in their insecticidal properties.

One of the putative receptors for *Bt* toxin, shares a sequence similarity with the cadherin superfamily of proteins (50). From the cDNA library of tobacco hornworm, *Manduca sexta*, a clone coding for receptor (*Bt*-R1) for an insecticidal *Bt* toxin was identified and expressed. The receptor was found to be a 210kDa membrane glycoprotein that specifically binds the CryIA(b) toxin of *Bt* var. *berliner* and leads to the death of the hornworm. *Bt*-R1 shows a 30-60% similarity and 20-40% identity to members of cadherin superfamily of proteins. Cadherins are membrane glycoproteins and are believed to mediate Ca++ dependent cell aggregation and sorting. *Bt*-R1 may be involved in membrane transport. Its function is similar to that of cadherin-like human intestinal peptide transport protein which channels peptide antibiotics through epithelial cells that line the small intestine.

The receptor for CryIA (c) delta-endotoxin in the midgut membrane of *M.sexta* larvae was purified by a combination of protoxin affinity chromatography and anion exchange chromatography. It was then microsequenced; N-terminal and internal amino acid were used to search the SWISS-PROT protein database and showed a high degree of similarity with the enzyme aminopeptidase N (AP-N) (28). AP-N is a highly glycosylated Zn metalloprotease that catalyses the removal of N-terminal, preferentially neutral amino acid. It is a common component of lepidopteran larval midgut. This ectoenzyme is commonly found in BBMV of alimentary tract in a variety of different organisms. In the mammalian intestinal epithelia, AP-N is involved in peptidyl bond cleavage releasing amino acids that are transported across the BBM. Comparison of the primary structure of *M.sexta* AP-N with that of other organisms (rat, rabbit, human) revealed a striking similarity to the functionally crucial Zn-binding motif. Though the physiological relevance and the factors responsible for the lack of toxin interaction with AP-N (human) could not be identified.

Receptors proteins in the BBMV are glycosylated containing N-acetyl galactosamine (GalNAc) residue (30). For this two evidences have been provided. One, the receptor binds to a GalNAc-specific lectin soybean agglutinin (SBA). Second, toxicity of CryIA(c) for a susceptible insect cell line was reduced by preincubation with GalNAc and that binding of CryIA(c) to BBMVs was blocked by GalNAc but not by N-acetyl glucosamine. This suggests that the toxin binding occurs on the carbohydrate moieties of the receptor protein.

If *Bt* toxins activity is indeed mediated predominantly by the carbohydrate moieties it is likely that the differing responses observed between insects is due to heterogeneity in glycosylation in addition to several other factors. A variety of glycoproteins could function as toxin binding proteins, hence facilitating toxin interaction with the midgut epithelium. The nature of carbohydrate moieties at the N- and O- glycosylation sites and the GPI anchor is not known. Their characterization should provide a better understanding of the selectivity of *Bt* toxin.

In the epithelial cells of mammalian kidney and intestine, AP-N is a type II membrane protein. In the lepidopteran BBMV, the C-terminal of AP N is 30 40 residues more extended than that of mammalian cell types. When the *M.sexta* BBMVs were treated with proteinase K, 100kDa soluble form of AP-N was released into the supernatant with the same termini amino acid sequence as the membrane bound form of the protein. This suggests that the *M.sexta* AP-N is a type I membrane protein.

To find out the moiety anchoring the AP-N, the BBMV from these larvae were incubated with bacterial phosphatidyl-inositol-specific phospholipaseC (PI-PLC), a fraction of AP-N was released from the cell membrane (39) thus indicating that some enzyme population is anchored by a GPI moiety. In order to corroborate that 120kDa AP-N glycoprotein functions as a CryIA(c) receptor *in vivo* partially purified AP was incorporated in liposomes. It was found that these vesicles required very less CryIA(c) toxin to induce a given amount of 86_{Rb}+ leakage compared with vesicles containing to BBM protein (43).

A number of possible roles for a *Bt* toxin receptor have been suggested:

(i) The receptor is itself a transmembrane channel and binding of the toxins opens it. This model suggests that there is a direct correlation of toxin binding with pore formation. But this doesn't hold true for all *Bt* toxins, some toxins display a high affinity binding to brush border membranes from insects that they do not kill, e.g., CryIA(c0 and *Spodoptera frugiperda* membranes (16).

(ii) The toxin and receptor together form a pore.

(iii) The receptor catalyses toxin association with / insertion into the insect membrane, but plays no further role in pore formation. This model has two variants :

(a) The receptor acts as a molecule to which the toxin binds, a raising its effective concentration at the membrane surface. In this case since the only role of the receptor is to concentrate the toxin at the cell surface, a high toxin concentration should be sufficient to form pores in a membrane lacking receptors. Some *Bt* toxins at high concentrations have been reported to form channels in phospholipid bilayers and liposomes in the absence of other proteins (12, 45). However, not all Cry toxins are able to form channels in lipid membranes in the absence of receptors.

(b) The toxin after binding to a receptor undergoes a conformational change which enables the insertion of the toxin into the membrane. This implies that receptor plays an important role in pore formation. The toxin structure offers the transmission of conformational change: helix alpha-7 and strand beta-1 form the interface between the putative binding and pore forming domains, so that a movement of beta-1 caused by receptor binding to domain II could be transmitted to alpha-7 and hence to the rest of the domain (33).

5.6 Ion Regulation

The lumen of lepidopteran midgut is electrically positive to both the cell and hemolymph sides as the epithelial apical membrane actively transport K^+ ions across the membrane. The uptake of amino acids across the midgut epithelium of the larvae depends on the existence of these transepithelial electrical potential differences. The uptake is carried out through a voltage-dependent symport mechanism, located in the apical border of the columnar eithelial cells, which allows the transport of K^+ ions and amino acids into the cytoplasm. Other monovalent cations, such as Na^+ and K^+ for amino acid transport *in vitro*. Six independent amino acid transport systems and a transporter with different affinities for Na^+ and K^+ have been identified in the lepidopteran midgut epithelium (21).

Activated *Bt* toxins apparently insert in the midgut membrane and form small pores. The creation of these pores leads to colloid-osmotic lysis (29); a net inflow of ions through the pore and an accompanying influx of water occurs resulting in swelling of the cell and eventual lysis. This leads to the disruption of the electrical, K^+ and pH gradients (41) which result in the depolarisation of the membrane and loss of the K^+ pump function.

5.7 Ion Pore Formation

Putative sequential mechanism by which the Cry toxins form a pore includes the following steps:

(i) The cell membrane-binding domain of the toxin binds to a high affinity receptor on the apical membrane of the insect midgut columnar cells.

(ii) The toxic domain inserts into the cell membrane after undergoing a change in the conformation.

(iii) Oligomerization of the toxin molecules in the cell membrane results in the formation of the pore leading to osmotic imbalance as a consequence of the influx of water and cations.

6. DEVELOPMENT OF RESISTANCE TO INSECTICIDAL TOXIN

The uncontrolled, nonjudicious use of chemical insecticides have posed a great problem of development of resistance in insects. Evolution of resistance to insecticides has occurred in more than 500 species of arthropods. Thus the use of chemical insecticides is being discouraged and as an alternative, biopesticide *Bt* toxin sprays are tried in fields. The success of biopesticides could be short-lived if pests developed resistance to them due to their uncontrolled use. So far, only one crop pest, the diamond back moth (*Plutella xylostella*) has evolved resistance to *Bt* in open field populations. Rapid degradation of *Bt* toxin due to UV light results in its limited persistence in the environment which is probably the major factor responsible for lowering selection for resistance to *Bt* toxin. The risk of evolution of resistance in pests has increased because of introduction of transgenic crops expressing *Bt* toxins. The insects will be under increased and prolonged selection pressure for resistance.

To delay the resistance to *Bt* toxin in pests, its important to understand the basis of resistance emergence. Resistance a defined as a genetically based decrease in susceptibility of a population to an insecticide. It is an evolutionary phenomenon.

6.1 Mechanisms of Resistance

McGaughey (36) demonstrated that laboratory selection of the Indian mealmoth (*Plodia interpunctella*), a lepidopteran pest of stored grains and grain products that had been exposed to *Btk* (Dipel) in the field resulted in development of 27-fold resistance in two generations and 97-fold resistance after 15 generations of selection. The resistance remained stable in the absence of selective pressure thus indicating that it was inherited.

Tabashnik *et al.* (49) analyzed resistance population of diamond back moth genetically and found that a single autosomal recessive gene conferred simultaneous resistance to four *Bt* toxins (CryIA(a), CryIA(b), CryIA (c) and CryIF / CryIIA) in diamond back moth. In another study laboratory selection with one *Bt* toxin produced cross-resistance (it occurs when selection with one toxin reduces susceptibility to other toxins) to other *Bt* toxins in a strain of tobacco budworm (*Heliothis virescens*) which suggests that one gene or set of genes conferred resistance to more than one toxin.

Reduced binding of *Bt* toxin to BBMV of the midgut epithelium has been identified as a primary mechanism of resistance in *P. interpunctella* (43) and *P. xylostella* (14 , 19). However, reduced toxin binding is not always associated with resistance to *Bt* (22). A second mechanism of resistance may involve mediation of gut proteinases that interact with Bt toxins.

Oppert *et al.* (39) demonstrated that some *Bt* toxin resistance strains of *P. interpunctella* lack a major gut tryspin-like proteinase due to which these strains have lower protoxin-activating abilities. This results in reduced levels of active toxin availability in the gut thereby not exposing insects to high levels of toxins.

Results from laboratory selection experiments show that evolution of resistance to *Bt* is possible in moths, beetles, mosquitoes and other flies. Even when many toxins are used simultaneously some pests are able to adapt themselves to a variety of *Bt* strains (35).

6.2 Delaying Development of Resistance to Insecticidal Toxin

Exposure of insect populations to single toxins can result in the rapid increase in resistance and the likely failure of *Bt* as an insect-control agent in the long run. To delay the onset of resistance, *Bt* formulations must contain toxins that interact either with multiple receptor sites or that have different modes of action. This is termed as multiple toxin approach. Due to the development of broad-spectrum cross-resistance in some insects (22), other tactics such as the use of synergists, temporal rotations of insecticides and providing untreated refuges in the field should be employed (47, 48).

Finally, total substitution of *Bt* for the use of chemicals may not be wise, thus judicious use of chemical insecticides in combination with *Bt* would probably delay the onset of insect resistance.

7. REFERENCES

1. Aronson, A.I., Beckman, W. and Dunn, P. 1986 *Bacillus thuringiensis* and related insect pathogens. *Microbiol Review 50*: 1-24.

2. Aronson, A.I., Wu, D. and Zhang, C. 1995. Mutagenesis of specificity and toxicity regions of a *Bacillus thuringiensis* protoxin gene. *J Bacteriol* **177**: 4059-4065.

3. Carroll, J., Li J., and Ellar, D.J. 1989. Functional mapping of an entomocidal delta-endotoxin. Single amino acid changes produced by site directed mutagenesis influence toxicity and specificity of the protein. *J Mol. Biol* **208**: 183-194.

4. Chen, X.J., Lee, M.K. and Dean, D.H. 1993 Site directed mutations in a highly conserved region of *Bacillus thuringiensis* delta-endotoxin affect inhibition of short circuit current across *Bombyx mori* midguts. *Proc Natl, Acid Sci* **90**: 9041-9045.

5. Choma, C.T., Surewicz, W.K., Carey, P.R., Pozsgay, M. and Raynor, T. 1990. Unusual proteolysis of the protoxin and toxin from *Bacillus thuringiensis* : structural implications. *Eur J. Biochem.* **189** : 523-527.

6. Choma, C.T., and Kaplan, H. 1990. Folding and unfoldihg of the protoxin from *Bacillus thuringiensis*: evidence that the toxic moeity is present in an active conformation. *Biochemistry* **29**: 10971-10977.

7. Cloffi, M. 1979. The morphology and the structure of the larval midgut of a moth (*Manduca sexta*) in relation to active ion transport. *Tissue Cell* **11** : 467-479.

8. Crickmore, N., Zeigler, D.R., Feitelson, J., Schnepf, E., Lambert, B., Lereclus, D., Gawron-Burke, C. and Dean, D.H. 1995. Revision of the nomenclature for *Bacillus thuringiensis* cry genes. In : Program and abstracts of the 28th annual meeting of the society for invertebrate pathology. Society for invertebrate pathology, Bethesda, MD. p.14.

9. De Maagd, R.A., Kwa, M.S.G. and Vander Klei, H. 1996. Domain III substitution in *Bacillus thuringiensis* delta-endotoxin CryIA(b) results in superior toxicity for *Spodoptera exigua* and altered membrane protein recognition. *Appl. Environ. Microbiol.* **62**: 1537-1543.

10. Dog, J.A.T. 1986. Insect midgut function. *Adv. Insect Physio.* **19**: 187-328.

11. Edomi, P., Morzari, R., Bhatnagar, R.K., Ahmed, S., Selvapandiyan and Bradbury A. 1997. Phage display of *Bacillus thuringiensis* CryIA(a) insecticidal toxin. *FEBS Lett* **411**: 27-31.

12. English, L.H., Readdy, T.R. and Bastian, A.L. 1991. Delta-endotoxin induced leakage of ^{86} RB-K and water from phospholipid vesicles is catalysed by reconstituted midgut membrane. *Insect Biochem* **21**: 177-184.

13. Feitelson, J.S., Payne, J., Kim and L. 1992. *Bacillus thuringiensis*: insects and beyond. *Biotechnology* **10**: 271-275.

14. Ferre, J., Real, M.D., Van Rie, J., Jansens, S. and Peferoen, M. 1991. Resistance to the *Bacillus thuringiensis*, bioinsecticide in a field population of *Plutella xylostella* is due to a change in a midgut receptor. *Proc. Natl. Acad. Sci. U.S.A.* **88**: 5119-5123.

15. Fiuza, L.M., Neilson-Leroux, C., Goze, E., Frutos, R. and Charles, J.F.1996. Binding of *Bacillus thuringiensis* Cry I toxins to the midgut BBMVs of *Chilo suppressalis* (Lepidoptera: Pyralidae): Evidence of shared binding sites. *Appl. Environ. Microbiol.* **62**: 1544-1549.

16. Garczynski, S.F. and Crim Adang, M.J. 1991. Identification of putative insect brush border membrane-binding molecules specific to *Bacillus thuringiensis* delta-endotoxin by protein blot analysis. *Appl. Environ. Microbiol.* **57**: 2816-2820.

17. Gazit, E., Bach, D., Sansom, I.D.K.M.S.P., Chejanovsky, N. and Shai, Y. 1994. The alpha-5 segment of *Bacillus thuringiensis* delta-endotoxin: *in vitro* activity, ion channel formation and molecular modelling. *Biochem. J.* **304**: 895-902.

18. Ge, A.Z., Shivarova, N.I. and Dean, D.H. 1989. Location if the *Bombyx mori* specificity domain on a *Bacillus thuringiensis* delta-endotoxn protein. *Proc. Natl. Acad. Sci. U.S.A.* **86**: 4037-4041.

19. Ge, A.Z., Rivers, D., Milne, R. and Dean, D.H. 1991. Functional domains of Bt insecticidal crystal proteins. *Biol. Chem.* **266**: 17954-17958.

20. Giordana, B., Sacchi, V.F. and Hanozet, G.M. 1982. Intestinal amino acid absorption in lepidopteran larvae. *Biochem Biophys Acta* **692**: 81-88.

21. Giordana, B., Sacchi, F.V., Parenti, P. and Hanozet, G.M. 1989. Amino acid transport systems in intestinal brush border membrane from lepidopteran larvae. *Am. J. Phys.* **257**: R494-R500.

22. Gould, F., Anderson, A., Reynolds, A., Bumgarner, L. and Moar, W.J. 1992. Broad spectrum resistance to *Bacillus thuringiensis* toxins in *Heliothis virescens. Proc. Natl. Acad. Sci. U.S.A.* **89**: 7986-7990.

23. Grochulski, P., Masson, L., Borisova, S., Pusztai-Carey, M., Schwartz, J.L., Brousseau, R. and Cygler, M. 1995.*Bacillus thuringiensis* CryIA(a) insecticidal toxin: crystal structure and channel formation. *J. Mol. Biol.* **254**: 447-464.

24. Haider, M.Z. and Ellar D.J. 1989. Functional mapping of an entomocidal delta-endotoxin. Single amino acid changes produced by site directed mutagenesis influence toxicity and specificity of the protein. *J. Mol. Biol.* **208**: 183-194.

25. Hofmann, C., Vanderbruggen, H., Hofte, H., Rie, J., Jansens, S. and Mellaert, H. 1988. Specificity of *Bacillus thuringiensis* delta-endotoxins is correlated with the presence of high-affinity binding sites in the brush border membrane of target insect midguts. *Proc. Natl. Acad. Sci. U.S.A.* **85**: 7844-7848.

26. Hofte, H., Greve, H., Seurinck, J., Jansens, H. and Mahillon, J. 1986. Structural and functional analysis of a cloned delta-endotoxin of *Bacillus thuringiensis* berliner 1715. *Eur. J. Biochem.* **161** : 273-280.

27. Hofte H. and Whiteley H.R. 1989. Insecticidal crystal proteins of *Bacillus thuringiensis*. *Microbiol. Rev.* **53**: 242-255.

28. Knight, P.J.K., Crickmore, N. and Ellar, D.J. 1994. The receptor for *Bacillus thuringiensis* CryIA(c) deta-endotoxin in the brush border membrane of the lepidopteran *Manduca sexta* is aminopeptidase N. *Mol. Microbiol.* **11**: 429-436.

29. Knowles, B.H. and Ellar, D.J. 1987. Colloid osmotic lysis is a general feature of the mechanism of action of *Bacillus thuringiensis* delta-endotoxin. *Biochem. Biophys. Acta.* **924**: 509-518.

30. Knowles, B.H., Knight, P.J.K. and Ellar, D.J. 1994. N-acetyl galactosamine is part of the receptor in insect gut epithelia that recognises an insecticidal protein from *Bacillus thuringiensis. Proc. R. Soc. Lond. B.* **254**: 31-35.

31. Knowles, B.H. 1994. Mechanism of action of *Bacillus thuringiensis* insecticidal delta-endotoxins. *Adv. Insect. Physiol.* **24**: 275-308 (1994).

32. Koller, C.N., Bauer, L.S. and Hollingworth R.M. 1992. Characterization of the pH-mediated solubility of *Bacillus thuringiensis* var. *san diego* native delta-endotoxin crystals. *Biochem. Biophys. Res. Commun* . **184**: 692-699.

33. Li, J., Beckman, W. and Dunn, P. 1986. *Bacillus thuringiensis* and related insect pathogens. *Microbiol. Rev.* **50**: 1-24.

34. Lu, H.L., Rajamohan, F. and Dean, D.H. 1994. Identification of amino acid residues of *Bacillus thuringiensis* delta-endotoxin CryIA(a) asociated with membrane binding and toxicity by *Bombyx mori. J. Bacteriol.* **176**: 5554-5559.

35. McGaughey, W.H. and Johnsonm D.E. 1992. Indian mealmoth resistance to different strains and mixture of *Bacillus thuringiensis. J. Econ. Entomol.* **85**: 1594-1600.

36. McGaughey, W.H. and Whalton, M.E. 1992. Managing insect resistance to *Bacillus thuringiensis* toxins. *Science* **258**: 1451-1455.

37. Munson, P.J. and Rodbard, D. 1980: LIGAND 1992. A versatile computerized approach for charaterization of ligand-binding systems. *Anal. Biochem.* **107**: 220-239.

38. Neilson-Leroux, C. and Charles, J.F. 1992. Binding of *Bacillus sphaericus* binary to a specific receptor on midgut brush border membrane from misquito larvae. *Eur. J. Biochem.* **210**: 585-590.

39. Oppert, B., Karmer, K.J., Beeman, R.W., Johnson, D. and McGaughey, W.H. 1997. Proteinase-mediated insect resistance to *Bacillus thuringensis* toxins. *J. Biol. Chem.* **272**: 23473-23476 (1997).

40. Rajamohan, F., Cotrill, J.A., Gould, F. and Dean, D.H. 1996. Role of domain II, loop 2 residues of *Bacillus thuringiensis* CryIA(b) delta-endotoxin in reversible and irreversible binding to Manduca sexta and *Heliothis virescens. J. Biol. Chem.* **271** : 2396.

41. Sacchi, V.F., Parenti, P., Hanozet, G.M., Giordana, B., Luthy, P., and Wolfersberger, M.G., 1986. *Bacilus thringiensis* toxin inhibits K-gradient-dependent amino acid transport across the brush border membrane of *Pieris brassicae* midgut cells. *FEBS Lett.* **204**: 213-218.

42. Sanchis, V., Lereclus, D., Menou, G.,, Chaufaux, J. and Lecadet, M.M. 1998. Multiplicity of delta-endotoxin genes with different insecticidal specificities in *Bt* aizawai 7.29. *Mol. Microbiol.* **2**: 393.

43. Sangadala, S., Walters, F.W., English, L.H. and Adang, M.J. 1994. A mixture of *Manduca sexta* aminopeptidase and phophatase enhances *Bacillus thuringiensis* insecticidal CryIA (c) toxin binding and 86Rb-K efflux *in vitro. J. Biol. Chem.* **269**: 10088-10092.

44. Schnepf, H.E., Tomezak, K., Ortega, J.P., and Whiteley, H.R. 1990. Specificity determining regions of a lepidopteran specific insecticidal protein produced by *Bacillus thuringiensis. J. Biol. Chem.* **265**: 20923-20930.

45. Slatin, S.L., Abrams, C.K. and English, L. 1990. Delta-endotoxins form cation-selective channels in plasma lipid bilayer. *Biochem. Biophys. Res.* Comm. **169**: 765-772.

46. Smith, G.P. and Ellar, D.J. 1994. Mutagenesis of two surface-=exposed loops of *Bacillus thuringiensis* CryIC delta-endotoxin affects insecticidal specificity. *Biochem. J.* **302**: 611.616.

47. Tabashnik, B.E. 1990. Modelling and evaluation of resistance management tactics. In : *Pesticide Resistance in Arthropods* (eds. B.E. Tabashnik and R.T. Roush). New York: Chapman and Hall, pp. 153-182.

48. Tabashnik, B.E. 1994. Evolution of resistance to *Bacillus thuringiensis. Annu. Rev. Entomol.* **39:** 47-79.

49. Tabashnik, B.E., Liu, Y.B., Finson, N., Masson, L. and Heckel, D.G. 1995. One gene in diamondback moth confers resistance to four *Bacillus thuringiensis* toxins. *Proc. Natl. Acad. Sci. USA.* **94:** 1640-1644.

50. Vadlamudi, R.K., Weber, B., Ji, I., Ji, T.H. and Bulla, L.A. 1995. Cloning and expression of a receptor for an insecticidal toxin of *Bacillus thuringiensis, J. Biol. Chem.* **270:** 5490-5494.

51. Van Rie, J., Jansens, S., Hofte, H. Degheele, D. and Van Mellaert, H. 1989. Specificity of *Bacillus thuringiensis* delta-endotoxins. *Eur. J. Biochem.* **186**: 239-247.

52. Van Rie, J., Jansens, S., Hofte, H., Degheele, D. and Van Mellaert, H. 1990. Receptors on the brush border membranes of the insect midgut as determinants of the specificity of *Bacillus thuringiensis* delta-endotoxins. *Appl. Environ. Microbiol.* **56**: 1378-1385.

53. Van Rie, J., McGaughey, W.H., Johnson, D.E., Barnett, B.D. and Van Mellaert, H. **1990b**. Mechanism of insect resistance to the microbial insecticide *Bacillus thuringiensis. Science* **247**: 72-74.

54. Von-Tersch, M.A., Slatin, S.L., Kulesza, C.A. and English, L.H. 1994. Membrane-permealising activities of *Bacillus thuringiensis* coleopteran-active toxin CryIII B2 and CryIII B2 domain I peptide. *Appl. Environ. Microbiol.* **56**: 1378-1385.

55. Walters, F.S., Slatin, S.L., Kulesza, C.A. and English, L.H. 1993. Ion channel activity of N-terminal fragments from CryIA(c) delta-endotoxin. *Biochem. Biophys. Res. Comm.* **196**: 921-926.

56. Whiteley, H.R., Schnepf and H.E. 1986. The molecular biology of parasporal crystal body formation in *Bacillus thuringiensis. Ann. Rev. Microbiol.* **40**: 549.

57. Wieczorck, H., Putzenlechner, M., Zeiske, W., Klein, U. 1991. A vacuolar type proton pump energizes K^+/H^+ antiport in an animal plasma membrane. *J. Biol. Chem.* **266**: 15340-15347.

58. Wu, D. and Aronson, A.I. 1992. Localised mutagenesis defines regions of the *Bacillus thuringiensis* delta-endotoxin involved in toxicity and specificity. *J. Biol. Chem.* **267**: 2311-2317.

3 Active Efflux - A Mechanism for Metal Ion Resistance in Bacteria

Sheela Srivastava, Ranginee Choudhury and Nalini Joshi

CONTENTS

ABSTRACT 26
1. INTRODUCTION 26
   1.1 Detoxification/Transformation 27
   1.2 Metal Binding 27
   1.3 Metal Efflux 27
   1.4 Bioprecipitation 27
2. EFFLUX AS A HEAVY METAL RESISTANCE MECHANISM 27
   2.1 P-type ATPase Based Heavy Metal Efflux Pumps 28
      2.1.1 Cadmium resistance 28
      2.1.2 Copper resistance 30
      2.1.3 Other metal resistances 32
   2.2 Chemiosmotic Antiport Ion Efflux Systems 32
      2.2.1 CZC system 32
      2.2.2 CNR & NCC systems 33
   2.3 Other Efflux Systems 33
      2.3.1 Arsenic resistance 33
      2.3.2 Copper resistance 34
      2.3.3 Silver resistance 36
      2.3.4 Zinc resistance 36
3. POST-EFFLUX MANAGEMENT OF METAL IONS 36
4. REFERENCES 37

Advances in Microbial Biotechnology
J.P. Tewari, T.N. Lakhanpal, Jagjit Singh, Rajni Gupta & B.P. Chamola (eds.)
APH Publishing Corporation, New Delhi - 110 002, India.

ABSTRACT

Essential inorganic cations and anions have corresponding genes and proteins that govern every movement of theirs in cells. There are highly specific membrane transport systems which bring about their regulated uptake. Once inside the cell, some inorganic nutrients are sequestered (e.g. with metalothionein) or enzymatically incorporated into specific protein (using enzyme ferrochelatase). Excess ions are either stored (e. g. Fe in ferritin) or extruded (by specific membrane transporters).

Microbes have developed cation and anion transport systems for heavy metals not only when they are limited but also when present in excess due to their constant exposure to polluted environments. The efflux or export of excess levels of an ion is a highly efficient strategy developed by microorganisms to resist toxic metals. Toxic metal resistance systems are usually found on plasmids, which facilitates their movement from cell to cell. However, heavy metal resistances encoded by chromosomal genes (e.g. Cd^{2+} resistance in *S. aureus*) as well as an interplay between chromosomal and plasmid genes are also frequently encountered.

The remarkable sequence conservation between Menkes and Wilson diseases (human diseases associated with copper metabolism) and bacterial Cd resistance proteins (Cad A) and copper transporter (Cop A & Cop B) illustrates the potential usefulness and relevance of studies on metal resistance bacteria. It also opens up an opportunity to study the homeostatic control of Cu in humans.

These studies not only give an insight into the evolutionary and academic aspects but also play a very important role in biorecovery of heavy metals. Post-efflux bioprecipitation can be used as a biotechnological technique to harness the heavy metals from the polluted sites.

1. INTRODUCTION

With the advent of industrial revolution, man's excessive use of metals seriously began to affect the environment. The multiplicity of industrial, agricultural and anthropogenic activities have enhanced the mobilisation of heavy metals above the rate manageable by biogeochemical cycles. Thus, resulting in an increased release of heavy metals into the environment. Nearly every index of microbial activity in aquatic and terrestrial ecosystems is affected by metal pollution including primary productivity, nitrogen fixation, biogeochemical cycling of C, N, S, P and other elements (including other metals), decomposition of organic matter, and enzyme synthesis and activity. Thus, the uncontrolled dissemination of heavy metals into the environment has made us aware of metal's inhibitory effect on living organisms, inspite of some of them being essential.

The toxicity of heavy metals is not only due to their presence in abundance, but also due to their non-biodegradable nature. Metal ions cannot be synthesized or degraded in accordance with current requirements of the cell. Presence of high levels of persistent heavy metals has forced the microorganism to adopt strategies to overcome their toxic effects, thus, resulting in the widespread appearance of metal resistance in microorganisms.

The different mechanisms of heavy metal resistance have evolved in microbes varying with the metal and the microbe. These are :

1.1 Detoxification/Transformation

This type of resistance mechanism involves enzyme mediated chemical transformation of metal to a less toxic form. Hg is an outstanding example of this mechanism. In this case, intracellular Hg(II) is reduced to Hg (0) by mercuric reductase, which subsequently volatilises from the cell. Since Hg(II) is highly toxic to certain intracellular targets, extracellular detoxification mechanisms also allow rapid induction of resistance, which would be favoured where there is a quick rise in the levels of the toxic metal ion (53).

1.2 Metal Binding

Cell can be rendered resistant via binding of metals either on the cell wall (biosorption) or intracellularly to specific binding components (sequestration). The selection of metals for biosorption as a resistance mechanism depends on its atomic weight and ionic size (74). Uranium tends to be biosorbed exceptionally well because of its large atomic weight in *Rhizopus arrhizus* (76).

Metal sequestering proteins are known to be present in many microorganisms. Metallothionein-like proteins have also been reported in several cases (24, 22).

1.3. Metal Efflux Systems

It includes exclusion by active export of metal from the cell via transport systems evolved for nutrient cell via transport systems evolved for nutrient cations or anions (61). Cd efflux in *Staphylococcus aureus* plasmid p1258 (64) and Cu efflux in *Enterococcus hirae* (48) have been well worked out. This is an exceptionally efficient mechanism to protect intracellular targets from metal poisoning.

The *czc* determinant from the Gram-negative multiple metal-resistant bacterium, *Alcaligenes eutrophus*, codes for regulated efflux of Co, Zn and Cd (40, 41).

1.4 Bioprecipitation

Bioprecipitation is a phenomenon in which metals precipitate in an insoluble form so that the microbial cell is rendered resistant to that particular metal. Macaskie *et al.* (34) have described enzyme mediated precipitation of metal. Phosphatase, produced by *Citrobacter* sp. facilitated precipitation of Cd and U as their phosphate derivatives. Similarly Cu can be precipitated as phosphate and hydroxide under carbon starvation conditions by *Pseudomonas putida* S4 (55).

Thus, a variety of mechanisms are involved in heavy metal resistances ranging from simple physico-chemical adsorption to more complex enzyme- mediated mechanisms. In this review, we are emphasizing chiefly on efflux and post-efflux mechanisms.

2. EFFLUX AS A HEAVY METAL RESISTANCE MECHANISM

Essential heavy metals are required in trace amounts to fulfil all physiological and metabolic requirements of living organisms. However, at high concentrations they disrupt the homeostatic balance of the cell there by, exerting toxicity. Therefore, the resistance mechanisms and their genetic controls have been responsive to the requirement so as to accumulate cations at trace levels and at the same time to reduce cytoplasmic concentrations, if potentially toxic levels are achieved, by efflux (8,9). In order to maintain metal homeostasis, there is every likelihood that this control inducible. The degree of expression of inducibility mirrors the external metal concentration (53). The regulation plays a crucial role so that a cell neither gets surfeited at high external concentration nor does it get starved at low concentration. The efflux genes may be located on the chromosome or on the plasmid. The interplay between these efflux systems is highly regulated by regulatory genes present on both the chromosome and the plasmid which respond by the metal concentrations present in the environment (8,9). This assures the meeting of normal cellular demands for low levels of metal, and at the same time accords protection from redox damage that free metal ions might cause from overly high intracellular metals levels present in resistant cells (65).

Relatively non-specific, non-energy dependent uptake systems function constitutively to maintain the physiological status of the cell. During metal ion excess, on the other hand, the cell gets induced to synthesise specific metal ion uptake efflux systems (40, 67) which are energy-dependent.

Two major groups of efflux pumps that forms the basis of plasmid resistance systems have been identified. These can either by ATPase (as the Cd^{2+} ATPase of Gram positive and the arsenite ATPase of Gram negative bacteria) or chemiosmotic (as the divalent cation efflux system of *Alcaligenes eutrophus* and the arsenite efflux system of *Bacillus* and *E. coli*) (40, 41, 59).

2.1 P-Type ATPase Based Heavy Metal Efflux Pumps

The integral membrane P-type ATPases are an important class of ion transport proteins that serve to maintain suitable ionic conditions. Heavy metal ATPases contain elements common to all P-type ATPases as well as several unique features. P-type ATPase consist of a single, large catalystic monomer of 70-200 kDa. The transfer of the r-phosphate of ATP to an aspartic acid residue results in the formation of an acylphosphate intermediate (49). Heavy metal P-types ATPases, for example, that of Cu and Cd, exhibit the following novel features(33) :

(i) putative heavy metal-binding sites in the polar amino-terminal region.

(ii) a conserved intramembranous CPC, CPH or CPS motif (CPx motif). Thus, they are called CPx type ATPases.

(iii) a conserved histidine-proline dipeptide (HP locus) 34 to 43 amino acids carboxy-terminal to the CPx motif.

(iv) a unique number and topology of the membrane-spanning domains.

Detailed studies of CPx type ATPases were carried out in cadium, copper, arsenic resistances in microorganisms. In this review, an insight into the efflux mediated by ATPases is dealt with.

2.1.1 Cadmium resistance

Cadmium ATPase has been reported from Staphylococcal plasmid pI258 (62, 63, 64), soil bacilli (23) and clinical isolate of *Listeria* (30).

In *S. aureus*, plasmid pI258 CadC resistance operon has been well worked out. Cloning and DNA sequence analysis identified two genes CadC and CadA in the operon (45). The Cad A gene product is a P-type ATPase, a member of the large family of related cation transport ATPase (64, 68). CadC is the product of the second gene of this operon. It is a DNA binding transcription regulatory repressor (20).

Cad A functions as a Cd^{2+} efflux ATPase (Fig. 1). It is a transmembrane polypeptide having 6 hydrophobic regions which form the cation channel. The initial Cd^{2+} recognition region, consisting of Cys23 and Cys26, is intracellularly located. In the first transmembrane hairpin structure there is an abundance of charged amino acids mostly positive at the inner surface and negative on the outer surface of the hairpins. Then comes a large domain of approximately 190 amino acids that functions as a transduction funnel (68) involved in moving the trapped Cd^{+2} ions from its initial binding sites to the membrane surface. Some authors also consider it to constitute a phosphatase domain (58). The conserved Thr-Gly-Glu-Ser tetrapeptide helps in removing the phosphate from its covalently bound position on aspartate 415 (57, 58). The second membrane hairpin structure acts as a membrane cation transport channel. The initially bound Cd^{2+} now binds to the Cys 371 and Cys 373 residues deriving its energy from the phosphate cleaved by bound ATP. Next is a large (240 amino acid) intracellular domain having ATP binding site (488-Lys-Gly-Ile-Val-492) and 7 amino acids kinase stretch (414-421). ATP is bound between the tetrapeptide listed above and a x-helical loop (618-626) which is predicted to make Mg-mediated salt bridge with r-phosphate of the ATP (58). It is followed by another membrane hairpin structure with an abundance of charged residues near the membrane surfaces and the sequence ends at Lys 727 (66).

It is proposed by Endo and Silver (20) that CadC sits as a dimer on the inverted repeat region forming retarded intermediate complex I. Subsequent addition of a second dimer of CadC (to form a tetramer bound to DNA) might result in retarded form II. Cd^{2+} does not cause complete dissociation of CadC from operator DNA but only an association is sufficiently reduced so that RNA polymerase can proceed or displace CadC. CadC is a member of a new sub-family of metal binding repressor, proteins having homology with the arsenic system repressor ArsR (25, 51, 79, 80, 81) and the cyanobacterial metallothionein repressor, SmtB (62, 75).

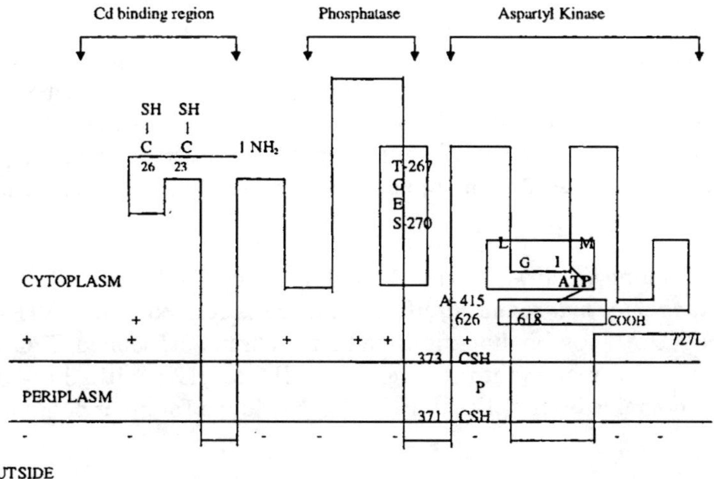

Fig.1. Cadmium resistance ATPase of *S. aureus* where C-SH stands for cys, T=thr, G=ghy, E=glu, S=ser, L=lys, I=ileu, V=val, A=asp, P=pro.

2.1.2 Copper resistance

The gram positive *Enterococcus hirae* possesses two P-type ATPases. CopA and CopB, that are involved in copper homeostasis. It is the first example of a P-type ATPase which has been shown to regulate copper transport. These enzymes are induced by extracellular copper concentrations that are either too low or too high for optimal growth. The two genes, CopA and CopB, determine uptake and efflux P-type ATPases, respectively. They are found in a single operon (48). During copper starvation Cop A uptake ATPase is needed whereas when copper is in excess CopB efflux ATPase is required (46).

Copper (I) and not copper (II) is the substrate of CopB and CopA. Copper which is not sequestered by enzymes occurs in copper (I) form either due to the reducing environment of the cytoplasm or through the action of cell-surface reductases. This situation is similar to eukaryotic cells where Cu^+ is thought to be the intracellular form (72). Cop B is an ATP-driven pump that can expel Cu^+ and Ag^+ from the cytoplasm. Cop B ATPase is a chromosomally-coded function having 745 amino acid residues (71). Cu^+ (Ag^+)-ATPase has one or two cysteines flanking proline residue that is believed to be the part of ion channel through the membrane, and has heavy metal binding motifs at the N-terminus. Cop B is the only ATPase known to possess the sequence CPH at the site of the proposed ion channel (72) whereas Cop A (727 amino acid residues) has the usual CPC motif. The conserved proline in one of the hairpin structure is surrounded by two cysteines in CopA ATPase and by cysteine and histidine in CopB, which is believed to form the ion transduction site (48). CopB possesses amino-terminal methionine-and histidine-rich sequences, which are the metal binding domains (71) CopA and CopB show 35% similarity to each other and Cop B ATPase have 3 putative canonical copper-binding sites, such as H.M. 6M on the cytoplasmic face of the membrane. Phosphatase and aspartyl kinase domain is conserved as in all P-type ATPases (48).

Enterococcus hirae contains a unique regulatory protein pair CopY CopZ of which the former determines a repressor protein and the latter an 'anti-repressor' (46). Under copper limiting conditions both CopY and CopZ have no copper bound and are free in the cytoplasm, allowing the expression of *cop* operon. When the cytoplasmic copper concentration is in the physiological range, CopY binds to Cu and represses transcription by binding to the operator. As the cytoplasmic copper concentration approaches toxic levels, CopZ also binds to Cu^+. This CopZ–Cu^+ complex then acts as an anti-repressor by binding to Cop Y–cu^+ and releasing it from the operator, hence activating transcription of the *cop* operon (46). Thus, there is simultaneous induction of synthesis of both uptake as well as efflux P-type ATPase by copper (Fig.2) (48). Copper transporting ATPase represent a novel mechanism for the control of cytoplasmic copper. The prevalence of similar enzymes in diverse species ranging from *E.hirae* to man suggests that ATPase-driven copper transport is a widely used mechanism for copper transport (48).

P-type ion pump (Cop A) having homology with Cd^{2+} and Cu^{2+}, ATPase have been reported in *Helicobacter pylori* and *H.felis* (2, 21). The predicted Cop A ATPases contained an N-terminal GMXCXXC ion binding motif and a membrane-associated CPC sequence. Cop A contains 4 pairs of transmembrane segments (H1 to HB) with the ATP binding and phosphorylation domains lying between H6 and H7 as found for other putative transition metal pumps (36).

A physiological role of copper P-type ATPase in maintaining copper equilibrium in divergent species and cell types became clear defects in putative copper P-type ATPase in two human disorders of copper metabolism, Menkes (X-linked disorder having defective Cu export in mostly all cells except hepatocytes which results in entrapment of copper in tissues like intestine and copper deficiency in most other tissues), and Wilson disease (defective copper export in liver cells causing liver damage with subsequently systemic spillover of copper) were uncovered (71). Studies have revealed that both of these disorders result from defects in the putative copper-transporting P-type ATPases, which are approximately 40% identical to CopA copper ATPase and 33% to CopB Cu ATPase of *E.hirae* (12, 37, 77).

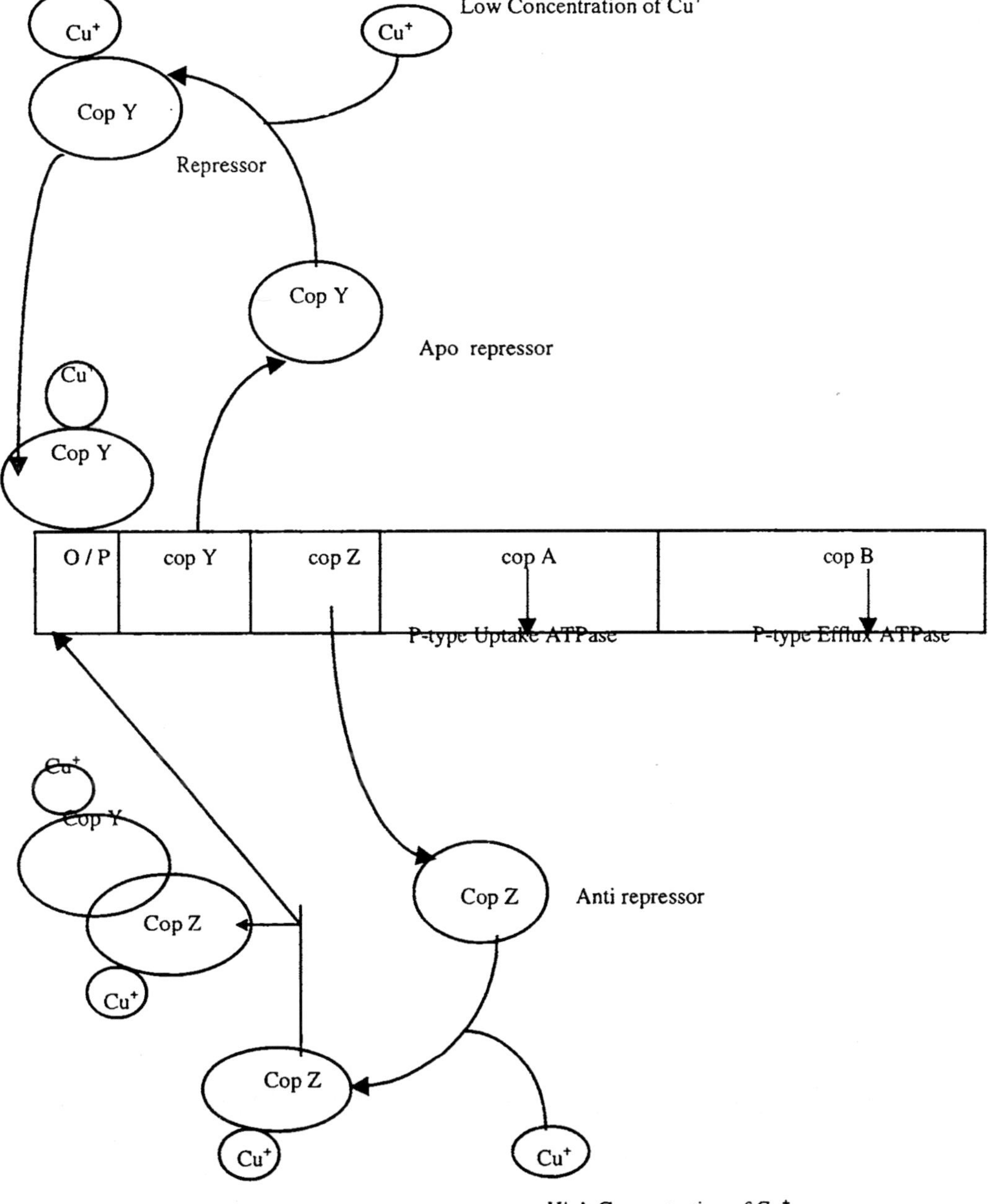

Fig. 2 : Regulation of *E. hirae* Copper resistance determinant.

2.1.3 Other metal resistances

Beard *et al.* (3) have identified *zntA* gene which is predicted to encode a protein of 732 amino acids similar to cation transport P-type ATPases which renders Zn resistance in *E.coli* K-12(3). Lead resistance on plasmids in *Alcaligenes* was also found to be P-type ATPase dependent (59).

2.2 Chemiosmotic Antiporter Ion Efflux Systems

Non-ATPase-dependent ion efflux systems involve chemiosmotic transporters. The best studied metal cation efflux system of this kind has been reported in *Alcaligenes eutrophus* CH34 for cadmium, zinc and cobalt resistance called CZC (39). Besides CZC other multi-protein non-ATPase efflux systems includes CNR (cobalt and nickel) in *A. eutropihus* (32) and NCC (nickel, cadmium and cobalt) in *A. xylooxidans* (56).

2.2.1 CZC system

The Czc system which confers resistance to Cd^{2+}, Zn^{2+} and Co^{2+} in *Alcaligenes eutrophus* CH34, a gram negative soil bacterium, function as a cation/proton antiporter effluxing cations from the cells (39, 40, 41, 42). The *czc* operon of pMOL30 consists of three structural genes *czcC, czcB* and *czcA* whose products form a complex cation efflux pump. Additional genes *CzcR* and *CzcD* are involved in the regulation of operon's expression (41,42) (Fig. 3).

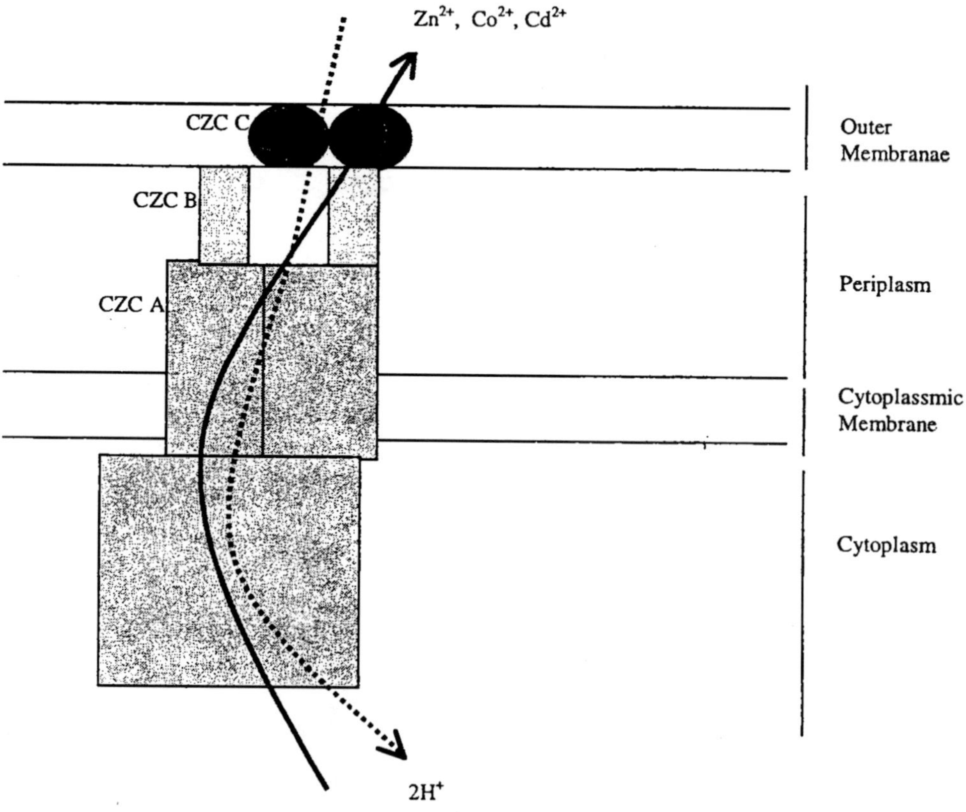

Fig. 3 : Model of CZC efflux system

Sequence homologies and topological similarities with outer membrane factors (OMF) suggest CzcC to to be an outer membrane protein (18) which leads to efficient transport across the outer membrane (24). CzcB shows some homology to membrane fusion proteins (19). It appears to span the periplasmic space and funnel cations across it thereby preventing the release of free cation (24). CzcB also displays some homology with calphotin which is a metal (Ca^{2+}) mobilizing proteins found in sponges (1). CzcB is thought to draw the outer membrane close to the CzcA antiporter, thereby connecting both the membrane in a flexible fashion (50).

CzcA is a chemiosmotic cation/H^+ antiporter. It is predicted to be made up of 12 transmembrane polypeptide with a symmetry between the first and second halves (18).

Deletion mutations in *czcA* result in the loss of resistances to all three ions, whereas deletion of *czcB* results in complete loss of resistances to Cd^{2+} and Zn^{2+} with only partial loss to Co^{2+}. When *czcC* gene was eliminated, cells retained full resistance to Zn^{2+}, partial resistance to Co^{2+} but were sensitive to Cd^{2+} (42). Thus, it can be summarized that CzcA is a primary pump protein inefficiently transporting Co^{2+}. Both CzcB and CzcC function as substrate range modulators. The addition of CzcC to the CzcAB complex extends the substrate range to include Cd^{2+} and leads to efficient transport across the outer membrane (24).

2.2.2 CNR & NCC systems

The *cnr* system is found on a different plasmid in *A.eutrophus* CH34 than the one that possesses the *czc* system. They differ in substrate specificity (15,42). The *cnr* system responds strongly to Ni^{2+} but only marginally to Co^{2+} (64). A mutant class adds zinc resistance to the *cnr* determinant. It also increases Co^{2+} and Ni^{2+} resistance (15).

Another related operon *ncc* from *Alcaligenes xylooxidans* located on plasmid confers resistance to Ni^{2+}, Cd^{2+} and Co^{2+}. It consists of seven genes having high homology with *cnr* gene products (56). The expression of *czc, cnr* and *ncc* operons is regulated but the total number of genes involved is not clear (24). The first regulatory gene identified was *czcD* (41,42). It is thought to be a cation sensor membrane protein. Nies (40, 41) proposed an additional regulatory gene called czcR which might act as a transcriptional activator(41). On further investigation into the same region, two potential reading frames upstream of *czcC* (*czcN* and *czcI*) were identified which may be involved in operon regulation. CzcN may be a transmembrane protein that could play a sensing role in regulatory response to Cd^{2+} and Co^{2+} (24). The *ncc* system contains an open reading frame closely homologous to CzcN. Cnr and Ncc systems possess unrelated (to Czc) regulatory genes called *cnrY*, *cnrX* and *cnrH* (or *nccY*, *nccX* and *nccH*) (22,56).

2.3 Other Efflux Systems Involved in Metal Resistance

Both arsenic and silver resistance mechanisms, which have been worked out, appear to have chemiosmotic as well as ATP-dependent efflux pump. Similarly, models of copper homeostasis and resistance involve the control of cellular uptake and efflux of copper by chromosomal genes as well as plasmid systems.

2.3.1 Arsenic resistance

Entry of arsenate is mediated by phosphate transporters (4). When phosphate is in abundance, Pit transport system (non-specific) is activated where both PO_4^{3-} and ASO_4^{3-} enter the cell. However, under Po_4^{3-} starvation, more specific Pst system is activated where entry of ASO_4^{3-} is reduced (Fig. 4).

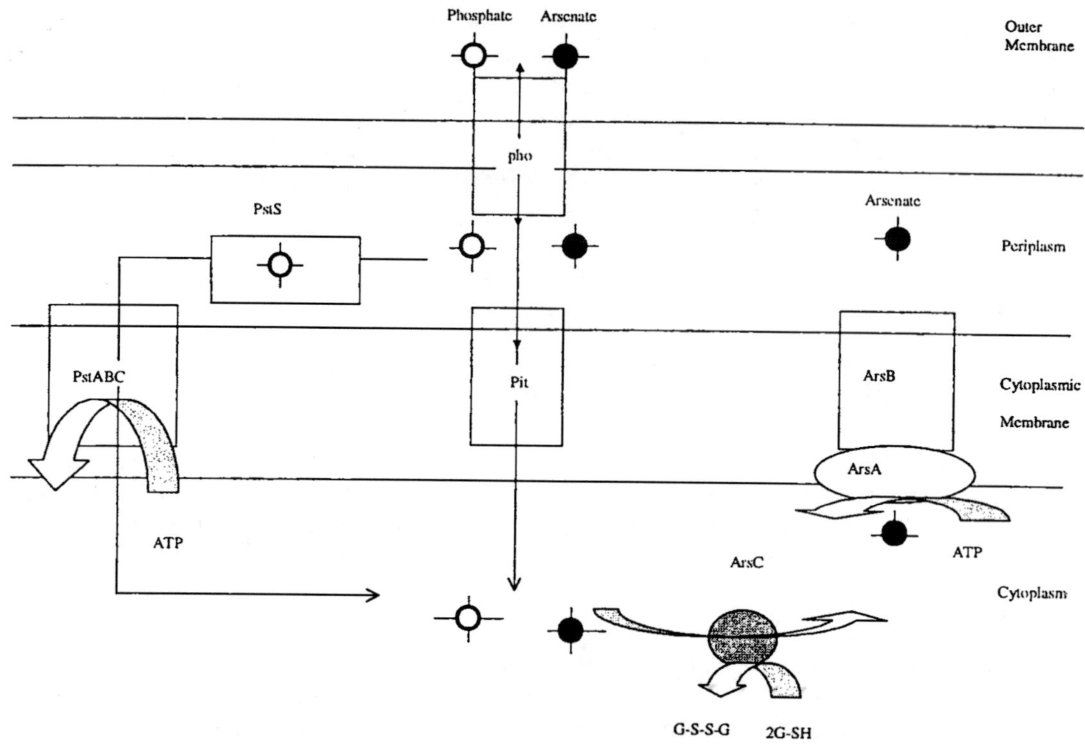

Fig. 4 : The mechanism of arsenate resistance in *E.coli*

Plasmid mediated arsenic resistance to arsenicals has been shown to result from the action of arsenical pump. This system is encoded by a single *ars* operon which is induced by ASO_4^{3-}, ASO_2^- or SbO^+ (70). The first two genes of the operon of plasmid R773, *ars RD* are regulatory genes and the last 3, *ars ABC*, determine proteins involved in the resistance mechanism (25).

The ArsC protein is an arsenate reductase that converts intracellular arsenite, the substrate of the Ars pump (10,26). The Ars A protein (429 amino acid) is an arsenite-stimulated ATPase that is part of a complex with the membrane-bound Ars B protein. The Ars B protein forms the transmembrane channel through which arsenicals are extruded from the cell (10). The Ars B protein can function alone as a chemiosmotic arsenite efflux transporter or together with Ars A as an ATP-driven primary pump (5, 17, 29).

ArsA homologs have recently been found in *Mycobacterium leprae* (27) and even in the human genome (28). Basically, the same *ars* operon conferring resistances to As (III), As (V) and Sb (III) occurs widely in Gram-negative and Gram-positive bacteria. The new *Yersinia* plasmid *ars* operon includes an additional gene *ars H*, which is not found in the other known systems, but which is required for arsenic resistance in this bacterium (38).

2.3.2 Copper resistance

Copper homeostasis in *E.coli* is determined by chromosomal and plasmid genes (Fig.5, 6) (31). The chromosomal counterparts have been designated as the *cut* operon. Two copper uptake systems Cut A and Cut B are present. Cut A is involved in zinc uptake as well as in copper transport (31). Cytoplasmic copper is believed to be reduced by glutathione to Cu (I) which can cause oxidative damage to the cell by producing hydroxyl radicals (7). Two intracellular copper storage/carrier proteins (Cut E and Cut F) are thought to protect the bacterial cell from Cu(I) toxicity by sequestering copper and by delivering copper to the sites of synthesis of copper metallo proteins (7,31). Cut F protein is responsible for intracellular delivery of copper to the export system (7). Structural genes *cutC* and *cutD* encode export proteins by which excess Cu is removed from the cytoplasm and are probably copper efflux ATPases (7). The cutC mutant shows increased sensitivity to both Cu and Zn suggesting its role in export of both the metals, while cutD is an efflux pump only for copper (7). Cut R controls the expression of the uptake and/or storage and/or export systems in order to maintain the intracellular concentration of copper within acceptable limits.

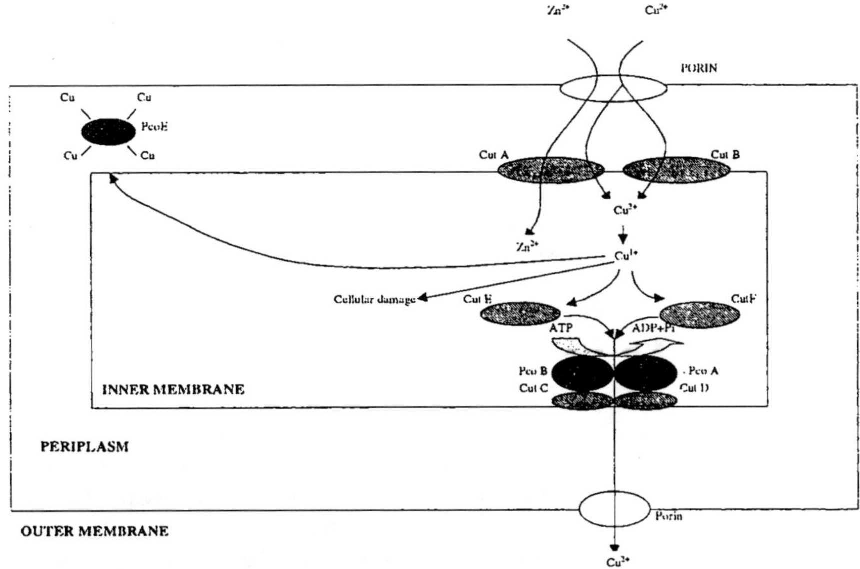

Fig.5. Relationship between chromosoma (*cuf*) and plasmid (*pco*) genes in the uptake and regulation of Copper ions in *E.coli*

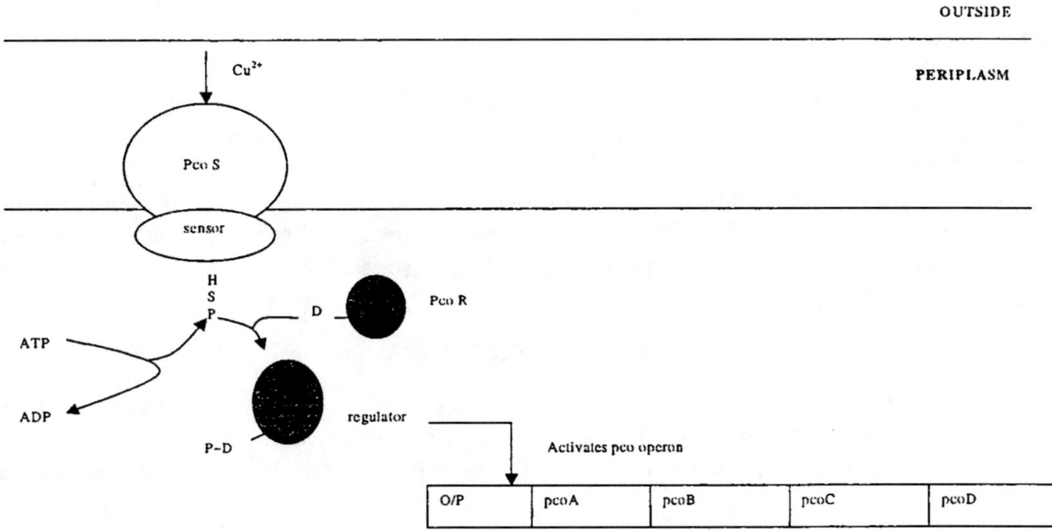

Fig. 6 : Model for the regulation of the pco genes of *E.coli* by pcoR and pcoS gene

The copper resistance determinant, pco on plasmid comprises seven genes, *pco* ABCDRSE (6, 31, 73). PcoA and C have copper binding sites. PcoD has potential transmembrane helices and conserved histidine and methionine groups predicted to be on the outer face of the cytoplasmic membrane (7).

Lee *et al.* (31) proposed that plasmid gene products of *pcoA* and *pcoB* interact with proteins CutC and CutD to form a 4-component efflux system which derives energy by hydrolysing ATP. PcoE protein is a periplasmic copper binding protein and appears to be transcribed in a Cu^{2+} inducible manner (52, 63). PcoR and S are the 2-component regulatory system involved in sensing environmental stimuli. The putative transmembrane protein Pco S senses high level of free Cu^{2+} in the periplasm and auto-phosphorylate via its kinase activity at a his 257 residue. Phosphate is subsequently transferred to asp 52 on Pco R protein present in the cytoplasm. Thereby converting it from an inactive to an active state to induce expression of the Pco operon (6, 16). Interestingly, *Pseudomonas synringae* plasmid mediate copper resistance genes, *CopABCDRS* are homologous to *pco* or *E.coli* through, the mechanism in the former is sequestration of copper (13, 14, 62). Copper resistance in *P. putida* strain S4 is also based on inducible efflux, located on the plasmid. Such cells are better prepared to face high external concentration of Cu^{2+} that provides the basis for resistance. From the inhibitor studies, it appears that efflux pump is a Cu^{2+} specific ATPase as both DCCD and valinomycin strongly inhibit the efflux (54 and unpublished results).

2.3.3 Silver resistance

The genetic basis of plasmid determined Ag^{2+} resistance is now actively under study. Silver resistance is encoded by 8 genes. *SilS* and *SilR* are homologous to *cop S* and *cop R* of *Pseudomonas syringae*. Sil E, a periplasmic silver binding protein, is homologus to Pco E. The remainder of the silver resistant genes are transcribed in the opposite direction unlike in *pco* system. They encode both a three component CBA system (Weakly homologous to *czc CBA*) and a P-type ATPase (marked by the absence of initial metal binding motif) (59).

2.3.4 Zinc Resistance

Not much work has been done on zinc resistance. A p-type ATPase for zinc resistance has been newly reported in *E.coli* (3). In *Pseudomonas* sp. strain UDG26 induction of ZN^{2+} resistance leads to enhanced accumulation of the metal, triggering an internal detoxification mechanism that results in the efflux of excess Zn^{2+}. It has been shown in strain UDG26 that both uptake and efflux are coordinaterly regulated. Thus, induced cells, which accumulate more zinc than uninduced has to involve an efflux mechanism to bring it to the same level as achieved by the latter. Efflux was inhibited by metabolic inhibitors like DNP and DCCD (35). Efflux was also found to be the main mechanism of Zn^{2+} resistance in a multiple metal resistant *Pseudomonas putida* strain S4. However, the effluxed metal is stored in the outer membrane for its sense in future. Further detailed study is in progress (Choudhury and Srivastava unpublished results).

3. POST-EFFLUX MANAGEMENT OF METAL IONS

Many investigators have addressed the question of management of effluxed ions so as to avoid the re-uptake of the same (9, 18). Post-efflux management of metal ions may involve bioprecipitation or binding and storage of metal ions to periplasmic proteins, or even the outer membrane.

Alcaligenes eutrophus CH34 shows an interesting phenomenon of bioprecipitaiton as a post-efflux management of metal ions. The pH of the growth medium increases as a function of increase in the metal ion concentration. It has been proposed that the progressive alkalinisation of growth medium is related to the proton influx during the czc-mediated efflux of cations. The presence of carbon sources like lactate or acetate facilitated such phenomenon. The effluxed metal is finally precipitated in the form of bicarbonates and hydroxides. Bioprecipitation seems to be strongly dependent on the availability of carbonates. The avoidance or the minimisation of recentry of toxic metals is crucial under oligotropic conditions where all czc bacteria are generally found(18).

In *P. putida* strain S4, effluxed copper ions bind to a periplasmic protein suggesting its reuse under copper limited conditions, whereas effluxed Zn is perhaps trapped in the outer membrane of the same strain (54).

ACKNOWLEDGEMENTS

The authors thank Dr. Rohit Mago and Dr. Deepa Saxena for allowing them to publish their unpublished results. The authors also thankfully acknowledge Mr. Rajiv Chawla for his expert assistance in the preparation of the manuscript.

4. REFERENCES

1. Ballinger, D.G.; Xue, N. and Maqrschman, K.D. 1993. A *Drosophila* photoreceptor cell specific protein. Calphotin, binds calcium and contains a leucine zipper. *Proc. Natl. Acad. Sci.* USA **90** : 1536-1540.

2. Bayle, D; Wangler, S; Weitzenegger, T; Steinhilber, W; Volz, J; Przybylski, M; Schafer, K.P; Scahs, G and Melchers, K. 1998. Properties of the P-type ATPases encoded by the *cop*AP operons of *Helicobacter pylori* and *Helicobacter felis*. *J.Bactenor.* **180** : 317-329.

3. Beard, S.J.; Hashim, R.; Menbrillo-Hernandez,J.; Hughes, M.N. and Poole, R.K. 1997, Zinc (11) tolerance in *Escherichia coli* K-12 : evidence that the *znt*A gene (0732) encodes a cation transport ATPase. *Molecular Microbiology*, **25** : 883-891.

4. Bennett, R.L. and Malamy, M.H. 1970. Arsenate-resistant mutants of *Escherichia coli* and phosphate transport. *Biochem. Biophys. Res. Comm.* **40** : 490-503.

5. Broer, S.; Ji, G.; Broer, A. and Silver, S. 1993. Arsenic effluxed governed by the arsenic resistance determinant of *Staphylococcus aureus* plasmid p1258. *J. Bacteriol.* **175** : 3480-3485.

6. Brown, N.L.; Barrett, S.R.; Camakaris, J.; Lee, B.T.O. and Rouch, D.A. 1995. Molecular genetics and transport analysis of the copper resistance determinant (pco) from *Escherichia coli* plasmid pRJ1004. *Mol. Microbiol.* **17** : 1153-1166.

7. Brown, N.L.; Lee, B.T.O. and Silver, S. 1994. Bacterial transport and resistance to copper. In: *Metal Ions in Biological Systems*, (eds. H. Sigel and A. Sigel) Vol. 30. Marcel Dekker Inc., Ny, USA, pp. 405-435.

8. Brown, N.L.; Rouch, D.A. and Lee, B.T.O. 1992. Copper resistance determinants in bacteria. *Plasmid* **27** : 41-51.

9. Brown, N.L., Camakaris, J.; Lee, B.T.O; Williams, T.; Morby, A.P.; Parkhill, J. and Rouch, D.A. 1991. Bacterial resistances to Mercury and Copper. *J. Cell Biochem.* **46** : 106-114.

10. Bruhn, D.F.; Li, J.; Silver, S.; Roberto, F. and Rosen, B.P. 1996. The arsenical resistance operon of Inc N plasmid R46. *FEMS Microbiol. Lett.* **139** : 149-153.

11. Bull, P.C. and Cox, D.W. 1994. Wilson disease and Menkes disease : new handles on heavy metal transport. *Trends Genet.* **10** : 246-252.

12. Bull, P.C.; Thomas, G.R.; Rommens, J.M.; Forbes, J.R. and Cox, D.W. 1993. The wilson diseae gene is a putative copper transporting P-type ATPase similar to Mewnks gene. *Nat. Genet.* **5** : 327-336.

13. Cha, J.S. and Cooksey, D.A. 1993. Copper hypersensitivity and uptake in *Pseudomonas syringae* containing cloned components of the copper resistance operon. *Appl. Environ. Microbiol.* **59** : 1671-1674.

14. Cha, J.S. and Cooksey, D.A. 1991. Copper resistance in *Pseudomonas syringae* mediated by periplasmic and outer membrane proteins. *Proc. Natl. Acad. Sci. USA* **88** : 8915-8919.

15. Collard, J.-M.; Provoost, A.; Taghair, S. and Mergeay, M. 1992. A new type of *Alcaligenes eutrphus* CH34 zinc resistance generated by mutations affecting regulation of the cnr cobalt-nickel resistance system. *J. Bacteriol.* **175** : 779-784.

16. Cooksey, D.A. 1994. Molecular mechanisms of copper resistance and accumulation in bacteria. FEMS *Microbiol. Rev.* **14** : 381-386.

17. Dey, S. and Rosen, B.P. 1995. Dual mode of energy coupling by the oxyanion-translocating Ars B protein. *J. Bacteriol.* **177** : 385-389.

18. Diels, L.; Dong, Q.; van der Llie, D.; B., Wilfnied and M. Mergeay. 1995. The *czc* operon of *Alcaligenes eutrophus* CH34 from resistance mechanism to the removal of heavy metals. *J. Ind. Microbiol.* **14** : 142-153.

19. Dinh, T.; I.T. Paulsen, and M.S. Saier. 1994. A family of extractyoplasmic proteins that allow transport of large molecules across the outer membranes of gram-negative bacteria. *J. Bacteriol.* **176** : 3825-3831.

20. Endo, G. and Silver, S. 1995. CadC, the transcriptional regulatory protein of the cadmium system of *Staphylococcus aureus* plasmid p1258. *J. Bacteriol.* **177** : 4437-441.

21. Ge, Z. and Taylor, D.E. 1996.*Helicobacter pylori* genes hpcop A and hp CopP constitute a cop operon involved in copper export. *FEMS Microbiol. Lett.* **145** : 181-188.

22. Gilotra, U. 1997. Molecular Genetic Characterization of Copper Resistance in *Pseudomonas pickettii* strain US321. Ph.D. Thesis. Delhi University.

23. Ivey, D.M.; Guffanti, A.; Shen, Z.; Kudyan, N. and Krulwich, T.A. 1992. The cadC gene product of alkaliphilic *Bacillus firmus* ORF4 prtially restores Na= resistance to an *Escherichia coli* strain lacking an Na^+/H^+ antiporter (NhaA). *J. Bacteriol.* **174** : 4878-4884.

24. Ji, G. and Silver, S. 1995. Bacterial resistance mechanisms for heavy metals of environmental concern. *J. Industrial Microbiol.* **14** : 61-75.

25. Ji, G. and Silver, S. 1992a. Regulation and expression of the Arsenic resistance operon from *Staphylococcus aureus* plasmid p1258. *Bacteriol.* **174** : 3684-3694.

26. Ji, G. and Silver S. 1992b. Reduction of arsenate to arsenite by ArsC protein of the arsenic resistance operon of *Staphylococcus aureus* plasmid, p1258. *Proc. Natl. Acad. Sci. USA* **89** : 7974-7978.

27. Koonin, E.V. 1993. A superfamily of ATPase with diverse funcitons containing either classical or deviant ATP-binding motif. *J. Mol. Biol.* **229** : 1165-1174.

28. Kurdi-Haider, B.; Aebi, D.; Heath, D.; Enns, R.E.; Naredi, P.; Mony, D.K. and Mowell, S.B. 1996. Isolation of the ATP-binding human homolog of the *arsA* component of the bacterial arsenite transporter. *Genomics* **36** : 486-491.

29. Kuroda, M., Dey, S.; Sanders, O.I. and Rosen, B.P. 1997. Alternate energy coupling of Ars B, the membrane sunbunit of the Arsanion-translocating ATPase. *J. Biol. Chem.* **272** : 326-331.

30. Lebrun, M.; Audurier, A. and Cossart, P. 1994. Plasmid-borne cadmium resistance genes in *Listeria monocytogenes* are present in Tn5422, a novel transposon closely related to Tn 917. *J. Bactriol.* **176** : 3049-3061.

31. Lee, B.T.O.; Brown, N.L.; Rogers, S.; Bergemann, A.; Camakaris, J. and Rouch, D.A. 1990. Bacterial. response to copper in the environment : copper resistance in *Escherichia coli* as a model system. *NATO ASI Series* **G23** : 625-632.

32. Liesegang, H.; Lemke, K.; Siddiqui, R.A. and Schlegel, H.G. 1993. Characterization of inducible nickel and cobalt resistance determinant *cnr* from pMOL28 of *Alcaligenes entrophus* CH34. *J. Bacteriol.* **175** : 767-778.

33. Lutsenko, S. and Kaplan, J.H. 1995. Organisation of P-tye ATPase : significance of structural diversity. *Biochemistry* **34** : 15607-15613.

34. Macaskie, L.E.; Blackmore, J.D. and Empson, R.M. 1988. Phosphatase overproduction and enhanced uranium accumulation by a stable *Citrobacter* sp. isolated by a novel method. *FEMS Microbiol. Lett.* **55** : 157-162.

35. Mago, R. and Srivastava, S. 1994. Uptake of zinc in *Pseudomonas* sp. strain UDG26. Appl. Environ, *Microbiol.* **60** : 2367-2370.

36. Melcher, K.; Weitzenegger, A.; Buhmann, A; Steinhilber, G.S. and Schafer, K.P. 1996. Cloning and membrane topology of a P-type ATPase from *Helicobacter pylori*. *J.Biol. Chem.* **271** : 446-475.

37. Mercer, J.F.; Livingston, J.; Mall, B.; Paynter, J.A.; Begy, C.; Chandrasekharappa, S.; Lockhart, P.; Grimes, A.; Bhava, M.; Siemieniak, D. and Glover, T.W. 1993. Isolation of a partial candidate gene for Menkes disease by positional cloning. *Nature Genet.* **3** : 20-25.

38. Neyt, C.; Iriarte, M.; Thi, V.H. and Cornelis, G.R. 1997. Virulence and arsenic resistance in Yersiniae. *J. Bacteriol.* **179** : 612-619.

39. Nies, D.H. 1995. The cobalt, zinc and cadmium efflux system Czc ABC from *Alcaligenes entrophus* function as a cation-proton antiporter in Escherichia coli. *J. Bacteriol.* **177** : 27-7-2712.

40. Nies, D.H. 1992a. Resistance to cadmium, cobalt zinc and nickel in microbes. *Plasmid* **27** : 17-28.

41. Nies, D. 1992b. Czc R and Czc D, gene products affecting regulation of resistance to cobalt, zinc and cadmium czc system in *Alcaligenes entrophus*. *J. Bacteriol.* **174** : 8102-8110.

42. Nies, D.H.; Nies, A.; Chu, L. and Silver, S. 1989. Expression and nucleotide sequence of a plasmid determined divalent cation efflux system from *Alcaligenes eutrophus*. *Proc. Natl. Acad. Sci. USA* **86** : 7351-7355.

43. Nies, D.H. and Silver, S. 1989. Plasmid-determined inducible efflux is responsible for resistance to cadmium, zinc and cobalt in *Alcaligenes eutrophus*. *J. Bacteriol.* **171** : 898-900.

44. Nies, D.; Mergeay, M.; Friedrich, B. and Schlegel, H.G. 1987. Cloning of plasmid genes encoding resistance to cobalt, zinc, and cadmium from *Alcaligenes entrophus* CH34. *J. Bacteriol.* **167** : 4865-4868.

45. Nucifora, G.; Chu, L.; Misra, T.K. and Silver, S. 1989. Cadmium resistance from *Staphylococcus* aureus plasmid p1258 cad A gene results from a cadmium-efflux ATPase. *Proc. Natl. Acad. Sci. USA* **86** : 3544-3548.

47. Odermatt, A.; Suter, H.; Krapf, R. and Solioz, M. 1994. Induction o the putative copper ATPases, Cop A and Cop B, of *Enterococcus hirae* by Ag^+ and Cu^{2+}, and Ag^+ xtrusion by Cop *B. Biochem. Biophys. Res. Commun.* **202** : 44-48.

48. Odermatt, A.; Suter, H.; Krapf, R. and Solioz, M. 1993. Primary structure of two P-type ATPases involved in copper homeostasis in *Enterococcus hirae*. *J. Biol. Chem.* **268** : 12775-12779.

49. Pederson, P.L. and Carafoli, E. 1987. Ion motive ATPass. I. Ubiguity, properties, and significance to cell function. *Trends Biochem. Sci.* **12** : 146-150.

50. Rensing, C.; Thomas, P. and Nies, D.H. 1997. New Functions for the Three subunits of the Czc CBA cation-Proton antiporter. J. Bacteriol. **179** : 6871-6879.

51. Rosentein, R.; Peschel, A.; Wieland, B. and Gotz, F. 1992. Expression and regulation of the antimonite, arsenite and arsenate resistance operon of *Staphylococcus xylosus* plasmid pSX267. *J. Bacteriol.* **174** : 3676-3683.

52. Rouch, D.A. and Brown, N.L. 1997. Copper-inducible transcriptional regulation at two promoters in *Escherichia coli* copper resistance determinant pco. *Microbiol.* **143** : 1191-1202.

53. Rouch, D.A.; Lee, B.T.O. and Morby, A.P. 1995. Understanding cellular responssen to toxic agents : a model for mechanism-cuoice in bacterial metal resistance. *J. Industrial Microbiol.* **14** : 132-141.

54. Saxena, D. 1998. Genetic analysis of copper management by *Pseudomonas putida* strain S4–A Copper Mine Isolate. Ph.D. Thesis. Delhi University.

55. Saxena, D. and Srivastava, S. 1998. Carbon source–starvation induced precipitation of copper by *Pseudomonas putida* strain S4. *World J. Microbiol. Biotechnol.* **14** : (In press).

56. Schmidt, t.a nd Schlegel, H.G. 1991. Combined nickel-cobalt-cadmium resistance encoded by the ncc locus of *Alcaligenes xylooxidans* 31A. *J. Bacteriol.* **176** : 7045-7054.

57. Serrano, R. 1988. Structure and function of proton translocation ATPase in plasma membranes of plants and fungi. *Biochem. Biophys. Acta* **947** : 1-28.

58. Serrano, R. and Portillo, F. 1990. Catalytic and regulatory sites of yeast plasma membrane H^+-ATPase studied by directed mutagenesis. *Biochem. Biophys. Acta* **1018** : 195-199.

59. Silver, s. 1998. Genes for all metals–a bacterial view of the periodic table the 1996 Thom Award Lecture. *J. Ind. Microbiol. Biotechnol.* **20** : 1-12.

60. Silver, S. 1996. Bacterial resistances to toxic metal ions–a review. *Gene* **179** : 9-19.

61. Silver, S. 1996. Transport of inorgnaic cations. In : *Escherichia coli and Salmonella typhimnrium*, (ed. F.C. Niedhardt) 2nd edn. ASM Press, Washington DC, pp. 1091-1102.

62. Silver, S. and Phung, L.T. 1996. Bacterial heavy metal resistance : new surprises. *Annu. Rev. Microbiol.* **50** : 753-789.

63. Silver, S. and Ji, G. 1994. Newer systems for bacterial resistance to toxic heavy metals. *Environ. Health Perspect.* **102** : 107-113.

64. Silver, S.; Nucifora, G. and Phung, L.T. 1993. Human Menkes X Chromosome disease and the staphylococcal cadmium resistance ATPase : a remarkable similarity in protein sequnces. *Mol. Microbiol.* **10** : 7-12.

65. Silver, S.; Lee, B.T.O.; Brown, N.L. and Cooksey, D.A. 1993. Bacterial plasmid resistances to Copper, Cadmium and Zinc. In : *Chemistry of Copper and Zinc Triads* (eds. A.J. Welch and S.K. Chapman) Royal Society of Chemistry, London, pp. 38-53.

66. Silver, S. and Walderhaug, M. 1992a. Gene regulation of plasmid–and chromosome-determined inorganic ion transport in bacteria. *Microbiol. Rev.* **56** : 195-228.

67. Silver, S. and Walderhaug, M. 1992b. *Ion Transport Encyclopedia of Microbiology* **2** : 549-560.

68. Silver, S.; Nucifora, G.; Chu, L. and Misra, T.K. 1989. Bacterial resistance ATPases : primary pumps for exporting toxic cations and anions. *Trends Biochem. Sci.* **14** : 76-80.

69. Silver, S. and Nies, D.H. 1989. Metal ion uptake by a plasmid-free metal-sensitive *Alcqligenes. eutrophus* strain. *J. Bacteriol.* **171**: 4073-4075.

70. Silver, S.; Budd, K.; Leahy, K.M.; Shaw, W.V.; Hammond. A.; Novick, R.P.; Willsky, G.R.; Malamy, M.H. and Rosenberg, H. 1981. Inducible plasid-determined resistance to arsenate, arsenite, and antimony (III) in *Escherichia coli* and *Staphyloccus aureus. J. Bacteriol.* **146**: 983-996.

71. Solioz, M. and Vulpe, C. 1996. CPx-type ATPases : a class of P-type ATPases that pump heavy metals. *TIBS* **21** : 237-241.

72. Solioz, M. and Odermatt, A. 1995. Copper and Silver Transport by Cop B–ATPase in membrane vesicles of *Enterococcus hirae. J. Biol. Chem.* **270** : 9217-9221.

73. Tetaz, T.J. and Luke, R.K.J. 1993. Plasmid controlled resistance in *Escherichia coli. J. Bacteriol.* **154** : 1263-1268.

74. Tobin, J.M.; Cooper, D.G. and Neufield, R.J. 1984. Uptake of metal ions by *Rhizopus arrhizus* biomass. *Appl. Environ. Microbiol.* **47** : 821-824.

75. Turner, J.S. and Robinson, N.J. 1995. Cyanobacterial metallothioneins : biochemistry and molecular genetics. *J. Ind. Microbiol.* **14** : 119-125.

76. Volesky, B. 1994. Advances in biosorption of metal. *FEMS Microbiol. Rev.* **14** : 291-302.

77. Vulpe, C.; Levinson, B; Whitney, S; Packman, S. and Gitschier, J. 1993. Isolation of a candidate gene for Menkes disease and evidence that it encodes a copper transporting ATPase. *Nat. Genet.* **3** : 7-13.

78. Willsky, G.R. and Malamy, M.H. 1980. Effect of arsenate on inorganic phosphate transport in E. coli, *J. Bacteriol.* **144** : 366-374.

79. Wu, J. and Rosen, B.P. 1993 a. Metalloregulated expression of ars operon. *J. Biol. Chem.* **268** : 52-58.

80. Wu, J. and Rosen, B.P. 1993b. The arsD gene encodes a second trans-acting regulatory protein of the plasmid-encoded arsenic operon. *Mol. Microbiol.* **8**: 615-623.

81. Xu, C. and Rosen, B.P. 1997. Dimerization is essential for DNA binding and repression by the Ars R metalloregulatory protein of *Escherichia coli. J. Biol. Chem.* **272**: 15734-15738.

4 Single - Cell Proteins

Inderjeet Singh Bakshi

CONTENTS

ABSTRACT 44

1. INTRODUCTION 44
2. MICROBIAL GROWTH PROCESSES 45
3. MICROBIAL PRODUCTS 46
   3.1 Advantages in the production of microbial proteins 48
4. BIOMASS PRODUCTION 48
   4.1 Algal biomass 48
   4.2 Bacterial and actinomycetous biomass 49
   4.3 Yeast biomass 49
   4.4 Fungal biomass (Mycoprotein) 50
5. PRODUCTION OF USEFUL PROTEINS 50
6. OTHER BIOTECHNOLOGICAL APPLICATIONS 51
7. REFERENCES 51

Advances in Microbial Biotechnology
J.P. Tewari, T.N. Lakhanpal, Jagjit Singh, Rajni Gupta & B.P. Chamola (eds.)
APH Publishing Corporation, New Delhi - 110 002, India.

ABSTRACT

Microorganisms are being extensively used for the production of a wide range of proteins, enzymes, antibiotics, pharmaceuticals, food supplements and steroids. For commercial production of these, microorganisms can be grown in a variety of mass culture systems. In addition to selection, mutation and screening processes, modern biotechnology uses specific genetic constructs and protoplast fusion techniques. Immobilized microbial cells/enzymes can be used to synthesize a wide range of microbial products. Use of engineered organisms to have microbial products of interest will be a major future development. The greatest interest will rest in the production of mammalian proteins by means of microorganisms.

1. INTRODUCTION

The biosphere provides us food, feed, construction material and energy. Population rise is the key driving force for development of new technologies for increasing biomass production. Since the ancient times, several microorganisms have been used as a component of diet. Usage of *Saccharomyces* sp. as a bread leavening agent dates back to as early as 2500 B.C., that of fermented milk and cheese (produced by *Lactobacillus* and *Streptococcus*) by Egyptians and Greeks to 50-100 B.C., that of edible mushrooms to 1 B.C. in Rome, and that of *Spirulina* as a source of protein to 16th century (8). Of late, the dried cells of bacteria, algae, actinomycetes and fungi are being used as food or feed. These are collectively termed as microbial proteins. At the First International Conference on Microbial Protein held at Massachusetts, USA in 1967, the term microbial protein was replaced by, a term single - cell protein based on single-cell nature of microorganisms used for food and feed. Since then some actinomycetes and many filamentous fungi have been found to produce proteins and terms fungal protein and microprotein are also in use.

Realizing the importance of microbial proteins, *Saccharomyces cerevisiae* was produced at a large scale in an aerated molasses medium containing ammonium salts during World-War-I and *Candida utilis* during World War-II in Germany (2). Since then, several microorganisms have been cultured for mass production of single - cell proteins through improved technologies. Modern Gene-cloning techniques also allow a range of alternatives.

Besides proteins, microorganisms are increasingly being put to use for industrial production of organic chemicals and antibiotics, for control of insects and pests, for the recovery of metals. A pre-historic technique of mutation and selection is now being implemented with an intentional transfer of genetic information. It is possible to combine sequences from different groups of microorganisms to have desired products (12). Microbial mass culture of such suitable microorganisms is followed to have required quantities of products of interest.

The desirable properties that an organism should possess to be most useful as a source of single - cell protein are (a) rapid growth; (b) simple nutrient requirement; (c) efficient utilization of energy source; (d) simple culture system; (e) simple processing and separation of cells; (f) non-pathogenic; (g) harmless when eaten; (h) good flavour; (i) high digestibility and (j) high nutrient content (4, 12).

The last property deserves further elaboration because the commercial value of single - cell protein depends on their nutritional performance. The mean crude protein in dry matter of algae and yeasts lies between 50% and 60% and for bacteria about 80%. Estimation of crude proteins is based on the total nitrogen content multiplied by the factor 6.25. The utility of single cell protein product as food for human beings and feed for animals differ. For human beings, protein digestibility, protein efficiency ratio and net protein utilisation are the parameters for food application, whereas for animals metabolisable energy, protein digestibility and feed conversion ratio are the measures of performance.

The role of microbes in agriculture, fermentation, and enzyme technology has been dealt with extensively in recent times (13, 14). Microbial biotechnology has a great potential to improve the welfare of mankind. In the following pages an attempt has been made to present the microbial growth processes, microbial products, biomass production and other biotechnological applications of microbes.

2. MICROBIAL GROWTH PROCESSES

Microbes of industrial application are grown in culture tubes, shake flasks and large scale mass culture unit or stirred fermenter under aseptic conditions (1). Other approaches include air-lift fermenters, solid-state fermenters, fixed bed reactors, fluidized bed reactors, dialysis culture systems and continuous culture techniques. A stirred - fermenter has motor-driven impellers to shake a microbial sample in an nutrient medium under defined pH and aseptic conditions.

The fermenters range in size from 3-4 liters to 100,000 liters depending on requirements of production. For precise control over the process and the production, constant monitoring is done for microbial biomass, levels of metabolic products, pH and gas composition at the inlet and outlet. These fermenters can be used both under aerobic and anaerobic conditions. The air-lift or lift-tube fermenters makes use of density differences of gas bubbles entrained in the medium to cause fluid circulation. The solid-state fermenter, as the name suggests involves growing the culture without added free water. In the bed reactors microbes are placed on surfaces of fixed support material (fixed-bed reactor) or on surfaces of particles suspended in liquid or gas flow (fluidized bed reactor) and the medium flows past the particles. The dialysis culture systems while allowing new substrates to diffuse into the culture, remove the toxic wastes from the culture. For such industrial processes where logarithmic phase of growth is not undesirable, continuous culture technique provides for improved cell out-puts.

The major components of growth media used for industrial processes include carbon, nitrogen, vitamins, iron, buffers and antifoam agents (11). The raw materials used for these major components are given in Table 1.

The levels and balance of minerals and growth factors are of key importance in production of desired metabolises. For example, biotin and thiamine influence biosynthetic reactions and that in turn controls product accumulation, and if key components of growth media become limiting it leads to a shift from growth to production of required metabolites. To overcome accumulation of waste products the carbon-source is added continuously such that the organism will not have excess of substrate at any given time. In addition, the physical environment plays a critical role, Precise control of agitation, cooling, pH changes and oxygenation is important for optimal microbial functioning in the mass culture system.

Table 1 : Major components of Growth media used for commercial culture of microbes.

Component	Source
Carbon	Molasses, Whey, Grains, Agricultural Wastes.
Nitrogen	Corn-steep Liquor, Soybean meal, Slaughter house products, Ammonia, Ammonium salts, Nitrates, Distiller's solubles.
Vitamins	Crude preparations of plant and animal products.
Iron, Trace Salts	Crude inorganic chemicals.
Buffers	Chalk or crude carbonates, fertilizer-grade phosphates.
Antifoam agents	Higher alcohols, silicones, natural esters, lard and vegetable oils.

3. MICROBIAL PRODUCTS

A wide range of compounds are produced by culturing microbes (5, 11). These include antibiotics, hormones, transformed steroids, solvents, organic acids, chemical feed stocks, amino acids, enzymes, fuels and energy related products (Table 2).

Table 2 : Major microbial products

Product	Source
Ethanol	*Saccharomyces cerevisiae*
Acetone	*Clostridium acetobutylicum*
Enzymes	*Aspergillus, Bacillus, Mucor, Trichoderma*
Gibberellins	*Gibberella fujikuroi*
Aminoacids	*Corynebacterium glutamicum*
Organic acids	*Aspergillus niger*
Nucleotides	*Corynebacterium glutamicum*
Vitamins	*Ashbya, Eremothecium, Blakeslea*
Polysaccharides	*Xanthomonas*
Antibiotics	*Penicillium, Streptomyces, Bacillus*
Alkaldids	*Claviceps purpurea*
Steroid Transformations	*Rhizopus, Arthrobacter*
Insulin, human growth hormone somatostatin, Interferon	*E.coli, S.cerevisiae*
Hydrogen	Photosynthetic microorganisms
Methane	*Methanobacterium*

Under conditions of balanced growth, the cells minimize accumulation of any particular cellular "building blocks" in amounts beyond those required for growth. In modern industrial microbiology microorganisms are "tricked" into producing large excesses of desired compounds, usually by selecting mutants that have lost the ability to control synthesis of a particular end product. Lysine, glutamic acid and monosodium glutamate (MSG) find use in the food industry as nutritional supplements and flavor-enhancing compounds. Commercial production of amino acids is carried out by means of regulatory mutants of *Corynebacterium glutamicum.* Low biotin level and presence of fatty acid derivatives aid in increased excretion of high concentrations of glutamic acid. *C. glutamicum* blocked in the synthesis of homoserine, accumulates lysine.

Unicellular bacteria, actinomycetes, yeasts and molds are being extensively used in bioconversions or microbial transformations or biotransformation (11). Such transformations have major advantages over chemical procedures because enzymes carry out these reactions under mild conditions and larger water-insoluble molecules can be transformed. Biocatalysts can be produced through continuous culture or batch culture techniques. After culturing cells are dried for direct use, or are processed to carry out desired bioconversions.

The microbial mass from freely suspended vegetative cells or spores is usually used only once and then discarded. For repeated use, cells are attached to ionexchange resins by ionic interactions or are immobilized in a polymeric material (11). Microbial cells, spores and enzymes are immobilized through ionic, covalent and physical entrapment approaches or on the inner-walls of fine tubes. The solution to be modified is passed through such immobilized microbial cells. Such immobilized microorganisms are extensively used in bioconversions of steroids, degradation of phenol and production of antibiotics, enzymes, amino acids and vitamins (Table 3).

Table 3 : Immobilized microbial cells and their products

Process/Product	Microorganisms used
Antibiotics	
Penicillin	*Penicillium chrysogenum*
Bacitracin	*Bacillus*
Cephalosporins	*Streptomyces clavuligerus*
Amino Acids	
L-alanine	*Corynebacterium dismutans*
L-glutamic acid	*Corynebacterium glutamicum*
L-tryptophan	*Eschericia coli*
L-Lysine	*Microbacterium ammoniaphila*
Enzymes	
Coenzyme A	*Brevibacterium ammoniagenes*
Protease	*Streptomyces fradiae*
Vitamins	
Pantothenic acid	*Eschericia coli*
Steroids	
Prednisolone	*Curvularia lunata*
Fermentation Products	
Ethanol	*Saccharomyces cerevisiae*
Lactic Acid	*Lactobacillus delbrueckii*
Environmental management	
Degradation of para nitrophenol	*Pseudomonas*.spp.
Degradation of phenol	*Candida tropicalis*
Denitrification	*Micrococcus* spp.
Heavy metal sorption	*Pseudomonas aeruginosa*

Biocatalysts are providing new opportunities for the industrial microbiologist and biotechnologist. Amylases, proteases, lipases, glucose isomerase, penicillin acylase are some of the enzymes being produced in large scale by microbial processes for commercial use (1). The potential for the production of useful enzymes has improved markedly because of the increased ease with which genes can be manipulated. Some extra cellular enzymes and their application is given in Table 4.

Table 4 : Microbial enzymes and their application

Enzyme	Source	Application
Amylase	Fungi	Bread, Syrup and glucose manufacturing, digestive aid
	Bacteria	Starch coatings, cold-swelling laundry starch, desizing.
Protease	Fungi	Bread
	Bacteria	Spot removal, meat tenderizing, wound cleansing, desizing, house hold detergent
Invertase	Yeast/fungi	Soft-center candies
Glucose-oxidase	Fungi	Glucose removal, oxygen removal, test paper for diabetes
Pectinase	Fungi	Pressing, clarification
Rennin	Fungi	Coagulation of milk

3.1 Advantages in the Production of Microbial Proteins

There are several advantages in the production of microbial protein. These include rapid succession of generations, easy genetic modification, high protein content (43-85%) in the dry mass, consistent quality of cultures, not dependent on climate, ecologically beneficial and high solar energy conversion efficiency per unit area. Succession of generations vary from a low of 0.5 hours in bacteria to upto 6 hours in algae. Genetic modifications have been resorted to especially for having desired amino acid complement (9).

4. BIOMASS PRODUCTION

4.1 Algal Biomass

Chlorella strains are being used for a variety of applications in biotechnology. *Chlorella* strains have high protein content. They are being used for production of animal protein. Strains of *Chlorella* are also being utilised for sewage oxidation and waste-water treatments. Mass cultivation depends on factors like illumination time, light intensity, CO_2 concentration, nitrogen sources and agitation of growing cells. Biomass yields ranges from 12 to 15 g/m^2/day on dry weight basis. Mass cultivation of algae is being undertaken in many countries. In India, research institutes at Nagpur, Lucknow, Hyderabad and Mysore are cultivating algae on a large scale for food and feed. For harvesting the algal biomass, the algal cells are recovered from ponds by concentration, dewatering and drying. Harvesting *Spirulina* is easier because its spiral filaments float on the surface of water. These algal mats are filtered and the algal suspension is dried with hot air to get a fine powder. A Mexican company named Sosa Texcoco is exporting *Spirulina* powder and marketing lozenges and capsules from the *Spirulina* powder supplemented with vitamin A and C. Biscuits and confectionery made of *Spirulina* powder are rich source of proteins.

4.2 Bacterial and Actinomycetous Biomass

Bacteria and actinomycetes with capacity to utilise a wide range of organic substrates and short life cycle are being widely used as a source of single-cell protein (4). Noteable genera are *Methylococcus, Pseudomonas, Hyphomicrobium, Acinetobacter, Flavobacterium, Brevibacterium, Achromobacter* and *Thermomonospora.* The substrates employed for growing bacteria include methane, ethane, propane, n-butane, isobutane, propylene, butylene, methanol, cellulosic wastes.

Imperial chemical industry has been engaged in production of pruteen from *Methylophilus methylotrophus* on methanol. Pruteen contains 72% protein, 86% liquid. Amino acid profile of pruteen reveals that it is rich in lysine and methionine. The conversion of methanol to single-cell protein occurs as under.

$$1.72\ CH_3OH + 0.23\ NH_3 + 1.51O_2 \rightarrow 1.0\ CH_{1.68}O_{0.36}N_{0.22} + 0.72\ CO_2 + 2.94\ H_2O$$

Production of biomass is influenced by strain of bacteria, genetic stability of the strain, pH of growth medium (most suitable being 5.7), temperature (15-35°C) according to strain used, availability of oxygen, organic substrate, absence of bacteriophage and aseptic conditions.

The cell density of bacteria is of the order of 10-20g/L. To cut down the high centrifugation cost, flocculents and decanter-type centrifuges are put to use.

4.3 Yeast Biomass

World production of yeast biomass met with an upswing since consumption of *S. cerevisiae* as food in Germany during World War-I. Of about 0.4 million metric tonnes per year production of yeast biomass, baker's yeast accounts for about 0.2 million tonnes.

Inorganic acids and sulphur salts together with organic wastes such as molasses, starchy materials, milk whey, fruit pulp, wood pulp are used for production of amino acids from yeast. As with bacteria chemical equations for yeast biomass production on carbohydrates and hydrocarbons are as under.

4.3.1 Carbohydrates

$$8n\ CHO + 0.8n\ O_2 + 0.19n\ NH_4^+ + \text{trace elements} \rightarrow$$

$$n(CH_{1.7}O_{9.5}N_{0.12}\ Ash) + 0.8n\ CO_{12} + 1.3n\ H_2O + 80{,}000\ n\ k\ Cal$$

4.3.2 Hydrocarbons

$$2n\ CH_2 + 2nO_2 + 0.19n\ NH_4^+ + \text{trace elements} \rightarrow$$

$$n\ (CH_{1.7}\ O_{0.5}\ N_{0.19}\ Ash) + nCO_{12} + 1.5\ H_2O + 2{,}00{,}000\ n\ k\ Cal$$

Factors affecting the yield include optimum C:N ratio between 7:1 and 10:1, pH in the range 3.5 to 4.5, temperature 30°C (varying with strain), oxygen (1-2 g/g dried Cells) and sterile conditions throughout the process (8).

The density of yeast cells reaches upto 1.1 g/ml. Cells are recovered by decantation centrifugation and drying treatment methods. Harvesting by rotary vaccum filter yields a cake containing 20-40% dry matter, which on drying gives a final harvesting with 6-10% water content.

4.4 Fungal Biomass (Mycoprotein)

Sporophytic fungi grow on complex organic compounds to produce high amount of fungal biomass. Some strains of *Aspergillus niger, A. fumigatus, Fusarium graminearum* are hazardous to humans. Use of such fungi as single - cell protein calls for toxicological evaluations (15). Mycoprotein popularity is because of following reasons (a) some filamentous fungi grow as fast as most of the single - celled organisms; (b) The product of filamentous fungi is fibrous and can be made into various textured foods; (c) these fungi have greater retention time in the digestive system; (d) their protein content varies between 35%-50%; (e) digestibility and net protein utilisation is higher than single -celled organisms; (f) the overall cost of protein production is more economical; (g) they have greater penetrating power into insoluble substrates; (h) their mushroom like odour and taste is readily acceptable and (j) the biomass produced by filamentous fungi can be used without any further processing (4).

For optimal growth of moulds the C:N should be 5:1 to 15:1 and ammonium salts be source of nitrogen. Required concentration of minerals, pH of growth medium ranging from 3.0 to 7.0, temperature ranging from 25°C to 3O°C and availability of oxygen are some other key requirements.

Cellulosic and lignocellulosic materials from agriculture, forestry and industry are favourable substrates for growth of fungi. The fungal mats or pellets are recovered from the growth medium by decantation and filtration.

Mushrooms, the members of higher fungi have been under cultivation since 1630 in France. After second World War, mushroom cultivation spread in several countries. Mushrooms are rich in protein. Their value as a food accessory can contribute in facing the challenge of world-wide food shortage. Mushrooms have the ability to grow on cheap carbohydrate materials and can transform various waste materials.

In India, lot of work has been done for mushroom cultivation at CMU, Solan (Himachal Pradesh); CFTRI, Hyderabad; RRL, Jammu; NBRI, Lucknow. On an average, the amount of protein varies from 34% to 44% of the total dry weight. Fresh mushrooms contain about 3% to 28% carbohydrates, several vitamins and mineral (4).

Mushroom cultivation is done either in gardens or fields or in tunnels in rocky areas or in ventilated houses with temperature and moisture control. To get pure cultures, mushrooms are isolated from nature and cultured on potato dextrose agar or malt extract medium under sterilised conditions.

5. PRODUCTION OF USEFUL PROTEINS

It is possible to engineer bacteria or other cells to produce the products of cloned genes, especially those proteins that are otherwise difficult to obtain because that are either present in only a few molecules per cell or that are produced in only a small number of cells or only in human cells (6). In principle, the method involves cloning a DNA sequence coding for the desired protein in a vector adjacent to a bacterial promoter (7, 10). cDNA with all the coding sequences spliced together in the right order is cloned in a high copy-number vector. That ensures many copies of the coding sequence in each bacterial cell to result in synthesis of the gene product at concentrations ranging from 1 to 5% of the total cellular protein.

The first cloned pharmaceutical product to be marketed was human insulin (16). The cloned genes assure an unlimited supply of insulin, not dependent on the vagaries of the livestock. Moreover, human insulin is not recognized as foreign, as has been the case with traditionally used pig and bovine insulin. Human growth hormone, is another example of a valuable protein. Children who inherit hypopituitary dwarfism make too little growth hormone. Treatment with human growth hormone enables them grow normally. Gene coding for human growth hormone has been cloned and the product is effective. The hormone thus obtained is free from contamination with the agent that causes a lethal brain condition called Creutzfeld - Jacob disease.

This idea is not confined to microbes. Recent experiments have shown that transgenic ewes can be made to produce human blood clotting protein called factor IX in their milk and transgenic tobacco plants can make antibodies (plantibodies), which might find use as supplements for infant formula (16). One can expect to harvest enormous quantity of foreign protein per acre from these green factories.

6. OTHER BIOTECHNOLOGICAL APPLICATIONS

Microorganisms can also be put to use as insecticides in biocontrol programs (11). For example, *Bacillus thuringiensis* is used on a wide variety of vegetable and field crops, fruits, shade trees and ornamentals. Among fungi, *Beauveria bassiana* and *Metarhizium anisopliae* are used for control of teh colorado potato beetle and the froghopper in sugar cane plantations, respectively. Nuclear polyhedrosis virus, granulosis virus and cytoplasmic polyhedrosis virus are pathogenic for specific insects.

Biopolymers, such as Dextran, Xanthan, Pullulan, Scleroglucan, microbial alginate and curdlan produced by a wide variety of microorganisms, find uses in the pharmaceutical and food industries because of their ability to modify flow characteristics of liquids and to serve as gelling agents (3).

Microorganisms are also proving their worth in a rapidly developing area of biosensor production. Microorganisms, or their enzymes or organelles are linked with electrodes and biological reactions are converted into electrical currents. These biosensors can find use in monitoring pollutants, detecting flavour compounds in food, clinical diagnosis, fermentation analysis, monitoring of industrial gases, measurement of toxic gas in mines and agricultural analysis.

This article has briefly described just a few examples of the many applications of microorganisms. One can look forward to many more significant contributions in the future.

7. REFERENCES

1. Brock, T.D. and Madigan, M.T. 1988. *Biology of Microorganisms*. Prentice Hall, New Jersey, USA.

2. Chahal, D.S. (ed.) 1991. *Food, Feed and Fuel from Biomass*, Oxford and IBH Publishing Company, New Delhi, India.

3. Crueger, W. and Crueger, A 1984. Biotechnology. In: *A Text Book of Industrial Microbiology* (ed: T.D. Brock). Sinauer Associates, Sunderland, Mass.

4. Dubey, R.C. 1993. *A Textbook of Biotechnology*. S.Chand and Company Ltd., New Delhi, India.

5. Eveleigh, D.E. 1981. The microbiological Production of Industrial Chemicals. *Scientific American* **245**: 154-178.

6. Gilbert, W. and Villa - Komaroff, L. 1980. Useful proteins from recombinant bacteria. *Scientific American* **242** : 74-94.

7. Hartl, D.L. 1995. *Essential Genetics*. Jones and Bartlett Publishers, Sudbury, Massachusetts.

8. Ignacimuthus S. 1996. *Applied Plant Biotechnology*. Tata Mc Graw-Hill Publishing Company Ltd., New Delhi, India.

9. Mantell, S.H. and Smith, H. (eds.) 1993. *Plant Biotechnology*. Combridge University Press, Combridge.

10. Prakash, J. and Pierik, R.L.M. (eds.) 1993. *Plant Biotechnology*. Oxford and IBH Publishing Company, New Delhi, India.

11. Prescott, L.M., Marley, J.P. and Klein, D.A. 1990. *Microbiology*. Wm C. Brown Publishers, U.S.A.

12. Purohit, S.S. and Mathur, S.K. 1990. *Fundamentals of Biotechnology*. Agrobotanical Publishers, Bikaner, India.

13. Sarwar, N. and Mukerji, K.G. 1996. New vistas in plant biotechnology. Pers. Bill. Sci. pp. 1-21

14. Singh, V.P. and Mukerji, K.G. 1989. Microbes in biotechnology In : *Frontiers in Applied Microbiology* Vol. III (eds. K.G. Mukerji, V.P. Singh and K.L. Garg), pp. 61-84.

15. Smith, J.C., Berry, D.R. and Kristiansen, B. 1980. *Fungal Biotechnology*. Academic Press, New York, U.S.A.

16. Weaver, R.F. and Hedrick, P.W. 1995. *Basic Genetics*. Wm. C. Brown Publishers, Kerper Boulevard, Dubuque.

5 Fungal Allelopathy - An Overview

Rajni Gupta

CONTENTS

ABSTRACT 54

1. INTRODUCTION 54
2. ALLELOPATHIC EFFECTS ON MYCORRHIZA AND OTHER MICROORGANISMS 56
3. CONCLUSION 61
4. REFERENCES 61

Advances in Microbial Biotechnology
J.P. Tewari, T.N. Lakhanpal, Jagjit Singh, Rajni Gupta & B.P. Chamola (eds.)
APH Publishing Corporation, New Delhi - 110 002, India.

ABSTRACT

Allelopathy is a process in which a plant releases, chemical compounds into the environment which inhibits the growth of another plant in the same or a neighbouring habitat. They play important roles in plant resistance to insects, nematodes and pathogen. Allelopathic crops and the chemicals have been proposed as an additional strategy for this exploitation in weed suppression in several agrosystems. Allelopathic inhibition typically results from a combination of allelochemicals, which interface with several physiological processes in the receiving plant or microorganisms.

1. INTRODUCTION

The phenomenon of allelopathy encompasses all types of chemical interactions among plants and microorganisms. Several hundred different organic compounds released from plants and microbes are known to affect the growth and function of the receiving species. Many new allelochemicals have been identified in recent years and it has become clear that the actions of allelochemicals are important features characterizing the inter-relationship among organisms. These compounds influence patterns in vegetational communities, plant succession, seed preservation, germination of fungal spores, the nitrogen cycle, mutualistic associations, crop productivity and plant defense (9). Allelopathy is tightly coupled with competition for resources and stress from diseases, temperature extremes, moisture deficit and herbicides. Allelopathic inhibition typically results from a combination of allelochemicals which interface with several physiological processes in the receiving plant or microorganism.

Molisch (55) coined the term allelopathy to refer to biochemical interactions between all types of plants including microorganisms. Khailov (36) demonstrated conclusively that the effect of any given compound may be inhibitory or stimulatory depending on the concentration of the compound in the surrounding medium.

Apparently, most of the organic compounds which are inhibitory at some concentrations are stimulatory to the same processes at lower concentrations Undoubtedly, many important ecological roles of allelopathy have been overlooked because of concern with only the detrimental effects of the added chemicals. Allelopathy is different from competition, which involves the removal or reduction of some factors from the environment, that is required by other plants sharing, the habitat. Factors which may be reduced include water, suggested the term interference to refer minerals, food and light. Muller (57) suggested the term interference to refer to the overall influence of one plant (or microorganism) on another. Interference would thus encompass both allelopathy and competition.

The confusion of competition and allelopathy has taken several forms. Competition is the process in which the reaction of a plant upon the habitat reduces the level of some necessary factor to the detriment of some other plant sharing the same habitat. Factors such as radiant energy, oxygen, carbon dioxide, mineral nutrients and water are all capable of depletion by plants. Competition occurs, however, only if the reaction involves a reduction demonstrably deleterious to other individual. Allelopathy is a process in which a plant releases chemical compound into the environment which inhibits the growth of another plant in the same or a neighbouring habitat. This process differs from competition in not involving the depletion of a necessary factor and in depending upon the addition of a deleterious factor. Both of these processes are reactions defining reaction as any change in a habitat resulting from the activity of an organism living in it.

Agronomists and horticulturists have long been more willing than ecologists to consider allelopathy as a significant phenomenon. This disparity is now decreasing as ecologists encounter increasingly effective evidence of allelopathy. A major influence in this development has been the recent availability of powerful analytical chemical techniques, such as gas chromatography, paper and thin layer chromatography, electrophoresis and spectrophotometry by means of which significant evidence of subtle chemical interactions could be revealed for the first time.

Evidences indicate that allelopathic compounds get out of plants by volatilization, exudation from roots, leachates from plants residues by rain or by decomposition of residues. Their toxic effects on the surrounding plants have been reported as stunting growth, inhibition of primary roots, increase of secondary roots, inadequate nutrient absorption, chlorosis, premature leaf abscission, slow maturation, delay or failure of reproduction and inhibition of seed germination.

There are some special terms suggested for chemical agents involved in allelopathy based on the type of plant producing the agent and type of plant affected. The term antibiotics is used for chemicals produced by microorganisms and effective against other microorganisms. In case the chemical is produced by higher plants and effective against a microorganism, the term 'phytoncide' is used. The term 'masrasnum' is used for a compound produced by a microorganism and active against a higher plant. Chemicals produced by higher plants and effective against higher plants are termed as 'kalines' (48).

Natural products play important roles in plant resistance to insects, nematodes and pathogens. Numerous phytotoxic secondary products have been isolated and identified from higher plants. Introduction of these compounds into the environment occurs by exudation of volatiles from living plants, leaching of water soluble toxins from aerial plant parts of subterranean tissues or by release of toxins from non-living plant materials. Allelopathic crops and the chemicals have been proposed as an additional strategy for this exploitation in weed suppression in several agrosystems. The concept that some crop plants may be allelopathic to common weeds of agricultural land is receiving greater attention as an alternative weed control strategy. Some allelopathic crop plants have already been used experimentally in weed control (63).

Root exudation of several plants contain nematicidal constituents which act as natural defense factors to ward off nematode infections. Asparagus root exudates contain a mixture of short chain alkaloids, acids and asparagustic acid which are nematicidal (63). Gilbert (38) reported that plantation of *Cymbopogon citratus* and *Tagetus miniuta* in nematode susceptible areas could successfully protect the plants from nematode infections. The allelochemicals affect the agriculture in several ways. Generally the allelochemicals produced by one plant species affect the other species. Chou and Lin (14) found that aqueous extracts of decomposing rice residues in soil retarded radicle growth of rice seedlings and growth of rice plants. Five inhibitory phenolic acids have been identified from decaying rice residues and several unidentified allelochemicals have been isolated.

There are several examples where the allelochemical of one crop affects other crops. When a leguminous crop is cultivated in the harvested paddy field, the yield of the legume is drastically reduced. However, there is a remarkable increase in the yield of the legumes when their plantation is done after burning the rice straws. Rice *et al.* (65) have reported that the decreased yields in unburned fields may result from the inhibition of the *Rhizobium* in the nodules of the leguminous crops. Some allelochemicals have been tried to find out the measure to overcome such harmful effects of allelochemicals. It has been reported by Burgos-Leon *et al.* (13) that inoculation of the decomposing roots, by *Trichoderma viride* has been found to reduce their allelopathic chemicals.

2. ALLELOPATHIC EFFECTS ON MYCORRHIZAE AND OTHER MICROORGANISMS

It has been well established that certain tree species, such as *Betula pendula* Roth. and spruce (*Picea abies*) frequently fall to develop in association with heather (*Calluna vulgaris*) (66). Handley (31) reported that heather inhibits growth of mycorrhizal in spruce and results in failure of spruce to establish among heather. Melin (52) pointed out that eco-mycorrhizal fungi are very sensitive to substances exuded from plant roots or leached from dead plant material. Robinson (66) employed experimental techniques which demonstrated conclusively that exudates run off from roots of living *Calluna vulgaris* and raw humus of this species contained a factor toxic to several mycorrhizal fungi. Robinson presented some evidence that the inhibitor may also prevent infection of *Calluna* by certain pathogenic fungi (66). Field observations indicated that the phytotoxin must be continuously produced, because the inhibitory effect disappears soon after the heather is removed. In prairie soils trees cannot grow normally unless the seedlings are inoculated with mycorrhizal fungi (59). These investigators decided, therefore, try to determine the causes which perclude the presence of free-living mycorrhizal fungi in prairie soils. They found that extracts of roots of prairie grasses did not affect oxygen uptake of mycorrhizal short roots of Monterey pine (*Pinus radiata*) and differently than extracts of roots of red pine (*P. resinosa*). On the other hand, water extracts of prairie soils significantly inhibited oxygen uptake by the same types of mycorrhizal roots compared with effects of extracts of forest soils. Theodorou and Bowen (73) reported that decomposition products of grass roots reduced the numbers of an ectotrophic mycorrhizal fungus of Monterey pine roots. Olsen *et al.* (58) found that aqueous extracts of aspen (*Populus tremula* L.) leaves strongly inhibited growth of several species of *Boletus*, a mycorrhizal fungus. They had a weak inhibitory effect on litter-decomposing species of *Marasmius*. The inhibitors were identified as catechol and benzoic acid and both had strong inhibitory effects of *Boletus* in bioassays. Shukla *et al.* (71) found that culture filtrates of *Mortierella substitissima* Ouedem, *Aspergillus candidus* Link, *A. flavus* link, *A. niger* V. Tiegh, *Penicillium rubrum*, *Papulaspora* sp., *Staphylococcus aureus* and *Bacillus subtills* Cohn contained antibiotics which are very inhibitory to the growth of various fungi isolated from leaf litter of *Shorea robusta* Gaetrn. Mallik and Zhu (47) found that aqueous extracts of fresh leaves and humus of *Kalmia* were inhibitory to primary root growth of black spruce. Further analysis leaf extract of *Kalmia* TLC and HPLC isolated and identified eight phenolic acids. These were m-conmaric, p-conmaric, ferulic, gentisic, p-hydroxy benzoic, o-hydroxyphenyl acetic, syringic and vanillic acid. All the compounds were inhibitory to the root growth of black spruce, with o-hydroxyphenylacetic acid being the most toxic and m-conmaric acid the least toxic. Toxicity of the compounds were greater with increasing concentration and decreasing pH. Mycelial growth of four ectomycorrhizal isolates NF4, GB45, GB23, GB12 was stimulated in presence of *Kalmia* leaf extract, Black spruce seedlings were inoculated with these four fungal isolates for subsequent experiments. Of these three fungal isolates NF4, GB45 and GB23 have potential in overcoming *Kalmia* growth inhibition in black spruce. Particularly important isolate is NF4 that caused 2-3 fold increase in seedling biomass compared to the control.

In addition to the unfavourable soil pH and nutrient conditions for tree seedling growth in *Kalmia* sites, it has been reported that leaf litter and soil of *Kalmia* contain allelopathic compounds that affect the primary root growth of black spruce (46). Similar organic accumulation and soil acidity increases have been reported for other ericaceous plants of cool, moist, temperate, climate, particularly *Calluna vulgaris* and *Erica cinerea*. Robinson demonstrated that root exudates of *Calluna vulgaris* contained allelopathic compounds that were inhibitory to some mycorrhizal fungi of conifer seedlings. Jalal *et. al.* (35) isolated and identified a number of phenolic compounds from *Calluna* plant and soil material. Some of these compounds were highly phytotoxic and were believed to contribute to the exclusion of other plants including trees from ericaceous heath (61). Aqueous extracts of leaves, roots, litter and soil of *Kalmia* were inhibitory to black spruce seedlings (45). The extracts affected the growth and development of primary roots of the germinants with no significant effects on percent seed germination and stem growth of seedlings. The primary root growth of other conifers such as red pine (*Pinus resinosa* Alt) and balsam fir (*Ables balsamea* L. Mill) was also inhibited by the water extracts of *Kalmia* (46). Allelopathic activity of the identified phenolic compounds was assayed on black spruce of four concentrations. All compounds were inhibitory to growth of black spruce seedlings, particularly to root growth and degree of inhibition varied depending upon the compound and its concentration.

Some mycorrhizal fungi that are capable of growing in acidic soil with *Kalmia* allelopathy can also form mycorrhizal association with black spruce. The effect of *Kalmia* leaf extracts on mycelial dry weight of the fungi was tested at different pH. The nineteen fungal isolates were also tested for their tolerance to eight phenolic acids that were identified from *Kalmia* leaves. This fungal bioassay was conducted in MMN agar containing pure substances of ethyl, acetic, ferulic, vanillic, syringic, o-hydroxy phenyl acetic, m-coumaric, p-hydroxy-benzoic, p-coumaric acid and gentisic acids. The effect of *Kalmia* leaf extracts on mycelial radial growth varied depending on the fungal isolate and concentration of the water extract. The mycelial growth of GB6, NF4 and GB4 was stimulated that of GB40, GB24, NFI, GH45, GH50, GH12 and GH8 remained unaffected, while in the root growth was severely inhibited at all the three concentrations. Similar growth stimulation and inhibition of mycorrhizal fungi in presence of plant residues, leaf leachates and soil extracts have been reported by others (15). Mallik and Zhu (47) inoculated the Black spruce germinants with NF4, GH12, GH23 and GH24 strains of mycorrhizal fungi. These isolates showed maximum tolerance to *Kalmia* leaf extracts. Mycorrhizal formation on the seedlings varied among the 19 isolates. Four isolates namely GH23, GH12, GH45 and NF4 formed mycorrhizae on 60-80% of the total short roots. The isolates GH56, GH50, GH25, GH20, ABI and NFI were unable to form mycorrhizae in presence of leaf extracts.

Increasing emphasis on sustainability in agriculture is redefining the role of vesicular arbuscular mycorrhizal fungi in the plant soil system (9). As a link between roots and the surrounding bulk soil, AM fungi integrate the interdependent functions of the plant-soil symbiosis (10). Abbott *et al.* (1) evaluated the effects of individual functions on the plant-soil system as a whole. In the majority of AM reports, however, plant growth response has been and still is the sole measure of AM-fungal input. A small number of reports during the past decade show a growing awareness of the need to integrate VAM research both into greater context of soil biology and into the concept of stable and sustainable plant soil system (26). Lindermann (43) related the interactions between VAM plants, VAM fungi and the soil fauna (33) with VAM soil and plant effects in mycorrhiza based soil food webs is still extant (74). Among the interactions between functional groups of the soil microflora that affect both plant and soil development, those between rhizobacteria and VAM fungi stand out because of their mutual influence on each other (24), they can influence in concert, the plant soil system in ways that depend on the specific composition of fungus bacterium populations (11). VAM fungi improve soil aggregation and inhibit some soil bacteria (3).

Root exudate mediated changes in the rhizosphere microflora in response to plant nutrient stress (34) fluctuations in soil aggregation with bacterial populations and nutrient allocation to above and below ground plant organs (39). Schreiner *et al.* (70) reported that VAM fungi influence both plant and soil functions and mediate their interactions between, above and below ground events. Inputs of different VAM isolates to plant and soil responses are similar in some aspects but distinct in others. Soil bacteria and VAM fungi may perform preferential associations.

The colonization of roots by VAM fungi induces biochemical changes into the host tissues. These include stimulation of the phenylpropanoid pathway (32). Change in levels of aliphatic polyamines (22), synthesis of proteins of unknown functions (20), activation of defense related genes (25) and enhancement of certain hydrolyse activities (19). Several studies have centered on elicitation of plant defense responses during ectomycorrhizal (49) and endomycorrhizal symbiosis (27). Increased chitinase activity have been reported in ectomycorrhizal symbiosis although no chitinase synthesis was found in *Picea abies* roots colonized by *Pisolithus tinctorius* (75).

Microbial growth in the rhizosphere is stimulated by the continual input of readily assimilable organic substrates from the root. In ecological terms, the substrates from the roots is the product of photosynthesis and is thus primary productivity and utilization of these organic substrates by rhizosphere microorganisms results in secondary productivity. In turn, these microorganism can influence the plant as primary producer. Microorganisms compete with each other for food and essential elements in the soil and around the rhizosphere (5). Vesicular arbuscular mycorrhizal fungi are wide spread in occurrence and because of their potential for crop improvement have been investigated extensively (56). Most soils contain spores or propagules of VAM fungi which effects the rhizosphere mycoflora of the plant (67). Gupta and Mukerji (30) reported the allelopathic effect of VAM on the rhizosphere of *Trigonella foenum graceum*. Mycoflora in the rhizosphere vary in the treated and control plants. *Trichoderma viride* was present frequently in the *Trigonella foenum graceum* inoculated with *Glomus macrocarpum*. *Cephalosporium acremonium* was frequently present in the treated plants while it was totally absent in control plants. *Trichoderma viride* and *Verticillium albo atrum* being good biocontrol agents were frequently present in the rhizosphere of treated *Trigonella foenum graceum*. They inhibit the growth of other fungi and bacteria. With the mycorrhizal fungi, plant exudates also effect the flora of rhizosphere.

When allelochemicals enter the soil, microbial transformations often occur or, their biological activity may be altered as they are adsorbed on soil particles. Impacts of these phenomena and other interactions in the soil on allelopathy are not well understood. Recently, Blum (12) reported that the amount of nitrates and other organic carbon sources in the soil modified the allelopathic action of p-coumaric acid. Higher levels of nitrate increased the amount of p-coumaric acid required to reduce growth of morning glory but elevated glucose or methionine in the soil reduced the concentration for growth inhibition. External hyphae of VAM fungi provide a physical or nutritional substrate for bacteria. Analysis of rhizosphere soil of VAM (*Glomus fasciculatum*) and non-VAM tomato plants showed greater population of bacteria and actinomycetes in the mycorrhizosphere, compared to non-inoculated control (4). Mc-Allister *et. al.* (50), while studying the interaction between saprophytic fungi (*Aspergillus niger, Trichoderma koningi, Fusarium solani* and *Glomus mosseae*, observed that

G. mosseae decreased the fungal population through its effect on the plant. Mycorrhizal association significantly changes the physiology and morphology of roots and plants in general which in turn cause a new microbial equilibrium to be established. These changes presumably involve the same type of organisms as were involved in the rhizosphere before formation of mycorrhizal but quantitative changes occur within these types as a result of direct metabolic interaction with the mycorrhizal fungal hyphae, spores or the indirect effect mediate by the host (42)

Studies on vesicular arbuscular mycorrhizal infections indicated that conditions which increase permeability of roots or exudation, increase infection (29). Once synthesized, the cost of maintaining VA mycorrhizae in terms of extra carbohydrate moved to the roots approximately 6-11% net fixed carbon (38).

Spores of most parasitic fungi are formed in dense populations in or on infected host tissue, and the spores usually remain ungerminated while located in their site of production (8). This can be due to several factors, one of which is the production by the spores of fungistatic agents which are excreted into the water around the spores. These self inhibitors generally assure dispersal of viable ungerminated spores. Endogenous germination stimulators that counteract inhibition of germination are produced by some spores (8). Stimulants most commonly introduced into the soil environment that alleviate fungistasis are components of seed and root exudates released during seed germination and root development. Since the majority of host parasite interaction in soil do not occur without the release of such stimuli (16) and understanding of the molecules involved in the activation of quiescent fungal propagules by seed and root exudates are of prime importance in understanding the initiation and control of seed and root infections by soil borne pathogens. Initiation of *Rhizobium*-Legume symbiosis involves a complex series of steps, where the bacteria and plant influence each other. *Rhizobium* are chemotactic toward plant roots and root colonization, growth and modulation by rhizobia inoculants are stimulated by seed and root exudates which include certain flavinoids (60). Recent studies show that both plant growth promoting rhizobacteria and vesicular arbuscular mycorrhizal fungi produce chemical signals that, mediate root growth. Volatile flavor compounds from fungal spores-seeds, bacteria and plant tissue have been implicated in a number of bioregulatory actions. Stem rust of wheat urediniospores (*Puccinia graminis* var. *tritici*) causal agent of wheat stem rust are stimulated to germinate by at levels as low as 0.01 ppm. These volatile compounds are not found alone and an array of compounds may be active stimulators of fungal spores.

Activity of allelochemicals against plants is often indirect through inhibition of the growth of microbial symbiosis. Suppression of fungal root colonization hinders water and nutrient absorption, slowing growth and perhaps contributing to delays in reforestation or decline problems in perennial crops (67). The sensitivity of *Rhizobium* sp. to allelochemicals leads to poor nodulation and subsequent reduction in nitrogen available to legumes growing in association with allelopathic plants or residue Allelochemicals from higher plants have been reported to alter microbial respiration, but for the most part their mode of action is unknown. Menzies and Gilbert (53) reported that decomposing plant residue produces volatile components which stimulate spore germination and fungal growth in soil. King and Coley-Smith (37) reported that a volatile principle evolved from onion and leek seedlings caused termination of dormant sclerotic of *Sclerotium cepivorum*.

Schenck and Stotzky (64)) found that volatile compounds released from germinating seeds have marked effects on several microorganisms, and they later surveyed several herbaceous and tree species for production of volatile compounds from their germinating seeds and seedlings (72). Allen and Newhook (2) demonstrated that ethanol suppress the chemotactic response of zoospores of *Phytophthora cinnamomi* Rands. Later, ethanol was found to be present in the rhizosphere of radicles of *Lupinus angustifolius* (77). Concentrations were commonly in the range of 1-5 mM, which were previously shown to exert a positively chemotactic influence on zoospores of *Phytophthora cinnamomi*.

Most propagules of pathogens do not survive to infect suitable host organisms, because of many adversities, one of which is antagonism by other organisms (8). One aspect of antagonism is allelopathy, and the evidence here is voluminous. This phenomenon merges into that of host-plant resistance to pathogens. Most research in this area has concerned production of antibiotics, both by other pathogens and by sarophytes (8). Lockwood (44) suggested that diffusible antibiotics produced by *Streptomyces* sp. might be an important cause. Mishra and Pandey (54) investigated numerous factors affecting soil fungistasis and concluded that their data supported "a biological origin of the inhibition of spore germination." A great many compounds known to inhibit spore germination and fungal growth, are produced by higher plants and enter the soil in various ways (62). Li *et al.* (41) presented much evidence concerning production by higher plants of compounds antagonistic to a root rot pathogen, *Poria weirii* Murr. Li *et al.* (40) described an excellent quantitative method for assaying soil for inhibitory fungi. These investigators found that many phenolics and other compounds which have been repeatedly isolated from soil are inhibitory to growth of *Poria weirii in vitro*. They found also that the combination of p-coumaric, syringic and ferulic acids, as found in the roots of Poria resistant red alder (*Alnus rubra* Bong) inhibited growth of *P. weirii* alone. Many species of plants which form an understory in red alder stands were demonstrated to produce several of the same phenolic compounds which are inhibitory to *P. weirii*.

The antagonistic activity of certain microorganisms can be used in biological control of plant diseases. *Fomes annosus* (Fr) cooke in a root pathogen of lobeally and slash pine and its basidiospores germinate on cut stumps, grow down through the stump and infect living roots through root grafts. McGrath (51) suggested the biological control of this disease by inoculating cut stumps with basidiospores of *Peniophora gigantea* (Fr.) Masse. Elliston *et al.*, (23) reported that bean plants (*Phaseolus vulgaris* L.) were protected locally and systemically against anthracnose caused by *Colletotrichum lindemuthianum* by *Colletotrichum* sp. non-pathogenic to bean.

Yakhontov (76) suggested the possibility of using the antagonistic activity of non host plants in controlling plant disease also. He found that water soluble excretions of numerous weeds and crop plants inhibited *Dactilospheara viticola* F. which causes the disease, phylloxera in grape plants.

There is a strong evidence for their host origin a doubt still exists that the noval isoforms could originate from the mycorrhizal fungi either as usual components playing a role in fungal morphogenesis and autolysis or as host inducible enzymes (68). Comparisons of different host plants inoculated with one AM fungus have shown that novel chitinase isoforms vary depending on the plant species, lending further support to the host origin of mycorrhiza induced chitinase isoforms (27). By comparing pea, tobacco and tomato roots infected by VAM or pathogenic fungi, it has also been demonstrated that different chitinase isoforms are activated by the two types of fungi (17), demonstrating that there is elicitation by AM fungi of specific root chitinase isoforms which differ from both constitutive and pathogen induced ones. In tomato roots which were first inoculated by the VAM fungus *G. mosseae* and two weeks later by the pathogenic fungus *Phytophthora nicotiana* var. *parasitica* all additional chitinase isoforms were detected. The roots of most plant species are natural hosts not only to pathogenic fungi, but also to beneficial mycorrhizal fungus symbionts. Dumas *et al.* (18) reported that mycorrhiza induced chitinase isoforms, firstly they have any antifungal activities against soil borne pathogens or even against other microorganisms and secondly if they are able to release elicitors from VAM fungi.

3. CONCLUSION

Allelochemicals are primarily secondary products of plant metabolism which have been an enigma to plant scientists, however, they under go a variety of reactions with plants, insects and animals that inhibit or stimulate the growth and development of such species. Increasing knowledge of allelopathy is helpful in under standing many ecological phenomenon. Besides the antibiotic and antitumour efficacy of some allelo-chemicals indicate their possible medicinal exploitation and there is a need of collaborative research between plant chemists and medical scientists to select some allelochemical models to explore their fruitfulness as antitumour agents. Thus allelopathy involves multiple disciplines. Its recent advances have been made through the use of modern instrumentations and it is moving from practical biology to basic molecular biology quite rapidly.

4. REFERENCES

1. Abbott, L.K. Robson, A.D. and Scheltema, M.A. 1995. Managing soils to enhance mycorrhizae benefits in mediterranean agriculture. *Crit Rev. Biotechnol.* **15** : 213-218.

2. Allen, R.N. and Newhook, F.J. 1974. Suppression by ethanol of spontaneous turning activity in zoospores of *Phytophthora cinnamoni. Trans. Brit. Mycol. Soci.* **63** : 383-385.

3. Andrade, G., Azcon, R. and Bethlenfalvay, G.J. 1995. A rhizobacterium modifier plant and soil responses to the mycorrhizal fungus *Glomus mosseae. Appl. Soil Ecol.* **2** : 195-202.

4. Bagyaraj. D.J. and Menge, J.A. 1978. Interactions between a VA mycorrhizae and *Azatobacter* and their effects on the rhizosphere microflora and plant growth. *New Phytol.* **80** : 567-573.

5. Baker, R. 1981. Eradictation of plant pathogens by adding organic amendments to sils. In: *Handbook of Pest Management in Agriculture*, (ed. D. Pimental). Chemical Rubber Company Press, Boca Raton, Florida, Vol. 2, pp. 137-157.

6. Bansal, M. and Mukerji, K.G. 1994a. Efficacy of root litter as a biofertilizer. *Biol. Fertel. Soils,* **18** : 228-230.

7. Bansal, M. and Mukerji, K.G. (1994b). Positive correlation between VAM induced changes in root exudation and mycorrhizosphere mycoflora. *Mycorrhizae,* **5** : 39-44.

8. Bell, A.A. 1977. Plant pathology as influenced by Allelopathy. In : *Report of the Research Planning Conference on the Role of Secondary Compounds in Plant Interactions* (Allelopathy), (eds. C.G. McWhorter, A.C. Thompson and E.W. Hanser) USDA, Agricultural Research Service. Tifton, Ga. pp. 64-99.

9. Bethlenfalvay, G.J. and Linderman, R.G. 1992. *Mycorrhizae in Sustainbale Agriculture.* American Society of Agromony Special Publication No. 54. American Society of Agronomy, Madison, Wisconsin, USA, p. 124.

10. Bthlenfalvay, G.J. and Schuepp 1994. Arbuscular mycorrhizae and agrosystem stability. In : *Impact of Arbuscular Mycorrhizae on Sustainable Agriculture and Natural Ecosystems* (eds. S. Gianinazzi and H. Schuepp). Birkhauser Verlag Basel, pp. 117-131.

11. Bethlenfalvay, G.J. Andrade, G., and Azlon, Aguilar, C. 1998. Plant and soil responses to mycorrhizae fungi and rhizobacteria in nodulated or nitrate fertilized peas. *Biol. Fert. Soils* (in Press).

12. Blum, Udo. 1993. The value of model plant microbe soil systems for understanding processes associated with allelopathic interaction. In : *Allelopathy Organisms Processes and Applications* (eds. Inderjit, K.M.M. Dakshini and Frank, A. Einhelling). American Chemical Society, Washington, DC. 1995, pp. 39-58.

13. Burgos Leon, W., Gangy, F., Nicou, R., Chopart, J.L., and Dommergues, Y. 1980. In : *Recent Advances in Phytochemistry Chemically Mediated Interactions Between Plants and other Organisms* (eds. Copper, Driver, Swain and Conu) Plenum Press, New York, pp. 546.

14. Chou, C.H. and Lin, H.J. 1976. *J. Chem. Ecol.* **2** : 353-367.

15. Cote, J., and Thibant, J. 1988. Allelopathic potential of Raspberry folia leachates on growth of ectomycorrhizal fungi associated with black spruce. *Amer. J. Bot.* **75** : 966-970.

16. Curl, E.A. and Truelove, B. 1986. *The Rhizosphere*. Springer, Verlag, Berlin, p. 288.

17. Dassi, B., Dumas Gaudot, E., Asselin, A., Richard, C. and Gianinazzi, S. 1996. Chitinase and β 1-3 glucanase isoforms expressed in pea roots inoculated with arbuscular mycorrhizae or pathogenic fungi. *EJPP* **102** : 105-108.

18. Dumas Gandot, E., Slezack, S., Bassi, B., Pozos M.J., Gianinazzi-Pearson, V. and Giannninazzi, S. 1996. Plant hydrolytic enzymes (Chitinase and β 1-3, glucanase) in root reactions to pathogenic and symbiotic microorganisms. *Plant and Soil* **185** : 211-227.

19. Dumas-Gandot, E., Grenier, J., Furlan, V. and Asselin, A. 1992. Chitinase and β 1-3 glucanase activities in *Allium* and *Pisum* roots colonized by *Glomus* species. *Plant Sci.* **84** : 17-24.

20. Dumas-Gandot, E., Guillaume, P., Tahiri Alaoni, A., Gianinazzi Pearson, V. and Gianinazzi, S. 1994. Changes in polypeptide patterns in tobacco roots colonized by two *Glomus* species. *Mycorrhizae* **4** : 215-221.

21. Einhelling, F.A. 1995. Allelopathy : current status and future goals. In : *Allelopathy Organisms, Processes and Applications* (eds. Inderjit, K.M.M. Dakshini and Frank, A. Einhellig). American Chemical Society, pp. 1-23.

22. El-Ghachtouli, N., Paynot, M., Morandi, D., Martin Tanguy J. and Gianinazzi, S. 1995. The effect of polyamines on endomycorrhizal infection of wild type *Pisum sativum*. cv. Fiersson (nod + myct) and two mutants (nod myc+ and non myc). *Mycorrhizae* **5** : 189-192.

23. Ellison, J., Kuc, J. and Williams, E.B. 1976. Protection of *Phaseolus vulgaris* against anthracnose by *Colletotrichum* species non pathogenic to bean. *Phytopathol. Z.* **6** : 117-126.

24. Garbaye, J. 1994. Les bacteres auxiliairesde la mycorrhization : une nouvella midension de la symbiose mycorrhizienne (mycorrhiza helper bacteria : a new dimension of the mycorrhizae symbiosis). *Acta Bot. Gallica,* **141** : 517-521.

25. Gianinazzi Pearson, V., Tahiri, Alaoui, A., Antoniw, J.F., Gianinazzi, S. and Dumas, E. 1992. Weak expression of the pathogenesis related. PR-b-1 gene and localization of related protein during symbiotic endomycorrhizae interactions in tobacco roots. *Endocytobiosis Cell Res.* **8** : 177-185.

26. Gianinazzi, S. and Schuepp, H. 1994. Impact of arbuscular mycorrhizas on sustainable agriculture and natural ecosystem. Birkhauser Verlag Basel, p. 226.

27. Gianinazzi, S., Gianinazzi-Pearson, V., Franken, P., Dumas Gandot, E., Van Thinen, D, Samara, A., Martin Lauraut, F. and Dassi, B. 1995. Molecules and genes involved in mycorrhiza

functioning. In : *Biotechnology of Ectomycorrhizae* (eds. P.V. Stocchi, P. Boofontane and M. Nuti). Plenum Press, New York, USA, pp. 67-76.

28. Gilbert, B. 1977. In : *Natural Products and Protection of Plants* (ed. Marini-Bettolo). Elsevier Scientific Publishing Company, New York, p. 209.

29. Graham, J.T., Leonard, R.T. and Menge, J.A. 1981. Membrane mediated decrease in root exudation responsible for phosphorus inhibition of vesicular arbuscular mycorrhiza formation. *Plant Physiol.* **68** : 548-552.

30. Gupta, R. and Mukerji, K.G. 1998. Mycorrhizal Allelopathy in *Trigonella foenum graceum.* In : *Ethnomycology to Fungal Biotechnology* - Exploiting Fungi from Natural Resources for Novel Products (eds. Jagjit Singh and K.R. Aneja). Plenum Press, New York, U.K. pp. 235-246.

31. Handley, W.R.C. 1963. Mycorrhizal associations and *Calluna* heathland afforestation. *Bull. For. Comm.* London. No. 36.

32. Harrison, M.J. and Dixon, R.A. 1993. Isoflavonoid accumulation and expression of defense gene transcripts during the establishment of vesicular arbuscular mycorrhizae associations in roots of *Medicago truncatula. MPMI* **6** : 643-654.

33. Ingham, R.E. 1992. Interactions between invertebrates and fungi : effects on nutrient availability. In : *The Fungal Community : Its Organization and Role in the Ecosystem* (eds. G.C. Carroll and D.T. Wicklow). Marcel Dekker, New York, pp. 669-690.

34. Jakobsen, I. and Rosendhal, T. 1990. Carbon flow into soil and external hyphae from roots of mycorrhizal cucumber plants. *New Phytol.* **115** : 77-83.

35. Jalal, M.F., Read, D.J. and Haslam, E. 1982. Phenolic contents and its seasonal variation in *Calluna vulgaris. J. Phytochem.* **21** : 1397-1401.

36. Khailov, K.M. 1974. Biochemical Trophodynamics in Marine Coastal Ecosystems, Naukova Dumka, Kiev.

37. King, J.E, and Coly Smith, J.R. 1968. Effects of volatile products of *Allium* species and their extracts on germination of sclerotia of *Sclerotium cepivorum* Berk. *Ann. Appl. Biol.* **61** : 407-414.

38. Koch, K.E. and Johnson, C.R. 1984. Photosynthate partitioning in split root citrus seedlings with mycorrhizae and non-mycorrhizae root systems. *Plant Physiol.* **75** : 26-30.

39. Kothari, S.K., Marschner, H. and George, E. 1990. Effects of VA mycorrhizae fungi and rhizosphere microorganisms on root and shoot morhphology, growth and water relations in maize. *New Phytol.* **116** : 303-311.

40. Li, C.Y., Lu, K.C., Trappe, J.M. and Bollen, W.B. 1969. A simple quantitative method of assaying soil for inhibitory fungi. U.S. Forest Service Res. Note. PNW-108.

41. Li, C.Y., Lu, K.C., Trappe, J.M. and Bollen, W.B. 1973. Formation of p-hydroxybenzoic acid from phenylacetic acid by *Poria weirii. Can. J. Bot.* **51** : 827-828.

42. Linderman, R.G. 1988. Mycorrhizal interactions with the rhizosphere microflora. The mycorrhizosphere effect. *Phytopathology* **78** : 366-371.

43. Linderman, R.G. 1994. Role of VAM fungi in biocontrol. In : *Mycorrhize and Plant Health* (eds. Pfleger and R.G. Linderman). The American Phytopathological Society Press, St. Paul, USA, pp. 1-27.

44. Lockwood, J.L. 1959. *Streptomyces* sp. as a cause of natural fungitoxicity in soil. *Phytopathology,* **49** : 327-334.

45. Mallik, A.V. 1987. *J. For. Ecol.* **20** : 43-51.

46. Mallik, A.V. 1992. In : *Allelopathy : Basic and Applied Aspects.* (eds. S.H.J. Rizvi and V. Rizvis) Chapman & Hall, London, pp. 321-340.

47. Mallik, A.V. and Zhu, H. 1995. Over coming allelopahtic growth inhibition by mycorrhizal inoculation. In : *Allelopathy Organisms : Processes and Applications.* (eds. Inderjit, K.M.M. Dakshini and Frank, A. Einhelling) American Chemical Society, Washington, DC, 1995, pp. 39-58.

48. Mandava, N.B. 1985. *Hand book of Natural Pesticides.* C.R.C. Press, Florida, pp. 163.

49. Martin, F., Lapeyrie, F. and Tagu, D. 1996. Altered gene expression during ectomycorrhiza development. In : *The Mycota* (eds. P. Lemke and G. Caroll). Springer Verlag, berlin, Germany.

50. McAlister, C.B., Garcia-Romera, I., Martein, J., Crodeas, A. and Ocampo, J.A. 1995. Interactions between *Aspergillus niger* var. *fiegh* and *Glomus mosseae* (Nical & Gred.) Gred. & Trappe. *New Phytol.* **129** : 309-316.

51. McGrath, W.T. 1972. Biological control of *Fomes unnosus* : a new possibility in the United States. *Consultant,* **17** : 94-96.

52. Melin, E. 1963. Some effects of forest tree roots on mycorrhizal Basidiomycetes. In : *Symbiotic Association* (eds. P.S. Nutman and B. Mosse, eds.). Cambridge University Press, London and New York, pp. 125-145.

53. Menzies, J.D. and Gilbert, R.G. 1967. Response of soil microflora to volatile components in plant residues. *Soil Sci. Soc. Amer. Proc.* **31** : 495-496.

54. Mishra, R.R. and Pandey, K.K. 1974. Studies on soil fungi : V. Effect of temperature, moisture content and incertation period. *Indian Phytopath.* **27** : 475-479.

55. Molisch H. 1937. Der Einfluss einer pflanze anf die andere-Allelopathie. Gustar Fischer Verlog, Jena.

56. Mukerji, K.G. 1995. Taxonomy of endomycorrhizal fungi In : *Advances in Botany* (eds. K.G. Mukerji, Binny Mathur, B.P. Chamola and P. Chitralekha) APH Publishing Corporation, New Delhi, pp. 212-218.

57. Muller, C.H. 1969. Allelopathy as a factor in ecological process. *Vegetatio* **18** : 348-357.

58. Olsen, R.A., Odham, G. and Lindeberg, G. 1971. Aromatic substances in leaves of *Populus tremula* as inhibitors of mycorrhizal fungi. *Plant Physiol.* **25** : 122-129.

59. Persidsky, D.J., Loewenstein, H. and Wilde, S.A. 1965. Effect of extracts of prairie soils and prairie grass roots on thé respiration of ecotrophic mycorrhizal. *Agron. J.* **57** : 311-312.

60. Peters, N.K. and Long, S.R. 1988. Alfalfa root exudates and compounds which promote or inhibit induction of *Rhizobium meliloti* nodulation genes. *Plant Physiol.* **88** : 396-400.

61. Read, D.J. 1984. In weed control and vegetation management in forests and amenity areas. *Asp. App. Biol.* No. 5, pp. 195-209.

62. Rice, E.L. 1974. *Allelopathy*. Academic Press. N.Y.

63. Rice, E.L., 1984. *Allelopathy* (2nd edition). Academic Press, New York, pp. 385.

64. Rice, E.L. 1992. In : *Allelopathy : Basic and Applied Aspect* (eds. S.J.H. Rizvi and V. Rizvi). Chapman and Hall, London, U.K. pp. 31-58.

65. Rice, E., Lin, C.Y. and Huang, C.Y. 1981. Effects of decomposing rice straw on growth of and nitrogen fixation by *Rhizobium. J. Chem. Ecol.* **7** : 333-334.

66. Robinson, R.K. 1972. The production of roots of *Calluna vulgaris* of a factor inhibitor to growth of some mycorrhizal fungi. *J. Ecol.* **60** : 219-224.

67. Rose, S.L., Perry, D.A., Pilz, D. and Schoenbergen, M.M.J. 1983. *J. Chem. Ecol.* **9** : 1153-1162.

68. Sahai, A.S. and Manocha, M.S. 1993. Chitinases of fungi and plants : their involvement in morphogenesis and host parasite interaction. *FEMS. Microbiol. Rev.* **11** : 317-338.

69. Schenck, S. and Stotzky, G. 1975. Effect of microorganisms of volatile compounds released from germinating seeds. *Can. J. Microbiol.* **21** : 1622-1634.

70. Schreiner, R.P., Mihara, K.L., McDaniel, H. and Bethlanfalvay, G.J. 1997. Mycorrhizal fungi influence plant and soil functions and interactions. *Plant Soil.* **188** : 199-209.

71. Shukla, A.N., Arora, D.K. and Dwivedi, R.S. 1977. Effect of microbial culture filtrates on the growth of sal (*Shorea robusta* Gaertru). Leaf litter fungi. *Soil. Biol. Biochem.* **9** : 217-219.

72. Stotzky, G. and Schenck, S. 1976. Observations on organic volatiles from germinating seeds and seedlings. *Amer. J. Bot.* **63** : 798-805.

73. Theodorou, C. and Bowen, G.D. 1971. Effects of non-host plants on growth of mycorrhizal fungi of radiata pine. *Aust. For.* **35** : 17-22.

74. Wardle, D.A. 1995. Impacts of disturbance on dieritus food webs in agroecosystems of contrasting tillage and weed management practices. *Ad. Ecol. Res.* **26** : 105-185.

75. Wiemken, V. and Ineichen, K. 1992. Effect of neutral and pathogenic fungi on mycorrhizal and non-mycorrhizal *Picea roots* : transpiration and accumulation of the stress aminocyclapropane carboxylic acid. *J. Plant Physiol.* **140** : 605-610.

76. Yakhontov, A.F. 1973. On the possibility of using the allelopathic action of some plants for controlling *Dactilosphera viticola* F. In : *Physiological-biochemicals Basis of Plant Interaction in Phytocenoses,* (ed. A.M. Grodzinsky) Nankova Damka Kiev. (In Russia, English Summary), Vol. 4, pp. 57-60.

77. Young, B.R., Newhook, F.J. and Allen, R.N. 1977. Ethanol in the rhizosphere of seedlings of *Lupinus angustifolius. L.N.Z. J. Bot.* **15** : 189-191

6 Involvement of Cellular Mechanisms in the Development of Drug Resistance in Bacteria

Kiran Singh, Protiti Bose and Mohd. Sarosh Khan

CONTENTS

ABSTRACT 68

1. INTRODUCTION 68
   1.1 Definition of Resistance 68
   1.2 Resistant and Insensitive Bacteria 69
      1.2.1 Pre-existant resistance 69
      1.2.2 Mutational resistance 69
      1.2.3 Transmissible resistance 69

2. HISTORY OF DISCOVERY 69

3. MODE OF ACTION OF ANTIBIOTICS 70
   3.1 Interference in Cell Wall Synthesis 71
   3.2 Interference with Protein Synthesis 71
   3.3 Interference with Cell Membrane Function 71
   3.4 Interference with Chemical Environment 72
   3.5 Interference with Nucleic Acid Synthesis 72

4. BASIS OF RESISTANCE IN BACTERIA 72
   4.1 Genetic Basis of Antibiotic Resistance 72
   4.2 Biochemical Basis of Antibiotic Resistance 73

5. SIGNIFICANCE 74

6. FUTURE PERSPECTIVES 75

7. REFERENCES 75

Advances in Microbial Biotechnology
J.P. Tewari, T.N. Lakhanpal, Jagjit Singh, Rajni Gupta & B.P. Chamola (eds.)
APH Publishing Corporation, New Delhi - 110 002, India.

ABSTRACT

The value of antimicrobial medications has been markedly impaired by the emergence and dissemination of drug resistant microbes. Studies of the mechanisms of drug action and the mechanisms of microbial resistance have aided in finding new antimicrobial medicines. Antibiotics and other antimicrobials have been of enormous value in controlling many serious infections, but their continued value depends largely on their judicious use, to minimise the incidence of organisms resistant to them. The establishment of such resistant strains in a community spread from person to person. The overcrowding and mobility of the human population, ventilation of houses and offices, sanitation and several other factors play a major role in the establishment of resistance. The administration of an antibiotic to a person markedly increases the probability that person will be colonized by extraneous antibiotic resistant strains of an organism presumably by the suppression, of sensitive strains in the person's normal flora.

This article finally highlights the impact of antibiotic resistance and the awareness we ought to bring in our population at large.

1. INTRODUCTION

The availability of clinically effective antimicrobial agents have regularly been followed by the rapid emergence of strains of resistant to them. This has seriously reduced the therapeutic value of many important antibiotics and have posed fresh challenges demanding new strategies in trying to circumvent this problem. Consequently, it is a major stimulus to the pharmaceutical industries in its constant search for newer and more effective antimicrobial drugs. This calls for a better understanding of bacterial resistance in different fields of biology, medicine and evolution. The spread of resistant genes in bacterial population should be taken upon, on a priority basis world-wide.

In general, resistance is stable and spreads rapidly within a host population for several generations. This is particularly important in case of hospitals, where the problem can become endemic and may infect many patients.

1.1 Definition of Resistance

Bacterial isolates have been labelled "SENSITIVE" or "RESISTANT" to antimicrobial agents ever since these agents were brought into use. The criteria, whether a bacterial isolate should be labelled "sensitive" or "resistant" ultimately depends on the likelihood that an infection with that organism can be expected to respond to treatment with a given drug. This may involve the relationship of the Minimum Inhibitory Concentration (MIC) of the antibiotic levels readily attainable in the body (This usually means, blood and tissue concentrations; an organism resistant to these may of course be sensitive to the higher concentrations; attainable in urine or by tropical applications), to the relationship of the sensitivity of the strain under examination or others of the same species. Over and above, in the way in which all these factors affect clinical responsiveness in also an important criteria (1).

1.2 Resistant and Insensitive Bacteria

In general bacterial species are not all intrinsically sensitive to all antibiotics. Species tend to vary widely in this respect. However, in considering the problems caused by resistance of bacteria to the action of antimicrobial drugs, it is usually the emergence of resistant in previously sensitive bacterial species that is of concern. At least these different situations, each producing its own problems can be distinguished (4).

1.2.1 Pre-existant resistance

In this type of bacterial species most of the strains are sensitive to a newly introduced antimicrobial, but a few strains are already resistant at the time the drug is introduced. The classic example here is *Staphylococcus aureus* and Penicillin.

1.2.2 Mutational resistance

This includes bacterial species fundamentally, all sensitive to the agent under consideration but which are able to mutate more or less readily to resistance. With these any large population of bacterial cells, a very few individual cells will be resistance at any one time. In absence of the agent, these cells will have no particular survival advantage, but once the antibiotic is brought into use, selective pressure again ensure replacement of sensitive ones by the resistant ones.

1.2.3 Transmissible resistance

Over the past two decades, the importance of these two situations have been overshadowed by the realization that genes conferring antibiotic resistance on bacterial cells can fairly readily pass from one bacterial cell to another both in the laboratory, and wild. In fact, transfer and selection is a much more fruitful source of antibiotic resistance genes in bacterial population than the development of resistance *de novo*. Development of resistant cells does not have to happen often or on a large scale. A single mutation or transfer event can happen in the appropriate selective pressures are operating, lead to the replacement of a selected population by a resistance one. Without selective pressure, antibiotic resistance may be a handicap rather than an asset to a bacterium.

2. HISTORY OF DISCOVERY

Bacterial resistance become much more of a problem as the widespread use of antibiotics led to the elimination of susceptible bacterium from the population with accompanying increase in the number of the resistant ones. The introduction of Penicillin and Sulphonamides in 1930's can be said to mark the beginning of the era of modern medicine. The development of bacterial resistant was rather infrequent during this time.

During the first few years after the introduction of the sulfonamides and penicillin, there was great hope that such agents would soon do away with most infectious diseases yet today, microbial resistance limits the usefulness of all known antimicrobial drugs.

In 1935, the dramatic increase in bacterial drug resistance to sulphonamides for the cure of the disease gonorrhoea brought in the first antibiotic Penicillin as "second - line drug". However, subsequent antibiotics continued bringing in similar problems. The extent has varied from one antibiotic to another with different bacteria.

Prior to the inception of antibiotics in clinical medicine, the prognosis for patient's with severe Staphylococcal infections was very poor. The use of Benzyl Penicillin in 1940's offered effective treatment to the infection caused by *Staphylococcus aureus*, unfortunately, by 1941 Penicillin resistant strains started appearing, although they occurred at a very low frequency of 1%. The WHO survey report showed that by 1946 the heavy use of Penicillin soon fostered their effectiveness, with the percentage of resistant strains going up to 14%. A year later, 38% were resistant and today resistant is found in more than 80% of all strains of *S. aureus*. Slowly multiple - resistant *S. aureus* strains were responsible for many outbreaks in hospital infections till the 1980's worldwide. This brought into use Vancomycin - a somewhat toxic antibiotic - the only drug effective against such staphylococci prevalent in hospitals.

In contrast, over the same period an equally important pathogen *Streptococcus pyogenes* has remained uniformly sensitive to penicillin. Penicillin resistance has also been reported recently in occasional strains of Pneumacocci. Staphylococci capable of destroying penicillin existed before penicillin become available, whereas no such strains of streptococci have been found. The importance of this has been underlined by the recent appearance and rapid spread of strains of *Neisseria gonorrhoea* capable of destroying penicillin; for many years this species was uniformly susceptible to penicillin and acquisition of the necessary genetic information for resistance apparently occurred as rare transfer event. So far, the closely related *N.meningitides* remain penicillin sensitive.

However, according to Greenwood (4) the possession of the genetic capacity of resistance in a species does not always explain its prevalence. Although about 80% of all strains of *S.aureus* are now penicillin resistant, the same has not happened to sulphonamide or ampicillin resistance in *Escherichia coli* under ostensibly similar selective pressure. At the present time only about 25% of *E. coli* strains are resistant to sulfonamides or ampicillin - plateau level of resistance which has existed more or less stably for many years.

Apart from these few case history reports many other cases *viz.* transfer of antibiotic resistance by conjugation, reported in 1958 by two Japanese Scientists, Akiba and Ochiai - they isolated both anti-sensitive and resistant organism of the same stereotype from patients with enteric infection being treated with a combination of drugs. According to them the resistant gene harboured in *E.coli* of the intestinal tract after being transferred to shigella dysentriae caused the infection. Antitiotic resistance has slowly become a seriuos problem worldwide and great efforts are being made by clinicians to understand its need (2).

3. MODE OF ACTION OF ANTIBIOTICS

A very large number of antibiotics (approx. 6000) have now been identified Gale (3) has classified those in current use on the basis of their mode of action. These include:

Group 1 those affecting the bacterial cell wall, such as penicillin and cephalosporins

Group 2 : agents affecting the cell membrane, such agents nystatin and amphotericin B

Group 3 : includes antibiotics interfering with protein synthesis - chloramphenicol, tetracycline and streptomycin

Group 4 : drugs which alter the biochemical environment - sulfonamides, isoniazid and para aminosalicylic acid, and

Group 5 griseofulvin which affects nucleic acid metabolism.

3.1 Interference with Cell Wall Synthesis

It follows from what has been said that bacterial and mammalian cells must differ in some fundamental respect, so that the principle of selective toxicity may function. One way in which cells differ is an anatomical one - bacterial cells possessing a thick cell wall (10-25 m.u.) while, the mammalian cells of the host do not. Inhibition of peptidoglycan synthesis in the bacterial cell takes place in there major stages.

(i) Synthesis of precursors in the cytoplasm.

(ii) Transfer of precursors to a lipid carrier molecule (undecaprenyl phosphate) which transports them a cross the cytoplasmic membrane.

(iii) Insertion of glycan units into the cell wall- attachment by transpeptidation and further final maturation steps. It has been found that certain antibiotics *viz.* Penicillin, cephalosporin, bacitracin, novobiocin, vancomycin and cycloserine act by chemical interference with the bacterial cell wall (3).

Penicillin acts by interfering with the mucopeptide synthesis which results in the production of deficient cell walls. This exposes the cells to osmotic action - the cell swells, rupture and die. It possible that differences in muco-peptide composition and its relation to the lipo-protein cell membrane are major factors in determining the response of Gram-positive and Gram-negative cells to the action of antibiotics of this group.

3.2 Interference with Protein Synthesis

Antibiotics affecting protein synthesis are chloramphenicol, streptomycin, tetracycline, kanamycin and neomycin which do not act on the bacterial cell wall, but inhibit protein synthesis within the cell. The antibiotics are thus selectively toxic towards bacteria (5). This selectivity of clinically useful inhibitors of bacterial protein synthesis arises principally from their ability to bind selectively to bacterial rather than mammalian ribosomes. In contrast to earlier views, a consensus is now emerging that ribosomal RNA is the primary target for a number of antibacterial drugs rather than ribosomal proteins.

3.3 Interference with Cell Membrane Function

All cells are bound by a cell membrane, but those of bacteria and fungus differ significantly from those of the animal cells to allow selective action of antimicrobial agents. Certain polypeptide antibiotics such as polymixins acts as detergents and dissolve bacterial cell membranes, probably by binding to phospholipids in the membrane. They are especially effective against Gram - negative bacteria, which have cell membranes phospholipids - phosphatidyl, ethanolamine, particularly sensitive to polymixins which explains why Gram-negative bacteria are more susceptible to polymixins than Gram-positive organisms. This is because membranes of the latter generally do not contain phosphatidylethanolamine.

Polymixins have only a minor place in medicine because they also have affinity for mammalian membranes. However, they appear to bind less readily to mammalian membranes than to bacterial membranes. The basis of this dissemination may relate to the presence of cholesterol in mammalian membranes.

3.4 Interference with the Chemical Envrionment

The sulphonamide drugs act by depriving susceptible bacteria of para-aminobenzoic acids (PABA) which is necessary for the formation of folic acid. Organisms which must synthesise folic acid, are susceptible to this drug action. However, organisms which use existing folic acid are not susceptible and bacteriostatic resulting from sulphonamides which can be reversed by providing PABA in the culture medium. Mammalian cells can use preformed folic acid - they are therefore in a similar position to insensitive microorganisms which do not synthesize folic acid (10).

3.5 Interference with Nucleic Acid Synthesis

This group includes antibiotic e.g. Quinolone, which acts by crossing both outer and cytoplasmic membrane located intracellularly and reaching the target (DNA gyrase), thus inhibiting nucleic acid synthesis. Interference with nucleic acid synthesis fall into three main catagories - compounds that interrupt nucleotide synthesis or nucleotide interconversion e.g. Sulphonamides; compounds that interfere with the role of DNA as a template in replication and transcription either by reacting with it directly to form a complex or by causing structural alternatives such as strand breakage or base removal; agents which directly inhibit enzyme processes in nucleic acid synthesis e.g. Rifampicia, Quinolones, etc.

4. BASIS OF RESISTANCE IN BACTERIA

Antibiotics have been widely available for the treatment of bacterial infections for over 40 years. The diversity of mechanisms used by bacteria to evade the action of antibiotics and the rapidity with which bacteria acquire resistance in depressing from the view point of acheiving successful therapy. Nevertheless, studies on the genetic and biochemical basis of antibiotic resistance have themselves suggested ways to circumvent it (6, 7).

In clinical setting, antibiotic resistant bacteria often emerge following heavy selective pressure with antibiotics. They are mainly put into two broad categories namely - intrinsic (inherent features of the cell, that is those expressed by chromosomal genes) and, acquired (mutation in chromosomal agents or by the acquisition of plasmids and transposons). Acquired rather than intrinsic antibiotic resistance usually pose the greatest threat to successful therapy. Both types of resistance can complicate the treatment of infections (8).

4.1 Genetic Basis of Antibiotic Resistance

The resistance of microorganisms to antimicrobial agents nearly always has its basis in genetic changes in the microbial cell. In only a few instances, have phenotypic changes been implicated in resistance. Consequently, a fundamental knowledge of microbial genetics is essential to the understanding of the development and spread of resistance to antimicrobial drugs. According to Russel and Chopra (9) resistance is determined by genes that reside in the host cell such as chromosomes, plasmids or on transposons. Genes conferring resistance have probably been spread from producer organisms to nonproducers via the agencies of transposons and plasmids, so that antibiotic resistance genes have been present in potential pathogens even before the advent of antibiotic chemotherapy. This has been confirmed by the examination of cultures lyophilized before the clininal use of antibiotics. These strains have been found to possess not only resistance genes, but plasmids which provide the vectors for their transmission and spread.

4.1.1 Resistance determined by chromosomal genes

Chromosomal mutation in acquired by the changes in base sequence of DNA. It can be categorised as *microlesion* in which a single base pair is altered by transition, transversion of frame shift mutation; whereas, *macrolesion* involves deletion, duplication, inversion and translocation (insertion). In principle each type of chromosal mutation might directly or indirectly confer antibiotic resistance on the cell.

Not all chromosomally - specified antibiotic resistance is due to mutation. For instance, the intrinsic resistance of the majority of Gram-negative bacteria to hydrophobic antibiotic is due to the expression of resident genes involved in synthesis and assembly of Lipopolysaccharide in the outer membrane, the macromolecule comprising an exclusion barrier to the penetration of hydrophobic compounds. Concerning this the chromosomally encoded beta - lactamases in Gram-negative bacteria that confer resistance to beta lactam antibiotics by hydrolysis of the beta - lactam ring.

4.1.2 Resistance determined by plasmids

It is well- known that plasmid, that is extra-chromosomal genetic element, that replicate independently of the chromosomes, frequently carry genes that confer antibiotic resistance. Plasmid determined antibiotic resistance is more common than chromosomal resistance in clinically important bacteria. The following reason explains the presence or absence of antibiotic selection pressure in majority of the cells in a particular population, deciding the maintenance of plasmid-carrying resistant progeny; genes encoded by plasmids are more mobile than chromosomal genes because plasmids can be transferred both within and between certain species; plasmids are frequently vectors of transposons.

4.1.3 Resistance determined by transposons

Transposons are mobile DNA sequences capable of transferring (transposing) themselves from one DNA molecule (the donor) to another (the recipient). Unlike plasmids, transposons are not able to replicate independently and must be maintained within a functional replicon (e.g. plasmid or chromosomes). Existence of such genetic element confer antibiotic resistance which account for the emergence of multiply - resistant bacteria. Since transposons do not require extensive DNA homology to insert into the recipient molecule, it is possible for the bacteria to acquire several different resistance genes by a series of transposon insertions in resident plasmids.

4.2 Biochemical Basis of Antibiotic Resistance

Initially it was usual to classify resistant bacteria into two groups - those able to grow in the presence of unchanged levels of antibiotic concentrations lethal to sensitive cells and those able to destroy the drugs. In the light of the recent discoveries, it is more useful to consider six mechanisms by which resistance can arise, as described in Table 1. However, resistance to particular antibiotic can result from the combined expression of several of the listed mechanisms. For example, the efficacy of a beta - lactam antibiotic against a Gram - negative organism will be determined by the diffusion rate of the antibiotic across the outer membrane, its affinity for and resistance towards any periplasmic beta - lactamase that may be present and finally, its affinity for the target penicillin binding proteins (PBPs) located in the cytoplasmic membrane.

Table 1 Mechanisms of antibiotic resistance

Mechanisms		Examples
1.	Alteration (antibiotic inactivation)	Acquired resistance to aminoglycosides, beta lactams and chloramphenicol. Intrinsic resistance to beta-lactams.
2.	Target site in the cell is insensitive to the inhibitor but still able to perform its normal physiological function	Acquired resistance to aminoglycosides, beta -lactarms, macrolides quinolones, rifampicin and trimethoprim.
		Intrinsic resistance to trimethoprim.
3.	Decreased antibiotic accumulation :	
	a. Impaired uptake	Intrinsic and acquired resistance to many antibiotics.
	b. Enhanced efflux	Acquired resistance to tetracycline.
4.	By-passed of antibiotic sensitive step by duplication of the target site, the second version being insusceptible to drug action.	Acquired resistance to methicilling, sulphonamides and trimethoprim.
5.	Over production of target so that antibiotic concentrations are needed to inhibit bacterial growth	Acquired resistance to trimethoprim.
6.	Absence of an enzyme/metabolic pathway	Intrinsic resistance to some antibiotics in certain species

5. SIGNIFICANCE

Although many communicable diseases have been effectively contained, bacterial infections remain a major cause of morbidity and mortality, particularly in developing countries. Drug resistance is a clinical problem because it limits, the number of therapeutically useful agents and puts constraints on those antibiotics which can be used. It sometimes forces the use of more expensive, more toxic and more difficult agents that could otherwise be chosen. It is important not to overstate the resident problem. However, most specific infections from anthraxs to yaws remains steadfastly susceptible to agents which have been available for many years, while our armoury is constantly being extended by the development of new and useful drugs.

In contrast to non - bacterial infections, therapeutic failure today is seldom due to the primary resistance of infecting bacteria to the drugs available for treatment. Until recently, antibiotic resistance was a serious problem with only three groups of bacteria the Staphylococci, "coliform" bacilli and *Mycobacterium tuberculosis.*

Nevertheless, the problem is a real one. In most situations in which antimicrobials need to be used treatment must be started and is often completed without the benefit of laboratory help This is true both in general practice where access to the lab may be limited and also in hospitals. In these circumstances, knowledge of local resistance trends may be an important factor. There are well - defined areas, particularly with hospital associated infections in patients with chronic illness or impaired body defences, where the rational choice of antimicrobial agents is increasingly being hampered by the emergence of bacterial drug resistance. In order to combat this situation, the nature of resistance should be well studied and understood.

6. FUTURE PERSPECTIVES

All antibiotic usage creates selective pressure which lead to the emergence of resistant bacterial strains. Antibiotics are vital to the practice of modern medicine; we can not abandon them or even unreasonably restrict their use. Resistant bacteria will continue to appear and to cause problems in the treatment of infections they cause. Surveillance of the use of antibiotics - locally, nationally and internationally coupled with the monitoring of emerging patterns of bacterial resistance will help influence prescribing practice and the development of meaningful antibiotic policies, as long as the people collecting the information ensure that it is passed onto the people who use the drugs.

Few will recklessly continue to prescribe antibiotic to which local resistance is common as long as they know that resistance is common. The dissemination of accurate, up-todate information is vital in establishing the appropriate use of antibiotics which is the best protection against the unrestricted spread of drug resistance problem, which are listed below:

(i) Drugs should not be prescribed to man or animals without valid indications of doing so.

(ii) Patients carrying resistance strains should be kept isolated from other patients.

(iii) Drugs should be given in adequate doses or drug combination should be used when appropriate.

(iv) Introducing an " antibiotic policy for a hospital or area". This involves designating certain drugs as available for general use but withdrawing others from circulation or permitting their use only on rare and special occasions (reserved antibiotic) (11).

ACKNOWLEDGEMENT

The authors are thankful to the Head, Institute of Microbiology and Biotechnology, Barkatullah University, Bhopal for the support rendered.

7. REFERENCES

1. Collins, C.H. and Lyne, P.M. 1990. In : *Medical Microbiology*. Butterwoths Groups publication, London and Boston.

2. Davies, J. 1992. Another look at antibiotic resistance. *J. Gen. Microbiol.* **138** : 1553-1559.

3. Gale, E.F. 1963. Mechanisms of antibiotic action. *Pharmacol. Rev.* **15** : 481.

4. Greenwood, D. 1983. *Antimicrobial Chemotherapy of Large Number of New Antibiotics*. Bailliere' Tindall publication, London. pp. 55-90.

5. Hillas, S. 1969. Chemotherapeutic Mechanisms. In : *Antibiotics in Clinical Practice*. Pitman Press Bath publication, UK, pp. 7-17.

6. Koch, A.L. 1981. Evolution of antibiotic resistance gene function. *Microbiol. Rev.* **45(2)**: 355-378.

7. Nester, E.W., Roberts, C.E., Pearsall, N.N. and Mc Carthy, B.J. 1978. In : *Microbiology*. Holt Rinehart and Winstone Publication, USA., pp 605-610.

8. Pedersons, P.L. 1995. Multidrug resistance - a fascinating, clinically relevant problem in bioenergetics. *J. Bioenerg. Biomembr.*, **27** (1) : 3-5.

9. Russell, A.D. and Chopra, I. 1990. In : *Understanding Antimicrobial Action Resistance*. Ellis Horwood series in Pharmaceutical technology, U.K., pp 39-90.

10. Weinstein, L. 1965. In : The pharmacological basis of therapeutics. L.S. Goodman and A. Gilman (Ed.). Collier-Macmillan Ltd., UK, pp 1148.

11. WHO 1995. The use of essential drugs - 6th report of the WHO expert committee; 850 series, Geneva, pp. 10-12.

7 Microbial Enzymes for Detergent Application

R.K. Saxena, W. Sheba Davidson and Rani Gupta

CONTENTS

ABSTRACT 78
1. INTRODUCTION 78
2. ADVANTAGES OF ENZYMES 79
3. COMPOSITION OF DETERGENTS 79
4. COST OF IMPLICATION 81
5. THE MICROBIAL "WASH" 81
6. ACTIVITY MEASUREMENT 84
7. WASHING PERFORMANCE 84
8. FUTURE TRENDS IN ENZYME-BASED DETERGENTS 84
9. CONCLUSIONS 90
10. REFERENCES 91

Advances in Microbial Biotechnology
J.P. Tewari, T.N. Lakhanpal, Jagjit Singh, Rajni Gupta & B.P. Chamola (eds.)
APH Publishing Corporation, New Delhi - 110 002, India.

ABSTRACT

Microbial enzyme - based detergents today occupy a place of prominence among various categories of detergents. Therefore, presently detergent - compatible enzymes have the largest share of the world market. The major enzymes in use are proteases, lipases, α- amylases and cellulases. These aid the removal of stubborn stains in a mild manner which cannot be achieved easily using the conventional chemical-based detergents. Extensive searches for novel thermostable, alkaline pH-stable and detergent- tolerant enzymes have today led to the widespread use of enzyme-based detergents. Being economical and eco-friendly, these enzymes are the choice of the environment- conscious modern society. Future trends in this direction envisage the use of consortium of natural protein- engineered enzymes whose demand will be on a rise for years to come.

1. INTRODUCTION

The term "enzyme" (literally "in enzyme") was coined by Kuhne in 1876. Yeast, because of the acknowledged importance of fermentation, was a favourite subject of research. A major controversy at that time associated most memorably with Liebig and Pasteur, was whether or not the process of fermentation was separable from the living cell. No belief in the necessity of vital forces, however, survived the demonstration by Buchner that alcoholic fermentation could be carried out by a cell-free yeast extract. The existence of extracellular enzymes had, for reasons of experimental accessibility, already been recognized. For example, as early as 1783, Spallanzani had demonstrated that gastric juice could digest meat *in vitro,* and Schwann called the active substance pepsin. Kuhne himself appears to have given trypsin its present name, although its existence in the intestine had been suspected since the early 19th century.

Till the beginning of the 20th century, however, enzymes were used only for a few purposes. It was in 1913 that pancreatic enzymes were used in a washing application, besides being of use as digestive aids. The introduction of enzymes into detergent formulations was a landmark event as the removal of stubborn stains which could earlier not be achieved using chemicals alone, was now possible biochemically. Enzyme-based detergents led to better, faster and easier wash results.

Enzymes are commonly added to detergent formulations to attack proteineous soils, which most often appear on clothes as stains. After drying and ageing, these stains strongly resist action of most low cost, high volume detergents, The enzymes are then added to attack hydrolytically a peptide or ester bond in the proteins, thereby either destroying it or making the fragments easy for a surfactant to remove. Without enzymes to act on the stains, the solidified particles of blood, egg and human sweat are not easily accessible to detergents (14).

The enzymes which are of interest to the detergent industry at present are proteases which act on protein to form amino acids, α-amylases which convert starches into dextrins, lipases which attack fats and oils and cellulases which act on cellulose.

The sources of these enzymes are today primarily microbial, due to their multifold properties, easy extraction procedures and unlimited supply (6, 17, 18, 19).

2. ADVANTAGES OF ENZYMES

i) The addition to enzymes in heavy and liquid laundry detergents provides a significant contribution to the protein-based stain removal of these detergents.

ii) All the stains can be removed by soaking at room temperature in enzyme-containing detergents, which saves a significant amount of energy required for heating water.

iii) Improved life of fabric by avoiding lengthy mixing and beating process, which is required for removal of the problem stains.

iv) Enzymes are totally biodegradable and hence, do not contribute to pollution.

3. COMPOSITION OF DETERGENTS

Detergent is a group of chemicals, which helps in the removal of soil from clothes during washing. Primary components which are responsible for cleaning action are surface active agents. These are molecules having both hydrophobic and hydrophilic ends. During washing, the hydrophobic ends of these molecules attach to the soil particles on the clothes and take them out into water (8). Apart from surfactants the other components of detergents, viz., dispersing agents, water softeners, bleaching agents, etc. help in improving the efficiency of washing. General composition of a detergent is given in Table 1.

Table 1 : General composition of a detergent

Low cost detergent		High cost detergent	
Components	%	Components	%
Anionics (LAS)	8	Anionics	15
Sodium carbonate (soda ash)	40	Sodium carbonate	40
Sodium sulphate	32	Nonionics	6
Salt	20	STPP	15
		Sodium perborate	10
		Sodium metasilicate	8
		CMC	2
		Salt	3
		Perfume	0.5
		Optical Brightener	0.5

Proximate composition of any detergent is :

(i) surface active agents

(ii) water softeners

(iii) dispersing agents

(iv) bleach agents/activators

(v) optical brighteners

(vi) perfume and colours

(vii) inorganic builders

3.1 Surface Active Agents

Four main classes of surface active agents have been reported for use, viz., anionics (e.g., alkyl benzene sulfonates), cationics (e.g., alkyl dimethyl benzyl ammonium chloride), non-ionics (e.g. octyl phenol) and amphoterics. However, among these four types, mostly combinations of anionics and non-anionics are used in detergent formulations for achieving better surfactancy.

3.2 Water Softeners

Hardness of water is a critical parameter which hinders the action of detergent. As invariably water in rural areas is quite hard, it becomes absolutely necessary to use softeners in detergent formulations. Water softeners are basically chelating agents which bind hardness causing ions like calcium and magnesium. Water softeners used are polyphosphates, EDTA, etc.

3.3 Dispersing Agents

It is very important to ensure that soil released on clothes during washing should not redeposit on clothes. In order to maintain these soil particles in water it is necessary to use a dispersing agent. Many dispersing or anti-redepositing agents are available. Commercially, however, most commonly used is carboxymethyl cellulose. Its level of usage is 1-2% in detergents.

3.4 Bleach Agents/Activators

Active oxygen present in water helps in giving bleaching effect, which aids the removal of many oxidizable stains. Bleaching agents like perborate, hypochlorites and hydrogen peroxide are well known. Low priced detergent formulations generally use hypochlorites while use of perborates is restricted to expensive detergents.

3.5 Optical Brighteners

They are used in detergents to complement any yellowishness generated in clothes (due to reuse) to make it look whiter and brighter. Compounds like Tinopal, Ultraphor, Fluorosal, etc. are popularly used.

3.6 Perfume and Colours

These are used at 0.1-0.5% level in detergents in order to improve its consumer appeal.

3.7 Builders

All the above mentioned components constitute less than 50% of detergent composition. Hence, other filler compounds are added as diluents in the detergent. Both organic and inorganic compounds are used as fillers or builders. Inorganic salts, viz., soda, sodium silicate and common salt are popular low cost builders.

4. COST IMPLICATION

The quality of detergents depends on the active matter present in it. Active matter (surfactant) at the level of 40% gives best result. But this high presence of active matter adds to the cost of detergents. Lowering the amount of active matter gives unsatisfactory performance. In order to increase the level of performance with a low amount of active matter, enzyme supplementation helps, as it hydrolyses big molecules like proteins to small fragments which are easier to remove. Enzymes which work at low levels of 0.2-0.5% are therefore, cost effective.

5. THE MICROBIAL "WASH"

As early as 1913, enzymes were used in laundry detergents by the German chemist Otto Rohm, who patented the application of pancreatic enzymes in a presoaking detergent composition. These enzymes, for example, trypsin and chymotrypsin, are active around pH 7-9, which was achieved by presoaking - in these days with sodium carbonate. The washing effect of trypsin and chymotrypsin would be negligible in modern detergents at higher pH (2).

The first detergent with a bacterial protease, Bio 40 did not appear on the market until 1959 and was produced by Gebruder Schnyder in Switzerland. The Danish company, Novo Nordisk, introduced Alcalase, an alkaline protease produced by *Bacillus licheniformis.* This was followed a few years later by Maxatase, made by Gist-brocades. Alcalase was formulated in a detergent product, still known as Biotex, which was marketed successfully and stimulated the development of many enzyme-based detergents (5).

The rapid growth of enzyme detergents was temporarily set back in the early 1970s, when workers in detergent factories developed allergies to the enzyme preparations. Enzyme manufacturers solved this problem by developing dust-free protease formulatiôns. In 1985, approximately 70%; of the heavy-duty laundry detergents (for domestic use) in Europe contained enzymes, compared to 15 %- in the United states. Presently, the major enzyme suppliers, produce a variety of differently formulated proteases for use in all types of liquid and powder detergents, with or without bleach and phosphates. In addition to proteases, other enzymes are also used in detergents (e.g., α-amylases), and the use of enzymes such as cellulases and lipases is being developed.

Following the allergy problems encountered in the 1970s, methods were developed to reduce the dusting properties of powdered enzymes in detergents involving enzyme encapsulation or granulation. These processes appear to improve the stability of the enzyme in the detergent powder. Non-ionic surfactants or sodium chloride/polyethylene glycol mixtures are used to encapsulate the enzymes. One process involves mixing the enzymes with melted non-ionic detergents and then spraying into a cold chamber so as to solidify the droplets as granules. It has been reported that cellulose aids the granulation process and produces particles with improved resistance to mechanical damage (2).

Composition of the wash liquor and laundering practices differ from country to country. In Europe, detergent concentrations of 4-10 g/L are used. The detergent formulation contains bleach and the washing process is characterized by a heating-up phase followed by a constant-temperature phase. In the United States and Japan, the detergent concentration is lower. Bleach and builders are often added separately or not at all. Washing is done at constant temperatures, significantly lower than those in Europe.

Enzyme suppliers have developed proteolytic enzymes that fulfill all the requirements of detergent formulation, stability and activity under washing conditions. The types of stains subjected to proteolytic attack are composed of protein and other soil. The protein acts as a binder and fixes the other soil components to the fabric. The protease acts synergistically with the detergent; limited proteolysis is sufficient to make the stain susceptible to removal by the detergent components. Detergent proteases are endoproteases which split the protein into oligopeptides (1).

Powdered laundry detergents, which constitute the largest part of the detergent market, function at a high pH. The relationship between temperature and activity is another important characteristic. The activity rises with increasing temperature until it decreases suddenly at higher temperatures due to enzyme inactivation. Because many powder detergents contain such sequestering agents as polyphosphates, nitrilotriacetic acid (NTA), or zeolites, enzymes whose activity depends on loosely bound metal ions like Mg^{2+} or Ca^{2+} cannot be used in detergents (5).

Table 2 : The average composition of a heavy-duty western european powder detergent and a liquid detergent

Ingredient	Powder (wt%)	Liquid (wt%)
Anionic surfactants	2 - 10	8 - 10
Nonionic surfactants	0.5 - 6	18 - 20
Soap	1 - 5	7 - 12
Sequestrants	30 - 50	-
Ethylenediamine tetraacetic acid	-	1
Bleach + activators	20 - 30	-
Triethanolamine	-	7 - 13
Ethanol	-	4 - 6
Propylene glycol	-	4 - 6
Enzymes	0.5	1
Perfume	0.2	0.4
Optical brighteners	0.4 - 0.8	0.2
Sodium sulfate	to make 100%	-
Water	-	to make 100%

The market for enzyme detergents has grown rapidly in the past three decades, despite a temporary set back in the early 1970s when detergent workers developed allergies. This enzyme market is today in excess of $100 million. The use of enzymes has gained widespread acceptance in a range of detergent types and the market share of such enzyme detergents is between 30 and 60% in most industrialized countries (8). Some of the major constituents are described below.

Table 3 : Examples of proteolytic enzyme detergent formulations

Constituent	Detergent		
	Pre-wash	Liquid	Heavy-duty
Anionics	5-15%	5-10%	5-15%
Nonionics	1-3%;	10-40%	1-6%
Sodium tripolyphosphate	25-50%	—	20-40%
Sodium silicate	0-5%-	—	5-10%
Sodium perborate	—	—	15-30%
Sodium carboxymethyl cellulose	0.5-1%	—	0.5-1%
Sequestering agents	—	0-15%	—
Optical brighteners/perfume	0.1-0.2%	0.1-0.5%	0.5-1%
Surfactants	—	5-15%	2-5%-
Water	—	35-50%	—
Sodium sulfate	35-60%	—	0-40%
pH	8-9.5	7-9.5	9.5-10.5
Alkaline protease	0.5-1%-	0.4-0.8%	0.4-0.8%

A close examination of these formulations, together with an appreciation of the types of temperature washing programmes used enables one to establish criteria related to the suitability of an enzyme preparation for use in detergents (9).

(i) Detergent enzymes must be stable during the shelf life of the detergent.

(ii) The enzyme must also have adequate stability and activity in the detergent wash solution and in presence of oxidizing agents.

(iii) It must be able to withstand high pH and temperature levels.

(iv) The enzyme should ideally have broad specificity.

(v) Enzymes whose activity depends on loosely bound metal ions like magnesium and calcium, can not be used in detergents because many detergents contain sequestering agents like polyphosphates, nitrilotriacetic acid (NTA), or zeolites.

(vi) Detergent enzymes should be protease-stable.

(vii) For efficient use in hard water, the activity of detergent enzymes should be stimulated by the presence of calcium ions.

(viii) Detergent enzymes must have high activity and specific activity.

(ix) The enzyme must also not be inhibited by chelating agents.

(x) The enzyme should be in free flowing powder or granular form.

The alkaline serine protease of *Bacillus licheniformis* is the predominant commercial detergent protease. There are reports of other sources of alkaline proteases suitable for detergents such as *B. licheniformis* (9). The use of fungal enzymes in detergent formulations is rather limited when compared to bacterial enzymes. In particular, bacterial proteases dominate the detergent scenario owing to their ability to fulfil most requirements for use in detergents. However, the compatibility of an alkaline protease from *Conidiobolus coronatus* was determined by Phadatare *et. al.* (11). Similarly, Samal *et. al.* (13) compared the laundry detergent compatibility of two proteases from the fungus, *Tritirachium album* Limber. They reported that protease *T. album* demonstrated superior stability in the presence of detergents at high temperature than protease R.

6. ACTIVITY MEASUREMENT

Formulated commercial enzyme preparations contain only a small percentage of enzyme protein (on a weight basis). They are sold on the basis of activity. The suppliers usually specify in the product sheets, the analytical method used to determine activity. In most cases, the enzyme is incubated at specified pH and temperature with large substrate molecules such as casein or haemoglobin. At the end of incubation, the amount of hydrolyzed and, therefore, solubilized protein is measured spectrophotometrically by UV absorption, usually preceded by acid precipitation of the larger fragments. Major detergent manufacturers perform their own tests of an enzyme's activity, stability, and performance before it is added to a detergent formulation (5).

7. WASHING PERFORMANCE

Laboratory model washing systems have been developed to screen, develop or evaluate enzyme detergents and these are designed to simulate standard washing processes. It is possible to test efficiently a range of parameters such as enzyme type and dose rate, physical, and chemical properties of the washing solution and types of stains/fabrics which can be treated.

The contribution of an enzyme to the washing performance of a detergent is determined in the laboratory by test washings in specially developed washing machines. The West European washing process is simulated in a so-called Launder-ometer; the American, in a Tergotometer.

For laboratory testing, different types of artificial stains such as milk, blood, cocoa, grass, egg, and grease are used. The stains are applied to different fabrics such as cotton or polyester. Specific stains are developed and produced in the laboratory. Other stains are commercially available. A well-known test fabric used to study the removal of protein is EMPA 116, a cotton cloth soiled with milk, blood and ink, the ink being glued to the fabric by milk and blood proteins.

Protein hydrolysis is required for the efficient removal of ink. The amount of ink left on the cloth after washing is determined by measuring the remission of the cloth. This is an indicator of the washing performance contributed by the enzyme to the detergent. Such laboratory tests are used to screen and study the fundamental characteristics of detergent enzymes. Since this type of testing on artificial stains in washing machines is only indicative of certain effects, detergent enzymes are tested in a later developmental phase on garments with real stains in real washing machines (5).

8. FUTURE TRENDS IN ENZYME - BASED DETERGENTS

The development of new detergent compositions will have an important impact on the development of new types of detergent enzymes (2). In several countries, environmental considerations led to the partial or complete substitution of phosphates by other sequestrants. As a result, detergent compositions were changed. The pH was increased to 10.0, and enzymes with altered pH activity profiles were required.

The tendency to reduce the amount of energy required for washing, i.e., the substitution of biochemical energy for thermal and mechanical energy, will affect the use of enzymes in detergents. Washing at lower temperatures requires the development of new bleaching systems. Traditional bleaching systems like perborate are effective only above 60^{0}C. Combinations of perborate with bleach activators such as tetraacetylethylenediamine or new bleaching compounds like magnesium monoperoxyphthalate, are active above 40-50^{0}C. In washing without a heating-up phase, traditional enzymes may not be fully compatible with the low-temperature bleaching systems. In the past, this was not a problem since the detergents used for constant-temperature processes did not contain a bleaching agent. At present, a strong tendency exists to incorporate a bleaching system into detergents for these washing processes. In that case, bleach-tolerant enzymes must be developed for specific markets (5).

The reduced activity of some of the nonbiological ingredients of the detergent is another consequence of low washing temperatures. In principle, enzymes other than proteases could compensate for the reduced efficacy of the detergent. For, example, removal of fatty stains at low temperatures may be improved by the addition of lipases. Lower wash temperatures are preferred as this not only saves energy but also maintains the quality and texture of the fabrics. As fat stains are difficult to remove under cold water, alkaline lipases which are active at low and moderate temperatures (20-45^{0}C) are needed in such a scenario (6, 12).

A tendency can be noted toward the development of multifunctional detergents, i.e., one product with the characteristics of several - the ideal being a product that contains a presoak function, a wash function, builders, a bleaching system, and a softener. One compatibility problem inherent in this type of formulation is the presence of anionic and cationic surfactants in one detergent. However, the cationic surfactants, which have a softening function, may be replaced by an alkaline cellulase which also acts as a softener, especially on cotton fabrics. In addition, the cellulose has soil-loosening and anti-redeposition effects (5).

Another trend is the increasing appearance of liquid detergents in the market, especially in the United States, where liquids comprise 25% of the laundry detergent market. In the following section, the raw materials found in liquid detergent compositions are listed to provide a convenient summary for further references (8). For the large-scale categories of light- and heavy-duty liquids, for automatic dishwasher detergents, and for fabric softeners, the list is fairly complete. For specialty liquids, shampoos and conditioners, and to a lesser extent for liquid soaps, only limited examples of special functional ingredients have been selected.

8.1 Light Duty Liquids

8.1.1 Surfactants

(i) Alkylbenzenesulfonate salts

(ii) Alkyl ether sulfate salts

(iii) Alkyl sulfate salts

(iv) Betaines

(v) Alkylpolyglycosides

8.1.2 Foam stabilizers

(i) Fatty acid alkanolamides (mono- and di-)

(ii) Alkyldimethylamine oxides

8.1.3 Hydrotropes

(i) Short-chain alkylbenzenesulfonates

(ii) Ethanol

8.2 Heavy Duty Liquids

8.2.1 Surfactants

(i) Alkylbenzenesulfonate salts

(ii) Alkyl ether sulfate salts

(iii) Alkyl sulfate salts

(iv) Alcohol ethoxylates

(v) N-methylglucamides

8.2.2 Builders

(i) Sodium citrate

(ii) Sodium salts of tartrate mono- and disuccinate mixture

8.2.3 Hydrotropes

(i) Salts of short-chain alkylbenzenesulfonates (xylenesulfonate, cumenesulfonate, toluenesulfonate)

(ii) Ethanol

8.2.4 Bases : Alkanolamines

8.2.5 Other special functional ingredients

(i) Enzymes (protease, amylase, lipase) : stain remover

(ii) Borax (cleaning aid)

(iii) Sodium formate, calcium chloride (enzyme stabilizing system)

(iv) Hydrogen peroxide (bleach)

(v) Propylene glycol (solvent)

8.3 Liquid Automatic Dishwater Detergents

8.3.1 Surfactants

(i) Alkyldiphenyl oxide disulfonate salts

(ii) Hydroxyfatty acid salts

8.3.2 Builders

(i) Pentasodium tripolyphosphate

(ii) Tetrasodium pyrophosphate

(iii) Sodium carbonate

(iv) Sodium silicate

8.3.3 Bases : Sodium and potassium hydroxide

8.3.4. Other special functional ingredients

(i) Sodium hypochlorite (bleach)

(ii) Clay (smectite, bentonite) viscosity regulator, suspending agent

(iii) Polyacrylate sodium salts (viscosity regulator)

(iv) Monostearyl acid phosphate (suds depressant)

8.4 Shampoo and Conditioners

8.4.1 Surfactants

(i) Alkyl sulfate salts

(ii) Alkyl ether sulfate salts

(iii) Acylaminopropyl betaines

(iv) α-Olefinsulfonate salts

8.4.2 Foam Stabilizers/viscosity regulators : Fatty acid alkanolamides

8.4.3 Other special functional ingredients

(i) Polyquaternium 7 (cationic conditioner)

(ii) Glycol monostearate (opacifier)

(iii) Aloe vera (luster promoter)

(iv) Jojoba (luster promoter)

(v) Derivatives of hydrolyzed keratin (skin and hair conditioner)

8.5 Liquid Soaps

8.5.1 Surfactants

(i) Alcohol sulfate salts

(ii) Alkyl ether sulfate salts

(iii) α-Olefinsulfonate salts

(iv) Fatty acid salts

8.5.2 Sequestrants

(i) EDTA

(ii) Sodium citrate

8.5.3 Foam stabilizers/viscosity regulators : Fatty acid alkanolamides

8.5.4 Other special functional ingredients

(i) Phenolic antimicrobials

(ii) Glycerin

8.6 Fabric Conditioners

8.6.1 Surfactants

(i) Dialkyldimethylammonium halide or methosulfate salts

(ii) Monoalkyltrimethylammonium halide salts

(iii) Alkylimidazolinium salts

(iv) Alcohol ethoxylates

(v) Alkyl (ethylene oxide/propylene oxide) nonionics

8.6.2. Solvents

(i) Isopropanol

(ii) Ethanol

8.6.3. Other special functional ingredients : Polydimethylsiloxanes (leveling agent)

8.7 Speciality Liquids

8.7.1 Surfactants

(i) Alkylbezenesulfonate salts

(ii) Alcohol,sulfate salts

(iii) Alkanesulfonate salts

(iv) Alkyl ether sulfate salts

(v) Alkylphenol ethoxylates

(vi) Alcohol ethoxylates

8.7.2 Builders/sequestrants

(i) Sodium carbonate

(ii) Sodium sesquicarbonate

(iii) Sodium citrate

(iv) EDTA

8.7.3 Acids/alkalis

(i) Hydrochloric acid

(ii) Phosphoric acid

(iii) Glycollic acid

(iv) Sodium hydroxide

(v) Sodium metasilicate

(vi) Alkanolamines

8.7.4 Hydrotropes : Short-chain alkylbenzenesulfonate salts (toluene-, cumene-)

8.7.5 Bases : Alkanolamines

8.7.6 Other special functional ingredients

(i) Pine oil (disinfectant)

(ii) Benzalkonium cationics (antimicrobials)

(iii) Sodium hypochlorite (bleach)

(iv) Calcium carbonate (soft abrasive)

(v) Alumina (suspending aid)

(vi) Alkyl glycol ethers (solvents)

(vii) Isopropanol (solvent)

Because the relatively low pH of enzyme-based liquid detergents may impair the removal of starch stains, α-amylases are today being used in some liquid detergents. The currently available bleaching systems are not stable in water-containing liquid detergents. A biological enzyme-containing bleaching system may represent an alternative, for example, methanol oxidase with methanol as substrate.

Enzyme sales for use in washing powders still remain the single largest market for industrial enzymes. At present, lipases do not play a very significant role in most household detergents, mainly because of the lack of enzymes that are sufficiently stable and active under alkaline conditions. However, the worldwide trend towards lower laundering temperatures has led to a much higher demand for such preparations. Understandably, fat stains are more difficult to remove in cold water than their hydrolysis products, which are more easily removed under alkaline conditions (16).

Recent intensive screening programmes, followed by genetic manipulation, have resulted in the introduction of several suitable preparations, for instance, Novo Nordisk's "Lipolase" (*Humicola* lipase expressed in *Aspergillus oryzae*) (7). This market is expected to grow, and undoubtedly we shall see the introduction of lipases from novel sources as well as genetically engineered enzymes with improved characteristics (3, 4).

At present, the largest market for lipases is due to their use as ingredients in heavy-duty laundry detergents. Lipases hydrolyse natural fat and oil stains that are difficult to remove at low wash temperatures with economical surfactants (10). However, a component in many laundry detergents is protease, and it is possible that the proteases will hydrolyse any added lipase, reducing the efficacy of the lipase. Consequently, workers at Unilever Research laboratories have identified the sites within a *Pseudomonas* species lipase that are hydrolysed by commercially available proteases used in detergents, and have engineered the lipase so as to increase its resistance to proteolysis. The engineering was accomplished by substituting or inserting amino acid residues at the site of proteolytic attack or its vicinity, resulting in a substantial increase in resistance of the lipase to proteolysis (16). A number of different approaches were shown to be effective in increasing stability toward proteolytic attack. Specifically, these were :

(i) the amino acid residue that upon cleavage would become the N terminus of the cleaved fragment is replaced by proline;

(ii) an amino acid residue two or four positions away from the cleavage site is replaced by a charged or more polar residue;

(iii) the electrostatic potential of the cleavage site is modified by insertion or substitution of positively charged residues, or by removal of negatively charged residues;

(iv) 1, 2 or 3 residues from the proteolytic cleavage site are deleted.

Another approach involves the use of compounds that bind even more tightly to the protease active sites and as a result inhibit this enzyme's activity in the product during shelf storage. Various boronic acids (for example, arylboronic acids and α–amino boronic acids), peptide aldehyde, peptide ketone, and aromatic borate ester compounds have been found that deliver this type of performance. It is believed that boronic acids inhibit proteolytic enzymes by attaching themselves at the active site. A boron-serine covalent bond and a hydrogen bond between histidine and a hydroxyl group on the boronic acid apparently are formed. The patent literature also describes methods of stabilizing the cellulose enzyme systems with hydrophobic amine compounds such as cyclohexylamine and *n*-hexylamine (8).

An important characteristic for enzymes destined for use in laundry detergents is the ability of the enzyme to withstand oxidative degradation caused by bleach components in the detergent formulation. Primarily, methionine residues in detergent enzymes are subject to oxidative attack, and oxidative stability can be increased by removing methionine residues that are suspectible to oxidative attack on the enzyme, or by replacing them with oxidatively stable residues. This approach to improve oxidative stability has been successfully applied in detergent protease enzymes, and has also been applied to potential detergent enzymes (16).

Another important characteristic of enzymes used in laundry detergents is the need for the enzyme to have an alkaline pH optimum, or to have an alkaline pI. Protein engineering can be used to increase the pI of the enzyme by increasing its net positive charge. This is achieved by the deletion of negatively charged residues or their replacement by neutral residues, by substituting positively charged residues for neutral or negatively charged residues, or by the insertion of positively charged residues (15, 16).

9. CONCLUSIONS

From the foregoing discussion on different hydrolytic enzymes being used for laundry formulation, it seems that although it started as early as in 1913, detergent development and process technology have taken more than eight decades to reach the present status. Even today, large scale efforts are on to develop detergents which not only clean clothes of dust particles, but are simultaneously efficient in removing stains of oils, proteins, tannins and other problem molecules. Although attempts for the utilization of proteases, lipases, α-amylases and cellulases in detergents have been made, great scope for further development of a novel detergent exists which will not only be economical but also be "complete" and efficient in all respects. Such a detergent can only be produced by using microbial biocatalysts.

10. REFERENCES

1. Aunstrup, K. and Andresen, 0. 1972. Proteolytic enzymes useful in washing agents and as depilatory agents. US Patent 3 : 827 938.

2. Barfoed, H.C. 1981. Detergents. In: *Industrial Enzymology* (eds. T. Godfrey and J.Reichelt) . Macmillan, London, Great Britain, pp. 284-293.

3. Bjorkling, F., Godtfredsen, S.E. and Kirk, 0.1991. The future impact of industrial lipases. *Trends Biotechnol.* **9** : 360-363.

4. Farin, F., Hazerwoude-Rijndijk, J., and Labout, N. 1990. Novel lipolytic enzymes and their use in detergent composition. US Patent 4 933 287.

5. Gerhartz, W. 1990. *Enzymes in Industry : Production and Applications.* VCH, Weinheim, Germany, pp.110-115.

6. Ghosh, P.K., Saxena, R.K., Gupta, R., Yadav, R.P. and Davidson, S. 1996. Microbial lipases : production and applications. *Sci. Prog.* **79** : 119-157.

7. Huge-Jensen, I.B., Gerspris, J. and Gormsen, E. 1989. Enzymatic detergent additive. US Patent 4 810 414.

8. Lai, K.Y. 1997. *Liquid Detergents.* Marcel Dekker, Inc., New York, U.S.A.

9. Moo-Young, M. 1985. Comprehensive biotechnology : the principles, applications and regulations of biotechnology in industry, agriculture and medicine. Pergamon Press, Oxford, Great Britain, **3** : 807-808.

10. Novotny, C. and Dolezalova, L. 1993. Effect of lipids and detergents on the production and solubilization of extracellular lipase in *Yarrowia lipolytica. Folia Microbiol.* **38(l)** :49-54.

11. Phadatare, S.U., Deshpande, V.V. and Srinivasan, M.C. 1993. High activity alkaline protease from *Conidiobolus coronatus* (NCL 86.8.20) : Enzyme production and compatibility with commercial detergents. *Enzyme Microb. Technol.* **15** : 72-76.

12. Saad, R.R. 1995. Production of lipase from *Aspergillus tamarii* and its compatibility with commercial detergents. *Folia Microbiol.* **40(3)** : 263-266.

13. Samal, B.B., Karan, B. and Stabinsky, Y. 1990. Stability of two novel serine proteinases in commercial laundry detergent formulations. *Biotechnol. Bioeng.* **35** : 650-652.

14. Ward, O.P. 1983. Proteinases. In : *Microbial Enzymes and Biotechnology* (ed. W.H. Fogarty). Applied Science, London, Great Britain, pp. 251-317.

15. Wolff, AM., Showell, M.S., Venegas, M.G., Barnett, B.L. and Wertz, W.C. 1996. Laundry performance of subtilisin proteases. In : *Subtilisin Enzymes : Practical Protein Engineering* (eds. R. Bott and C.Betzel). Plenum Press, New York, U.S.A., pp.113-120.

16. Woolley, P. and Petersen, S.B. 1994. Lipases : their structure, biochemistry and application. Cambridge University Press, Great Britain.

17. Yadav, R.P., Saxena, R.K., Gupta, R. and Davidson, S. 1997. Production of biosurfactant from sugar alcohols and natural triglycerides by *Aspegillus terreus. J. Sci. Indus. Res.* **56** : 479-482.

18. Yadav, R.P., Saxena, R.K., Gupta, R. and Davidson, S. 1998. Purification and characterization of a regiospecific lipase from *Aspergillus terreus. Biotechnol. Appl. Biochem.* **28(3)** : 243-249.

19. Yadav, R.P., Saxena, R.K., Gupta, R. and Davidson, S. 1998. Lipase production by *Aspergillus* and *Penicillium* species. *Folia Microbiol.* **43(4)** : 373-378.

8 Microbial Phytases in Food Biotechnology

Ashima Bali and T. Satyanarayana

CONTENTS

ABSTRACT 94
1. INTRODUCTION 94
2. PHYTIC ACID 96
2.1 Importance of Phytic Acid and Hydrolysis Products 96
2.1.1 Role in the plants 96
2.1.2 Role in nutrition 96
2.2 Methods of Phytic Acid Removal 97
3. PHYTASES 97
3.1 Sources of phytases 97
3.1.1 Plant sources 97
3.1.2 Bacterial sources 98
3.1.3 Fungal sources 98
4. REGULATION OF PHYTASE SYNTHESIS 100
4.1 Effect of Phosphorus Concentration 100
4.2 Effect of Medium Ingredients, Inoculum Size on Pellet Formation on Phytase Yields 101
4.3 Effect of Surfactants 101
5. PURIFICATION AND CHARACTERIZATION OF PHYTASE 101
6. SEQUENCING AND CLONING OF PHYTASE 102
7. ACTIVE SITE DETERMINATION 102
8. APPLICATIONS OF PHYTASES 102
8.1 Animal Feed Industry 102
8.1.1 Feed trials 103
8.1.2 Economics and Potential effect of phytase on pollution abatement 104
8.2 Food Industry 104
8.3 Antioxidant 105
9. FUTURE PERSPECTIVES 106
10. REFERENCES 106

Advances in Microbial Biotechnology
J.P. Tewari, T.N. Lakhanpal, Jagjit Singh, Rajni Gupta & B.P. Chamola (eds.)
APH Publishing Corporation, New Delhi - 110 002, India.

ABSTRACT

Phytic acid (myoinositol hexakisdihydrogen phosphate) is a major storage form of organic phosphorus in legumes, cereals, oilseed, pollen and nuts. It is an anti-nutritional factor in monogastrics and humans due to its chelation of various cations, reduction in the digestibility of proteins starch and lipids and inhibition of the action of certain enzymes. Microbial phytases can improve the nutritional quality of food by hydrolyzing phytic acid and releasing phosphate groups.

1. INTRODUCTION

The food industry plays an important role in our present day society. As we know without the food industry, urbanization would not have been possible. Presently, the food industry employs biobased processes which account for about 95% of the industrial applications of living systems. There can be little doubt that the new biotechnologies are having and will continue to have significant impact on the practice of food processing, agriculture and related research and development. From plant and animal improvement to any number of food processing techniques, biotechnology holds the promise of contributing to the provision of more efficient production techniques and superior food products. Moreover, the new biotechnologies have the present and future potential not only for changing food production and processing, but also for altering the institutional structure of agriculture.

The nutritional quality of food depends on the total amount of nutrients present in the food consumed, as well as the levels of antinutritive factors. Nutritional stress factors are defined as elements which increase human nutritional requirements by destroying essential nutrients, making essential nutrients unavailable, interfering with the digestion of food and/or decreasing food intake. Some nutritional stress factors present in foods are presented in Table 1. Traditional processing methods can improve nutritional quality by increasing food palatability and digestibility or increasing nutrient availability by destroying toxic factors or minimizing their effects (30). These involve mechanical, physical, chemical and biological modes of operation (33).

Biological processes traditionally have not been included as unit operations in food processing and there is a limited information on the efforts of these processes on nutritional quality (76).

Enzymes are routinely employed in a myriad of industrial applications. Traditional applications include their use in food processing as well as in the brewing, baking and leather baiting industries.

An increased understanding of enzymes and their properties now suggests several novel applications. One such application is the use of selective enzyme activities in the food and feed industry. Enzymatic treatment can reduce the level of nutritional stress factors in food. For example, raffinose and stachyose present in soybeans cause flatulence in humans and animals because there is no α-galactosidase activity in the mammalian intestine (48), thus decreasing the intake of soy products. Treatment of soybean meal with invertase and α-galactosidase resulted in the conversion of these oligosaccharides to monosaccharides (12).

Table 1 : Nutritional stress factors in food

Stress factor	Chemical nature	Occurrence	Action	Dietary effect
Phytates	Organic acid	Cereals, legumes	Chelate metals	Reduce mineral availability
Oxalates	Organic acid	Spinach, amaranth	Chelate cations	Calcium and iron unavailable
Gossypol	Polyphenol	Cottonseed	Chelate metals	Anaemia Potential poisoning
Tannins	Polyphenol	Beans Sorghum	Bind proteins	Proteins insoluble, inactive enzymes
Avidin	Protein	Egg white	Binds biotin	Biotin unavailable
Trypsin inhibitors	Protein	Legumes, cereals	Inhibit proteolysis	Reduce protein digestibility
Solanine	Alkaloid	Potato	Inhibits cholinesterase	Gastrointestinal or neurological disorders, potential poisoning
Goitrin	Glucosinolate	Rapeseed	Goiterogenic	Reduces iodine uptake
Limarin	Cyanogenic glucoside	Cassava	Releases cyanide	Potential poisoning
Thiaminase	Protein	Fish	Destroy thiamine	Possible thiamine deficiency

Source : Teutonico (67).

Phytase is an enzyme that hydrolyzes phytic acid to myoinositol and phosphoric acid in a stepwise manner forming myoinositol phosphate intermediates. Phytic acid, a major storage form of phosphorus in the mature seeds of both monocot and dicot plants, acts as an anti-nutritional factor. The research on phytase spans 87 years from its discovery by Suzuki *et al.* (64) until its commercialization in Europe in 1993-1994 by Gist-Brocades. The commercialization required not only a practical use and delivery of the enzyme, but also the ability to produce the enzyme economically. The milestones in the discovery and commercialization of phytase have been described by Wodzinski *et al.* (77).

The International Union of Biochemistry (1979) listed two phytases: (i) 3-phytase, EC 3.1.3.8 which hydrolyzes the ester bond at the 3 position of myoinositol hexakisphosphate to D-myoinositol 1, 2, 4, 5, 6 - pentakisphosphate and orthophosphate and (ii) 6-phytase, EC 3.1.3.26 which hydrolyzes the 6-position of myotinositol hexakisphosphate to D-myionositol 1, 2, 3, 4, 5- pentakisphosphate and orthophosphate.

Subsequent ester bonds in the substrate are hydrolyzed at different rates. The 6-phytase dephosphorylates phytic acid completely. However, at least one of the 3-phytases described does not hydrolyze the phosphomonoester. In this chapter the use of phytase in mitigating anti-nutritional risks posed by phytates is discussed.

2. PHYTIC ACID

Postenak described phytic acid in 1903 (46) for the first time. It is an organic form of phosphorus, which is chemically a myoinositol hexakis-dihydrogenphosphate (IP6). The molecular formula of phytic acid is $C_6H_{18}O_{24}P_6$ and its molecular weight is 659.86. Salts of phytic acid are called phytates. Phytin is the calcium or magnesium salt of phytic acid.

2.1 Importance of Phytic Acid and Hydrolysis Products

2.1.1 Role in the plant

The presence of phytates in plant food stuffs is well documented. It is the primary source of inositol and storage form of phosphorus in plant seeds that are used as animal feed ingredients (oilseed meals, cereal grains and legumes) (38). Approximately 75% of the total phosphorus in cereals, legumes and seeds (11) exists as phytic acid phosphorus. Cotton seed meal, wheat, corn, rice, oats, soybean meal and other plant-derived feedstuffs contain varied amounts of phytate (40).

The role of phytin-P in the plant was earlier speculated as a storage product. It was believed that large amounts of phosphorus was stored in the seed and it was liberated on germination and incorporated into ATP. Recent studies have established the role of inositol phosphate intermediates in the transport of materials into the cell. Their role in transport as secondary messengers and in signal transduction in plant and animal cells is a very active area of research (3). Graf (21) reviewed and listed 59 different applications of phytic acid including rust removal, prevention of dental caries and its use as a hypocholesterolaemic agent.

All the applications centre around the properties of phytic acid to chelate metals, interact electrostatically with proteins, its high affinity for hydroxyapatite and interact with miscellaneous targets (21). Phytic acid or inositol intermediates have been implicated in starch digestibility and blood glucose response (68), as an antioxidant (22), in the lowering of cholesterol and triglycerides in tumour formation in the treatment of Alzheimers disease (50, 51), in the treatment of Parkinson's disease (52) and in the treatment of multiple sclerosis (53).

2.1.2 Role in nutrition

Ruminant animals sustain the microflora that enzymatically release inorganic phosphorus from phytic acid. However, phytic acid phosphorus is essentially unavailable to monogastric animals like humans, chickens and pigs (40, 41). They produce little or no phytase in the intestine and therefore the phytin - P is excreted. An external source of phosphorus must be supplied in sufficient quantity to meet their daily mineral requirements (40).

Phytic acid that is present in the manure of these animals is enzymatically hydroylzed by soil and water-borne microorganisms. The released phosphorus is transported into rivers and lakes causing eutrophication in aquatic environments in which phosphorus is limiting. When it is introduced in high quantities, excessive algal growth and oxygen depletion ensue.

Phytic acid acts as an anti-nutritional factor (ANF) in several ways as described below:

(i) Six reactive groups in the molecule of IP6 make a strong chelating agent which binds cations such as Ca^{2+}, Mg^{2+}, Fe^{2+} and Zn^{2+}. Under gastrointestinal pH conditions insoluble metal phytate complexes are formed which make the metal unavailable for absorption in the intestinal tract of animals and humans (17, 38).

(ii) Phytates reduce digestibility of proteins, starch and lipids. Phytate complexes with proteins making them less soluble. There is an evidence for the fact that phytate protein complexes are less subject to proteolytic digestion than the same protein alone (28).

(iii) The action of certain enzymes has been found to be inhibited by phytic acid and also by inositol pentaphosphate (IP5). This effect was confirmed in relation to amylases, trypsin, acid phosphatase, tyrosinase and alpha amylase (28).

2.2 Methods of Phytic Acid Removal

Several treatments (autoclaving, cooking, ion exchange techniques, elution with buffer, germination of seeds, chemical technique) have been tested for removing phytates from food products. The use of these methods in most cases lead to a loss of nutritional constituents such as proteins and minerals.

One method for phytic acid removal is enzymatic treatment with phytase. The phytate phosphorus is decreased by 81% in a mixture of ground maize, soybean meal and wheat bran when treated with the native plant phytases. Although some feed ingredients contain native phytase, it is not a consistent source of the enzyme. The pH and temperature during feed processing may inactivate the enzyme (38, 40). The feeding of a phytase produced by a microorganism may ameliorate these problems. A natural lactic acid fermentation was found to decrease phytate content in corn meal. Mixed cultures of yeast and lactobacilli were effective for pearl millet.

3. PHYTASES

Phytases (myoinositol hexaphosphate phosphohydrolase) belong to a special group of acid phosphatases which are capable of hydrolyzing phytate to a series of lower phosphate esters of myoinositol and phosphate.

3.1 Sources of Phytase

3.1.1 Plant sources

Suzuki *et al.* (64) were the first investigators to make a preparation of phytase. They detected activity in rice and wheat bran. They also isolated inositol as a product of the reaction. The occurrence of phytase in germinating plants has been exhaustively reviewed by many workers (49). They reported that phytase has been isolated and/or characterized in cereals such as triticale, wheat, corn, barley, rice and beans such as mung beans dwarf beans and California small white beans.

3.1.2 Bacterial sources

Phytase activity has been detected in *Aerobacter aerogens (Enterobacter aerogens) (23).* Examination of the activity in the bacteria showed that most of the enzyme is bound to the insoluble debris obtained after ultrasonic disintegration of washed cells. The activity of the debris was shown to be optimal between pH 4 and 5 and to be inhibited by high substrate concentration and by inorganic orthophosphate. With myonositol hexa-phosphate as a substrate the Michaelis constant was 0.135 mM. The rate of hydrolysis of various inositol phosphates is affected by the number of substituent phosphate radicals and their spatial arrangements. The dephosphorylation of myonositol polyphosphates occurs in a step wise manner resulting in the appearance of several intermediate inositol polyphosphates which were separated by ion exchange chromatography (23).

Two distinct molecular forms of phytase have been detected in *Klebsiella aerogenes* (65). The cell free extract from this organism could be resolved into two distinct fractions on DE-52 by iso-electro focusing technique and on Sephadex G-200, one eluting in the void volume and the other at the end of the column run.

The molecular weight of the second peak was 10-13 kDa and could possibly be a fragment of the native enzyme. The two species of the enzyme thus isolated differed in their Km, pH optima, pH values and temperature sensitivity. Evidence is provided to show that low molecular weight enzyme is generated during isolation. This is one of the rare instances where such a small fragment of enzyme peptide retained a full complement of enzyme activity.

Phytase has also been shown to be present in the culture filtrates of *Bacillus subtilis* and has been purified 50 fold. The enzyme differed from other previously known phytases in its metal requirement and specificity. It has a specific requirement for Ca^{2+} for its activity. The enzyme hydrolyzes only phytate and has no action on several other phosphate esters which were tested (47).

A bacterial strain that produced extracellular phytase was isolated from soil near the roots of the leguminous plants and identified to be a genus of *Enterobacter* sp.4. The optimum condition for phytase production in phytase screening medium was pH of 5.5 and 3 days of cultivation at 37^{0}C. In the bacteria 81.7% of the enzyme activity was found in the extracellular fraction, 4.4% of the activity was detected in the periplasmic space and the rest was in the intracellular and cell bound fractions. The maximal phytase activity was observed at pH 7.0-7.5 and it was most stable at pH 6.0-8.0. The optimum temperature of this enzyme activity was 50^{0}C. The crude enzyme was inhibited by 1mM Zn^{2+}, Ba^{2+}, Cu^{2+}, Al^{2+} and EDTA (79).

Nineteen strains of lactic acid producing bacteria of the genera *Lactobacillus* and *Streptococcus* were screened for the production of extracellular phytase (63). Among them *Lactobacillus amylovorus* B4552 secreted the maximum amounts of phytase ranging from 125 to 146 units per ml secreted in submerged cultivations. *Lactobacillus amylovorus* has potential in improving nutritional qualities of cereal and pulse based food fermentations (63).

3.1.3 Fungal sources

Howson and Davis (29) surveyed 84 fungi from 25 species for phytase production. The incidence of phytase production was highest in aspergilli (Table 2). The yields of phytase produced extracellularly are shown in Table 3. Of all the organisms surveyed, *A.niger* NRRL 3135 produced the most active extracellular phytase in corn starch (57) and semisynthetic (3) media. It produced 110 nKat P/ml. *Aspergillus niger* NRRL 3135 produced two different phytases, one with pH optima at 5.5 and 2.5 and the other with a optimum pH of 2.0. Later these enzymes were designated as phy A and B respectively.

Table 2 : Production of extracellular phytase by microorganisms

Organism	No. of cultures tested	No. producing phytase	Reference
Fungi			
Aspergillus spp. 3	3	3	(57)
A.niger (syn) *A.ficuum*	1	1	(29)
A.flavus	10	2	(57)
A.niger	7	3	(29)
A.terreus	1	1	(78)
A.versicolor	5	2	(29)
A.wentii	1	1	(29)
Botrytis cinerea	5	2	(29)
Geotrichum candidum	1	1	(29)
Mucor spp.	37	1	(57)
Mucor piriformis	4	2	(29)
Penicillium spp.	58	1	(57)
Rhizopus oryzae	5	2	(29)
R. oligosporus	1	1	(29)
Saccharomyces cerevisiae	8	6	(29)
Schwanniomyces occidentalis (syn *S.castelli)*	1	1	(54)
Bacteria			
Bacillus subtilis	1	1	(47)
B.subtilis (natto) *N-77*	1	1	(59)
Lactic acid bacteria	19	14	(63)

Table 3 : Yields of extracellular phytase produced in culture filtrates of supernates

Organism	I.U. (ml-1)	References
Fungi		
A. awamorii ATCC 11382	1.20	(57)
A. awamorii ATCC 11358	1.72	(57)
A. carbonarius NRRL 368	0.99	(57)
A. carbonarius PCC 1040	0.78	(57)
A.niger syn *A. ficuum* NRRL 3135	6.62	(57)
A.niger syn *A. ficuum* WB 320	0.52	(57)
A.niger syn *A. ficuum* WB 3 64	0.78	(57)
A.niger syn *A. ficuum* WB 4016	0.57	(57)
A.niger syn *A. ficuum* WB 4541	0.52	(57)
Bacteria		
Bacillus subtilis	0.24	(47)
Bacillus subtilis (natto)	0.11	(59)
Lactobacillus amylovorous	125	(63)

The production of phytase during solid state fermentation using *A.niger* NRRL 3135 in canola meal has been studied. Optimum moisture content of media for these processes was 64%. The rate of phytase production increased with an increase in inoculum age between 2 and 5 days (15).

Solid state fermentation (SSF) using *Aspergillus carbonarius* with canola meal as a substrate showed that the production of phytase was associated with growth; maximum activity was achieved after 72 h. Apparent 25% and 10% increases in the protein content of the canola meal were recorded after 48h and 72h respectively. The rate of decrease in phytic acid content was optimum with a moisture content between 53% and 60% (21). A variety of organisms produce phytase intracellularly. Casida (6) has reported the production of phytase in whole cells of *Aspergillus* spp., *Penicillium* spp. and *Pseudomonas* spp. A crude preparation of intracellular acid phosphatase (EC 3132) from a waste mycelium of *A.niger* was utilized for dephosphorylation, of phytate compounds present in food and feed ingredients. The hydrolysis of phytate phosphorus in wheat bran. soyabean meal was carried out at 40°C and a pH value of 4.5.

Moulds commonly used in oriental food fermentations have been examined for their ability to produce phytase. Except for *Mucor dispersus* NRRL 3103 and *Actinomucor elegans*, all other molds tested produced both extra- and intracellular phytase. The pH optimum of 5.3 characterizes the enzyme as an acid phosphohydrolase and its maximum activity at 50°C suggests a relatively high theremostability. The *A.oryzae* phytase is active with either phytic acid or glycerophosphate as substrate. The Km value of the enzyme was 0.47 mM and Vmax was 11.9μ moles Pi per min per mg protein (75). A comparison of the properties of phytases that have been highly purified is shown in Table 4.

Table 4 : Physiochemical and kinetic parameters of microbial phytase and acid phosphatases

Enzyme	Mw (kDa)	Pi	Kcat	Km (μM)	Kcat/Km	Optimum pH	Optimum temperature	Reference
A. ficuum PhyA[a]	85	4.5	348	27	1.29×10^7	2.5	58	(70)
A. ficuum PhyB[a]	68	4.0	628	103	6.1×10^6	2.5	63	(70)
A. ficuum APase[b]	85	4.9	260	200	1.3×10^6	6.0	63	(71)
E. coli phytase P2[c]	42	6	209	130	4.77×10^7	4.5	35	(24)
E. coli APase[d]	45	6.3	2698	5000	5.4×10^5	2.5	37	(9)
B. subtilis[e]	38	6.3	5.5	500	1.1×10^4	6.0 60		(59)

4. REGULATION OF PHYTASE SYNTHESIS

4.1 Effect of Phosphorus Concentration

Ware and Shieh (76) discovered that the available inorganic phosphorus content of the medium controlled the synthesis of the enzyme. Maximum yields (113 nkat P/ml in shake flasks in 5 days) *in A.niger* NRRL 3135 were obtained if the inorganic phosphorus content was controlled in the range of 0.0001% to about 0.005% by weight determined as phosphorus. They reported that the repression of phytase synthesis at high levels of P was a general phenomenon because it was observed with all moulds and yeasts that produced phytase. Han and Gallagher (26) also confirmed that high P concentrations inhibited phytase synthesis by *A.niger* NRRL 3135. Gibson (19) also confirmed that synthesis of phytase was controlled by the levels of phosphorus in the medium.

4.2. Effect of Medium Ingredients, Inoculum Size on Pellet Formation on Phytase Yields

When a simple sugar such as glucose or fructose is used as a sole source of carbon for phytase production by *A.niger* NRRL 3135, mycelial pellets were formed and the enzyme was produced in low yields (57). However, if they used a medium containing a surfactant (sodium oleate, 0.5% v/v), growth was dispersed and phy A yields were 4.7 - fold higher than in controls in a 17-day incubation. It was observed that glucose concentration upto 5.2% caused increase in biomass, production of phytase and the rate of phytic acid hydrolysis, while systems with 9.8 and 17.8% of glucose had an adverse effect on these processes.

When the inoculum size was too small, or a relatively low viscosity medium (one derived of corn starch) was used, the organism formed pellets. Pellet formation could be minimised by the shearing action of the agitation system that separated aggregates of hyphae (62). When pellets were formed, a decline in extracellular enzyme yields was observed in *A.niger* NRRL 3135.

4.3 Effect of Surfactants

During the growth of *Aspergillus carbonarius*, the rate of growth, phytase production and reduction of phytic acid content in canola meal media during solid state fermentation were higher in the presence of Na-oleate or Tween 80 than in the control medium which was not supplemented with these surfactants. Addition of Triton X-100 had a negative effect on phytase production. The optimum concentration of Na-oleate in solid state culture media was 1% (1).

5. PURIFICATION AND CHARACTERIZATION OF PHYTASE

Phytase from *Aspergillus carneus* was purified about 43-fold from the culture filtrate by precipitation with acetone, gel filtration through sephadex G-75 and ion exchange chromatography on DEAE-cellulose. Activity of both crude and purified phytase was influenced greatly by changing pH and reaction temperature; maximum activity of the crude enzyme occurred at 35^{0}C and pH 5.6, whereas that of the purified preparation at 40^{0}C and pH 5.6. The pure enzyme was found stable between pH 5.6-6.2 and it retained its activity over a long period when stored at 4^{0}C (19).

An extracellular phytase and an extracellular acid phosphatase were purified from *Aspergillus oryzae* K1. The molecular masses of *A.oryzae* phytase and acid phosphatase were 60 and 70 kDa respectively. An extracellular phytase from *B.subtilis* N-77 was purified 322-fold by gel filteration and DEAE chromatography (59). The molecular mass of the active monomeric form was judged to be 36 kDa by SDS-PAGE. Two periplasmic phytases, PI and P2, were purified from *E.coli* to near homogeneity (24). The active species were judged to be a monomer with a molecular mass of 42 kDa. Both enzymes were very specific for phytate. A phytase was purified from yeast, but purity and molecular data were not reported. The pH and temperature optima were 4.6 and 45^{0}C respectively.

A cytoplasmic phytase was purified about 410-fold to apparent homogeneity with a recovery of 28% from *Klebsiella terrigena* (26). The enzyme was inducible under carbon limitation in the presence of phytate. It behaved as a monomeric protein of a molecular mass of about 40 kDa. The phytase was rather specific for phytate and it exhibited optimal conditions for phytate degradation at pH 5.0 and 58^{0}C.

Phytase from the yeast *Schwanniomyces casteillii* was purified by anion exchange and gel filteration chromatography. The enzyme has a molecular weight of 490,000 with a glycosylation rate around 31%. The structure of the deglycosylated protein was tetrameric, with one large subunit (M.W. 125, 000) and three identical small subunits (M.W. 70, 000). The enzyme exhibited optimum activity at 77°C and at a pH of 4.4. Phytate was completely dephosphorylated by the phytase and its Km was 38 µM (54).

6. SEQUENCING AND CLONING OF PHYTASE

A gene (phy A) of *Aspergillus niger* NRRL 3135 coding for extracellular, glycosylated phytase was isolated using degenerate oligodeoxy ribonucleotides deduced from phytase amino acid (aa) sequences. Nucleotide sequence analysis of the cloned region revealed the presence of an open reading frame coding for 467 aa and interrupted once by an intron of 102 bp in the 5' part of the gene. Expression of phyA was shown to be controlled at the level of mRNA accumulation in response to inorganic phosphate levels (74). In transformants of *A.niger,* the expression of multiple copies of phyA resulted in upto more than ten fold higher phytase levels than in the wild type strain.

Phytase sequence has also been deduced from the cloned DNA of *A.niger* van Tieghem (74). A third phytase sequence was obtained from *A.niger* var. *awamori* (GenBank Accession No. L02421) and revealed 97.2% homology to *A.niger* NRRL 3135 phytase. The primary structure of *A.niger* NRRL 3136 phy B and *A.niger* var. *awamori* phy B was elucidated from the cloned DNA and a partial sequence was verified by chemical sequencing (16, 45). The phyb from *A.niger* NRRL 3135 showed 99% homology to the correspnding protein from *A.niger* var. *awamori.* The primary structure of bacterial phytases has not been deduced chemically; however, the primary structure of pH 2.5 optimum acid phosphatase from *E.coli* was deduced from the cloned DNA (10).

7. ACTIVE SITE DETERMINATIONS

The active site of phytases showed remarkable homology to the active site residues of the members of a particular class of acid phosphatase termed "histidine phosphatase". Chemical probing at the active site of human prostatic acid phosphatase suggested that an arginine residue is involved in catalysis. A similar observation was also made in *A.niger* NRRL 3135. A conserved tripeptidic region with the sequence RHG has been found to be present in N terminal region of phyA and phyB *in A.niger* NRRL 3135. Further chemical probing of the fungal phytase also suggested a sensitive histidine at the active site (73). On closer examination of the active site residues of phyA and phyB in *A.niger* NRRL 3135, pH 2.5 optimum acid phosphatase in *E.coli,* pho3 and pho5 gene products in yeast, human prostatic and lysosomal acid phosphatase, it was observed that the most conserved sequence is RHGXRXP. The acid phosphatases and phytases containing the active site motif in the N terminal segment of the protein are grouped under "histidine phosphatase". The positive charge of guanido group of arginine is responsible for the recognition and anchoring of the negatively charged phosphate group to the proximity of a histidyl residue in the active site. The phosphate group is transiently transferred to the histidine group to form an unstable phosphoenzyme complex before hydrolytic cleavage to form orthophosphate.

8. APPLICATION OF PHYTASES

8.1 Animal Feed Industry

The bulk of animal-feed constituents are vegetable in nature. As a consequence, many feed stuffs contain components that may be considered anti-nutritional. Such anti-nutritional factors (ANFS) interfere with the digestibility, absorption or utilization of nutrients and hence, adversely affect animal performance.

The major protein supplements in poultry feeds are soyabean and cottonseed meals. In cooler climates however, rapeseed is a major oil seed crop (82). Several anti-nutritional factors prevent this ingredient from being extensively used in poultry nutrition (34). The presence of phytates in rapeseed causes zinc, calcium and magnesium deficiency syndrone in chickens (43). On the other hand, two thirds of rapeseed meal phosphorus is bound as phytate and unavailable for poultry, making it necessary to supplement the diet with the calcium phosphate (42).

Canadian rapeseed varieties, which are relatively low, in glucosinolates and erucic acid, are the source of canola meal and it is a byproduct of canola oil production from canola seeds. It contains 37-40% protein and it is used as a feedstuff for livestock and poultry (8). Canola meal contains 4-6% phytic acid which reduces the nutritional value of the meal. It has been shown that the phytic acid in canola meal binds to multivalent cations such as Zn^{2+}, Ca^{2+} and Fe^{3+} and so reduces their bioavailability. It has also been reported that phytic acid inihibits enzymes such as α-amylase (56), trypsin, tyrosinase and pepsin. Hydrolysis of phytic acid prior to animal consumption would increase the availability of inorganic phosphorus and other minerals in the animal diet (27). This can be achieved enzymatically in the plants whose seeds contain endogeneous phytase, such as beans (7), cotton seeds, soybean (27), wheat (44) and mung beans (39). There are also plants, such as canola, that do not contain this enzyme. In such cases, the possibility for the reduction of phytic acid using microbial phytases should be considered.

8.1.1. Feed trials

Studies on the enzymatic treatment of feeds using microbial phytase sources have demonstrated that this method increases the bioavailability of proteins and essential minerals and provides levels of growth performance as good as or better than those with phosphate supplementation. Nelson *et al.* (40) were the first to pretreat corn-soya diet with culture filtrate containing phytase of *A.niger* NRRL 3135 and fed it to 1 day old chicks. The chicks showed increases in bone ash due to the phytin-P released from the dietary substances by the action of phytases.

Nelson *et al.* (41) supplemented a commercial corn soybean meal diet with solvent precipitated *A.niger* NRRL 3135 phytase and fed it to chicks. They measured bone ash and feed-to-gain ratios. The enzyme was active in the animals and they were able to incorporate the released phytin-P into bones. They assessed the effect of phytase fed, on calcium requirements. The amount of calcium from natural ingredients required in the diet was at least one third less than 90% of the phytin-P hydrolyzed by phytase. The phytase of *Schwanniomyces occidentalis* could be used to produce protein rich feed stuff free of phytic acid.

Wheat bran and cotton flour had phytic acid contents of 45 μmol/g and 61 μmol/g, respectively. Phytates were removed from these materials by incubation with the phytase of *S.castellii* at pH 4.4 and 70°C (55).

A 21-d experiment was conducted with one-day old male broilers to evaluate the effectiveness of the supplemented phytase for improving the availability of phytate P in soybean meal when varying levels of P were fed. The results clearly showed that phytate bound P in soybean meal was made more available to broilers (13).

Simons *et al.* (60) were the first to demonstrate the efficacy of feeding phytase to swine. They concluded that the addition of enzyme (100μMP/hr/ml of phytase/kg of diet) was sufficient to obtain levels of performance equal to or better than that attained by supplementing broiler diet with inorganic phosphate. When the amount of enzyme was increased to 1500µm/hr/ml of phytase/kg of diet, an improved performance as compared to birds fed with control diets was recorded.

Royal Gist-brocades, Delft (Netherlands) has supplied phytase (Natuphos produced by cloning phyA from *A.niger* NRRL 3135 in *A.niger* CBS 513.8) to various researchers for studies to find out whether the phytin-P in feed ingredients supplemented with phytase is utilized by monogastric animals (18). The studies have been extended to layers, ducks and ducklings. Lei *et al.* (36) demonstrated that the addition of phytase to the diets of weaning pigs improved the bioavailablity of zinc and phytate phosphorus.

8.1.2 Economics and potential effect of phytase on pollution abatement

One of the driving forces in the development of phytase as a product is the awareness of the environmental effects of the release of concentrated pollutants, especially in intensive agricultural production of animals. Although some feed ingredients contain phytase, it is not a reliable source of the enzyme.

In mongastric animals, a large portion of the phytic acid phosphorus is excreted, creating phosphorus pollution problems in areas of intensive livestock production. Pretreatment of feeds with microbial phytases may help in circumventing these problems.

When microbial phytase was fed to low-P diets for broilers, the availability of P increased to 60% and the amount of P in the droppings decreased by 50% (60). The potential benefits accrued due to the use of phytase in commercial diets in abating phosphate pollution in soil and water have been established.

Plant phosphorus which accounts for the major part (2/3 to 3/4) of the available phosphorus in monogastric animals, consists mainly of phytates. The availability of phosphorus may be improved by adding microbial phytase to the feed or by using phytase-rich cereal diet. This enzyme optimises the utilization of plant phosphorus in poultry and pigs and minimizes the need to add inorganic phosphorus, thus markedly reducing the excretion of phosphorus in manure.

8.2 Food Industry

The presence of phytates in plant foodstuffs is well documented (11). This organic form of phosphorus is poorly utilized by man and by monogastric animals because they lack or have low phytase activity to catalyze the hydrolysis of phytates to inositol and phosphoric acid in their intestines (4, 40).

Moulds commonly used in oriental food fermentation have been examined for their ability to produce phytase. Tempeh is a popular oriental fermented food made from soyabeans inoculated by moulds *(Rhizopus oligosporus)* in the koji process. Convincing evidence has accumulated which shows that not only the digestibility, vitamin content and flavour of soyabean are improved by this mould fermentation, but the phytate content was also reduced to one third. The pH optimum of this enzyme was 5.6 and the Km of the phytase-phytate reaction was 0.28×10^{-3} M(81).

Lactic acid bacteria are widely used all over the world to improve the shelf life and nutritive qualities of perishable and other foods such as milk and vegetables, meat, fish, legumes and cereals. Cereal and pulse based food products are a major staple food worldwide. Lactic fermentations have been in practice since time immemorial in most countries for preparing sour bread, pancake, idli, dosa, dhokla, porridges, alcoholic and non alocholic beverages. The majority of cereals and pulses are rich in phytate which complexes phosphate, calcium, iron and zinc and also proteins, thus making them non-available to monogastric animals and humans. A number of lactic acid bacteria exhibit enzyme activity in the fermentation medium, but *Lactobacillus amylovorus* B4552 produces the maximum amounts of phytase ranging from 125 to 146 units ml^{-1} extracellularly under the submerged cultivation conditions (63). *Lactobacillus amylovorus* thus has a potential in improving the nutritional qualities of cereal and pulse based food fermentations.

There are several reports on the hydrolysis of seed phytates by means of soaking and heating. Dry small white beans *(Phaseolus vulgaris)* lost 90% of their phytate when presoaked and heated at 60^0C for 10h (7). Germination can improve autohydrolysis of seeds by endogenous phytases, but such processing is questionable from an economical point of view. One possible solution seems to be the hydrolysis of food phytates by exogenous phytase preparations (81).

Phytate dephosphorylation in legume products was first attempted in the 1960s when Shieh *et al.* (58) screened for moulds producing extracellular phytate degrading enzyme and found that the *A.ficuum* NRRL 3135 strain was the most active.

A 78% phytate loss resulted when a preparation from *A. ficuum* was mixed with soyabean meal and incubated for 15h (27). A preparation of phytase from wheat was considerably more active towards phytate rapeseed meal than one from *A. ficuum* but the reaction was slow (29). A crude preparation of intracellular acid phosphatase rich in phytase activity, obtained from *A.niger* caused nearly complete dephoshorylation of protein isolates from soyabean in the course of a few hours (80). In commercial wholewheat bread, phytic acid is present at levels of 0.29 to 1.05% (w/w). By adding mould phytases during bread making dough phytates could be almost completely eliminated. A preparation of phytate degrading enzymes, that might be recommended for baking, would have to be safe, highly active, Ca^{2+} independent, with a pH optimum of 4.5 to 5.0 and have a high reaction rate at about 30^0C. The possibility of using mould phytases during bread making has been explored by Knorr *et al.* (32) who found that by adding phytase, dough phytates could be almost completely eliminated.

Steeping is a process required in wet milling of corn for softening the kernel of the corn seed and breaking the cell walls as well as obtaining corn steep liquor. Caransa *et al.* (5) reported that microbial phytase can accelerate this process, improve the separation of germ and starch from fibre and give a higher yield of starch and gluten, as well as improve the properties of corn steep liquor (81).

8.3 Antioxidant

The catalysis of radical formation by iron and subsequent oxidative damage has been well documented. Although many iron- chelating agents potentiate reactive oxygen formation and lipid peroxidation, phytic acid (abundant in edible legumes, cereals and seeds) forms an iron chelate which greatly accelerates Fe^{2+} mediated oxygen reduction yet blocks iron driven hydroxyl radical generation and suppresses lipid peroxidation.

Furthermore, high concentrations of phytic acid prevent browning of various fruits and vegetables by inhibiting polyphenol oxidase. These observations indicate an important anti-oxidant function for phytate in seeds during dormancy and suggest that phytate may be a substitute for presently employed preservatives, many of which pose potential health hazards.

9. FUTURE PERSPECTIVES

In view of a wider implications of microbial phytases in animal and human nutrition and environmental pollution abatement, further investigations are needed to look for novel phytases which can liberate phosphorus from phytates rapidly.

10. REFERENCES

1. Al-Asheh, S. and Duvnjak, Z. 1994. The effect of surfactants on the phytase production and the reduction of the phytic acid content in canola meal by *Aspergillus carbonarius* during a solid state fermentation process. *Biotechnology Letters* **16(2)**: 183-188.

2. Al-Asheh, S. and Duvnjak, Z. 1995. Phytase production and decrease of phytic acid content in canola meal by *Aspergillus carbonarius* in solid-state fermentation. *World Journal of Microbiology & Biotechnology* **11**: 228-231.

3. Berridge, M.J. and Irvine, R.F., 1989. Inositol phosphates and cell signalling. *Nature* **341**: 197-205.

4. Bitar, K. and Reinhold, J.G. 1 972. Identification and properties of phytase in cereal grains and oilseed products. *Biochem. Biophys. Acta.* **208**: 442-452.

5. Caransa, A., Simell, M., Lehmussari, M., Vaara, M. and Vaara, T. 1988. A novel enzyme application in corn wet milling. *Starch* **40** : 409-411.

6. Casida, L.E., Jr. 1959. Phosphatase activity of some common soil fungi. *Soil Sci.* **87** : 305-310.

7. Chang, R., Schwinmmer, S. and Burr, H.K. 1977. Phytate removal from whole dry beans by enzymatic hydrolysis and diffusion. *J. Food Sci.* **42**: 1098-1101.

8. Clandinin, D.R. 1986. Canola meal for livestock and poultry. Winnipeg, Manitoba Canola Council of Canada.

9. Dassa, E., Tetu, C. and Boquet, P.L. 1980. Identification of the acid phosphatase (optimum pH 2.5) of *Escherichia coli. FEBS Lett.* **113**: 275-278.

10. Dassa, E., Marek, C. and Boquet, P.L. 1990. The complete nucleotide sequence of the *E. coli* gene app A reveals significant homology between pH 2.5 optimum acid phosphatase and glucose-l-phosphatase. *J. Bacteriol.* **172**: 5497-5500

11. De Boland, A.R., Garner, G.B. and O'Dell, B.L. 1975. Identification and properties of "phytate"in cereal grains and oilseed products. *J.Agri. Food Chem.* 23: 1186-1189.

12. Delente, J. and Ladenburg, K. 1972. Quantitative determination of the oligosaccharides in deflated soybean meal by gas-liquid chomatography. *J. Food Sci.* **37**: 372-374.

13. Denbow, D.M., Ravindran, V., Kornegay, E.T., Yi-Z., Hulet, R.M. 1995. Improving phosphorus availability in soybean meal for broilers by supplemental phytase. *Poultry Science* **74(11)** : 1831- 1842.

14. Dox, A.W. and Golden, R. 1911. Phytase in lower fungi. *J.Biol. Chem.* **10**: 183-186.

15. Ebune A., Al-Asheh, S. and Duvnjak, Z. 1995. Effect of phosphate, surfactants and glucose on phytase production and hydrolysis of phytic acid in canola meal by *Aspergillus ficuum* during solid state fermentation. *Bioresource Technology* **54** : 241-247.

16. Ehrlich, K.C., Montalbano, B.G., Mullaney, E.J., Dischinger H.C., Jr. and Ullah, A.H.J. 1993. Purification and cloning of a second phytase gene (Phy B) from *Aspergillus niger (A. ficuum). Biochem. Biophys. Res. Commun.* **195**: 53-57.

17. Erdman, J.W. Jr. 1979. Oilseed phytates: Nutritional Implications. *J. Amer. Oil Chem. Soc.* **56**: 736-741.

18. Farrell, D.J., Martzin, E., Du Preez, J.J., Bongarts, M., Betts, M., Sudaman, A. and Thomson, E. 1993. The beneficial effect of a microbial feed phytase in diets of broiler chickens and ducklings. *J. Anim. Physiol. Anim. Nutr.* **69**: 278-283.

19. Ghareib, M. 1990. Biosynthesis, purification and some properties of extracellular phytase from *Aspergillus carneus. Acta-Microbiol-Hung.* 37: 159-164.

20. Gibson, D.M. 1987. Production of extracellular phytase from *Aspergillus ficuum* on starch media. *Biotechnol. Lett.* **5** : 305-310.

21. Graf, E. 1986. (ed. E. Graf.) In : *Phytic Acid Chemistry and Applications.* Pilatus Press, Minnealpolis, pp.1-21.

22. Graf, E., Empson, K.L. and Eaton, J.W. 1987. Phytic acid - A natural antioxidant. *J.Biol. Chem.* **262**: 11647-11650.

23. Greaves, M.P. anderson, G. and Webley, D.M. 1967. The hydrolysis of inositol phosphates by *Aerobacter aerogenes. Biochem. Biophys. Acta* **132**: 412-418.

24. Greiner, R., Konitzny, U. and Jany, K.D. 1993. Purification and characterization of two phytases from *Escherchia coli. Arch. Biochem. Biophys.* **303**: 107-113.

25. Greiner, R., Haller, E., Konitzny, U.and Jany, K.D. 1997. Purification and characterization of a phytase from *Klebsiella terigena.* Arch.Biochem. Biophys.341, 201-206.

26. Han, Y.W. and Gallagher, D.J. 1987. Phytase production by *Aspergillus ficuum* on semi-solid substrate. *J.Ind. Microbiol.* **2**: 195-200.

27. Han, Y.W. 1988. Removal of phytic acid from soybean and cotton seed meals. *J.Agri. Food Chem.* **36**: 1181-1183.

28. Harland, B.F and Morris, E.R. 1995.Phytate:A good or a bad food component. *Nutrition Research* **15(5)** : 733-754.

29. Howson, S.J. and Davis, R.P. 1983. Production of phytate- hydrolysing enzyme by some fungi. *Enzyme Microb. Technol.* **5**: 377-382.

30. Hudson, H.J.F. 1980. Detoxification of foods in food processing. In : *Food and Health: Science and Technology* (eds. G.G.Birch and K.J. Parker). Applied Science, London, pp. 305-317.

31. Irving, G.C.J. and Cosgrove, D.J. 1971. Inositol phosphate phosphatases of microbiological origin: some properties of a partially purified bacterial *(Pseudomonas* sp.) phytase. *Aust. J. Biol. Sci.* **24**: 547-557.

32. Knorr, D., Watkins, T.R. and Carlson, B.L. 1981 Enzymatic reduction of phytate in whole wheat breads. *Journal of Food Science* **46**: 1866-1869.

33. Knorr, D. and Sinskey, A.J. 1985. Biotechnology in food production and processing. *Science* **239**: 1224-1229.

34. Koreleski, J., Rys, R., Mlodkowski, M., Ombach, A. and Kubicz, M. 1986. Effect of feeding of oilmeal from traditional or low glucosinolate rapeseeds on performance and vitamin mineral supply of broiler chickens and growth of rats. *Rooz Nauk Zoot-Monogr i Rozpr* **24**: 281-297.

35. Kuenzi, M., Assi, F., Chmiel, A.K., Collins, C.H., Donikan, M., Domiguez, J.B., Financsek, L., Fogarty, L.M., Frommer, W., Pasko, F., Hovland, J., Houwink, E.H., Mahler, J.L., Sandkvist, A., Sargent, K., Sloover, C. and Tuijnenburg Mujis, G., 1985. Safe biotechnology: General considertions. *Appl. Microbiol. Biotechnol.* **21**: 1-6.

36. Lei, X., Ku, P.K., Miller, E. R., Vllrey, D.E. and Yokoyama, M.J. 1993. Supplemented microbial phytase improves bioavailability of dietary zinc to weanling pigs. *J. Nutr.* **123**: 1117-1123.

37. Lolas, G.M., Poamidis, N. and Markakis, P. 1976. The phytic acid total phosphorus relationship in barley, oats, soybeans and wheat. *Cereal Chem.* **53**: 867.

38. Maga, J.A. 1982. Phytate: Its chemistry, occurence, food interaction, nutritional significance and methods of analysis. *J. Agic. Food Chem.* **30**: 1-9.

39. Mandal, N.C., Burman, S. and Biswas, B.B. 1972. Isolation, purification and chracterization of phytase from germinating mung beans. *Phytochemistry* **11**: 495-502.

40. Nelson, T. S. 1967. The utilization of phytate phosphorus by poultry. *Poultry Sci.* 46: 862.

41. Nelson, T.S., Shieh, T.R., Wodzinski, R.J. and Ware, J.H. 1971. Effect of supplemental phytase on the utilizaion of phytate phosphorus by chicks. *Journal of Nutrition*, 101: 1289-1294.

42. Nelson, T.S. 1976. The hydrolysis of phytate phosphorous by chicks and laying hens. Poultry *Science*, **55**: 2262-2264.

43. Nwokolo, E.N. and Bragg, D.B., 1977. Influence of phytic acid and crude fiber on the availability of minerals from four protein supplements in growing chicks. *Can J. Anim Sc.* **57**: 475-477.

44. Pears, F.G. 1953. The phytase of wheat. *Biochemical Journal* **53**: 102-109.

45. Piddington, C.S., Houston, C.S., Palobeimo, M., Cantrell, M., Micttinen - Oinonen, A., Nevalinen, H. and Ram Bosek, J. 1993. The cloning and sequencing of the genes encoding phytase (phyA) and pH 2.5 optimum acid phosphatase(aph) from *Aspergillus niger* var. *awamori. Gene.* **133**: 55-62.

46. Postenak 1903. Compt. Rend. 137, 202. Phytase. In:*Advances in Applied Microbiology*, 1996. **42**: 263-302

47. Powar, V.K. and Jagannathan, V. 1982. Purification and properties of phytate specific phosphatase from *Bacillus subtilis. J. Bacteriol.* **151**: 1102- 1108.

48. Rackis, J.J. 1977. Enzymes in soybean processing and quality control. In: *Enzymes in Food and Beverage Processing* (eds. R.L.Ory and A.J.St.Angelo) American Chemical Society, Washington, D.C., pp.244-265.

49. Reddy, N.R. and Salunkhe, D.K. 1981. Interaction between phytate, protein and minearals in whey fractions of black gram. *J. Food Sci.* **46**: 564

50. Sabin, R. 1988. U.S. Patent No.4, 758, 430. July 19.

51. Sabin, R. 1989. U.S. Patent No.4, 847, 082. July 11.

52. Sabin, R. 1992. U.S. Patent No.5, 112, 814. May 12.

53. Sabin, R. 1993. U.S. Patent No.5, 217, 959, June.

54. Segueilha, L., Lambrechts, C., Boze, H., Moulin, G. and Galzy, P. 1992. Purification and properties of phytase from *Schwanniomyces castellii. Journal of Fermentation and Bioengineering* **74** (**1**): 7-11.

55. Segueilha, L., Moulin, G. and Galzy, P., 1993. Reduction of phytate contents in wheat bran and glandless cotton flour by *Schwanniomyces castellii. J.Agric. Food Chem.* **41** (**12**): 2451-2454.

56. Sharma, C.B., Goel, M. and Irshad, M. 1978. Myoinositol hexaphosphate as a potential inhibitor of α-amylase of different origins. *Phytochemistry* **17**: 201-204.

57. Shieh, T.R. and Ware, J.H. 1968. Survey of microorganisms for the production of extra cellular phytase. *Appl. Microbiol.* **16(9)** 1348-1351.

58. Shieh, T.R., Wodzinski, R.J. and Ware, J.H. 1969. Regulation of the formation of acid phosphatase by inorganic phosphate in *Aspergillus ficuum. J. Bacteriol.* **100**: 1161-1165.

59. Shimizu, M. 1992. Purification and characterization of phytase from *Bacillus subtilis* (natto) N-77. *Biosci. Biotechnol. Biochem.* **56(8)**: 1266-1269.

60. Simons, P.C.M., Versteegh, H.J.A., Jongbloed, A.W., Keeme, P.A., Slump, P., Bos. K.D., Wolters, M.G.E., Beudeker, R.F. and Verschoor, G.J. 1990. Improvement of phosphorus availability by microbial phytase in broiler and pigs. *Brit. J. of Nutr.* **64**: 525-540.

61. Skowronaski, T. 1978. Some properties of apartially purified phytase from *Aspergillus niger. Acta Micro. Pol.* **27**: 41-48.

62. Smith, J.E. and Beny, D.R. 1974. *Biochemistry of Fungal Development.* Academic Press, London. pp. 282-308.

63. Sreeramulu, G., Srinivasa, D.S., Nand, K. and Joseph, R. 1996. *Lactobacillus amylovorus* as a phytase producer in submerged culture. *Letters in Applied Microbiology* **23**: 385-388.

64. Suzuki, U., Yoshimura, K. and Takaishi, M. 1907. Tokyo Imp. Univ. Coll. Agr. Bull. 7: 503-512. (Adapted from Wodzinski, J. and Ullah A.H.J. 1996).

65. Tambe, S.M., Kakli, S.G., Kelkar, S.M. and Parekh, L.J. 1994. Two distinct molecular forms of phytase from *Klebseilla aerogenes;* Evidence for unusually small active enzyme peptide. *J. Ferm. Bioeng.* **77(1)**: 23-27.

66. Teutonico, R.A. and Knorr, D. 1985. Impact of biotechology on the nutritional quality of food plants. *Food Technol.* **39**: 127- 134.

67. Teutonico, R.A. 1987. Impact of biotechnology on the nutritional quality of Foods. In: *Food Biotechnology* (es. Dietrich Knoff) Marcel Dekker, Inc., New York and Basel.

68. Thompson, L.U. 1986. In : *Phytic Acid: Chemistry and Applications* (ed. E. Graf), Pilatus Press, Minneapolis, pp. 173-194.

69. Ullah, A.H.J. and Gibson, D.M. 1987. Extracellular phytase (E.C.3.1.3.8) from *Aspergillus ficuum* NRRL 3135: purification and characterization. *Prep. Biochem.* **17**: 63-91.

70. Ullah, A.H.J. and Cummins B.J. 1987. Purification, N terminal amino acid sequence and characterization of pH 2.5 optimum acid phosphatase (E.C.3.1.3.2) from *Aspergillus ficuum. Prep. Biochem.* **17**: 397-422.

71. Ullah, A.H.J. and Cummins B.J. 1988. *Aspergillus ficuum* extracellular pH 6.0 optimum acid phosphatase: Purification, N terminal amino acid sequence and biochemical characterizaion. *Prep. Biochem.* **18**: 37-65.

72. Ullah, A.H.J. and Dischinger, H.C. Jr. 1993a. *Aspergillus ficuum* phytase: Complete primary structure determination by chemical sequencing. *Biochem. Biophys. Res. Commun.* **192**: 747-753.

73. Ullah, A.H.J. and Dischinger, H.C., Jr. 1993b. Identification of active site residuse in *Aspergilllus ficuum* extracelllular pH 2.5 optimum acid phosphatase. *Biochem. Biophys. Res. Commun.* **192**: 754-759.

74. Van Hartingsveldt, W., Van Zeijl, C.M.J., Harteveld, G.M., Gouka, R.J., Sukerbuyk, M.E.G., Luiten, R.G.M., Van Paridon, P.A., Selten, G.C.M., Veenstra, A.E., Van Gorcom, R.F.M. and Van den Hondel, C.A.A.M.J.J. 1993. Cloning characterization and overexpression of the phytase encoding gene(phyA) of *Aspergillus niger. Gene* **127**: 87-94.

75. Wang, H.L., Savain. W. and Hesseltine, C.W. 1980. Phytase of molds used in oriental food fermentation. *Journal of Food Science* **45**: 1261-1266.

76. Ware, J.H. and Shieh, T.R. 1967. U.S. Patent No.3.297, 548.

77. Wodzinski, R.J. and Ullah, A.H.J. 1996. Phytase. In: *Advances in Applied Microbiology*. **42**: 263-302.

78. Yamada, K., Minoda, Y., Kobayashi, T., Hidaka, Y., Matuo, H. and Kobayashi, M. 1968.Phytase from *Aspergillus terreus. J. Ferment. Technol.* **46**: 857-862.

79. Yoon, Seong Jun, Choi, Yun Jaie, Min, Hae Ki, Cho, Kwang Keun, Kim, Jin Wook, Zee Sang Cheol and Jung, Yeon Hoo. 1996. Isolation and identification of phytase producing bacterium, *Enterobacter* sp.4 and enzymatic properties of phytase enzyme. *Enz. Microbiol. Technol.* **18**: 449-454.

80. Zyta, K., Koreleski, J. and Kujawski, M. 1989. Dephosphorylation of phytate compounds by means of acid phosphatase from *Aspergillus niger. Journal of the Science of Food and Agriculture,* **49**: 315-324.

81. Zyta, K.1992. Mould phytases and their application in the food industry. *World J. Microbiol. Biotechnol.* **8**: 467-472.

82. Zyta, K. 1993. *In vitro* and *in vivo* dephosphorylation of rapeseed meal by means of phytate degrading enzymes derived from *Aspergillus niger. J. Sci. Food Agric.* **61**: 1-6.

9 Molecular Probes in Detection and Monitoring of Fluorescent Pseudomonads

Bhavdish N. Johri

CONTENTS

ABSTRACT ... 112
1. INTRODUCTION ... 112
2. FLUORESCENT PSEUDOMONADS ... 113
2.1 General Characteristics ... 113
2.2 Methods in Identification and Characterization ... 113
2.3 Taxonomic and Phylogenetic Tools ... 113
2.4 Methods Based on Protein Finger Printing ... 115
3. GROWTH PROMOTORY FUNCTION ... 115
3.1 ACC Deaminase ... 116
3.2 Other Tools ... 116
4. MOLECULAR METHODS ... 117
4.1 PCR/RFLP ... 117
4.2 Newer Tools in Diversity and Heterogeneity ... 117
4.3 Phenotypic Characterization vs PCR Finger Printing ... 117
4.4 PCR/RFLP for Plant Pathogenic Pseudomonads ... 118
4.5 Siderophore Typing of *P. aeruginos* ... 118
5. MARKERS IN RHIZOSPHERE COLONIZAITON, SPREAD AND SUSTENANCE ... 118
5.1 lac z System ... 119
5.2 *lux* reporter System ... 120
5.3 inaz System ... 120
5.4 xyle System ... 120
5.5 RNA-based System ... 121
5.6 Immunotyping ... 121
6. MARKERS FOR *IN-SITU* ECOLOGY, DETECTION AND MONITORING ... 121
6.1 Direct Soil DNA Extraction ... 121
6.2 Siderophore-mediated Disease Suppression ... 123
6.3 Antibiotic mediated disease suppression ... 123
6.4 Use of Genetically Engineered Pseudomonads. ... 123
7. REFERENCES ... 124

Advances in Microbial Biotechnology
J.P. Tewari, T.N. Lakhanpal, Jagjit Singh, Rajni Gupta & B.P. Chamola (eds.)
APH Publishing Corporation, New Delhi - 110 002, India.

ABSTRACT

Fluorescent pseudomonads comprise a group of Gram-negative bacteria that are characterized by secretion of yellow-green diffusible pigment in King's agar and are currently being studied extensively as potentially exploitable forms for promotion of plant productivity. Two major species that have attracted most attention are, *P. fluorescens* and *P. putida*. In view of their variability and heterogeneity, these species have been further demarcated into biovarieties and identification of such forms therefore, is not an easy task. In the context of applicability in natural ecosystem comprising largely of bulk soil and rhizosphere there are serious problems in detecting the introduced forms on laboratory media on account of existence of non-culturable cells and the very nature of the selectivity of the techniques. Recent years have seen application of both, protein and DNA fingerprinting techniques in detection and monitoring as also in evaluating *in situ* metabolic functions of fluorescent pseudomonads. In addition, immunotyping and introduction of various molecular markers has permitted detailed investigation of root colonization, *in situ* interactions and sustenance of such bacterial inoculants. These recent developments are presented in this article.

1. INTRODUCTION

Bacteria that efficiently and effectively colonize plant roots have been termed "rhizobacteria" by Kloepper and Schroth (42) and Suslow *et al.* (92). The purpose of coining this term was to delineate aggressively root colonization forms from those whose natural habitats occur elsewhere. To this term has added the epithet "plant-growth promoting rhizobacteria" (PGPR) to include all forms that bring about promotion of plant growth irrespective of the mechanism in question. For more specific plant growth responses more discrete terms have been coined such as nodulation promoting rhizobacteria (NPR), emergence promoting rhizobacteria (EMR) , deleterious rhizobacteria etc. The net result of this conglomeration of free-living and symbiotic forms has attracted increased scientific attention on account of their intimate role in especially improved and sustained crop productivity. Bacterial forms that have received most attention as PGPR's include *Acetobacter, Azotobacter, Arthrobacter, Azospirillum, Bacillus, Flavobacter, Pseudomonas* and *Rhizobium.*

Detailed analysis of soil microbial community structure shows that short, Gram (-) rods predominate in the rhizosphere milleu among which genus *Pseudomonas* plays a key role in biodegradation, biotransformation, biocontrol and direct plant growth promotion. While large pseudomonad population exerts stimulatory response on root elongation and shoot growth, isolates that are deleterious or neutral are a normal component of the population structure. In the previous decade, fluorescent pseudomonads have received considerable attention as potentially promising bioinoculants of future on account of their ability to release IAA resulting in growth promotion as also by virtue of the inhibition of root and soil borne phytopathogens through either siderophore-mediated iron chelation and/ or by release of antibiotics such as 2, 4-acetyphloroglucinol, pyrolnitrin and phenazines (39, 100). Researches undertaken during the last years have resulted in a new dimension to the disease control potential i.e., that of induced systemic resistance (56, 97). This has enlarged the scope of application of fluorescent pseudomonads from root and soil borne diseases to foliar pathogens (57) and as a means of long-term control possibly through singnal molecules (42). However, inspite of these developments, the question of population heterogeneity, strain variability and monitoring of fluorescent pseudomonads has been of paramount concern. This however, has received recent attention in order to devise tools that would help check inconsistences by way of root colonization and spread, stability in soil milleu and quality control of the end product. This article will focus on recent developments including use of molecular tools in diversity, systematics and monitoring of fluorescent pseudomonads in natural ecosystems including plant growth promotion and biocontrol.

2 FLUORESCENT PSEUDOMONADS

2.1 General Characteristics

Within the genus *Pseudomonas*, several species are able to synthesize under iron deficient conditions, yellow-green fluorescent, water-soluble pigments which are components of iron-chelating bacterial siderophores (72), two major structural groups are the pyoverdives and pseudobactin (67). Siderophores are responsible for iron transport to the bacteria and are often highly specific for a strain (61).

Species within the fluorescent pseudomonas have been studied extensively from an ecological viewpoint on account of the diversity of the metabolites secreted in the environment besides their influence on plant and animal systems (11). For example, *P. aeruginosa* is pathogenic to animals and has been investigated in detail (65). Several species, viz., *P. corrugata, P. fuscovaginae, P. syringae* and *P. viridiflava* cause disease in plants (37, 101). *Pseudomonas tolaasii* has been associated with diseases of mushrooms. However, several isolates within the species, *P. fluorescens* and *P. putida* have been found to promote plant growth or control soil-borne diseases (25). Since this requires use of diverse population of the fluorescent pseudomonads, identification and characterization are important for which various tools have now been developed.

2.2 Methods in Identification and Characterization

Fluorescent pseudomonads are placed in rRNA group I by Palleroni *et al.* (76) based on DNA/DNA homology. Included in this group are the heterogenous species *P. fluorescens* and *P. putida* which have been sub-divided into biovarieties by Palleroni (74); the former has bv I to V and the latter, bv A and B. Plant pathogenic *P. syringae*, on the other hand, is subdivided into pathovars based on host range. However, to tide over the problems associated with the characterization of multitude of isolates usually recovered in ecological investigations, Laguerre *et.al.* (49) developed a method based on polymerase chain reaction (PCR) and 16S rDNA genes for differentiation of fluorescent and non-fluorescent pseudomonads. Other methods employed in identification and characterization are listed in Table 1.

2.3 Taxonomic and Phylogenetic Tools

In a wide array of other tools used in taxonomy and phylogeny of fluorescent pseudomonads, Hildebrand (33) utilized simple pectate and pectin-containing gels for the differentiation of plant pathogenic pseudomonads (Table 2). On the other hand, sub-grouping of *P. solanacearum* isolates was find out based on tRNA consensus primers utilizing PCR amplification (83). Lipololysaccharide patterns have helped characterize a variety of Gram-negative forms and pseudomonads are no exception either (86). However, more specific methods in determining strain variability immunotyping, epitope mapping and use of monoclonal antibodies (36, 69, 70, 80). On the phylogeny front Lie *et.al.* (55) utilized 16S-rRNA sequencing to determine relationships withing group II pseudomonads. Therefore, a whole array of tools and techniques are available to characterize the heterogenous fluorescent pseudomonads that today utilize both, catabolic property on a host of substrates and combine it with one or the other molecular methods among which PCR/RFLP is gaining significance in the context of determine heterogeneity and variability and 16S sequencing for specific identification.

Table 1 : Major methods used in characterization and phylogeny of fluorescent pseudomonads

Characteristic Employed	References
Phenotypic Characteristics	(4, 32)
DNA homology	(23, 75, 76)
DNA-rRNA hybridization	(20, 76)
Partial 16S rDNA sequence	(98, 99)
Macrorestriction fragment fingerprinting of genomic DNA	(29)
rRNA cistron similarities	(19)
Whole cell fatty acid profile	(66, 86, 87, 91)
PCR/RFLP of 16S-rDNA	(49)
16S-rRNA sequences	(55)
Whole-cell protein pattern	(88)
Outer membrane protein heterogeneity	(46)
23S r-DNA sequences	(14)
Siderophore typing	(65)

Table 2 : Markers used in determining taxonomic and phylogenetic relationship

Probe/Gene/Marker	Function/Assessed	Reference
Antibodies based on synthetic peptide representing protein F	Immunotyping of heterologous strains of *P. aeruginosa*	(36)
Epitope mapping	Characterization of major outer membrane porin protein Opr F	(80)
tRNA consensus primers	sub-grouping of *P. solanacearum* by PCR	(83)
Peetate and pectin gels	Differentiation of plant pathogenic pseudomonads	(33)
LPS pattern	Serological variability in *P. corrugata*	(86)
16S-r RNA sequences	Phylogeny of r RNA group II pseudomonads	(55)
Monoclonal antibody	Specific to outer membrane lipoprotein of *P. fluorescens*	(69)

2.4 Methods Based on Protein Fingerprinting

Three species makeup the central group within the fluorescent pseudomonads, viz., *P. aeruginosa*, *P. fluorescens* and *P. putida*; each species is characterized by biotypes which makes their identification difficult. However, in view of the difficulties encountered in distinguishing various forms based on biochemical features, proteini based characterization protocol was developed by Sorensen *et al.* (88). Both, whole cell and membrane-bound proteins have been found to be useful because of the conserved nature of the molecules. Buss *et al.* (12) applied PAGE analysis of soluble proteins for characterization of pseudomonads at the species and subspecies level. On the other hand, heterogeneity of membrane proteins was utilized in characterization of environmental isolates of *P. aeruginosa* by Hancock and Chan (31). Zyl and Steyn (103) developed a PAGE-based numerical taxonomy for differentiating phytopathogenic isolates of *Pseudomonas* and *Xanthomonas*. Thus, cell and membrane proteins provide distinct features for characterization of species and biovarieties within the fluorescent pseudomonads. Lambert *et al.* (50, 51) utilized this variability to characterize rhizobacteria of maize and sugarbeet plants; they defined "electrotype" as an assemblage of strains with nearly identical whole cell protein pattern. Employing this technique with 350 isolates of fluorescent pseudomonads from maize roots, these workers identified 43 "electrotypes".

A major problem with whole-cell protein patterning however is variability on account of modification in cultural conditions. Sorensen *et al.* (88) therefore utilized proteinase-K resistant protein-fragments in identification of environmental isolates of fluorescent pseudomonads and coupled it to biochemical tests which included growth at 42°C, phenazine pigment formation, denitrification and growth on meso-inositol, sorbitol, mannitol and phenylacetate. The major protein fragment bands utilized for identification purpose were located at approx. 53 Kda, 45Kda, 18 Kda and 16 Kda. For *P.aeruginosa*, a 53 Kda band was distinct whereas 45 Kda fragment was common in *P. fluorescens* and *P. putida*: These two species could however be distinguished based on bands in 15-20 Kda range. In a subsequent study Kragelund *et al.* (46) examined outer membrane protein heterogeneity within *P. fluorescens* and *P. putida* biovarieties and found that the patterns were not useful in assigning strains to a specific biovar even though several dominant proteins migrating at 42 to 46 KDa, 33 to 38 KDa, and 20 to 22 KDa were conserved among the strains. Tentatively these proteins were identified as Opr E (44 KDa), Opr F (38 KDa), Opr H (2.1 KDa) and Opr L (20.5 KDa), known proteins of *P. aeruginosa*. An immunoblot assay was developed based on polyclonal antibody raised against Opr I (37 KDa) protein which was found to be specific for other strains belonging to *Pseudomonas* rRNA homology group I. Application of this antibody in a colony blot assay for determine fluorescent pseudomonads in the rhizosphere soil of barley resulted in about 2 to 14 times greater CFU counts on 10% tryptic soy agar compared to the values obtained in more specific King's agar and Gould S1 agar; apparently not all fluorescent pseudomonads are recovered on the commonly used selective media. Authors further confirmed the identify of rRNA homology group I pseudomonads employing API 20 NE system which utilizes a variety of biochemical tests. Therefore, while protein profiling provides a rapid means for determining the heterogeneity of fluorescent pseudomonads, additional tests including biochemical profiles and immunotyping help confirm identity.

3. GROWTH PROMOTORY FUNCTIONS

Promotion of plant growth is one of the major activity for which fluorescent pseudomonads are the most sought after bacteria (25). A normal protocol to search for promising isolate(s) revolves around recovery of large number of pseudomonads from bulk soil, rhizoplane, rhizosphere and endorhizosphere milleu on selective media such as King's agar and evaluation of growth promotory potential employing seed germination and root elongation test on plant species of choice; also evaluated are the siderophore producing ability, antagonistic potential, HCN secretion and others. Therefore, the whole exercise is a long drawn process and no specific short protocols are available.

3.1 ACC Deaminase

Glick *et al.* (25) described a novel isolation method for plant growth promoting pseudomonad based on the ability of soil bacteria to utilize 1-aminocyclopropane-1-carboxylate (ACC) as sole N source (Table 3). The basic premise of this method is that the enzyme ACC deaminase hydrolyzes ACC which is the immediate precursor of ethylene and would thus lower ethylene levels in seed, resulting in growth promotion. Therefore, any bacterium that contains ACC deaminase signifies that after seed bacterization it would promote, root elongation. Employing mutants of various kind these workers confirmed this hypothesis. Based on this hypothesis and its subsequent testing Glick *et al.* (26) have recently proposed a model based on lowering of ethylene concentration by PGPR to explain improved plant growth and productivity. According to the proposed model, biological activity of PGPR is a reflection of the relative amounts of ACC deaminase and ACC carboxylate oxidase; to lower plant ethylene levels to the desired critical limit, levels of deaminase should be 100-to 1000-fold greater than the oxidase.

Table 3 : Probes and markers utilized in assessment of plant growth-related functions

Probe/Gene/Marker	Function Assessed	Reference
IAA-overproducing mutants	Growth modulation by *P. putida* GR 12-12	(100)
1-Aminocyclopropane-1-carboxylic acid deaminase mutants	Non-elongation of canola roots by *P. putida* GR12-2	(24)
lux AB-tagged reporter system	(i) Metabolic activity and phosphorus starvation response of *P. fluorescens* in rhizosphere	(45, 47)
	(ii) Detection of bacteria in rhizosphere and soil	(22, 63, 64)
Phosphate-reporter system	Phosphate limitation in the rhizosphere and bulk soil	(21)
Use of exotic carbon source	Selective increase in metabolic activity and growth of *P. putida* in soil	(15)

3.2 Other Tools

In other marker assisted functions related to growth promotion, Xie *et al.* (100) utilized Tn5 mutagenesis for developing IAA-overproducing mutants of *P. putida* GR12-2 to confirm the role of auxin in root elongation of canola (*Brassica campestris* CV Tobin). In the same bacterial strain, Glick *et.al.* (24) had earlier tested the operation of ACC deaminase through aminase-negative mutants. Other markers employed in assessment of metabolic and growth-related functions are listed in Table 3. Among these, *lux* reporter system is discussed elsewhere in the context of fluorescent pseudomonads.

4. MOLECULAR METHODS

In recent years several molecular methods have been developed. These are briefly presented here.

4.1 Polymerase Chain Reaction/Restriction Fragment Analysis

Rapid methods of microbial identification and monitoring have developed for various groups on account of the pace and often, one step confirmation of identify of the environmental isolate(s). Woese *et al.* (98) showed through rRNA cataloguing that 16S rRNA in case of fluorescent and non-fluorescent pseudomonads differed in their nucleotide sequence information. Based on this premise, Laguerre *et al.* (49) have developed a technique for differentiation of fluorescent pseudomonads which utilizes restriction fragment analysis of PCR-amplified 16S rDNA. These workers used oligonucleotide sequences f D1 and rD1 as primers since these are derived from conserved regions of 16S rRNA genes. The two primers have the sequence- fD1,5' CCG AATT CGA CAA GA GAG TTT GAT CCT GGC TCAG-3'; rD1, 5'-CCC GGG ATC CAA GC TTA AGG AGG T GAT CCA GC-3'. A total of 13 restriction enzymes were used by these workers; they were , AluI, CfoI, DdeI, HaeIII, HinfI, MseI, MspI, NciI, NdeII, RsaI, TaqI, Scr FI, and Sau96I. In an analysis of ten species of *Pseudomonas* and biovars spread over 31 strains, 17 different 16S rDNA genotypes could be characterized based on the data for restriction patterns. This technique thus permitted differentiation of various species except *P. aureofaciens* and *P. chloroaphis*; good correlation existed between PCR/ RFLP data and the information obtained from DNA/DNA homology data.

4.2 Newer Tools in Diversity and Heterogeneity or Pseudomonads

Natural populations of fluorescent pseudomonads from rhizoplane and rhizosphere have been reported to exhibit a great deal of heterogeneity and improper understanding of this natural indigenous consortia has often resulted in inconsistencies associated with the use of PGPR's in improved plant growth and biocontrol action. An evident reason for this lacuna has been non-availability of rapid and reliable identification protocols for such a large number of isolates. However, the situation has changed totally with the availability of suitable molecular tools.

4.3 Phenotypic Characterization vs PCR Fingerprinting

Lemanceau *et al.* (53) carried out detailed study of the influence of plant species, flax (*Linum usitatissimum* L.) and tomato (*Lycopersicon esculentum* Mill.) on the bacterial traits that governed the diversity of soil borne populations of fluorescent pseudomonads. A total of 176 isolates recovered from the uncultivated soil, the rhizosphere, the rhizoplane, and the root tissue of flax and tomato alongwith 30 reference strains were characterized assimilation of 147 different organic compounds; based on numerical analysis and clustering of isolates, 87.8% similarity was found. Based on phenotypic characters it could be inferred that the soil bacterial population was different from plant-associated isolates. That plant species were different was particularly evident in flax where isolates recovered from uncultivated soil and root tissue were markedly different. However, the selection was to an extent plant specific since it was possible to discriminate between isolates from tomato and flax based on only 10 different substrates. The phenotypic data was found to correlate well with genotypic characterization based on repetitive extragenic pallindronic PCR fingerprinting. According to these workers, the ability of fluorescent pseudomonads to assimilate varied organic substrates would be an important trait that would confer selective advantage in soil and rhizosphere colonization and sustenance.

With the above base data about the phenotypic heterogeneity. Latour *et al.* (52) carried out investigation of fluorescent pseudomonads from two uncultivated soils (Chateaurenard and Dijon) and compared it with that from the roots of flax and tomato in order to understand the reasons for the inconsistencies associated with the PGPR or biocontrol action. Alongwith the substrate assimilation pattern, the genotypic diversity in this case was studied employing, (i) types of 16S gene coding for rRNA (rDNA), (ii) their repetitive extragenic pallindronic patterns by PCR, and (iii) plasmid profiles. Numerical analysis and clustering permitted the conclusion that both plant and soil type influenced the pseudomonad diversity. Among the two, soil had distinctly more prominent role, whereas the plant selection was more plant-specific. The genotypic characterization and phenotypic clustering exhibited high correlation substantiating the usefulness of the two techniques in functional categorization of fluorescent pseudomonads. According to these workers REP-PCR and fingerprinting was more valuable than PCR-RFLP analysis of 16S rDNA in distinguishing fluorescent pseudomonads. This could not be inferred for plasmid DNA because few isolates of fluorescent pseudomonas harbor plasmids.

4.4 PCR/RFLP for Plant Pathogenic Pseudomonads

Application of PCR-RFLP has helped define diversity of some plant pathogenic fluorescent pseudomonas besides deciphering the heterogeneity of geographically varied isolates. In one such study, Jaunet *et al.* (37) analysed the diversity of *P.fuscovaginae* responsible for sheath brown rot in low and high elevations in temperate countries. viz, Madagascar, Nepal, Montpellier (France), Philippines, Indonesia and Japan. On the basis of biochemical characters 23 different nutritional groups were identified whereas employing PCR-RFLP, 25 composite 16S-rDNA haplotypes were obtained; truly pathogenic and opportunistic isolates could be distinguished. In a recent investigation of fluorescent pseudomonad strains causing tomato pith necrosis (TPN), they were studied by numerical taxonomy based on biochemical characteristics, LPS pattern, fatty acid profiles and DNA relatedness (93). Based on the dendrograms drawn, strains of authentic *P. corrugata* could be placed in one phenon whereas 17 of the 33 TPN strains clustered in a different phenon; DNA: DNA hybridization data confirmed the true character of *P. corrugata* however this was not true for TPN strains. Authors have utilized this composite information to designate the distinict TPN strains as *Pseudomonas* FP 1, FP2, and FP3.

4.5 Siderophore Typing of *Pseudomonas aeruginosa*

P. aeruginosa is an important fluorescent pseudomonad for which siderophore mediated detection and monitoring system has been proposed by Meyer *et al.* (65). This species operates pyoverdine mediated iron incorporation system which was evaluated by, (i) isoelectric focussing, (ii) growth stimulation, (iii) immunoblot detection of outer-membrane receptor protein, and (iv) pyoverdine-facilitated iron uptake. Employing these methods, three different siderophore types could be distinguished.

5. MARKERS IN RHIZOSPHERE COLONIZATION, SPREAD AND SUSTENANCE

A major area of use for fluorescent pseudomonads comprises of field application for biocontrol of soil and root-borne diseases and growth promotion of crops. Detection of introduced strain(s) has undergone a sea change from the days of antibiotic marking to *lux*-tag and others. These developments are briefly summarized below in the context of biocontrol, PGPR action and behaviour and tracking in the rhizosphere (Table 4).

Table 4: Markers used in colonization, monitoring and ecology of fluorescent pseudomonads

Probe/Gene/Marker	Function Assessed	References
Plasmid and chromosomally encoded luminescence marker system	(i) Detection of *P.fluorescens* in soil (ii) Plant root colonization	(24) (7)
pvd-ina Z fucsion-based biosensor for iron	Detection of engineered *P. fluorescens* and *P. syringae* in bean rhizosphere and phyllosphere	(59)
Charge coupled device-enhanced microscopy	Detection of single genetically modified bacterial cell	(85)
Root exudate-induced promoter activity	Detection of *P. fluorescens* mutants in wheat rhizosphere	(96)
Root adhesin	Homology with porin F or *P. aeruginosa* and *P. syringae*	(18)
Sandwich Elisa	Detection of *P. fluorescens* and related psychrotrophs in milk	(28)
Pho E-caa construct	PCR-and immunologically detectable marker gene for *P. putida*	(102)
RFLP with DNA probe	Genetic diversity of *P. solanacearum*	(16)
MAB against surface determinants	Tracking genetically engineered *P. putida* 2440	(81)
Tn7-*lux* system based on *E. coli* rRNA promotor	Direct detection of rhizosphere-colonizing *Pseudomonas* sp.	(10)

5.1 lac Z system

Incorporation of lac Z and lac Y genes from *E. coli* K-12 into fluoroscent pseudomonads has been an early development among the genetic markers. Barry (65) described the insertion of these two genes into the chromosome of a rifampicin and nalidixic acid-resistant strain of *P. aureofaciens* through Tn7 vector system. As a result it was possible to select fluorescent pseudomonads on lactose as a sole carbon and energy source; selection as blue colonies was possible on the chromogenic dye X-Gal. In a different approach Hofte *et al.* (35) inserted lac Z gene into *P. aeruginosa* alongwith the gene for kanamycin resistance, permitting selection on the minimal medium containing lactose and kanamycin; a detection limit of 10 cfug-1 soil was observed.

Kluepfel (41) reported nearly similar detection limit for *P. aureofaciens* strains marked with Tn7: lac ZY. The only limitation to the use of this marker is the necessity of dilution plating and selective media which permit only culturable cells to be counted. In our laboratory

Khokar *et al.* (40) successfully tagged *P. fluorescens* with lac marker, system for *in situ* detection of a growth promotory isolate.

5.2 *lux* Reporter System

Bioluminescence observed through the introduction of *lux* genes from *Photobacterium* or *Vibrio* spp. has found wide ranging application in detection and monitoring of a variety of microorganisms. Employing *lux* reporter system, Shaw *et al.* (84) have reported a detection limit of 50 cfu g^{-1} soil and 1.5×10^4 cfu $leaf^{-1}$ in cabbage (*Brassica oleracea*). However, for bioluminescent *P. syringae* pv. *phaseolicola*, Silcock *et al.* (85) have reported detection of even single cells employing charge coupling derive-enhanced microscopy. The greatest advantage of this device, has been the non-destructive nature of the technique. DeWeger *et al.* (22) have utilized *lux* system to monitor fluorescent pseudomonads in the rhizosphere of plants; these workers have reported detection limits of 10^3 to 10^4 cfu cm^{-1} of root tissue. In a more recent use of *lux*, Kragelund *et al.* (47) have studied the distribution of metabolic activity and availability of phosphate for *P. fluorescens* in the barley rhizosphere. The expression of *lux* AB fusion in strain DF57-4-E7 was constitutive and thus permitted detection of bioluminescence under defined assay conditions. The *lux* AB gene fusion was controlled by regulatory sequences similar to phosphate (pho) regulon. According to these workers while P-limitation was characteristic under the gnotobiotic rhizosphere conditions, natural rhizosphere situation was different on account of indigenous microbiota. Addition of indigenous strains to sterilized soil resulted in amelioration of P-limitation.

5.3 ina Z System

The property of the gene ina Z to code for an outer membrane protein in *P. syringae* and other bacteria for bringing about ice nucleation is known for a long time (60). However, application of ice-nucleation property as a reporter system for *in situ* expression of pseudomonads in the rhizosphere was reported by Loper *et al.* (58). These authors suggested use of ina Z as a good measure of bacterial population since ice-nucleation activity is quantitatively related to the amount of ina Z protein; the numbers of ice nuclei were representative of pseudomonad population and through this bacterial population as low as 10^4 cells $root^{-1}$ could be counted. The gene expression did not require culturing of bacteria. Lindgren *et.al.* (54) have utilized this reporter system for the identification of inducible pathogenicity genes in *P. syringae* pv. *phaseolicola*. In another interesting use of this reporter system, Loper and Lindow (59) developed a biosensor for iron based on the siderophore pyoverdine-ina Z fusion product, for detection and monitoring of *P. fluorescens* and *P. syringae* in bean rhizosphere and phyllosphere. Presence of iron (Fe-III) resulted in reduced number of ice nuclei on leaves and in soil confirming reduced transcription of the pvd-ina Z reporter gene. This reporter system thus permitted monitoring of iron availability in the natural environment, an element crucial in regulation of siderophores.

5.4 xyl E

This interesting chromogenic marker system has been cloned and expressed in various species of *Pseudomonas* and can be easily detected by spraying catechol on the plate when the modified bacterial colonies appear bright yellow on account of formation of 2-hydroxymuconic semi-aldehyde. This marker has been used for tracking *P. putida* in lake water and *P. syringae* in the phyllosphere however it has been used only infrequently (41).

5.5 rRNA Based Systems

Microbial classification and phylogeny has undergone a complete change ever since sequencing of rRNA's (5s, 16s, 23s) came on the scene. Also, the presence of conserved sequences unique to species/ subspecies has helped development of oligonucleotide probes specific for detection of various taxa including pseudomonads. Application of these methodologies in the context of systematics has been discussed in an earlier section. Other applications in the context of pseudomonads shall be briefly highlighted here.

Applying rRNA hybridization coupled to epifluorescence microscopy, Hahn *et al.* (30) were able to detect *P. aeruginosa* in soil; a theoretical sensitivity of 10^2 to 10^3 cell g^{-1} soil was possible *in situ* 16S rRNA hybridization technique. In order to utilize this technique at non-destructive scale and for enumeration of quantitative and qualitative change in indigenous microflora, Amann *et al.* (1) have described coupling of this technique to flow cytometry.

5.6 Immunotyping

Besides utilization of outer-membrane protein based polygonal and monoclonal antibodies, several other detection systems based on immunological methods have been used with fluorescent pseudomonads. Hughes *et al.* (36) developed an immunotyping protocol for heterologous strains of *P. aeruginosa* which utilized antibodies based on a synthetic peptide representing protein F. On the other hand, Rawling *et al.* (80) utilized epitope mapping system for characterization of this major outer membrane porin protein. In alternate approaches, a monoclonal antibody specific to outer membrane lipoprotein of *P. fluorescens* group was developed by Mutharia *et al.* (70) whereas a Sandwich ELISA system was developed for detection of *P. fluorescens* and related psychrotrophs in refrigerated milk (28). Distribution and monitoring of rRNA homology group I pseudomonads which includes the fluorescent species has been possible through polyclonal antibody developed against Opr F protein (46). However, for tracking the genetically engineered *P. putida* 2440,? utilized a MAB active against the cell surface determinants. In another approach, a Pho E-caa construct was used for PCR and immunological detection of *P.putida* (102). In this species, Cebolla *et.al.* (13) have used an antigenic surface reporter system to detect the activity of the catabolic promoters (Table 4).

6. MARKERS FOR *IN SITU* SOIL ECOLOGY, DETECTION AND MONITORING

6.1 Direct Soil DNA Extraction

In order to assess *in situ* soil bacterial population and its ecological behaviour, methods for direct DNA extraction from soil have come to fore for further molecular dissection employing restriction endonucleases, PCR amplification and other protocols. Trevors *et al.* (95) had used this technique to investigate transportation of a genetically engineered *P. fluorescens* through a soil microcosm. Employing suitable probe(s) against the DNA extracted directly from soil, a sensitivity level close to 10^3 bacteria g^{-1} soil has been reported which does not make it comparable with the more sensitive dilution plate techniques but permits *in situ* detection of both, culturable and non-culturable forms.

In a continuing programme on growth promotory fluorescent pseudomonads in our laboratories, direct soil DNA extraction protocol for strains PRS_9 and GRP_3 was standardized by Kumar (48). Restriction analysis of DNA with Bgl II revealed RFLP patterns that were distinct for the two strains and different from an unsterile unseeded control soil. For PRS^9, 8 to 10 bands were observed in 22-4 kb range whereas in GRP^3, 4 to 6 bands of 22-8 kb were detected; a soil co-seeded with PRS^9 and GRP^3 produced 6 bands of 22-12 kb (Kumar and Goel, unpublished observations).

Sensitivity of direct soil DNA technique has been considerably imvroved by use of PCR to amplify and detect specific gene sequences. Steffan and Atlas (90) applied this technique to a genetically engineered strain of *P. cepacia* (now *Burkholderia cepacia*) and detected as little as one bacterial cell g^{-1} soil; this was a 1000-fold improvement over a similar protocol with nonamplified soil samples. Klucpfel (41) isolated restrictable DNA from a field site where a lacZY- marked pseudomonad isolate had been released six months earlier. Using the standard dilution plating protocols the isolate remained undetected for nearly three months. In the amplification of directly extracted soil DNA, these workers used two 19-base-pai (bp) primers which were specific for a unique 503-bp fragment of the Tn7R segment of the transposon Tn7. While diluting plating gave negative results, Tn7R probe was able to hybridize to the amplified sample; a population level of $10cfug^{-1}$ soil was estimated.

Among the mechanisms of plant growth promotion exhibited by fluorescent pseudomonads, siderophore-mediated and antibiotic-mediated inhibition of soil-brone phytopathogens is of prime significance. While methods to assess *in situ* antagonism abound it is not always easy to corroborate this information with *in situ* behaviour of introduced strain(s). To this end various approaches have been used; some salient features are discussed here (Table 5).

Table 5: Markers used in evaluating biocontrol and environmental applications of fluorescent psuedomonads

Probe/Gene/Marker	Function Assessed	References
pub B gene	Inducible ferric-pseudobactin receptor of *P. putida* wcs 358	(44)
Salicylate utilization	Activity of genetically engineered biocontrol pseudomonad	(15)
Cloning of pyrrolnmitrin genes	Biocontro potential of a *P. fluorescens*	(34)
Tn7-based chia A gene	Control of phytopathogens in soil by *P. fluorescens*	(43)
PhzR gene	Biosynthesis of phenazine antibiotic in *P. aureofaciens* strain 30-84	(78)
Housekeeping sigma factor	enhanced antibiotic production in biocontrol strain *P. fluorescens* CHAO	(81)
pqq genes	overproduction of pyluteorin in *P. fluorescens* CHAO	(82)
Antigenic surface reporter system	Detection and activity of catabolic promoters of *P. putida*	(13)
bph gene construct	Degradation of polychlorinated biphenyl in a rhizosphere pseudomonad	(9)
Ice nucleation reporter gene system	Identification of inducible pathogenicity genes in *P.syringae* pv. *phaseolicola*	(54)
Lux modified constructs	Bioavailability of heavy metals for *P. fluorescens*	(77)
Opr F antibody	Distribution and ecology of rRNA homology Group I pseudomonads	(46)

6.2 Siderophore-mediated Disease Suppression

The early approach to confirm the involvement of siderophores in biocontrol action by *P. fluorescens* strains 3551 and B224 were derived from mutants obtained through chemical, Tn5 insertion and UV mutagenesis; sid⁻ mutants provided much less control than sid⁺ parental strains against *Pythium* causing damping-off of cotton and wheat (3, 8). In *P. fluorescens* strain B10, genes for siderophore (pyoverdine) production were cloned by Moores, *et. al.* (68), mutant complement analysis showed that 12 genes were arranged in four clusters. For pseudobactin system, on the other hand, receptors located on the membrane have been cloned (62, 73). Detailed characterization of cloned receptors has shown that they are required for early steps involved in binding of ferric-pseudobactin complex.

6.3 Antibiotic-mediated Disease Suppression

While siderophore-mediated disease suppressiveness has been analyzed to understand the mechanisms involved it is the antimicrobial compounds that have received most attention.

Among the molecules analyzed at molecular level, phenazine-1-carboxylate (PCA) from *P. fluorescens* 2-79 has been shown to be responsible for 50-90% reduction in root disease of wheat known as 'take-all' (*Gaeumannomyces graminis* var. *tritici*) (94). Confirmation for this role came from phenazinedeficient (Phz) mutants created through Tn5 mutagenesis, use of similar tools has confirmed the role for three other phenazines in *P. chloroaphis* 30-84 and for the antibiotic, 2,4-diacetylphloroglucinol (phl). In another interesting system, *P.fluorescens* Hv 37a secretes oomycin A which requires glucose for its biosynthesis. Detailed investigation of genetic regulation shows multilevel controls with several gene clusters in operation. For Phz system of *P. chloroaphis* 30-84 as well at least four clustered genes are required for antibiotic biosynthesis, i.e., phz B, phzC, phzR and phzI. On the other hand, PCA production by *P. fluorescens* strain 2-79 requires operation of at least two loci (17). One of the other genetically dissected system is represented by *P. fluorescens* strain CHAO; this strain exhibits strong antifungal action through release of HCN-and the antibiotics pyrolnitrin, pyoluteorin and Phl. Cook *et al.* (17) have presented evidence to show that antibiotic production in fluorescent pseudomonads is a conserved character because the antibiotic Phl has been shown to be active against *F. oxysporum, Thielaviopsis basicola, G. graminis* var. *tritici, Pythium ultimum* and *Rhizoctonia solani* recovered from globally different regions of the world. Therefore, antibiotic production provides ecological advantage to the producer strain(s). On the other hand, production of the same antibiotic by genetically distinct fluorescent pseudomonads would suggest that this character has helped them to variety of soil and plant habitats. Based on this and other related information of molecular analysis of antibiotic production, regulation and monitoring Cook has suggested use of antibiotic biosynthesis genes as transgenes which if kept under control of a root-specific promoter would not result in problems of toxic residue at the later stage of plant development.

6.4 Use of Genetically Engineered Pseudomonads

In other developments of potentially useful bicontrol strains of fluorescent pseudomonads other molecular markers have also been used. Colbert *et al.* (15) monitored the activity of a genetically engineered pseudomonad based on salicylate utilization pattern whereas Koley *et al.* (43) developed a Tn 7-based chia A gene system to monitor the biocontrol potential of *P. fluorescens* in soil. On the antibiotic related control systems, pyrolintrin, Phz R genes, pqq genes and a house keeping sigma factor have been utilized (34, 78, 81, 82).

A major area of environmental application of fluorescent pseudomonads is in the realm of degradation of environmental pollution and in bioremediation. While this subject is not the main theme of this article on account of vast published literature, it is of interest to highlights two interesting systems. In one such study Brazil *et.al.* (9) used a polychlorinated biphenyl (bph) gene construct to determine the pollutant removal capacity of a pseudomonad recovered from the rhizosphere. As opposed to this, Cebolla *et.al.* (13) utilized a totally different approach wherein they developed an antigenic surface reporter system for detecting *in situ* presence of *P. putida* and the functional activity of the catabolic promoters of this bacterium. Other systems utilized for environmental applications include, *lux* modified constructs for *P. fluorescens*, icenucleation reporter system for *P. syringae* pv. *phaseolicola* and Opr F antibody for ecological profiling of rRNA homology of group I pseudomonads (Table 5).

7. REFERENCES

1. Amann, R.I., Binder, J.B., Olson, R.J., Chisholm, S.W., Devereux, R. 1990. Combination of 16s rRNA-targeted of igonucleotide probes with flow cytometry for analyzing mixed microbial populations. *Appl. Environ. Microbiol.* **56**: 1919-1925.

2. Amin-Hanjani, S., Meikle, A., Glover, L.A., Prosser, J.I. and Killham, K. 1993. Plasmid and chromosomally encoded luminescence marker systems for detection of *Pseudomonas fluorescens* in soil. *Mol. Ecol.* **2**: 47-54.

3. Bakker, P.A.H.M., Bakker, A.W., Marugg, J. D., Weisberg, P.J. and Schippers, B. 1987. A bioassay for studying the role of siderophores in potato growth stimulation by *Pseudomonas* spp. in short potato rotation. *Soil. Biol. Biochem.* **19**:443-450

4. Barrett, E.J., Solanes, R.E., Tang, J.S. and Palleroni, N.J. 1986. *Pseudomonas fluorescens* biovar. V: its resolution into distinct groups and the relationships of these groups to other *P. fluorescens* biovars, to *P. putida* and to *Psychrotrophic pseudomonas* associate with food spoilage. *J. Gen. Microbiol.* **132**: 2709-2721.

5. Barry, G.F. 1986. Permanent insertion of foreign genes into the chromosomes of soil bacteria. *Biotechnology*. **4**: 446-449.

6. Barry, G.F. 1988. A broad-host-range shuttle system for gene insertion into the chromosome of gram negative bacteria. *Gene* **71**: 75-84.

7. Beauchamp, C.J., Kloepper, J.W. and Lamke, P.A. 1993. Luminometric analysis of plant root colonization by bio-lumninescent pseudomonads. *Can. J. Microbiol.* **39**: .434-441.

8. Beacker, J.O. and Cook, R.J. 1988. Role of siderophores in suppression of *Pythium* species and production of increased growth response of wheat by fluorescent pseudomonads. *Phytopathology*. **78**: 778-782.

9. Brazil, G.M., Kenefick, L., Hara, C M., deLorenzo, V., Dowling, D.N. and O'Gara, F. 1995. Construction of a rhizosphere pseudomonad with potential to degrade polychlorinated biphenyb and detection of bph' gene expression in the rhizosphere. *Appl. Environ. Microbiol.* **61**: 1946-1952.

10. Brennerova, M. and Coley, D.E. 1994. Direct detection of rhizosphere - colonizing *Pseudomonas* sp. using an *E. coli.* rRna promotor in a Tn-7-lux system. *FEMS Microbiol. Ecol.* **14**: 319-330.

11. Budzikiewicz, H. 1993. Secondary metabolites from fluorescent pseudomonads. *FEMS Microbiol Rev.* **104**: 209-228.

12. Busse, H.J., El-Banna, T. and Auling. G. 1989. Evaluation of different approaches for identification of xenobiotic-degrading pseudomonads. *Appl. Environ. Microbiol.* **55**: 1578-1583.

13. Cebola, A., Guzman, C. and Lorenzo, V. De 1996. Nondisruptive detection of activity of catabolic promoters of *Pseudomonas putida* with an antigenic surface reporter system. *Appl. Environ. Microbiol.* **62** : 214-220.

14. Christensen, H., Boye, M., Poulsen, L.K and Rasmussen, O.F. 1994. Analysis of fluorescent pseudomonads based on 23S fibosomal DNA sequences. *Appl. Environ. Microbiol.* **62**: 2196-2199.

15. Colbert, S.F., Hendson, M., Ferri, M. and Schroth, M.N. 1993. Enhanced growht and activity of a biocontrol bacterium genetically engineered to utilize salicylate. *Appl. Envrion. Microbiol.* **59**: 2071-2076.

16. Cook, D., Barlow, E. and Sequeria, L. 1989. Genetic diversity of *Pseudomonas Solanacearum* detection or restriction fragment length polymorphism with DNA probex that specify virulence and hypersensitive response. *Mol. Plant Microbe Intract.* **2**: 113-121.

17. Cook, R.J., Thonmshow, L.S., Weller, D.M., Fujimoto, D., Mazzola, M., Bangera, G. and Kim, D. 1995. *Proc. Nati. Acad. Sci. USA* **92**: 4197-4201.

18. De Mot, R., Proost, P., Van Damme, J. and Vanderleyden, J. 1992. Homology of the root adhesin of *Pseudomonas fluorescens* OE 28.3 with potin F of *P. aeruginosa* and *P. syringae*. *Mol. Gen. Genet.* **231**: 489-493.

19. De Vos, P., Goor, M., Gillis, M. and De Ley, J. 1985, Ribosomal ribonucleic acid cistron similarities of phytopathogenic *Pseudomonas* species. *International Journal of Systematic. Bacteriology* **35**: 169-184.

20. DeVos, P., Van Lands Choot, A. and Segers, P. 1989. Genotype relationships and taxonomic localization of unclssified *Pseudononas* and *Pseudomonas*-like strains by deoxyribonucleic acid: ribosomal ribonucleic acid hybridizations. *International Journal of Systematic Bacteriology.* **39** : 35-49.

21. De Weger, L.A., Dunbar, P., Mahejee, W.F. and Lugtenber, B.J.J., 1990. Use of reporter bacteria for studying the availability or phosphate in rhizosphere and soil (abst.). *Int. Symp. Mol. Genl. Plant Microbe Intract.* **2**: 113-121.

22. DeWeger, L.A., Dunbar, P., Mahafee, W.F., Lugtenberg, B.J.J. and Sayler, G.S. 1991. Use of bioluminescence markers to detect *Pseudomonas* spp. in the rhizosphere. *Appl. Environ. Microbial.* **57**: 3641-3644.

23. Gardan, L., Bollet, C., Abu Ghorrah, M., Grimont, F. and Grimont, P.A.D. 1992. DNA relatedness among the pathover strain of *Pseudonomas syringae* subsp. *savastonoi* Janse (1982) and proposal of *Pseudomonas savastonoi* sp. nov. *International Journal of Systematic Bacteiology* **42** : 606-612.

24. Glick, B.R., Jacobson, C.B., Schwarze, M.M.K. and Pasternak, J.J. 1994. 1-Aminocyclopropane-1-carboxylic acid deaminase mutants of the plant growth promoting rhizobacterium *Pseudomonas putida* GR 12-2 do not stimulate canola root elongation. *Can. J. Microbiol.* **40**: 606-612.

25. Glick, B.R., Karaturovic, D.M. and Newell, P.C. 1955, A novel procedure for rapid isolation of plant growth promoting pseudomonads. *Can. J. Microbiol.* **41**: 533-536.

26. Glick, B.R., Penrose, D.M. and Li, J. 1998. A model for the lowering of plant ethylene concentrations of plant growth promoting bacteria. *J. Theor. Biol.* **190**: 63-68.

27. Gonzalez, M.I., Raiz-Cabello, F., Bretter, I., Garrido, F. and Raman, J.L. 1992. Tracking genetically engineered bacteria: monoctonal antibodies against surface determinants of soil bacterium *Pseudomonas Pudida* 2440. *J. BActeriol.* **174**: 2978-2985.

28. Gonzalez, I., Martin, R., Garcia, T., Morales, P., Sanz, B. and Hernadez, P.E. 1993. A sandwich enzyme-linked immunosorbent assay (ELISA) for detection of *Pseudomonas fluorescens* and related psychrotrophic bacteria in refrigerated milk. *J. Appl. Bacteriol.* **74** : 394-401.

29. Grothues, B. and Tununler, B. 1991. New approaches in genome analysis by pulse-field gel electrophoresis: application to analysis of *Pseudomonas* species. *Molecular Microbiology* **5**: 2763-2776.

30. Hahn, D., Amann, R.I., Ludwig, W., Akkermans, A.D.L and Schleifer, K.H. 1992. Detection of microorganisms in soil after *in situ* hybridization with rRNA targeted, fluorescently labelled oligonucleotides. *J. Gen. Microbiol.* **138**: 879-887.

31. Hancock, R.E.W. and Chan, L.W. 1988. Outer membranes of environmental isolates of *Pseudomonas aeruginosa. J. Clin. Microbiol.* **26** : 2423-2424

32. Haynes, W.C. and Burkholder, W.H. 1957. Genus I. *Pseudomonas migula* 1984. In : *Bergey's Manual of Determinative Bacteriology.* (eds. R.S., Breed, F.G.D., Murray, N.R. Smith) 7th edn. Williams and Wilkins Co, Baltimore pp. 89-152.

33. Hildebrand, D.C. 1971. Pectate and Pectine gels for differentiation of *Pseudomonas* sp. and other bacterial plant pathogens. *Phytopathology* **61** : 1430-1436.

34. Hill, D.S., Stein, J.L.D., Torkewitz, N.R., Morse, A.M., Howell, C.R., Pachlatko, J.P., Becker, J.O. and Ligon, J.M. 1994. Cloning of genes involved in the synthesis of pyrrolnitrin from *Pseudononas fluorescens* and role of pyrrolnitrin synthesis in biological control of plant disease. *Appl. Env. Microbiol.* **60**: 78-85.

35. Hofte, M., Mergeay, M. and Vestraete, W. 1990. Marking the rhizopseudomonas strain $7NSK_2$ with a Mu d (lac) element for ecological studies. *Appl. Env. Microbiol.* **56**: 1046-1052.

36. Hughes, E.E., Gilleland, L.B. and Gilleland, H.E. 1992. Synthetic peptides representing epitopes of outer membrane protein F of *Pseudomonas aeuginosa* that elict antibodies reactive with whole cells of heterologous immunotypes strains of *P. aeruginosa. Infect. Immumn.* **60**: 3497-3503.

37. Jaunet, T., Laguerre, G., Lemanceau, P., Frutos, R. and Notteghem, J.L. 1995. Diversity of *Pseudomonas fuscovaginae* and other fluorescent pseudomonas isolated from diseased rice. *Phytopathology* **85(12)** : 1534-1541.

38. Johnson, J.L. and Palleroni, N.J. 1989. Deoxyribonycleic acid similerities among *Pseudomonas* sp. *Int. J. Syot. Bacteriol.* **39**: 230-235.

39. Johri, B.N., Rao Ch. V.S. and Goel, R. 1997. Fluorescent pseudononads in plant disease management. In : *Biotechnological Approaches in Soil Microorganisms for Sustainable Crop production* (ed. K.R. Dadarwal) Scientific Publishers. Jodhpur, India pp. 193-121.

40. Khokar, A., Sinha, R. and Goel, R. 1996. A novel approach to tag *Pseudomonas fluorescens* with tac marker for environmental monitoring. *Ind. J. Microbiol.* **36**: 163-164.

41. Kluepfel, D.A. 1993. The behaviour and tracking of bacteria in the Rhizosphere. *Annu. Rev. Phytopathol.* **31**: 441-472.

42. Kleopper, J.W. and Shroth, M.N. 1978. Plant growth-promoting trhizobacteria on radish. In : *Proc., 4th Int. Conf. Plant Pathogenic Bact.* Gillbert-Clarey, Tours, france, pp. 879-882.

43. Koley, S., Schickler, H., Chet, J. and Oppenheim, A.B. 1994. The chitinas encoding Tn 7-based Chi A gene endows *Pseudomonas fluorescens* with the capacity to control plant pathogens in soil. *Gene* **147** : 147 : 81-83.

44. Koster, M., Van de Vossenverg, Leong, J. and Weisbeek, P.J. 1993. Identification and characterization of the pup B gene encoding in inducible ferric-pseudobactin receptor of *Pseudomonas putida* WCS 358. *Mol. Microbiol.* **8**: 591-601.

45. Kragelund, L., Leopold, K. and Nybroe, O. 1996. Culturability and expression of outer membrane proteins during carbon, nitrogen, or phosphorous stravation fo *Pseudomonas fluorescens* DF57 and *Pseudomonas putida* DF14. *Appl. Environ. Microbiol.* **60(8)** : 2944-2948.

46. Kragelund, L., Leopold, K. and Nybroe, O. 1996. Outer membrane protein heterogeneity within *Pseudomonas fluorescens* and *P. putida* and use of Opr F antibody as a probe for r RNA homology group I pseudomonads. *Appl. Envrion. microbiol.* **62 (2)** : 480-485.

47. Kragelund, L; Hosbond, C. and Nybroe, O. 1997. Distribution of metabolic activity and phosphate starvation response of lux-t agged *Pseudomonas fluorescens* reporter bacteria in the barley rhizosphere. *Appl. Environ. Microbiol.* **63 (12)** : 4920-4928.

48. Kumar, C. 1997. DNA extraction from soil samples for DNA finger-print analysis, M.S. Thesis.

49. Laguerre, G., Rigottier-gois, L. and Lemanceau, P. 1994. Fluorescent *Pseudomonas* species categorized by using polymerase chain reaction (PCR) restriction fragment analysis of 16s rDNA. *Molecular Ecology* **3** : 479-487.

50. Lambert, B. Leynp, F., Van Yvoyen, L. Gossele, F. Papon, Y. and Swings, J. 1987. Rhizobacteria of maize and their antifungal activities. *Appl. Environ. Microbiol.* **53**:? 1866-1871.

51. Lambert, B., Meire, P., Joos, H., Lens, P. and Swings, J. 1990. Fast-growing, aerobic, heterotrophic bacteria from the rhizosphere of young sugar beet plants. *Appl. Environ. Microbiol.* **56**: 3375-3381.

52. Latour, X., Coberand, J. Lagurre, G., Allerd, F. and Lamanceau, P. 1996. *Appl. Environ. Microbiol.* **62**: 2449-2451.

53. Lemanceau, P., Corberand, T., Gardan, L., Latour, X, Laguerre, G., Boeufgras, J.M. and Alabouvette, C. 1995. Effect of two plant species, flax (*Linum suitatissinum* L.) and tomato (*Lycopersicon esculentum* Mill.), on the dirersity of soil born populations of fluorescent pseudomonads. *Appl. Environ. Microbiol.* **61 (3)**: 1004-1012.

54. Lindgren, P.B., Frederick, R., Govidarajan, A.G., Panopoulos, N.J., Staskawicz, B.J. and Lindow, S.E. 1989. An ice nucleation reporter gene system; identification of inducible pathogenicity genes in *Pseudomonas syringae* pv. *phaseolicola. EMBOJ* **8**: 1291-1301.

55. Li, X., Dorsch, M., Del Dot, T., Sly, L.I., Stackebrandt, E. and hayward, A.C. 1993. Phylogenetic studies of the rRNA group II pseudomonads based on 16s rRNA sequences. *J. Appl. Bacteriol.* **74**: 324-329.

56. Liu, L., Kleopper, J.W. and Tuzun, S. 1995a. Induction fo systemic resistance in cucumber against *Fusarium* wilt by plant growth promoting rhizobacteria. *Phytopathology* **85 (6)** : 695-698.

57. Liu, L., Kloepper, J. W. and Tuzun, S. 1995b. Induction of systemic resistance in cucumber against bacterial angular leaf spot by plant growth-promoting rhizobacteria. *Phytopathology* **85 (8)** : 843-47.

58. Loper, J.E. and Henkels, M.D. 1992. Ice-nucleation activity as a reporter of *in situ* gene expression by rhizosphere pseudomonads. *Ann. Meet. Am. Phytopathol.* Portlant.

59. Loper, J.E. and Lindow, S.E. 1994. A biological sensor for iron available to bacteria in their habitats on plant surfaces. *Appl. Environ. Microbiol.* **60**: 1934-1941.

60. Maki, R.L., Galyon, E.L., Chang, Chein, M. and Caldwell, D.R. 1974. Ice nucleation induced by *Pseudomonas syringae. Appl. Env. Microbiol.* **28**: 456-460.

61. Marugg, J.D., Neilander, H.B., Horresvoits, A.J.G., Van Megan, I,, Van Genderson, J. and Weisbeek, P.J. 1988. Genetic organization and transcriptional analysis of a major gene cluster involved in siderophore biosynthesis in *Pseudomonas putida* WCS358. *J. Bacteriol.* **170**: 1812-1819.

62. Marugg, J. D, de Weger, L.A., Neilander, H.B., Oorthvizen, N., Recourt, R., Lugtenberg, B, van der Hofstad, M. and Weisbeek, P.J. 1989. Cloning and characterization of gene encoding an outer-membrane protein required for siderophore mediated Fe^{3+} in *Pseudomonas* putida WC S 538. *J. Bacteriol.* **171**: 2819-2926.

63. Meikle, A., Killham, K., Prosser, J.I. and Glover, L.A. 1992. Luminometric measurement of population activity of genetically modified *Pseudomonas flurescens* in the soil. *FEMS Microbiol. Lett.* **99**: 217-220.

64. Meikle, Al., Glover, L.A., Killham, K. and Prosser, J.I. 1994. Potential luminescence as an indicator of activation of genetically modified *Pseudomonas fluorescens* in liquid culture and in soil. *Soil Biol. Biochem.* **26**: 747-755.

65. Meyer, J.M., Stintzi, A., DeVos, D., Cornelis, P., Tappe, R. Taraz, K. and Budzikiewicz, H. 1997. Use of siderophores to type pseudomonads: the three *Pseudomonas aeruginosa* pyoverdine systems. *Microbiology* **143**: 35-42.

66. Miller, L.T. and Berger, T. 1985. Bacterial identification by gas chromatography of whole cell fatty acids. Hewlett-Packard application note 228-241. Hewlett-Packard Co., Palo Alto, Calif.

67. Mohn, G., Taxar, K. and Budzikiewicz, H. 1990. New Pyoverdin-type siderophores from *Pseudomonas fluorescens*. (1). *Z. Naturforsch* **45b**: 1437-1450.

68. Moores, J.C., Magazine, M., Ditta, G.S. and Leong, J. 1984. Clonong of genes involved in the biosynthesis of pseudobactin, a high affinity iron transport agent of a plant growth promoting *Pseudomonas* strain. *J. Bacteriol.* **157** : 53-58.

69. Mutharia, L.M. and Hancock, R.E.W. 1983. Surface localization of *Pseudomonas aeruginosa* outer membrane porin protein F by using monoclonal antibodies. *Infec. Immuno.* **42**: 1027-1033.

70. Murtharia, L.M. and Hancock, R.E.W. 1985. Characterization of two surface-localized antigenic sites on porin protein F of *Pseudomonas aeruginosa. Can. J. Microbiol.* **31**: 381-386.

71. Mutharia, L.M., Nicas, T.I. and Hancock, R.E.W. 1988. Outer membrane proteins of *Pseudomonas aeruginosa* serotype strains. *J. Infect. Dis.* **146**: 770-779.

72. Neilands, J.B. 1981. Microbial iron compounds. *Annu. Rev. Biochem.* **50**: 715-731.

73. O'Sullivan, D.J. and O'Gara, F. 1990. Iron regulation of ferric iron uptake in fluorescent pseudomonads: Cloning of a regulatory gene. *Mol. Plant-microbe Interact.* **3**: 86-93.

74. Palleroni, N.J. 1984. *Pseudomonas.* In : *Manual of Systematic Bacteriology*. Vol. 1. (ed. N.R. N.R. Krieg). Williams and Wilkins Co. Baltimore, pp. 141-199.

75. Palleroni, N.J., Ballard, R.W., Ralston, E. and Doudroff, M. 1972. Deoxyribonucleic acid homologies among some *Pseudomonas* species. *J. Bact.* **110**: 1-11.

76. Palleroni, N.J., Kunisawa, R., Contopolou, R. and Doudroff, M. 1973. Nucleic acid homologies in the genus *Pseudomonas International Journal of Systematic Bacteriology* **23**: 111-121.

77. Paton, G.I., Campbell, C.D., Glover, L.A. and Killham, K. 1995. Assessment of bioavailability of heavy metals using *lux* modified constructs of *Pseudomonas fluorescens* . *Lett. Appl. Microbiol.* **20**: 52-56.

78. Pierson, L.S. III, Keppenne, V.D. and Wood, D.W. 1994. Phenazine antibiotic biosynthesis in *Pseudomonas aureofaciens* 30-84 is regulated by $ph_{,}$ R in response to cell density. *J. Bacteriol.* **176**: 3996-3974.

79. Romos-Gonzalex, M.I., Ruiz-Cabello, F., Brettar, I., Garrido, F. and Ramos, J.L. 1992. Tracking genetically engineered bacteria: monoclonal antibodies against surface determinants of the soil bacterium *Pseudomonas putida* 2440. *J. Bacteriol.* **147**: 2978-2985.

80. Rawling, E.G., Martin, N.L. and Hancock, R.E.W. 1995. Epitope mapping of the *Pseudomonas aeruginosa* major outer membrane porin protein Opr F. *Infect. Innun.* **15**: 617-623.

81. Schnider, U., Keel, C., Blumer, C, Troxler, J., Defago, G. and Hass, D. 1995a. Amplification of the housekeeping sigma factor in *Pseudomonas fluorescens* CHAO enhances antibiotic production and improves biocontrol abilities. *J. Bacteriol.* **177**: 5387-5392.

82. Schnider, U., Keel. C., Voisard, C., Defago, G. and Hass, D. 1995b. Tn5-directed cloning of pqq genes from *Pseudomonas fluorescens* CHAO : mutational inactivation of the genes results in over production of the antibiotic pyluteorin. *Appl. Environ. Microbiol.* **61**: 3856-3864.

83. Seal, S.F., Jackson, L.A. and Daniels, M.J. 1992. Use of tRNA consensus primers to indicate subgroups of *Pseudomonas solanacearum* by polymerase chain reaction amplification. *Appl. Environ. Microbiol.* **58**: 3759-3761.

84. Shaw, J.J., Dane, R., Geiger, D. and Kloepper, J.W., 1992. Use of bioluminescence for detection of genetically engineered microorganisms released into the environment. *Appl. Environ. Microbiol.* **58(1)**: 267-273.

85. Silcock, D.J., Waterhouse, R.N., Glover, L.A., Prosser, J.J. and Killham, K. 1992. Detection of a single genetically modified bacterial cell in soil by using charge coupled device-enhanced microscopy. *Appl. Envrion. Microbiol.* **58**: 2444-2448.

86. Siverio, F., Cambra, M., Gorris, M.T. Corzo, J. and Lopez, M.M. 1993. Lipololysaccharides as determinants of serological variability in *Pseudomonas corrugata. Appl. Environ. Microbiol.* **59**: 1805-1812.

87. Siverio, F., Carbonell, A.A., Garcia, F. and Lopez, M.M., 1996. Characteristics of the whole cell fatty acid profiles of *Pseudomonas corrugata. Eur. J. Plant. Pathol.* **102**: 519-526.

88. Sorensen, J., Skouv, J., Jorgensen, A. and Nybroe, O. 1992. Rapid identification of environmental isolates of *Pseudomonas aeruginosa, P. fluorescens* and *P. putida* by SDS-PAGE analysis of whole-cell protein patterns. *FEMS Microbiol. Ecol.* **101**: 41-50.

89. Stanier, R.Y., Palleroni, N.J. and Doudroff, M. 1996. The aerobic pseudomonads, a taxonomic study, *J. Gen. Microbil.* **43**: 159-271.

90. Steffan, R.J. and Atlas, R.M. 1988. DNA amplification to enhance detection fo genetically engineered bacteria in environmental samples. *Appl. Environ. Microbiol.* **45**: 137-61.

91. Stead, D.E. 1992. Grouping of plant-pathogenic and some other *Pseudomonas* spp. by using cellular fatty acid profiesl. *Int. J. syst. Bacteriol.* **42**: 281-295.

92. Suslow, T.V. and Schrotch, M.M. 1982. Rhizobacteria of sugar beats: Effect of seed application and not colonization on yield. *Phytopathology* **72**: 199-203.

93. Sutra, L., Siverio, F., Lopez, M.M., Hunault, G., Bollet, C. and Gardan, L. 1997. Taxamony of *Pseudomonas* strains isolated from tomato pith necrosis: emended description of *Pseudomonas corrugata* and proposal of three unnamed flurescent *Pseudomonas* Genomospecies. *International Journal of Systematic Bacteriloogy* **47(8)**: 1020-1033.

94. Thomashow, L.S. and Weller, D.M. 1988. Role of antibiotics in root disease suppresion by fluorescent pseudomonds. Abstr. 5th ICPP Congress, Kyoto, Japan. II, 7-4, 94.

95. Trevors, J.T., Van Elsas, J.D. Van Overbeek, L.S. and Starodub, M.E. 1990. Transport of a genetically engineered *Pseudomonas fluorescens* strain through a soil microcosm. *Appl. Environ. Microbiol.* **56**: 401-408.

96. Van Obverbeek, L.S. and Van Elsas, J.D. 1995. Root exudate-induced promoter activity in *Pseudomonas fluorescens* mutants in the wheat rhizosphere. *Appl. Environ. Microbiol.* **61**: 890-898.

97. Wei. G., Kloepper, W. and Tuzun, S. 1996. Induced systemic resistance to cucumber diseases and increased plant growth by plant growth promoting rhizobacteria under field conditions. *Phytopathology* **86** **(2)** : 221-224.

98. Woese, C.R., Weisburg, W.G. and Hahn, C.M. 1985. The phylogeny of purple bacteria : the gamma subdivision. *Systematic Applied Microbiology* **6**: 25-33.

99. Woese, C.R., Blanz, P. and Hahn, C.M. 1984. What is not a pseudomonad: the importance of nomenclature in bacterial classification *Systematic Applied Microbiology* **5**: 179-195.

100. Xie, H., J. and Glick, B.R. 1996. Isolation and characterization of mutants of the plant growth-promoting rhizobacterium *Pseudomonas putida* GR12-2 that overproduced indoleacetic acid. *Current Microbiol.* **32**: 67-71.

101. Young, J.M., Dye, D.W., Bradbury, J.F., Panagopoulos, C.G. and Robb, C.F. 1978. A proposed nomenclature and classification for plant pathogenic bacteria. *New Zealand Journal of Agricultural Research* **21**: 153-177.

102. Zaat, S.A.J., Slegtenhorst-Eedgeman, K., Tommassen, J., Geli, V., Wijffelman, C.A. and Lugtenberg, B.J.J. 1994. Construction of pho E-caa, a novel PCR-and immunologically detectable marker gene for *Pseudomonas putida. Appl. Environ. Microbiol.* **60**: 3965-3973.

103. Zyl, E.V. and Steyn, P.L. 1990. Differentiation of phytopathogenic *Pseudomonas* and *Xanthomonas* species and pathovars by numerical taxonomy and protein gel electrophoresis. *J. Syst. Appl. Microbiol.* **13**: 60-71.

10 Bacterial Thermostable Cellulase-Free Xylanases in Environment Friendly Paper Pulp Bleaching Technology

A. Archana and T. Satyanarayana

CONTENTS

ABSTRACT 132

1. INTRODUCTION 132
2. STRUCTURE AND LOCATION OF XYLAN IN PLANTS 132
3. HEMICELLULOSE IN PULPING OF WOOD 133
   - 3.1 Xylanolytic Enzymes and their Mode of Action 133
4. OCCURENCE OF XYLANASES 133
   - 4.1 Cellulase-free Nature 134
   - 4.2 Suitability of Xylanases in Prebleaching of Paper Pulp 134
     - 4.2.1 Thermostability 134
     - 4.2.2 Alkalostability 134
5. APPLICATION OF XYLANASES IN BLEACHING 135
6. ENVIRONMENTAL HAZARDS OF ORGANOCHLORINES 135
   - 6.1 Overcoming the Environmental Pollution 136
7. COMMERCIALISATION OF XYLANASES 136
8. CONCLUSIONS 137
9. REFERENCES 138

Advances in Microbial Biotechnology
J.P. Tewari, T.N. Lakhanpal, Jagjit Singh, Rajni Gupta & B.P. Chamola (eds.)
APH Publishing Corporation, New Delhi - 110 002, India.

ABSTRACT

There is an enormous demand for paper because it is useful in education, civilization and spread of knowledge around the world. But paper manufacturing industries are among the worst offenders for deteriorating environment health. The organochlorines produced during pulping process have been a matter of concern in the paper industry for the past two decades. These compounds are produced mainly by the reactions between residual lignin present in wood fibres and the chlorine used for bleaching. Some of the organochlorines are found to be toxic, mutagenic, persistent, bioaccumulating and cause harm to biological systems. Therefore, the possibility of xylanases to replace and/or supplement chemical methods is gaining interest. The prerequisites of xylanases to be useful in the paper industry are their cellulase-free and thermoalkalostable nature. Thermophilic bacteria are known producers of such xylanases. Furthermore, if one of the aforesaid feature are naturally present in the microbe, the other may be acquired by enzyme engineering. Cellulase coproduced with xylanase activity may be chemically inactivated, eliminated by purification methods, or expressed in isolation by cloning in xylanase encoding gene from xylanolytic organism in heterologous non-cellulolytic bacterial mutants. It is important to knock down the contaminating cellulase activity from xylanase preparation, so as to protect cellulose of the pulp from being degraded, that maintains the quality and quantity of thus produced paper very high.

The enzyme-aided bleaching of kraft pulps leads to reductions in chemical consumption and costs, and maintains product quality. This ecofriendly and unpolluting approach has already been put to practice in mill scale in several countries, and its cost effectiveness is being worked out.

1. INTRODUCTION

The pulp and paper industry is emerging as one of the largest markets for enzyme applications in the world. The demand for paper is increasing globally as standards of living are rising and the need for clean, efficient processing is also rising ever greater. Increased pulp yield, improved fibre properties, enhanced paper recycling, and reduced processing and environmental problems are all consequences of enzyme applications in the pulp and paper industry. As knowledge and expertise of these is increasing, the acceptance of these technologies is also growing.

2. STRUCTURE AND LOCATION OF XYLAN IN PLANTS

Xylan is the major component of hemicellulose, which includes, in general the non-cellulosic polysaccharides of plant cell walls (41). The location and structure of hemicellulases in the fibres affect the delignification of pulps as well as the technical properties of the fibre products. The removal of residual lignin seems to be both physically and chemically restricted by hemicelluloses in the fibre matrix. Lignin carbohydrate linkages have been suggested to restrict the chemical removal of residual lignin from pulp fibres (44). Recent developments in pulp chemistry have emphasised on the fundamentals of hemicellulose-aided bleaching. In natural softwood and hardwood and also during solubilization and relocation caused during pulp process, hemicellulose and lignin are found physically and/or chemically interlinked. In the natural woods and grasses, hemicelluloses appear to be the cementing material between cellulose and lignin. Hence, hemicellulases such as xylanase, mannanase etc. serve to degrade this cementing layer, rendering the cellulose free of lignin. The latter may be eliminated by specifically designed protocols. This enzymatic step is termed prebleaching, and alongwith a chemical extraction of lignin, it reduces the extent of hazardous chlorine used in the bleaching step of paper making.

3. HEMICELLULASES IN PULPING OF WOOD

Hemicellulases comprise a group of enzymes, which act synergistically on different types of hemicelluloses. In softwood (pine, spruce), the main hemicelluloses are galactoglucomannans and arabinoglucuronoxylans, whereas in hardwoods (birch) the main hemicellulose is acetylglucuronoxylan. The enzymes necessary for the hydrolysis of softwoods and hardwoods have been well characterised (11). During kraft pulping most of the glucomannan of wood is degraded and the relative amount of xylan compared to glucomannan is increased. Also, most of the side groups of the xylan polymer are cleaved off due to alkaline conditions during pulping. Therefore, endoxylanases are very crucial enzymes in the biobleaching of kraft pulps. Even partial hydrolysis of hemicelluloses in paper pulps have been found to enhance the extractability of lignins in the conventional chemical bleaching step (10).

3.1 Xylanolytic Enzymes and their Mode of Action

Xylan, being a heterogeneous polymer, requires the concerted action of a complex enzyme system composed of the enzymes degrading the xylan backbone as well as the side chains, all of which act cooperatively to convert xylan to its constituent sugars (6). Xylanases (1,4-β-D-xylan xylanohydrolase, E.C.3.2.1.8) catalyse the hydrolysis of xylan to xylooligosaccharides and xylose. The use of xylanase for pulp bleaching technology is based on the understanding of the structural relation of lignin that binds to cell wall polysaccharides to form lignin-carbohydrate complexes, and thus it is difficult to release lignin from wood. Hemicellulose cements cellulose to lignin and thus degradation of hemicellulose leads to mutual dissociation of the two, consequently facilitating enhanced lignin removal. Experiments have indicated that endoxylanase is the most important enzyme in hemicellulose hydrolysis (11, 43). In most of the investigations, crude enzyme mixtures have been employed. The culture filtrates, however, contained predominantly xylanases as the main component. Hemicellulose - rich culture filtrates from different microorganisms have been shown to have similar effects on kappa number reduction (a measure of lignin removal) in subsequent delignification steps when the same xylanase dose is used. Increased delignification during enzymatic pretreatment of pulp may be due to hydrolysis of reprecipitated xylan in inner layers. If xylan is attached to lignin molecules through side groups, hydrolysis of linkages between side groups and xylan backbone may enhance delignification. Side group cleaving enzymes have been shown to further enhance the delignification processes (22).

4. OCCURRENCE OF XYLANASES

Xylanases are produced by a vast variety of prokaryotes and eukaryotes including bacteria, fungi, protozoa, insects, snails and germinating plant seeds (12). The extensively studied bacterial producers include the species of *Aeromonas, Agrobacterium, Bacillus, Dictyoglomus, Nocardia, Pseudomonas, Streptomyces, Thermotoga*, and *Xanthomonas* species (14, 15, 24, 29, 32). Among these *Bacillus* spp. are the most common and significant producers of xylanase (6).

4.1 Cellulase-free Nature

A prerequisite for utilizing microbial xylanases in the, paper and pulp industry with a target to produce superior quality dissolving pulps is their freedom from contaminating cellulase activity (16, 39). Due to lack of cellulase activity in such a preparation, selective removal of only the hemicellulose component with minimum damage to cellulose pulp is facilitated. This translates into prevention of yield loss and brittleness of thus formed paper (42). Some organisms exhibiting this type of xylanase activity include *Chainia* sp. (6) *Streptomyces* sp. (45) and *Cephalosporium* sp. (5). High levels of cellulase-free xylanases may be produced when certain special physiological conditions are tailor-made. Glucose limited fed-batch cultures of a thermophilic *Bacillus* strain was reported to secrete high titres of xylanases with no detectable contaminating cellulase production (37). However, in many instances where cellulase is coproduced alongwith xylanases, the xylanase titres are generally reported to be high (18). Inactivation of this accompanying cellulase activity in crude culture, filtrate has been attempted using mercurial compounds (30), separation of contaminating cellulase by bulk scale purification (40), and sequencing and cloning of xylanase genes to obtain selective expression in heterologous non-cellulolytic bacterial hosts (33) for its successful commercial usage, but with partial success (39). The gene encoding xylanase of the extreme thermophile *Thermotoga* strain Fj SS 3B.1 was cloned and expressed in *Kluyveromyces lactis* (46).

4.2 Suitability of Xylanases in Prebleaching of Paper Pulp

Xylanases are useful in a number of applications in the paper and pulp industry (16, 39, 43). Since the most common pulping reactions and recycled fibre processes are alkaline, and the bioreactor temperature rises during pulping operations, the obvious need of the hour is procurement or isolation of thermophilic microbe producing thermoalkalostable xylanases, that are more robust in terms of temperature and pH tolerance.

4.2.1 Thermostability

The advantages of employing a thermostable enzyme are immense (8). Although fungi produce higher quantities of xylanases than bacteria (6, 34), the xylanases from thermophilic eubacteria and archaebacteria have requisite half-lives at relatively higher temperatures. Mostly thermophilic microbes have been found suitable for treatment of pulp for the simple reason that xylanases secreted by them are frequently quite thermostable (9). Horikoshi and Atsukawa (21); Ratto *et al.* (36); Namamura *et al.* (29); Okazaki *et al.* (32) and Rajaram and Verma (35) have reported various thermophilic *Bacillus* spp. that could be useful in paper pulp treatment. High values of temperature and pH optima have been reported for *Bacillus* xylanases by Devy *et al.* (13); Ratto *et al.* (36) and Nakamura *et al.* (29). Xylanases with a half life of a few minutes upto 90 min at 80°C have been detected in *Bacillus thermophilus* (19); *B. licheniformis* (1); *Caldocellum saccharolyticum* (25); *Clostridium stercorarium* (7) and from thermophilic actinomycetes *Thermomonospora fusca, T. curvata* and *T. chromogena* (27). The xylanase of an extremely thermophilic *Thermotoga* sp. was reported to have a half life of more than 20 min at 105°C.

4.2.2 Alkalstability

Many alkali stable xylanases from alkalophiles have also been reported (20, 29, 32). Morales *et al.* (28) studied a bacterial xylanase from *Bacillus polymyxa,* whereas Nakamura *et al.* (29) reported the production of an alkaline xylanase from an alkalophilic *Bacillus* sp. Ohkoshi *et al.* (31) studied xylanase of *Aeromonas* sp., that exhibited activity in a broad pH range. Mathrani and Ahring (26) reported a xylanase of a strictly anaerobic, non-sporulating bacterium,

Dictyoglomus sp. R146B1 that showed a strong activity at alkaline conditions. A good alkaline xylanase producing thermophilic *Bacillus* sp. was studied by Dey *et al.* (13). An alkaline xylanase was produced by *Bacillus* sp. isolated from an alkaline lake (17). Out of thermostability and alkalstability, even if one kind of tolerance is naturally present in the microbe, the other one may be acquired by enzyme engineering.

5. APPLICATION OF XYLANASES IN BLEACHING

Xylanases form an ecofriendly alternative of prebleaching of paper pulp due to the selective removal of hemicellulose caused by them in the pulp, without affecting cellulose degradation, thus maintaining the quality of paper produced quite high (11). This helps to reduce the dependence of pulp brightening process on chlorine. The conventional chemical bleaching process generates large quantities of chlorine compounds measured as TOCl (total organically bound chlorine) or AOX (adsorbable organic halogen) which cause serious environmental pollution. The employment of xylanolytic enzymes in biobleaching processes enhances the brightness of bleached pulps, and reduces the amount of Cl_2/ClO_2 used in the bleaching stage. In the pollution conscious developed world, application of xylanase in pulp industry along with other bleach reagents such as oxygen and hydrogen peroxide is an environment friendly alternative to the use of toxic chlorine compounds. This has been accepted and projection of totally cellulase-free xylanase pulp bleaching technology has been postulated in the present decade (42).

Xylanases can be used at different stages of the pulping process, and their single most important involvement is in the prebleaching of kraft pulp. The process of kraft pulping consists of cooking the wood chips at 170°C with sodium sulphide and NaOH; this leads to solubilisation of most of the lignin. However, 3-5% of the residual lignin still remains in hardwood as well as softwood pulps, which imparts an undesirable brown colour to the pulp. Bleaching with Cl_2 or ClO_2 in conventional pulping process removes the brown coloured alkali extract of chlorinated derivatives of lignin comprising of dioxin and related compounds. These are toxic, carcinogenic, recalcitrant to biodegradation and pose serious hazards as environmental pollutants. Xylanase pretreatment of kraft pulp has been consistently shown to reduce the brown colour of the pulp with concomitant reduction in the amounts of Cl_2 compounds. Viikari *et al.* (42) observed that xylanase pretreatment of pulp caused substantial reduction in residual lignin content, bleach chemicals, AOX (adsorbable organic halogens) and dioxin levels.

6. ENVIRONMENTAL HAZARDS OF ORGANOCHLORINES

Due to ever increasing demand for paper, the paper pulp industry is rapidly growing and in the process it is becoming worst offender to the environment. Not only the huge forests are being consumed for paper making, the process of making paper is itself a highly polluting one. During the past few years, the toxic organic chlorine compounds arising from this process contributing to water pollution has attracted public attention (44). The organochlorines produced mainly by reactions between residual lignin present in the pulp and chlorine used for bleaching in the paper mills are recalcitrant and bioaccumulating. They are responsible for increasing oxygen demand (BOD and COD), effluent colour, toxicity, mutagenicity and carcinogenicity (4). About 300 different organochlorines including chlorinated resin acids, chlorinated phenolics and dioxins have been reported to be released by pulping industries, which employ conventional chemical methods for pulp bleaching.

6.1 Overcoming the Environmental Pollution

This acute toxicity problem may be tackled successfully by purifying waste bleach water or, better still by avoiding the hazardous chlorine compounds in newer and modified pulp bleaching techniques. The latter may be accomplished by replacement of chlorine by less toxic chlorine dioxide, which decreases effluent colour, and drastically reduces total organic chlorine (TOCl), adsorbable organic halogens (AOX), amount of dioxin and also chlorinated phenolics. Some chlorates, however, are produced that are herbicidal killing certain aquatic plants and brown algae when chlorine dioxide is used. Hydrogen peroxide and ozone are not suitable bleaching agents since they are very expensive at such a large scale (34). Other biochemical methods like employing lignolytic enzymes (lignin peroxidases, manganese peroxidases and laccases) have been found to be effective but their practical utility has turned out to be limited primarily due to lack of understanding of enzyme degradation systems (44). Hemicellulases are a relatively better understood enzyme consortium and could be utilised better for pulp bleaching. Among these, xylanases are very important in enhancing paper pulp brightness, that calls for lesser amounts of chlorine in final bleaching step (43). Optimization of microbial xylanase production, development of hyperxylanolytic strains, isolation of cellulase-free xylanase producing microbes, isolating thermoalkalostable xylanases or modifying the pre-existing ones by means of enzyme engineering and enzyme stabilisation techniques may enhance their feasibility and usefulness towards the objective of turning paper technology, non-polluting in true sense.

7. COMMERCIALISATION OF XYLANASES

Pretreatment of lignocellulosics can be combined with the pulping process using different bleaching sequences. The combinations of a hemicellulase pretreatment with a conventional bleaching sequence results in a gain of saving about 25% of active chlorine in the prebleaching stage. The quality analysis of the pulp revealed that employing cellulase-free xylanases in enzyme-aided bleaching protocols did not adversely affect the quality of paper sheets produced (10). Recent reports from literature indicate that enzymatic prebleaching has been successfully demonstrated on mill scale in which a pulp with 88% ISO brightness was achieved when used together with ClO_2 and H_2O_2. The pulp similarly treated without the enzyme component had only 82% ISO brightness. This is a very favourable and encouraging observation which has stimulated sustained research and technology for a totally chlorine free (TCF) paper technology towards the turn of the century. Pulpzyme introduced by NOVO Nordisk A/S was the first commercially available xylanase for use in biobleaching of wood pulp. Several multinational companies have entered the field at present and commercial xylanases such as Irgazyme 10 and Irgazyme 40 (Genencor International), Cartazyme HS and HT (Sandoz), Ecopulp (Alko Ltd.), VAI Xylanase (Voest-Alpine) have been under large scale evaluation for commercial application in different technologically advanced countries. Of these, Pulpzyme and Cartazyme are manufactured from *Bacillus* strains. As an illustration the following data published from Finland of a mill trial on large scale would enable a better appreciation of the enzymatic prebleaching with xylanase: 35 tons of Albazyme 10 was used to treat and produce 35,000 tons of fully bleached pulp from hardwoods as well as softwoods. The enzyme was added to a kraft cooked pulp after suitably adjusting the pH and temperature to suit the optimum for enzyme activity. An overall reduction of 12% Cl_2 usage could be achieved (23). Productivities have been reported to be increased by employing optimised production methods, efficient screening methods, developing hyperproducing mutants and by cloning (2, 33, 44).

With the successful demonstration of the efficacy of enzymatic bleaching in large scale trials, current efforts are aimed at process optimization, simplification and cost reduction for application. Using the xylanases available at present, a pH adjustment of the pulp from 10-11 to 6-8 is necessary for optimum activity. From an industrial point of view it is more difficult to adjust the temperature whereas pH is relatively easier to control. Thus screening criteria for xylanases with better thermostability and a possibly, increased pH optima (so that pH adjustment step could be eliminated) have come into greater attention and focus. The identification of a thermostable, alkalistable xylanase from an anaerobic bacterium, *Dictyoglomus* has been one such development, Perttula *et al.* (34) have also studied xylanase of thermophilic bacteria from Iceland hotsprings. The xylanolytic *Dictyoglomus* strains were isolated from sludge from a paper pulp cooling tank in a paper board factory in Finland. The rod shaped anaerobe grew only on xylan while no growth occurred with soluble sugars, other polysaccharides, peptone or yeast extract. The enzyme was secreted at 68°C, and had a half-life of 80 min at 90°C. This alongwith its alkalotolerance makes it a possible attractive source as a contributor of desirable gene pools for cloning and expression in heterologous host systems in future. Several substantial reports on alkalophilic and thermophilic xylanolytic organisms are available, including a *Thermotoga* strain producing the most thermostable xylanase with a half-life of 90 min at 95°C.

Increasing public concern about the environment has been the driving force for dramatic changes in pulp bleaching. Papermaking processes that move towards total elimination of chlorine requirement is the ultimate goal in this direction. Increasing stringent regulations are causing bleach plant operators to modify and have the currently existing bleaching schemes. The transition from laboratory to industrial scale has been aided by a relative ease in achieving this target, and due to the fact dut intermediate pilot stages are not needed for scale up. It is usually observed that higher brightness values can be obtained in full scale than in laboratory scale due to more efficient mixing systems and higher pulp consistencies. In terms of fresh capital inputs in existing industries, the requirement usually is of pH adjustments. In elementally chlorine free (ECF) bleaching sequences, specially where the utilisation of chlorine gas has been abandoned, the use of enzymes increases the productivity of the bleaching plant when the production capacity of chlorine dioxide is the limiting factor. The addition of enzymes in totally chlorine free (TCF) bleaching sequences, increases the final brightness value and also leads to savings in the TCF bleaching chemicals, which is important both in terms of cost and the strength properties of the pulp. Several alternative TCF bleaching methods based on different oxygen chemicals have been developed. These chemicals are oxygen, ozone and peroxide. Frequently enzymes are added to oxygen delignified pulps to increase the brightness. Present day TCF technologies are usually based on bleaching of oxygen delignified pulps with enzymes and hydrogen peroxide. Finnish forest companies have been pioneers in implementing the enzymatic pretreatment in pulp treatment with mill scale trials conducted way back in 1988. By 1992, about one hundred mill trials had been carried out with half of them conducted in Europe. With the development of more efficient production strains and technologies, the prices of the enzymes are expected to decrease, and consequently bring down the price of enzymatic treatment. Though the price of the enzyme is usually compared with that of the competing bleaching chemicals, factors such as decreased AOX loadings and retention of viscosity confer additional advantages in case of enzyme-aided technology.

8. CONCLUSIONS

Paper pulp bleaching using microbial xylanases is a promising and environment friendly alternative to the conventional chemical bleaching. The requirement of such a replacement is becoming mandatory, due to the pollution and long-term hazards caused by the organochlorines. Xylanases used for this purpose must be free from cellulase activity, so as to obtain high quality paper and to avoid yield loss. Thermophilic bacteria are known producers of cellulase-free xylanases. Tolerance of such xylanases to elevated temperature and pH make them very suitable candidates for kraft pulping because such conditions are typical of the pulping protocols. Microbial xylanases thus offer an opportunity as well as a challenge to microbiologists, biotechnologists and industrialists in the quest for global competitiveness, and the future outcome would depend upon intelligent planning and approaches to meet these challenges in biotechnological alternative implementation.

9. REFERENCES

1. Archana, A. 1996. Xylanolytic enzymes of a moderate thermophile *Bacillus licheniformis* A99. Ph.D. Thesis, University of Delhi South Campus, New Delhi, India.

2. Bailey, M.J., Buchert, J. and Viikari, L. 1993. Effect of pH on production xylanase by *Trichoderma reesei* on xylan and cellulose based media. *Appl. Micrbiol. Biotechnol.* **40**: 224-229.

3. Bajpai, P. and Bajpai, P.K. 1994. Biological colour removal of pulp and paper mill wastewaters. *J. Biotechnol.* **33**: 221-220.

4. Bajpai, P. and Bajpai, P.K. 1997. Reduction of organochlorine compounds in bleach plant effluents. In : *Biotechnology in the Pulp and Paper Industry* (ed. K.E.L. Eriksson) Springer Verlag, Berlin, Heidelberg, pp. 213-259.

5. Bansod, S.M., Dutta-Choudhry, M., Srinivasan, M.C. and Rele, M.V. 1993. Xylanase active at high pH from an alkotolerant *Cephalosporium* species. *Biotechnol. Lett.* **15**: 965-970.

6. Bastawde, K.B. 1992. Xylan structure, microbial xylanase and their mode of action. *World J. Microbiol. Biotechnol.* **8**: 353-368.

7. Berenger, J.F., Frixon, C., Bigliardi, J. and Crenzet, N. 1985. Production, purification and properties of thermostable xylanase from *Clostridium stercorarium*. *Can. J. Microbiol.* **31**: 635-643.

8. Bragger, J.M., Daniel, R.M., Coolbear, T. and Morgan, H.W. 1989. Very stable enzymes from extremely thermophilic archaebacteria and eubacteria. *Appl. Microbiol. Biotechnol.* **31**: 556-561.

9. Brodel, B., Samain, E. and Debeire, P. 1990. Regulation and optimization of xylanase production from *Clostridium thermolacticum*. *Biotechnol. Lett.* **12**: 65-70.

10. Buchert, J., Kantelinen, A. and Viikari, L. 1991. Enzymatic bleaching of pulp. In : *National Conference of Biophysics and Biotechnology*, University of Kuopio, Research Report No. 4.

11. Coughlan, M.P. and Hazlewood, G.P. 1993. β-1, 4-D-Xylan-degrading enzyme systems : biochemistry, molecular biology and applications. *Biotechnol. Appl. Biochem.* **17**: 259-289.

12. Dekker, R.F.H. and Richards, G.N. 1976. Hemicellulases: Their occurrence, purification, properties, and mode of action. *Adv. Carbohyd. Chem. Biochem.* **32**: 277-352.

13. Dey, D., Hinge, J., Shendye, A. and Rao, M. 1992. Purification and properties of extracellular endoxylanases from alkalophilic thermophilic *Bacillus* sp. *Can. J. Microbiol.* **38**: 436-442.

14. Dung, N.V., Vetayasuporn, S., Kamio, Y., Abe, N., Kaneko, J. and Izaki, K. 1993. Purification and properties of β-1, 4-xylanases 2 and 3 from *Aeromonas caviae* W-61. *Biosci. Biotech. Biochem.* **57**: 1708-1712.

15. Elegir, G., Szakacs, G. and Jeffries, T.W. 1994. Purification, characterization and substrate specificities of multiple xylanases from *Streptomyces* sp. strain B-12-2. *Appl. Environ. Microbiol.* **60**: 2609-2615.

16. Eriksson, K.E.L., Blanchette, R.A. and Ander, P. 1990. Microbial and enzymatic degradation of wood and wood components. Springer Verlag, Berlin, Heidelberg, New York.

17. Gessess, A. and Gashe, B.A. 1997. Production of alkaline xylanase by an alkalophilic *Bacillus* sp. isolated from an alkaline lake. *J. Appl. Microbiol.* **83**: 402-406.

18. Gilbert, H.J. and Hazlewood, G.P. 1993. Bacterial cellulases and xylanases. *J. Gen. Microbiol.* **139**: 187-194.

19. Gruninger, H. and Fiechter, A. 1986. A novel, highly termostable D-xylanase. *Enz. Microb. Technol.* **8**: 309-314.

20. Honda, H., Kudo, T., Ikura, Y. and Horikoshi, K. 1985. Two types of xylanases of alkalophilic *Bacillus* sp. No. C-125. *Can. J. Microbiol.* **31**: 538-542.

21. Horikoshi, K. and Atsukawa, Y. 1973. Xylanase produced by alkalophilic *Bacillus* C-59-2. *Agricul. Biol. Chem.* **37**: 2097-2013.

22. Kantelinen, A., Ranua, M. and Ratto, M. 1988. International Pulp Bleaching Conference Proceedings. TAPPI Press, Atlanta, p. 1.

23. Loponen, R. 1971. Enzyme systems prove their potential. *Pulp and Paper Internat.* **33**: 20-21.

24. Kyu, K.L., Ratanaknanokehi, K., Uttapap, D. and Tanticharoen, M. 1994. Induction of xylanase in *Bacillus circulans. Bioresource Technol.* **48**: 163-167.

25. Luthi, E., Jasmat, N.B. and Bergquist, P.L. 1990. Xylanase from the extremely thermophilic bacterium *Caldocellum saccharolyticum* : overexpression of the gene in *Escherichia coli* and characterization of the gene product. *Appl. Environ. Microbiol.* **56**: 2677-2683.

26. Mathrani, I.M. and Ahring, B.K. 1992. Thermophilic and alkalophilic xylanases from several *Dictyoglomus* isolates. *Appl. Microbiol. Biotechnol.* **38**: 23-27.

27. McCarthy, A.J., Peace, E. and Broda, P. 1985. Studies on the extracellular xylanase activity of some thermophilic actinomycetes. *Appl. Microbiol. Biotechnol.* **21**: 238-244.

28. Morales, P., Madarro, A., Gonzalez, J.A.P., Sendra J.M., Pinaga, F. and Flors, A. 1993. Purification and characterization of alkaline xylanases from *Bacillus polymyxa. Appl. Environ. Microbiol.* **59**: 1376-1382.

29. Nakamura, S., Wakabayashi, K., Nakai, R., Aono. R. and Horikoshi, K. 1993. Production of alkaline xylanase by a newly isolated alkalophilic *Bacillus* sp. strain 41 M-1. *World J. Microbiol. Biotechnol.* **9**: 221-224.

30. Noe, P., Chevalier, J., Mora, F. and Comtat, J. 1986. Action of xylanases on chemical pulp fibres. Part II, enzymatic beating. *Jour. Wood Sci. Technol.* **6**: 167-184.

31. Ohkoshi, A., Kudo, T., Mase, T. and Horikoshi, K. 1985. Purification of three types of xylanases from an alkalophilic *Aeromonas* sp. *Agric. Biol. Chem.* **49**: 3037-3038.

32. Okazaki, W., Akiba, T., Horikoshi, K. and Akahoshi, R. 1985. Purification and characterization of xylanases from alkalophilic thermophilic *Bacillus* sp. *Agric. Biol. Chem.* **49**: 2033-2039.

33. Paice, M., Bermerm, M. and Jurasek, L. 1988. Viscosity enhancing bleaching of hardwood kraft pulp with xylanase from a cloned gene. *Biotechnol. Bioengin.* **32**: 235-239.

34. Perttula, M., Ratto, M., Kondradsdothir, M., Kristjansson, J.K. and Viikari, L. 1993. Xylanases of thermophilic bacteria from icelandic hot springs. *Appl. Microbiol. Biotechnol.* **38**: 592-595.

35. Rajaram, S. and Varma, A. 1990. Production and characterisation of xylanase from *Bacillus thermoalkalophilus* grown on a agricultural wastes. *Appl. Microbiol. Biotechnol.* **34**: 141-144.

36. Ratto, M., Mathrani, I.M., Ahring, B. and Viikari, L. 1994. Application of thermostable xylanase of *Dictyoglomus* sp. in enzymatic treatment of kraft pulps. *Appl. Microbiol. Biotechnol.* **41**: 130-133.

37. Samain, E., Debeire, P. and Touzel, J.P. 1997. High level production of cellulase free xylanase in glucose limited fed batch cultures of a thermophilic *Bacillus* strain. *J. Biotechnol.* **58**: 71-78.

38. Simpson, H.D., Haufler, V.R. and Daniel, R.M., 1991. An extremely thermostable xylanase from the thermophilic eubacterium *Thermotoga. Biochem. J.* **277**: 413-417

39. Srinivasan, M.C. and Rele, M.V. 1995. Cellulase-free xylanases from microorganisms and their application to pulp and paper biotechnology: an overview. *Ind. Jour. Microbiol.* **35**: 93-101.

40. Tan, L.U.L., Yu, E.K.C., Louis-Seize, G.W. and Saddler, J.N. 1987. Inexpensive, rapid procedure for bulk purification of cellulase-free β-1-4-Dylanase of high specific activity. *Biotechnol. Bioengin.* **3**: 96-100.

41. Timell, T.E. 1967. Recent progress in the chemistry of wood hemicelluloses. *Wood Sci. Technol.* **1** : 45-70.

42. Viikari, L., Kantelinen A., Sundquist, J. and Link, M. 1994. Xylanases in biobleaching: from an idea to the industry. *FEMS Microbiol. Rev.* **13**: 335-350.

43. Viikari, L., Ranua, M., Kantelinen, A., Linko, M. and Sundquist, J. 1987. Application of enzymes in bleaching. In : *Proceedings of 4th International Symposium on Wood and Pulping Chemistry.* Vol. I. pp. 69-90.

44. Viikari, L., Tenakanen, M., Buchert, J., Ratto, M., Bailey, M., Siika-aho, M and Linko, M. 1993. Hemicellulases for industrial applications. In : *Bioconversion of Forest aid Agricultural Plant Residues.* (ed. L. Saddler) CAB International, Wallingford, UK, pp. 131-182.

45. Vyas, P., Chauthaiwale, V., Phadatare, S., Deshpande, V. and Srinivasan, M.C. 1990. Studies on the alkalophilic *Streptomyces* with extracellular xylanolytic activity, *Biotechnol. Lett.* **12**: 225-228.

46. Walsch, D.J., Gibbs, M.D. and Bergquist, P.L. 1998. Expression and secretion of a xylanase from the extreme thermophile *Thermotoga* strain Fj SS3B.1 in *Kluyveromyces lactis. Extermophiles* **2**: 9-14.

11 Aflatoxins: Knowing and Combating a Major Health Hazard

Chitreshwar Sen, Saswati Bardhan and B. Dasgupta

CONTENTS

ABSTRACT 142

1. INTRODUCTION 142
2. WHAT ARE AFLATOXINS ? 143
3. CHEMISTRY OF AFLATOXINS AND THEIR BIOSYNTHESIS 149
4. DETECTION & QUANTIFICATION OF AFLATOXINS 152
5. BIOLOGY OF ASPERGILLI, GROWTH AND AFLATOXIN PRODUCTION 154
6. TOXIC MODE OF ACTION OF AFLATOXINS 162
7. MANAGEMENT APPROACHES-PREVENTING AFLATOXIN PRODUCTION 165
8. CONCLUSION 173
9. REFERENCES 174

Advances in Microbial Biotechnology
J.P. Tewari, T.N. Lakhanpal, Jagjit Singh, Rajni Gupta & B.P. Chamola (eds.)
APH Publishing Corporation, New Delhi - 110 002, India.

ABSTRACT

Of the numerous devious means of survival adopted by microbes, one may be its ability to produce toxic metabolites. The hazards of such toxic metabolites produced by fungi, whether by design of leftovers of metabolism, is known to humanity since ages and are grouped under general terms, mycotoxins and mycotoxicoses. Of these, afflatoxins and researches relating to them, dominated the scenario for several decades since early 1960s till Scientists understood largely the action mechanisms of these toxophores and approaches to reduce the hazards. The present review sums up the existing knowledge on sources of afflatoxins, their chemistry and biosynthetic pathways and their detection and quantification methods.

Produced in the producer system or on processed food through contamination by *Aspergillus flavus* and *A. parasiticus*, their consumption even at low dosage, may cause a variety of malaise and symptoms, the chief concern among which is the carcinoma. The factors affecting production by the involved fungi, both *in vitro* and *in situ*, are therefore detailed along with symptom expression; the biochemical malfunctions and effects on nucleic acid metabolism leading to carcinoma are briefly described.

Since the management, a major component of which is awareness specially in the third World around the tropical and subtropical belts, more detailed and critical assessment, both of existing knowledge and tools are presented with the hope that these will be brought to focus among the relevant authorities through the existing mass communication media.

1. INTRODUCTION

The highly biodiverse filamentous fungi thriving on exotic food resources, often produce a variety of secondary metabolites, many of which have been exploited by man fruitfully while others have been a source of health hazard to various domestic animals and man. Those that are poisonous are called mycotoxins derived from the Greek words *Mykes* = fungus ; *toxicum* = poison. These residues appear in food and feed through natural events over which man has little control. Most such mycotoxins have carcinogenic properties adding to the concern of man although the general syndrome or malaise is described as mycotoxicosis usually caused by consumption or ingestion of mycotoxin(s)-contaminated food or feed.

Since mycotoxins are associated with plants in the field and stored plant produced, one could easily visualise that mycotoxicosis has co-existed with agricultural history and perhaps the first record was that by the Greek Physician Galen who stated that people forced to eat stored grains of wheat and barley often suffer from a putrid and pestilential fever, others are seized by a scabby and leprosy like condition. Ergotism of a convulsive nature was known from middle ages, caused by consumption of cereals and breads containing the sclerotia of *Claviceps purpurea*–fortunately of rare occurrence now. The beginning of this century saw severe outbreaks of mycotoxicosis in Russia caused by *Fusarium* spp. in cereals. People eating rye bread in 1923 summer suffered from weakness, vertigo, headache, nausea and vomiting.

Similar problems were caused by linseed oil in Ukraine region. *Fusarium* contaminated grains caused alimentary toxic aleukia (ATA), that caused death sporadically in early parts of 20th century but became explosively hazardous causing death by hundreds of thousands following the 2nd World War, both in Russian and East Germany. The symptoms were haemorrhage, necrotic angina, sepsis and exhaustion of the bone marrow (94.)

The major mycotoxins identified, included among others aspergillic acid, aspertoxin, citrinin, gliotixins, kojic acid, ochratoxin, oxalic acid, patulin, penicillic acid, sterigmatocystin and a group of secondary metabolites called aflatoxins.

The aflatoxins and the resultant aflatoxicosis was brought to focus by epidemic fatality among 100,000 turkey poults in England in 1960 showing symptoms of acute necrosis of the liver and hyperplasia of the bile duct (The turkey X disease). The disease was correlated with a groundnut feed imported from Brazil. The meal was found to be contaminated by the fungus *Aspergillus flavus* and hence the toxic metabolite(s) produced by it was called Aflatoxins (A=*Aspergillus* ; FLA=*flavus*; Toxin = poison (51). Characteristic symptoms of aflatoxicosis was reported before the turn of the 20th century (41, 120). Mayo suggested that toxicosis was a disease unique to horses since other farm animals did not seem to be affected and the syndrome was called "mouldy corn poisoning". This group of toxin was soon revealed to cause widespread, world wide problems and was also found to be carcinogenic leading to intensification of researches. Its presence in wide range of commodities was demonstrated in the tropics and the presence in cotton seed, maizé and groundnuts in USA and led to an explosive research agenda till the problem was brought under partial control.

The present article attempts to highlight the cardinal findings that led to the understanding of both the producer and the product–from the source to the sink–to its management and the existing problems primarily of awareness that needs to be tackled in the developing world.

2. WHAT ARE AFLATOXINS?

It has already been pointed out that aflatoxins are secondary metabolites of *Aspergillus flavus*. Subsequently it was also shown to be produced by *A. parasiticus*.

2.1 Source

The genus *Aspergillus* belongs to class Hyphomycetes, a subdivision of Deuteromycotina. This subdivision includes conidial stages of organisms whose perfect stages are unknown. These produce conidia directly on the mycelium or on single or clustered conidiophores (24). Apart from these two species stated above, *A. nominus*, a new species similar to *A. flavus*, seems to be involved (110).

These fungi are found in soil, in or on living or dead plants and animals and their air-borne spores can contaminate food and other agricultural commodities. Theoretically, all processed foods are susceptible to contamination during the stages of production, processing, transportation and storage.

These fungi have a world wide distribution and are particularly prevalent in tropical and subtropical regions where they may colonize plant organs both in the field and under poor storage conditions after harvest.

2.2 Occurrence in Food and Feed

The aflatoxin producing fungi are so ubiquitous that in them natural aflatoxins may be universally distributed in all foods and feeds even if undetected by the currently available sensitive techniques. Levels that fail to produce aflatoxicosis symptoms are transmitted unreacted into the edible products obtained from these animals–and finally to the sink-man.

The USDA, USDI, US-PL-480, FAO and WHO have poured in a lot of funds following reports by Salamon and Newborne (169) of presence of aflatoxin in domestic groundnut meal and considerable information is available from their data banks.

2.3 In Agricultural Commodities from Plant Sources

The common approach used was survey–testing and reviewed by Armbrecht (3). The commodities from which aflatoxicogenic moulds have been isolated include : Cassava, cornmeal, cocoa, cottonseed, Brazil nuts, flour, oilseeds, pumpkin seeds, groundnuts, pistachio nuts, rice, copra, oats, sorghum, soybean, wheat, etc. (64).

2.3.1 Groundnut

Groundnut (peanut) contamination has received the highest attention and has been reviewed through a status report presented by ICRISAT (75). Infestation of groundnuts ranged from 35-100% on an area basis and toxin production ranged from nil (no toxin) to about 300 mg kg^{-1} of peanut substrate with a medium yield of 1-5 mg kg^{-1} (200). In shelled ground, getting a representative sample was a critical problem (204).

The heterogenous distribution is largely decreased in producing meal by a process of grinding and milling. Particle size is much smaller in peanut butter. In a random sample survey of groundnut in Denmark important from Brazil, Argentina, Nigeria, Sudan, Senegal, Indonesia, Kenya, Ghana, Congo and Uganda. Out of 52 samples, none were detected in seven, two contained 100 μgkg^{-1}, 27 contained 100-1000 μgkg^{-1} and 16 contained over 1000 μgkg^{-1}. Highest contamination was 3.5 mg Kg^{-1} during the period from June to October, 1968. Largely similar results were obtained for imports from Ireland (212), Formosa (121). Siwela and Caley (191) analysed 441 stored groundnut samples over a period from 1982-83 to 1986-87 in Zembabwe that belonged to seven varieties. Sixty eight percent had contamination level upto 25 μgKg^{-1}. Similar observations in Zambia (101) showed that 6.3% of the 28, 410 samples had contamination level above 5 μgKg^{-1}. In Mozambique (9) 87-100% samples were contaminated, the aflatoxin content ranging over 3-5500 Kg^{-1}.

In People's Republic of China (225), out of a total 1172 samples spread over 24 provinces, 26.3% and 47.3% were contaminated for kernels and oils respectively. Contamination was higher in southern than in northern China. In Phillipines the survey conducted by Quitco *et al.* (166) was more comprehensive. At harvest groundnut contained, on an average, 3-16μg Kg^{-1} aflatoxin; at wholesaler's level it reached 188 μgKg^{-1} and those stored for over 3 months contained 120.6 μgKg^{-1}. Consumption of roasted peanut is common in Asia. In Pakistan roasted peanut contamination ranged over 24-800 μgKg^{-1} (169) all samples from Khuzdar area, the main growing area, were contaminated. Lahore samples showed higher content (373 μgKg^{-1}) although only 15% of the samples were contaminated (11): In India Sharma *et al.* (186) estimated aflatoxin B_1 content in deoiled / oil cakes and grains during different seasons. Among 97 samples, the B^1 content was 200 and 2000 ppb in deoiled and oil cakes in 51 and 10% of the samples respectively. In samples from southern and western parts of India the aflatoxin B_1 content ranged over 400-800 ppb. In Brazilean samples, from where the story of aflatoxin B_1 Kg^{-1}.

Considering the fact that permissible levels of aflatoxin contamination is very low in most countries (Netherland) 5 µgKg^{-1} for feed, USA 20 µgKg^{-1} it is obvious that more stringent measures need to be adopted in most developing countries to promote groundnut export.

The overall data as surveyed by FAO/WHO/UNEP monitoring programmes (1977-83) for groundnut and peanut-butter (popular in U.S.A.) showed that 90th percentile (level below which 90 µgKg^{-1}), Ireland, 300-4000 µgKg^{-1} ; Mexico, 7090 µgKg^{-1} ; Switzerland, 338 µgKg^{-1} ; UK, 75 µgKg^{-1} ; U.S., 24 µgKg^{-1} ; Soviet Union, 329 µgKg^{-1}. These levels are higher than the theoretically permissible level of 20 µgKg^{-1}.

2.3.2 Other nuts and seeds

Among these cotton seed and meal has received special attention as *A. flavus* is responsible for a boll rot disease that causes serious economic losses. However Whitten's (213) report showed low levels, if at all, of aflatoxin contamination in seeds, meal or hull from seeds or even the processed oil. Subsequent analysis of imported seeds (74) from Brazil, Columbia, Guatemala, Nicaragua, El Salvador, Syria, Turkey and Russia gave the following picture :

Not detected	66.7%
<30 µgKg^{-1}	19.2%
30-100 µg Kg^{-1}	11.6%
>100 µg Kg^{-1}	2.5%

Permissible limit : 100 µg Kg^{-1}. concentrations of aflatoxins recorded in other nuts and seeds by FAO/WHO/UNEP monitoring Programme (1976-1981) revealed as follows :

Pistachios (Canada)	110 µg Kg^{-1}
Pistachios (US)	15-67 µg kg^{-1}
Pumpkin Seeds (Canada) :	63 µg kg^{-1}
Brazil nuts (UK)	65 µg Kg^{-1}
Brazil nuts (US)	40-65 µg Kg^{-1}

Linseed does not appear to be implicated in field episodes. Coconut, by the very nature of the crop, should not present a serious problem. Any invasion will quickly impart an "off-flavour" and a characteristic odour change in the milk. Once, however, the protective shell is broken the copra becomes vulnerable. Copra has actually been found to be an excellent substrate for aflatoxin production (6).

2.3.3 Grains

Among the grains maize (corn), rice, oats, sorghum, soybean and wheat have received serious attention.

The literature on aflatoxin production in corn grain has been reviewed in depth (217): Normally other grains such as rye, wheat, oats, buckwheat, barley, millet and rice have not been implicated with serious aflatoxicosis. However, given favourable conditions (as prevails in many of our grainaries), they are capable of supporting aflatoxicogenic strains of *A. flavus* and produce toxic levels of aflatoxins. In sample surveys in USA (189) contamination between levels ranged 0.5%-7.1% in corn and 3% or lower in others. Aflatoxin content in μg Kg^{-1} ranged from 3-27 μg Kg^{-1} in corn and 6-19 μg Kg^{-1} in others.

Under favourable conditions of higher than 16% moisture and temperature over a range of 24-35°C, corn grains may have as much as 700 μg Kg^{-1} in storage. Boller and Schroeder (20) found that average contamination in rough rice was about 50 μg Kg-1 but after the bran was removed. Ninety five percent of the toxins originally present in rough rice was found in the bran. Rice bran oil, therefore, would need special monitoring and handling to prevent occurrence of human carcinoma. On the other hand a report from CFTRI, Mysore (127) stated that aflatoxins production was more in polished rice than on unpolished rice. Aflatoxin production was far less in presence of bran oil.

2.3.4 Other crops

In an Ugandan family, three children were fed cassava preparation that had an aflatoxin content of 1.7 mg kg^{-1}. One child died of abdominal pain, edema of both legs and a prolonged PR interval with a partial blockage of the right bundle branch of his heart (183). Another became ill with similar symptoms but survived. Cassava is a tuberous root plant cultivated in tropical America and Africa, yielding a starchy product from which bread, porridge or a tapioca is made and extensively consumed.

Similarly feeding of mouldy sweet potatoes to livestock show toxic signs such as pulmonary edema, hepatic damage and cellular changes in the kidneys. Were these due to aflatoxins? No proper analysis has been made.

Aflatoxin contamination of green grasses and forage crops has not been reported. Similarly mouldy coffee pulp samples did not yield significant levels of aflatoxins and thus identifies these as low risk-prone crops.

2.4 In Animal Tissues–The First Order Sink

Although kinetics of aflatoxins in the food chain has not been studied properly, as will be seen later, many undergo metabolic changes in animal systems to produce new toxophores. The earlier work has been reviewed by Armbrecht (3). Often signs of aflatoxicosis in animals is not symptomatically visible. They are recognisable after slaughter. Time dosage relationship studies do not show quantal relationships as residues get distributed in different organs and go below the 2 μg Kg^{-1} in milk and eggs. Acute single dosage toxicity of aflatoxin B1 as well as lower multiple dosage has been tabulated (3). The dosage range from less than 1 mg Kg^{-1} in ducks (Khaki Cambell), rabbit (Dutch belted), cats and pigs to above 5 mg Kg^{-1} in adult rats, salmon, coho and catfish channel. In the fishes, dosage were administered over 5 successive days. Ducklings were unusually sensitive and provided a tool for bioassay. Turkeys, were shown to develop tolerance following graded exposure and age. Pigs fed with aflatoxin B_1 diet at the high rate of 400 μg Kg^{-1} showed 0.83/0.19 (liver), 0.62/0.30 (kidney), 0.36/0.08 (blood) and 0.27/0.06 (ham) of aflatoxin B_1 and M_1 in the tissues stated (22). No clinical signs or impaired conversion efficiency was noted in Yorkshire swines.

Chronic exposure over a period of 100-180 days led to changes in liver progressing through steatosis, bile duct profile rations, pericellular fibrosis to parenchymal karomegaly and terminally a nodular hyperplasia (79).

Two important domestic animals other than those stated above are diary cattle and fish. In dairy cattle it was found that aflatoxin B_1 was metabolised into M_1 in the liver and excreted in the urine, faeces and in lactating animals, in milk. The least (0.5%) was excrerted in the milk as most passed out through urine (9). In a market survey of freeze dried milk in South Africa (165), aflatoxin M_1 was detected at levels of 0.05 nm L^{-1} on whole milk and negative in skimmed milk.

Among fishes, trout and cat fishes have been cultured in the orient since ages. Feed conversion in warm blooded homeotherms generally exceeds 2 Kg for 1 Kg body weight. In the poikilothermic fishes, such conversions are 0.8 to 1.5 as no energy is spent to maintain the thermal body temperature. A no response diet for rainbow trout appeared to be 1 μg Kg^{-1} (193.) for 12 months of continuous exposure. However the trout hepatoma has been reviewed (76). Very little is known about the high temperature tolerant carps and cat fish.

The acute and chronic toxicity in animals may be summed up as follows:

Acute aflatoxicosis:

Characteristic symptoms	:	Liver enlargement, fat deposition, necrosis Hyperplasia of bile duct.
Less specific effects		Loss of appetite, lethargy, wing weakness in turkey X disease, Gastrointestinal haemorrhage in poultry.

In mammals considerable variation in resistance was found to be related to sex, age and size; LD50 values ranging from 1.0 - 1.4 mg AFB, Kg^{-1} in guinea pigs to 17.9 mg Kg^{-1} in 150g female rats.

Other symptoms noted are congestion of lungs and intestinal submucosa, haemorrhagic necrosis of the adrenals and spleen, necrosis of kidneys and death reabsorption or reduced growth in foetuses.

2.5 In Processed Foods–The Second Order Sink

Normally the house wife's practice is to cut away mould damaged parts and use apparently healthy ones for food. However, the aflatoxins excreted from the fungal mycelium diffuses through processed food like cheese, (114) butter or more so through the more porous bread (81).

On the other hand cured meat, fresh meat and cheese products that often acquire an extensive mould cover while ageing in cool storage rooms, along with fruit juices kept under refrigeration are not favourable pastures for *A. flavus* or for aflatoxin production. In mould ripened Roquefort or Camembert cheese no aflatoxin were detected.

2.6 In Humans–The Terminal Sink

Acute hepatitis in humans due to aflatoxicosis has been reported from China, India, Senegal and Uganda (214). The fatalities were more than 100 in Western India where the cause was consumption of maize containing 6.25-15.60 mg AFB Kg^{-1}. The daily consumption of those affected was calculated at 2-6 mg (106). Kwashiorkar syndrome, a deficiency disease reported from tropical areas of the world was shown to be due to aflatoxin induced liver damage (85). The disease occurred among those who consumed starchy staples such as maize, rice or

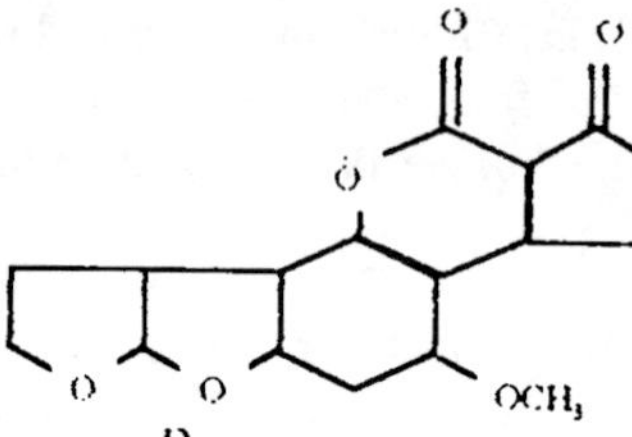

B1 B2 M1 M2 G1 G2 AFLH1 AFL(aflatoxicol) B2a AFQ1 AFP1 AFB1 2,3-oxide

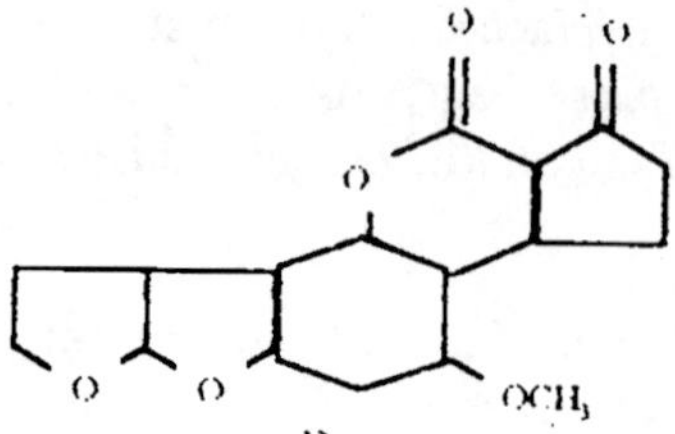

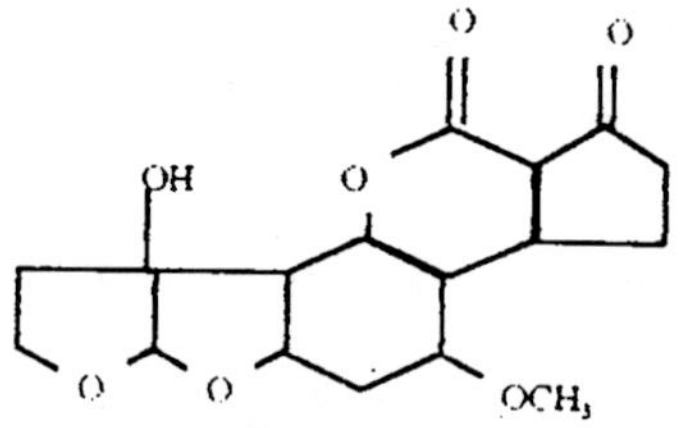

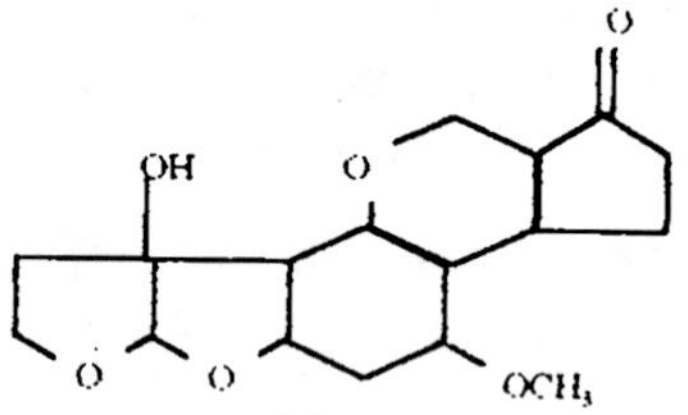

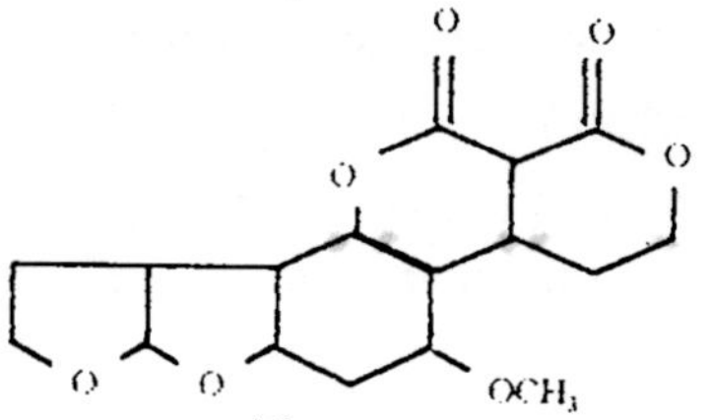

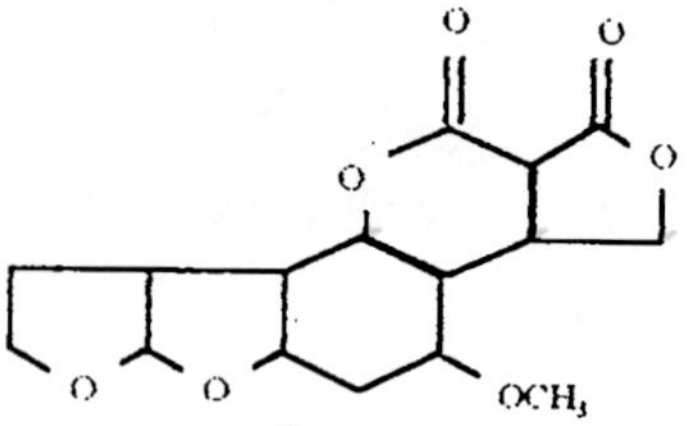

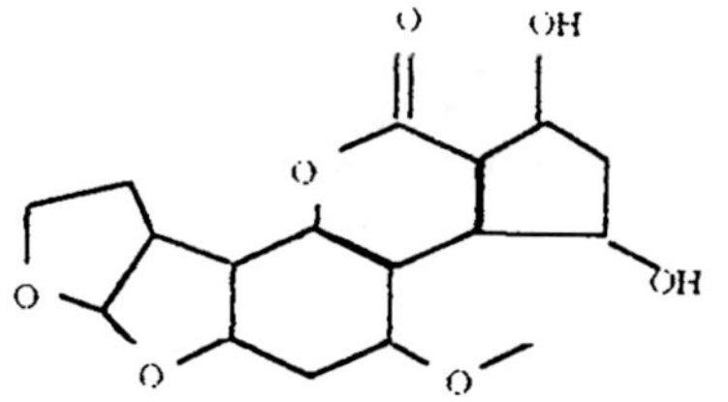

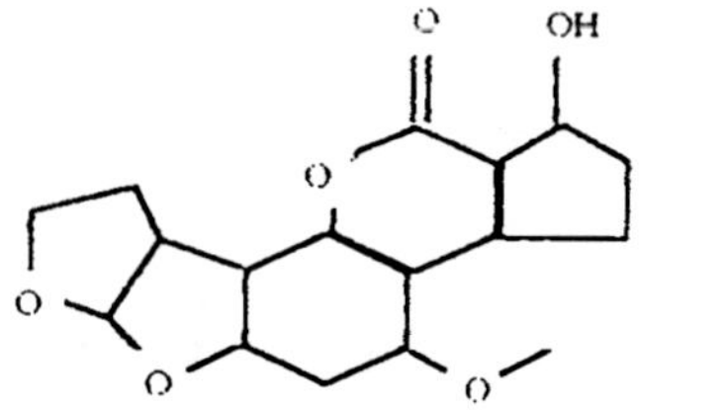

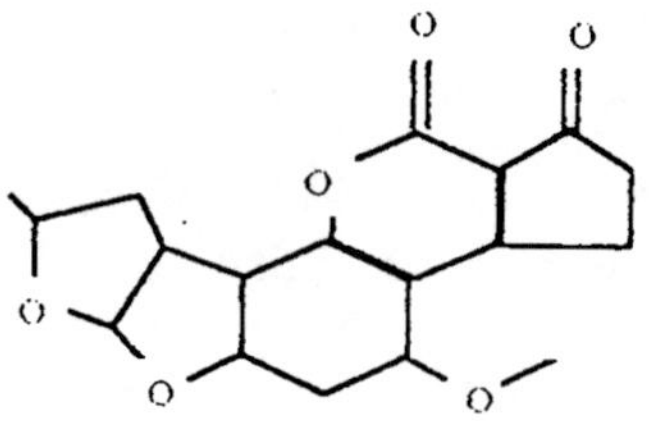

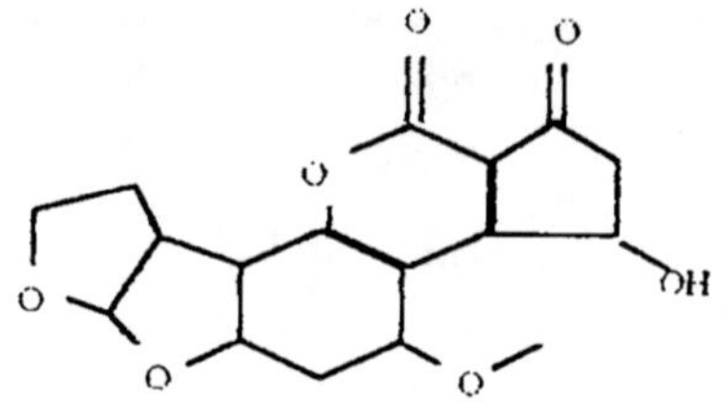

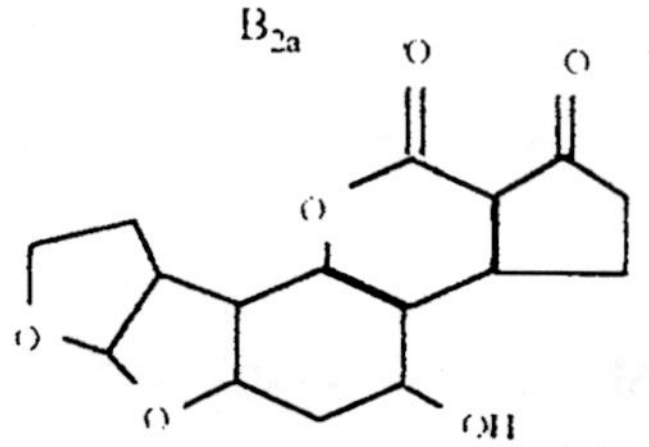

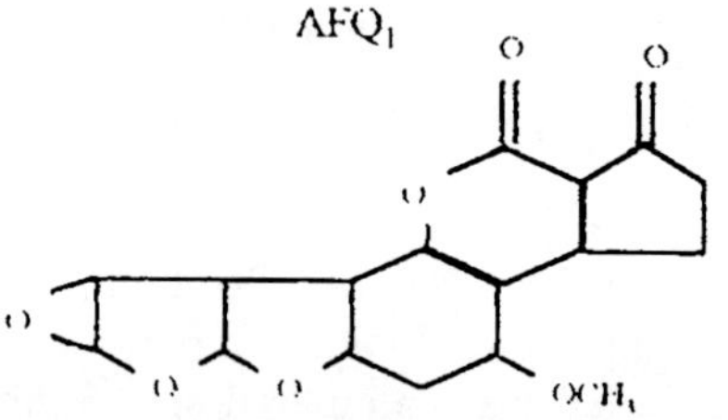

Fig. 1. Aflatoxins reported

plantains. In humans, aflatoxins have been shown to accumulate in body fluids (47) and are only slowly eliminated. Reye's syndrome, a disease common among children and characterized by acute encephalopathy, hepatic enlargement and serum transeliminase level has been reported from Czechoslovakia, New Zealand, Thailand and U.S.A. The disease has been ascribed to aflatoxins although blood levels among affected and unaffected do not warrant such conclusion. Overriding factors may contribute to Reye's syndrome that include viral infection and a range of xenobiotic compounds including aflatoxins. Evidence of an association between aflatoxins in food and primary hepatocellular carcinoma in humans living in Central and South Africa, Thailand and Indonesia was evaluated by (214). The incidence of the disease was highest where aflatoxin in food was greatest (mean daily uptake 3.5-221.1 ng Kg-1 body weight). Such studies are however flawed as (i) population and not individual daily uptake was taken into account and (ii) no consideration was given to high incidence of Hepatitis B virus and alcohol consumption in these areas.

3. CHEMISTRY OF AFLATOXINS AND THEIR BIOSYNTHESIS

3.1 Structure and Chemistry of Aflatoxins

Chemically, aflatoxins are difurano coumarin derivatives (25) Presently 18 different types of aflatoxins have been identified with aflatoxins B_1, B_2, G_1, G_2, M_1 and M_2 being the most common (Fig. 1). Of these B_1 and G_1 occurs most frequently, B_1 being the most potent. The letters B and G refer to the fluorescent colours observed under long wave length UV light and the subscript 1 and 2 to the separation patterns of these compounds in TLC plates (26).

The aflatoxin molecule contains a coumarin nucleus linked to a bifuran and either a pentanone (AFB_1) or a six membered lactone as in AFG (Fig. 1). *A. parasiticus* produces all four of these aflatoxins but toxicogenic strains of *A. flavus* produce only AFB_1 and AFB_2. Two 4-hydroxylated derivatives of these last toxins have been found in groundnut and maize. These derivatives AFM_1 and AFM_2 (Fig. 1) were isolated from milk of cows fed on aflatoxin contaminated rations (100, 113). They have also been recovered from the meat, liver, kidneys and urine of various animals, and from eggs. Various other minor aflatoxins are produced by *A. flavus* in culture (144). Some are also produced in liver and other organs by reduction, oxidative hydroxylation, o-demethylation and epoxidation of B_1 (107).

3.2 Biosynthesis of Aflatoxins

It has already been stated that aflatoxins are secondary metabolites whose synthesis normally occurs through a polyketide pathway–a major pathway for moulds (16, 17). Aflatoxins are formed by a polyketide process that involves the condensation of an acetyl unit with three or more malonyl units with the loss of CO_2 and production of polyketide intermediates probably built up by cyclic processes similar to the fatty acid biosynthesis (181). The difference is that the reduction–elimination–reduction sequence is eliminated resulting in the loss of acetate oxygen (89) and thus giving a paraffin chain structure. This highly reactive structure undergo a series of condensation leading to formation of polyketides (136) ranging from triketides onwards. Most aflatoxins are nonaketides but their biosynthesis occurs through decaketide intermediates. This precursor then undergoes a direct cyclization and aromatization to form an anthrone–which is then oxidised to give norsolorinic acid.

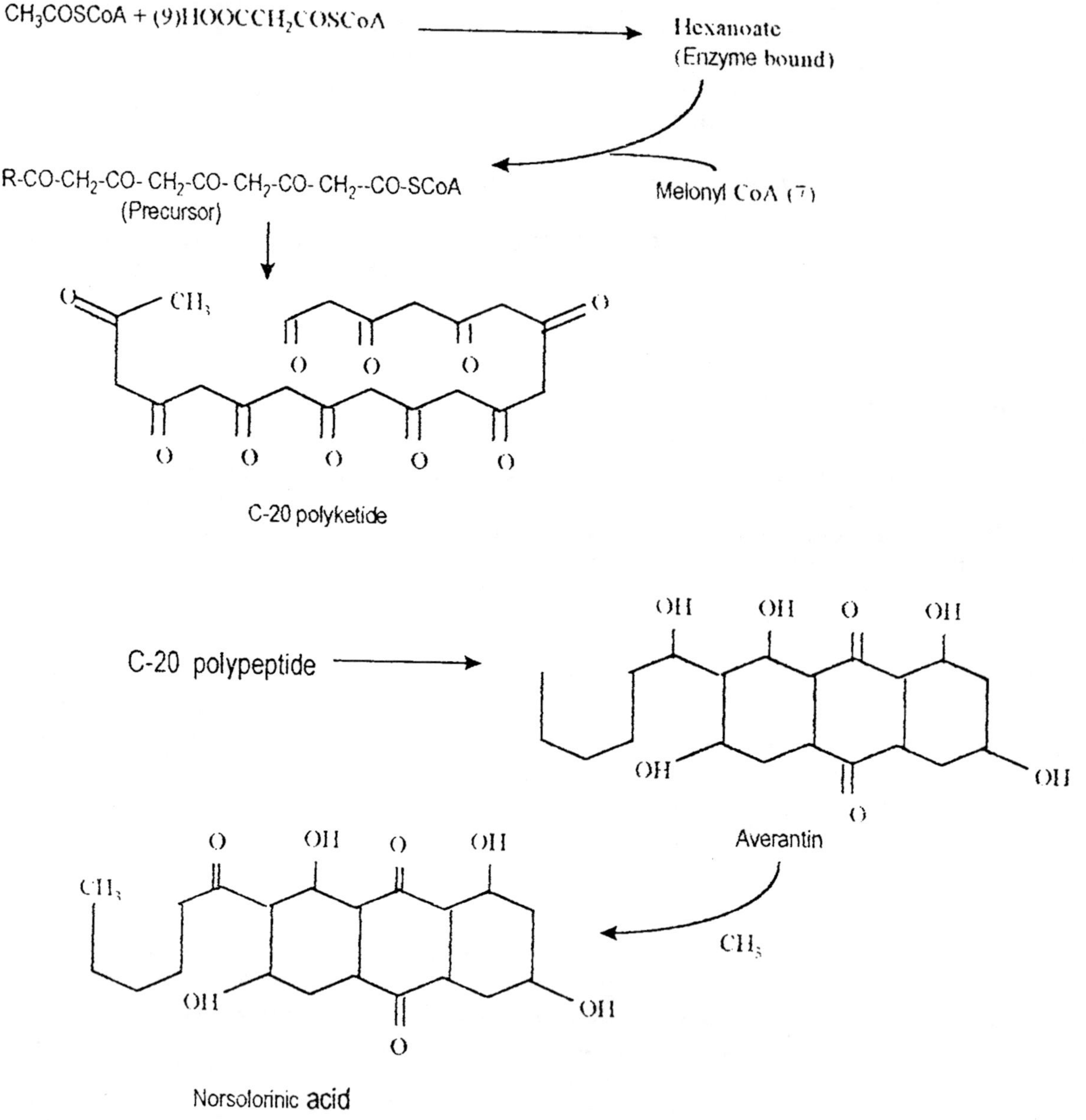

$CH_3COSCoA + (9)HOOCCH_2COSCoA$ → Hexanoate (Enzyme bound)

Melonyl CoA (7)

$R\text{-}CO\text{-}CH_2\text{-}CO\text{-}CH_2\text{-}CO\text{-}CH_2\text{-}CO\text{-}CH_2\text{--}CO\text{-}SCoA$
(Precursor)

CH_3

C-20 polyketide

C-20 polypeptide →

Averantin

CH_3

CH_3

Norsolorinic acid

The norsolorinic acid to formed undergoes several metabolic conversions involving several decaketide intermediates (averufin), versiconal hemiacetal acetate, (Versicorlin A etc.) and forms aflatoxin B_1. Most intermediates are enzyme-bound while the processes of cyclization, alkylation and reduction occur. The pattern of folding during polyketide formation being specific, they occur on specific enzyme surfaces for specific moulds. Divalent metal ions like Ca^{++}, Mg^{++} or Zn^{++} plays a key role in forming bridge bonding.

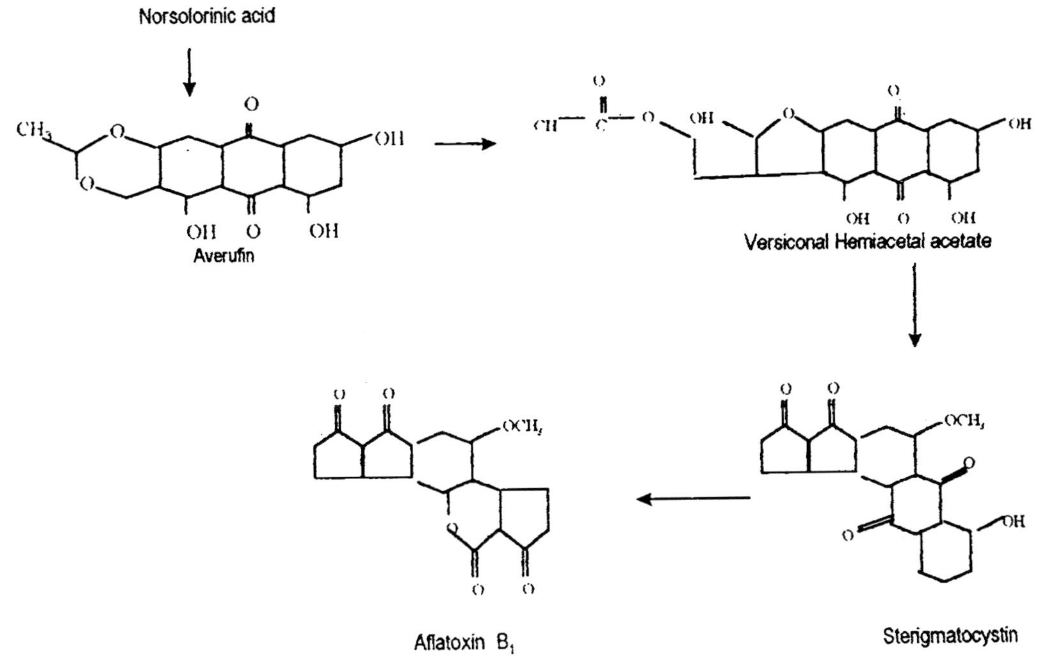

Norsolorinic acid

Averufin

Versiconal Hemiacetal acetate

Sterigmatocystin

Aflatoxin B_1

Once aflatoxin B_1 is formed, interconversions to B_2, G_1 and M_1 can occur (84).

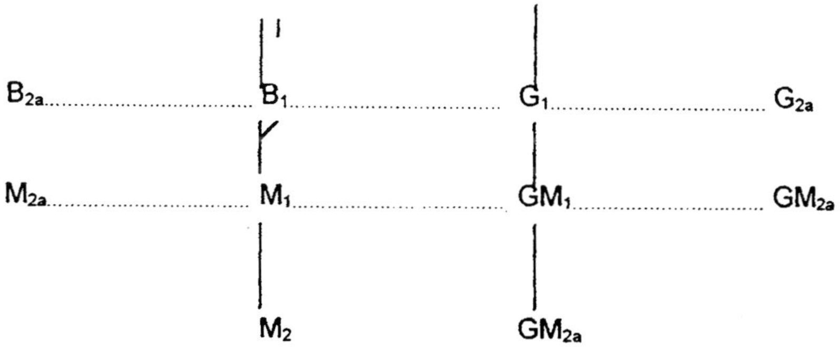

B_{2a} .... B_1 .... G_1 .... G_{2a}

M_{2a} .... M_1 .... GM_1 .... GM_{2a}

M_2 GM_{2a}

(From Ellis *et. al.*, 1991. *Cril. Rev. Food Sci. Nutrition*, 30 : 403 - 439).

Reference (64).

These pathways, some confirmed and some deduced, were largely traced using mutant strains of *A. parasiticus* suitably labelled ^{14}C intermediates.

4. DETECTION AND QUANTIFICATION OF AFLATOXINS

A large number of techniques have been developed with increasing degree of sensitivity to lower levels of aflatoxin content. It can now be estimated to picagram levels. The earlier methods of extraction and extract purification are reviewed by Stoloff (198) depending upon the material being analysed. The American Association of Cereal Chemists (AACC) joined hands with AOAC and International union of Pure and Applied Chemistry (IUPAC) and examined four methods, viz., Contaminats Branch (CB) method (66), the celite method (140), Best Food Method (BF) (209) and the SRRL method (160), having minimum detection levels of one to five Kg^{-1} sample.

Third generation methods capable of detecting levels below one $\mu g\ Kg^{-1}$ were described by Jacobson *et al.* (92)–the so-called FDA Beltsville method using whole milk. Since these early times more sophisticated methods have been developed, some for rapidity and ease of handling large samples, others for qualitatively and quantitatively detecting presence of traces in samples and these can be briefly summed up under the following heads:

4.1 Physicochemical method

4.1.1 Chromatographic methods

(i) Thin layer chromatography (TLC)

(ii) High performance liquid chromatography (HPLC)

(iii) Capillary gas chromatography (CGC)

4.1.2 Instrumental methods

(i) Fluorodensimetry

(ii) Spectrophotometry

4.1.3 Rapid Methods

(i) Blue green yellow fluorescence (BGYFO)

(ii) Minicolumn detection

4.2 BIOLOGICAL METHODS

4.2.1 Bioassays

(i) Cell and tissue cultures

(ii) Brine and shrimp larvae

(iii) 1-d old ducklings

(iv) Trout

(v) Microorganisms

42.2 Immunoassays

(i) Radioimmunoassay (RIA)

(ii) Enzyme-linked immunosorbent assay (ELISA)

(iii) Polymerase chain reaction (PCR)

Whatever be the method used the key features that determine analytical precision include:

(i) Sampling

(ii) Sample preparation

(iii) Extraction

(iv) Purification / Cleanup

(v) Development / Separation

(vi) Quantification and confirmation.

Samples have to be representative. Besides, analysis of cereals, nuts and fruits need reduction in particle size by grinding or milling prior to extraction (27). Extraction involves removal of aflatoxin from the sample for quantification without altering its properties. The solvents used may vary with the nature of the samples. Thus:

Sample	**Solvent**	
Seeds, Copra,	Acetone	: Water (85 : 15)
Coconut etc.	Chloroform	: water (91 : 9)
Nuts (pea, Pistachio)	Chloroform	: Water (91 : 9)
	Methanol	: Water (55 : 45)
Powdered milk	Acetone	: Water (70 : 30)

Particularly for precise quantifications, purifications is necessary. This may be achieved through liquid-liquid partition system. Precipitation may be achieved using lead acetate (162), silver nitrate (179) and silica gel column chromatography (65). Others used include aluminium oxide, silicic acid, sephadex, floriset celite, liquid column chromatography, aluminium oxide and silica gel TLC.

Mayura and Murthy (128) used all four aflatoxins from 1 X 1 neutral alumina column using chloroform : acetone : methyl alcohol (6 : 3 : 1). Mehan *et al.* (131) used the BF, CB, SRRL methods for groundnut and also its meal and found that BF and SRRL gave better efficiency. Distinct approach is used for HPLC estimation (86, 206). Concentration is achieved by any of the several process available like :

(i) Rotary evaporation under reduced pressure

(ii) Steam bath evaporation

(iii) Use of an enclosed hot plate to evaporate solvent.

Once these steps are completed the final analytical tools to be used is determined by several factors. Thus, two dimensional TLC is good were there are high amount of contaminants (coffee, cocoa). Detection is based on the fluorescent properties of the aflatoxins. Quantification by visual or other colour intensity measurements has only a 20% accuracy. Better results are now obtained by fluorodensitometers. HPLC is more commonly used now for cotton seeds (5), peanut products, pea and corn (161); milk and milk products (136) and blood of manmals (203). Low levels may be detected by attaching sensitive detection and sophisticated data retrieval equipments to the HPLC.

The CGC technique was good for B_1 and B_2 but was far less sensitive for G_1 and G_2 (71).

Among the rapid methods BGYF test is based on fluorescence of cotton seed or corn colonized by *A. flavus* when viewed under UV light. It is used as a rapid detection test for aflatoxin contamination of cereals. The fluorescence is due to Kojic acids produced by mouls converted to a fluorescing substance by plant tissue peroxidases. That is why BGYF is only a presumptive test for aflatoxin presence. Minicolumn detection technique is usually resorted to for detection of total aflatoxins, thus screening a large number of samples.

The several bioassay methods available today can also be used as toxicity test for aflatoxins. The one day old ducklings have been successfully used for detection of aflatoxin B_1, B_2, G_1, G_2 and M_1 (136). Use of trouts is expensive.

Commonly sensitive bacteria used for bioassay are *Bacillus megaterium* and *B. brevis* (28). A standard curve may be created similar to that used for antibiotic assay. Other microbes used include *Escherichia coli, B. myucoides,* etc.

Aflatoxins are non-antigenic but are able to cause antibody response in test animals (136). RIA technique has been successfully used by Chu (34) and Thean *et al.* (202) and that of ELISA by Lawellin *et al.* (112) and Pestka *et al.* (125).

All bioassay methods have some limitations. Thus, cell and tissue culture methods and use of shrimps, ducklings or microbes are non-specific and less accurate, use of animals specially trouts, as also immune response techniques are expensive and needs expert handling.

Recently polymerase chain reaction technique has been used by Shapira *et al.* (183)for detecting aflatoxicogenic moulds in grains. They identified 3 key genes, ver-1, omt-1 and apa-2, coding for key enzymes, as also a regulatory factor in aflatoxin biosynthesis. Three primer pairs, each complementing the coding portion of one of the genes, were generated. DNA extracted from mycelia of 5 *Aspergillus* spp., among other, were used as PCR templates. DNA extracted from groundnut, maize and 3 insect spp. were also tested. DNA amplification was achieved with *A. parasiticus* and *A. flavus* in all 3 primer pairs. These workers concluded that genes involved in the aflatoxin biosynthetic pathway may form the basis for an accurate, sensitive and specific detection system, using PCR, for aflatoxicogenic strains in grains and foods.

5. BIOLOGY OF ASPERGILLI, GROWTH AND AFLATOXIN PRODUCTION

The question that needed convincingly documented answers were (i) Is good growth of the producer a prerequisite to high aflatoxin production? In other words, are the parameters of optional growth same as those of aflatoxin production? (ii) Are controlled condition experiments reflected in their *in vivo* functioning? (iii) Do over riding factors of strains, strain-hot interaction or microbe-microbe interaction play a determining role in aflatoxin production? Answer to all these questions have been intensively investigated and reviewed (16, 64). The salient observations are presented briefly.

5.1 Taxonomy of Aflatoxin Producers

There appears to be some confusion in relation to identity of *A. flavus* and *A. parasiticus*. Thus to quote Diener (55) "*Researchers have frequently failed to distinguish between the two species* .... Therefore throughout this review we use *Aspergillus flavus* to denote both species." Hence a taxnomic description largely taken from Domsch *et al.* (57) will be in order :

Aspergillus flavus group included 10 species. Within them following description accounts for *A. flavus* and *A. parasiticus* :

Conidia 5-6 μm or more in diameter; old colonies more or less brown; conidial mass on CzA media +0.5% anisic acid regularly pink

(*A. oryzae* subgroup) -------------------------- 2

Conidia 4-5 μm in diameter, old colonies remaining yellow-green or greenish brown. Conidial mass sometimes turning pink on anisic acid

(*A. flavus* ----- group) --------------------------4

Metulae absent; conidia prominently echinulate; sclerotia absent—*A. parasiticus.*

Metulae present in many conidiophores; conidia finely echinulate; brown irregularly shaped sclerotia sometimes present—*A. flavus.*

5.1.1 Primary taxonomic descriptions of *Aspergillus flavus*

Features: Often treated as *A. flavus-oryzae* because they are difficult to distinguish. Several subdivisions are of toxicological significance. Colonies reach 30-70 mm in ten days at 24-26 °C on CzA or 60 -70 mm in MEA; characteristically yellow green. Conidiophores hyaline, rough walled. On larger conidiophores a layer of metulae supports phailides. No such metulae to be seen on smaller ones. Conidia globose to subglobose, finely roughened to echinulate; SEM shows irregularly curved short ridges on conidia (63). Some isolates produce light to dark red-brown sclerotia, 400-700 μm in diameter and formed at 34°C especially with 3% sucrose and 0.5% $NaNO_3$ in the medium (139). Seven linkage groups have been described through parasexual cycle. The sterols cholesterol, ergosterol and 5, 7-ergostadienol were found in aflatoxin-producing and non-producing isolates (168). Analysis of soluble protein fractions did not give support to the assumption of a close relationship with *A. parasiticus* (109). A diagnostic medium consisting of tryptone, yeast extract and ferric citrate has been described (23).

A. flavus has a relatively high competitive ability (95) and reported from rhizosphere of many crops, vegetables, spices, nuts and seeds of many plants including groundnut, soybean, green coffee beans, corn, wheat, barley, sorghum, cotton, grasses, olives, tomato, sweet potato, etc. It has also been found in alimentary canal of man and commonly in insects (78).

An exogenous carbon source is required for germination; a mixture of amino acids was more suitable than glucose and NH_4Cl (153). Conidia germinates between 12 -37°C at a minimum of -270 bars and do not survive-350 bars at 29°C (201). The growth limiting water potential is -310 bars and optimum-30 bars. Best pH for growth is 7.5 and for conidia production 6.5

Aflatoxins are produced in the range of 20-40°C (50) with specific optima for different compounds and derivatives. Aeration favours aflatoxin production (73). Phenolic amino acids and tryptophan acts as precursors in amino acid synthesis (1). Fungus has a high tolerance for allyl alcohol and formalin (137). *A. flavus* is sensitive to γ-radiation (138, 159) depending on the age of the conidia or mycelium (134).

5.1.2 Primary taxonomic discritation of *Aspergillus parasiticus*

Colonies reach 2.5-4 cm in 8-10 days with more pronounced green tints than in others of this group. Conidiophores are coarsely roughened and metulae absent; conidia 3-4 nucleate with prominently echinulate walls, discoloured pinkish in anisic acid fortified medium. SEM of conidia shows them to be spiny (63).

A. parasiticus has been isolated occasionally from the cultivated soils of India (61, 73, 132), Japan, Italy, Argentina, USA and Czechoslovakia. It was found in rice, groundnut, pecans and in the rhizosphere of wilted pineapple plants. Occurs also on insects.

Optimum growth temperature is 30^{0}C while maximum aflatoxin production occurs at 25^{0}C. Ten percent glucose is good enough for growth but for good aflatoxin production upto 30% glucose may be required. Identified metabolites include the volatile flavour compounds 1-octan-3-ol, 2-octan-1-ol, 3-octanol, 3-methyl butanol (102), kojic acid, norsolorinic acid (117), aflatoxins B_1, B_2, G_1, G_2, the B_1 related parasiticol (139) and parasiticolide (77).

Infected rice stored at -140 bars up showed presence of aflatoxins. Invasion of seeds however occurred at much lower water potentials (-350 bars) (20). Production of aflatoxin is lowered under near anaerobic conditions (187). Presence of cholesterol, ergosterol or 5,7-ergostadienol was not related to production of aflatoxins. Growth was restricted in presence of 10% NaCl. (108).

For a monographic account of such asperigilli one is referred to Raper and Fennal (171) and Smith and Paterman (192).

5.2 Growth and Aflatoxin Production in Culture

The cardinal environmental factors that affect both growth and aflatoxin production are temperature, light pH, atmospheric gases and water activity (Aw.).

5.2.1 Temperature effects

Generally speaking the optimal temperature for fungal growth and aflatoxin production are 25-30^{0}C. The optimum temperature for growth of *A. flavus* has been reported to be 29-35^{0}C (167) and that of *A. parasiticus* is 25-30^{0}C (143), on the other hand optimum temperature for aflatoxin production for *A. parasiticus* to be same as growth.

Both low and high temperatures affect production of aflatoxin ca. low$<$ 13^{0}C and high $>$ 42^{0}C (207). Generally B_1 is produced at higher temperatures than G_1 whose production is optimal at 25^{0}C (48). When one recalls that B_1 is the primary aflatoxin from which G_1 is derived it becomes obvious that enzyme system responsible for this transformation is affected at higher temperatures. In nature, however, there usually significant variations in temperature compared to controlled condition experiments and hence the yield of aflatoxin is likely to vary. Usually at lower temperatures yield of B and G are in equal amounts but at higher temperatures B production becomes dominant. Thus, Dinner and Davis (51) showed that B_1 and G_1 are produced in equal amounts at 15-18^{0}C while they are produced in a ratio of about 12 : 1 at 32^{0}C.

5.2.2. Light effects

The few studies on effect of light shows that aflatoxin production increases five folds in dark (94). On the other hand, Om Prakash and Siradhana (143) reported that B_1 production in hybrid Ganga-5 maize was normal in daylight compared to darkness.

Light in general, influences production of aflatoxins both in liquid and semisolid medium; the level of influence largely depending on photochemical effects on the medium (31). A detailed investigation by Bennet *et al.* (13, 14) showed that conidiospores were more abundant in both the species when exposed to light. Fungal photoresponses were highest in the blue region. The effect of light was not as significant at intermediate temperatures as it was at lower and higher (20 - 25°C) side of the range. Interestingly Nkana *et al.* (141) showed increase rate of B_1 destruction in light.

5.2.3 pH effects

Generally aflatoxin producing moulds can grow over a wide range of pH (1.7 - 9.3) with an optimum of around 3.5 - 6.0 (116). However, aflatoxin production varied at different pH levels and observations do not complement. Thus, it has been recorded that 26 times more toxin was produced when the moulds were grown at an initial pH of 4.0 rather than at 7.4 (95). Aflatoxin production by *A. parasiticus* increased with decreasing pH of a synthetic medium up to around 6.0 favoured production of B_1 while higher pH favoured production of G_1. Types of medium is an over riding factor.

5.3 Effects of Atmospheric Gases

CO_2 has been shown to catalyse the malonyl CoA involved in fatty acid and ring compound synthesis and hence one would expect an increase in aflatoxin synthesis at high CO_2 concentration. However, concentrations of CO_2 above 20% inhibits mould germination while more than 10% CO_2 suppresses toxin synthesis (199). A decrease in atmosphere in atmospheric O_2 to less than 20% or its increase above 90% inhibits aflatoxin production (187). Increase in gaseous N also reduced its production (67).

The concensus is that maximum aflatoxin production not only depends on mould strains but also an appropriate concentrations of head space CO_2 and O_2. This is due to the fact that aflatoxin production is neither associated with maximal concentration of O_2 present, not with maximal growth (187).

5.4 Effects of Relative Humidity, Moisture and Water Activity

Relative humidity (RH) and moisture content seems to play a key role in aflatoxin production. Thus, though a RH of 85% was optimal for growth (8) a higher level of around 95-100% was required to maximise aflatoxin production. Higher RH increased aflatoxin production in peanuts while optimum toxin production in Indian red pepper occurred at 95% RH while none was produced below 70% RH (182).

Similarly a moisture content of 15-30% in seed grains stimulated abundant aflatoxin production but below 13% none was produced as fungus failed to grow (12).

Water activity (Aw) is the ratio of water vapour pressure of the substrate to that of pure water at the same temperature and pressure. At low Aw water is bound by salts, sugars, proteins, etc. and is not available for fungal growth. Aflatoxin production ceases below an Aw of 0.85 although a certain amount of fungal growth still occurs at Aw of 0.78-0.80. Minimum Aw for *A. flavus* has been shown to be 0.78-0.84 while for *A. parasiticus* it is 0.84 for growth and 0.87 for aflatoxin production (49). Optimum production in both the moulds occur at 0.95-0.99 (51).

5.5 Effects of Chemical Factors

The three major chemical interfaces that modulate aflatoxin production are the substrate, nutrients and anti fungal agents.

5.5.1 Substrates

Generally speaking aflatoxin producing microbes prefer carbohydrate rich solid substrates such as flour, cocoa, corn meal, cheese, groundnut, cotton seed, cassava, nuts, rice, oilseeds, pumpkin seeds, coconut (specially its copra), sausage, meat pie, bread, milk, macaroni, cooked meat, frozen patties etc. This is specially true for *A. parasiticus. A. flavus*, however, has been reported to utilize low amounts of carbohydrates and still produce substantial amounts of aflatoxin (145). Prolonged incubation at temperatures around 25°C degrades proteins to its component amino acids which serves as source for both N and C. However, release of large amounts of NH_3 usually hinders aflatoxin production (56).

5.5.2 Nutrients

It is the nutrient composition of the substrates that affect aflatoxin production. Thus, preferred C sources for such production has been reported to be simple sugars like glucose, fructose and sucrose (*A. flavus*) while for *A. parasiticus* it is xylose and mannose. Amino acids like alanine, aspartate, glycine, glutamate, glutamine and proline along with some bivalent metal ions like Ca^{++}, Mg^{++} and Zn^{++} stimulate aflatoxin production (174). Zinc possibly acts through regulation of formation of the intermediates in aflatoxin B_1 biosynthesis, i.e. versicolin A and C. Some of the B group vitamins other than riboflavin has been shown to stimulate aflatoxin production (10).

5.5.3 Pesticidal agents

Generally growth appears to go hand in hand in aflatoxin production and many of the fungistats like propionic acid and sorbic acid (a food preservative) act as inhibitors. Low concentrations may, however, stimulate such production through effects on inhibition of TCA cycle leading to accumulation of acetyl CoA, a precursor to aflatoxin synthesis (70) Dichlorvos acted as an inhibitor to toxin production in rice, wheat, maize and ground nut (170). Devi and Polassa (45) reported inhibitor effects of carbendazim on growth and toxin production. Glyphosate herbicide is used extensively today as a broad specturm, safe (?) weedicide. It has been reported that this herbicide may increase aflatoxin synthesis (80) upto 500 ppm level in wheat straw and couch grass (113, 122).

5.6 Effects of Biological Factors

Three biological factors play a major role in aflatoxin production : (a) Genetic strain variability; (b) Inoculum size and (c) Associated microflora.

5.6.1 Genetic strain variability

It has been convincingly and repeatedly shown that aflatoxin production is not only species dependent but also strain dependent. Typically aflatoxin B_1 and G_1 are present in more amounts than their dihydroxy derivatives B_2 and G_2 while *A. flavus* primarily produces B_1 and B_2. *A. parasiticus* produces B_1, B_2 along with G_1 and G_2.

Although very little study have been conducted on identifying specific genes for aflatoxin production, the differences are attributed to the presence of diploids and subsequent diploid and haploid mitotic recombinants.

Papa (146, 147, 148, 149, 150, 151, 152) in a series of papers made interesting genetic observations. Thus, he found a mutant that produces more B_2 than B_1. Further analysis showed that this is mediated by a single diallelic gene locus in the mutant in the 8th linkage group. On the other hand, parasexual cycle of *A. parasiticus* involved 6 diploids with 12 genetic markers illiciting independent segregation. The results revealed that aflatoxin-less mutant (afl-1) was dominant but all others were recessive in diploids. Their pairing considerably reduced aflatoxin production.

These results revealed the need for clear understanding of strain differentiation in order to develop technologies for prevention, restriction or propagation of hazardous strains.

5.6.2 Incoculum size

It has been observed that in a synthetic medium the amount of aflatoxin produced is dependent upon mycelial branching and differentiation. Thus, nutrient depletion or staling products increase lateral branching which in turn leads to production of higher amount of toxins (72, 93). Maximum aflatoxin in cultures have been reported with an inoculum load of 10^3 spores/ml (103), both higher and lower dosage resulting in lower aflatoxin yield.

5.6.3 Associated microflora

Moss and Frank (135) stated that the competing mould microflora may affect aflatoxin production by

(i) Metabolisation of aflatoxin produced

(ii) Alteration of the metabolism of the producer

(iii) Compete for substrate necessary for aflatoxin formation

(iv) Rendering conditions unfavourable for aflatoxin production.

Nocardia and *Scopulariopsis brevicaulis* have been shown to detoxify aflatoxins when they grow with the producers. Critical studies with maize showed that *A. flavus* failed to produce aflatoxins in presence of either *A. niger* or *Trichoderma viride*. *Rhizopus nigricans* and *Saccharomyces cerevisiae* were reported to stimulate both growth and aflatoxin production by *A. parasiticus*. *Acetobacter aceti*, however, stimulated both the processes in *A. flavus*. In maize kernels, interference with toxin production by associated microbiota has been recorded in India (17, 32, 33, 40, 46). These results have future practical implications.

Thus in rice, inspite of normal growth of *A. parasiticus*, presence of *A. chevalieri* of *A. candidus* reduced aflatoxin production probably through competition. On the other hand presence of *A. niger* led to high amounts of citric acid produciton that reduced to pH to <3.0, low enough to inhibit aflatoxin production. Horn and Wicklow (87) isolated an inhibitory factor from *A. niger* that reduced aflatoxin formation. Restriction of aflatoxin producers by *A. niger* and *Rhizoctonia solani* has been recorded.

5.7 Epidemiological Features of Aflatoxin Production

Both *A. flavus* and *A. parasiticus* are closely related and are found contaminating seeds and plant debris in field, during harvest, in storage or during processing. They are also pathogenic to some insects (215). The former is adapted to aerial and foliar environment (corn, cotton seed, tree fruits) while the latter is to soil environment (groundnut). Clearcut taxonomic differences, as discussed earlier, has not been defined so far although the former produces aflatoxin B^1 and B^2 only while the latter produces these along with G_1, G_2 and M_1. The major thrust area of research has been on groundnut (53) maize (54) and cotton (37, 190) and their epidemiology has been adequately reviewed (55, 96).

The preharvest inoculum during spring appears to be the spores in soil, mycelium overwintering in plant debris and litter in soil, insects or as sclerotia of *A. flavus*, in soil. A population upto 165 propagules per g occurred in soils adjacent to rye and groundnut fruits, 110-200 in infested debris cotton fields. The primary inoculum may be carried on insects like *Chorochoa saye* or *Lygus herperus* in cotton fields (197).

Sources of secondary inoculum in cotton appeared to be floral bracts (fimbriate regions), floral nectaries and bracteeoles (190), on leaves damaged by leaf feeding insects. In aflatoxin-high fields more spores of producers were trapped (116) although aerial flora showed occasional fluctuations both in cotton and maize. In maize, spore production increased during the tillage and harvesting and was double in non-irrigated plots over irrigated ones.

Apparently dispersal is primarily through air-borne conidia. In cotton it has been shown that wind driven soil can disperse propagules to developing cotton bolls and infect when followed by rain. Further epidemiological details in these three crops are phytotrons.

5.7.1 Groundnut

Groundnut flowers inoculated with washed conidia initially were restricted primarily to the stigma and pollen adhering to it. With time it grew down the style and eventually reached the top of the ovary. In some flowers the stamens were colonised (211). Under natural conditions the anthers could form a relatively large area for colonization and these are well placed to provide secondary inoculum for other flowers. Insects may play a role in infection of the stigmata. Infection could also occur in the aerial pegs. Direct invasion of the fruit in the geocarposphere led to eventual colonization of the kernels.

The major factors that epidemiologically modulate infection (and consequently aflatoxin production) are temperature, moisture and time. The colonization was inversely proportional to maturity and was greatest in groundnuts grown under drought conditions with elevated geocarposphere temperatures (177).

However, excessively high temperature in the geocarphosphere could prevent aflatoxin production even when the fungus was present.

Temperature limits were identified as :

Undamaged kernels-lowers limits: $<25.7^0C$; Higher limits : $30.5-31.3^0C$

Damaged kernels-Mean : $26.3-29.6^0C$

None : $> 31.3^0C$

Obviously high temperature inhibited growth of afla fungi (91).

Drought effect were only apparent as drought heated and cooled soils showed no aflatoxin above certain temperatures. However, water stress during last 40-75 days before harvest contributed to aflatoxin in sound mature kernels (221). The threshold of aflatoxin contamination was between 20-30 days before harvest at optimum temperatures (178). Neither drought, nor elevated temperatures alone resulted in aflatoxin production in sound, mature kernels (19).

Other factors that affected infection and aflatoxin production was insect damage by the lesser corn stalk borer (*Elasmopalus lignosellus*) specially during periods of drought (120). Wounding by insects may provide infection courts and also allow kernels to dry down to levels more favourable for the growth of the fungus and aflatoxin production.

On the other hand, *A. niger*, a common competitor for *A. flavus*, was also favoured at high temperatures and more prevalent in irrigated plots (which incidentally hardly showed aflatoxin production) antagonised *A. flavus*.

5.7.2 Maize

A. flavus invades maize ears through the silks. Even in uninoculated ears, the emerged silks carried the fungus and the extent increased with degree of emergence and time (123). Both natural and green house observations revealed that infection proceeded from tip of the ears downwards towards the base of silks, then the glumes and by the milk stage of the kernels, their surfaces were colonised. It, however, reached the pith of the cob. Phytotron studies revealed a day-night rhythm of 34/30°C to be favourable. In fields it took 4-13 days to reach the base. SEM showed green silks to be poor substrate for colonisation, either due to inhibitors present or lack of nutrients. Yellow brown silks were susceptible to infection especially near the pollen grains (123). Colonisation was restricted to the parenchyma and oriented parallel to the axis of the silks. Penetration usually occurred through cracks and intercellular gaps. The nutrient-rich pollens played a key role in infection of silks and subsequent spread. From silks the fungus enters the kernels through the stylar canal. The presence of BGYF in small area of the crown shows the possibility of its additional entrance mode through scars where the silk was attached (96, 123). It is thus possible that the fungal mycelium grows from the silk attachment site onto the adjacent pericarp and only colonises the kernel surface and does not actually penetrate the stylar abscission layer into the kernel. However, regular association of the fungus at the top of the kernel suggests that *A. flavus* enter through this region. SEM revealed presence of fungus at the tip cap (pedicel), its intercellular spaces and across the surfaces of tip cap parenchyma cells, sclereids and protoxylem elements (123). Plating such kernels revealed sporulation in the region of the germ (68%) and tip (48%) and least over endosperm (12%) (69). Initial growth, however, was confined to the germ region. High temperature appears to be a basic requirement for *Aspergillus* contamination and for maize it ranged between 32-38°C. Phytotron studies revealed need for a night temperature above 30°C. A good day-night regime noted was 34/30°C. Infected kernels were distributed evenly all over the ears perhaps a little more at the tip. Critical time was perhaps 16-24 days after silking when kernel moisture was approximately 30% (154) Aflatoxin levels were also higher in ears inoculated these weeks after silking and in kernels growing at a temperature regime of 30/26°C.

As in groundnut, maize exposed to drought were more susceptible to infection (44). Water stress before and after inoculation increased the level of aflatoxin by ten times in the kernels. Aflatoxin production may occur at RH 85% but is more significant at increasingly higher levels upto 93% (119). Although infection occurred at early milk stage, aflatoxin production was highest at late milk stage (4) when kernel moisture is about 30%.

Other factors influencing aflatoxin production in corn are:

(i) Fertility stress contributes to aflatoxin contamination

(ii) Low levels of N promoted aflatoxin levels

(iii) So did dense population of plants and reduced fertilization

(iv) Weed canopy promoted aflatoxin contamination.

5.7.3 Cotton

Main point of entry for cotton were the flower buds, specially the corolla and the involucral nectaries (105). Unlike groundnut, stigmatal inoculation was of little or no significance. When inoculated on the nectaries at staggered post anthesis dates, the harvested bolls showed higher infection upto 25 days post anthesis but not later (104). This is possibly because of the degeneration of the funiculus. Physical and biochemical blockage may hinder nectary invasion. Then how were cotton infected? It was shown that cotton plants could be infected through the fresh cotyledonary scars of seedlings of 4-8 leaf stage. Seventy eight percent of the bolls with seeds contaminated also had the fungus in subtending stems and peduncles. Fungus was shown not to grow downwards readily but preferred a upward movement. Other avenues of infection could be the fibres on partial opening as indicated by the presence of BGYF in the lint (7). Inoculating through sutures did not lead to aflatoxin detection in the seeds. Aflatoxin levels and the ratio of toxic to non-toxic seed were similar in naturally infected bolls and bolls wound inoculated 33 days after anthesis. Bolls collected from lower most 1/3 of cotton plants represented most of the cotton plant infection and almost all the aflatoxic seed (7). Symptoms of infection was rare in fully fluffed bolls and was associated with partially opened ones.

Again temperature appeared to be a key factor in preharvest aflatoxin contamination of cotton (174). Daily minimum temperature above 24^{0}C in combination with precipitation exceeding 20-30 mm resulted in high levels of aflatoxin in cotton seed. High moisture and/or high RH was essential for spore germination and fungal proliferation. Lack of infection till bolls opened to about 1-3 mm was considered to be a moisture related phenomenon. Moisture is more critical for boll opening than temperature. Areas lacking in aflatoxin in cotton seeds resulted possibly from higher humidity leading to high pH (8.0 - 10.2) in aqueous extracts of cotton fibres, where as favourable pH for aflatoxin production is 5-7.8 (124). The highest incidence of seed infection occurred with water potentials between -1.6 and -1.9 MPa which are associated with drought.

The relationship between pink boll worm and other insects to *A. flavus* boll rot and aflatoxin contamination revealed that exit holes made by the mature larvae predispose bolls to infection and production of aflatoxin contaminated seed (175). Controlling the pink boll worm by insecticides decreased the number of BGYF cotton seed and aflatoxin contamination (129).

Other factors affecting aflatoxin production in cotton are:

(i) Unknown physical or chemical factors seem to mediate high aflatoxin levels at a fixed time;

(ii) Aflatoxin being concentrated in one to two seeds in the apex half of the boll, even though the fungus is present in the locks, shows that aflatoxin formation in seed within infected locks and bolls is highly selective, about which little is known.

6. TOXIC MODE OF ACTION OF AFLATOXINS

We have seen that the aflatoxin producing asperigilli, particularly *A. flavus*, can cause expression of symptoms in plants. But the primary thrust in aflatoxin research came from the knowledge of its ability to cause toxicoses, particularly liver cancer in domestic animals and man.

6.1 Biological Mode

The mode of action has been studied from biological, biochemical and genetic point of view. Of these we have already seen albeit briefly, the biological effects under section II, the major effects being:

(i) Carcinogenic

(ii) Mutagenic

(iii) Teratogenic

(iv) Hepatotoxic

(v) Chronic and acute aflatoxicosis.

6.1.1 Afltoxicosis

The last named needs a little more elaboration here: Acute toxicity occurs due to either administration and chance intake of very high dosage of aflatoxin as was seen in the turkey back in 1960, leading almost invariably to mortality. On the other hand chronic aflatoxicosis occurred when prolonged consumption of low to moderate level of aflatoxin occurred unknowingly. The primary symptoms of these two types of aflatoxicosis were listed by Ellis *et al.* (64). Acute aflatoxicosis symptoms included:

(i) Enlarged, fatty, pale decolorised livers

(ii) Failure of blood clotting system leading to haemorrhages

(iii) Reduction in total serum proteins of the liver

(iv) Increase in specific serum enzymes of the liver

(v) Accumulation of blood in the gastrointestinal canal

(vi) Hyperplasia of the bile duct.

Other symptoms included glumerural nephirits of the kidney and congestion of lungs.

The primary symptoms of chronic aflatoxicosis were listed as:

(i) Congested liver with haemorrhagic and nectrotic zones

(ii) Proliferation of the hepatic parenchyma and epithelial cells of the duct

(iii) Congestion of kidneys that show occasional haemorrhagic enteritis.

The net result of chronic aflatoxicosis was reduced growth rate and reproductive efficiency.

A 3rd type of aflatoxicosis may occur due to secondary effects resulting from regular consumption of low to very low levels of aflatoxins. The salient indicators are :

(i) Impairment of native resistance

(a) Reduction in the ability of macrophages to function as efficient phagocytes

(b) Reduction in production of non specific humoral substances such as complements.

(ii) Impairment of inmunogenisis. This involves

(a) Affection of the cell-mediated immune system that impairs the effectiveness of vaccination set up in animals.

Such damge to defence mechanisms of the body would contribute to the observed reduction in resistance of animals to viral, bacterial and *Candida* infections.

6.2 Biochemical Mode

Of the biochemical mechanisms, the most plausible, DNA binding as a possible mode of action (195), is considered separately.

Other biochemical effects at the cellular level include the disruption of energy production by B_1, B_2, and M_1 through inhibition of the cytochrome electron transport. B_1 and M_1 also uncouple oxidative phosphorylation. B_1 has been shown to affect the mitochondrial permeability (90). These result in inhibition of the enzyme ATPase leading to reduced formation of ATP (136).

Reduction in hepatic glycogen due to aflatoxicosis has been ascribed to:

(i) inhibition of glycogenesis

(ii) depression of glucose transport in liver cells

(iii) acceleration of glycogenolytic processes.

6.2.1 Carcinogenic effects

Before getting into the biochemical mechanisms it will perhaps be fruitful at this juncture to summarise the carcinogenic effects on animals and humans.

6.2.1.1 Animals

Experimental simulation of carcinoma in trouts, ducks, rats and monkeys through dietary uptake of aflatoxins is documented, the primary target being liver and secondary targets the kidney, pancreas and other organs. The potency of B_1 and M_1 varied with the type of animal, rats being much more sensitive to B_1 than trouts when compared with M_1. Generally, the most prominent, B_1 causes:

(i) multicellular hepatocentric carcinoma

(ii) haemorrhagic and necrotic tumourogenesis

(iii) chronic cirrhosis of the liver

(iv) reduced weight gain

(v) decreased resistance to infectious disease

6.2.1.2 In humans beings

Primary hepatocellular carcinorma (PHC) has been recorded in central and southern Africa, Thailand and Indonesia.

Once aflatoxin B_1 is taken in, it reacts with a specific cytochrome P-450 mono-oxygenase to form as C_2-C_9 epoxide which is believed to be the active form of B_1. Electrophilic attack of the N-7 position of guanyl residues in DNA by the epoxide is likely to be the major cause of lesions which appear to account for the mutagenic and carcinogenic nature of B_1. B_1 mutagenetic effects have been noted in bacteria, algae, fungi and insect species. Point mutations are common. Frameshit mutation is also possible. The hydration of epoxide forms a dihydrodiol which may form Schiff's bases by reacting with the amino groups of bases or, as seen *in vitro*, the dihydrodiol binds to DNA and the product acts as a strong mutagen (196).

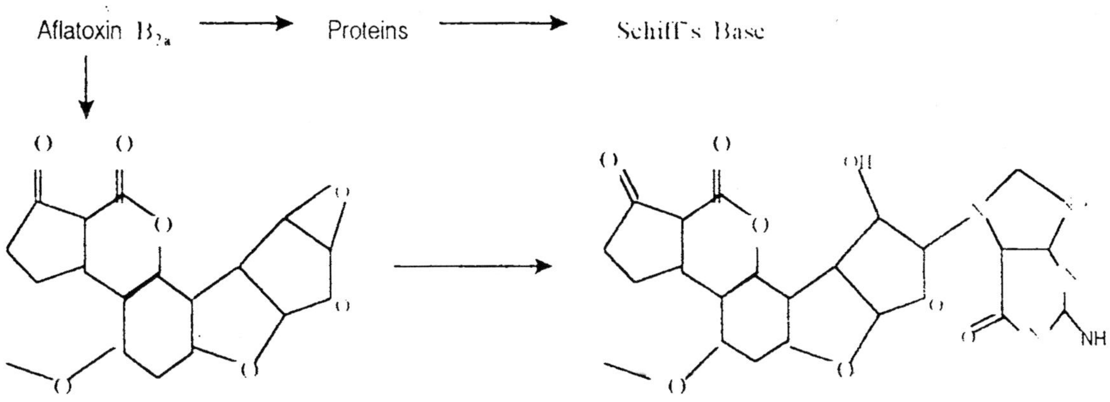

Aflatoxin B_{2a} ⟶ Proteins ⟶ Schiff's Base

AFB$_1$2,3-oxide

However, it is likely that such a reaction, recorded *in vitro*, may not occur at all in animal system because of the high reactivity of the product which may lead to immediate sequestration by proteins and thus detoxifying them. B_1 and its derivatives are also known to bind histones. ribosomal RNA and proteins. The toxicosis of B_1 to liver is because it affects DNA synthesis at much lower concentration than required for affecting protein or RNA synthesis. Binding of B_1 or its metabolites to DNA and/or proteins causes a modification of DNA protein template activity and inactivation of DNA polymerases and other necessary enzymes for DNA synthesis (88). On the other hand, inhibition of RNA synthesis affects synthesis of several key inducible enzymes and components of immune system. Deficiency of complement fixation, delayed production of interferon, reduced levels of IgA and IgG has been noted in chickens. Phagocytic activity of reticuloendothelial cells are also reduced.

7. MANAGEMENT APPROACHES-PREVENTING AFLATOXIN PRODUCTION

We have seen that aflatoxin is produced in crop plants by fungi that attack them during the growing phase or colonise the economically important parts during harvest, drying, storage or processing. During any of these phases, given suitable environment, aflatoxin(2) are produced by the producer. From the producer it passes into the first order sink–processed food and feed and there may undergo changes that produce stronger or weaker secondary metabolites. Finally, it reaches the terminal sink–the consumers where, depending upon the nature of the metabolite and its concentration, it may cause various disease syndrome and death mostly related to mutations and carcinogenicity of the liver tissues. It is obvious that earlier the producer proliferation is blocked in this chain, greater will be effectiveness of management. We therefore briefly summarise the various approaches being developed and tested for reducing/eliminating aflatoxin contamination in order of effective priority:

Pre plant–Incorporating resistance

(i) In Plant — Improved farm management

At harvest : Improved sampling, screening and biological treatment techniques

(ii) During storage — Improved environmental condition

(iii) During food and feed processing : detoxification/elimination.

7.1 Breeding Resistant Hybrids–Pre-plant Management

Breeding for resistance would involve essentially:

(i) Breeding for resistance to infection

(ii) Structural and biochemical defense markers to reduce growth of aflatoxin producers

(iii) Triggering aflatoxin detoxifying mechanisms

(iv) Developing suitable and rapid selection techniques.

7.1.1 Breeding for resistance to infection

The plant tissues penetrated and time of penetration are important. Thus, Patil *et al.* (156) observed that *Aspergillus* invading cotyledonary cells of groundnut may be a primary source of infection. They further observed that the developing shells are easily invaded but it is the penetration through the shell into the pod cavity that varies from cultivar to cultivar. Pods that formed lignified sclerenchyma bans early in the development were less susceptible to hyphal penetration than those without such bans. Kernel invasion is influenced by features of the hilum and the seed coat. Resistance was more in cultivars having small, covered hila and compact seed coats. Walliyar and Bocklee-Morovan (208) reported that genotypes with seed resistance to *A. flavus* had a lower proportion of it in the rhizosphere than those whose seeds were susceptible to fungal invasion. Zambeltakus (226) reported two cultivars *Daron IV* and *Shulamit* to illicit lower levels of pod infection in Senegal. Kushalappa *et al.* (111) concluded that resistance to pod infection was highly variable and seed infections were seen even where there were no visible signs of pod infection. Natural seed infection was shown to be lower in IVSCAF-resistant cultivars than IVSCAF-susceptible cultivars. Mehan (130) identified the problems in screening for resistance on groundnuts. They pointed out that genotypes resistant to seed colonization by aflatoxicogenic fungi may prove good substrates for aflatoxin production (163). This largely affects the utility of the observation that infection and contamination process are under genetic control (229).

For corn, the infection process, preimposes a different screening procedure. Initial work of Widstrom *et al.* (218) i.e., wound inoculation to the ear 20 days after silking, gave differing results (30). Subsequently Widstrom *et al.* (219) showed that waiting until grain is mature before sampling was most effective. Initial results did not indicate presence of resistance. Darrah *et al.* (42) confirmed that resistance to infection and contamination were genetically controlled and recent selection (29) provides several new inbred lines. However, Davis *et al.* (43) reported that no differences for resistance existed among the hybrids. In recent years the following observations are noteworthy:

(i) Scott (180) reported that the number of fluorescing particles can be used to select maize genotypes with low aflatoxin contamination.

(ii) Zhung *et al.* (228) reported that, based on data on genetics of resistance to kernel infection of maize by *A. flavus*, use of inbred lines like L-729 and B-73 as female parents should be avoided as they showed significant, positive maternal reciprocal effects.

(iii) Woloshuk *et al.* (223) suggested that the best inducers of aflatoxins biosynthesis are carbon sources readily metabolised by glycolysis. They also suggested that L- amylase produced by *A. flavus* has a role in the induction of aflatoxin biosynthesis in infected maize kernels.

(iv) *In vitro* experiments conducted by Zeringue *et al.* (227) using several maize genotypes showed that volatiles released (C6-C12 alkanal and alkenal contents) through decay of polyunsaturated fatty acids impart resistance to diseases and aflatoxin producers through a lipoxygenase pathway.

7.1.2 Structural and biochemical defence markers

Petit *et al.* (156) working with groundnut seed coats found certain tannin-like compounds, that included umbelliferonol and methyl catechol, inhibited growth of aflatoxin producers. They suggested that isolation of various plant constituents to detect the presence of specific proteins, tannin like compounds, lignins, phytoalexins and other compounds may correlate with levels of resistance and should be helpful in screening cultivars. Recently Cleveland *et al.* (36) identified antifungal peptide genes that are being used to engineer cotton to prevent invasion by *A. flavus*. A gene cluster of 70 kb was identified on which resided several genes governing AFB_1 pathway. This gene cluster is now being targeted in attempts to interfere with the aflatoxin production in plants.

7.1.3 Aflatoxin detoxifying mechanisms in seed

It has been observed that the yellow varieties of maize, richer in carotenes and xanthophylls in the endosperm are likely to have less amounts of aflatoxin inspite of the fungal contamination (142). *In vitro* it was shown that β-carotene, lulein and Zeaxanthin inhibit of AFB1. α-ionone type of compounds like α-carotene, lutein, etc. caused over 90% inhibition of aflatoxin production by *A. parasiticus*.

Huang *et al.* (90) isolated two maize seed proteins from the seeds of *Tex 6*, one a growth inhibitor having a molecular mass of 28 KDa and the other an aflatoxin biosynthesis inhibitor having a molecular mass of >100 KDa. *Tex 6* incidentally is resistant to aflatoxin accumulation.

7.2 Preventing Mould Contamination in Field

We have already seen that considerable amount of contamination occurs through infection in field and the significant role of temperature and moisture in epidemic development. The obvious management strategies to reduce such contamination will be through such steps that will avoid drought and favourable temperature during the susceptible phases of crop growth. We can look at this aspect taking corn as a model other than selection of suitable varieties.

7.2.1 Manipulating environmental factors

A common recommendation in risk prone areas will be to drought stress during production of crop. However, it is to be realised that complex interrelationships exist among moisture, insect damage, temperature and plant resistance to aflatoxin production. Irrigation regimes may also be useful in modifying soil temperature.

Other than weather parameter edaphic factors like adequate fertilisation, tillage, and soil pH (pre plant application of lime) are considerations that may ameliorate aflatoxin production in given situations (98). Destruction of crop residues is important as sclerotia produced on crop debris may survive for long periods (215). A note of caution was sounded by Wilson *et al.* (220) when they showed N over fertilisation resulted in increase in incidence and contamination.

Indirect methods of managing the environment and reducing contamination include manipulation of plant population, planting time and insect and disease management.

7.2.1 Plant population

High plant population impose a distress by competition for water and nutrients on one hand and reduced exposure to high inoculum load of air-borne spores through a dense canopy. Hence exceeding the optimum population recommended for a variety will not be desirable.

High densities increase the diurnal duration of free water on ears and silks during kernel maturation thus contributing to infection. On the other hand Bilgrami *et al.* (18) reported from India that lowest plant population sustained the highest aflatoxin contamination in a crop grown during monsoon when rainfall ranged from 50-100 mm wk-1. Temperature might have been a more important consideration in this study. However, Cole *et al.* (38) observed that pre harvest contamination with aflatoxin originates mainly from the soil and not via floral invasion.

7.2.1.2 Planting date

In North Carolina in USA, Jones *et al.* (99) reported that late planting produced high aflatoxin contamination and recommended early planting. Obviously such recommendations are location specific. Delayed planting recommendations in coastal areas (216) needs to give weightage to cocomitantly reduced yields. The basic tenet of late planting is the time to planting so that the critical grain filling period beginning at 20 days after silking, occurs after the highest seasonal temperatures and periods of net evaporation.

7.2.1.3 Biotic pest management

We have seen that in general several insects, pathogens and weed infestation predispose plants to infection by aflatoxin producers although Bilgrami *et al.* (18) from India reported no difference from one, two or three cultivation for weed control.

In general highest level of aflatoxin contamination was associated with insect damage (222) insects acting as carriers of aspergilli. Husk tightness was an important component of plant resistance to most ear feeding insects and also reduced the amount of aflatoxin contamination.

The predisposition of seed to aflatoxin contamination by ear rotters of maize like *Helminthosporium maydis* was first reported from Georgia, USA in 1971 (58).

Managing such biotic stress components provide obvious options for reducing aflatoxin contamination.

7.3 Management at Harvest and Transit

Harvest time and transit delays may considerably modify aflatoxin contamination and this has been specially demonstrated for maize. The basic tenet is to harvest the crop as soon as possible after physiological maturity to maintain grain quality and minimise other losses.

7.3.1 Harvest time

Early and prompt harvest at maturity appears to be critical in obtaining a produce with minimal aflatoxin contamination. Delay increases such contamination in a cumulative fashion. Thus a month's delay can cause additional field losses of 5-10% in yield and more losses due to increased aflatoxin contamination.

Since corn matures when moisture in the grain reaches 300-320 Kg^{-1} moisture, it will be wise to harvest the crop in aflatoxin-sensitive area when contamination is below 20 g Kg^{-1} even if it causes some losses due to higher moisture. Harvesting at 260-280 g Kg^{-1} will minimise harvest losses due to cracking, lodging, shelling etc. (118). The cracked kernels are risk prone for aflatoxin accumulation during storage.

The infected kernels in the ear are usually widely distributed and their isolation and elimination will require some modification of the harvesting methods (97).

Thus, in choosing an appropriate harvest date, the grower must assess the level of contamination present in the field against yield losses due to early harvest at higher moisture and cost of drying the harvested crop.

7.3.2 Transit time

The extent of care and speed of action required in transit for market or storage will largely depend on presence and level of *Aspergillus* contamination at harvest. Meticulous phytosanitary measures become imperative for all pre-harvest infected maize crops and includes.

(i) Cleaning of augur everyday

(ii) Complete clean out of bins prior to dumping

(iii) Maintaining clean trucks, trailers, vans, etc.

When a railway wagon carrying contaminated high moisture maize are made to wait for 4-6 hr contamination builds up to alarming levels of anywhere between two to ten folds.

7.4 Management During Storage

It is known that high moisture (Aw >0.86) and temperature (> 30^{0}C) favour rapid growth of *A. flavus* and a concomitant build up of aflatoxins. Hence, to reduce contamination and aflatoxin synthesis both pre-storage and in-storage measures are essential and both require a system of efficient monitoring at regular intervals.

7.4.1 Management of storage environment

Since moisture and temperature are cardinal both for the growth of aspergilli and aflatoxin(s) production which builds up as a function of time, manipulating these parameters in storage are of paramount importance.

7.4.1 Moisture

At humidity level of 70% or less RH, moisture content of wheat is about 13% and that of oil-rich seeds (groundnut, cotton, maize) 7-10%. This moisture keeps the aflatoxin production in check. Any rise in RH may increase the aflatoxin level in corn by as much as 10 folds. However, such increases in moisture level in localised pockets in the stored commodities is likely to occur due to inadequate aeration, leading to temperature variations. So called "wet spots" may develop in cooler areas (83). Proper, sustained aeration and moisture monitoring may be key to reduce aflatoxin levels during storage.

7.4.1.2 Temperature

Although favourable temperature for growth of aflatoxin producing fungi are reported to be 30°C or above, aflatoxin is known to be produced even at 7.5-10°C. Hence the storage temperature of particularly the perishable commodities should be lowered to 50°C as quickly as possible. The modern practice of packaging perishables. In sealed plastic containers provides a potentially hazardous situation by permitting accumulation of moisture even at refrigeration temperatures. Such packages need to be maintained at 0°C to avoid build up of aflatoxin. Such storage, for large–scale agricultural commodities is economically extravagant.

7.4.1.3 Modified gaseous atmosphere

The two approaches of atmospheric modification in storage are controlled atmosphere storage (CAS) and modified atmosphere packaging (MAP). In CAS a correct gas mixture is applied and maintained all through the storage time. It is good for bulk materials like maize or groundnut (52). Map is applied to packaged food products where the gaseous environment is changed to slow down respiration rate, microbial growth rate and enzymatic degradation by reducing O_2 and increasing N_2 and CO_2 in the head space of the package (188). Both these methods ought to be practicable in controlling mould growth and aflatoxin production.

7.5 Feed and Food Decontamination

The various approaches that have been tested experimentally and/or put to practice may be considered under following heads:

(i) Blending, sorting and cleaning

(ii) Extraction and detoxification using chemicals

(iii) Using heat and/or irradiation

(iv) Adsorption

(v) Using biologicals

7.5.1 Blending, sorting and cleaning

Processors of maize kernels often use blending or dilution process to achieve grade or moisture requirements (210). However, this practice is now believed to be hazardous and self defeating. It should neither be recommended nor practised.

Cleaning by mechanical removal of contaminated kernels has been largely ineffective. The methods relate to density differences between affected and healthy kernels when placed in sucrose solution. Non-buoyants were hardly contaminated. The method is as yet impractical but the principle is sound and provides basis for designing other separation systems.

In Australia aflatoxin containing groundnut kernels were separated by a process of selective segregation and sorting (172). The method however, is yet to be standardised and even with high standards of sampling and analysis in experimental systems, uncertainly in aflatoxin control was significant.

7.5.2 Extraction and detoxification using chemicals

A wide variety to chemical approaches have been tested for detoxification of aflatoxins in foods and feeds and some have yielded very promising results.

7.5.2.1 Reagents

Reagents used include acids, bases, oxidising agents, aldehydes, several gases and bisulphites (136). Acids convert B_1 and G_1 to their hemiacetal forms B_2 and G_2 through incorporation of water. However, acids have little effect on B_2 or G_2.

7.5.2.2 Volatiles

Sodium hydroxide, a base, has been used in refining oils to reduce aflatoxin to very low levels. Similarly $Ca(OH)_2$ treatment has been used to reduce aflatoxin content of groundnut and cotton seed meals, where formaldehyde was also claimed to be affective.

7.5.2.13 Volatiles

Of the bases, ammonia gas seems to be most effective, economically feasible and put to practice. Used as anhydrous gas under pressure and high temperature, a 98% reduction in groundnut meal has been reported and a pilot plant set up in India in 1988 (39). It is believed that ammonia opens the lectone ring of B_1,. forming an ammonium salt with the resulting hydroxy acid. Decarboxylation of --°-keto acid leads to formation of AFD_1. This technique is good for animal feed stuff and is now in use in Senegal, France and USA. Unfortunately, this approach reduces the protein efficiency of the product and causes off-flavours. Other volatile bases like methyl and ethyl amines have been found to reduce aflatoxin levels.

7.5.2.4 Oxidizing Agents

Dwarkanath *et al.* (62) compared O_3, O_2 and air as oxidising agents for aflatoxin degradation in cottonseed and groundnut meals at 100°C for 2 hr (22% moisture). Storage in O_3 resulted in complete detoxification of B_1 and G_1 within one hour but had no effect on B_2. The method also reduced the protein efficiency ratio (PER) level as well as lysine content.

Hydrogen peroxide also destroyed 97% of aflatoxins in groundnut meal while having no effect on PER of wheat and other feeds. In conjunction with riboflavin it reduced the AFM1 in milk by 98% possibly riboflavin induces production of singlet oxygen which reacts with the double bond of the furanofuran ring of M_1 (194).

7.5.2.5 Bisulphites and others

Bisulphetes are usually used as food preservatives and acting as antioxidants, it prevents browning and inhibits microbial growth. It was show to degrade B_1 and G_1 resulting in loss of fluorescence. Moerck *et al.* (133) found that 0.5% and 1% sodium bisulphite was more effective than aqueous NH_3 solution.

Shanthala (184) reported that aflatoxins are present in the finely suspended solids in groundnut oil and most of it can be removed by filtration or by extraction with 10% NaCl.

7.5.2.6 Solvent extraction

Solvents fund useful in removing aflatoxin from seeds include 95% ethanol, 90% aq acetone, 80% isopropyl alcohol, hexane-ethanol and hexane-methanol. They remove all traces of aflatoxins without affecting the nutritional quality of the meal. The logistic problems of additional processing, solvent removal and disposal restricts the use of this approach.

7.5.3 Using heat and/or irradiation

Generally aflatoxins are heat stable (ça. 250^{0}C) but if the product is moist, thermal inactivation temperature is considerably reduced. Thus cottonseed meal containing 30% moisture heated for 2 hr at 100^{0}C eliminated 85% of the aflatoxin present. When the meal contained 6.6% moisture, only 50% degradation occurred. Roasting has been claimed to eliminate aflatoxins in groundnut pecan and corn to having to effect in Pakistan (169). Such treatment on the other hand, reduce the nutritive value of the product.

Aflatoxins are sensitive to UV light but in practice it has doubtful effectiveness (68). Results with gamma rays were similar. An increase in aflatoxin production in cereals and root vegetables due to irradiation has been reported (164).

Apparently, both heat and radiation as an approach to reduce aflatoxin contamination in food and feed have doubtful credibility.

7.5.4 Chemosorbent additives

Two additives of promise are bentonite and hydrated sodium calcium aluminosilicate (HSCAS).

The ability of the clay, bentonite, to adsorb and retain aflatoxin was dependent on temperature and particle size (125). Further research is needed to standardise this approach.

On the other hand HSCAS selectively binds aflatoxins (157). When added in proportion of 0.5%, 90% of the aflatoxin got bound. It has also been shown to decrease AFM_1 in milk from lactating dairy cows. This is a promising arena (158) that requires further and immediate attention. It has been suggested the HSCAS could modify the toxicity of AFB_1 in chickens through a sequestering effect resulting reduced viability of AFB_1.

7.5.5 Biologicals

Many microbes have been shown to degrade aflatoxins of which most promising appear to be *Flavobacterium aurantiacum* NRRL B-184 which degrades B_1, G_1 and M_1 (35, 79) from peanut milk, be it defatted or partially defatted. Others shown to be of promise include *Absidia repens, Aspergillus niger, Dactylium denroides, Mucor ambigus, M. griseo-cyanus, M. alternaus, Rhizopus arrhizus, R. stolonifer, R. oryzae, Trichoderma viride, Tetrahymena pyriformis, Corynebacterium rubrum, Lactobacillus acidophilus, L. bulgaricus, L. plantarum.* Doyle (60) attributed the efficiency of this approach using moulds to hinge upon:

(i) Age of mycelia-optimum 8-10 day

(ii) Disrupted mycelia

(iii) Substrate type

(iv) Mould strain

(v) Amount of mycelia in media

(vi) Temperature (around 28^{0}C)

(vii) pH (between 5-6.5).

Basically this is a slow approach, requiring about 3-4 days to transform 60% of AFB_1 to aflatoxicol, given the best of conditions. Production of peroxidases by these moulds possibly leads to release of free radicals that react with the aflatoxins (59).

8. CONCLUSION

This paper overviewing aflatoxins and the concomitant problems of contamination leading to aflatoxicoses shows that aflatoxicoses, known in the 19th century, did not receive the attention it deserved because it did not occur in areas where innovative scientific research was is vogue. With the advent of the problem in the Western World, its existing and potential problems and the producer-consumer equation was furiously worked upon over a wide research agenda and today we stand on a platform where the Western World can breath a sigh of relief as they feel confident that they have the tools and motivation to restrict aflatoxicosis and in near future can even reduce aflatoxion production. The head-space atmosphere in packages required to repress aflatoxin production is a case in point in a consumer world going for packaged, ready-touse fast food.

The knowledge and technology both are available to the third world. Can the third World also have its sigh of relief? The answer is no. Yet those in the administrative helm of food, health and hygeine conveniently turn their minds off problems that they consider to be non-existent or minimal. Aflatoxins may have been the major cause of terminal carcinoma in this part of the World where almost all the agricultural substrates are consumed without knowledge of the concomitant hazards in absence of any semblance of monitoring. Monitoring in a vast country like India with the common populace consuming roasted groundnut, maize and various oils from groundnut, cotton seed, rice husk, and the elite consuming other more high priced nuts, cheese, butter, packaged fish, ducks etc. may be difficult. But awareness can certainly be generated, making use of the fairly well developed mass communication media to make people conscious of hazards of consuming products that present serious health risks. Carcinoma is easy to diagnose but can the same be said of the root cause of carcinoma?

Today, not only tools are available for producing varieties resistant to the aflatoxin producers and production in plants but at the interface of the biotechnology, it may be possible through its tools to largely replace from the ecosystem the aflatoxin producers by aflatoxin (-) strains or to produce varieties that will prevent aflatoxin production both in plant and their storage products. Thus, the first step in the genetic transformation system for *A. flavus* taken by Wolshuk *et al.* (224) helped not only in understanding biosynthesis of aflatoxins better but the genes for the biosynthetic pathogens was identified. Cloning of subsequently identified *apa-2* gene associated with regulation of aflatoxin biosynthesis has brought us nearer to identifying molecular approaches to control pre-harvest aflatoxin production. Again a gene cluster of over 70kb was identified (36) both in *A. flavus* and *A. parasiticus* on which many of the genes of aflatoxin biosynthesis appeared to reside. This will naturally lead to rational designing of non-aflatoxicogenic bio competitor strains of *A. flavus* group (through use of gene disruption techniques), which being conjugative may transfer the aflatoxin (-) character to existing aflatoxin formers, thus reducing the latter's population. Fortunately not all strains of *A. flavus* produce aflatoxins (173).

Another approach emerges from the negative observation that weevils (*Sitophilus zeamais*) that often act as biological synergists enhance aflatoxic B_1 content in stored maize (15). Such risks can be obviated by the potential tool presented through the observation of existence of antifungal peptide genes that can cause inhibition of growth and aflatoxin production by the producers. Such genes, properly transferred may reduce the invasion by *A. flavus* (36).

Even though options are many, they do not as yet provide controls to eliminate aflatoxin(s) production but do provide multiple options for their containment. Further researches on quantifying the impact of multiple environmental parameters through mathematical modelling, gaseous requirements of aflatoxicogenic moulds in packaged foods and irrdatiaon effects may provide a more refined protocol for reducing the problem below a threshold level. In the meanwhile, for the impoverished third world awareness through mass communication media by experts provides the major key to unknown or unidentified health hazards posed by these aflatoxicogenic moulds.

ACKNOWLEDGEMENT

The authors recognize the encouragement from Dr. D.C. Khatua and Dr. A.K. Chowdhury for providing inspiration for writing this article. They also recognize assistance received from Dr. Manoj Kumar Nanda of Department of Agronomy and Mr. Jayanta Kumar Roy, Research Scholar, Department of Plant Pathology for assistance in composing the simulation formulae.

9. REFERENCES

1. Adye, J. and Mateles, R.I. 1964. Incorporation of labelled compounds into *aflatoxin. Biochiim. Biophys. Acta.* **86** : 418 -420.

2. Allcroft, R., Roberts, B.A. and Loyd, M.K. 1968. Excretion of aflatoxin in lactating cow. *Food Cosmet. Toxicol.* **6**: 619-624.

3. Ambrecht, B.H. 1971. Aflatoxins residues in food and feed derived from plant and animal sources. *Res. Rev.* : 13-54.

4. Anderson, H.W., Nehring, E.W. and Wichser, W.R. 1975. Aflatoxin contamination of corn in the field. *J. Agric. Food Chem.* **23** : 775-82.

5. AOAC. 1984. Association of Official Analytical Chemists, *Official Methods of Analysis,* (14th ed.), AOAC, Washington, D.C. USA.

6. Arseculeratne, S.N., De Silva, L.M., Wijesundera, S. and Bandunatha, C.H.S.R. 1969. Coconut as a medium for the experimental production of aflatoxin. *Appl. Microbiol.* **18**: 88-91.

7. Ashworth, L.J.Jr., McMeans, J.L. and Brown, E.N. 1969. Infection of cotton by *Aspergillus flavus* : Time of infection and the influence of fiber moisture. *Phytopathology.* **59** : 383-85.

8. Austwick, P.K.C. and Ayurerst, G. 1963. Toxin products in groundnuts, groundnut microflora, toxicity. *Chem. Indi. (Lond.)* **2**:55-61.

9. Baquete, E.F. and Freie, M.J. 1989. Present status and perspectives of aflatoxin research in Mozambyque, pp. 93-94. In : *Aflatoxin Contamination of Groundnuts.* (ed. S.D. Hall), ICRISAT: India.

10. Basappa, S.C., Jayarman, A., Srenivasamurthy, V. and Parpia, H.A.B. 1967. Effects of B vitamins and ethyl alcohol on aflatoxin production by *Aspergillus oryzae. Indian J. Exp. Biol.* **5** : 262.

11. Begum, N., Adil, R. and Shah, F.H. 1985. Contamination of ground nuts with aflatoxins. *Pakistan J. Med. Res.* **24** : 129-131.

12. Behere A.G., Sharma, A., Pâdwal-Desai, S.R. and Mandakini, G.B. 1978. *J. Food Sci.* **43** : 1102-1103.

13. Bennett, J.W., Ferrhoiz, F.A. and Lee, L.S. 1978. Effect of light on aflatoxins, anthraquinones and sclerotia in *Aspergillus flavus* and *Aspergillus parasiticus. Mycologia* **70** : 104.

14. Bennett, J.W., Dunn, J.J. and Goldman, C.I 1981. Influence of white light on production of aflatoxin and anthraquinones in *Aspergillus parasiticus. Appl. Environ. Microbiol.* **41** : 488.

15. Beti, J.A., Phillips, T.W. and Smalley, E.B. 1995. Effects of maize weevils (Coleoptera: Cucurculionidae) on production of aflatoxin B_1 by *Aspergillus flavus* in stored corn. *J. Econ. Entomol.* **88**: 1776-1782.

16. Bhatnagar, D., Enrlich, K.C. and Cleveland, T.E. 1992. Oxidation-reduction reaction in biosynthesis of secondary metaboliles, In : *Handbook of Applied Mycology.* Vol. 5. (ed. D. Bhatnagar, E.B. Lillehoj and D.K. Arora). Dekker, N.Y. pp. 255-286.

17. Bilgrami, K.S. and Misra, R.S. 1981. Aflatoxin production by *Aspergillus flavus* in storage and standing maize crops. *Adv. Fract. Mycol. Plant Pathol,* **26** : 67-78.

18. Bilgrami, K.S. Ranjan, K.S. and Masood, A. 1992. Influence of cropping pattern on aflatoxin contamination in preharvest kharif (monsoon) maize crop (*Zea mays*). *J. Sci. Food Agric.* **58**: 101-106.

19. Blankenship, P.D., Cole, R.J., Sanders, T.H. and Hill, R.A. 1984. Effect of geocarposphere temperature on preharvest colonization of drought-stressed peanuts by *Aspergillus flavus* and subsequent aflatoxin contamination. *Mycopathologia* **85** : 69-74.

20. Boller, R.A. and Schroeder, H.W. 1966. Aflatoxin producing potential of *Aspergillus flavus-oryzae* isolates from rice. *Cereal Sci. Today,* **11** : 342-346.

21. Boller, R.A. and Schroeder, H.W. 1974. Influence of *Aspergillus candidus* on production of aflatoxin in rice by *Aspergillus parasiticus. Phytopathology* **64** : 121-134.

22. Booth, A.N. 1969. Review of the biological effects of aflatoxin on swvine, cattle and poultry. *J Amer. Oil Chem. Soc.* **46**: 154 (Abs.)

23. Bothast, J.J. and Fennell, D.I. 1974. A medium for rapid identification and enumeration of *Aspergillus flavus* and related organisms. *Mycologia* **66** : 363-369.

24. Bouchart, L.R. 1987. *Food and Beverage Mycology*, 2nd (ed. Van B. Nostrand), Reinhold, New York, 28 pp.

25. Buehl, G. and Rae, I.D. 1969. *The Structure and Chemistry of the Aflatoxins*. In : *Aflatoxins*, (ed. L.A. Goldblatt), Academic Press, New York, pp.55.

26. Bullerman, L.B. 1979. Significance of mycotoxins to food safety and human health. *J. Food Prot.* **42** : 65.

27. Bullerman, L.B. 1987. Methods of detecting mycotoxins in foods and beverages. In : *Food and Beverage Mycology*, (ed. L.R. Beuchat), 2nd ed. AVI, New York,. 7, pp 571.

28. Burmeister, H.R. and Hasseltine, C.W. 1966. Survey of the sensitivity of microorganisms to aflatoxins. *App. Microbiol.* **20** : 437-440.

29. Campbell, K.W., White, D.G. and Toman, J. 1993. Sources of resistance in F_1 corn hybrids to ear rot caused by *Aspergillus flavus. Plant Dis.* **77** : 1169.

30. Campbell, K.W. and White, D.G. 1994. An inoculation device to evaluate maize for resistance to carrot and aflatoxin production by *Aspergillus flavus. Plant Dis.* **78**. 778-781.

31. Carlile, M.J. 1970. The photoresponses of fungi. In : *Photobiology of Microoorganisms*, (ed. P. Halldal). John Wiley and Sons, New York, pp. 310.

32. Chowdhary. A.K. 1992. Influence microbial co-inhabitants of aflatoxin synthesis of *Aspergillus flavus* on maize kernels. *Lett. Appl. Microbiol.* **14**: 143-147.

33. Chowdhary, A.K. and Sinha, K.K. (1993). Competition between toxigenic *Aspergillus flavuys* strain and other fungi on stored maize kernels. *J. Stored Prod. Res.* **289**: 75-80.

34. Chu, F.S. 1984, Immunoassays for analysis of mycotoxins. *J. Food Prot.* **47**: 562-569.

35. Ciglar A., Lillehoj, E.B., Peterson, R.E. and Hall, H.W. 1966. Microbial detoxification of aflatoxin. *Appl Microbiol.* 14 : 934-939.

36. Cleveland, T.E., Cary, J.W., Brown, R.L., Bhatnagar, D., Yu, J., Chang, P.K., Chlan, C.A. and Rajasekaran, K. 1997. Use of biotechnology to eliminate aflatoxin in perharvest crops. *Bull. Inst. Compr. Agric. Sci.* **5** : 75-90.

37. Cole, R.J., Blankenship, P.D., Hill, R.A. and Sanders, T.H. 1984. Effect of geocarposphere temperature on preharvest colonization of drought stressed peanuts by *Aspergillus flavus* and subsequent aflatoxin contamination. In : *Toxigenic Fungi: Their Toxins and Health Hazard*, (ed. H. Kurala and Y. Ueno). Elsevier, Amesterdam, pp. 44-51.

38. Cole, R.J., Sanders, T.H., Doriner, J. W. and Blankanship, P.D. 1989. Environmental conditions of groundnuts: summary of six years research. In : *Aflatoxin Contamination of Groundnuts* (ed. S.D, Hall), ICRISAT India, pp. 279-287

39. Coker, R.D. 1989. Control of aflatoxin in groundnut products with emphasis on sampling analysis and detoxification. In : *Aflatoxin Contamination of Groundnuts* (ed. S.D. Hall). ICRISAT, India, pp. 125-132.

40. Cotty, P.J. and Bhatnagar, D. 1994. Variability among atoxigenic *Aspergillus flavus* strains in ability to prevent aflatoxin contamination and production of aflatoxin biosynthetic pathway enzymes. *Appl. Environ. Microbiol.* **60**: 2248-2251.

41. Dalrymple, W.H. 1983. Report of the veterinarian. *La. Agric. Exp. Stn. Bull.* **22**: 724-730.

42. Darrah, L.L., Lillehoj, E.B., Zuber, M.S., Scott, G.E., Thompson, D., West, D.R., Widstrom, N.W. and Fortnun, B.A. 1987. Inheritance of aflatoxin B_1 levels in maize krnels under modified natural inoculation with *Aspergillus flavus. Crop Sci.* **27**: 869-872.

43. Davis, N.D., Currier, C.G. and Diener, U.L. 1985. Response of corn hybrids to aflatoxin formation by *Aspergillus flavus. Alabama Agric. Exp. Stn. Bull.* p. 575.

44. Davis, N.D., Currier, C.G., and Diener, U.L. 1986. Aflatoxin contamination of corn hybrids in Alabama. *Cereal Chem.* **63**: 467-70.

45. Devi, G.R. and Polasa, H. 1984. Inhibitory effects of Bavistin (carbendazim) on growth and mycotoxin production by *Asperigillus* grown on maize. *Pesticides* **18** : 32.34.

46. Devi, G.R. and Polas, H. 1987. Interference in toxin production among toxigenic *Asperigillus* spp. *J. Stored Prod. Res.* **23**: 149-150

47. De Vris, H.R., Maxwell, S.M. and Hendrickse, R.G. 1990. *Mylopathologia* **110**: 1.

48. Diener, U.L. and Davies, W.D. 1966. Aflatoxin production by isolates of *Aspergillus flavus. Phytopathology* **56**: 1390-1393.

49. Diener, U.L. and Davies, N.D. 1967. Limiting temperature and relative humidity for growth and production of aflatoxin and free fatty acids by *Aspergillus flavus* in sterile peanuts. *J.Am. Oil Chem. Soc.* **44**: 259.

50. Diener, U.L. and Davies, N.D. 1968. Effect of environment an aflatoxin production in freshly dug peanuts. *Trop. Sci.* **10**: 22-28.

51. Diener, U.L. and Davis, N.D. 1970. Limiting temperature and relative humidity for aflatoxin production by *Aspergillus flavus* in stored peanuts. *J. Am. Oil Chem. Soc.* **47**: 347.

52. Diener, U.L. and Davies, N.D. 1972 Highlights. *Agric. Res.* p. 19.

53. Diener, U. L., Pettit, R.E. and Cote, R.J. 1982. Aflatoxins and other mycotoxins in peanuts. In : *Peanut Science and Technology*, (ed. H.E. Pattee, C.T. Young), *Am Peanut Res. Edu. Soc*, USA, pp. 486-519.

54. Diener, U.L., Asquith, R.L. and Dickens, J.W., (eds.) 1983. Aflatoxin and *Aspergillus flavus* in Corn. *Soc. Corp. Ser. Bull.* 279. Auburn University, Ala. Ala Agric. Exp. Stn. 112 pp.

55. Diener, U.L., Cole, R.J., Sanker, T.H., Payne G.A., Lee, L.S. and Klich, M. 1987. Epidemiology of aflatoxin formation by *Aspergillus flavus. Annu. Rev. Phytopathol.* **25**: 249-270.

56. Diew, S.W. and Demain, A.L. 1977. Effects of primary metabolism on secondary metabolism. *Ann. Rev. Microbiol.* **31**: 343

57. Domsch, K.H., Gams, W. and Anderson, T.H. 1980. *Compendimum of Soil Fungi*, Vol. 1. Academic Press, London, 859 pp.

58. Doupnik, B. 1972. Maize seed predisposed to fungal invasion and aflatoxin contamination by *Helminthosporium maydis* ear rot. *Phytopathology* **62**: 1367-1368.

59. Doyle, M. P. and Marth, E.W. 1979. Peroxidase activity in mycelia of *Aspergillus parasiticus* that degrades aflatoxin. *Eur. J. Appl. Microbiol. Biotech.* **7**: 211.

60. Doyle, M.P., Applebaum, R.S., Brackett, R.E. and Marth, E.H. 1982 Physical, chemical and biological degradation of mycotoxins in foods and agricultural commodities. *J. Food Prot.* **45**: 964.

61. Dutta, B. G. and Ghosh, G.R. 1964. Soil fungi of paddy fields. *Mycopath. Mycol. Appl.* **25**: 316-322.

62. Dwarakanath, C.T., Rayner, E.T., Mann, G.F. and Dollear, F.G. 1968. Reduction of aflatoxin levels in cotton seed and peanut meals by ozonization. *J. Am. Oil Chem. Soil.* **45**: 93.

63. Ellis, J.J., Bulla, L.A., Julian, G. and Hesseltine, C.W. 1970. Scanning electron microscopy of fungal and bacterial spores. Proc. 3rd A. Scanning el. Mier. Symp. p. 145-152. IIT Res. Inst. Chicago.

64. Ellis, W.U., Smith, J. P. and Simpson, B.K. 1994. Aflatoxins in food occurrence effects on organisms, detection and methods of control. *Crit. Rev. Food Sci. Nutr.* **30**: 403-439.

65. Eppley, R.M. 1966. A versatile procedure for assay and preparatory separation of aflatoxins from peanut products. *J. Assoc. off. Anal. Chem.* **49**: 1218-1223.

66. Eppley, R.M., Staloff, L. and Campbell, A.D. 1968. Collaborative study of "a versatile procedure for assay of aflatoxins in peanut products," including preparatory separation and confirmation of identity. *J. Assoc. Off. Anal. Chem.* **51**: 67.

67. Epstein, E., Steinberg, M. P., Nelson, A.I. and Wei, L.S. 1970. Aflatoxin production as affected by environmental conditions. *J. Food Sci.* **35**: 389.

68. Fennel, A.J. 1966. Aflatoxins in groundnuts. IV. Problems of detoxification. *Trop. Sci.* **8**: 61.

69. Fennell, D.I., Bothast, R.J., Lillehoj, E.B., Peterson, R.E. 1973. Bright greenish-yellow fluorescence and associated fungi in white corn naturally contaminated with aflatoxin. *Cereal Chem.* **50**: 404-414.

70. Gareis, M., Bauer, J., Von Montgales, A. and Gedek, B. 1984. Stimulation of aflatoxin B_1 and T-2 toxin production by sorbate. *Appl. Environ. Microbiol.* **47**: 416.

71. Goto, T., Matsui, M. and Kitsuwa, T. 1980. Determination of aflatoxins by capillary column gas chromatography. *J. Chromatogr.* **447**: 410-412.

72. Gottlieb, D. 1971. Limited growth in fungi. *Mycologia* **63**: 619.

73. Gupta, M.L. 1975. Studies on microfungi from different soils of Gorakhpur with special reference to ecology and taxonomy of aspergilli. Ph. D. Dissertation, Univ. Gorakhpur.

74. Hald, B. and Krogh, P. 1970. Occurrence of aflatoxins in imported cotton seed products. *Nord. Vet. Med.* **22**: 39.

75. Hall, S.D. (ed). 1989. *Aflatoxin Contamination of Groundnuts.* ICRISAT, Hyderabad, India, pp: 423.

76. Halver, J.E. 1969. Aflatoxicosis and trout hepatoma. In : *Aflatoxins.* (ed. L.A. Goldblatt), Academic Press, N.Y.

77. Hamasaki, T., Kuwano, H., Isono, K., Hatsuda, Y., Fukuyama, K., Tsukihara, T., and Katsube, Y. 1975. A new metabolite, Parasiticolide-A, from *Aspergillus parasiticus. Agric Biol. Chem.*, Tokyo **39**: 749-751.

78. Hara, S., Fennel, D.I., and Hesseltine, C.W. (1973). Studies on the mycological characters of aflatoxin-producing strains belonging to the *Aspergillus flavus* group *Rep. Res. Inst. Brewing,* **145**: 8-12.

79. Harding, T.D.J., Done, J.P., Lewis, G. and Allcroft, R. 1963. Experimental groundnut poisoning in pigs. *Res. Vet. Sci.* **4**: 217.

80. Hasan, H.A.H. 1994. Effect of glyphosate herbicide on aflatoxin production on wheat straw and couch grass. *Rostlinna Vyroba* **40**: 189-192.

81. Hauseen, E. and Hagedom, G. 1969. Occurrence and penetration of aflatoxin B in food stuffs and the effect of some processing on it. *Z. Lebensm. Untersuch.* **141**: 129-138.

82. Hayes, A.W., Davis, N.D. and Diener, U.L. 1966. Effect of aeration on growth and aflatoxin production by *Aspergillus flavus* in submerged culture. *Appl. Microbiol.* **14**: 1019-1021.

83. Heathcote, J.G. and Hibbert, J. R. (eds.). 1978. *Aflatoxins: Chemical and Biological Aspects.* Isevier, New York.

84. Heathcote, J.G., Dutton, M. F. and Hibbert, J.R. 1976. Biosynthesis of aflatoxins II. *Chem. Ind.*: 270-273.

85. Hendreickse, R.G. 1984. The influence of aflatoxins in child health in the tropics with particular reference to Kwashiokor. *Trans Roy. Soc. Trop. Med. Hyg.* **78**: 427-435.

86. Hetmanski, M. T. and Scudemore, K.A. 1989. A simple quantitative assay of animal foodstuffs using gel permeation chromatography cleanup. *Food Additives Contr.* **6**: 35-48.

87. Horn, B.W. and Wicklow, D.T. 1983. Factors influencing the inhibition of aflatoxin production in corn by *Aspergillus niger. Can. J. Microbiol.* **29**: 1087.

88. Hsieh, D.P.H. 1987. The role of antibiotics in human cancer. In : *Mycotoxins and Phycotoxins.* (eds, P.S. Vleggaar and R. Vleggaar,). Elsevier, Netherlands, pp. 447-486.

89. Hsieh, D.P. H., Lin, M. T., Yao, R.C. and Singh, R. 1976. Biosynthesis of aflatoxin: conversion of norsolorinic acid and other hypothetical intermediates into aflatoxin B_1. *J. Agric. Food Chem.* **24**: 167-170.

90. Huang, Zheng Yu., White, D.G., Payne, G.A., Payne, G.A. and Huang, Z.Y. 1997. Corn seed proteins inhibtory to *Aspergillus flavus* and aflatoxin biosynthesis. *Phytopathology* **87**: 622-627.

91. Hussein, A.M., Sommer, N.F. and Fortlage, R.J. 1986. Suppression of *Aspergillus flavus* in raisins by solar heating during sun drying. *Phytopathology* **76**: 335-38.

92. Jacobson, W.C., Harmeyer, W.C. and Wiseman, H. G. 1971. Determination of aflatoxins B_1 and M_1 in milk. *J. Dairy Sci.* **54**: 21-26.

93. Jinks, J.L. 1969. Selection for adaptibility to new environment in *Aspergillus glaucus. J. Gen. Microbiol.,* **20**: 223.

94. Joffe, A.Z. and Liskern, N. 1969. Effect of light, temperature and pH value on aflatoxin production in vitro. *Appl. Microbiol.* **18**: 517-518.

95. Johri, K., Johri, B.N. and Saksena, S.B. (1975). Colonization capacity of the soil microorganisms in relation tofungistasis. *Plant Soil* **43**: 347-354.

96. Jones, R.K. 1979. The epidemiology and management of aflatoxins and other mycotoxins. In *Plant Disease.* Vol. **4**. (eds. Hortsfall, J.G. and Cowling, E.B.). Academic Press, N.Y. pp. 381-392.

97. Jones, R.K. 1987. The influence of cultural practices on minimizing the development of aflatoxin in field maize. In : *Aflatoxin in Maize, Proc. Workshop.* (eds. M. S. Zuber, L.B. Lilehoy and B.L. Renfro). CIMMYT, Mexico, pp. 136-144.

98. Jones, R.K. and Duncan, H.E. 1981. Effect of nitrogen fertilizer, planting date, and harvest date on aflatoxin production in corn inoculated with *Aspegillus flavus. Plant Dis.* **65**: 741-744.

99. Jones, R.K., Duncan, H.E. and Hamilton, P.B. 1981. Planting date, harvest date and irrigation effects on infection and aflatoxin production by *Aspergillus flavus* in field corn. *Phytopathology* **171**: 810-816.

100. Kadian, S.K., Monga, P. and Goel, M.C. 1988. *Mycopathologia* **104**: 33.

101. Kainnaiyan, S., Sandhu, R.S. and Phiri, A.L. 1989. Aflatoxin and *Aspergillus flavus* contamination problems of groundnut in Zambia. In : *Aflatoxin Contamination of Groundnuts*, (ed. S.D. Hall). ICRISAT, India, pp. 65-70.

102. Kaminski, E., Stawicki, S. and Wasowicz, E. 1974. Volatile flavour compounds produced by molds of *Aspergillus, Penicillum* and Fungi imperfecti. *Appl. Microbiol.* **27**: 1001-1004.

103. Karunaratne, A. and Bullerman, L.B. 1990. Interactive effects of spore load and temperature on aflatoxin production. *J. Food Prot.* **53**: 227.

104. Klich, M. A. and Chmielewski, M. A. 1985. Nectaries as entry sites for *Aspergillus flavus* in developing cotton bolls. *Appl Environ. Microbiol.* **50**: 602-604.

105. Klich, M.A., Thomas, S.H. and Mellon, J. E. 1984. Field studies on the mode of entry of *Aspergllus flavus* into cotton seeds. *Mycologia* **76**: 665-79.

106. Krishnachari, K.A.V.R., Bhat, R.V., Nagarajan, V. and Tilak, T.B.G. 1975. Hepatitis due to aflatoxicosis, an outbreak in Wcstern India. *Lancer* **1**: 1061-1063.

107. Krogh, P. (ed.). 1987. *Mycotoxins in Food.* Academic Press, London.

108. Kulik, M. M. and Harlin, R.T. 1968. Osmophilic strains of some Aspergillus species. *Mycologia* **60**: 961-964.

109. Kulik, M.M. and Brooks, A.G. 1970. Electrophoretic studies of soluble proteins from *Aspergillus* spp. *Mycologia* **62**: 365-376.

110. Kurtzman, C.P., Horn, B.W. and Hesseltine, C.W. 1987. Antonie von Leeuwen. *J. Microbiol Serol.* **53**: 147.

111. Kushalappa, A.C., Bartz, J. A. and Norden, A.J. 1979. Susceptibility of pods of different peanut genotypes to *Aspergillus flavus* group fungi. *Phytopathology* **69**: 159-162.

112. Lawellin, D.W., Grant, D.W. and Joyce, B. K. 1977. Enzyme-linked immunosorbent analysis of aflatoxin B_1. *Appl. Environ. Microbiol.* **34**: 94-96.

113. Leach, L.L. and Paha, K.E. 1974. *Mycopath. Mycol. Appl.* **52** : 223-230.

114. Lee, J.H. and Mark, E.H. 1967. Formation of aflatoxin in cheddar cheese by *Aspergillus flavus* and *Aspergillus parasiticus*. *J. Dairy Sci.* **50** : 1708-1711.

115. Lee, L. and Marth, E.H. 1968. Aflatoxin formation by *Aspergillus flavus* and *Aspergillus parasiticus* in a casein substrate at different pH values, *J. Dairy Sci.* **51**: 1743-1747.

116. Lec, L. S., Goynes, W.H. and Lacey, P.E, 1986. Aflatoxin in Arizona cotton seed; Simulation of insect vectored infection of cotton bolls by *Aspergillus flavus*. *J. Am. Oil Chem. Soc.* **63**: 468.

117. Lee, L.S., Bennett, J. W., Goldblatt, L. A. and Lundin, R.E.. 1970. Norsolorinic acid from a mutant strain of *Aspergillus parasiticus*. *J. Am. Oil Chem. Soc.* **48**: 93-94.

118. Lee, R.D. 1994. Irrigated corn production in Georgia. Ga. *Crop. Ext. Serv. Bull.* p. 891.

119. Lillehoj, E.B. 1983. Effect of environmental and cultural factors on aflatoxin contamination of developing corn kernels. In. *Corn. So. Corp. So Bull.* **279**. Aubrn Univ., Alabama, pp. 27-34.

120. Lillehoj, E.B., Kivolek, W.F., Horner, E.S., Widstrom, N.W., Josephson, L.M. 1980 Aflatoxin contamination of preharvest corn: Role of *Aspergillus flavus* inoculum and insect damage. *Cereal Chem.* **57**: 255-257.

121. Ling, K. H., Tung, C.M., Sheh, I, Wang, J. J., Wang Jung, T. 1968. Aflatoxin B_1 in unrefined peanut oil and peanut in Taiwan. *J. Formosan Med. Assoc.* **67**: 309.

122. Maggon, K.K., Gupta, S.K. and Venkatasubramanian, T.A. 1977. *Bact Rev.* **41**: 822-885.

123. Marsh, S.F. and Payne, G.A. 1984. Preharvest infection of corn silks and kernels by *Aspergillus flavus*. *Phytopathology* **74**: 1?84-1289.

124. Marsh, P. B., Simpson, M.E. and Filsinger, E.C. 1973. *Aspergillus flavus* boll rot in the U.S. cotton belt in relation to high aqueous-extract pH of fiber, an indicator of exposure to humid conditions. *Plant Dis. Rep.* **57**: 664-668.

125. Masimango, N., Remacle, J. and Renault, J. H. 1978. The rol of absorption in the elimination of aflatoxin B_1 from contaminated media. *Eur. J. Appl. Microbiol. Biotech.* **6**: 101.

126. Mayo, N. S. 1891. Enzootic cerebritis of horses. *Kan. Agric. Expt. Stn. Bull.* **24**: 1-12.

127. Mayura, K., Basappa, S.C. and Murthy, V. 1985. Studies on some factors affecting aflatoxin production by *Aspergillus flavus* and *A. parasiticus*. *J. Food Sci. Technol.* **22**: 126-129.

128. Mayurya, K. and Murthy, Sreenivasa V. 1969. A quantitative method for estimation of aflatoxins in peanuts. *J. Assoc. Off. Anal. Chem.* **52**: 77-81.

129. McMeans, J. L. and Brown, C.M. 1975. Aflatoxins in cottonseed as affected by the pink boll worm. *Crop Sci.* **15**: 865-866.

130. Mehan, V.K. 1989. Sereening groundnut for resistance to seed invasions by Aspergillus flavus and to aflatoxin production. In : *Aflatoxin Cintamination in Groundnuts*, (ed. S.D. Hall). ICRISAT, India, pp 323-334.

131. Mehan, V.K., Bhavanishankar, T.N. and Bedi, J.S. 1985. Comparison of different methods of extractions and estimation of aflatoxin B_1 in groundnut. *J. Food Sci. Technol.* **22**: 123-125.

132. Mehrotra, B. R. and Kakkar, R.K. 1972. Ecological study of soil fungi of an agricultural field in Allahabad. *Mycopath. Mycol. Appl.* **47**: 41-58.

133. Moerck, K.E., McElfresh, P., Wohiman, A. and Hilton, B. W. 1980. Aflatoxin degradation in corn using sodium bisulphate, sodium hydroxide and aqueous ammonia. *J. Food Prot.* **43**: 571-574.

134. Mohyddin, M., Osman, N. and Skoropad, W. 1972. Inactivation of coniospores and mycelia of *Aspergillus flavus* by gamma-radiation. *Radiat. Bull.* **12**: 427-431.

135. Moss, M. O. and Frank, J. M. 1985. The influence on mycotoxins production of interactions between fungi and their environment. In : *Trichothecenes and other Mycotoxins*, (ed. J. Lacey). John Wiley and Sons, Chichester, pp. 257.

136. Moss, M. O. and Smith, J. E. 1985. *Mycotoxins Formation: Analysis and Significance*. John Wiley and Sons, Chichester, p.7.

137. Moubasher, A.H. and Mazer, M. B. 1971. Selective effects of three fumugants on Egyptian soil fungi. *Trans. Br. Mycol. Soc.* **57**: 447-454.

138. Muenzner, R. 1969. Ueber Einige Die Strahlewenpfindilch keit Von Schimmel-Kilzen Beeinflussende faktoren. *Arch. Mikrobiol.* **64**: 349-356.

139. Murakami, H. 1971. Taxonomic studies of the Japanese industrial strains of *Aspergillus* 10. Molds other than the Koji mold and mycological characters of the *Aspergillus* strains. *Rep. Res. Inst. Brew.* Japan **143**: 16-29.

140. Nesheim, S., Banes, D.D., Stoloff, L. and Campbell, A.D. 1964. Note on aflatoxin analysis in peanuts products. *J. Assoc. Offi. Agr. Chem.* **47**: 586-588.

141. Nkana, I., Nobbs, J. H. and Muller, H. G. 1987. Destruction of aflatoxin B_1 in rice exposed to light. *J. Cereal Sci.* **5**: 167.

142. Norton, R.A. 1997. Effect of carotenoids on aflatoxin B_1 synthesis by *Aspertillus flavus. Phytopathology* **87**: 8: 814-821.

143. Om Prakash and Sirdhana, B.S. 1978. Aflatoxin B_1 formation by *Aspergillis flavus* on the grains of some common maize cultivars. *Curr. Sci.* **47**: 695-696.

144. Palmgren, M. S. and Heyes, A.W. 1987. In : *Mycotoxins in Food*, (ed. P. Krogh). Academic press, N.Y. pp. 149.

145. Park, K.Y. and Bullerman, L. B. 1983. Effect of substrate and temperature on aflatoxin production by *Aspergillus parasiticus* and *Aspergillus flavus. J. Food Prot.* **46**: 178.

146. Papa, K. E. 1973. *Mycologia* **65** : 1201-1205.

147. Papa, K. E. 1976. *Mycologia* **68** : 159-165.

148. Papa, K. E. 1977a. *Appl. Environ. Microbiol.* **33** : 206.

149. Papa, K. E. 1977b. *Mycologia* **69**: 556-562.

150. Papa, K. E. 1977. *Mycologia* **69** : 1185-1190.

151. Papa, K. E. 1979. *Genet. Res* **34** : 1-10

152. Papa, K. E. 1980. *J. Gen. Microbiol.* **118** : 279-282.

153. Pass, T. and Griffin, F. J. 1972. Exogenous carbon and nitrogen requirements for conidial germination by *Aspergillus flavus*. *Can. J. Microbiol.* **18**: 1453-1461.

154. Payne, G.A. 1983. Nature of field infection of corn by Aspergillus flavus. In : *Aflatoxin and Aspergillus flavus in Corn*, (eds. U. L. Deimer. R.L. Asquith and J.W. Dickens). *Corn Soc. Copt. Bull.* **279**. Aurburn Univ., Ala, USA, pp-16-19.

155. Pestka, J.S., Li, Y.K. and Chu, F.S. 1983. Determination of aflatoxin B_1-DNA adducts and guanyl-aflatoxin B_1 prepared *in vitro* enzyme-linked immunosorbent assay. *Appl. Environ. Microbiol.* **44**: 1159-1161.

156. Petit, R.E., Araizeh, H.D., Taber, R. A. and Smith, O.D. 1989. Screening groundnut cultivar for resistance to *Aspergillus flavus, A. parasiticus* and aflatoxin contamination In : *Aflatoxin Contamination of Groundnuts*, (ed. S.P. Hall). ICRISAT, India, pp. 291-304.

157. Philips, T.D., Clement, B.A., Kubena, L. F. and Harvey, R.B. 1991. Selective chemisorption of aflatoxicosis in animals and reduction of aflatoxn residues in food of animal origin. In : Aflatoxin in Corn: New perspectives, (eds. O.K. Shotwell and C.R. Hurburg, Jr.). N. Cen. Reg. Pub. 329, iowa Agric. Home Econ. Sta. Ames. Iowa, pp. 359-368.

158. Philips, T.D., Clement, B.A. and Park, D.L. 1994. Approaches to reduction of aflatoxin in foods and feeds. In : *The Toxicology of Aflatoxins*. (eds. D.L. Eaton and J.D. Groopman). Academic Press, San Diego, CA, pp. 383-406.

159. Pollock, R.T. 1973. Environmental factors affecting the pattern of perithecium development in *Sordaria fimicola*, on agar medium. *Bull. Torrey Bot. Club.* **100**: 78-83.

160. Pons, W.A., Jr. 1969. Collaborative study on the determination of aflatoxins in cottonseed products. *J. Assoc. Official Anal. Chem.* **52**: 61-72.

161. Pons, W.A. and Franz, A.O. 1977. High performance liquid chromatography of aflatoxins in cotton products. *J. Assoc. Off. Anal. Chem.* **60**: 89-95.

162. Pons, W.A., Cucullu, A.E., Lee, L.S., Franz, A.O, Goldblatt, L.A. 1966. Determination of aflatoxins in agricultural products: use of aqueous acetone for extraction. *J. Asoc. Off. Anal. Chem.* **49**: 554-562.

163. Priyadarshini, E. and Tulpule, P.G. 1978. *Food Cosmet. Toxicol.* **14**: 293.

164. Priyadarshini, E. and Tulpule, P.G. 1978. Relationship between fungal growth and aflatoxin production in varieties of maize and groundnut. *Agric. Food Chem.* **26**: 249-252.

165. Purchase, I.F.H. and Vorster, L. J. 1968. Aflatoxin in commercial milk samples. *S. Afr. Med. J.* **42**: [illegible]

166. Quitco, R., Bautista, L. and Bautista, C. 1989. Aflatoxin contamination of groundnuts at the post production level of operation in the Philippines. In : *Aflatoxin Contamination of Groundnuts*, (ed. S.D. Hall.). ICRASAT. India pp. 101-110.

167. Rabie, C.J. and Smalley, E.B. 1965. In : *Symposium on Mycotoxins in Foodstuffs*. Agricultural Aspects, Pretoria, pp 18-29.

168. Rambo, G.W., Tuite, J. and Crane, P. 1974. Preharvest inoculation and infection of dent corn ears with *Aspergillus flavus* and *A. parasiticus*. *Phytopathology* **64**: 797-800.

169. Rana, A. 1989. Aflatoxin contamination of groundnuts in Pakistan. In : *Aflatoxin Contamination of Groundnuts*, (ed. S.D. Hall). ICRISAT. India, pp. 111-114.

170. Rao, H.R.G. and Harein, P.K. 1972. Dichlorvos as an inhibitor of aflatoxin production on wheat, corn, rice, and peanuts. *J. Econ. Entomol.* **65**: 988-989.

171. Raper, K.B. and Fenel, D.I. 1965. *The Genus Aspergillus,* Williams and Wilking Co. Baltimore.

172. Read, J.D. and Kasali, O.B. 1989. Hazards to livestock of consuming aflatoxin contaminated groundnut meal in Africa. In : *Aflatoxin Contamination of Groundnuts*, (ed.S.D. Hall). ICRISAT, India, pp. 31-36.

173. Ruiz, J.A., Bentabol, A., Gallego, C., Angulo, R. and Jodral, M. 1996. Aflatoxin producing strains of *Aspergillus flavus* in the mold flora of the different greenhouse substrates for the cultivation of cucumber (*Cucumis sativus L.) J. Sci. Food Agric.* **74**: 64-68.

174. Russel, T.E. 1980. Aflatoxin contamination of cottonseed-climate, insect, cultural practices and segregation. *Proc. Belt. Cotton. Prod. Conf. Spec. Sess. Aflatoxin.* St. Louis, Mo. Memphis: Natl. Cotton Counc. pp. 53-55.

175. Russel, T.E., Watson, T.F. and Ryan, G.F. 1976. Field accumulation of aflatoxin in cotton seed as influenced by irrigation termination dates and pink boll worm infestation. *Appl. Environ. Microbiol.* **31**: 711-713.

176. Salmon, W.D. and Newborne, P.M. 1963. Occurrence of hepatomas in rats fed diets containing peanut meal as a major source of protein. *Cancer Res.* **23**: 571.

177. Sanders. T.H., Hill, R.A., Cole, A.J. and Blankenship, P.D. 1981. Effect of drought on occurrence of *Aspergillus flavus* in maturing peanuts. *J. Am. Oil Chem. Soc.* **58**: 966A-970A.

178. Sanders, T.H. , Cole, R.J., Blankenship, P.D. and Hill, P.A. 1985. Relation of environmental stress duration to *Aspergilius flavus* invasion and aflatoxin production in perharvest peanuts. *Peanut Sci.* **12**: 90-93.

179. Scott, P.M. 1969. Analysis of beans for aflatoxins. *J. Assoc. Off. Anal. Chem.* **56**: 827-831.

180. Scott, G.E. 1996. Association of bright greenish yellow fluorescence and aflatoxin in grain of maize inbreds. *Maydica* **41**: 41-48.

181. Sedwick, B. and Morris, C. 1980. Stereo chemical course of hydrogen transfer catalysed by the enoyl reductase enzyme of the yeast fatty acid synthetase. *J. Chem. So.* **D**: 96-98.

182. Seenappa, M. and Kempton, A.G. 1980. *Mycopathologia* **70**: 135-138.

183. Sercek-Hanssen, A. 1970. Aflatoxin-induced fatal hepatitis. *Arch. Environ. Health* **20**: 729.

184. Shanthla. T. 1994. Methods of estimation of aflatoxins: A critical appraisal. *J. Food Sc. Technol.* **31**: 91-103.

185. Shapira, R., Paster, N., Eyal, O, Menasherov, N., Melt, A. and Salomon, R. 1996. Detection of aflatoxigence molds in grains by PCR. *Appl. Environ. Microbiol.* **62**: 3270-3273.

186. Sharma, R.S., Trivedi, K. R., Wadodbar, U.R, Murthi, T.N. and Punjrath. J.S. 1994. Aflatoxin B_1 content in deoiled oil cakes, cattle feed and damaged grains during different seasons in India. *J. Food Sci. Technol.* **31**: 244-246.

187. Shih, C.N. and Marth, E.H. 1973. Aflatoxin produced *Aspergillus parasiticus* when incubated in the presence of different gases. *J Milk Food Tech.* **36**: 421.

188. Shotwell, O.L. 1986. Chemical survey methods for mycotoxins. In : *Modern Methods in the Analysis and Structural Elucidation of Mycotoxins*, (ed. R.J. Cole). Academic Press, New York, pp. 60.

189. Shotwell, O.L., Hessettine, C.W., Burmeister, H.R., Kwolck, W.F., Shamon, G.M. and Hall, H.H. 1969. Survey of cereal grains and soybeans for the presence of aflatoxin I wheat grains, sorghum and oats. *Cereal Chem.* **58**: 124-127.

190. Simpson, M.E. and Batra, L.R. 1984. Ecological relations in respect to boll rot of cotton caused by *Aspergillus flavus*. In : *Toxicogenic Fungi; Their Toxins and Health Hazards*, (eds. H. Kurata and Y. Ueno). Elsevier, Amsterdam, pp. 24-32.

191. Simela, A.H. and Caley, A.D. 1989. Aflatoxin contamination of stored groundnuts in Zimbabwe,. In (Ed. S.D. Hall) *Aflatoxin Contamination of Groundnuts*. ICRISAT, India, pp 59-63.

192. Smith, J. E. and Paterman, J. A. (Eds). 1977. *Genetics and Physiology of Aspergillus. Symp. Br. Mycol. Soc.* 1. Academic Press, London, 552 pp.

193. Sinnhuber, R.O. 1967. Aflatoxin in cotton seed meal and liver cancer in rainbow trout. *Trout Hepatoma Res. Conf. Res. Rept.* **70**: 48-50.

194. Sreenivasamurthi, V., Parbia, H.A.B, Srikanta, S. and Shankarmurti, A. L. 1967. Detoxification of aflatoxin in peanut meal by hydrogen peroxide. *J. Assoc. Off. Anal. Chem.* **50**: 350-354.

195. Stark, A.A. 1980. Mycotoxins as mutagens and carcinogems. *Ann. Rev. Microbiol.* **34**: 82.

196. Stark, A. A. 1980. Mutagenicity and carcinogenity of mycotoxins: DNA binding as a possible mode of actions. *Ann Rev. Microbiol.* **34**: 235-262.

197. Stephenson, L.W. and Russel, T.E. 1974. The association of *Aspergillus flavus* with hemepterous and other insects infesting cotton bracts and foliage. *Phytopathology* **64** : 1502-1506.

198. Stoloff, L. 1971. Report on mycotoxins (for 1970). *J. Assoc. Offi. Anal. Chemists* **54**: 305.

199. Tabak, H. H. and Cooke, W. B. 1968. The effect of gaseous environment on the growth and metabolism of fungi. *Bot. Rev.* **34**: 124.

200. Taber, R.A. and Schroeder, H. W. 1967. Aflatoxin producing potential of isolates of the *Aspergillus flavus oryzae* groups from peanuts (*Arachis hypogea*). *Appl. Microbiol.* **15** : 140-150.

201. Teitell, L. 1958. Effect of relative humidity on viability of conidia of Aspergilli. *Am. J. Bot.* **45**: 748-753.

202. Thean, J. E., Lorenz, D.L., Wilson, D.M., Rodgers, K. and Gueldner, R.C. 1980. Extraction, clean up and quantitative determination of aflatoxin in corn. *J. Assoc. Off. Anal. Chem.* **63**: 631-633.

203. Thiel, P.G. 1986. HPLC determination of aflatoxin and mammalian aflatoxin metabolites. In: *Mycotoxins and Phytotoxins*, (eds. P.S. Steyn, and R. Vleggar). Elsevier, New York, 329 p.

204. Tienistra, P. J. 1969. A study of the variability associated with sampling peanuts for aflatoxins. *J. Am. Oil. Chem. Soc* **46**: 667.

205. Tulpule, P.G., Bhat, R.V., Nagarjuna, V. and Priyadarshini, E. 1977. Variations in aflatoxin production due to fungal isolates and crop genotypes and their scopes in production of aflatoxin production. *Arch. Instt.* Pasteur. Tunis **54**: 187-193.

206. Tutelyan, V.A., Eller, K.J. and Sobolev, V.S. 1989. A survey using normal phase high performance liquid chromatography of aflatoxins in domestic and imported foods in the USSR. *Food Additives Cont.* **6**: 459-465.

207. Van Walbeek, W., Clademenos, T. and Thatcher, F.S. 1969. Influence of refrigeration on aflatoxin production by strains of *Apergillus flavus. Can. J. Microbiol.* **15**: 629-633.

208. Walliyar, F. and Bockley-Morovan, A. 1989. Resistance of groundnut varieties to *Aspergillus flavus* in Senegal. In: *Aflatoxin Contamination of Groundnuts,* (ed. S.D. Hall). *ICRISAT*, pp. 305-310.

209. Waltking, A.E. 1907. Collaborative study of three methods for determination of aflatoxin in peanuts and peanut products. *J. Assoc. Off Anal. Chem.* **53**: 104-108.

210. Watson, S.A. 1987. Measurement and maintenance of quality. In : *Corn: Chemistry and Technology,* (eds. S.A. Watson and P.E. Ranistad). Am. Assoc. Cereal Chemists, St. Paul, MN. pp. 125-183.

211. Wells, T.R. and Kreutzer, W.A. 1972. Aerial invasion of peanut flower tissue by *Aspergillus flavus* under gnotobiotic conditions. *Phytopathology* **62**: 797(Abs)

212. Wheeler, B. 1969. An examination of cereals and feeds for aflatoxin. *Irish J. Agr. Res.* **8**: 172.

213. Whitten, M.E. 1967. Occurrence of aflatoxins in cotton seed and cotton seed products *Proc. 1967 Mycotoxin Research Seminar, U.S.D.A.* Washington, D.C., June 8-9.

214. WHO, 1979. Environmental health criteria **11**: *Mycotoxins.* World Health Organisation, Geneva, Switzerland.

215. Wicklow, D.T. and Donahue, J.E. 1984. Sporogenic germination of sclerotia in *Aspergillus flavus* and *A. parasiticus. Trans. Br. Mycol. Soc.* **82**: 621-624.

216. Widstrom, N. W. 1992. Aflatoxin in developing maize: Interactions among involved biota and pertinent econice factors. In : *Handbook of Applied Mycology*, Vol. 5. (eds. D. Bhatnagar, E.B. Lilliehoj and D.K. Arora). Marcel Dekker, New York, pp 23-58.

217. Widstrom, N.W. 1996. The aflatoxin problem with corn grain. *Adv. Agron.* **56**: 219-280.

218. Widstrom, N.W., Wilson, D.M. and McMillan, W.W. 1981. Aflatoxin contamination of preharvest corn as influenced by timing and method of inoculation. *Appl. Environ, Microbiol.* **42**: 249-251.

219. Widstrom, N.W., Wilson, D.M. and McMillian, W.W. 1986. Differentiation of maize genotypes for aflatoxin concentration in developing kernels. *Crop. Sci.* **26**: 935-937.

220. Wilson, D.M., Walker, M.E. and Gascho, G.J. 1989. Some effects of mineral nutriton on aflatoxin contamination of corn and peanuts. In : *Soil-borne Plant Pathogens : Management of Diseases with Macro-and Microelements*, (ed. A.W. Englehard). APS Press, St. Paul, MN. pp. 137-151.

221. Wilson, D.M. and Stansell, J. R. 1983. Effect of irrigation regimes on aflatoxin contamination of peanut pods. *Peanut Sci.* **10**: 54-56.

222. Wilson, D.M., Widstrom, N.W., Marti, L.R. and Evans, B.D. 1981. *Aspergillus flavus* group, aflatoxin, and bright greenish yellow fluorescence in insect-damaged corn in Georgia. *Cereal Chem.* **58**: 40-42.

223. Woloshuk, C.P., Cavaletto, J. R. and Cleveland, T.E. 1997. Inducers of aflatoxin biosynthesis from colonized maize kernels are generated by an amylase activity from *Aspergillus flavus. Phytopathology* **87(2)**: 164-169.

224. Woloshuk, C.P., Seip, E.R., Payne, G.A. and Adkins, C.R. 1989. Genetic transformation system for the aflatoxin producing fungus *Aspergillus flavus. Appl. Environ, Microbiol.* **55**: 86-89

225. Xiao Daren, 1987. Research on aflatoxin contamination of groundnut in the Peoples Republic of China. In : *Contamination of Aflatoxins in Groundnuts*, (ed. S.D. Hall) ICRISAT, India, pp. 95-100.

226. Zambettakis, C. 1975. Etude de la contamination de quelques varieties d' arachide parl' *Aspergillus flavus* (In Fr.). *Oleagineus* **30**: 161-167

227. Zeringue, H. J. Jr., Brown, R.L., Neucere, J. N. and Cleveland, T.E. 1996. Relationships between C_6-C_{12} alkanal and alkenal volatile contents and resistance of maizegenotypes to *Aspergillus flavus* and aflatoxin production. *J. Agril. Food. Chem.* **44**: 403-407.

228. Zhung, Y., Kang, M.S. Magari, R. 1997. Genetics of resistance to kernel infection by *Aspergillus flavus* in maize. *Plant Breed.* **116**: 146-152.

229. Zuber, M. S. and Lilehoj, E.B. 1979. Status of aflatoxin problems in corn. *J. Environ. Qual.* **8**: 1-5.

12 Protoplast Fusion for Improvement of Fungal Strains

Sumeet

CONTENTS

ABSTRACT 188
1. INTRODUCTION 188
2. PROTOPLAST ISOLATION 189
2.1 Lytic Enzymes 189
2.2 Osmotic Stabilizer 190
2.3 The Organisms 190
2.4 pH and Buffer 191
2.5 Temperature and Time of Contact with the Lytic Enzyme 191
3. PROTOPLAST AS TOOL FOR GENETIC MANIPULATION 191
3.1 Protoplast Fusion 191
4. APPLICATIONS OF PROTOPLAST FUSION 192
4.1 Conventional Genetic Studies 192
4.2 Strain Improvement 194
4.3 Protoplast Fusion and Fungal Biotechnology 195
5. CONCLUSIONS 196
6. REFERENCES 196

Advances in Microbial Biotechnology
J.P. Tewari, T.N. Lakhanpal, Jagjit Singh, Rajni Gupta & B.P. Chamola (eds.)
APH Publishing Corporation, New Delhi - 110 002, India.

ABSTRACT

The use of protoplast fusion has aroused as a tool in the genetic improvement of fungi. This application of protoplast technology stems from a long and detailed study of protoplast system relating to their isolation and culture. Although protoplasts have been isolated from all the taxonomic groups of fungi but due to the diverse chemical nature of cell wall, conditions for protoplast isolation should be standardized for each and every strain. These aspects have great significance in the context for protoplast fusion. The fusion technique by which genetic recombination can be achieved along with the different factors for isolation of protoplasts are discussed.

1. INTRODUCTION

The genetic improvement of fungi has traditionally involved induced mutagenesis and selection of strains with the appropriate enhanced characteristics. The maintenance of fitness character and ability to proliferate in a natural environment is crucial. In addition to desired genetic changes, the mutagenic programmes result in the introduction of deleterious mutations into the strains. Moreover, random screening is most laborious because large number of isolates need to be processed in order to detect the rare improved strains. Whereas, recombination is a rational approach to strain improvement which despite the power of mutation and selection approach, provides a worthwhile alternative strategy. Various natural methods of recombination including heterokaryosis and sexual cycle in fungi have some application in the improvement of strains but the most widely used method of gene transfer for strain improvement are those which are unlikely to occur in nature *viz*., parasexual breeding, protoplast fusion and genetic engineering. Parasexuality could be initiated by hyphal anastomosis but protoplast fusion circumvents the incompatibility barriers to hyphal anastomosis and may allow genetic recombination between genomes of fungi from different species and genera. The use of protoplast fusion has aroused a great deal of interest in improvement of fungal strains.

The term protoplast is normally used to describe the osmotically fragile cells completely devoid of cell wall material, while sphaeroplast is a cell with the some ramnants of cell wall material (12). Although, the work of Giaja (31) identified the lytic activity of gastric juice of snail, *(Helix pomatia)* against yeast cells, but the work of Weibull (99) on isolation of protoplasts from bacteria provided the impetus for large scale production of protoplasts from fungi. Initially the protoplasts were isolated from *Saccharomyces cerevisiae, S. carlsbergenesis* (26) and *Neurospora crassa* (6). Since than protoplasts have been isolated from all the taxonomic groups of fungi by enzymatic digestion of cell wall. Protoplasts have been used to study the regeneration of protoplasts to vegetative mycelium, structure and biochemical composition of cell wall, fusion, gene transfer, antibiotic action, metabolic production and ultrastructure. Before, the large scale production of protoplasts, the studies on the above mentioned areas may be restricted by the presence of rigid cell wall. Drastic mechanical treatments are often required to produce cell fractions which may lead to loss and damage of cellular components. The formation of protoplasts limits the need of such harsh treatments.

2. PROTOPLAST ISOLATION

The isolation of protoplasts from different taxonomic groups of fungi is a routine process and large quantities of this are required for use in physiological and biochemical studies. Also, isolated protoplasts must retain their ability to maintain normal cellular functions in the absence of cell wall. Several factors govern the success of technique *viz.* nature and concentration of lytic enzyme, osmotic stabilizer, cultural conditions and growth phase of organism, use of disulphide bond reducing agents, pH and concentration of buffer system, temperature and time of protoplasting, etc. These factors may interact and influence the isolation and regeneration of protoplasts.

2.1 Lytic Enzyme

Protoplasts have been isolated from fungi by enzymatic digestion of entire or part of the cell wall. Chemically, fungal cell wall is composed of 80-90% polysaccharides with the remainder consisting of protein and lipids (7). The arrangement of these cell wall components has marked effect on the efficiency of the lytic enzymes used for protoplast production from various species.

Enzymes suitable for protoplast release from fungi are now readily available commercially including cellulase, driselase, β-glucanase, glusulase, β-glucuronidase, helicase, Novozym 234, Rhozym HP150, zymolyase and sulphatase. Most of them are crude or partially purified preparations with multiple activities, the main component being chitinase, α and β-glucanase and protease.

Commercially available enzymes are rarely used singly in the isolation of protoplasts from fungi. In combinations they are found to be more effective than used alone. Exception including the use of commercial preparation of enzyme mutanase also known as Novozym 234 to obtain high yield of protoplasts from number of fungi. The higher activities of protease, chitinase and β-1,3-glucanase make it suitable for releasing protoplasts from yeasts and filamentous fungi (2, 24, 68, 76, 77, 83, 95, 100).

Although commercial enzymes are convenient to use, many workers have found that the lytic enzymes produced from microorganism are more effective in obtaining protoplasts. These mycolytic enzymes are produced by suitable microorganism or derived from autolytic cultures of the specific fungus (74, 79). Number of inducible substrates such as cellulose, chitin, laminarin and purified fungal cell wall as the sole carbon source have been used to promote the synthesis of lytic enzyme complexes (43). Number of fungi have been used successfully for preparation of lytic enzyme including *Penicillium purpurogenum (67), P. italicum* (9) and *Trichoderma* spp. (22,69).

Enzymes present in a incubation mixture exerts a profound effect on release and viability of protoplasts. Therefore enzyme concentration must be adjusted to reduce the incubation time, as longer incubation causes lysis of protoplasts thereby lowering the protoplast yield. Also amount of enzyme required for protoplast release may vary with the size of organism.

The analysis of various enzyme activities present in the lytic enzyme throw light on their contribution to wall lysis and protoplast formation. However, such detailed studies were carried out in only a few instances. The activities of endo-β-1,4-glucanase, cellobiase, β-1, 3-glucanase, protease, α–mannase, α-mannosidase and chitinase are present in *T. harzianum* enzyme (58). The studies of Kitamato *et al.* (53) suggested that *T. harzianum* enzymes used alone for protoplasting is superior to combination of commercial enzymes. The enzymes production from *T. harzianum* showed a high proteinase activity which on prolonged incubation caused bursting of protoplasts, therefore, proteinase activity was decreased using bentonite in the culture medium. The enzyme preparation with reduced proteinase activity resulted in increased protoplast yield after prolonged incubation periods as compared to the once without treated with bentonite.

Number of commercial enzymes *viz.*, cellulase (Merck), cellulase CP, cellulase CT, chitinase, β–D-glucanase, helicase, lytic enzyme LI (BDH), Novozym 234, β-D-glucuronidase were investigated for their lytic activity against different polysaccharides known to be components of fungal cell wall. The high level of β-D-glucanase and chitinase were associated with Novozym 234 and cellulase CP which gave the best yield of protoplasts. The Novozym 234 was also found to contain considerable proteolytic activity (36). The most prominent activities of endo β-1,4-glucanase, aryl-β-glucosidase and β-1, 3-glucanase were associated with driselase, whereas no chitinase activity was detected (78).

2.2 Osmotic Stabilizer

Isolated protoplasts are osmotically fragile therefore osmotic stabilizer is necessary to prevent lysis. A wide range of osmotic stabilizers including sugar, sugar alcohols and inorganic salts such as KCI, NaCl, $MgCI_2$, NH_4NO_3, $(NH_4)_2SO_4$ are most frequently used for protoplast stabilization. These factors also depend on the organism and there is no universal stabilizer for all groups of fungi. The choice of osmotic stabilizer greatly influences the protoplast yield. They are generally effective only at optimum concentration and any change in concentration may decrease the protoplast yield. At low concentration protoplasts began to swell and rupture after a short time while at higher concentration the protoplast will shrink (21).

Complete separation of protoplasts from hyphal debris may require density gradient centrifugation. Hashiba and Yamada (39) described the liquid-liquid two phase system for obtaining homogenous suspension of protoplasts from crude preparation of protoplasts. The optimum stabilizer suitable for protoplast release may not be suitable for subsequent stabilization of the isolated protoplasts. It may be necessary to change the stabilizer after digestion to prevent the gradual loss of protoplasts through lysis(5):

2.3 The Organism

Both filamentous fungi and single celled yeasts respond to different environmental conditions in a variety of ways therefore the nature of growth medium and cultural conditions are of prime importance in considering protoplasts isolation. The presence of sulphur containing amino acids or 2-deoxyglucose (2DG) in the growth medium has been shown to increase the protoplast yield from log phase cells of *Shizosaccharomyces pombe (46), Rhodotorula rubra* (27) and *Trichoderma* spp. (88). The use of sulphur containing amino acids in the growth medium enables enrichment of the cell wall proteins and addition of 2DG further interferes with the constitution of the cell wall by partially inhibiting the cell wall synthesis. Although, in the presence of 2DG culture grew slowly but the old cultures were also protoplasted.

The influence of physiological age of the culture on protoplast formation is better documented. In general protoplast yield is much greater from mycelium in the exponential phase of growth (73). Deutch and Parry (23) demonstrated that the cell wall build up resistance to lysis in the transitional growth period between exponential and stationary growth phases. The maximum yield of protoplasts is obtained during the period of growth coinciding with the occurrence of maximum branching and cell wall synthesis (43). The region of cell wall most susceptible to lysis is near the hyphal tip where wall is thinnest and wall synthesis is actively taking place.

Pretreatment of yeast cells or fungal mycelium with appropriate thiol compound or detergent accelerates the rate of protoplast release. The effectiveness of pretreatment on protoplast isolation is influenced by the age and growth of the culture, strain of an organism, the pretreatment medium, the subsequent treatment conditions and the length of time for pretreatment and treatment. The reducing agents cysteine, mercaptoethanol and dithiothreitol increased the rate of lysis from the cells in exponential phase of growth in *Candida apicola* and are essential for forming protoplasts from stationary phase cells grown in YEPG medium (80). Pretreatmnt with Na-thioglycolate resulted in 70-80% conversion of midstationary cells of *C. albicans* to protoplasts (90).

2.4 pH and Buffer

The choice of buffer in the incubation mixture can affect the rate of protoplast release. There is no universal buffer used for protoplast isolation. Affect of molarity of buffer on the protoplast yield was observed in *T. reesei*. The gradual increase in protoplast yield was observed between lOmM and 25mM phosphate buffer, above that protoplast yield was gradually decreased and was minimum at l5OmM (58).

Depending upon the nature of cell wall the optimum pH of incubation mixture is different for all groups of fungi. Maximum number of protoplasts are obtained at optimum pH which is evident from the experiments with *T. harzianum* (82), *T. reesei* (58) and *Aspergillus sojae* (93).

2.5 Temperature and Time of Contact with the Lytic Enzyme

Effect of temperature is related to the activities of lytic enzyme system and higher temperature caused a significant decrease in protoplast count. Kovac and Subik (55) observed that unfavourably high temperature may cause agglutination of certain organelles in isolated protoplasts and low temperature may affect stability of protoplast membrane.

The prolonged contact with lytic enzyme than optimum time of contact results in a decrease of protoplast count, which may be due to the lysis of already formed protoplasts (15, 58).

3. PROTOPLAST AS TOOL FOR GENETIC MANIPULATION

Protoplasts are important in both classical and molecular genetics. Its use in genetic manipulation of fungi has aroused as a major area of development in the past decade or so. Different methods like protoplast fusion and transformation have been employed to develop the new strains with improved characters.

3.1 Protoplast Fusion

The possibility of fusing isolated protoplasts was first demonstrated in the early 1970s (17), but significant progress was made following the discovery of the fusogenic property of the polyethylene glycol (PEG) (47) and was soon applied to induce fusion in fungal protoplasts. The report using PEG/Ca^{2+} to induce fusion in fungal protoplasts showed 1000 times higher fusion frequencies than previously achieved (3, 39). Over a period of few years this combination has proved universal to produce intraspecific, interspecific and intergeneric hybrids in both yeasts and filamentous fungi. In the presence of PEG, the membranes of two or more protoplasts merge into one large structure. The time required for complete fusion of fungal protoplasts depends upon the reaction of the protoplast membrane with fusogenic factor and osmotic stabilizers present in the incubation mixture. Generally 25-40% of PEG (mw 4000-8000) is used in the incubation mixture. The Ca^{2+} ions promote the fusion of fungal protoplasts at concentration ranging form 0.01 M to 0.1 M and negligible fusion frequency was observed in the absence of Ca^{2+} (28, 29).

The mechanism by which PEG promotes fusion is still unresolved. It is assumed that PEG acts as a molecular bridge between adjacent membranes either directly by hydrogen bond or indirectly by Ca^{2+} (18), further Ca^{2+} potentiate fusion by creating areas of membrane destability at the points of closest contact thus fascilitating the coalescence of adjacent cytoplasm. Subsequent resuspension of the protoplasts in hypertonic buffer ensures rehydration which results in the expanding of the inter protoplast cytoplasmic continuities. Following protoplast fusion, karyogamy may proceed or nuclei may co-exist in the cytoplasm as heterokaryon (49).

In 1982, another method of inducing fusion previously used for animal and plant cells was applied to *Saccharomyces cerevisiae* (35). This technique is known as electrofusion. It is two step process (i) the establishment of strong intermembranous contact (ii) optimum membrane destabilization. In this, protoplasts are subjected to a high frequency alternating electric field which causes the adhesion between protoplasts and formation of pearl like chain of protoplasts. It ensure a close membrane contact essential for fusion, follows by application of one to several short duration direct current pulses which, momentarily destabilize the protoplast membrane at site of membrane contact (63). This two step process can be followed microscopically and also controlled the strength and duration of dielectrophoretic field and pulses. Uniform size of the protoplasts in incubation mixture ensures the high frequency electrofusion. Large disparity in the protoplast size or presence of high proportion of cell debris results in their dielectrophoretic separation with a corresponding reduction in fusion frequency (43, 75).

Selection procedure for the recovery of fusion products must be employed to ensure that only fusion products are able to regenerate into colonies. Identification of fusants would be easier if the parental strain possess suitable markers. For such studies mutants with auxotrophic, antibiotic resistance, temperature sensitive and respiration deficient marker are used. Among these complementary auxotrophic mutants have been used frequently. Direct visual identification of fusion products may be possible if the parental strain carry colour markers i.e., either by spore colour (69) or by release of pigment in the medium (25). Pigmentation markers are also used along with auxotrophic markers (76, 94). Resistance to antimicrobial compounds or respiratory deficient markers can also be used as a method of selection of fusion products (44, 61).

4. APPLICATIONS OF PROTOPLAST FUSION

The essence of protoplast fusion is the bringing together of nuclear and cytoplasmic genomes of two fungal strains. It has great potential value in species where sexual mechanism is lacking or difficult to achieve and also to overcome species and genus barrier.

4.1 Conventional Genetic Studies

In fungi, vegetative incompatibility is widely spread and constitute a major barrier to strain improvement via hyphal anastomosis. It can be overcome by protoplast fusion, indicating that the expression of incompatibility occurs primarily at the level of cell wall since by fusing protoplasts of relatively incompatible strains heterokaryons were formed (20, 47, 91). In *Verticillium* spp. number of incompatible factors preventing the formation of heterokaryons via normal process. However, both heterokaryons and diploids in *Verticillium albo-atrum, V. dahliae* (91) and *V. lecanii* (47) were formed by using protoplast fusion technique. In *Pseudocercosporella herpotrichoides* protoplast fusion has also been used to overcome natural vegetative incompatibility (44). It is a valuable tool to force formation of hybrids in incompatible isolates of *Phoma tracheiphila* (13). In *Cryphonectria parasitica,* protoplast fusion resulted in the occurrence of a genetic rearrangement between 2 genomes, one of which appeared dominant. PEG-fused protoplast showed their ability to grow in the presence of both PCNB and $CoCl_2.6H_20$ (97). Both heterokaryons and diploids were recovered in *Aspergillus nidulans* with protoplasts from different heterokaryon compatibility (h-c) groups. A linkage map of genes responsible for h-c group in *A. nidulans* was made by treatment of diploids with benomyl (20).

In *Gibberella zeae,* heterokaryotic growth between two incompatible strains could not be sustained because the hyphal bridges essential for the maintenance of the heterokaryons did not form. The slow growing hybrid colonies obtained following protoplast fusion were considered to be the result of nuclear fusion at an early stage and consequently polyploid (1).

Protoplast fusion can also be used to overcome species and genus barrier producing interspecific and intergeneric hybrids. Generally, interspecific and intergeneric fusion frequency is below that of intraspecific fusion and it indicates the amount of genetic relatedness between the parent species involved in the fusion. Fusion products between *Aspergillus nidulans* and *A. rugulosus* grow as stable colonies on minimal medium and then on complete medium colony segregated. The pattern of segregation and DNA estimation suggested that the hybrids originated from the fusion of two parental nuclei and these fusants were diploid and/or aneuploid in nature (11, 50, 51). In case of *Plueurotus ostreatus* and *P. sajor-caju,* paired culture tests and isozyme analysis revealed that the fusant produced by using protoplast fusion method was a strain distinctly different from the two parents (66). Strains obtained in the *P. ostreatus* + *P. pulmonarius* and *P. pulmonarius* + *P. cornucopiae* fusions formed fruiting bodies on sawdust-wheat bran medium (89).

The protoplast fusion between cycloheximide - resistant mutant of *Fusarium oxysporum* f.sp. *lycopersici* and the mycostatin-resistant mutant of *F. oxysporum* f.sp. *radicis-lycopersici* produced hybrids which differed significantly from the parents in their expression of pathogenicity, growth and the electrophoretic pattern of their proteins enzymes and iso-enzymes. It indicate a transformed genetic basis exists for this altered expression and using protoplast fusion technology, it is feasible for examining the biology of pathogenicity genes and for elucidating the disease and virulence potential of new races from within hybridizable taxa of *Fusarium* spp. (64).

In intergeneric protoplast fusion between *Pleurotus ostreatus* and *Schizophyllum commune,* most of the fusants were monokaryotic and sterile and no heterokaryons occurred. These fusants showed a significantly higher nuclear DNA content when compared to parental strains. Arbitarily-primed PCR fingerprints revealed that most of the fusants were recombinant hybrids (102). Similarly protoplast fusion between *P. sapidus* and *Lentinus edodes* yielded hybrids. The results of random amplified polymorphic DNA analysis showed that the fusants were true hybrids but the amount of genetic information from the parents were not equal (62).

In stable fusion products by UV irradiation the frequency of mitotic segregation was enhanced above the spontaneous level in *Candida albicans.* Differences in the mitotic segregation pattern was established in one and the same fusion combinations. The chromosome rearrangement may be due to the processes connected with the hybrid formation after protoplast fusion (54).

In *Saccharomyces diastaticus*, the cell size in hybrid was larger than the haploid ones and the average cell volume being as large as the sums of the parental strain. The DNA content of fusants were variable (81). Increase in conidial diameter and DNA content was reported in *Aspergillus awamori* var. *kawachi (70)*, *A. sojae* (93) and *Candida albicans* (60). In *Pseudocercosporella herpotrichoides* and *P. anguioides* the hybrids produced spores with size inbetween the parents (42).

The electrophoretic karyotype of the fusant is a useful tool for characterization of fusion products and for assessing the interaction of different genomes in hybrids. Such an attempt was made by Kobori *et al.* (55) in studying the karyotype of fusants resulting from fusion between *Candida tropicalis* and C. *biodinii.* Some of the fusants yielded noval bands due to rearrangement of chromosomes of both parents while some of the fusants showed bands resembling to either of the parents whereas one fusant showed band of both the parents.

Isozyme analysis is also used for characterization of fusants. Some of the fusants resulting from the cross between *Pseudocercosporella herpotrichoides* and *P. anguioides* showed segregation of malate dehydrogenase and glucose phosphate isomerase (42). In *Candida albicans* and *C. tropicalis* isozyme analysis of hybrids showed either of the two parent partners (57). The hybrids derived from cross between *Penicillium chrysogenum and P. baarnense* showed an entirely new band for glucose-6-phosphate dehydrogenase (4).

4.2 Strain Improvement

In addition to its use in genetic analysis, protoplast fusion is an invaluable aid for strain improvement in relation to higher productivities of commercial enzymes and strains with suitable mycoparasitic and antagonistic attributes for potential as biocontrol organisms.

The diploid strain resulting from fusion between *Aspergillus awamori* var. *kawachii* exhibited high citric acid and amylase productivities corresponding to those of original strain, (70) and a protoprophic haploid segregance of the diploid strain from the cross between *A. usamii* mut. *shirousamii* and *A. niger* showed higher citric acid and amylase activities than those of diploid strain (71). A diploid strain resulted from fusion between *A. niger* and *A. kawachii* with high cellulose productivity and higher level of citric acid and it would be very useful for the degradation of cellulosic substances in mash from shochu fermentation (72). A recombinant strain of *A. niger* was isolated with 2.5 fold increase in β-glucosidase activity than the parental strain (45). A stable fusant from protoplasts fusion between *A. terreus* (an itaconic acid producer) and *A. usamii* (a glucoamylase producer) produced 5 times higher productivity of itaconic acid from soluble starch than *A. terreus* and 70% of *A. terreus* from glucose (52). Four prototrophic hybrids of *Aspergillus* sp. and *A. flaviceps* exhibited enhanced production of endo-pectinase and pectin lyase than those produced by the wild type strain (87).

Fusants obtained by *Saccharomyces diastaticus* produced more ethanol than the parental strains. The ethanol productivity changed with the carbon source used. Some fusants produced more ethanol in glucose medium than the parental strain but not always in starch medium (81). Whereas fusants obtained from cross between *S. cerevisiae* and *S. diastaticus* produced 45% more ethanol than the parental strain (32). Cross between *Saccharomycopsis fibuligera* and *S. cerevisiae* produced hybrids with 2.5 times higher amylase activity than the parental strain

(16). Hybrids from *Pachysolen tannophilus* and *S. cervisiae* showed capability of utilizing D-xylose. The hybrids resembled S. *cerevisiae* parent in morphology and sugar assimilation, the genes for utilization of xylose were transferred from *P. tannophilus* (40). A stable hybrid strain from protoplast fusion of *Malbranchea sulfurea* mutants resisted catabolite repression and produced an α-amylase, that digested raw starch, showed twice the activity of either parents (34).

Protopast fusion is achieved in large number of fungi but only few studies are related to fungal biocontrol agents. Using the technology it is possible to combine different traits into single superior, strain. The protoplast fusion has been used to improve biocontrol ability of *Trichoderma* sp. The diversity created by this process can provide a source of improved biocontrol strains. Recently, few strains with improved biocontrol efficacy have been obtained by using protoplast fusion. Pe'er and Chet (76) found only one strain from fusion between mutants of *T. harzianum,* which gave significant biocontrol of *Rhizoctonia solani.* Similarly only 2-4% of the fusants from fusion between *T. harzianum* strain T12 and T95, showed higher biocontrol ability. One strain 1295-22 was effective against an array of plant pathogenic fungi *viz., Botrytis cinerea, Fusarium graminearum, Pythium ultimum, R. solani, Sclerotinia homoeocarpa* and *Sclerotium rolfsii.* This strain is found to be more rhizosphere competent than the parental strains, maintained a constant high level along with root and increased the root elongation rate in cotton, maize, pea and cucumber (38, 84).

Two fusants resulted from the fusion between *T. harzianum* and *T. longibracheatum* showed improved antagonistic potential against *Bipolaris oryzae, Curvularia lunata, F. oxysporum* f.sp., *lycopersici, R. solani* and *Venturia inaequalis* and also tolerant to copper sulphate and carbendazim (59). Likewise, protoplast fusion between di-auxotrophic mutants of *Beauveria barssiana* resulted in the recovery of some stable fusants with enhanced biocontrol efficacy of *Ostrinia nubilalis* and *Lweptinotarsa decemlineate* (19).

4.3 Protoplast Fusion and Fungal Biotechnology

Protoplasts have also proved as a centre of development of fungal biotechnology. Transfer of isolated cytoplasmic genetic elements *viz.*, nuclei, mitochondria or plasmids into protoplasts may provide a novel means for complementation of genetic information.

The recombination of nuclei into protoplasts eliminates the recombination of extranuclear components and genetic analysis of the progeny resulted from nuclear transfer becomes less complex. Nuclei from *Saccharomyces cerevisiae* were transferred into protoplasts of the same species (29). In filamentous fungi protoplast of one strain was hybridized by using isolated nuclei of other strain e.g. *T. harzianum* (85) and *A. nidulans* (95). Hybridization of *Pichia stipitis* (xylose fermenting yeast) with nuclei isolated from *Fusarium moniliforme* and *T. ressei* was achieved using PEG. Resultants hybrids showed capability to hydrolyse xylan and were morphologically similar to *P. stipitis.* One of the hybrids showed higher xylanase activity than the parental strains (98). Similarly several other studies have demonstrated that mitochondria can be isolated and transferred to protoplasts of *S. cerevisiae* (33, 101).

In another field, namely transformation, using protoplasts it is possible to introduce genes or construction involving parts of different genes into a cell using a DNA molecule or vector which can be maintained in that cell. DNA mediated transformation was first described in *S. cerevisiae* in 1978 (8, 41) and shortly afterwards in *Neurospora crassa* (14). For isolation of transformed products, resistance to benomyl and hygromycin B was used in *Pleurotus ostreatus* (77), *Pseudocercosporella herpotrichoides* (10) and *T. harzianum* (65, 86, 92).

Number of the filamentous fungi actively secrete enzymes into their environment and because of this, they are important in the cycling of nutrients in natural ecosystem. Using this approach it is possible to manipulate the production of particular gene product. In one such approach protoplasts of *T. harzianum* were co-transformed using two plasmids containing bacterial chitinase gene from *Serratia marcescens* and other encoding acetamidase from *A. nidulans*. Two transformants showed increased constitutive chitinase activity and higher antagonistic activity against *Sclerotium rolfsii* than wild parents (37).

The major advantage of such genetic manipulations is the ability to introduce genes from one strain to other strain thus enhancing the potency of biocontrol agents and making a superior single strain which is effective, stable and consistent against an array of plant pathogenic fungi without the hazardous effect of chemical pesticides.

5. CONCLUSIONS

It is now technically possible to isolate and culture protoplasts from wide range of fungi. The diversity of fungal cell wall is a limiting factor in the application of particular enzyme. Therefore, the procedure can be improved particularly with respect to the enzyme system.

With the advent of protoplast fusion it is possible to develop constitutive high yielding strains and superior biocontrol agents with desirable properties. The protoplast fusion has great potential value to overcome incompatibility barriers and allow characteristics from even distantly related isolates to be combined. Furthermore it opens the door to hybridization between distant relatives with possibilities of novel products derived from resultant progeny. Protoplast fusion will undoubtedly play an important part in the exploitation of fungi.

6. REFERENCES

1. Adams, G., Johnson, N., Leslie, J.F.and Hart, L.P. 1987. Heterokaryons of *Gibberella zeae* formed following hyphal anastomosis or protoplast fusion. *Exp. Mycol.* **11** : 339-353.

2. Andrade, C.M.M., Oliveira, F.M. and Valter, R.L. 1992. *Candida fennica* : enhancement of protoplast formation and fusion. *Can. J. Microbiol.* **38** : 807-810.

3. Anne, J. and Peberdy, J.F. 1976. Induced fusion of fungal protoplasts following treatment with polyethylene glycol. *J. Gen. Microbiol.* **92** : 413-417.

4. Anne, J. and Peberdy, J.F. 1985. Protoplast fusion and interspecies hybridization in *Penicillium*. In : *Fungal Protoplasts : Application in Biochemistry and Genetics* (eds. J.F. Peberdy and L. Ferenczy), Marcel Dekker Inc., New York. pp. 259-277.

5. Arnold, W.N. and Garrison, R.G. 1979. Isolation and characterization of protoplasts from *Saccharomyces rouxii. J. Bacteriol.* **137**: 1386-1394.

6. Bachmann, B.J. and Bonner, D.M. 1959. Protoplasts from *Neurospora crassa. J. Bacteriol.* **78**: 550-556.

7. Bartinicki-Garcia, S. 1968. Cell wall chemistry, morphogenesis and taxonomy of fungi. *Ann. Rev. Microbiol.* **22**: 87-108.

8. Beggs, J. 1978. Transformation of yeast by a replicating hybrid plasmid. *Nature*, **283** : 835-840.

9. Benitez, T. , Villa, T.G. and Garcia Acha, I. 1975. Chemical and structural differences in mycelial and regeneration walls of *Trichoderma viride. Arch. Microbiol.* **105**: 277-282.

10. Blakemore, E.J.A., Dobson, M.J., Hocart, M.J., Lucas, J.A. and Peberdy, J.F. 1989. Transformation of *Pseudocercosporella herpotrichoides* using two heterologous genes. *Curr. Genet.* **16** : 177-180.

11. Bradshaw, R.E., Lee, K. and Peberdy, J.F. 1983. Aspects of genetic interaction in hybrids of *Aspergillus nidulans* and *Aspergillus rugulosus* obtained by protoplast fusion. *J. Gen. Microbiol.* **129**: 3525-3533.

12. Brenner, S., Dark, F.A., Gerhardt, P., Jeynes, M.H., Kandler, O., Kellenberger, E., Klienberger-Nobel, E., McQuillen, K., Rubio-Huertos, M., Salton, M.R.J., Strange, R.E., Tomcsik, J. and Weibull, C. 1958. Bacterial Protoplasts. *Nature* **181** : 1713-1715.

13. Cacciola, S.O. and Faedda, R. 1996. Production and fusion of protoplasts of genetically marked strains of *Phoma tracheiphila. Petria* **6** : 159-175.

14. Case,M.E., Schwiezer, M., Kushner, S.R. and Giles, N.H. 1979. Efficient transformation of *Neurospora crassa* by utilizing hybrid plasmid DNA. *Proceedings of the National Academy of Science*, U.S.A. **76**: 5259-5263.

15. Chadegani, M., Brink, J.J., Shehata, A. and Ahmadjian, V. 1989. Optimization of protoplasts formation, regeneration and viability in *Microsporum gypseum. Mycopathologia* **107**: 33-50.

16. Choi, S.H., Sung, C., Oh, M.J. and Kim. C.J. 1997. Intergeneric protoplast fusion in *Saccharomycopsis fibuligera and Saccharomyces cerevisiae. J. Ferment. Bioengg.* **84**: 156-161.

17. Cocking, E.C. 1973. Isolation, fusion and development of protoplasts of higher plants. In *: Yeast, Mould and Plant Protoplasts* (ed. J.R. Villaneuva, I. Garcia-Acha, S. Gascon and F. Uruburu) Academic Press, London, pp.309-318.

18. Constable, F. and Kao, K.N. 1974. Agglutination and fusion of plant protoplasts by polyethylene glycol. *Can. J. Bot.* **52**: 1603-1606.

19. Couteaudier, Y., Viaud, M. and Riba, G. 1996. Genetic nature, stability and improved virulence of hybrids from protoplast fusion in *Beauveria. Microbial Ecology* **32** : 1-10.

20. Croft, J.H. 1985. Protoplast fusion and incompatibility in *Aspergillus*. In *: Fungal Protoplasts : Applications in Biochemistry and Genetics* (eds. J.F. Peberdy and L. Ferenczy). Marcel Dekker Inc.,New York. pp.225-240.

21. Davis, B., d'Avillez Paixao, M.T., Deans, S.G. and Smith, J.E. 1977. Protoplast formation from giant cells of *Aspergillus niger. Trans. Br. Mycol. Soc.* **69** : 207-212.

22. de Vries, O.M.H. and Wessels, J.G.H. 1972. Release of protoplasts from *Schizophyllum commune* by a lytic enzyme preparation from *Trichoderma viride. J. Gen. Microbiol.* **73**: 13-22.

23. Deutch,C.E. and Parry, J.M. 1974. Sphaeroplast formation in yeast during the transition from exponential phase to stationary phase. *J. Gen. Microbiol.* **80**: 259-268.

24. Dias, E.S., Araujo, E.F., Guimaraes, W.V., Coelho, J.L.C. and Silva, D.O. 1997. Production and regeneration of *Penicillium expansum* and *Penicillium grisoroseum* protoplasts. *Revista de Microbiologia* **28** : 116-120.

25. Didek-Brumec, M., Gaberc-Porekar, V., Alacevic, M. and Socic, H. 1993. Strain improvement of *Claviceps purpurea* by protoplast fusion without introducing auxotrophic markers. *Appl. Microbiol. Biotechnol.* **38**: 746-749.

26. Eddy, A.A. and Williamson, D.H. 1957. A method of isolating protoplasts from yeast. *Nature* **179** : 1252-1253.

27. Evans, C.T. and Conrad, D. 1987. An improved method for protoplast formation and its application in the fusion of *Rhodotorula rubra* with *Saccharomyces cerevisiae*. *Arch. Microbiol.* **148** : 77-82.

28. Ferenczy, L. 1976. Some characteristics of intra- and interspecific protoplast fusion products of *Aspergillus nidulans* and *Aspergillus fumigatus*. In : *Cell Genetics in Higher Plants* (eds. D. Dudits, G.L. Farkas and P. Maliga,), Akademiai Kiado, Budapest, pp. 171-181.

29. Ferenczy, L. and Pesti, M. 1982. Transfer of isolated nuclei into protoplasts of *Saccharomyces cerevisiae*. *Curr. Microbiol.* **7**: 157-160.

30. Ferenczy,L., Kevei, F. and Szegedi, M. 1975. High frequency fusion of fungal protoplasts. *Experimentia* **31** : 1028-1030.

31. Giaja, J. 1919. Emploi des ferments dans less etudes de physiologie cellulaire : Le globule de levure depouille de sa membrane. *C.R. Soc. Biol. Fil. Paris* **82** : 719-720.

32. Grewal, H., Dilbaghi,N. and Sharma, S. 1994. Protoplast fusion between *Saccharomyces cerevisiae* and *Saccharomyces diastaticus* for ethanol production from starch. *Annals of Biol.* **10**: 80-85.

33. Gunge, N. and Sakaguchi, K. 1979. Fusion of mitochondria with protoplasts in *Saccharomyces cerevisiae*. *Mole. Gen. Genet.* **170**: 243-247.

34. Gupta, A.K. and Gautam, S.P. 1995. Improved production of extracellular alpha-amylase, by the thermophilic fungus *Malbranchea sulfurea,* following protoplast fusion. *World J. Microbiol. Biotech.* **11**: 193-195.

35. Halfmann, H.J., Rocken, W., Emeis, C.C. and Zimmermann, U. 1982. Transfer of mitochondrial function into a cytoplasmic respiratory deficient mutant of *Saccharomyces* yeast by electrofusion. *Curr. Genet.* **6**: 25-28.

36. Hamlyn, P.F., Bradshaw, R.E., Mellon, F.M., Santiago, C.M., Wilson, J.M. and Peberdy, J.F. 1981. Efficient protoplast isolation from fungi using commercial enzymes. *Enz. Microb. Technol.* **3** :321-325.

37. Haran, S., Schickler, H., Pe'er, S., Logemann, S., Oppenheim, A. and Chet, I. 1993. Increased constitutive chitinase activity in transformed *Trichoderma harzianum*. *Bio. Control* **3**: 101-108.

38. Harman, G.E., Taylor, A.G. and Stasz, T.E. 1989. Combining effective strains of *Trichoderma harzianum* and solid matrix priming to improve biological seed treatment. *Plant Dis.* **73** : 631-637.

39. Hashiba, T. and Yamada, M. 1982. Formation and purification of protoplasts from *Rhizoctonia solani*. *Phytopathology* **72** : 849-853.

40. Heluane, H., Spencer, J.F.T. , Spencer, D., de Figueroa, L. and Callieri, D.A.S. 1993. Characterization of hybrids obtained by protoplast fusion, between *Pachysolen tannophilus* and *Saccharomyces cerevisiae*. *Appl. Microbiol. Biotechnol.* **40** : 98-100.

41. Hinnen, A., Hicks, J.B. and Fink, G.R. 1978. Transformation of yeast. *Proceedings of the National Academy of Sciences* U.S.A. **75**: 1929-1933.

42. Hocart, M.J. and McNaugton, J.E. 1994. Interspecific hybridization between *Pseudocercosporella herpotrichoides* and *P. anguioides* achieved through protoplast fusion. *Mycol. Res.* 98: 47-56.

43. Hocart, M.J. and Peberdy, J.F. 1989. Protoplast technology and strain selection. In : *Biotechnology of Fungi for Improving Plant Growth* (eds. J.M. Whipps and R.D. Lumsden), Cambridge University Press, Cambridge, pp.235-258.

44. Hocart, M.J., Lucas, J.A. and Peberdy, J.F. 1993. Parasexual recombination between W and R pathotype of *Pseudocercosporella herpotrichoides* through protoplast fusion. *Mycol. Res* **97**: 977-983.

45. Hoh, K.Y., Tan, T.K. and Yoeh, H. 1992. Protoplast fusion of β-glucosidase producing *Aspergillus niger* strain. *Appl. Biochem. Biotechnol.* **37**: 81-88.

46. Housset, P., Nagy, M. and Schwencke, J. 1975. Protoplasts of *Schizosaccharomyces pombe* : an improved method for their preparation and the study of their guanine uptake. *J. Gen. Microbiol.* **90**: 260-264.

47. Jackson, C.W. and Heale, J.B. 1987. Parasexual crosses by hyphal anastomosis and protoplast fusion in the entomopathogen *Verticillium lecani. J. Gen. Microbiol.* **133** : 3537-3547.

48. Kao, K.N. and Michayluk, M.R. 1974. A method for high frequency intergeneric fusion of plant protoplasts. *Planta* **115** : 355-367.

49. Kavanagh, K., Walsh, M. and Whittaker, P.A. 1991. Enhanced intraspecific protoplast fusion in yeast. *FEMS Microbiol Lett.* **81** : 283-286.

50. Kevei,F. and Peberdy, J.F. 1984. Further studies on protoplast fusion and interspecific hybridization within the *Aspergillus nidulans* group. *J. Gen. Microbiol.* **130**: 2229-2236.

51. Kevei, F. and Peberdy, J.F. 1985. Interspecific hybridization after protoplast fusion in *Aspergillus*. In : *Fungal Protoplasts : Applications in Biochemistry and Genetics* (eds. J.F. Peberdy and L. Ferenczy), Marcel Dekker Inc., New York. pp.241-257.

52. Kirimura, K., Imura, M., Lee, S.P., Kato, Y. and Usami, S. 1989. Intergeneric hybridization between *Aspergillus niger* and *Trichoderma viride* by protoplast fusion. *Agric. Biol. Chem.* **53**: 1589-1596.

53. Kitamoto, Y., Mori, N., Yamamoto, M., Ohiwa, T. and Ichikawa, Y. 1988. A simple method for protoplast formation and improvement of protoplast regeneration from various fungi using an enzyme from *Trichoderma harzianum. Appl. Microbiol. Biotechnol.* **28**: 445-450.

54. Klinner, U. and Bottcher, F. 1985. Chromosomal rearrangements after protoplast fusion in the yeast *Candida maltosa. Curr. Genet.* **9**: 619-621.

55. Kobori, H. Takata, Y. and Osumi, M. 1991. Interspecific protoplast fusion between *Candida tropicalis* and *Candida boidinii* : characterization of the fusants. *J. Ferment. Bioengg.* **72**: 439-444.

56. Kovac, L. and Subik, J. 1970. Some aspects of the preparation of yeast protoplasts and isolation of mitochondria. *Acta. Fac. Med. Univ. Brun.* **37**: 91-94.

57. Kucsera, J. and Ferenczy, L. 1986. Interspecific hybridization between *Candida albicans and Candida tropicalis. FEMS Microbiol. Lett.* **36**: 315-318.

58. Kumari,J.A. and Panda, T. 1992. Studies on critical analysis of factors influencing improved production of protoplasts from *Trichoderma reesei* mycelium. *Enz. Microb. Technol.* **14**: 241-284.

59. Lalithakumari, D., Mrinalini, C., Chandra, A.B. and Anamalai, P. 1996. Strain improvement by protoplast fusion for enhancement of biocontrol potential integrated with fungicide tolerance in *Trichoderma* spp. *Zeitschrift fur Pflanzenkrankheiten und Pflanzenschutz* **103** : 206-212.

60. Law, C., Kavanagh, K. and Whittaker, P. 1994. Analysis of hybrids of *Candida albicans* formed by protoplast fusion. *FEMS Microbiol. Lett.* **115**: 77-82.

61. Layton,A.C. and Kuhn, D.N. 1988. Heterokaryon formation by protoplast fusion of drug resistant mutants in *Phytophthora megasperma* f.sp. *glycinea*. *Exp. Mycol.* **12** : 180-194.

62. Liu, G., Lui, Z., Lia, J., Liu, L., Tai, L., Li, X., Zhu, H., Zhu, L., Liu, G.Z., Lia, J.H., Liu, L.J., Tai, L.F., Li, X.B., Zhu, H. and Zhu, L.H. 1995. Studies on fusants derived from intergeneric protoplast fusion of *Pleurotus sapidus* and *Lentinus edodes* by RAPD analysis. *Heridatas*, Beijing. **17**: 37-40.

63. Lynch, P.T., Issac, S. and Collin, H.A. 1989. Electrofusion of protoplasts from *Aspergillus nidulans. FEMS Microbiol. Lett.* 59: 225-228.

64. Madosingh, C. 1994. Production of intraspecific hybrids of *Fusarium oxysporum* f.p. *radic-lycopersici* and *Fusarium oxysporum* f.sp. *lycopersici* by protoplast fusion. *J. Phytopath.* **142**: 301-309.

65. Manczinger, L., Antal, Z. and Ferenczy, L. 1995. Isolation of uracil auxotrophic mutants of *Trichoderma harzianum* and their transformation with heterologus vectors. *FEMS Microbiol. Lett.* **130** :59-62.

66. Matsumoto, M., Eguchi, F. and Hiqaki, M. 1997. Polyethylene glycol centrifugal fusion between *Pleurotus ostreatus* and *Pleurotus sajor-caju.* Mokuzai-Gakkaishi - *J. Jpn. Wood. Res. Soc.* **43** 215-219.

67. Musiikova, M., Fencl, Z. and Seichertova, O. 1969. Release of *Aspergillus niger* protoplasts by *Penicillium purpurogenum* enzymes. *Folia Microbiol.* **14**: 47-50.

68. Nga, B.H., Abu Baker, F.D., Loh, G.H., Chiu, L.L., Harashima, S., Oshima, Y. and Heslot, H. 1992. Intergeneric hybrids between *Saccharomycopsis fibuligera* and *Yarrowia lipolytica. J. Gen. Microbiol.* 138: 223-227.

69. Ogawa, K., Brown, J.A. and Wood, T.M. 1987. Intraspecific hybridization of *Trichoderma reesei* QM 9414 by protoplast fusion using colour mutants. *Enz. Microb. Technol.* **9** : 229-232.

70. Ogawa, K., Ohara, H. and Toyama, N. 1988. Intraspecific hybridization of *Aspergillus awamori* var. *kawachi* by protoplast fusion. *Agric. Biol. Chem.* **52**: 337-342.

71. Ogawa, K., Tsuchimochi, M., Taniguchi, K. and Nakatsu, S. 1989 Interspecific hybridization of *Aspergillus usamii* mut. *shirousamii* and *Aspergillus niger* by protoplast fusion. *Agric. Biol. Chem.* **53** : 2873-2880.

72. Ogawa, K., Fujii, N., Furukawa, K., Ashadi, R.W., Thayanauphat, P. and Tanaka, K. 1993. Transfer of cellulolytic character of *Aspergillus niger* into *Aspergillus kawachi* by their protoplast fusion. *J. Gen. Appl. Microbiol.* **39**: 443-452.

73. Peberdy, J.F. 1979. Fungal protoplasts : isolation, reversion and fusion. *Ann. Rev. Microbiol.* **33** 21-39.

74. Peberdy, J.F. 1985. Mycolytic enzymes. In : *Fungal Protoplasts: Applications in Biochemistry and Genetics* (eds. J.F. Peberdy and L. Ferenczy), Marcel Dekker Inc., New York. pp. 31-44.

75. Peberdy, J.F. 1989. Fungi without coats - protoplasts as tools for mycological research. *Mycol. Res.* **93** : 1-20.

76. Pe'er, S. and Chet, I. 1990. *Trichoderma* protoplast fusion : a tool for improving biocontrol agents. *Can J. Microbiol.* **36**: 6-9.

77. Peng, M., Lemke, P.A. and Shaw, J.J. 1993. Improved conditions for protoplast formation and transformation of *Pleurotus ostreatus. Appl. Microbiol. Biotechnol.* **40**: 101-106.

78. Picataggio, S.K., Schamhart, D.M.J., Montenecourt, B.S. and Eveleigh, D.E. 1983. Sphaeroplast formation and regeneration in *Trichoderma reesei*. *Eur. J. Appl. Microbiol. Biotechol.* **17**: 121-128.

79. Reyes, F., Perez-Leblic, M.I., Martinez, M.J. and Lahoz, R. 1984. Protoplast production from filamentous fungi with their own autolytic enzymes. *FEMS Microbiol. Lett.* **24** : 281-283.

80. Rlike, O., Baum, A., Weiss, J., Hommel, R. and Kleber, H.P. 1992. Kinetics of enzymatic lysis, formation and regeneration of protoplasts of *Candida (Torulopsis) apicola*. *World J. Microbiol. Biotechnol.* **8** : 14-20.

81. Sakai, T., Koo, K., Saitoh, K. and Katsuragi, T. 1986. Use of protoplast fusion for the development of the rapid starch fermenting strains of *Saccharomyces diastaticus*. *Agric. Biol. Chem.* **50**: 297-306.

82. Sandhu, D.K., Bawa, S. and Bagga, P.S. 1989. Optimization of protoplast production from *Trichoderma harzianum*. *Ind. J. Exp. Biol.* **27**: 180-181.

83. Shabana, Y.M. and Charudattan, R. 1997. Preparation and regeneration of mycelial protoplasts of *Alternaria eichhorniae*. *J. Phytopath.* **145**: 335-338.

84. Sivan, A. and Harman, G.E. 1991. Improved rhizosphere competence in a protoplast fusion progeny of *Trichoderma harzianum*. *J. Gen. Microbiol.* **137**: 23-29.

85. Sivan, A., Harman, G.E. and Stasz, T.E. 1990. Transfer of isolated nuclei into protoplasts of *Trichoderma harzianum*. *Appl. Environ. Microbiol.* **56**: 2404-2409.

86. Sivan, A., Stasz, T.E., Hemmat, M., Hayes, C.K. and Harman, G.E. 1992. Transformation of *Trichoderma* spp. with plasmids conferring hygromycin B resistance. *Mycologia* **84**: 687-694.

87. Solis, S., Flores, M.E. and Huitron, C. 1997. Improvement of pectinase production by interspecific hybrids of *Aspergillus* strain. *Lett. Appl. Microbiol.* **24** : 77-81.

88. Stasz, T. E., Harman, G. and Gullino, M.L. 1989. Limited vegetative compatibility following intra- and interspecific protoplast fusion in *Trichoderma*. *Exp. Mycol.* **13** : 364-371.

89. Takehara, T. , Kumata, A. and Aono, S. 1993. Interspecific protoplast fusion between some *Pleurotus* species using auxotrophic mutants. *J. Jpn. Wood Res. Soc.* **39** : 855-859.

90. Torres-Bauza, L.J. and Riggsby, W.S. 1980. Protoplasts from yeast and mycelial forms of *Candida albicans*. *J. Gen. Microbiol.* **119**: 341-349.

91. Typas, M.A. 1983. Hetrokaryon incompatibility and interspecific hybridization between *Verticillium albo-atrum* and *Verticillium dahliae* following protoplast fusion and microinjection. *J. Gen. Microbiol.* **129**: 3043-3056.

92. Ulhoa, C.J., Vainstein, M.H. and Peberdy, J.F., 1992. Transformation of *Trichoderma species* with dominant selectable markers. *Curr. Genet.* **21**: 23-26.

93. Ushijima, S. and Nakadai, T. 1987. Breeding by protoplast fusion of koji mold, *Aspergillus sojae*. *Agric. Biol. Chem.* **51** :1051-1057.

94. Ushijima, S., Nakadai, T. and Uchida, K. 1990. Further evidence on the interspecific protoplast fusion between *Aspergillus oryzae* and *Aspergillus sojae* and subsequent haploidization, with special reference to their production of some hydrolyzing enzymes. *Agric. Biol. Chem.* **54**: 2393-2399.

95. Ushijima, S., Nakadai, T. and Uchida, K. 1991. Interspecific electrofusion of protoplasts between *Aspergillus oryzae* and *Aspergillus sojae*. *Agric. Biol. Chem.* **55** : 129-136.

96. Vagvolgyi, Cs. and Ferenczy, L. 1991. Isolation of nuclei from *Aspergillus nidulans* protoplasts, *FEMS Microbiol. Lett.* **48**: 247-252.

97. Vannacci, G., Pecchia, S. and Fanti, S. 1997. Protoplast fusion of spontaneous drug resistant mutants from vegetatively incompatible isolates in *Cryphonectria parasitica. J. Plant Pathology* **79**: 51-59.

98. Vazquez, F.G., Heluane, H., Spencer, J.F.T. , Spencer, D.M. and DeFigueroa, L.I.C. 1997. Fusion between protoplasts of *Pichia stipitis* and isolated filamentous fungi nuclei. *Enz. Microbiol. Technol.* **21** :32-38.

99. Weibull, C. 1953. The isolation of protoplasts from *Bacillus megaterium* by controlled treatment with lysozyme. *J. Bacteriol.* **66**: 688-695.

100. Yang, H.A., Tommerup, I.C., Sivasithamparam, K. and O'Brien, P.A. 1992. Heterokaryon formation with homokaryons derived from protoplasts of *Rhizoctonia solani* anastomosis group eight. *Exp. Mycol.* **16**: 268-278.

101. Yoshida, K. and Takeuchi, I.K. 198O. Cytological studies on mitochondria induced cytoplasmic transformation in yeasts. *Plant Cell Physiol.* **21**: 497-509.

102. Zhao, J. and Change, S.T. 1996. Intergeneric hybridization between *Pleurotus ostreatus* and *Schizophyllum commune* by PEG-induced protoplast fusion. *World J. Microbiol. Biotech.* **12**: 573-578.

13 Health Problems in Buildings - A Holistic Approach

Jagjit Singh and Douglas Oughton

CONTENTS

ABSTRACT 204

1. INTRODUCTION 204
2. HEALTH PROBLEMS IN BUILDINGS 204
3. CAUSES OF THE PROBLEM 205
4. CAUSAL AGENTS OF ILLNESS AND STRESS 205
5. BUILDING ENVIRONMENTS 205
6. INDOOR ENVIRONMENTS 206
7. MICROBIAL POLLUTANTS 207
8. ALLERGIC LOAD AND COCKTAIL EFFECT 207
9. OCCUPATIONAL HAZARDS 208
10. AIRBORNE EXPOSURE 208
11. FOOD CONTAMINANTS 208
12. COMFORT AND HEALTH 208
13. ENVIRONMENTAL ASSESSMENT 209
14. REGULATIONS AND STANDARD 210
15. CONCLUSION 210
16. REFERENCES 210

Advances in Microbial Biotechnology
J.P. Tewari, T.N. Lakhanpal, Jagjit Singh, Rajni Gupta & B.P. Chamola (eds.)
APH Publishing Corporation, New Delhi - 110 002, India.

ABSTRACT

The interrelationship of building materials, construction, services and spatial arrangements with their environments, occupants and contents is complex and can influence our health and comfort in buildings. The causal agents of illness and stress can be of chemical, physical or biological origin and have a measurable impact on productivity. These illnesses may occur due to 'allergic load' and can be particularly aggressive to immuno-suppressed individuals. There are a number of environmental, design--use, management and construction factors which influence allergic components, for example, flora and fauna, materials and structures, design of ventilation systems, thermal insulation, tightness, air change and energy. This article will describe how scientific investigations combined with environmental engineering skills can provide a better understanding of building related illnesses, and can provide exemplary indoor environments with little cast penalty.

1. INTRODUCTION

The house of Commons Committee on Indoor Pollution states that the annual cost in loss of productivity and absenteeism due to indoor air quality problems is at least -$330 million (14). Among the influencing factors are pollution (ozone, formaldehyde, volatile organic, compounds etc.), inert fibres, microbiological growth (fungi, viruses, bacteria, mites, algae and other allergens), radon and electromagnetic fields (28).

Microbial contamination of indoor environments has received increasing attention in recent years as a possible cause of indoor-air-related illness (1). People spend long periods indoors, at home and work, and so they breathe mostly indoor air. The indoor air as well as the temperature, light and sound conditions in our dwellings, offices, schools and other premises is of considerable importance for the health, comfort, morale, productivity and well-being of the occupants (3, 8). The main reason for the neglect of indoor air quality issues was the lack of awareness of the problem, because the effects are mostly chronic and long-term and are not directly and immediately life-threatening (4, 9). The cost of one sick day due to an indoor air quality problem can equal the cost of energy for that one person's space for up to six months (12).

2. HEALTH PROBLEMS IN BUILDINGS

Indoor environment may influence the health of occupants in buildings in the following different ways:

2.1 Allergy and Environmental Hypersensitivity

Allergies such as rhinitis and asthma can be caused by diverse allergens, for example house dust mites, pollen, cat dander and moulds(7).

2.2 Building Related Illness

Building Related Illness and building associated illness are the terms used to cover the range of ailments which commonly affect occupants in buildings. For example Legionnaire's disease, radon, asbestos etc. (23, 24, 25, 26, 27, 28).

2.3 Sick Building Syndrome (SBS)

The World Health Organisation (WHO) defines health as 'a state of complete physical, mental and social well being, not merely the absence of disease and infirmity'. SBS is the name given to a condition in which the occupants of a building experience symptoms which disappear soon after the affected people leave the building. Other terms used are Tight Building Syndrome and Stuffy Office Syndrome (25).

The WHO identifies the following typical symptoms :

Stuffy nose	Blocked, runny or itchy nose
Dry chest	Dry skin
Chest tightness	Watering or itchy eyes
Lethargy	Headache
Loss of concentration	

3. CAUSES OF THE PROBLEM

The following categories broadly illustrate the causes of the SBS problem. These categories operate cumulatively and their cocktail effect is an additional risk factor contributing to SBS.

(i) **Design and construction factors** eg. office design and layout, poor lighting scheme, ergonomics.

(ii) **Environmental factors** eg. odour, lighting, noise, outdoor and indoor environment.

(iii) **Perceptual psychological factors** eg. hysteria and stress due to lack of privacy, or because of lack of control or claustrophobic effects due to sealed construction.

(iv) **Cultural and organisational factors** eg. maintenance and management and their relationships with occupants.

4. CAUSAL AGENTS OF ILLNESS AND STRESS

The causal agents of illness and stress may be categorised as chemical, biological, physical, psychosomatic and psychogenic (for example depression, anxiety, overwork and frustration).

5. BUILDING ENVIRONMENTS

The health of the occupants in buildings is influenced by the internal building environment which has a varied microclimate depending upon structural and services aspects.

Buildings can be likened to living organisms and their useful life depends on internal and external environments, both in terms of longevity of materials and as on appropriate habitat for its occupants (2, 5, 21, 29). Buildings work as spatial environmental ecosystems and provide ecological riches and microclimates for a wide variety of biological agents and must be understand as a whole. These biological agents can cause damage to the building structure and can raise concerns for the indoor air quality and the health of the building occupants (27, 31).

The interrelationships of building structures and materials with their environments and the living organisms within them are very complex (10, 11, 25). These interrelationships must be analysed through a multidisciplinary scientific study of the building in order to improve the role of the building environment from a total health perspective. The building must function in close correspondence with the processes and biorhythms of the body, for example regulation of moisture, breathing and heat balance(5, 21, 29).

6. INDOOR ENVIRONMENT-SOURCES OF INDOOR CONTAMINATION

Indoor air contaminants come from two sources; those from the outdoor environment moving inside and those from sources originating inside the buildings.

(i) **Materials**-formaldehyde, solvents, mineral fibres, radon gas and pesticides

(ii) **Construction**-airtightness and energy conservation versus ventilation for occupants and fabric

(iii) **Interior furnishings**-volatile organic compounds

(iv) **Services and controls**-thermal comfort, air conditioning, and control of indoor microclimate

(v) **Spatial arrangement**-building layouts, ceiling heights and volume of space per occupant

(vi) **Occupants**-occupant activities, moisture and introduction of pollutants, tobacco smoking

(vii) **Lighting levels**

(viii) **Humidity and mould growth**

(ix) **Noise**

(x) **Maintenance**-poorly maintained building fabric and controls

(xi) **Radon**

(xii) **Photocopying, cleaning and other office activities**-ozone, organic compounds, particulates

(xiii) **Odour and irritation**

(xiv) **Contaminated land**-emission of gases

Indoor environments can allow the accumulation and proliferation of microorganisms and their metabolites (i.e. endotoxins and mycotoxins) as well as other volatile organic compounds (VOC), and their circulation within the indoor air. Within the health care environment, treatment such as respiratory therapy, which forces room air into the patient's lungs, and invasive procedures that allow airborne pathogens to bypass the normal body defences, place a further strain on the immune response ability of already compromised (sick) people.

Illnesses in buildings usually occur as a response to microbial antigens that become aerosolised via heating, ventilation and air conditioning (HVAC) equipment that may contaminated, thereby causing, for example, legionnaire's disease.

7. MICROBIAL POLLUTANTS

Microbial pollution is a risk to health and is associated with allergic illnesses (19). Published results indicate that 20% of the population can be sensitised by airborne fungal spores in the UK. While 40% of the inspected houses in Germany suffer from mould-related problems (31). The medical consequences of immune response, allergic reactions, endotoxins, mycotoxins and epidemiology have been extensively studied by Miller (18), Morey *et al.* (20), Gravensen *et al.* (13) and Burge *et al.* (8). Similarly, Legionnaire's disease and Pontiac fever are associated with wet cooling towers and domestic hot-water systems in complex buildings.

According to the official published figures, some 560,000 people need treatment because of indoor pollution due to mites and mould in damp houses (14). Indoor airborne allergic components come from two sources: outdoor air-borne spores moving inside and allergic components originating inside the dwelling (23, 26). Biological growth within buildings is associated with moisture and the formation of microclimates; it also depends upon the type of buildings and their ventilation (30). Mould fungi thrive on surfaces on which there is nourishment and suitable humidity, for example on damp water pipes, windows and walls in kitchens and bathrooms, in central air-conditioning systems, circulation pumps, blowers, ventilation ductwork and air filters, central dehumidifiers, and inside damp structures. Allergenic substances can be airborne and inhaled, such as pollen, fungus and dust, or digested with mouldy food or drink. Investigations suggest that airborne allergies cause more problems throughout the world than all other allergies combined. Additionally cross-infection from patient to patient is of great concern in hospitals. The medical field that treats allergies recognises the following allergenic diseases: asthma, allergic rhinitis, serous otitis media, bronchopulmonary aspergillosis and hypersensitivity pneumonitis (23, 24, 25, 26, 27, 28).

8. ALLERGIC LOAD AND COCKTAIL EFFECT

For some people, an allergic reaction in the indoor environment may be friggered by non-biological factors, such as chemicals or other indoor air pollutants, emotional stress, fatigue or changes in the weather. These factors burden allergic people further if they are suffering from allergic reactions to biological contaminants. This combination is known as the 'allergic load'. Microbial contaminants propagated within the health care establishment are particularly aggressive to patients due to reduced immune system resistance (28).

Recently, attention has been focused on the cocktail effect of chemicals present in indoor air (21, 29). Volatile organic compounds may be produced from the use of wood preservatives and remedial timber treatment chemicals, moth-proof carpets, fungicides, mouldicide-treated paints, furnishing materials such as particle board and foamed insulation which may emit formaldehyde (2). Biological pollutants alone or in synergistic effect with any of the above-mentioned volatile organic compounds may produce symptoms such as stuffy nose, dry throat, chest tightness, lethargy, loss of concentration, blocked, runny or itchy nose, dry skin, watering or itchy eyes or headache in sensitive people. The 'sick building syndrome' (SBS) or tight building syndrome may arise from a variety of causes. Because of the uncertainties about the causes of SBS and the rising levels of health problems in buildings there is an increasing use of term building-related illness (BRI) to cover a range of ailments which commonly affect building occupants.

9. OCCUPATIONAL HAZARDS

Fungal, viral, mycoplasmic and bacterial diseases in man and animals are world-wide in distribution (7). Many diseases in people and their pets are closely related and in a few cases some species can infect both, for example fleas, mites, etc. In such cases, pets act as a reservoir of the inoculum and increase the chance that exposure will lead to infection in humans. Fungal infection of man and animals can be split into four categories:

(i) **Superficial**—where the surface of the skin is invaded.

(ii) **Cutaneous**—where the keratinized tissue of the body (outer epidermal cells of the skin, hair and nails) are invaded.

(iii) **Subcutaneous**-where invasion occurs in the deeper layers of the skin.

(iv) **Systemic**-where infection of the lung takes place by direct inhalation of fungal spores. From the lungs infection may be disseminated to all organs of the body and may cause death in humans and their pets.

10. AIRBORNE EXPOSURE

The cutaneous and subcutaneous mycoses are not considered to be airborne diseases. Common airborne systemic mycoses include: histoplasmosis, coccidiodomycosis, blastomycosis and paracocccidiodomycosis. Studies suggest that mycotoxins may play a role in the symptoms experienced by those building occupants who are heavily contaminated by certain fungal species. Mycotoxins are secondary metabolites produced by fungi. The most common mycotoxigenic fungi found in buildings are *Stachybotrys* spp., *Aspergillus* spp. and *Penicillium* spp. (17).

11. FOOD CONTAMINANTS

There are several species of fungi, bacteria and viruses which can cause contamination of food in hospitals. The most common species belong to the genera *Bacillus, Lactobacillus, Clostridium, Botrytis, Penicillium* and *Aspergillus*. Among food poisoning organisms the most common are *Salmonella, Staphylococcus, Clostridium* and *Bacillus*.

In some building environments, insects, rodents and birds act as reservoirs of infection or vectors of diseases to humans and their pets. An example of this is the association of bird roosting with the infection of school children by *Histoplasma capsulatum*. Pigeons are known to carry a mild form of psittacosis of *Salmonella*, which can be a potential health hazard. Rodents are known to transmit leptospirosis, frichinosis, rat bite fever, murine typhus fever, plague and salmonellosis.

12. COMFORT AND HEALTH

The quality of the built environment is associated with the health, comfort and productivity of building occupants. Perception of an odour is a comfort effect, whereas irritation is usually defined as an acute health effect. Comfort has been defined as that condition of mind which expresses satisfaction with the environment. Of the many days of work lost through absenteeism a notable number is caused by SBS and building-related illnesses which can also lead to low morale, inability to concentrate, eye strain and poor productivity. Careful environmental assessment of building plans and monitoring of the built environment can alert managers to problems before they arise (6). An independent investigation is required to recommend ways of improving standards and to suggest strategies for improving the quality of the work-place and minimising the impact of buildings on the environment. Most comfort standards are based on acceptable level of dissatisfaction normally taken as 20%.

Occupational exposure limits do not include the synergistic or cocktail effect of pollutants or their effects on compromised individuals. For example, some immuno-compromised individuals may experience allergic reactions to which normally healthy persons may not react. Sick building syndrome (or tight building syndrome) symptoms are sometimes associated with inadequate ventilation and result in loss of productivity and absenteeism (23, 26).

Health and comfort issues should be addressed as below:

(i) advice and guidance on environmental design, control and maintenance

(ii) environmental assessment and monitoring

(iii) air and water quality control

(iv) building services design and review

(v) energy efficiency assessment

(vi) stimulation-based problem solving

Health and comfort in the built environment is a cross-disciplinary issue requiring engineers and scientists expert in health and comfort, air conditioning and environmental control.

13. ENVIRONMENTAL ASSESSMENT

13.1 Assessment of Indoor Airborne Fungi

Biological aerosols are viable contaminants occurring as solid or liquid particulates in the air. These particulates vary in size and are measured in terms of microns (a micron is equivalent to 1/25,000 of an inch). They can occur as single, unattached organisms or can combine to form colonies and complex colloidal forms.

There are several methods for trapping bioaerosols and fungal spores in the built environment for quantitative and qualitative analysis. One technique is to collect spores on a microscope slide coated with a sticky substance such as Vaseline, or on a Petri dish coated with fungal nutrient media. A more reliable method for studying airborne fungal spores is through the use of suction pumps for example Andersen Sampler, SAS or the RCS sampler. Recently scientists have used biological systems and bioassays for monitoring indoor air quality (3).

A visual examination of the source, for example fungi growing on the edges of bathtubs and sinks, on wallpaper and on items stored in damp cupboards and basements, could lead to diagnosis of the fungi involved. Mould odours can be assumed to represent fungal contamination (16). This can be assessed by using animals, for example trained dogs, or complex biochemical analysis for volatile compounds.

The most common areas in the built environment which allow fungal growth and development are those related to moisture sources and moisture reservoirs. These reservoirs can be assessed for fungal contamination. They include humidifiers, dehumidifiers, wet carpeting, files, walls, wallpaper, sinks, bathtubs, windowsills and basements. Damp organic materials in the indoor air, including stored food, wood chips, bird droppings, and skin from pets, provide a food source for mites and for fungal growth and proliferation.

13.2 Risk Assessment

There are no official guidelines, standards or regulations for fungal or bacterial aerosols. The World Health Organisation suggests that:

- the presence of certain fungal pathogens such as *Aspergillus fumigatus* and certain toxigenic fungi such as *Stachybotrys atra* should be considered unacceptable.
- more than 50 colony forming units (CFU) of fungi from indoor sources per cubic metre of air should prompt investigation if there is only one species present.
- up to 150 CFU of fungi form indoor sources per cubic metre of air should be considered acceptable if there is a mixture of species.
- up to 500 CFU of fungi from indoor sources per cubic metre of air should be considered acceptable if the species present are primarily *Cladosporium* or other common phylloplane fungi; higher counts should be investigated to ensure there is no indoor source.

14. REGULATIONS AND STANDARDS

The workplace and the environment are now regulated by several Acts of Parliament, Control of Substances Hazardous to Health Regulations and various Building Regulations, HSE, EPA and CIBSE guidelines. The recent HSE-approved Code of Practice for Legionella, for example, requires employers and others to

(i) identify and assess the sources of risk

(ii) prepare a scheme for preventing or controlling the risk

(iii) keep records of its implementation.

Employers now have to show that they have exercised due diligence in the operation of their offices and buildings.

15. CONCLUSION

Buildings can be likened to living organisms and require an understanding of the interrelationships of architectural and engineering design, construction and installation, usage and operational procedures and preventative maintenance. These interrelationships must be analysed through a multidisciplinary scientific study of the building in order to improve the ongoing comfort and well-being needs of the occupants.

16. REFERENCES

1. Anderson, J. 1990. Fungal growth in damp dwellings: Assessment of prevalent moulds and their pathogenic potential. *Building Pathology* **90** : 121-126.
2. Anon. 1988. *Toxic Treatment*, London Hazards Centre, London.
3. Baird, J.C., Berglund, B. and Jackson, W.T. (eds) 1991. *Indoor Air Quality for People and Plants*, Swedish Council for Building Research, Stockholm.
4. Berglund, B., Lindvall. t. and Mansson, L. 1988. *Healthy Building* '88, Swedish Council of Building Research, Stockholm, p.445.
5. Bordass, W. 1989. The effect-for good or ill-of building services and their controls. *Building Pathology* 89 : 9-68.

6. BREEAM 1991. New Homes Version 3/91. An Environmental Assessment of New Homes, Building Research Establishment, Garston, Watord.

7. Burge, P. 1990. Building sickness-a medical approach to causes. *Indoor Air* **90(5)**: 3-14.

8. Burge, P.S., Jones P. and Robertson, A.S. 1990. Sick Building Syndrome: environmental comparisons of sick and healthy buildings. *Indoor Air* **90 (1)** 479-484.

9. Curwell, S., March, C. and Venables, R. 1990. *Buildings and Health*: The *Rosehaugh Guide* RIBA, London.

10. Fielden, B. 1970. *Conservation of Historic Buildings*. Butterworth Scientific, London.

11. Garratt, J. and Nowak, F. 1991. Trackling condensation, a guide to causes and remedies for surface condensation and mould in traditional housing. *Building Research Establishment*. Garston, Watford, p. 100.

12. Gaynor, R.P. 1993. Developing an IAQ management plan for Commercial Buildings in hearting, piping and air conditioning, pp.55-58.

13. Gravensen, S., Larsen, L., Gyntelberg, F. and Skov, P. 1990. The role of potential immunogenic components of dust (MOD) in the sick building syndrome. *Indoor Air* **90(1)** : 9-14.

14. House of Commons Environment Committee, 1991. *Indoor Pollution*, 6th report, HMSO, London.

15. Hutton, T.C., Lloyd, H., and Singh, J. 1991. The environmental control of timber decay. *Structural Survey* **10** : 5-21.

16. Hyppel, A. 1994. Mould odour in buildings. In : *Allergy Problems in Buildings*. (eds. Singh and Walker), Quay Publişhing, Lancaster.

17. Kuehn, K.A., Garrison, R., Robertson, L., Koehn, R.D., Johnson, A.L. and Rea, W.J. 1992. Identification of airborne microfungal populations from home environments within the Dallas Fort Work (Texas) region. *Indoor Environment* **1** : 285-92.

18. Miller, J. 1990. Fungi as contaminants in indoor air. *Indoor Environment* **90 (5)** : 51-64.

19. Monroe, J. 1990. Allergy and environmental mediine, Breakspear Hospital, St Albans.

20. Morey, A. (ed.) 1990. *Biological Contaminants in the Indoor Environment* ASTM.

21. Schimmelschmidt, M. 1990. Breathing life into housing, *RIBA Journal*, November, 57-58.

22. Singh, J. 1986. Biological Control of Late Blight Fungus, PhD thesis, University of London, London.

23. Singh, J. 1993. Biological contaminants in the built environment and their health implications. *Building Res. Inform.* **21(4)** : 216-224.

24. Singh, J. 1994a. *Building Mycology: Management of Health and Decay in Buildings*. E & FN Spon, London.

25. Singh, J. 1994b. Indoor air quality in buildings. Office Health and Safety Briefing, Crona.

26. Singh, J. 1996. Impact of indoor air pollution on health, comfort and productivity of the occupants. *Aerobiologia* **12** : 121-127.

27. Singh, J. 1996. Health, comfort and productivity in the indoor and built environment. Jan-Feb 1996. Karger Publications. pp. 22-34.

28. Singh, J. and Walker, B. 1994. *Allergy Problems in Buildings*. Quay Publishing, Lancaster.

29. Walker, B. 1990. A building aware of our needs, *Building Services*, December, 35-36.

30. Wathes, C. 1989. An aerobiology of animal houses. *Building Pathology* **89**: 96-107.

31. Waubke, N.V. and Kusterle, W. 1990. Mould infestations in residential buildings. Paper presented at 1st Symposium on Mould infestations, Innsbruck, January 1990.

14 Plant Pathogens and Microbial Products as Agents for Biological Weed Control

Robert E. Hoagland

CONTENTS

ABSTRACT 214
1. INTRODUCTION AND BACKGROUND 214
2. WEED CONTROL - SOME HISTORICAL PERSPECTIVE 216
3. WEED CONTROL WITH PATHOGENS 216
3.1 Classical Approach 217
3.2 Inundative Approach 218
4. DISCOVERY AND SCREENING OF MICROBIAL BIOHERBICIDES 220
5. PLANT DEFENSE 221
5.1 Physical/Chemical Defenses 221
5.2 Biochemical Defenses 223
6. PATHOGEN INTERACTIONS WITH CHEMICALS 228
6.1 Synergistic or Antagonistic Chemicals 229
6.2 Formulations 231
7. PHYTOTOXINS FROM MICROBES AS BIOHERBICIDES 232
8. LIMITATIONS AND CONSTRAINTS 234
8.1 Biological Limitations 234
8.2 Environmental Limitations 234
8.3 Technological Limitations 235
8.4 Commercial/Economic Limitations 235
9. RECENT STRATEGIES FOR THE DEVELOPMENT AND USE OF BIOHERBICIDES 235
9.1 Phytopathogenic Bacteria for Weed Control 236
9.2 Rhizobacteria as Bioherbicide Agents 236
9.3 Genetic Manipulation of Bacterial and Fungal Weed Pathogens 236
10. SUMMARY AND CONCLUSIONS 237
11. REFERENCES 238

Advances in Microbial Biotechnology
J.P. Tewari, T.N. Lakhanpal, Jagjit Singh, Rajni Gupta & B.P. Chamola (eds.)
APH Publishing Corporation, New Delhi - 110 002, India.

ABSTRACT

Some microbes (plant pathogens) and secondary microbial products (phytotoxins) exhibit potential as weed control agents. The initiative for using pathogens and phytotoxins from pathogens and other microorganisms as biological weed control agents (bioherbicides) began about three decades ago. Interest in these organisms (either directly or as sources of naturally-occurring phytotoxins) has increased recently, due to the search for less persistent, more selective, and more environmentally benign herbicides. Pathogens, and in some cases microbial phytotoxins, may be used directly on target weed species; or these phytotoxins may provide unique chemical templates for the synthesis of new herbicide classes with novel molecular modes of action. Pathogens have potential for use in integrated weed management programs, if the organisms can tolerate various agricultural chemicals. They may also interact synergistically with agrochemicals. Genetic engineering and microbial strain selection can be used to increase pathogen virulence, alter host range, and enhance interactions with other chemical regulators or synergists. Recent advances in the isolation and identification of novel microbial chemistries with phytotoxic properties, and perspectives on other aspects and problems of the use of microbes for weed control are presented.

1. INTRODUCTION AND BACKGROUND

It is estimated that agricultural output will need to increase nearly 35% by the year 2025 to support the projected world population of 8 billion (279). Modern agricultural practices rely heavily on synthetic chemicals to maintain insect pests and weed populations below economic thresholds. Although insects are often assumed to be the greatest threat to crop production, weeds cause greater economic losses. However, of the 250,000 species of plants identified to date, those weed species causing serious economic losses in cultivated world crops number less than 250 (156). Notable advances in weed control technologies have been achieved, yet weeds continue to cause major agricultural losses. Reasons for this include the expansion of weed ranges, the introduction and spread of invasive exotic weeds, and increased incidences of herbicide resistance. The 1995/96 International Survey of Herbicide-Resistant Weeds recorded 183 herbicide-resistant weed biotypes, representing 124 different species in 42 countries (133). The rate of increase in herbicide-resistant weeds has remained relatively constant over the past 20 years, averaging nine new cases per year worldwide.

In the U.S. alone, weeds were estimated to cause reductions in productivity resulting in monetary losses in field crops and vegetables of over $7 billion per year in 1975-79. In addition to losses in productivity are the cost of the pesticides and their application. From the late 1960's through 1987, the use of herbicides in the U.S. exceeded that of insecticides and fungicides combined (10). The world pesticide market grew by about 3.6% in 1996, the third annual increase, to approximately $31.25 billion U.S. (1). Very recent global data shows that herbicides accounted for 47% of total pesticide sales in 1997, again exceeding the combined sales of insecticides (27%) and fungicides (19%) (Figure 1).

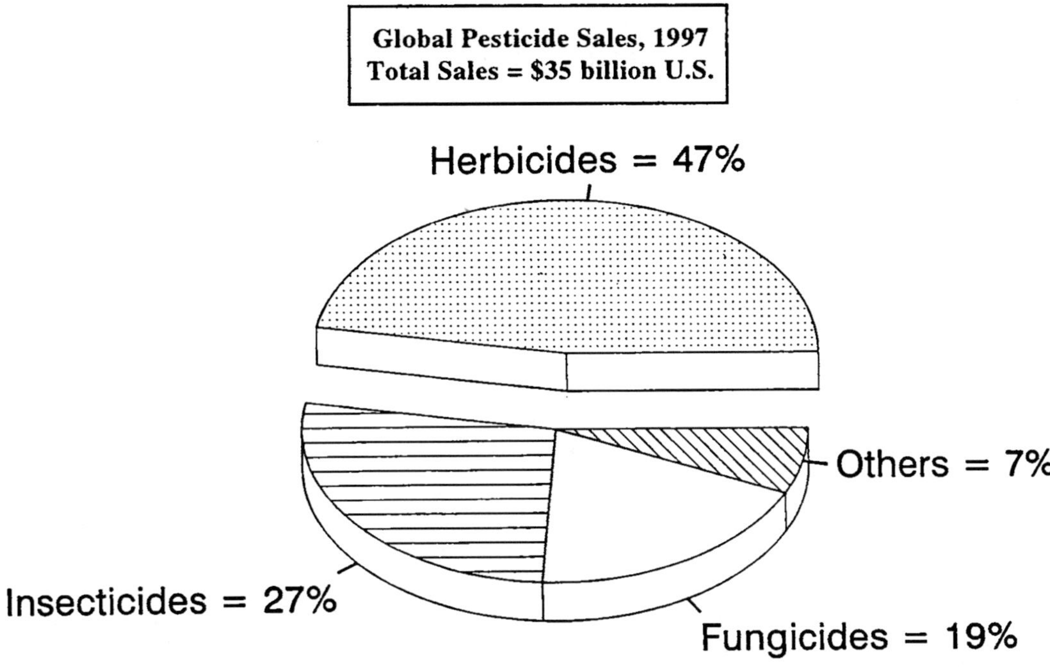

Fig. 1 : Global pesticide sales for 1997 (157).

Although synthetic chemical pesticides play a major role in providing the food and fiber required by the world's population, their use is not without negative aspects. Real and perceived public concern about food safety, adverse environmental impact, and pest resistance have caused increased scrutiny of synthetic pesticide use. For example, in Ontario, Canada, a government initiative, "Food Systems 2002," has been proposed to reduce the use of all pesticides by 50% by the year 2002 (266). These restrictive trends, coupled with the rising costs of developing, testing, and registering conventional synthetic materials with the U.S. Environmental Protection Agency and the European Community, have provided impetus to the research and development of pesticides with reduced risk, enhanced environmental safety, increased selectivity, and less persistence (93, 280). Generally, the motivation for developing mycoherbicides was for use against major weeds in agronomic crops. However, there has recently been increased interest in developing these products for areas where synthetic herbicides are not or cannot be registered for use. The increasing interest in biologically-based pesticides, including bioherbicides, has resulted in an estimated sales increase of 10-25% for these products; but generally agrochemical markets have plateaued or are declining slightly (240). Although various biological pest control systems have been studied and used with some success for over 100 years, many of these biological methods have not been, or cannot easily be, adapted to modern agricultural management practices. However, the developments outlined above should continue to promote research in the area of biological pest control, including the use of biological agents for weed control. The sustained interest in this area is evidenced by the publication of three major books on biological weed control using microbes and microbial products (27, 147, 270) in less than 10 years, and a continuing stream of journal reports on various aspects of this subject.

This chapter will examine, discuss, and assess the rationale, progress, potential, problems, and challenges related to the use of plant pathogens and the novel phytotoxins derived from microorganisms for weed control. Emphasis will be placed on the biochemistry of pathogen-weed and pathogen-chemical interactions, and on the inundative use of pathogens for weed control (defined below). Hopefully, this presentation will stimulate further discussion and interest in the use of microbial products as biological weed control agents, a strategy whose time may now be at hand.

2. WEED CONTROL - SOME HISTORICAL PERSPECTIVES

Weed control has gone through many phases since man shifted from his nomadic hunter-gatherer lifestyle to form agrarian communities. Hand-weeding and cultivation were the main methods of weed control until the development of synthetic herbicides beginning in the early- to mid- 1940's. There are currently about 150 herbicides comprising about 15 major chemical classes registered for use in the U.S. (302, 303). A chronology of some weed control developments, with particular emphasis on biological weed control with pathogens, is presented in Table 1. Biocontrol has been recognized as a feasible method of weed control for some time. However, the advent (and success) of highly efficacious synthetic chemical herbicides has made biological control a less popular weed control option. Most weed biocontrol successes have involved the use of introduced phytophagous insects rather than phytotoxic microorganisms. A major exception was the introduction of *Puccinia chondrilla* for control of rust skeletonweed (*Chondrilla juncea*) and skeletonweed (*Lygodesmia juncea*) in Australia (298). The use of naturally-occurring diseases and their causative agents is a rather new approach to an age-old problem. Plant diseases have been recognized and have concerned farmers and interested researchers for years, and many pathogens with bioherbicidal potential have been discovered. However, most of these pathogens lack sufficient virulence to adequately overcome weed defenses under normal environmental conditions; and there have been few attempts to formally apply the concept of the selective use of diseases for weed control. Other important conceptual and technological milestones in the development of mycoherbicides for weed control have been reviewed elsewhere (29, 50, 122).

Much research has been conducted in the area of biological control of weeds with plant pathogens since these initial successes. However, much of what is needed to identify, establish, and market a new individual successful biocontrol agent remains to be determined. In my view, a rational and environmentally beneficial approach to the weed control problem is integrated control; i.e., using synthetic chemicals, together with biologically-based components such as phytophagous insects (51), plant pathogens (131, 169, 274, 298), allelopathic cover crops, and/or chemicals derived from plants and microbes (88, 103, 262).

3. WEED CONTROL WITH PATHOGENS

The use of biotic agents to reduce or suppress pest populations is referred to as biological control. Bioherbicides can be defined as microorganisms (fungi, bacteria, viruses, algae, etc.) or their secondary products that are used, or have potential for use, in weed control when inundatively applied. Fungal pathogens used in this manner for weed control are often referred to in the literature as "mycoherbicides." The development, marketing, and use of living organisms as bioherbicides is considerably more complex than the developmental protocols for synthetic phytotoxins (herbicides). There are two generally-accepted approaches to biological weed control, i.e., classical (inoculative) and inundative or augmentative (49, 50).

Table 1 : Chronology of some important events related to Weed Control with Herbicides, Biocontrol Agents, and Microbial Products

1941:	Chemical synthesis of 2,4-D (234)
1942:	2,4-D reported to be a plant growth regulator (315)
1944:	2,4-D used for field weed control (127)
1945:	Principle of preemergence soil herbicides established
1954:	Role of diseases in weed control recognized (106)
1968:	*Principles of Plant and Animal Pest Control. Vol. 2: Weed Control* (publication 1597) published by the National Academy of Sciences, confirmed importance of weed science, including biological control, as a discipline
1969:	Discovery of the endemic disease, *Colletotrichum gloeosporioides* (Penz.) Sacc. f. sp. aeschynomeme on the weed jointvetch [*Aeschynomeme virginica* (L.) B.S.P.] (257)
1971:	Successful establishment of *Puccinia chondrilla* against rush skeletonweed (*Chondrilla juncea*) and skeletonweed [*Lygodesmia juncea* (Pursh.) D. Don] (298)
1973:	First successful use of augmentative approach for biocontrol of northern jointvetch (76)
1976	Phosphinothricin (glufosinate) herbicidal properties patented (152)
1977:	First microbial herbicide (bialaphos) patented (211)
1981:	DeVine®, a formulation of *Phytophthora palmivora* (Butler) Butler, registered for strangelvine (*Morrenia odorata* Lindl.) control in citrus (272)
1982:	*Colletotrichum gloeosporioides* (Penz.) Sacc. f. sp. *aeschynomeme* registered as Collego®, a formulation for selective control of northern jointvetch (272)
1985:	Development of glyphosate-tolerant tobacco via genetic transformation (60)
1989:	Microbial strain selection for bialaphos production to give 500-fold increase above wild type (267)

3.1 Classical Approach

Epidemics of late blight of potato (*Solanum tuberosum* L.) in Northern Europe in the 1840s, which resulted in the death of hundreds of thousands of people, illustrate the impact plant diseases can have on crops. Plagues of plant pathogens in crops have occurred throughout history; but the concept of using pathogens for weed control only occurred around the turn of the 20th century. Initial attempts utilized the classical approach of biological weed control; i.e., the selection and introduction of an exotic organism that is self-perpetuating and can thus provide long-term control of an introduced pest. This approach is considered to be ecological since the organism moves, propagates, and controls its host, as environmental conditions permit. Several biocontrol microorganisms have been successfully used in the classical approach to control a variety of weed pests (Table 2). At least 23 exotic plant pathogens have been investigated for classical biological weed control (221). A landmark example demonstrating success of this approach was the control of skeletonweeds by *Puccinia chondrilla* in Australia (Tables 1 and 2).

3.2 Inundative Approach

The inundative approach involves applications of large quantities of a mass-produced biological control agent over a localized area to achieve short-term pest control. Repeated applications can be made as necessary during a growing season, or from season to season. Although most agents used in this approach are indigenous, exotic plant pathogens have also been used (209). The first weed control success using this approach was the use of *Colletotrichum gloeosporioides* (Penz.) Sacc. f. sp. *aeschynomene* against northern jointvetch [*Aeschynomene virginica* (L.) B.S.P.] in rice (*Oryza sativa*) in 1973 (76). Another approach is the augmentative approach, in which inoculum is also introduced into the field, though it is not mass-produced or inundatively applied over large areas. In this case, weed control results from the natural increase of the disease through many disease cycles (45). Many fungal pathogens have been isolated and identified as having potential for bioherbicidal activity against weeds (Table 3). Some of these pathogens have very narrow host ranges and may only have potential to control one weed species. However, several fungal pathogens can utilize multiple weed species as hosts (Table 4). About 70 weeds have been targeted for biocontrol using over 100 mycoherbicidal pathogens (50). To date, three pathogens have been registered and marketed as bioherbicides in the U.S. or Canada for control of weeds in agricultural situations. These are Collego® (*Colletotrichum gloeosporioides* (Penz.) Sacc. f. sp. *aeschynomene*), a foliar pathogen for northern jointvetch control; DeVine® [*Phytophthora palmivora* (Butler) Butler], a soil-borne fungus for stranglervine (*Morrenia odorata*) control; and BioMal® (*Colletotrichum gloeosporioides* f. sp. *malvae*), a foliar pathogen for round-leaved mallow (*Malva pusilla* Sm.) control (13, 271, 300).

Table 2. Examples of weed control using plant pathogens in the classical biological control approach.

Biocontrol organism	Target weed: Scientific name	Target weed: Common name	Reference
Puccinia chondrilla Bub. & Syd.	*Chondrilla juncea* L.	skeletonweed	(130)
Phragmidium violaceum (Schultz) Winter	*Rubus* sp.	blackberry	(131)
Uromycladium tepperianum (Sacc.) McAlp.	*Acacia saligna* (Labill.) Wendl.	acacia tree	(216, 217)
Puccinia carduorum Jacky	*Carduus thoermeri* Weinm.	musk thistle	(235)
Puccinia jaceae Otth.	*Centaurea diffusa* Lam.	diffuse knapweed	(129, 222, 301)
Entyloma ageratinae Barretto and Evans	*Ageratina riparia* (Regel) K. & R.	mistflower	(216, 278)

Table 3. Examples of fungal bioherbicides for economically important terrestrial and aquatic weeds.

Weed	Pathogen	Reference
Velvetleaf (*Abutilon theophrasti*)	*Colletotrichum coccodes*	(150)
	Fusarium lateritium	(289)
Giant ragweed (*Ambrosia trifida*)	*Fusarium lateritium*	(11)
Wild oat (*Avena fasuca*)	*Septoria tritici* f.sp. *avenae* Desm.	(200)
Common lambsquarters (*Chenopodium album*)	*Ascochyta caulina* Sacc.	(244)
	Cercospora chenopodii Fres.	(244)
	C. dubia (Riess) Wint.	(244)
Field bindweed (*Convolvulus arvensis*)	*Phomopsis convolvulus*	(11) (287)
Yellow nutsedge (*Cyperus esculentus*)	*Cercospora caricis* Oud.	(11)
Purple nutsedge (*C. rotundus*)	*Phyllachora cyperi* Rehm.	(11)
Large crabgrass (*Digitaria sanguinalis*)	*Pyricularia grisea* (Cke.) Sacc.	(11)
Barnyardgrass (*Echinochloa crus-galli*)	*Cochliobolus lunatus* Nelson & Haasis	(243)
Waterhyacinth (*Eichhornia crassipes*)	*Alternaria eichhorniae* Nag Raj & Ponnappa	(246)
Goosegrass (*Eleusine indica*)	*Bipolaris setariae* (Saw.) Shoemaker	(11)
Common purslane (*Portulaca oleracea*)	*Dichotomophthora indica* Rao *D. portulacae* Mehrlich & Fitzpatrick ex M.B. Ellis	(17) (180)
Itchgrass (*Rottboellia cochinchinensis*)	*Curvularia* sp.	(99)
	Phaeoseptoria sp.	(99)
Johnsongrass (*Sorghum halepense*)	*Sphacelotheca holci* Jack.	(205)
	Bipolaris halepense Chiang, Leonard & Van Dyke	(54)
	B. sorghicola (Lefebvre & Sherwin) Alcorn	(309)
	Colletotrichum graminicola (Ces.) G.W. Wils.	(11)
	Gloeocercospora sorghi D. Hain & Edg.	(11)
Sicklepod (*Cassia obtusifolia*)	*Pseudocercospora nigricans*	(153)

Adapted from Charudattan (49)

Table 4. Examples of broad-spectrum fungal bioherbicides that have been tested for control of various weeds.

Pathogen	Weed species or family	Reference
Alternaria cassiae	sicklepod (*Cassia obtusifolia*)	(31)
	coffee senna (*Cassia occidentalis*)	(53)
	showy crotalaria (*Crotalaria spectabilis*)	(290, 291)
Amphobotrys ricini (Buchw.) Hennebert	Members of Euphorbiaceae	(155, 307)
		(307)
Colletotrichum gloeosporioides	Members of Leguminosae, Malvaceae,	(76)
various f. spp.	Convolvulaceae (dodders)	(219)
C. dematium (Fr.) Grove f. spp.	Members of Leguminosae	(41)
C. graminicola	Members of Gramineae	(11)
Exserohilum monoceras	Echinochloa species; Rottboellia cochinchinensis	(313)
Fusarium lateritium	velvetleaf (*Abutilon theophrasti*)	(289)
	giant ragweed (*Ambrosia trifida*)	(11)
	spurred anoda (*Anoda cristata*)	(289)
	prickly sida (*Sida spinosa*)	(289)
	Potamogeton spp.	(22)
Myrothecium verrucaria	sicklepod; various species of other plant families	(295)
Pyricularia grisea	large crabgrass (*Digitaria sanguinalis*)	(11)
	goosegrass (*Eleusine indica*)	(102)
Sclerotinia sclerotiorum (Lib.) de Bary	Multiple genera and species	(37)

Adapted from Charudattan (49)

4. DISCOVERY AND SCREENING OF MICROBIAL BIOHERBICIDES AND MICROBIAL PHYTOTOXINS

Most commercial synthetic herbicides have been discovered through random screening programs, rather than via biorational approaches for herbicide design. Over the past 50+ years, perhaps about a million synthetic compounds have been screened for herbicidal activity. The number of compounds screened has continually increased and, estimates in the 1980's suggested that 12 to 15 thousand compounds per year were synthesized and screened for each successfully-marketed compound. When potential herbicidal candidates are discovered, derivative synthesis is followed by phytotoxicity studies. Screening of prospective compounds typically involves greenhouse testing on up to a dozen or more important crop and weed species. However, screening strategies have recently undergone a change, and specific target sites are also being considered in order to improve chances of discovering potent herbicides that have minimal effect on non-target sites. Simple model species, cell free assays, and specific enzyme assays may provide additional advantages in such screening programs (25). This emphasis on synthetic compounds has obviously resulted in some excellent products that have saved producers much time, labor, and money, allowing them to provide food and fiber to consumers at reasonable costs. Research efforts on the study of pathogens for weed control have been dwarfed by the massive synthesis, screening, and evaluation efforts carried out in the search for synthetic herbicides.

As with synthetic herbicides, it is important that bioherbicidal pathogens exhibit sufficient host specificity and efficacy with minimal impact on non-target organisms. However, this narrow host specificity can also have a negative aspect; a broad weed-host range can make a particular organism more attractive commercially. The two main purposes of host-range studies are to determine the range of weed species controlled and to identify non-host plants (crops, ornamentals, etc.) that are susceptible to the weed control agent. Host-range tests of a pathogenic organism are typically conducted in the greenhouse or growth chamber prior to field testing (192, 297), using optimal disease-development environments, and the use of both susceptible and resistant plants. Host-range tests include all known biotypes of the host and other plants closely related to the known host (16, 297). Even if preliminary tests indicate that a pathogen has a desirable host range, scrutiny of the epidemiology is a critical factor in determining the potential for further development as a bioherbicide. Although there is a perception by some that naturally-derived weed control agents (pathogens or phytotoxins) are more environmentally benign than their synthetic counterparts, they are not without risk. Several safety factors need to be taken into account when considering these bioherbicidal weed control options. These include displacement of non-target organisms, toxigenicity or pathogenicity to non-target organisms, and allergic reactions in humans and other animals (64). More information is needed to understand the mechanisms of weed defense and the virulence factors of pathogens (recognition, phytotoxin production, etc.), as related to host range. Attempts to alter host range via formulation, synergistic interactions, and biotechnological approaches are discussed later.

5. PLANT DEFENSE

Before any pathogen or phytotoxin can have the desired effect on a target weed population, several host defense parameters must be addressed. In the case of a pathogen, the natural defenses of the intended host must be overcome. For a phytotoxin to be successful, there must be a lack of a detoxification mechanism in the target species. A thorough knowledge and understanding of plant defense systems is critical to this success. Therefore, a rather in-depth examination of these defense mechanisms and processes will be presented. Plant defenses (physical/chemical barriers and biochemical responses) serve to protect plants against attack from essentially all microorganisms and other stressors. But some pathogens possess mechanisms that allow them to either evade or breakdown plant defenses, causing infection that can lead to injury or death. Resistance to pathogens has been a major focus of crop germplasm improvement programs. The opposite is true for weed control; i.e., it is desirable to increase pathogenicity or weaken weed defenses. Thus, there is rationale for the exploitation of synergistic or additive interactions of chemicals and bioherbicides for weed control. Herbicides that affect many of the physical and biochemical barriers have been reported. Thus, if weed defenses can be lowered via chemical regulators, weeds can be made more susceptible to pathogen attack, allowing reductions in the concentration of bioherbicides, chemicals, or both. The host range of a given pathogen might then be expanded to other weeds by using chemical synergists.

5.1 Physical/Chemical Defenses

5.1.1 Callose

Callose is composed mostly of β-1-3 glucans situated at the surface of some intact plant cells. This material is synthesized by UDP-glucose:glucan synthase, and cellular damage via pathogen attack or other stresses can potentiate increased callose production (144). Callose production has also been related to limiting viral spread in hypersensitive responses (251). Callose synthesis and deposition can be regulated chemically. Activation of callose synthesis is associated with increased levels of Ca^{+2} (172), and UDP-glucose:glucan synthase is synergistically activated by both Ca^{+2} and a β-glucoside (38, 160). Syringomycin (a bacterial phytotoxin) elicits Ca^{+2}-dependent callose synthesis in cell cultures of Madagascar periwinkle (*Catharanthus roseus*) at very low concentrations (173). β-Furfuryl-β-glucoside (FG), a specific endogenous activator of higher plant callose synthase (Ohana et al. 1992), may be a component of the elicitation signal of callose synthase *in vivo* (226). Glucans also have other roles in plant defense processes, such as elicitation of antiviral properties (265) and phytoalexins (antimicrobial compounds) (15). 2-Deoxyglucose can reduce formation of callose-containing papillae in ml-o barley (*Hordeum vulgare* L.) tissues, enhancing pathogenicity of powdery mildew (18). Ca^{+2} chelators (or ionophores) or other compounds that interfere with callose synthase have been suggested as candidates to test for interactions with bioherbicides (144).

5.1.2 Wax-lipid-herbicide interactions

Epicuticular waxes are physical barriers to pathogen attack, and can reduce spore germination and germ tube number e.g., *Alternaria brassicae* (Berk.) Sacc. conidia (61). Several herbicides can affect wax biosynthesis, composition, or deposition in crop plants and weeds. For example, seven herbicides inhibited and four (auxin-type) stimulated lipid synthesis in hypocotyls of the weed hemp sesbania [*Sesbania exaltata* (Raf.) Cory] (203). Effects of various herbicides on lipid or wax accumulation in several plant species are summarized in Table 5. Not all studies used pure compounds; thus adjuvants, surfactants, and other formulation additives could have contributed to these effects. Although these studies were designed to determine herbicide penetration and mode of action, and herbicide effects on transpiration, they are among the few research efforts that demonstrate chemical effects on a defense barrier of weeds.

Table 5. Herbicides that alter wax or lipid biosynthesis in crop or weed species.

Herbicide	Plant species	Effect	Reference
Amitrole[1]	*Carissa* sp., *Maba* sp., *Gymnosporia* sp., & *Randia* sp.	reduced epicuticular waxes & lipids	(83)
CDEC[2]	Pea (*Pisum sativium*)	reduced/altered leaf waxes	(261)
CIPC[3]	Pea	little effect	(261)
Diallate[4]	Pea	reduced epicuticular lipids	(261)
Diuron[5]	*Brassica Oleracae*	inhibited *de novo* fatty acid synthesis	(183)
EPTC[6]	Pea Sicklepod (*Cassia obtusifolia*)	reduced/altered leaf waxes reduced petiole cuticle thickness	(261) (308)
Metolachlor[7]	Sorghum (*Sorghum bicolor*)	inhibited alcohols & fatty acids	(95)
Propanil[8]	*Carissa* sp., *Maba* sp., *Gymnosporia* sp., & *Randia* sp.	reduced epicuticular waxes & lipids	(83)
TCA[10]	Pea, *Brassica Oleracae*	reduced epicuticular waxes	(168, 182)
2,4,5-T[11]	*Carissa* sp., *Maba* sp., *Gymnosporia* sp., & *Randia* sp.	reduced epicuticular waxes & lipids	(83)

[1] 1*H*-1,2,4-triazol-3-amine
[2] 2-chloroallyl diethyldithiocarbamate
[3] 1-methylethyl 3-chlorophenylcarbamate
[4] *S*-(2,3-dichloro-2-propenyl) bis(1-methylethyl) carbamothioate
[5] *N'*-(3,4-dichlorophenyl)-*N,N*-mimethylurea
[6] *N'*-(3,4-dichlorophenyl)-*N,N*-dimethylurea
[7] *S*-ethyl dipropyl carbamothioate
[8] *N*-(3,4-dichlorophenyl) propanamide
[10] trichloroacetic acid
[11] 2,4,5-trichlorophenoxyacetic acid

5.1.3 Hydroxyproline-rich glycoproteins (HRGPs)

Defense proteins can be categorized according to role (28). Accumulation of HRGPs (also called extensins) is reported to be a dicotyledonous plant defense reaction to pathogens and stress (43, 208). These compounds exist in low amounts in plant cell walls and were first shown to accumulate via de novo synthesis in response to infection (208). They may be involved in pathogen recognition, and their defensive action is via physical agglutination of negatively charged fungal and/or bacterial cells or cell components (190, 212). The plant growth regulator ethylene causes accumulation of HRGPs in several plant species (98), and aminovinylglycine (inhibits ethylene-synthesis via 1-aminocyclopropane-1-carboxylic acid synthase inhibition) reduces ethylene and HRGP levels (239). Several other compounds can inhibit ethylene production in plants (6), but apparently have not been tested for effects on HRGP production.

5.2 Biochemical Defenses

Numerous reviews have discussed plant biochemical responses to pathogens, but few have emphasized defense mechanisms in relation to biological weed control (140, 196, 197). Nearly all literature on plant defense mechanisms focuses on crop plants and their associated pathogens; there are few reports on weed defenses against pathogens. Many of the plant-pathogen interaction studies also deal with the role of secondary plant metabolism in defense response mechanisms, which will be presented in greater detail later in this chapter.

5.2.1 Free radicals

Oxygen radicals can be generated in response to pathogenic attack. These radicals are highly reactive and attack the double bonds in unsaturated membrane fatty acids (125). This reaction leads to autocatalytic lipid peroxidation and subsequent loss of membrane function leading to cell death, and is documented for many plant/pathogen interactions (85, 86, 177, 178). Free radical scavengers (e.g., glutathione, ascorbate, α-tocopherol, etc.) might be combined with bioherbicides to lower free radical defense systems in weeds. Some of these scavengers have lowered defense responses of potato (86). Some herbicides that generate active oxygen species in plants can be synergized using specific chelators that inhibit metallo-enzymes that detoxify these oxygen species. However, it is unknown if such chelators can act synergistically with pathogens for weed control.

5.2.2 Pathogenesis-related (PR-) proteins

The term PR-protein (or stress protein) is applied to proteins induced by infection or specific environmental stress (26, 40). They are generally of low molecular weight and are different from heat shock proteins. At least 6 classes of PR-proteins have been identified (see 144). PR-proteins are associated with resistance in infected plants including bean (*Phaseolus vulgaris* L.) (296), cucumber (*Cucumis sativus* L.) (9), cowpea [*Vigna ungiculata* (L.) Walp.] and tobacco (*Nicotiana tobacum*) (65). PR-proteins can be produced by viruses, bacteria, or fungi that cause hypersensitivity in hosts during induced resistance by pathogens (108), or by chemicals such as ethylene (19), benzoic and polyacrylic acids (12), or salicylic acid (306). Other compounds (aspirin, kinetin and abscisic acid) altered protein patterns without inducing resistance to *Phytophthora infestans* (Mont.) de Bary infections in tomato (*Lycopersicon esculentum*); but indole acetic acid, ethephon, and fusicoccin induced PR-proteins and/or resistance (55). An elicitor from P. *megasperma* also induced PR-proteins and both local and systemic resistance (55).

Even though PR-proteins may be induced in stressed plants, evidence of their involvement in resistance mechanism(s) remains unclear (28). PR-proteins have not been examined in weeds, and little is known of their function, role, or regulation in other plants. Thus, it will be a formidable task to obtain specific inhibitors of their synthesis or inhibition which would be useful in biological weed control.

5.2.3 Suppressors

Suppressors are non-phytotoxic compounds secreted by pathogens that delay or prevent elicitation of host defenses (228, 253). Suppressor compounds may render host cells susceptible to avirulent or non-pathogenic organisms (181, 229, 230, 252). Suppressors produced by phytopathogenic fungi suppress various defense-related parameters, including phytoalexins, PR-proteins, superoxides, and hypersensitivity (253). However, in many cases, little is known about suppressor identity, characterization, or site(s) of action. The host plant series examined thus far again indicates a lack of investigations of this defense response (suppressors) in weed hosts.

5.2.4 Elicitors of phytoalexins

Elicitors are generally glucans, glycoproteins, and other high molecular weight cell wall components of pathogenic fungi (112), and were first discovered when a protein was found to elicit phaseolin production in beans (69). Elicitors have been characterized from only three fungal genera (*Phytophthora*, *Colletotrichum*, and *Fusarium*). Elicitors induce the development of necrotic symptoms, as well as the production and accumulation of phenolics and phytoalexins. Phytoalexins are secondary plant compounds, produced post-infection, that exhibit anti-microbial activity (104). They can be grouped into three chemical categories: phenolics, isoprenes, and polyacetylenes (284). Phenolics are derived from the shikimate pathway, isoprenes (mono-, di-, and sesquiterpenes) from the acetate-mevalonic acid pathway (68), and polyacetylenes from the condensation of acetate (237). Only about 60 phytoalexins have been isolated and identified from about 100 plants encompassing 20 families (112).

Some elicitors have been characterized and synthesized; and the use of elicitors of phytoalexins as herbicides has been suggested (174, 232). Since many of these materials are high molecular weight chemicals, they may not be readily absorbed or recognized as an elicitor by the plant. Formulation materials could aid in the absorption of elicitors by weeds, or elicitors might enter weeds at infection site of via mechanical wounding. This could theoretically provide additive or synergistic effects of a given elicitor and pathogen.

Bacteria, fungi, and plants can enzymatically degrade phytoalexins (283). Environmental factors and chemicals can also influence phytoalexin accumulation. Thus, there are potential regulatory sites in weeds that may be sensitive to chemicals that either decrease phytoalexin synthesis or activate degradation enzymes. Since many phytoalexins are phytotoxic (256), potentially synergistic combinations of phytoalexins and bioherbicides are possible.

5.2.5 Plant enzymes related to defense

Various enzymes have been implicated in host plant resistance to disease. Among the more important of these enzymes are chitinase, glucanase, glycosidase, NADPH oxidase, and several enzymes of secondary plant metabolism, including phenylalanine ammonia-lyase (PAL), peroxidase, and polyphenoloxidase (140, 144). Since most reports are related to enzymes of secondary plant metabolism and defense, these latter three enzymes will be examined here. Of these three, PAL has received by far the most in-depth attention. Several aspects will be examined here, including its role in disease, its inhibition by specific inhibitors, and the effects of herbicides on PAL activity.

Many reactions are involved in channeling carbohydrate metabolism into the shikimic acid pathway and the subsequent formation of products of secondary plant metabolism. PAL is a pivotal regulatory enzyme connecting these two pathways by transforming phenylalanine to t-cinnamate. PAL activity increases dramatically in response to various stimuli, including elicitors and challenge by pathogens (39). Increased enzyme activity can, in turn, lead to increased production of phenolic products. PAL activity is under feedback control regulation by various phenolic products in some systems (39). The phenolic pathway is responsible for a variety of phenolic compounds, some of which are toxic to bacteria and fungi, and some (phytoalexins) that are produced only in response to the above stimuli.

Results from studies on enzymatic resistance to pathogen attack indicate the lack of a universal resistance mechanism(s). That is, specific enzyme activities in various plant species have been both positively and negatively correlated with infection by various pathogens. Nearly all enzymatic defense information is based on crop-pathogen interactions; few studies have examined these responses in weeds. Hence, there is a need to examine resistance mechanisms in pathogen-weed situations if the full potential of such interactions for biological control are to be realized.

5.2.6 Pathogen challenge : effects on PAL

Numerous examples correlate PAL activity with pathogen challenge and plant defense, but only a few will be presented here. PAL activity increased rapidly in potato leaves within hours after treatment with zoospores or an elicitor from *Phytophthora infestans* (Mont) de Bary (105). PAL activity was examined in soybean (*Glycine max* Merr.) cultivars possessing differential resistance to *P. megasperma* Drechs. f. sp. *glycinea* at different temperatures. PAL activity was lower at 33°C than at 25°C in each of six cultivars, but was lowest in two cultivars that became susceptible to the pathogen at 33°C (23). PAL activity increased rapidly 2 h after *P. megasperma* f. sp. *glycinea* inoculation of a resistant soybean cultivar, but not in a susceptible cultivar (24). Glyceollin increases were concomitant with increases in enzyme activity (24). Abscisic acid decreased PAL activity and glyceollin synthesis, causing increased susceptibility of a resistant soybean cultivar to *P. megasperma* f. sp. *glycinea* (299). PAL increased 2 h after inoculation of barley leaves with *Erysiphe graminis* DC, reached a peak after 4 h, and then declined (254). PAL activity increased 12-fold above control in tomato cell cultures inoculated with *Verticillium albo-atrum* Reinke et Berthold (21). Inoculation of blast fungus conidia induced PAL activity in rice leaves (123). Fungal cell wall elicitors from *Colletotrichum lindemuthianum* (Sacc. & Magnus) Lams.-Scrib. also stimulated PAL activity in bean (66, 70), and fungal or elicitor challenge induced PAL activity in several plant species (124). Several graminaceous plants infected by various pathogens responded with enhanced PAL activity and changes in phenolic metabolism (116, 207, 312).

As pointed out previously, most research on pathogen-plant interactions pertains to crops and their associated pathogens, but there are some reports of weed defense against pathogens. The fungal pathogen *Alternaria cassiae* Jurair and Kahn, which has bioherbicidal activity against sicklepod (*Cassia obtusifolia*) (290, 292, 294), alters phenylpropanoid metabolism in this weed (142). PAL activity levels increased prior to visual symptomology of pathogenesis when *A. cassiae* spores were applied to seedlings. PAL activity was elevated about 3-fold above uninoculated plants, and remained high for several days. Methanol-soluble hydroxyphenolic compounds were also increased in stems and leaves of infected plants compared to untreated tissues. *A. cassiae* also had bioherbicidal activity against *Cassia alata* (ringworm bush, closely related to the weed sicklepod), and caused elevated PAL activity (143). *Alternaria crassa* (Sacc.) Rands is a fungal pathogen with bioherbicidal potential for jimsonweed (*Datura stramonium* L.) control (30). Examination of defense responses of red-stemmed and green-stemmed biotypes of jimsonweed during *A. crassa* challenge showed that PAL activity was elevated 2- to 3-fold above untreated control levels in both biotypes (146). Anthocyanin content (~11-fold higher in red- than green-stemmed plants) also increased after infection. In these examples of weed-pathogen interactions, phenylpropanoid metabolism appears to be altered via increased PAL activity. However, other defense mechanisms may also be operative in these plants; and other pathogens might trigger other defense mechanisms.

5.2.7 Specific PAL inhibitors

Some synthetic compounds are potent PAL inhibitors *in vitro* and *in vivo*, including aminooxyacetic acid (AOA), α-aminooxy-β-phenylpropionic acid (AOPP), (1-amino-2-phenylethyl)phosphonic acid (APEP), and O-benzylhydroxylamine (OBHA). AOA is the least specific [also inhibits transaminase(s)] (4), causes injury to plant tissues (148), and has been registered as a herbicide (152). AOPP is a more specific and potent PAL inhibitor that can be used in plant to alter secondary metabolism without substantial effects on growth (3, 5, 164). One of the most potent PAL inhibitors is APEP (188). OBHA inhibits PAL and accumulation of some secondary plant products (136). All of these PAL inhibitors lowered levels of some phenolic compounds, some of which are altered by a variety of herbicides and plant growth regulators, such as ethylene (144). Generally, levels of ethylene are increased in infected plants, causing activation of defense enzymes and resistance, in some instances (140, 285). AOA, AOPP, and OBHA also inhibit ethylene production in some plants (6).

PAL inhibitors have been used to test the role of PAL in disease resistance. Resistance expression and glyceollin accumulation in soybean tissue treated with an incompatible race of P. *megasperma* f. sp. *glycinea* were not blocked by AOA or AOPP (154, 175). AOPP reduced glyceollin levels and increased soybean susceptibility to P. *megasperma* f. sp. *glycinea* (120, 215). AOA and AOPP reduced phenolic metabolism via PAL inhibition and decreased hypersensitive reaction of tobacco mosaic virus (TMV) in tobacco (231). AOA caused PAL activity inhibition in vivo, but not its synthesis, resulting in increased extractable PAL activity from maize (*Zea mays* L.) mesocotyls which prevented resistance expression (231). This again suggests PAL involvement in resistance. Phosphite increased PAL activity and inhibited growth and infectivity of *Phytophthora cryptogea* Pethybr. & Lafferty in cowpea (241). Pretreatment of leaves with AOA, prior to phosphite and P. *cryptogea*, lowered PAL activity, reduced kievitone and phaseollidin accumulation, and increased symptomology (241). AOPP increased successful fungal infection of appropriate (but not inappropriate) *Erysiphe graminis* in cereal plants (42). AOA increased disease severity in tomato plants resistant to *Fusarium oxysporum* (35), increased infection by *Phytophthora parasitica* Dastur var. *nicotianae* in tobacco (Fenn and Coffey 1989), and induced susceptibility of potato tuber tissue to non-pathogenic *Cladosporium cucumerinum* Ellis & Auth. (126). Inhibition of avenalumin [phytoalexin in oat (*Avena sativa* L.) leaves] accumulation by AOA was correlated with increased susceptibility of oats to the rust pathogen *Puccinia coronata* Corda f. sp. avenae (206). AOA and other compounds, including the herbicide dinitramine [N3,N3-diethyl-2,4-dinitro-6-(trifluoromethyl)-1,3-benzenediamine] lowered ethylene biosynthesis and induced resistance of melon (*Cucumis melo* L.) to *Fusarium* wilt (59). APEP prevented glyceollin induction and transformed an incompatible pathogen interaction into a compatible one, suggesting a role for PAL and toxic phenolic compounds in resistance (288). R-APEP caused soybean to loose resistance to *P.*

megasperma (288). The use of PAL inhibitors allowed infection of reed canarygrass (*Phalaris arundinacea* L.) by *Helminthosporium avenae*, an oat pathogen which normally doesn't infect reed canarygrass (250). Contrary results with some PAL inhibitors have been reported; i.e., PAL inhibitors were unable to increase susceptibility in wheat (*Triticum aestivum* L.) (236) or bean leaves (233).

It is evident that PAL inhibitors or compounds that alter secondary plant metabolism can weaken plant defense, thereby increasing disease severity and/or extending pathogen host range. In biological weed control, such compounds might be useful as pathogen synergists, but there are apparently no successful reported results using these PAL inhibitors in weed-pathogen interactions.

5.2.8 Agrochemical effects on PAL and secondary plant metabolism

In the search for chemical synergists of weed pathogens, herbicides may be the most probably group to examine, since they are designed to cause plant injury. Plant growth regulators (synthetic and naturally-occurring) are compounds that can also contribute to synergistic interactions with pathogens. PAL inhibition was suggested as a possible target for herbicide action (165), but attempts to design herbicides around this mode of action were unsuccessful. Although not their primary mode of action, many herbicides can vastly alter extractable PAL activity in plants. Herbicides from 14 chemical classes altered total hydroxyphenolic compound levels in soybean seedlings, and effects on extractable PAL activity levels ranged from none to substantial decreases or increases in activity (147, 149). Herbicides also increased PAL levels in other plant species. Glyphosate [N-(phosphonomethyl)-glycine] substantially increased PAL levels in other plants [maize (Duke and Hoagland 1978) and cotton (*Gossypium hirsutum* L.) (Duke and 136)]. Other herbicides [dichlobenil (2,6-dichlorobenzonitrile) (97), amitrole (1H-1,2,4-triazol-3-amine) (147), acifluorfen {5-[2-chloro-4-(trifluoromethyl)phenoxy]-2-nitrobenzoic acid} (138, 184), and paraquat (1,1_-dimethyl-4,4_-bipyridinium ion) (147)] also elevated PAL activity levels in certain plants.

Herbicides can also decrease PAL activity in *planta*. Herbicides that substantially lower PAL activity include diuron [N'-(3,4-dichlorophenyl)-N,N-dimethylurea] in buckwheat (*Fagopyrum esculentum* Moench.) (7) and strawberry [*Fragaria chiloensis* (L.) Duch.] (67). MCPA [(4-chloro-2-methylphenoxy)acetic acid] had no effect on PAL in wheat, which is resistant, but decreased PAL in susceptible tobacco (282). Atrazine [6-chloro-N-ethyl-N_-(1-methylethyl)-1,3,5-triazine-2,4-diamine] reduced PAL activity in both wheat and tobacco (282). When supplied alone, glyphosate and atrazine dramatically elevated and lowered PAL activity in soybean seedings, respectively, but these herbicides produced an antagonistic interaction when supplied together (137). PAL activity in vivo is also inhibited by 2,4-D [(2,4-dichlorophenoxy)acetic acid] in several plant systems (79, 147). These herbicides had no inhibitory effect on PAL *in vitro*.

There are many reports on the influence of various xenobiotic compounds on secondary plant metabolism. Some xenobiotics with diverse chemistries and their potential for manipulation of and interaction with this pathway for agricultural purposes have been reviewed (87, 140, 196). Pesticides (herbicides, insecticides, and fungicides) and plant growth regulators produce important effects on a variety of secondary compound levels (144, 199). Some secondary plant compounds have potent phytotoxicity (88, 199), and their levels in plant tissues can be altered by agrochemicals (87, 195, 196). Agrochemicals can influence susceptibility or resistance by altering secondary compound levels. This can occur in field situations where plants receive sublethal doses from residues in soils, from spray drift, or when a weed is resistant to the pesticide. For example, glyphosate and the herbicide lactofen {(±)-2-ethoxy-1-methyl-2-oxoethyl 5-[2-chloro-4-(trifluoromethyl)phenoxy]-2-nitrobenzoate} interactions resulted in glyphosate-stimulated shikimate accumulation in little mallow (*Malva parviflora* L.) (305). Some of these secondary products are phytoalexins (e.g., glyceollins, phaseolin, and pisatin). Several herbicides alter acetate and mevalonate incorporation into terpenoids by a key enzyme of this pathway, 3-hydroxy-3-methylglutaryl-coenzyme A-reductase (121).

Herbicide mode of action at the molecular level (the primary action) is separated from action on PAL activity and secondary metabolism (a secondary or tertiary effect). Nevertheless, these and other secondary or tertiary effects could be of critical importance in weakening weed defenses and/or in aiding pathogen infection. Major parameters involved in achieving successful synergism (or avoiding antagonism) are concentration and timing of application ingredient, and plant growth stage.

Some herbicides, plant growth regulators, specific enzyme inhibitors, and other chemicals can alter these defenses. Secondary plant metabolism (phenylpropanoid pathway) is a major biochemical pathway related to several defense processes in weeds, as previously discussed. Increased activity of PAL, a key enzyme of this pathway, is a response to pathogen attack, as demonstrated in three weeds and their associated bioherbicidal pathogens: *Alternaria cassiae* on sicklepod (142) and C. alata (143), and *A. crassa* on jimsonweed (146).

5.2.9 Other defense enzymes of secondary metabolism

Peroxidase (PO) and polyphenol oxidase (PPO), enzymes associated with secondary plant metabolism, are also implicated in plant defense. PO is important in the oxidation and polymerization of hydroxycinnamic alcohols into lignin. PPO converts hydroxylated phenolics into antimicrobial quinones (140) which react with each other to form insoluble phenolic polymers. PO and PPO activities increase dramatically in many infected plants (140, 284, 286). Compounds that inhibit PAL, PO, or PPO activities have been discovered, and their use to suppress disease resistance has been discussed (140). An analog of the triazine herbicide metribuzin [4-amino-6-(1,1-dimethylethyl)-3-(methylthio)-1,2,4-triazin-5(4H)-one] strongly inhibited lignification catalyzed by PO in lupin (*Lupinus albus*) (223). Such inhibitory compounds have the potential to lower weed resistance and, hence interact synergistically with pathogens.

6. PATHOGEN INTERACTIONS WITH CHEMICALS

Several tactics have been examined to increase pathogen efficacy for weed control. Weakening of physical and biochemical defenses, and lowering of resistance to pathogen attack, may result from reduced production of phenolics, lignin, and phytoalexins caused by herbicides and other chemicals that affect cuticular component biosynthesis and/or key aspects of secondary plant metabolism. Various synthetic and naturally-occurring compounds (herbicides and plant growth regulators) alter secondary plant product levels in a variety of plant species (144). In many cases, these compounds lowered secondary plant metabolite levels in plant tissues. However, acifluorfen increased phytoalexin accumulation in several species (185), and 2,4-D increased scopolin in sunflower (*Helianthus annuus*) (84). This indicates that some herbicides can increase secondary plant metabolites which may increase disease resistance, an undesirable effect when attempting to synergize pathogen infectivity with chemicals for use in weed control.

6.1 Synergistic or Antagonistic Chemicals

Since the beginning of pesticide use in agriculture, combinations of two or more chemicals for sequential or simultaneous application on a given crop have been used. Sometimes these combinations produced interactions that either increased or decreased efficacy. Terms commonly used are "interaction", "synergism", "antagonism", and "additive effect". Definitions, experimental design, and statistical analysis related to these terms have been reviewed (118, 119, 132).

The use of synergistic synthetic herbicide interactions to reduce application rates has been discussed (119, 132). There are some analogous results using chemical interactions with bioherbicides that have provided additive or synergistic enhancement of bioherbicide activity. When synergistic interactions are found that expand bioherbicide host range, options are available to maximize this host regulation potential. For example, directed spray applications of fungicides might be used to manage the plant-control spectrum of broad host-range fungal weed pathogens, allowing use of these pathogens in situations where the crop may also be a host. There are also promising results using various microbes to control pathogens (63, 198), and these might be used to protect crops when a broad-spectrum weed pathogen is used as a bioherbicide. Increased infectivity and virulence of pathogens have been achieved by optimizing formulations, chemical additives, and application of microbial pathogens for weed control (34, 74). Some weed pathogens are incompatible with agrochemicals, but this can be overcome by careful integration with such compounds (48). Incompatiblity may be especially high when fungicides are used in cropping situations together with fungal bioherbicides. This problem can be circum 'ented using proper timing for integrated management (271).

Interactions between herbicides and plant pathogens (2, 171, 258), herbicide induction of microbial invasion of plant roots (115), and the interactions of sublethal herbicide doses on root pathogens (193) have been examined. Some of the earliest reports indicating that herbicides could block resistance to pathogens involved the use of glyphosate to increase infection of an incompatible race of *Phytophthora megasperma* f. sp. *glycinea* in soybean (175). Sicklepod produced a fungitoxic flavonoid phytoalexin (247, 249), and glyphosate suppressed its production and acted synergistically with*Alternaria cassiae* (248). This phytoalexin was derived from phenylalanine (249), which supported the fact that elevated PAL levels occurred with this weed was challenged by this pathogen (142).

A soil-borne fungus, *Fusarium solani* (Mart.) App. & Wr. f. sp. cucurbitae that could infect the weed Texas gourd [*Curcurbita texana* (Scheele) Gray] was tolerant to the herbicide trifluralin [2,6-nitro-N,N-dipropyl-4-(trifluoromethyl) benzenamine] and the combined herbicide and pathogen produced greater control of this weed than either treatment alone (304). *Cochliobolus lunatus*, a pathogen of barnyardgrass, combined with the herbicide atrazine, resulted in synergistic action against the weed (243). A synertistic interaction was also reported when the fungus D. monceras was combined with the herbicide pyrazosulfuron-ethyl against barnyardgrass (111).

Other interactions involved pathogens of aquatic weeds. The herbicides 2,4-D and diquat (6,7-dihydrodipyrido[1,2-a:2_,1_-c]pyrazinediium ion) interacted with the pathogen *Cercospora rodmonii* in waterhyacinth (*Eichhornia crassipes*) (47). The herbicide endothall (7-oxabicyclo[2.2.1]heptane-2,3-dicarboxylic acid) and the pathogen *Colletotrichum gloeosporioides* interacted in Eurasian water milfoil (*Myriophyllum spicatum* L.) (259). Several other plant pathogens have also been investigated for weed control in aquatic situations, as summarized elsewhere (271). Interactions such as these in aquatic situations could help provide improved bioherbicidal weed control and reduce the amount of herbicides introduced directly into surface water systems. The fact that synergistic interactions have been found for a diverse group of herbicides, pathogens, and weed hosts suggests promising areas for future research.

Synergistic interactions of several other chemicals (herbicides and plant growth regulators) and four fungal pathogens have also been discovered and patented (44). The herbicides acifluorfen and bentazon [3-(1-methylethyl)-(1H)-2,1,3-benzothiadiazin-4(3H)-one 2,2-dioxide] were the most effective synergists, providing increased control of the following weedy hosts by their respective bioherbicides: sicklepod by *Alternaria cassiae*, northern jointvetch by *Colletotrichum gloeosporioides*, hemp sesbania by C. *truncatum*, and Florida beggarweed (*Desmodium tortuosum*) by *Fusarium lateritium* (56, 144). The plant growth regulator thidiazuron (N-phenyl-N_-1,2,3-thidiazol-5-yl-urea) acted synergistically with C. coccodes, and increased mortality of velvetleaf compared to the fungal bioherbicide alone (311). These interactions also point out that there is no universal chemical synergist for all pathogens. These examples (and those mentioned previously) show that many herbicides with vastly different chemistries and molecular modes of action can act synergistically with microbes to improve weed control. Tests with these chemical-pathogen combinations were, however, not able to broaden the weed host range of the pathogens.

Attempts to mix bioherbicides and agricultural chemicals are sometimes detrimental to the living microbes, and the potential toxicity of herbicides and other chemical pesticides may hinder the integration of bioherbicides into intensive crop production systems (275). Collego® mixed with chemicals used in rice production {propanil [N-(3,4-dichlorophenyl)propanamide}, molinate (S-ethyl hexahydro-1-H-azepine-1-carbothioate), 2,4,5-T [(2,4,5-trichlorophenoxy) acetic acid], and the fungicide benomyl [methyl-1-(butylcarbamoyl)-2-benzimidazolecarbamate]} resulted in loss of bioherbicide efficacy (258). Interactions detrimental to the pathogen *Phytophthora palmivora* were found when chlamydospores were applied simultaneously with herbicides such as bromacil [5-bromo-6-methyl-3-(1-methylpropyl)-2,4(1H,3H)pyrimidinedione], diuron, glyphosate, paraquat, and simazine (6-chloro-N,N_-diethyl-1,3,5-triazine-2,4-diamine) (46). Sequential application of herbicides and pathogen resulted in no inhibition of pathogen growth or activity. This points out that timing (and perhaps concentration) is a major factor in integrating pathogens and other chemicals in attempts to find synergy.

Combinations of bacteria and chemical agents for enhanced weed control (termed X-tend bioherbicide systems) have recently been studied (56) in field and greenhouse weed control tests, summarized as follows. The herbicide sulfosate (trimethylsulfonium-carboxymethylaminomethyl phosphonate) causes varying degrees of plant injury to a wide variety of weeds. Bacterial strains alone caused little or no plant injury, but sulfonate plus bacteria resulted in greater injury than from the herbicide alone. Similar synergistic results were obtained in field tests with two bacterial preparations plus glufosinate [2-amino-4-(hydroxy-methylphosphinyl) butanoic acid]. Although this synergy between synthetic herbicides and bacterial agents could significantly reduce amounts of herbicides used to control a broad spectrum of weeds, it was concluded that improved bacterial strains and/or formulations would be needed before successful commercialization of these products for weed control (56). Although some of these chemical and pathogens combinations were synergistic, the modes of these interactions at the molecular and physiological levels are unknown and have received little or no attention. Synergism of pathogens with other pathogens or with insects can also occur (271). In one example cited, the mortality of spurred anoda (*Anoda cristata*) was increased by the synergistic interaction that occurred following sequential applications of *Alternaria macrospora* and *Fusarium lateritium* (a weak pathogen of spurred anoda). Other strategies useful to increase pathogenicity are strain selection, mutant selection, genetic transformation, and nutrient availability.

Some non-pathogenic organisms can alter herbicide efficacy, and some herbicides influence disease development in plants. Research has shown some synergistic interactions of microbes and chemicals with relevance to weed control. Further research on pathogen interactions with agrochemicals (or other chemicals/regulators), other pathogens, insects, and formulation ingredients could result in increased efficacy of pathogen-herbicide combinations, reduction of herbicide and pathogen levels required for weed control, and expanded pathogen host range.

6.2 Formulations

Most attempts at using pathogens for weed control have involved fungi. The spores of many fungal species require several hours of free moisture or dew to germinate and infect the intended host. Formulations are extremely important for the successful application of pathogens for weed control. In addition, the availabity of bioherbicides in familiar forms that are easy to apply will aid in consumer acceptance. Formulation materials are generally a liquid carrier plus other components (such as adjuvants or surfactants) that, when combined with pathogen spores or propagules, produce a suspension/solution that can be efficiently applied to target weeds. Numerous wettable powders and solid granular formulations have been also examined and developed to meet the specific survival or growth requirements of certain pathogens. Many reviews of the formulation of microorganisms for biological control of weeds have been published (33, 34, 62, 75, 117).

Studies of *Colletotrichum truncatum* grown under various C:N ratios (5:1 to 40:1) showed that the 5:1 ratio produced more efficacious spores than did the other ratios, resulting in greater growth reduction of the weedy host, hemp sesbania (245). Nitrogen nutrition was also found to be important in the development of liquid media to produce conidia of C. *truncatum* (163). Complex or defined nitrogen sources which provide a complete compliment of amino acids but are low in cysteine, methionine, lysine, tryptophan, leucine, and isoleucine promoted high conidial yield in liquid cultures. These media formulation factors are useful in economic production of this fungus.

Abundant biomass of chlamydospores obtained by liquid formulation of mycoherbicidal strains of *Fusarium oxysporum* was incorporated in alginate with various substrates (134). Self-life was significantly affected by fungal strain, formulation method, and interactions of these two parameters. Research indicates most of these formulations meet several critera for production of effective biological control agents -- low viability losses during formulation, satisfactory shelf life at room temperature, abundant chlamidia production, and rhizosphere colonization (134).

Inoculum thresholds needed to control sicklepod and jimsonweed with *Alternaria cassiae* and *A. crassa*, respectively, were reduced to one conidium per droplet when an invert emulsion formulation was used for the inoculum. However, invert emulsions also had potential for crop damage (8).

Biological control of aquatic weeds with plant pathogens has been reviews and summarized (167). Hundreds of microbes have been examined for potential to control a number of economically important aquatic weeds. Several issues need to be addressed to realize the efficient use of plant pathogens in aquatic situations, including the relatively high inocula levels needed, appropriate formulations, and the development of delivery systems.

7. PHYTOTOXINS FROM MICROBES AS BIOHERBICIDES

Co-evolution of toxin-producing pathogens with their plant host species has resulted in the production of many host-specific toxins. Presently, host-specific phytotoxins have only been characterized from crop pathogens; studies on such phytotoxins from weed pathogens have been generally neglected. Although many microbial metabolites have excellent phytotoxic activity, only a few have been developed into commercial herbicides (Table 6). Isolation and chemical characterization of host-specific and non-specific phytotoxins from pathogens and non-pathogens may provide templates for analog synthesis and lead to the development of herbicides with even more desirable characteristics than the synthetic compounds presently marketed. Anisomycin was used as a template for herbicidal derivatives, one of which was methoxyphenone (4-methoxy-3-3_-dimethylbenzophenone) in 1974 (103). Thus, anisomycin became the first microbial metabolite to lead to a biologically-based synthetic commercial herbicide. Many synthetic analogs of the fungal phytotoxin moniliformin have been produced and screened as potential herbicides; but none have reached the commercialization stage, due mainly to toxicological concerns (103). Irpexil [methy-2-(acetamidooxy)acetate] is a phytotoxin discovered in a basidiomycete (103). A corresponding acid [2-(acetamidooxy)acetate] was synthesized to cover derivatives of the herbicide benzadox (benzamidooxyacetate). Benzadox breaks down in plants to aminooxyacetate (AOA), a compound closely related to irpexil. AOA (discussed earlier as a PAL inhibitor) has been registered for use as a herbicide (281). But, despite structural similarities, the modes of action of AOA and irpexil appear to be different (103). The most successful of the microbial products used as commercial herbicides are bialaphos (L-2-amino-4-[hydroxy(methyl)phosphinyl]butyryl-L-alanyl-L-alanine) and phosphinothricin (L-2-amino-4-[hydroxy(methyl)phosphinyl]butyric acid). These are the first two microbial metabolites that have lead directly to commercialized herbicide production. Phosphinothricin is marketed as the synthetic ammonium salt, called glufosinate (145, 151). Bialaphos is marketed in Japan as Herbiace® (211), and is the only commercial herbicide produced by fermentation. The compound is cleaved by plant peptidases to yield phosphinothricin as the active phytotoxin. Many biotechnological advances have been made with this chemistry, including elucidation of its biosynthetic pathway, isolation of genes responsible for synthesis and resistance to these compounds, and the development of transgenic crops of many species that are resistant to these herbicides. Both compounds are potent non-selective herbicides with rapid dissipation in the environment and relatively low toxicity to non-target species (145, 151).

Table 6. Attempts to Commercialize herbicides based on compounds produced by microbes.

Microbial compound	Microbial source	Herbicide (common name)	Reference
Anisomycin	*Streptomyces* sp.	NK-046; methoxyphenone	(103)
Phosphinothricin	*S. hygroscopicus* *S. viridochromogenes*	phosphinothricin (glufosinate) bialaphos	(152) (211)
Moniliformin	*Fusarium moniliforme*	numerous analogs examined, but no commercial products	(103)
Irpexil[1]	*Irpex pachyodon*, a basidiomycete	many herbicidal derivatives synthesized, but no commercial products	(103)

Because plant pathogens can be devastating to plant tissues, pathogenic organisms may be the first to come to mind when contemplating phytotoxin-producing agents for bioherbicides. However, many non-pathogenic organisms also produce phytotoxins. Indeed, some of the antibiotics isolated from non-pathogenic organisms possess very potent phytotoxic properties. The genus Streptomyces has only a couple of pathogenic species, but this genus has been a rich source of phytotoxins (82, 262, 269). Some progress has been made in the areas of isolation and/or screening of microbial phytotoxins as herbicides, as evidenced in many reviews (71, 72, 73, 88, 89, 103, 139, 264). Screening of compounds produced by soil microorganisms has also resulted in the discovery of many bioherbicidal compounds (72, 80, 82, 135, 176, 242). Table 7 outlines some phytotoxins produced by a variety of bacterial and fungal organisms and indicates their phytotoxic effects. These, and numerous other microbial phytotoxins (fungal and bacterial pathogens and non-pathogens) have been isolated, identified, and screened in molecular, tissue, and seed germination bioassays, or on whole plants. However, most have not received extensive testing for selectivity or physiological/ biochemical effects and molecular mode of action on weeds. The need to control weeds that have become resistant to some routinely-used herbicides has demonstrated to industry the need to find phytotoxic compounds that have molecular sites of action different from previously-discovered herbicides. These discoveries are very important, since there are only about a dozen known sites of synthetic herbicide action (90); and the known sites of action of microbial phytotoxins differ from those of the synthetic herbicides (89). Table 8 shows several microbial phytotoxins and their known molecular sites of action. The fact that these microbial phytotoxins have modes of action different from those of synthetic herbicides strongly suggests that the continued search for other microbial phytotoxins will be a very worthwhile endeavor. The diverse chemistry of phytotoxic compounds from microbes that have been, and will be, discovered can provide a rich source of compounds

Table 7. Examples of non-host-selective phytotoxins produced by microorganisms.

Phytotoxin	Source	Effect	Reference
Acetylaranotin	*Aspergillis terreus*	Growth inhibition	(170)
Brefeldin A	*Alternaria carthami*	Chlorosis	(276)
Coronatine	*Pseudomonas syringae*	Chlorosis	(213)
Dehydrocurvularin	*Alternaria macrospora*	Necrosis	(238)
Gougerotin	*Streptomyces* sp.	Growth inhibition	(225)
Naramycin B	*Streptomyces griseus*	Growth inhibition	(20)
Ophiolbolin	*Helminthosporium oryzae*	Membrane disruption	(58)
Phosphinothricin	*Streptomyces viridochromogenes* *S. hygroscopicus*	Chlorosis, wilting, necrosis	(145)
Radicinin	*Alternaria helianthi*	Necrosis	(268)
Viridiol	*Gliocladium virens*	Necrosis	(159)

with these new and sensitive molecular sites of action. Most microbial phytotoxins have been isolated from non-phytopathogenic species; little has been done on the exploitation of phytotoxins from pathogens as herbicide leads. The chemical complexity of natural products from microbes transverses a wide range, from simple to very complex. It will be difficult or impossible to develop economic synthesis of many large molecules due to their complexity and the fact that they may possess multiple chiral centers. However, complex phytotoxins may be produced by large-scale fermentation, as with bialaphos (211), and there are many phytotoxins of low molecular weight that can be easily synthesized; e.g., phosphinothricin (glufosinate) (145, 151). Studies of phytotoxins from pathogens are also important to provide information on the virulence and pathogenicity of these microorganisms, as has been demonstrated in several cases (158, 214).

Table 8. Known sites of action of selected microbial toxins.

Affected site (enzyme)	Phytotoxin	Reference
Methionine synthesis (β-cystathionase)	Rhizobitoxine	(110)
Arginine synthesis (Ornithine carbamoyl transferase)	Phaseolotoxin	(109)
Glutamine synthesis (glutamine synthetase)	Bialaphos, phosphinothricin	(103) (145)
Energy transfer uncoupling (CF_1)	Tentoxin	(260)
Membrane potential	Fusaric acid	(77)
Singlet oxygen production	Cercosporin	(78)

8. LIMITATIONS AND CONSTRAINTS TO BIOHERBICIDE DEVELOPMENT

Several levels of constraints or limitations govern the success of a given bioherbicide, as outlined in recent reviews (13, 221). These include biological, environmental, technological, and economic or commercial constraints.

8.1 Biological Limitations

It is desirable for a bioherbicide microorganism to act relatively quickly and have sufficient efficacy to control one or more weeds. Unfortunately, many of the weed pathogens discovered may provide only partial control of only one weed species, even under ideal conditions. This host specificity is related to the basic biology of the pathogen in question and to host variability (192, 192). However, several means of expanding the host range of weed pathogens have been investigated. These include the use of several pathogens in a combined application (14, 57); addition of plant extract to inoculum (32); and combinations of synthetic herbicides with pathogens (113, 114, 144, 258). The use of insects to deliver weeds pathogens to host sites has also been studied (47, 81, 186), and may have some utility.

8.2 Environmental Limitations

The most important environmental factors that limit efficacy of foliarly-applied bioherbicides are moisture (dew) and sub-optimal temperature (202, 210). Timing of bioherbicide application to coincide with or take advantage of dew periods or field irrigation can be utilized to meet the necessary moisture requirements for spore germination, mycelial growth, infection, and disease development (202, Walker and 30). Formulations that retain water have also been developed to aid germination, growth, and infectivity of some pathogen propagules (74, 75, 310). As with synthetic herbicides, various factors such as low humidity, UV effects, and rain events can be effect-limiting (191).

8.3 Technological Limitations

The mass production of genetically-stable, viable, and efficacious inoculum with an adequate shelf-life is a paramount obstacle in the development of microbes useful as bioherbicides (162). Such formulated living materials must also be easily applied to weed targets (33, 161). Various surfactants or adjuvants can be included in formulations to enhance propagule growth, spore germination, appressorial development, and to aid in propagule distribution and adhesion on leaf surfaces (96, 224). Thus, formulation is critical to the successful technological development of bioherbicides as presented and discussed previously (34, 74, 162).

8.4 Commercial/Economic Limitations

As with any successful product, there must be sufficient market potential, and the product must have an economically-sound basis. Phytotoxins must be easily and economically mass-produced via synthesis and/or fermentation. Several promising microbial bioherbicides have not been commercially developed by industry because they were deficient in one or more of these critical areas (218, 220). Even though bioherbicide agents are less expensive (< \$3 million) to develop and register than synthetic herbicides (> \$30 million) (221), these investments still require patent protection to safeguard the recovery of research and development costs (13). The lack of market demand for many bioherbicides is also a major deterrent to the involvement of large industry in the development of these products. However, weed control in settings other than agronomic crops (e.g., forests, rights-of-way, lawns, gardens, etc.) may be areas where specific pathogens and microbial products can be developed to fill these niche markets.

Another important criterion for the commercial development of a pathogen for agronomic weed control is its interaction and compatibility with other agrochemicals, which allows implementation of an integrated pest management strategy. Compatibility of bioherbicides with other chemicals has been previously discussed.

Bioherbicides may have advantages over other methods of pest control, but still must compete in the market against highly effective synthetic herbicide products. Ecological vs economic comparisons result in complicated cost:benefit ratio analyses which further compromises the development of biologically-based products (277).

9. RECENT STRATEGIES FOR THE DEVELOPMENT AND USE OF BIOHERBICIDES

It is evident that, in the future, many more weed pathogens and microbial phytotoxins will be discovered, isolated, identified, and characterized. Studies on the biological control of weeds with pathogens has resulted in a deeper understanding of the efficacy, host range, difficulties in pathogen mass production, length of biological activity on the shelf, and biochemical interactions of plants with pathogens. Information on all of these factors is necessary to in order to fully understand and utilize pathogens for weed control on a commercial scale. As with many biological control systems, many parameters must be identified and evaluated to ascertain if the biological system is or can be adapted to changing agricultural practices or to varied and changing environmental conditions. Some relatively new aspects of the use of bioherbicides for weed control are emerging, and several of these areas are discussed below.

9.1 Phytopathogenic Bacteria for Weed Control

. Some bacterial weed pathogens may have significant potential as biological control agents, due to several traits. They generally grow rapidly in liquid culture, viability can be preserved in frozen or dried formulations, and they can be genetically manipulated and screened for mutant selection with relative ease. They may be very effective in soils to attack germinating weed seeds and seedlings. One drawback to the use of these bacteria in foliarly-applied weed control is their lack of penetration into plant cells. However, there has been some progress to increase bacterial penetration by wounding of plant tissues and by using surfactants (166, 1992). For example, combination of a polyalkyleneoxide-modified polydimethlysiloxane surfactant with *Pseudomonas syringae* pv. *tagetis* improved the percentage of seedlings [Canada thistle (*Cirsium arvense*), common ragweed (*Ambrosia artemisiifolia*), sunflower, and marigold (*Tagetes* sp.)] infected by this pathogen from 0-5% to 100%.

9.2 Rhizobacteria as Bioherbicide Agents

The use of rhizobacteria for weed control has recently been reviewed (187). Various deleterious rhizobacteria have recently been discovered that can colonize root surfaces of weeds and effectively suppress weed growth. Some of these bacteria can reduce weed density, biomass, and seed production, allowing crops to be more competitive with weeds.

A recent survey identified rhizobacteria from seven genera being studied for their potential as biological weed control agents (187). The projects included nearly 20 economically-important weeds and covered diverse ecosystems including rangelands, cereal crops, forests, turfgrasses, row crops, and pastures. Techniques for improving the survival of rhizobacteria to suppress downy brome (*Bromus tectorum*) has been investigated (255). The mechanisms by which these organisms reduce weed growth are presently unknown, but are being investigated. Although this research area is in the early development stages, it is one that offers promise in weed control with microbes.

9.3 Genetic Manipulation of Pathogens to Improve Bioherbicide Efficacy

Although the main objective in studying the molecular genetics of pathogens has been to clarify their role in pathogenicity, the information obtained may also be used to develop biologically based herbicides (194). In bacterial phytopathogens in genera such as *Erwinia*, *Pseudomonas*, and *Xanthomonas*, several types of gene clusters appear to be involved in phytotoxicity, virulence, hypersensitivity, and host range. Work on the exploration of the use of bialaphos genes is being performed by transforming *Xanthomonas campestris* pv. *campestris* using bialaphos genes (*campestris*). Alteration of such bacterial genes could produce overall increases in virulence, host range, or other related traits. Transfer of genes controlling production of toxins or specific enzymes to improve the performance of potential mycoherbicides is also a developing area (273). Transformation of *Colletotrichum gloeosporioides* f. sp. *aeschynomene* (Gga) (*Glomerella cingulata*) with the bar gene encoding resistance to the herbicide bialaphos has recently been achieved (36). Pathogen host range was not affected by this transformation. However, coapplication of the herbicide bialaphos and the transformed bioherbicide resulted in a significant synergistic effect of disease severity on Indian jointvetch (*Aeschynomene indica*). This suggests that virulence and host range may be increased if a transformed bioherbicide (with phytotoxin resistance) is used with a phytotoxin. Perspectives on biological engineering of prokaryotes for biological control of weeds (189), plant pathogenic fungi for weed control (179), and techniques such as protoplast fusion (128) have been reviewed. It should be possible to genetically enhance fungal bioherbicides to increase virulence and efficacy and to modify host range (191). However, prerequisites to significant progress in this area are the identification and isolation of genes responsible for these factors and the ability to perform stable genetic transfers without loss of virulence or fitness.

10. SUMMARY AND CONCLUSIONS

It is apparent from the amount of research presently being conducted, and the interest in this area, that many more fungal and bacterial weed pathogens will be discovered that possess useful bioherbicidal activity. However, many of the major problems still exist; i.e., most mycoherbicides require free moisture or dew periods that cannot realistically and predictably be found in Nature or be provided for by current formulations. Providing optimum temperature for spore germination and infectivity under field conditions is also problematic. Other problems include finding efficacious pathogens that control more than one weed, and the ability to mass-produce, package, and store these products for reasonable periods prior to field application. It has been suggested that research efforts must be shifted from the discovery of bioherbicide candidates to solving the universal bioherbicidal product problems of production, storage, and efficay. Economics also are a major consideration, not only in the many developmental aspects of the bioherbicide itself, but also in terms of competition with highly effective synthetic herbicidal products. It is clear from the recent developments in formulation and genetic engineering, the increased interest in rhizobacteria and other phytopathogenic bacteria, and the discovery of important phytotoxins from pathogenic and non-pathogenic microbes that major advances may be made within the next several years. The interactions of pathogens and weeds are complex, and our understanding of these interactions is only in its infancy. There is a need to more fully understand the many facets of the biochemistry of weed defense, pathogen virulence, infectivity mechanisms, and host specificity (host recognition, phytotoxin production, etc.), and to discover synergists. All of these parameters are critical to developing bioherbicidal products that can be competitive in the marketplace. To achieve this, a multi-disciplinary approach, utilizing scientists with expertise in microbiology, plant pathology, biochemistry, physical chemistry, and molecular biology is needed. The proficiency of industry in screening, mass-production, formulation, synthesis, risk and economic assessment could also play an important role in the development of pathogens and microbial products as herbicides. The interactions of adjuvant and surfactant ingredients on weed defense mechanisms and microbial propagules, and the development of improved formulations should also be studied.

Although most bioherbicidal products have been targeted at agronomic weeds, the development of herbicide-resistant crops may render the use of bioherbicides impractical in agronomic crops that are genetically transformed for resistance to non-selective herbicides that can control nearly all weeds. For example, over 20 transgenic crops are currently being used or developed with resistance to glufosinate or bialaphos (145). There are also many crops being developed that are resistant to other selective and non-selective herbicides such as glyphosate, various sulfonylurea herbicides, 2,4-D, and others (95). Since glufosinate and bialaphos are non-selective herbicides that also possess potent antifungal and antibacterial activities, the concurrent use of bioherbicidal organisms would not only be unnecessary, but inappropriate in these herbicide-resistant crops. However, bioherbicides may have an important role to play in weed control in non-agronomic areas such as recreational areas, forests, rights-of-way, lawns, gardens, etc., where synthetic herbicides are either not registered or their use is cost-prohibitive.

Weeds are big business. Pathogens can play a critical role in integrated weed management systems, but most likely in situations outside of herbicide-resistant crops. As pointed out, some of the bioherbicides studied are compatible with various synthetic agrochemicals. Some compounds have even been shown to synergize pathogen activity against weeds; and it is likely that even more synergistic combinations will be found as research in this area continues. These bioherbicides, and others yet to be discovered, could play important roles in the integrated pest management of non-transformed cropping situations, or in crops transformed to the more selective herbicide types. Bioherbicides may have utility in controlling weeds previously controlled by herbicides that are no longer registered due to environmental concerns, or whose re-registrations have not been pursued because of diminished market size. Bioherbicides may also play an important role in controlling weeds that have become resistant to one or more of the synthetic herbicides.

As with synthetic herbicides, the concern of host resistance from overuse must be considered in a weed management strategy utilizing bioherbicides. Bioherbicides are not a "silver bullet" that will solve all environmental and pest management problems associated with synthetic herbicides; they will not replace the current arsenal of synthetic herbicides. Rather, their role will be as complimentary components in successful integrated weed management systems, and in the discovery of novel phytotoxins with new molecular sites of action.

11. REFERENCES

1. Agrow, World Crop Protection News, July 11, 1997.

2. Altman, J., Neate, S. and Rovira, A.D. 1990. Herbicide-pathogen interactions and mycoherbicides as alternative strategies for weed control. In: *Microbes and Microbial Products as Herbicides* (ed. R.E. Hoagland) Amer. Chem. Soc. Symp. Ser. No. 439. ACS Books, Washington, DC, pp. 240-295.

3. Amrhein, N. and Gödeke, K.H. 1977. a-Aminooxy-b-phenylpropionic acid - a potent inhibitor of phenylalanine ammonia-lyase (PAL) *in vitro* and *in vivo*. *Plant Sci. Lett.* **8**:313-317.

4. Amrhein, N., Gödeke, K.H. and Gerhardt, J. 1976. The estimation of phenylalanine ammonia-lyase (PAL) activity in intact cells of higher plant tissue. I. Parameters of the assay. *Planta* **131**:33-40.

5. Amrhein, N. and Holländer, H. 1979. Inhibition of anthocyanin formation in seedlings and flowers by the enantiomers of a-aminooxy-b-phenylpropionic acid and their N-benzyloxycarbonyl derivatives. *Planta* **144**:385-389.

6. Amrhein, N. and Wenker, D. 1979. Novel inhibitors of ethylene production in higher plants. *Plant Cell Physiol.* **20**:1635-1642.

7. Amrhein, N. and Zenk, M.H. 1971. Untersuchungen zur rolle der phenylalanin ammonium-lyase (PAL) bei der regulation der flavonoid-synthese in buckweizen (*Fagopyrum esculentum* Moench.) *Z. Planzenphysiol.* **64**:145-168.

8. Amsellem, Z., Sharon, A. and Gressel, J. 1991. Abolition of selectivity of two mycoherbicidal organisms and enhanced virulence of avirulent fungi by an invert emulsion. *Phytopathology* **81**:985-988.

9. Andebrhan, T., Coutts, R.H.A., Wagih, E.E. and R.K.S. Wood. 1980. Induced resistance and changes in the soluble protein fraction of cucumber leaves locally infected with *Colletotrichum lagenarium* or TMV. *Phytopathol. Z.* **98**:47-52.

10. Anonymous, 1989a. *Alternative Agriculture*, National Academic Press.

11. Anonymous, 1989b. Discovery and Development of Plant Pathogens for Biological Control of Weeds. In: *Plant Pathol. Reg. Res. Proj. SRCS* 8801 (S-136). Dept., Univ. Florida: Gainesville, p. 44

12. Antoniw, J.F. and White., R F. 1980. The effect of aspirin and polyacrylic acid on soluble leaf proteins and resistance to virus infection in five cultivars of tobacco. *Phytopathol.* Z. **98**:331-341.

13. Auld, B.A. and Morin, L. 1995. Constraints in the development of bioherbicides. *Weed Technol.* **9**:638-652.

14. Auld, B.A., Schrauwen, J.M.A., Talbot, H.E. and Raburn, K.B. 1994. Interaction between *Colletotrichum obiculare* and *Alternaria zinniae* or a *Phomopsis* sp. on *Xanthium spinosum. Plant Protect. Quart.* **9**:86-87.

15. Ayers, A.R., Ebel, J., Valent, B. and Albersheim, P. 1976. Host-pathogen interactions. X. Fractionation and biological acitivity of an elicitor isolated from the mycelial walls of *Phytophthora megasperma* var. *sojae. Plant Physiol.* **57**:760-765.

16. Barrett, S.C.H. 1982. Genetic variation in weeds. In: *Biological Control of Weeds with Plant Pathogens*, (eds. R. Charudattan and H.L. Walker) John Wiley and Sons, New York, pp. 73-98.

17. Baudoin, A.B.A.M. 1986. First report on *Dichotomophthora indica* on common purslane in Virginia. *Plant Dis.* **70**:352.

18. Bayles, C.J., Ghemawat, M.S. and Aist, A.R. 1990. Inhibition by 2-deoxyglucose of callose formation, papillae deposition and resistance to powdery mildew in a ml-o barley mutant. *Physiol. Mol. Plant Pathol.* **36**:63-72.

19. Bellés, J.M., Carbonell, J. and Conejero, V. 1991. Polyamines in plants infected by citrus exocortis viroid or treated with silver ions and ethephon. *Plant Physiol.* **96**:1053-1059.

20. Berg, D., Schedel, M., Schmidt, R.R., Ditgens, K. and Weyland, H. 1982. Naramycin B, an antibiotic from Streptomyces griseus strain 587 with herbicidal properties: Fermentation, isolation, and identification. Z. *Naturforsch.* **37C**:1100-1106.

21. Bernards, M.A. and Ellis. B.E. 1991. Phenylalanine ammonia-lyase from tomato cell cultures inoculated with *Verticillium albo-atrum. Plant Physiol.* **97**:1494-1500.

22. Bernhardt, E.A. and Duniway, J.M. 1986. Decay of pondweed and *Hydrilla hybernacula* by fungi. *J. Aquat. Plant Manage.* **24**:20-24.

23. Bhattacharyya, M.K. and Ward, E.W.B. 1987. Temperature-induced susceptibility of soybeans to *Phytophthora megasperma* f. sp. *glycinea*: phenylalanine ammonia-lyase and glyceollin in the host; growth and glyceollin I sensitivity of the pathogen. *Physiol. Mol. Plant Pathol.* **31**:407-419.

24. Bhattacharyya, M.K. and Ward, E.W.B. 1988. Phenylalanine ammonia-lyase activity in soybean hypocotyls and leaves following infection with *Phytophthora megasperma* f. sp. *glycinea. Can. J. Bot.* **66**:18-23.

25. Böger, P and Sandmann, G. (eds.) 1989. *Target Sites of Herbicide Action.* CRC Press, Boca Raton, FL.

26. Bol, J.F., Linthorst, H.J.M. and Cornelissen, B.J.C. 1990. Plant pathogenesis-related proteins induced by virus infection. *Annu. Rev. Phytopathol.* **28**:113-138.

27. Boland, G.J. and Kuykendall, L.D. (eds.) 1998. *Plant-Microbe Interactions and Biological Control.* Marcel Dekker, Inc., New York.

28. Bowles, D.J. 1990. Defense-related proteins in higher plants. *Annu. Rev. Biochem.* 59:873- 907.

29. Boyetohko, S.M. 1999, Innovative application of microbial agents for biological weed control. In : *Biotechnological Approach in Biocontrol of Plant Pathogens* (eds. K.G. Mukerji, B.P. Chamola. and R.K. Upadhyay) Plenum Publishers, New York, pp. 73-98.

30. Boyette, C.D. 1986. Evaluation of *Alternaria crassa* for biological control of jimsonweed: Host range and virulence. *Plant Sci.* **45**:223-228.

31. Boyette, D.C. 1988. Biocontrol of three leguminous weed species with *Alternaria cassiae. Weed Technol.* **2**:414-417.

32. Boyette, C.D. and Abbas, H.K. 1994. Host range alteration of the bioherbicidal fungus *Alternaria crassa* with fruit pectin and plant filtrates. *Weed Sci.* **42**:487-491.

33. Boyette, C.D., Quimby, P.C. Jr., Caesar, A.J., Birdsall, J.L., Connick, W.J. Jr., Daigle, D.J., Jackson, M.A., Egley, G.H. and Abbas,H.K. 1996. Adjuvants, formulations, and spraying systems for improvement of mycoherbicides. *Weed Technol.* **10**:637-644.

34. Boyette, C.D.,Quimby, P.C. Jr., Connick, W.J. Jr., Daigle, D.J. and Fulgham F.E. 1991. Progress in the production, formulation, and application of mycoherbicides. In: *Microbial Control of Weeds.* (ed. D.O. TeBeest) Chapman and Hall, New York, pp. 209-222.

35. Brammall, R.A. and Higgins, V.J. 1988. The effect of glyphosate on resistance of tomato to *Fusarium* crown and root rot disease and on the formation of host structural defensive barriers. *Can. J. Bot.* **66**:1547-1555.

36. Brooker, N.L., Mischke, C.F., Patterson, C.D., Mischke, S., Bruckart,W.L. and Lydon, J. 1996. Pathogenicity of bar-transformed *Colletotrichum gloeosporioides* f. sp. *aeschynomene. Biol. Control* **7**:159-166.

37. Brosten, B.S. and Sands, D.C. 1986. Field trials of *Sclerotinia sclerotiorum* to control Canada thistle (*Cirsium arvense*). *Weed Sci.* **34**:377-380.

38. Callaghan, T., Ross, R., Weinberger-Ohana, P. and Benziman, M. 1988. β-Glucoside activators of mung bean UDP-glucose:β-glucan synthase. II. Comparison of effects of an endogenous β-linked gluco-lipid with synthetic *n*-alkyl-D-monoglucopyranosides. *Plant Physiol.* **86**:1104-1107.

39. Camm, E.L. and Towers, G.H.N. 1973. Phenylalanine ammonia-lyase. *Phytochemistry* **12**:961-973.

40. Carr, J.P. and Klessing, O.F. 1990. The pathogenesis-related proteins of plants. In: *Genetic Engineering, Principles and Methods*, Vol. 11. (ed. J.K. Setlow), Plenum Press, New York, pp. 65-109.

41. Cardina, J., Littrell, R.H. and Hanlin, R.T. 1988. Anthracnose of Florida beggarweed (*Desmodium tortuosum*) caused by *Colletotrichum truncatum. Weed Sci.* **36**:329-334.

42. Carver, T.L.W., Zeyen, R.J., Robbins, M.P. and Dearne, G.A. 1992. Effects of the PAL inhibitor AOPP on oat, barley, and wheat cell responses to appropriate and inappropriate formae speciales of *Erysiphe graminis* DC. *Physiol. Mol. Plant Pathol.* **41**:397-409.

43. Cassab, G.I. and Varner, J.A. 1988. Cell wall proteins. *Annu. Rev. Plant Physiol. Plant Mol. Biol.* **39**:321-353.

44. Caulder, J.D. and Stowell, L. 1988. U.S. patent #4,766,873.

45. Charudattan, R. 1984. Microbial control of plant pathogens and weeds. *J. Georgia Entomol. Soc.* **19**:40-62.

46. Charudattan, R. 1985. The use of natural and genetically altered strains of pathogens for weed control. In: *Biological Control in Agricultural IPM Systems.* (eds. M.A. Hoy and D.C. Herzog) Academic Press, New York, pp. 347-372.

47. Charudattan, R. 1986. Integrated control of waterhyacinth (*Eichhornia crassipes*) with a pathogen, insects, and herbicides. *Weed Sci.* **34 (Suppl. 1)**:26-30.

48. Charudattan, R. 1988. Inundative control of weeds with indigenous fungal pathogens. In: *Fungi in Biological Control Systems* (ed. M.N. Burge), Manchester University Press, New York, pp. 86-110.

49. Charudattan, R. 1990. Pathogens with potential for weed control. In: *Microbes and Microbial Products as Herbicides* (ed. R.E. Hoagland) Amer. Chem. Soc. Symp. Ser. No. 439, ACS Books, Washington, DC, pp. 132-154.

50. Charudattan, R. 1991. The mycoherbicide approach with plant pathogens. In: *Microbial Control of Weeds* (ed. D.O. TeBeest), Chapman Hall, New York, pp. 24-57.

51. Charudattan, R. and DeLoach ,C.J. Jr. 1988. Management of pathogens and insects for weed control in agroecosystems. In: *Weed Management in Agroecosystems: Ecolological Approaches.* (eds. M.A. Altier and M. Liebman), CRC Press, Boca Raton, FL, pp. 245-264.

52. Charudattan, R., Prange, V.J. and Devalerio, J.T. 1996. Exploration of the use of the "bialaphos genes" for improving bioherbicide efficacy. *Weed Technol.* **10**:625-636.

53. Charudattan, R., Walker, H.L. , Boyette, C.D., Ridings, W. H., TeBeest, D.O., Van Dyke, C.G. and Worsham, A.D. 1986. Evaluation of *Alternaria cassiae* as a mycoherbicide for sicklepod (*Cassia obtusifolia*) in regional field test. In: *South. Coop. Ser. Bull.* 317. Alabama Agric. Exp. Sta., Auburn, AL. p. 19.

54. Chiang, MY., Leonard, K.J. and Van Dyke, C.G. 1989. Bipolaris halepense: a new species from *Sorghum halepense* (johnsongrass). *Mycologia* **81**:532-538.

55. Christ, U. and Mösinger, E. 1989. Pathogenesis-related proteins of tomato: I. Induction by *Phytophthora infestans* and other biotic and abiotic inducers and correlations with resistance. *Physiol. Mol. Plant Pathol.* **35**:53-65.

56. Christy, A.L., Herbst, K.A., Kostka, S.J., Mullen, J.P. and Carlson, P.S. 1993. Synergizing weed biocontrol agents with chemical herbicides. In: *Pest Control with Enhanced Environmental Safety.* (eds. S.O. Duke, J.J. Menn and J.R. Plimmer) Amer. Chem. Soc. Symp. Ser. No. 524. ACS Books, Washington, DC, pp. 87-100.

57. Crowley, D.K., Walker, H.L. and Riley, J.A. 1985. Interaction of *Alternaria macrospora* and *Fusarium lateritium* on spurred anoda. *Plant Dis.* **69**:977-979.

58. Cocucci, S.M., Morgotti, S., Cocucci, M. and Gianani, L. 1983. Effects of ophiobolin A on potassium permeability, transmembrane potential and proton exstrusion in maize roots. *Plant Sci. Lett.* **32**:9-16.

59. Cohen, R., Riov, J., Lisker, N. and Katan, J. 1986. Involvement of ethylene in herbicide-induced resistance to *Fusarium oxysporum* f. sp. *melonis. Phytopathology* **76**:1281-1285.

60. Comai, L., Facciotti, D., Hiatt, W.R., Thompson, G., Rose, R.E. and Stalker, D.M. 1985. Expression in plants of a mutant aroA gene from *Salmonella typhimurium* confers tolerance to glyphosate. *Nature* **317**:741-744.

61. Conn, K.L. and Tenari, J.P. 1989. Interactions of *Alternaria brassicae* conidia with leaf epicuticular wax of canola. *Mycol. Res.* **93**:240-242.

62. Connick, W.J., Jr., Lewis, J.A. and Quimby, P.C. Jr. 1990. Formulation of biocontrol agents for use in plant pathology. In: *New Directions in Biological Control: Alternatives for Suppessing Agricultural Pests and Diseases.* (eds. R.R. Baker and P.E. Dunn) Alan R. Liss, Inc., New York, pp. 345-372.

63. Cook, R.J. and Baker, K.F. 1983. *The Nature and Practice of Biological Control of Plant Pathogens.* American Phytopathology Society, St. Paul, MN.

64. Cook, R.J., Bruckart, W.R., Coulson, J.R., Goettel, M.S., Humber, R.A., Lumsden, R.D., Maddox, J.V., McManus, M.L., Moore, L., Meyer, S.F., Quimby, P.C. Jr., Stack, J.P. and Vaughn, J.L. 1996. Safety of microorganisms intended for pest and plant disease control: A framework for scientific evaluation. *Biol. Control* **7**:333-351.

65. Coutts. R.H.A. 1978. Alterations in the soluble protein patterns of tobacco and cowpea leaves following inoculation with tobacco necrosis virus. *Plant Sci. Lett.* **12**:189-197.

66. Cramer, C.L., Ryder, T.B., Bell, J.N. and Lamb, C.J. 1985. Rapid switching of plant gene expression induced by fungal elicitor. *Science* **227**:1240-1243.

67. Creasy, L.L. 1968. The increase in phenylalanine ammonia-lyase activity in strawberry leaf discs and its correlation with flavonoid synthesis. *Phytochemistry* **7**:441-446.

68. Croteau, R. and Johnson, A. 1984. Biosynthesis of terpenoids in glandular trichomes. In: *Biology and Chemistry of Plant Trichomes.* (eds. E. Rodriguez, P. Healey and I. Mehta) Plenum Press, New York, pp.133-186.

69. Cruickshank, I.A.M. and Perrin, D.R. 1968. The isolation and partial characterization of monilicolin A, a polypeptide with phaseollin-inducing activity from *Monilinia fructicola. Life Sci.* **7**:449-458.

70. Cunha da, A. 1987. The estimation of L-phenylalanine ammonia-lyase shows phenylpropanoid biosynthesis to be regulated by L-phenylalanine supply and availability. *Phytochemistry.* **26**:2723-2727.

71. Cutler, H.G. 1986. Isolating, characterizing, and screening mycotoxins for herbicidal activity. In: *The Science of Allelopathy*, (eds. A.R. Putnam and D.S. Tang) Wiley-Interscience, New York, pp. 147-170.

72. Cutler, H.G. 1988. Perspectives on the discovery of microbial phytotoxins with herbicidal activity. *Weed Technol.* **2**:525-532.

73. Cutler, H.G. 1991. Phytotoxins of microbial origin. In: *Handbook of Natural Toxins*, Vol. 6, *Toxicology of Plant and Fungal Compounds*, (eds. R.F. Keeler and A.T. Tu) Marcel Dekker, New York, pp. 411-438.

74. Daigle, D.J. and Connick, W.J. Jr. 1990. Formulation and application technology for microbial weed control. In: *Microbes and Microbial Products as Herbicides*, (eds. R.E. Hoagland) Amer. Chem. Soc. Symp. Ser. No. 439. ACS Books, Washington, DC, pp. 288-304.

75. Daigle, D.J., Connick, W.J., Quimby, P.C. Jr., Evans, J., Trask-Morrell, B. and Fulgham F.E. 1990. Invert emulsion: carrier and water source for the mycoherbicide *Alternaria cassiae. Weed Technol.* **4**:327-331.

76. Daniel, J.T., Templeton, G.E., Smith, R.J. and Fox, W.T. 1973. Biological control of northern jointvetch in rice with an endemic fungal disease. *Weed Sci.* **21**:303-307.

77. D'Alton, A. and Etherton, B. 1984. Effects of fusaric acid on tomato root hair membrane potentials and ATP levels. *Plant Physiol.* **74**:39-42.

78. Daub, M.E. and Ehrenshaft, M. 1993. The photoactivated toxin cercosporin as a tool in fungal photobiology. *Physiol. Plant.* **89**:227-236.

79. Davies, M.E. 1972. Effects of auxin on polyphenol accumulation and the development of phenylalanine ammonia-lyase activity in dark-grown suspension cultures of Paul's Scarlet Rose. *Planta.* **104**:66-77.

80. DeFrank, J. and Putnam, A.R. 1985. Screening procedures to identify soil-borne Actinomycetes that can produce herbicidal compounds. *Weed Sci.* **33**:271-274.

81. de Nooij, M.P. 1988. The role of weevils in the infection process of the fungus *Phomopsis subordinaria* in *Plantago lanceolata. Oikos* **52**:51-58.

82. Deshpande, B.S., Ambedkar, S.S. and Shewale, J.G. 1988. Biologically active secondary metabolites from Streptomyces. *Enzyme Microb. Technol.* **10**:455-473.

83. Devi, M.S., Rao, J.V.S. and Das, V.S.R. 1976. Herbicide induced changes in the levels of epicuticular waxes and cuticle. *Ind. J. Plant Physiol.* **19**:249-253.

84. Dieterman, L.J., Lin, C.Y., Rohrbaugh, L.M. and Wender, S.H. 1964. Accumulation of ayapin and scopolin in sunflower plants treated with 2,4-dichlorophenoxyacetic acid. *Arch. Biochem. Biophys.* **106**:275-279.

85. Doke, N. 1983. Involvement of superoxide anion generation in the hypersensitive response of potato tuber tissues to infection with an incompatible race of *Phytophthora infestans* and to the hyphal wall components. *Physiol. Plant Pathol.* **23**:345-357.

86. Doke, N., Miura, Y., Sanchez, L. and Kawakita, K. 1994. Involvement of superoxide in signal transduction: Responses to attack by pathogens, physical and chemical shocks, and UV irradiation. In: *Causes of Photooxidative Stress and Amelioration of Defense Systems in Plants*, (eds. C.H. Foyer and P.M. Mullineaux) CRC Press, Boca Raton, FL, pp. 177-197.

87. Duke, S.O. 1985. Biosynthesis of phenolic compounds: Chemical manipulation in higher plants. In: *The Chemistry of Allelopathy*, (ed. A.C. Thompson) Amer. Chem. Soc. Symp. Ser. No. 268. ACS Books, Washington, DC, pp. 113-131.

88. Duke, S.O. 1986a. Naturally occurring chemical compounds as herbicides. *Rev. Weed Sci.* **2**:15-44.

89. Duke, S.O. 1986b. Microbially produced phytotoxins as herbicides - A perspective. In: *The Science of Allelopathy*, (eds. A.R. Putnam and D.S. Tang) Wiley-Interscience, New York, pp. 287-304.

90. Duke, S.O. 1990. Overview of herbicide mechanisms of action. *Environ. Health Perspect.* **87**:262-271.

91. Duke, S.O. and Hoagland, R.E. 1978. Effects of glyphosate on metabolism of phenolic compounds. I. Induction of phenylalanine ammonia-lyase activity in dark-grown maize roots. *Plant Sci. Lett.* **11**:185-190.

92. Duke, S.O. and Hoagland, R.E. 1985. Effects of glyphosate on metabolism of phenolic compounds. In: *The Herbicide Glyphosate*. (eds. E. Grossbard and D. Atkinson) Butterworths, London, pp. 75-91.

93. Duke, S.O., Menn, J.J. and Plimmer, J.R. 1993. Challenges of pest control with enhanced toxicological and environmental safety: An overview. In: *Pest Control with Enhanced Environmental Safety*, (eds. S.O. Duke, J.J. Menn, and J.R.Plimmer) Amer. Chem. Soc. Symp. Ser. No. 524. ACS Books, Washington, DC, pp. 1-13.

94. Dyer, W.E. 1996. Techniques for producing herbicide-resistant crops. In: *Herbicide-Resistant Crops*, (ed. S.O. Duke) CRC Press, Boca Raton, FL, pp. 37-51.

95. Ebert, E. and Ramsteiner, K. 1984. Influence of metolachlor and the metolachlor protectant CGA43089 on the biosynthesis of epicuticular waxes on the primary leaves of *Sorghum bicolor* Moench. *Weed Res.* **24**:383-389.

96. Egley, G.H. and Boyette, C.D. 1995. Water-corn oil emulsion enhances conidia germination and mycoherbicidal activity of *Colletotrichum truncatum*. *Weed Sci.* **43**:312-317.

97. Engelsma, G. 1973. Induction of phenylalanine ammonia-lyase by dichlobenil in gherkin seedlings. *Acta Bot. Neerl.* **22**:49-54.

98. Esquerré-Tugayé, M.T. and Lamport, D.T. 1979. Cell surfaces in plant-microorganism interactions. I. A structural investigation of cell wall hydroxyproline-rich glycoprotein which accumulates in fungus-infected plants. *Plant Physiol.* **64**:314-319.

99. Evans, H. and Ellison, C. 1988. Preliminary work on the development of a mycoherbicide to control *Rottboellia cochinchinensis*. In: *Proc. VII Int. Symp. Biol. Control Weeds*. (ed. E.S. DelFosse) 1st Sper. Patol. Veg. (MAF), Rome, p. 76.

100. Farkas, G.L. and Király, Z. 1962. Role of phenolic compounds in the physiology of plant disease and disease resistance. *Phytopathol.* Z. **44**:105-150.

101. Fenn, M E. and Coffey, M.D. 1989. Quantification of phosphonate and ethyl phosphonate in tobacco and tomato tissues and significance for the mode of action of two phosphonate fungicides. *Phytopathology* **79**:76-82.

102. Figliola, S.S., Camper, N.D. and Ridings, W.H. 1988. Potential biocontrol agents for goosegrass (*Eleucine indica*). *Weed Sci.* **36**:830-835.

103. Fischer, H.P. and Bellus, D. 1983. Phytotoxicants from microorganisms and related compounds. *Pestic. Sci.* **14**:334-346.

104. Friend, J. 1981. Plant phenolics, lignification, and plant disease. *Prog. Phytochemistry* **7**:197-261.

105. Fritzmeier, K.-H., Cretin, C., Kombrink, E., Rohwer, F., Taylor, J., Scheel, C. and Hahlbrock, K. 1987. Transient induction of phenylalanine ammonia-lyase and 4-coumarate:CoA ligase mRNAs in potato leaves infected with virulent or avirulent races of *Phytophthora infestans*. *Plant Physiol.* **85**:34-41.

106. Fullaway, D.T. 1954. Biological control of cactus in Hawaii. *J. Econ. Entomol.* **47**:696-700.

107. Gabriel, D.W. 1991. Parasitism, host species specificity, and gene-specific host cell death. In: *Microbial Control of Weeds*. (ed. D.O. TeBeest) Chapman Hall, New York pp. 115-131.

108. Gianinazzi, S. 1982. Antiviral agents and inducers of virus resistance: Analogies with interferon. In: *NATO Advanced Study Institutes Series*, Vol. 37 (ed. R.K.S. Wood). Plenum Press, New York, pp. 275-298.

109. Gilchrist, D.G. 1983. Molecular modes of action. In: *Toxins and Plant Pathogenesis*. (eds. J.M. Daly and B.J. Deverall) Academic Press, New York, pp. 81-136.

110. Giovanelli, J., Owens, L. and Mudd, S. 1971. Mechanism of inhibition of b cystathionase by rhizobitoxin. *Biochem.* Biophys. Acta **227**:671-684.

111. Gohbara, M. and Yamaguchi, K. 1993. Biological agents for the control of paddy weeds in Japan. *Exp. Bull. No.* 369. Food and Fertilizer Technol. Ctr., Taipai City, Republic of China on Taiwan.

112. Goodman, R.N., Király, Z and Wood, K.R. 1986. Secondary metabolites. In: *The Biochemistry and Physiology of Plant Disease*. Univ. Missouri Press, Columbia, MO, pp. 211-244.

113. Grant, N.T., Prusinkiewicz, E., Makowski, R.M.D., Holmström-Ruddick, B. and Mortensen, K. 1990a. Effect of selected pesticides on survival of *Colletotrichum gloeosporioides* f.sp. *malvae*, a bioherbicide for round-leaved mallow (*Malva pusilla*). *Weed Technol.* **4**:701-715.

114. Grant, N.T., Prusinkiewicz, E., Mortensen, K. and Makowski, R.M.D. 1990b. Herbicide interactions with *Colletotrichum gloeosporioides* f.sp. *malvae*, a bioherbicide for round-leaf mallow (*Malva pusilla*). *Weed Technol.* **4**:716-723.

115. Greaves, M.P. and Sargent, J.A. 1986. Herbicide induced microbial invasion of plant roots. *Weed Sci.* **34 (Suppl. 1)**:50-53.

116. Green, N., Hadwiger, L.A. and Graham, S.O. 1975. Phenylalanine ammonia-lyase, tyrosine ammonia-lyase, and lignin in wheat inoculated with *Erysiphe graminis* f. sp. *tritici*. *Phytopathology* **65**:1071-1074.

117. Green, S., Stewart-Wade, S.M., Boland, G.J., Teshler, M.P. and Liu, S.H. 1998. Formulating micro-organisms for biological control of weeds. In: *Plant-Microbe Interactions and Biological Control*, (eds. G.J. Boland and L.D. Kuykendall) Marcel Dekker, New York, pp. 249-281.

118. Gressel, J. 1990. Synergizing herbicides. *Rev. Weed Sci.* **5**:49-82.

119. Gressel, J. 1993. Synergizing pesticides to reduce use rates. In: *Pest Control with Enhanced Environmental Safety*, (eds. S.O. Duke, J. J. Menn, and J. R. Plimmer) Amer. Chem. Soc. Symp. Ser. No. 524. ACS Books, Washington, DC, pp. 48-61.

120. Grisebach, H., Börner, H. and Moesta, P. 1982. Induction of phytoalexin synthesis in soybean and its significance for the resistance against *Phytophthora megasperma* f. sp. *glycinea*. *Ber. Deutsch. Bot. Ges.* **95**:619-642.

121. Grumbach, K H. and Bach, T.J. 1979. The effect of PS II herbicides amitrole and San 6706 on the activity of 3-hydroxy-3-methylglutaryl-coenzyme-A-reductase and the incorporation of [2-14C] acetate and [2-3H}mevalonate into chloroplast pigments of radish seedlings. Z. *Naturforsch* **34C**: 941-943.

122. Gupta, R., Mukerji, K.G. and Upadhyay, R.K. 1998. Mycoherbicides : An overview. In : *Integrated Pest and Disease Management,* (eds. R.K. Upadhyay, K.G. Mukerji, B.P. Chamola, and O.P. Dubey), APH Publishing House, New Delhi pp. 515-532.

123. Haga, M., Haruyama, T., Kano, H., Sekizawa, Y., Urushizaki, S. and Matsumoto, K. 1988. Dependence on ethylene of the induction of phenylalanine ammonia-lyase activity in rice leaf infected with blast fungus. *Agric. Biol. Chem.* **52**:943-950.

124. Hahlbrock, K. and Scheel, D. 1989. Physiology and molecular biology of phenylpropanoid metabolism. *Annu. Rev. Plant Physiol. Molec. Biol.* **40**:347-369.

125. Halliwell, B. 1978. Biochemical mechanisms accounting for the toxic action of oxygen on living organisms: The key role of superoxide dismutase. *Cell Biol. Int. Rep.* **2**:113-128.

126. Hammerschmidt, R. 1984. Rapid deposition of lignin in potato tuber tissue as a response to fungi non-pathogenic on potato. *Physiol. Plant Pathol.* **24**:33-42.

127. Hamner, C.L. and Tukey, H.B. 1944. The herbicidal action of 2,4-(dichlorophenoxy)acetic acid and 2,4,5-(trichlorophenoxy)acetic acid on bindweed. *Science* **100**:154-155.

128. Harman, G.E. and Stasz, T.E. 1991. Protoplast fusion for the production of superior biocontrol fungi. In: *Microbial Control of Weeds* (ed. D.O. TeBeest) Chapman Hall, New York, pp. 171-186.

129. Harris, P. 1993. Effects, constraints, and the future of weed biocontrol. **Agric**. *Ecosyst. and Environ.* 46:289-303.

130. Hasan, S. 1972. Specificity and host specialization of *Puccinia chondrillina*. *Ann. Appl. Biol.* **72**:257-263.

131. Hasan, S. 1988. Biocontrol of weeds with microbes. In: *Biocontrol of Plant Diseases*. (eds. K.G. Mukerji and K.L. Garg) CRC Press, Boca Raton, FL, pp. 129-151.

132. Hatzios, K.K. and Penner, D. 1985. Interactions of herbicides with other agrochemicals in higher plants. *Rev. Weed Sci.* **1**:1-63.

133. Heap, I.M. 1997. The occurrence of herbicide-resistant weeds worldwide. *Pestic. Sci.* **51**:235-243.

134. Hebbar, K.P., Lumsden, R.D., Lewis, J.A., Poch, S.M. and Bailey, B.A. 1998. Formulation of mycoherbicidal strains of *Fusarium oxysporum*. *Weed Sci.* **46**:501-507.

135. Heisey, R.M., DeFrank, J. and Putnam, A.R. 1985. A survey of soil microorganisms for herbicidal activity. In: *The Chemistry of Allelopathy* (ed. A.C. Thompson) Amer. Chem. Soc. Symp. Ser. No. 268. ACS Books, Washington, DE, pp. 337-349.

136. Hoagland, R.E. 1985. O-Benzylhydroxylamine: An inhibitor of phenylpropanoid metabolism in plants. *Plant Cell Physiol.* **26**:1353-1359.

137. Hoagland, R.E. 1989a. Biochemical interactions of atrazine and glyphosate in soybean (*Glycine max*) seedlings. *Weed Sci.* **37**:491-497.

138. Hoagland, R E. 1989b. Acifluorfen action on growth and phenolic metabolism in soybean (*Glycine max*) seedlings. *Weed Sci.* **37**:743-747.

139. Hoagland, R.E. 1990a. Microbes and microbial products as herbicides: An overview. In: *Microbes and Microbial Products as Herbicides* (ed. R.E. Hoagland) Amer. Chem. Soc. Symp. Ser. No. 439, ACS Books, Washington, DC, pp. 2-52.

140. Hoagland, R.E. 1990b. Biochemical responses of plants to pathogens. In: *Microbes and Microbial Products as Herbicides* (ed. R. E. Hoagland) Amer. Chem. Soc. Symp. Ser. No. 439. ACS Books, Washington, DC, p. 87-113.

141. Hoagland, R.E. (ed.) 1990c. *Microbes and Microbial Products as Herbicides*. Amer. Chem. Soc. Symp. Ser. No. 439. ACS Books, Washington, DC.

142. Hoagland, R.E. 1990d. *Alternaria cassiae* alters phenylpropanoid metabolism in sicklepod (*Cassia obtusifolia*). J. *Phytopathol.* **130**:177-187.

143. Hoagland, R.E. 1995. The mycoherbicide *Alternaria cassiae* infects and alters phenolic metabolism of *Cassia alata* seedlings. *Phytoprotection* **78**:67-74.

144. Hoagland, R.E. 1996. Chemical interactions with bioherbicides to improve efficacy. *Weed Technol.* **10**:651-674.

145. Hoagland, R.E. 1999. Biochemical interactions of the microbial phytotoxin phosphinothricin and analogs with plants and microbes. In: *Biologically Active Natural Products in Agriculture and Pharmaceuticals* (eds. H.G. Cutler and S.J. Cutler) CRC Press, Boca Raton, FL (in press).

146. Hoagland, R.E. and Boyette, C.D. 1994. Pathogenic interactions of *Alternaria crassa* and phenolic metabolism in jimsonweed (*Datura stramonium* L.) varieties. *Weed Sci.* **42**:44-49.

147. Hoagland, R.E. and Duke, S.O. 1981. Effects of herbicides on extractable phenylalanine ammonia-lyase activity in light- and dark-grown *Glycine max* (L.) Merr. seedlings. *Weed Sci.* **29**:433-439.

148. Hoagland, R.E. and Duke, S.O. 1982. Effects of glyphosate on metabolism of phenolic compounds. VIII. Comparison of the effects of aminooxyacetate and glyphosate. *Plant Cell Physiol.* **23**:1081-1088.

149. Hoagland, R.E. and Duke, S.O. 1983. Relationship between phenylalanine ammonia-lyase activity and physiological responses of soybean (*Glycine max*) seedlings to herbicides. *Weed Sci.* **31**:845-852.

150. Hodgson, R.H., Wymore, L.A., Watson, A.K., Snyder, R.H. and Collette, A. 1988. Efficacy of *Colletotrichum coccodes* and thidiazuron for velvelleaf (*Abutilon theophrasti*) control in soybean (*Glycine max*). *Weed Technol.* **2**:473-480.

151. Hoerlein, G. 1994. Glufosinate (phosphinothricin), a natural amino acid with unexpected herbicidal properties. *Rev. Environ. Contam. Toxicol.* **138**:73-145.

152. Hoescht, A.G. 1977. (German patent) DOS 2 717 440.

153. Hofmeister, F.M. and Charudattan, R. 1987. *Pseudocercospora nigricans*, a pathogen of sicklepod (*Cassia obtusifolia*) with biocontrol potential. *Plant Dis.* **71**:44-46.

154. Holliday, M.J. and Keen, N.T. 1982. The role of phytoalexins in the resistance of soybean leaves to bacteria: Effect of glyphosate on glyceollin accumulation. *Phytopathology* **72**:1470-1474.

155. Holcomb, G.E., Jones, J.P. and Wells, D.W. 1989. Blight of prostrate spurge and cultivated poinsettia, caused by Amphobotrys ricini. *Plant Dis.* **73**:74-75.

156. Holm, L.G., Plucknett, D.L., Pancho, J.V. and Herberger, J.P. 1991. *The World's Worst Weeds.* Krieger Publ. Co., Malabar, FL.

157. Hopkins, W.L. 1997. *Global Herbicide Directory*, (2nd ed.) Ag Chem Information Services, Indianapolis, IN.

158. Hoppe, H.H. 1997. Fungal phytotoxins. In: *Resistance of Crop Plants to Fungi* (eds. H. Hartleb, R. Heitefuss, and H.-H. Hoppe) Fischer, Jena, Germany, pp. 58-83.

159. Howell, C.R. and R.D. Stipanovic. 1984. Phytotoxicity to crop plants and herbicidal effects on weeds of viridiol produced by *Gliocladium virens*. *Phytopathology* **74**:1346-1349.

160. Hyashi, T., Read, S.M., Bussell, J., Thelen, M.T., Lin, F.C., Brown, R.M. Jr. and Delmer, D.P. 1987. UDP-glucose:(1®3)-b-glucan synthases from mung bean and cotton. Differential effects of Ca^{+2} and Mg^{+2} on enzyme properties and on macromolecular structure of the glucan product. *Plant Physiol.* **83**:1054-1062.

161. Jackson, M.A., Schisler, D.A., Slininger, P.J., Boyette, C.D., Silman, R.W. and Bothast, R.J. 1996a. Fermentation strategies for improving the fitness of a bioherbicide. *Weed Technol.* **10**:645-650.

162. Jackson, M.A., Shasha, B.S. and Schisler, D.A. 1996b. Formulation of *Colletotrichum truncatum* microsclerotia for improved biocontrol of the weed hemp sesbania (*Sesbania exaltata*). *Biol. Control* **7**:107-113.

163. Jackson, M.A. and Slininger, P.J. 1993. Submerged culture conidial germination and conidiation of the bioherbicide *Colletotrichum truncatum* are influenced by the amion acid composition of the medium. *J. Indust. Microbiol.* **12**:417-422.

164. Janas, K.M., Filipiak, A., Kowalik, J., Mastalerz, P. and Knypl, J.S. 1985. 1-Amino-2-phenylethylphosphonic acid: an inhibitor of L-phenylalanine ammonia-lyase *in vitro*. *Acta Biochim.Polonica* **32**:131-143.

165. Jangaard, N.O. 1974. The effect of herbicides, plant growth regulators and other compounds on phenylalanine ammonia-lyase activity. *Phytochemistry* **13**:1769-1775.

166. Johnson, D.R., Wyse, D.L. and Jones, K.J. 1996. Controlling weeds with phytopathogenic bacteria. *Weed Technol.* **10**:621-24.

167. Joye, G.F. 1990. Biological control of aquatic weeds with plant pathogens. In: *Microbes and Microbial Products as Herbicides* (ed. R.E. Hoagland) Amer. Chem. Soc. Symp. Ser. No. 439, ACS Books, Washington, DC, pp. 155-174.

168. Juniper, B.E. 1957. The effect of pre-emergent treatment of peas with trichloroacetic acid on the submicroscopic strucure of the leaf surface. *New Phytol.* **58**:1-5.

169. Jutsum, A.R. 1988. Commercial application of biological control: Status and prospects. Phil. *Trans. Royal Soc. London* B **318**:357-373.

170. Kamata, S., Sakai, H. and Hirota, A. 1983. Isolation of acetylaranotin, bisdithiode(methylthio)-acetyl-aranotin and terrein as plant growth inhibitors from a strain of *Aspergillus terreus*. *Agric. Biol. Chem.* 47:2637-2638.

171. Katan, J. and Eshel, Y. 1973. Interactions between herbicides and plant pathogens. *Residue Rev.* **45**:145-177.

172. Kauss, H. 1990. Role of the plasma membrane in host/pathogen interactions. In: *The Plant Plasma Membrane: Structure, Function, and Molecular Biology*. (eds. C. Larsson and I. M. Meller) Springer-Verlag, Berlin, pp. 320-350.

173. Kauss, H., Waldmann, T., Jeblick, W. and Takemoto, T.Y. 1991. The phytotoxin syringomycin elicits Ca^{2+}-dependant callose synthesis in suspension-cultured cells of *Catharanthus roseus*. *Physiol. Plant.* **81**:134-138.

174. Keen, N.T. 1990. Phytoalexins and the elicitors. In: *Microbes and Microbial Products as Herbicides* (ed. R.E. Hoagland) Amer. Chem. Soc. Symp. Ser. No. 439. ACS Books, Washington, DC, pp. 114-129.

175. Keen, N.T., Holliday, M.J. and Yashikawa, Y. 1982. Effects of glyphosate on glyceollin production and the expression of resistance to *Phytophthora megasperma* f. sp. *glycinea* in soybean. *Phytopathology* **72**:1467-1470.

176. Kenfield, D., Bunkers, G., Strobel, G.S. and Sugawara, F. 1988. Potential new herbicides - Phytotoxins from plant pathogens. *Weed Technol.* **2**:519-524.

177. Keppler, L.D. and Baker, C.J. 1989. O_2-Inhibited lipid peroxidation in a bacteria-induced hypersensitive reaction in tobacco cell suspensions. *Phytopathology* **79**:555-562.

178. Keppler, L.D., Baker, C.J. and Atkinson, M.M. 1989. Active oxygen production during a bacteria-induced hypersensitive reaction in tobacco suspension cells. *Phytopathology* **79**:974-978.

179. Kistler, H.C. 1991. Genetic manipulation of plant pathogenic fungi. In: *Microbial Control of Weeds* (ed. D.O. TeBeest) Chapman Hall, New York, pp. 152-170.

180. Klisiewicz, J.M. 1985. Growth and reproduction of *Dichotomophthora portulacae* and its biological activity on purslane. *Plant Dis.* **69**: 761-762.

181. Kodama, M., Kajiware, K., Otani, H. and Kohmoto, K. 1989. A host-recognition factor from *Botrytis* affecting scallion. In: *Host-Specific Toxins, Recognition and Specificity Factors in Plant Disease*. (eds. K. Kohmoto and R. D. Durbin) Tottori Univ. Press, Tottori, Japan, pp. 33-44.

182. Kolattukudy, P.E. 1965. Biosynthesis of wax in *Brassica oleracea*. *Biochemistry* **4**:1844-1855.

183. Kolattukudy, P.E. 1967. Biosynthesis of paraffins in *Brassica oleracae*: Fatty acid elongation-decarboxylation as a plausible pathway. *Phytochemistry* **6**:963-975.

184. Kömives, T. and Casida, J.E. 1982. Diphenylether herbicides: Effects of acifluorfen on phenylpropanoid biosynthesis and phenylalanine ammonia-lyase activity in spinach. *Pestic. Biochem. Physiol.* **18**:191-196.

185. Kömives, T. and Casida, J.E. 1983. Acifluorfen increases the leaf content of phytoalexins and stress metabolites in several crops. *J. Agric. Food Chem.* **31**:751-755.

186. Kremer, R.J. 1995. Integration of a seed-feeding insect and fungi for management of velvetleaf (*Abutilon theophrasti*) seed production. In: *Proc. VIII Int. Symp. Biol. Contr. Weeds.* (eds. E.S. DelFosse and R.R. Scott) DSIR/CSIRO, Melbourne, pp. 627-631.

187. Kremer, R.J. and Kennedy, A.C. 1996. Rhizobacteria as biocontrol agents of weeds. *Weed Technol.* **10**:601-609.

188. Laber, B., Kiltz, H.H. and Amrhein, N. 1986. Inhibition of phenylalanine ammonia-lyase in *vitro* and in vivo by (1-amino-2-phenylethyl)phosphonic acid, the phosphonic analogue of phenylalanine. Z. *Naturforsch.* **41C**:49-55.

189. Lacy, G.H. 1991. Perspectives for biological engineering of prokaryotes of biological control of weeds. In: *Microbial Control of Weeds* (ed. D.O. TeBeest) Chapman Hall, New York, pp. 135-151.

190. Leach, J. E., Cantrell, M.A. and Sequeira, L. 1982. Hydroxyproline-rich bacterial agglutinin from potato. Extraction, purification, and characterization. *Plant Physiol.* **70**:1353-1358.

191. Leathers, T.D., Gupta, S.C. and Alexander, N.J. 1993. Mycopesticides: status, challenges and potential. *J. Indust. Microbiol.* **12**:69-75.

192. Leonard, K.J. 1982. The benefits and potential hazards of genetic heterogeneity in plant pathogens. In: *Biological Control of Weeds with Plant Pathogens* (eds. R. Charudattan and H.L. Walker) John Wiley and Sons, New York, pp. 99-112.

193. Lévesque, C.A. and Rahe, J.E. 1992. Herbicide interctions with fungal root pathogens, with special references to glyphosate. *Annu. Rev. Phytopathol.* **30**:579-602.

194. Lydon, J. 1995. The molecular genetics of bacterial phytotoxins. *Proc. 22nd Annual Plant Growth Regul. Soc. Amer.* pp. 98-101.

195. Lydon, J. and Duke, S.O. 1988. Glyphosate induction of elevated levels of hydroxybenzoic acids in higher plants. *J. Agric. Food Chem.* **36**:813-818.

196. Lydon, J. and Duke, S.O. 1989. Pesticide effects on secondary metabolism of higher plants. *Pestic. Sci.* **25**:361-373.

197. Lydon, J. and Duke, S.O. 1993. The role of pesticides on host allelopathy and their effects on allelopathic compounds. In: *Pesticide Interactions in Crop Production, Beneficial and Deleterious Effects* (ed. J. Altman) CRC Press, Boca Raton, FL, pp. 37-56.

198. Lynch, J.M. and Ebben, M.H. 1986. The use of micro-organisms to control plant disease. *J. Appl. Bact. Symp. Suppl.* **15**:115S-126S.

199. Macias, F.A., Castellano, D., Oliva, R.M., Cross, P. and Torres, A. 1997. Potential use of allelopathic agents as natural agrochemicals. *Brighton Crop Protect. Conf. - Weeds*, **2**:33-38.

200. Madariaga, R.B. and Scharen, A.L. 1985. *Septoria tritici* blotch in Chilean wild oat. *Plant Dis.* **69**:126-127.

201. Makowski, R.M.D. 1993. Effect of inoculum concentration, temperature, dew period, and plant growth stage on disease of round-leaved mallow and velvetleaf by *Colletotrichum gloeosporioides* f. sp. *malvae. Phytopathology* **83**:1229-1234.

202. Makowski, R.M.D. and Mortensen, K. 1990. *Colletotrichum gloeosporioides* f. sp. *malvae* as a bioherbicide for round-leaved mallow (*Malva pusilla*): conditions for successful control in the field. In: *Proc. VII Int. Symp. Biol. Contr. Weeds.* (ed. E.S. DelFosse) 1st Sper. Patol. Veg. (MAF), Rome, pp. 513-522 .

203. Mann, J.D. and Pu, M. 1968. Inhibition of lipid synthesis by certain herbicides. *Weed Sci.* **16**:197-198.

204. Massala, R., Legrand, M. and Fritig, B. 1987. Comparative effects of two competitive inhibitors of phenylalanine ammonia-lyase on the hypersensitive resistance of tobacco to tobacco mosaic virus. *Plant Physiol. Biochem.* **25**:217-225.

205. Massion, C.L. and Lindow, S.E. 1986. Effects of *Sphacelotheca holci* infection on morphology and competetiveness of johnsongrass (*Sorghum halepense*). *Weed Sci.* **34**:883-888.

206. Mayama, S., Hayashi, S., Yamamoto, R., Tani, T., Ueno, T. and Fukami, H. 1982. Effects of elevated temperature and a-amino-oxyacetate on the accumulation of avenalumins in oat leaves infected with *Puccinia coronata* f. sp. *avenae*. *Physiol. Plant Pathol.* **20**:305-312.

207. Mayama, S., Tani, T. and Matsuura, Y. 1981. The production of phytoalexins by oat in response to crown rust, *Puccinia coronata* f. sp. *avenae*. *Physiol. Plant Pathol.* **19**:217-226.

208. Mazau, D., Rumeau, D. and Esquerré-Tugayé, M.T. 1987. Molelcular approaches to understanding cell surface interactions between plants and fungal pathogens. *Plant Physiol. Biochem.* **25**:337-343.

209. McRae, C.F. 1988. Classical and inudative approaches to biological weed control compared. *Plant Protect. Quart.* **3**:124-127.

210. McRae, C.F. and Auld, B.A. 1988. The influence of environmental factors on anthracnose of *Xanthium spinosum*. *Phytopathology* **78**:1182-1186.

211. Meiji Seika Kaisha. J 5 4092 628 (1979; priority 29.12.1977).

212. Mellon, J.E. and Halgeson, J.P. 1982. Interaction of a hydroxyproline-rich glycoprotein from tobacco callus with potential pathogens. *Plant Physiol.* **70**:401-405.

213. Mitchell, R.E. 1981. Structure: Bacterial. In: *Toxins in Plant Disease* (ed. R.D. Durbin) Academic Pess, New York, pp. 259-293.

214. Mitchell, R.E. 1984. The relevance of non-host-specific toxins in the expression of virulence by pathogens. *Annu. Rev. Phytopathol.* **22**:215-124.

215. Moesta, P. and Grisebach, H. 1982. L-a-Aminooxy-b-phenylpropionic acid inhibits phytoalexin accumulation in soybean with concomitant loss of resistance against *Phytophthora megasperma* f. sp. *glycinea*. *Physiol. Plant Pathol.* **21**:65-70.

216. Morris, M.J. 1991. The use of plant pathogens for biological weed control in South Africa. *Agric. Ecosyst. Environ.* **37**:239-255.

217. Morris, M.J. 1996a. Impact of a gall-forming rust fungus, *Uromycladium tepperioanum*, on populations of an invasive tree, *Acacia saligna*, in South Africa. In: *Proc. IX Int. Symp. Biol. Contr. Weeds* (eds. V.V. Moran and J.H. Hoffman) Univ. of Capetown, Capetown South Africa, p. 509.

218. Morris, M.J. 1996b. The development of mycoherbicides for an invasive shrub, *Hakea sericea*, and a tree, *Acacia mearnsii*, in South Africa. In: *Proc. IX Int. Symp. Biol. Contr. Weeds* (eds. V.V. Moran and J.H. Hoffman) Univ. of Capetown, Capetown South Africa, p. 547.

219. Mortensen, K. 1988. The potential of an endemic fungus, *Colletotrichum gloeosporioides*, forbiological control of roundleaf mallow (*Malva pusilla*) and velvetleaf (*Abutilon theophrasti*). *Weed Sci.* **36**:472-478.

220. Mortensen, K. 1996. Constraints in development and commercialization of a plant pathogen, *Colletotrichum gloeosporioides* f. sp. *malvae*, for biological weed control. In: *Proc. 2nd Int. Weed Control Congress* (eds. H. Brown, G. Cussans, M.Devine, S. Duke, C. Fernandez-Quintanilla, A. Helweg, R. Labrada, M. Landes, P. Kudsk and J. Streibig) Dept. of Weed Control, Pesticides, and Ecology, Flakkebjerg, Denmark, pp. 1297-1300.

221. Mortensen, K. 1998. Biological control of weeds using microorganisms. In: *Plant-Microbe Interactions and Biological Control* (eds. G.J. Boland and L.D. Kuykendall) Marcel Dekker, New York, pp. 223-248.

222. Mortensen, K., Harris, P. and Kimm, W.K. 1991. Host ranges of *Puccinia jaceae, P. centaureae, P. acroptili,* and *P. carthami*, and the potential value of P. jaceae as a biological control agent for diffuse knapweed (*Centaurea diffusa*) in North America. *Can. J. Plant Pathol.* **11**:322-324.

223. Muñoz, R., Martinez-Martinez, A., Ros-Barceló, A. and Pedreño, M.A. 1990. Effect of 4- amino-6-methyl-3-phenylamino-1,2,4-triazin-5(4H)-one on the lignification process catalysed by peroxidase from lupin (*Lupinus albus*). *Pestic. Sci.* **28**:283-288.

224. Munyaradzi, S.T., Campbell, M. and Burge, M.N. 1990. The potential for bracken control with mycoherbicidal formulations. *Aspects Appl. Biol.* **24**:169-177.

225. Murao, S. and Hayashi, H. 1983. Gougerotin, as a plant growth inhibitor from *Streptomyces* sp. No. 179. *Agric. Biol. Chem.* **47**:1135-1136.

226. Ohana, P., Benziman, M. and Delmer, D.P. 1993. Stimulation of callose synthesis *in vivo* correlates with changes in intracellular distribution of the callose synthase activitor b-furfuryl-b-glucoside. *Plant Physiol.* **101**:187-191.

227. Ohana, P., Delmer, D.P., Volman, G., Steffens, J.C., Matthews, D.E. and Benziman, M. 1992. b-Furfuryl-b-glucoside: an endogenous activator of higher plant UDP-glucose:(1-3)-b-glucan synthase. Biological activity,distribution, and *in vitro* synthesis. *Plant Physiol.* **98**:708-715.

228. Oku, H., Shiraishi, T. and Ouchi, S. 1977. Suppression of induction of phytoalexin, pisatin, by low- molecular-weight substances from spore germination of fluid of pea pathogen, *Mycosphaerella pinodes. Naturwissenschaften* **64**:643-644.

229. Oku, H., Shiraishi, T. and Ouchi, S. 1987. Role of specific suppressors in pathogenesis of *Mycosphaerella* species. In: *Molecular Determinants of Plant Diseases* (eds. S. Nishimura, C. P. Vance and N. Doke) Japan Weed Sci. Soc. Press/Springer-Verlag, Tokyo, pp. 145-156.

230. Oku, H., Shiraishi, T., Ouchi, S. and Ishiura, M. 1980. A new determinant of pathogenicity in fungal plant disease. *Naturwissenschaften* **67**:310-311.

231. Pascholati, S.F., Helm, D. and Nicholson, R.L. 1985. Phenylalanine ammonia-lyase and susceptibility of the maize mesocotyl to *Helminthosporium maydis*. *Physiol. Plant Pathol.* **27**:345-356.

232. Paxton, J.D. 1988. Fungal elicitors of phytoalexins and their potential use in agriculture. In: *Biolgically Active Natural Products: Potential Use in Agriculture* (ed. H.G. Cutler) Amer. Chem. Soc. Symp. Ser. No. 380. ACS Books, Washington, DC, pp. 109-119.

233. Permulla, C.J. and Heath, M.C. 1991. The effects of inhibitors of various cellular processes on the wall modification induced in bean leaves by the cowpea rust fungus. *Physiol. Mol. Plant Pathol.* **38**:293-300.

234. Pokorny, R. 1941. Some chlorophenoxyacetic acids. *J. Amer. Chem. Soc.* **63**:1768.

235. Politis, D.J., Watson, A.K. and Bruckart, W.L. 1984. Susceptibility of musk thistle and related composites to *Puccinia carduorum*. *Phytopathology* **74**:687-691.

236. Ride, J.P. and Barber, M.S. 1987. The effects of various treatments on induced lignification and the resistance of wheat to fungi. *Physiol. Mol. Plant Pathol.* **31**:349-360.

237. Robinson, T. 1980. Miscellaneous unsaponifiable lipids. In: *The Organic Constituents of Higher Plants*. Cordus Press, N. Amherst, MA, pp. 110-132.

238. Robeson, D.J. and Strobel, G.A. 1985. The identification of a major phytotoxic component from *Alternaria macrospora* as b-dehydrocurvularin. *J. Nat. Prod.* **48**:139-141.

239. Roby, D., Toppan, A. and Esquerré-Tugayé, M.T. 1985. Cell-surfaces in plant-microorganism interactions. V. Elicitors of fungal and of plant origin trigger the synthesis of ethylene and of cell-wall hydroxyproline-rich glycoprotein in plants. *Plant Physiol.* **77**:700-704.

240. Rodgers, P.B. 1993. Potential of biopesticides in agriculture. *Pestic. Sci.* **39**:117-129.

241. Saindrenan, P., Barchietto, T. and Bompeix, G. 1988. Modification of the phosphite induced resistance response in leaves of cowpea infected with *Phytophthora cryptogea* by a-aminooxy-acetate. *Plant Sci.* **58**:245-252.

242. Sakamura, S., Ichihara, A. and Yoshihara, T. 1988. Toxins of phytopathogenic microorganisms: Structural diversity and physiological activity. In: *Biologically Active Natural Products* (ed. H.G. Cutler) Amer. Chem. Symp. Ser. No. 380, ACS Books, Washington, DC. pp. 57-64.

243. Scheepens, P.C. 1987. Joint action of *Cochliobolus lunatus* and atrazine on *Echinochloa crus-galli* (L.) Beauv. *Weed Res.* **27**:43-47.

244. Scheepens, P.C. and van Zon, H.C.J. 1982 Microbial herbicides. In: *Microbial and Viral Pesticides* (ed. E. Kurstak) Marcel Dekker, New York, pp. 623-641.

245. Schisler, D.A., Jackson, M.A. and Bothast, R.J. 1991. Influence of nutrition during conidiation of *Colletotrichum truncatum* on conidial germination and efficacy in inciting disease in *Sesbania exaltata*. *Phytopathology* **81**:458-461.

246. Shabana, Y.M.N.E. 1987. Biological Control of Water Weeds by Using Plant Pathogens. Dissertation, Mansoura University, El-Mansoura, Egypt.

247. Sharon, A. and Gressel, J. 1991. Elicitation of flavonoid phytoalexin accumulation in *Cassia obtusifolia* by a mycoherbicide: Estimation by AlCl3-spectrofluoimetry. *Pestic. Biochem. Physiol.* **41**:142-149.

248. Sharon, A. and Gressel, J. 1992. Glyphosate suppression of an elicited defense response. Increased susceptibility of *Cassia obtusifolia* to a mycoherbicide. *Plant Physiol.* **98**:654-659.

249. Sharon, A., Ghirlando, R. and Gressel, J. 1992. Isolation, purification, and identification of 2-(p-hydroxyphenoxy)-5,6-dihydroxychrome: a fungal induced phytoalexin. *Plant Physiol.* **98**:303-308.

250. Sherwood, R.I. and Vance, C.P. 1976. Histochemistry of papillae formed in reed canarygrass leaves in response to noninfecting pathogenic fungi. *Phytopathology* **66**:503-510.

251. Shimomura, T. 1971. Necrosis and localization of infection in local lesion hosts. *Phytopathol. Z.* **70**:185-196.

252. Shiraishi, T., Oku, H., Yamashita, M. and Ouchi, S. 1978. Elicitor and suppressor of pisatin induction in spore germination fluid of pea pathogen, *Mycosphaerella pinodes*. *Ann. Phytopathol. Soc. Japan* **44**:659-665.

253. Shiraishi, T., Yamada, T., Saitoh, K., Kato, T., Toyoda, K., Yoshioka, H., Kim, H.M., Ichinose, Y., Tahara, M. and Oku, H. 1994. Suppressors: determinants of specificity produced by plant pathogens. *Plant Cell Physiol.* **35**:1107-1119.

254. Shiraishi, T., Yamaoka, N. and Kunoh, H. 1989. Association between increased phenylalanine ammonia-lyase activity and cinnamic acid synthesis and the induction of temporary inaccessibility caused by *Erysiphe graminis* primary germ tube penetration of the barley leaf. *Physiol. Mol. Plant Pathol.* **34**:75-83.

255. Skipper, H.D., Ogg, A.G. and Kennedy, A.C. 1996. Root biology of grasses and ecology of rhizobacteria for biological control. *Weed Technol.* **10**:610-620.

256. Smith, D.A. 1982. Toxicity of phytoalexins. In: *Phytoalexins.* (eds. J. A. Bailey and J. W. Manfield) John Wiley and Sons, New York, pp. 218-252.

257. Smith, R.J., Jr. 1986. Biological control of northern jointvetch (*Aeschynomene virginica*) in rice (*Oryza sativa*) and soybean (*Glycine max*) -- a researcher's view. *Weed Sci.* **34 (Suppl. 1)**:17-23.

258. Smith, J.R., Jr. 1991. Integration of biological control agents with chemical pesticides. In: *Microbial Control of Weeds.* (ed. D.O. TeBeest) Chapman Hall, New York, pp. 189-208.

259. Sorsa, K.K., Nordheim, E.V. and Andrews, J.H. 1988. Integrated control of Eurasian water milfoil, *Myriophyllum spicatum*, by a fungal pathogen and a herbicide. *J. Aquat. Plant Manage.* **26**:12-17.

260. Steele, J.A., Uchytil, T.F., Durbin, R.D., Bhatnagar, P. and Rich, D.H. 1976. Chloroplast coupling factor 1: A species-specific receptor for tentoxin. *Proc. Nat. Acad. Sci. U.S.A.* **73**:2245-2248.

261. Still, G.G., Davis, D.G. and Zander, G.L. 1970. Plant epicuticular lipids: Alteration by herbicidal carbamates. *Plant Physiol.* **46**:307-314.

262. Strobel, G., Kenfield, D., Bunkers, G., Sugawara, F. and Clardy, J. 1991. Phytotoxins as potential herbicides. *Experientia* **47**:819-826.

263. Strobel, G.A., Sugawara, F. and Clardy, J. 1987. Phytotoxins from plant pathogens of weedy plants. In: *Allelochemicals: Role in Agriculture and Forestry* (ed. G.R. Waller) Amer. Chem. Soc. Symp. Ser. No. 330; ACS Books, Washington, DC, pp. 516-523.

264. Stonard, R.J. and Miller-Wideman, M.A. 1995. Herbicides and plant growth regulators. In: *Agrochemicals from Natural Products* (ed. C.R.A. Godfrey) Marcel Dekker, New York, pp. 285-310.

265. Stübler, D. and Buchenauer, H. 1993. Antiviral properties of lichenan (b{1-3, 1-4}D-glucan) in tobacco. In: *Mechanisms of Plant Defense Responses* (eds. B. Fritig and M. Legrand) Kluwer Acad. Publishers, Boston, MA, p. 455.

266. Swanton, C.J., Harker, K.N. and Anderson, R.L. 1993. Crop losses due to weeds in Canada. *Weed Technol.* **7**:537-542.

267. Takebe, H., Imai, S., Ogawa, H., Satho, A. and Tanaka, H. 1989. Studies on the production of bialaphos. 1. Breeding of bialaphos-producing strains from a biochemical engineering viewpoint. *J. Ferm. Bioengineering* **67**:226-232.

268. Tal, B., Robeson, D.J., Burke, B.A. and Aasen, A.J. 1985. Phytotoxins from *Alternaria helianthi*: radicinin, and the strucutrues of deoxyradicinol and radianthin. *Phytochemistry* **24**:729-731.

269. Tanaka, Y. and Omura, S. 1993. Agroactive compounds of microbial origin. *Annu. Rev. Microbiol.* **47**:57-87.

270. TeBeest, D.O. (ed.) 1991. *Microbial Control of Weeds.* Chapman and Hall, New York.

271. TeBeest, D.O. 1996. Biological control of weeds with plant pathogens and microbial pesticides. *Adv. Agron.* **56**:115-137.

272. TeBeest, D.O. and Templeton, G.E. 1985. Mycoherbicides: Progress in the biological control of weeds. *Plant Dis.* **69**:6-10.

273. TeBeest, D.O., Yang, Z.B. and Cisar, C.R. 1992. The status of biological control of weeds with fungal pathogens. *Annu. Rev. Phytopathol.* **30**:637-657.

274. Templeton, G.E. 1982. Biological herbicides: Discovery, development, deployment. *Weed Sci.* **30**:430-433.

275. Templeton, G.E., Smith, R.J. and TeBeest, D.O. 1986. Progress and potential of weed control with mycoherbicides. *Rev. Weed Sci.* **2**:2-14.

276. Tietjen, J.G., Schaller, E. and Matern, U. 1983. Phytotoxins from *Alternaria carthami* Chowdhury: Structural identification and physiological significance. *Physiol. Plant Pathol.* **23**:387-400.

277. Tisdell, C. 1987. Economic evaluation of biological weed control. *Plant Prot. Quart.* **2**:10-11.

278. Trujillo, E.E. 1985. Biological control of hamakua pa-makani with Cercosporella sp. in Hawaii. In: *Proc. VI Internatl. Symp. Biol. Contr. Weeds* (ed. E.S. DelFosse) Agriculture Canada, Ottawa, pp. 661-671.

279. U.S. Bureau of the Census. 1998. http://www.census.gov/ipc/www/worldpop.html

280. U.S. Environmental Protection Agency. Notice to manufacturers, formulators, producers and registrants of pesticide products. Pesticide Regulation (PR) Notice 97-3, Sept. 4, 1997.

281. U.S. Patent Office No. 3 162 525 (1964).

282. Vallee, J. C., Paynot, M., Martin, C., Vansuyt, G. and Prevost, J. 1975. Action de molecules et proprietes hormonales sur l'activite phenylalanine ammoniac-lyase. *Phytochemistry* **14**:2147-2151.

283. Vidhyasekaran, P. 1988a. Phytoalexin and disease resistance. In: *Physiology of Disease Resistance in Plants*, Vol. I. CRC Press, Boca Raton, FL pp. 83-107.

284. Vidhyasekaran, P. 1988b. Host enzymes and disease resistance. In: *Physiology of Disease Resistance in Plants*, Vol. II. CRC Press, Boca Raton, FL, pp. 5-18.

285. Vidhyasekaran, P. 1988c. Plant growth regulators and disease resistance. In: *Physiology of Disease Resistance in Plants*, Vol. II. CRC Press, Boca Raton, FL, pp. 19-32.

286. Vidhyasekaran, P. 1988d. Physiology of resistance against virus diseases. In: *Physiology of Disease Resistance in Plants*, Vol. II. CRC Press, Boca Raton, FL, pp. 83-117.

287. Vogelgsang S., Watson, A.K., Ditommaso, *A.* and Hurle, K. 1998. Effect of the pre-emergence bioherbicide *Phomopsis convolvulus* on seedling and established plant growth of *Convolvulus arvensis. Weed Res.* **38**:175-182.

288. Waldmüller, T. and Grisebach, H. 1987. Effect of R-(1-amino-2-phenylethyl)- phosphonic acid on glyceollin accumulation and expression of resistance in *Phytophthora megasperma* f. sp. *glycinea* in soybean. *Planta* **172**: 424-430.

289. Walker, H.L. 1981. *Fusarium lateritium*: a pathogen of spurred anoda (*Anoda cristata*), prickly sida (*Sida spinosa*), and velvetleaf (*Abutilon theophrasti*). *Weed Sci.* **29**:629-631.

290. Walker, H.L. 1982. A seedling blight of sicklepod caused by *Alternaria cassiae*. *Plant Dis.* **66**:426-428.

291. Walker, H.L. 1983. Control of sicklepod, showy crotalaria, and coffee senna with a fungal pathogen. U.S. Patent 4,390,360.

292. Walker, H.L. and Boyette, C.D. 1985. Biocontrol of sicklepod (*Cassia obtusifolia*) in soybean (*Glycine max*) with *Alternaria cassiae. Weed Sci.* **33**:212-215.

293. Walker, H.L. and Boyette, C.D. 1986. Influence of sequential dew periods on biocontrol of sicklepod (*Cassia obtusifolia*) by *Alternaria cassiae. Plant Dis.* **70**:962-963.

294. Walker, H.L. and Riley, J.A 1982. Evaluation of *Alternaria cassiae* for the biocontrol of sicklepod (*Cassia obtusifolia*). *Weed Sci.* **30**:651-654.

295. Walker, H.L. and Tilley, A.M. 1997. Evaluation of an isolate of *Myrothecium verrucaria* from sicklepod (*Senna obtusifolia*) as a potential mycoherbicide agent. *Biol. Control* **10**:104-112.

296. Walter, M.H., Liue, J.W., Grand, C., Lamb, C.J. and Hess, D. 1990. Bean pathogenesis-related (PR)proteins induced from elicitor-induced transcripts are members of a ubiquitous new class of conserved PR proteins inducing pollen allergens. *Mol. Gen. Genet.* **222**:353-360.

297. Wapshere, A.J. 1974. A strategy for evaluating the safety of organisms for biological weed control. *Ann. Appl. Biol.* **77**:201-211.

298. Wapshere, A.J., DelFosse, E.S. and Cullen, J.M. 1989. Recent developments in biological control of weeds. *Crop Protect.* **8**:227-250.

299. Ward, E.W.B., Cahill, D.M. and Bhattacharyya, M.K. 1989. Abscisic acid suppression of phenylalanine ammonia-lyase activity and mRNA, and resistance of soybeans to *Phytophthora megasperma* f. sp. *glycinea*. *Plant Physiol.* **91**:23-27.

300. Watson, A.K. (ed.) 1993. *Biological Control of Weeds Handbook*. Weed Sci. Soc. Amer., Champaign, IL.

301. Watson, A.K., Schreoder, D. and Alkhoury, I. 1981. Collection of *Puccinia* species from diffuse knapweed in eastern Europe. *Can. J. Plant Pathol.* **3**:6-8.

302. Weed Science Society of America. 1994. *Herbicide Handbook*, 7th Edition. Weed Sci. Soc. Amer., Champaign, IL.

303. Weed Science Society of America. 1998. *Herbicide Handbook*, Supplement to 7th edition. Allen Press, Lawrence, KS.

304. Weidemann, G.J., and Templeton, G.E. 1988. Control of Texas gourd, *Cucurbita texana*, with *Fusarium* solani f. sp. *cucurbitae*. *Weed Technol.* **2**:271-274.

305. Wells, B.H. and Appleby, A.P. 1992. Lactofen increases glyphosate-stimulated shikimate production in little mallow (*Malva parviflora*). *Weed Sci.* **40**:171-173.

306. White, R.F. 1979. Acetylsalicylic acid (aspirin) induces resistance to tobacco mosaic virus in tobacco. *Virology* **99**:410-412.

15 Microbial Herbicides : Factors in Development

Surinder Kaur

CONTENTS

ABSTRACT 258

1. INTRODUCTION 258
2. THE MICROBIAL HERBICIDE STRATEGY 259
3. DEVELOPMENT OF MICROBIAL HERBICIDE 260
4. MICROBIAL HERBICIDE PATENTS 262
5. COMMERCIAL CONSIDERATIONS 264
6. PRODUCT REGISTRATION 265
7. PRODUCT SELECTIVITY AND REHABILITY 265
8. FORMULATION TO IMPROVE HELD RELIABILITY 266
9. GENETIC MANUPULATON OF MICROBIAL HERBICIDES 267
10. CONCLUSIONS 271
11. REFERENCES 271

Advances in Microbial Biotechnology
J.P. Tewari, T.N. Lakhanpal, Jagjit Singh, Rajni Gupta & B.P. Chamola (eds.)
APH Publishing Corporation, New Delhi - 110 002, India.

ABSTRACT

Key to the development of biologically - based agents such as mycoherbicides and phytotoxins as effective and practical components of weed management systems are the advancement of practical, reliable, cost-effective methods for their production, stabilization, formulation and application. Some of the advantages of mycoherbicides over traditional chemical herbicides are their specificity for the target weed; absence of adverse effects on humans, wildlife or domestic animals; rapid degradation and absence of residues in surface or ground water, crops, soil or food chains. However, there are several intrinsic initiations which are common to nearly all biological agents which must be over come before they will be widely acceptable for practical use.

Microherbicide research and development over the past three decades has led to commercial use of several indigenous plant pathogens for weed control and several others are in development that will be available for commercial use within next few years.

An array of constraints, biological, economic or regulatory, have been overcome to achieve success with microbial herbicides. Many other barriers to commercialization have been identified and appear to be surmountane by research on disease and pathogen biology and by application of recent advances in biotechnology. In this paper, no discuss the status of mycoherbicide development, the barriers to their development and some anticipated benefits of biotechnology for improvement of microbes to enhance microherbicide potential.

1. INTRODUCTION

Weeds are a major constraint to the realization of optimum yield in agricultural systems. Unwanted plant species reduce efficiency of plant establishment in range and forestry systems as well (22, 64). Synthetic chemical herbicides account for approximately 60 to 70% of all the pesticides used on cropland (43, 70). Agricultural practices such as weed control with synthetic chemical herbicides are a major cause of nonpoont source surface and ground water pollution (66). The increased use of pesticides has intensified concern and public debate on public health and environmental consequences of current agricultural practices. Synthetic herbicides are coming under fire as contributors to problems of ground water pollution (20), herbicide persistence (38) and weed resistance (49). Although these warnings were needed to some extent, and research in a number of disciplines aimed at developing an integrated approach to Pest Management was initiated, both the development of new herbicides and the concerns about their use continued to increase to the late 1980s. Since then, increasing stringency of pesticide regulations and, in a sense fortuitously, world recession, have reduced the number of herbicides available for use. In particular, these herbicides perceived as offering a serious risks of contamination of ground-water supplies have been withdrawn from use.

Despite this concerns about environmental damage are still prevalent. Undoubtedly, many of the concerns are exaggerated, either deliberately for political reasons or through ignorance. Nonetheless, the concerns are generally based on a genuine desire to avoid damage to the environment and, as such, deserve a genuine response in attempts to establish alternative approaches to weed management in modern agriculture. This response is clearly evident in the expanding interest in developing so-called 'sustainable agricultural systems', which depend heavily on integrated pest management.

2. THE MICROBIAL HERBICIDE STRATEGY

Biological weed control arises from the observation that biotic factors can have a significant effect on the presences and growth of plant species (36, 71, 76). With that in mind, microorganisms that selectively and suppress weed species may alter competition among plants (47, 58). These plant pathogens may be used to regulate the growth of unwanted plant species growing simultaneously with more desirable plants. (93)

It is possible that biological control of weeds could include the use of other microbial groups, although, at this time, research specifically in this area is lacking.

It is the host/pathogen/environment interaction which determines the success of a biocontrol program. Biocontrol requires that a susceptible host weed a favourable environment and a virulent pathogen be present to cause weed suppression. Alteration in any one of the three components may change the disease, thus altering the biocontrol potential (Fig. 1). The environment plays a key role by influencing the survival of the pathogen and the susceptibility of the weed to the negative effects.

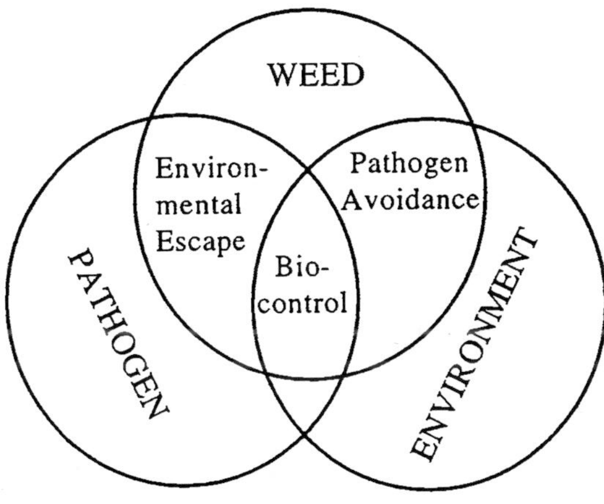

Fig. 1

The first demonstration that an endemic, indegenous plant pathogenic microorganism might be exploited as a biological control agent in agriculture (i.e., as a microbial herbicide) was by Daniel *et al.* (32) They showed that the natural constraints operating to restrict disease development in the field could be overcome by inundating the target weed, at an appropriate time, with an inoculum of a phytopathogenic fungus. The approach allowed exploitation of any conditions required to optimize disease development (32) summarized the requirements, for a successful microbial herbicide (Mycoherbicide if the agent is a fungus) as :

(i) Being capable of producing abundant and durable inoculum in artificial culture.

(ii) Being genetically stable and specific to the target weed.

(iii) Being infective and able to kill the target weed in a wide range of environments.

Subsequently, the microbial herbicide concept has been expanded and refined and has been described comprehensively in many excellent reviews (2, 4, 24, 27, 48, 52, 69)

In the course of this expansion the original definitions have been redefined variously. Thus, the term (mycoherbicide, coined at the start of the research in this area, was defined as 'plant pathogenic fungi' developed and used in the inundative strategy to control weeds in the way chemical herbicides are used. (92) Later, Templeton *et al.* (95) repharased the definition as 'having products that control weeds in agriculture as effectively as chemicals. This rephrasing is welcome in that it expands the concept to include organisms other than fungi. Certainly, there is every reason to expect that many organisms, including bacteria and viruses can be exploited as weed control agents to be used in the inundative strategy. For this reason, the term 'mycoherbicide' is slowly being replaced by more generic terms such as 'bioherbicide' or 'microbial herbicide'.

As the first microbial herbicides to reach the market were plant pathogenic fungi, it is understandable that the resulting expansion in research focussed on this group. Following the early work of Daniel *et al.*, (32) which established the microbial herbicide strategy, interest developed in two other forms of biological weed control for use in agriculture as opposed to stable ecosystems where the classical biological control strategy is appropriate. (102) These two approaches are the augmentative strategy and use of necrogenic micro-organisms. In the former, microbial inoculum is introduced in localized areas within larger areas that are already infected at a low level by the disease, thus producing an augmentation effect. This differs from the microbial herbicide approach in that the inoculum is not necessarily mass-produced, a few grams of inoculum per hectare is usually sufficient and it is not inoculated inundatively over the whole of the weed-infested crop. To date there has been relatively little development of this approach, probably because it generally involves biotrophic fungi that are difficult or even impossible to mass produce on artificial media. However, one potential product has been developed. This is based on *Puccinia canaliculata* schw a rust fungus that is native to the USA and that is effective against yellow nut-sedge, *Cyperus esculantus* L., to be called Pr. Biosedge, is a preparation of the uredospores which are obtained from infected plants especially grown to produce the fungus. Such fungi, used in the augmentative approach, are usually host-density dependent for dissemination from inoculum sites to infect the weed throughout the infested area.

The use of necrogenic organisms for weed control relies on their ability to produce phytotoxins in the soil environment. They are generally not pathogenic organisms. Several fungi have been shown to offer potential as weed control agents in this approach. *Gliocladium virens* Miller, Giddens & Foster (50, 51, 54) is perhaps the best known example of a phytotoxin-producing non-pathogenic fungus with potential fast weed control. The fungus can produce sufficient viridiol to control red root, pig weed, *Amaranthus retroflexus* L. in cotton (54) if inoculum is prepared in a suitable nutrient base. Control was achieved by causing root necrosis and was effective against several annual composites and some monocots. Crop phytotoxicity was avoided by placement of the fungus away from the crop root zone.

3. DEVELOPMENT OF MICROBIAL HERBICIDE

The earlier work of Daniel *et al.* (32) was the start of an intensive study of the potential for *Colletotrichum gloeosporioides* (Penz) & Sacc. f. sp. *aescheynomene* to control the legume weed northern jointvetch *Aeschynomene virginica* (L.) B.S.P. This work which has been published in detail (9, 10, 86, 87, 90, 91, 94, 97) resulted in the formulation and marketing of the fungus as COLLEGO, which is used widely against its weed target in rice and soybean in

Arkansas and adjacent states in the U.S.A. At the same time, a second microbial herbicide, ultimately marketed as De Vine, was also developed (19, 55, 76, 77, 108). This is a preparation of a suspension of chlamydospores of a pathotype of *Phytophthora palmivora* (Butl.) Butl., which is native to Florida and infects strangler vine *Morrenia odarata* (HBA) Lind., a vine weed of citrus in Florida. This product is unique in the sense that it has no appreciable shelf life and must, therefore, be custom prepared for each purchaser, supplied and stored in refrigerated containers, and used soon after receipt. These restrictions are now seen as inimicable to the development of a commercially viable microbial herbicide which should ideally have a shelf-life in normal conditions (5-50°C) of 18 months or more.

Besides these, the only other registered product in the commercial sense is Biomal, a preparation of *Colletotrichum gloeosporiodes* f. sp. *malvae* for the control of sound-leaved malon *Malva pusilla* sm. in Canada and the northern USA. This was registered for practical use in 1992, ten years after the first products Registration was achieved only after exhaustive tests to ensure that even though safflower is also a host for the fungus, use of the microbial herbicides were put into use in the intervening period, although not on a truly commercial basis. For example, a wilt-causing fungus *Cephalosporium diospyr* crandall is one in use in Oklahoma against persimmon *Diospyros virginiana* L. (30, 46, 108, 109). This fungus, which is supplied by a charitable foundation, is wound-inoculated into the trunks of the persimmon trees, which it then kills quite rapidly.

As well as the few microbial herbicides in practical use, there are several which are at an advanced stage of commercial development of these. CASST, a wettable powder preparation of *Alternaria cassiae* Jurair & Khan for the control of sickle pod, *Cassia obtusifolia* L. is perhaps the best known example (3, 8, 101). This organism appears to have an advantage over other microbial herbicides in that, rather than bring totally specific to one weed, it can control three economically important leguminous species. In addition to sicklepod, it is effective against coffee senna (*Cassia occidentalis* L.) and showy crotalaria (*Crotalaria spectabilis* Roth.) (11). Despite this advantage, CASST has yet to appear in the market. Another promising microbial herbicide is *Colletotrichum coccodes* (Wallr.) Hughes, proposed trade name VELGO, for the control of velvet leaf *Abutilon theophrasti* Medic (110, 111). Like CASST, this agent has been under commercial development for some time, but has not yet reached the market.

Scheepens and Hoogerbrugge (84) and De Jong (37) have developed a somewhat revolutionary herbicide for the control of black berry (*Prunus serotina* Enrl.), a serious woody weed of forestry plantations in Holland. They have selected *Chondrosternum purpurreum* (Pers.) Pouzar as the control agent, a fungus which is the causal agent of serious disease in commercial fruit tree orchards. Through exhaustive epidemiological studies they have been able to show that the organism does not disseminate widely from the treated areas. Thus, if its use is restricted to infestations of black cherry which are distant (by 1 km) from *Prunus* orchards there is no risk of disease spreading to the crop. All infected *P. seratina* material must be burned in the treatment area and not removed. On the basis of these restrictions, the agent was given clearance for experimental field use by the Dutch regulatory authorities and it is now under commercial development.

The most recent example of a microbial herbicide undergoing exhaustive development for commercial use is that of *Colletotrichum orbiculare* (Berk & Mort) V. Arx for the control of Bathurst burr or spiny cockelbur, *Xanthium spinosum* L., in Australia (5, 66, 67). Field trials in both dry land pasture and irrigated soybean crops gave levels of kill ranging from 50-100%, with the best results (98-100%) occurring in the dryland pasture.

Many other fungi have been the subjects of exhaustive research, as pre-requirements for commercial development. These include *Cercospora rodmanii* Conway for the control of water hyacinth, *Eichhornia crassipes* (Meert.) Solms', (23), *Colletotrichum coccodes* (Walls) Hughes for velvet leaf, *A. theophrasthi* (113), *Colletotrichum gloeosporioides* f. sp. *jussiaceae* for winged water primrose, *Jussiaea decurrens* (Wait.) DC (13); *Colletotrichem malvacum* (Braun & Caspar) Southworth for prickky sida, *Sida spinosa* L. (58). *Fusarium solani* f. sp. *cucurbitae* Sonyd. & Hane for Texas gourd *Cucurbita texana* Gray (12, 101, 102, 103) and *Phomapsis convolvulus* Ormeno for field bind weed *Convolvulus arvensis* L.

In view of the extent of research over the past two decades, it is perhaps surprising that the number of products developed is so small. It appears that the identification of potential value as a microbial herbicide may only be in the eye of the researchers and be exaggerated by a desire to stay with a fascinating area of research, while this may be so in some cases, objective analysts of the literature shows clearly that more tangible reasons are operating.

4. MICROBIAL HERBICIDE PATENTS

A convenient means of reviewing progress in microbial herbicide research is to examine the disclosures found in patent specifications. Much of the information presented in patents has a unique nature in that it is, often not published elsewhere. A recent monography published by the UK Patent Office (1992) presents a valuable synthesis of the state of the art in microbial herbicides. While not claiming to be totally comprehensive, it does, through the judicious selection of 43 sets of patent publications or related disclosures, present a valuable focus on the strengths and weaknesses of microbial herbicides and of the associated research and development.

The two earliest patents identified originate, interestingly, from Russia in 1970 and 1971. Thus, they provide the most significant patents which related to COLLEGO and other mycoherbicide products in the USA. Surprisingly, this patent list does not include any reference to *Phytophthora palmivora* (De Vine). The only agent for *Morrenia odorata* which is referred to is the *Arauijio* mosaic virus. The patents refer to one other virus (TMV) for control of *Solanum carolinense* L. in Russia (Patent No. SU 663357, 1977) and to three bacteria. Otherwise, all the control agents disclosed are 32 species of 14 genera of fungi. The three genera represented most frequently are *Alternaria, Colletotrichum* and *Fusarium*, which are mentioned in 15, 13 and 10 patents respectively.

Many of the patents refer to methods of formulation and delivery but only five refer to these specifically. These cover alginate granules (two patents), synergistic formulations with crop oils and surfactants, encapsulation in polymers, and oil-in-water invert emulsion. A further six patents refer to combinations of fungi with chemical herbicides and six others refer to combinations of two fungal pathogens to broaden the spectrum of weed control. Broad-spectrum control is the subject of two patent applications which introduce the important concept of pathogens, namely *Alternaria cassiae* and *Fusarium lateraitium* Nees, being able to control several species of weed.

The pathogens, weed targets and relevant patents are listed in Table 1. However, this progress in the development of microbial herbicides contrasts with the lack of practical achievements, as demonstrated by products on the market. Although effective means of formulating and delivering the large numbers of candidate microbial herbicides are available (Table 2), but there are significant factors that inhibit the translation of promise into practice. Some of the important reasons for failure are discussed below.

Table 1 : Microbial herbicide patents (15)

Pathogen	Target Weed	Patent No.*	Date**
Albugo tragopognis (Pers.) S.F. Gray	*Ambrosia artemisiifolia* L.	SU343671	1970
Alternaria alternata (Fr.) keissler	*Xanthium* spp.	JP62278978	1987
Alternata cassiae Jurair & Khan	*Cassia occidentalis* L. *Cassia obtusifolia* L. *Croalaria spectabilis* Roth.	US4390360	1983
Alternaria crassa (Sacc.) Rands	*Datura stramonium* L.	US7092100	1987
Alternaria euphorbiicola Simmons & Engelhard	*Euphorbia* spp.	US4755208 US4871386	1988
Alternaria (Zinniae ?)	*Carduus tenuiflorus* Curt.	US4636386	1987
Amphobotrys ricini (Buchwald) Hennebert	*Caperonia palustrsi* St.-Hil	US4909826	1990
Arauijio mosaic Virus	*Morrenia odorata* (H & A) Lindl.	US4162912	1979
Ascochyta hyalospora Lib.	*Chenopodium* spp.	EP296057	1988
Bipolaris sorghicola (Lefebure Scherwin) Alcorn	*Sorghum halepense* (L.) Pers.	US4606757	1986
Cercospora rodmanii Conway	*Eichhornia crassipes* Solms.	US4097261	1978
Colletotrichum coccodes (Wallr.) Hughes	*Abutilon theophrasti* Medic.	CA1224055	1987
Colletotrichum coccodes (Wallr.) Hughes	*Solanum ptycanthum* L.	US4715881	1987
Colletotrichum gloeosporioides (Penz.) Sacc. f. sp. *aeschynomene*	*Aeschynomene virginica* (L.) B.S.P.	US3849104	1974
Colletotrichum gloeosporiodes (Penz.) Sacc. f. sp. *malvae*	*Malva pusilla* Sm. *Abutilon theophrasti*	EP218386	1987
Colletotrichum malvarum (Braun & Caspary) Southwarth	*Sida spinosa* L.	US3999973	1976
Colletotrichum orbiculare (Berk. & Mont.) Arx.	*Xanthium spinosum* L.	AU8818454	1989
Colletotrichum truncatum (Schw.) Andrus & Moore	*Desmodium tortuosum* DC. *Serbania exaltata* (Raf.) Rydb.	US4643756 US(NTIS)7338680	1987 1989
Colletotrichum sp.	*Cyperus rotundus* L.	WO90/06056	1990
Drechslera spp.	*Echinochloa crus-galli* Beauv.	EP374499	1990
Fusarium lateritium Nees	*Abutilon theophrasti* Medic *Sida spinosa* L. *Anoda cristata* Schlecht	US4419120	1983
Fusarium orobanches fom.	*Orobanche* spp.	SU387689	1973
Fusarium oxysporum Schl.	*Echinochlao* sp.	UP02013367	1990
Fusarium roseum Lk.	*Hydrilla verticillata* Prestl.	US4263036	1981
Fusarium tricinctum (Corda) Sacc.	*Cuscuta* spp.	US4915726	1990
Fusarium sp.	'Arrowroot'	JP53099321	1978
Hyphomycetes sp.	*Eleocharis kuroganwai* L.	JP01238507	1989
Lactic acid bacteria	General weeds	JP01100106	1989
Phomopsis cirsii (Sacc.) Sacc.	Compositae	EP136850	1985
Pseudomonas spp.	*Bromus tectoreim* L.	WO89/12691	1989
Puccinia canaliculata Schw.	*Cyperus esculentus* L.	US4731104	1988
Septoria cirsii	Compositae	EP136850	1985
Tobacco mosaic Virus	*Solanum carolinense* L.	SU663357	1979
Xanthomonas compestris	*Poa annua* L.	WO88/01172	1988

* Letters prefacing patent numbers; AU, Australia; CA, Canada; JP, Japan; SU, Russia; US, United States of America; EP, European Patent Office; WO, World Patent (Patent Co-operation Treaty).

** Dates refer to date of publication of the patent application.

Table 2 : Microbial herbicide technology patents (15)

Technology	Patent*	Date**
Microbial herbicides containing chemical herbicides and plant growth regulators	EP 207653	1987
Alginate granule formulation	US 4718935 US 4767441	1988
Synergistic formulation with crop oil	US 4755207	1988
Encapsulation with non-ionic polymer	EP 320483	1989
Invert emulsion (oil-in-water) formulation	US 4902333	1990
Mixed microbial herbicide inoculum	WO 90/06056	1990
Chemical herbicide mixed with pathogenic bacterium	WO 91/03161	1991

* Letters prefacing patent numbers : US, United States of America; EP, European Patent Office; WO, World Patent (Patent Co-operation Treaty).

** Dates refers to publication of the patent application.

5. COMMERCIAL CONSIDERATIONS

Auld[4] has identified three main areas to be considered before developing a microbial herbicide (Table 3). Of these considerations, one which seems to be ignored in many research projects is that of market size. Obviously, where the research is commercially funded, market size has been carefully researched before the project commences. In contrast, many publically funded projects address weeds that appear to offer little or no attraction for commercial development investment. Understandably, public funders, especially those at a local level, may be anxious to provide control strategies for weeds that are a significant problem at a local level. A case in point in the development of a microbial herbicide to control *Aeschynomene virginica*. This leguimenous weed is a serious problem in rice and soybean in Arkansas and a few neighbouring states in the USA. However, this weed does not appear in a listing of 224 species of weeds selected by U.S. weed scientists as prevalent in croplands, grazing lands, non croplands and aquatic sites in the USA (USDA-ARS 1971). In practice, this microbial agent did achieve commercial product status, as did COLLEGO, and is used successfully to control the weed. This is despite the fact that the market is small. Indeed, only a part of this market has been penetrated and, although farmers are making repeat purchases, it would not be considered as commercially viable by many companies. the fact that the principal chemical herbicide to give satisfactory control of the weed was 2,4,5-T, which was withdrawn, obviously helped sales but even so, they remain small by pesticide market standards.

Table 3 : Commercial consideration in development of microbial herbicides (4)

1. **Market Size and Stability :**
   a) Does the weed cause a similar level of problem each year?
   b) What area of use is likely?
   c) What is the value per unit area of affected crops?
   d) What is the cost of alternative effective controls?
2. Is the technology protected by patent or secrecy?
3. **Product development costs.**

It would be foolish to imagine that chemical herbicides are going to be anything other than the dominant means of weed control in the foreseeable future. Therefore, microbial herbicides must be capable of withstanding the competition these chemicals present and these microbial herbicides must be developed commercially. That being so, they must compete with chemicals in the market. A major factor in establishing a viable competitive position is to target an appropriate market before initiating the research.

6. PRODUCT REGISTRATION

The advantage of registration of biological agents over chemicals is that they are cheaper. However, there are major differences in national attitudes to registration requirements and these are exaggerated when the question of genetic modification of the agent is raised. In extreme cases, some countries have refused to consider biological agents through fears of mutation leading to epidemics of disease that affect crop species. While there is no evidence to show that mutation is likely in conditions of practical use, neither is there evidence to prove that it will not and so, allay the concerns.

Despite some efforts to harmonize registration requirements, notably in the European community, differences are bound to persist. These will provide hindrance, perhaps seriously to widespread use of a product - a prerequisite in many instances, for commercial development. For example, any country proposing to permit widespread use of a microbial herbicide may face objections from countries with whom they share a frontier and who may anticipate spread into their territory of what they view as an alien species. Countering such objections would, at least, involve considerable expense. In, for example, proving that the exact strain in question was already endemic in these neighbouring countries, that it would not spread from the use sites, as that it was sufficiently selective not to affect native species. Such expense, of course, erodes the cost advantage that the microbial herbicide may have over chemical herbicides in the country of registration. Until this method is resolved, there is bound to be some reluctance by potential developers to make the necessary financial investment.

7. PRODUCT SELECTIVITY AND RELIABILITY

The majority of microbial herbicide products are claimed to be totally specific to their selected host weed. Indeed, this is a major factor in their environmental safety. The absolute specificity claimed may, in some cases, be an artifact of limited host-range testing. This was fo far for COLLEGO which, since its introduction to the market, has been shown to infect, but not kill some leguminous crops.[89] Under field use there is unlikely to be spread to such crops, even if they are grown close to crops treated with COLLEGO, as the agent's spores are sticky and it spreads poorly. A similar analysis, based on detailed epidemiological studies was made before commercial development of *Chondrostereum purpureum*, a serious pathogen of fruit trees, as a control agent for *S. serotina* was initiated in Holland.

In this case also, release of the agent was approved as its spread by natural means was limited.[33,84] These few examples of lack of absolute specificity do not seriously impair the perceived environmental safety of the agents. However, specificity to one weed target does seriously impair marketability in most instances. It is axiomatic that farmers and growers rarely face a problem from a monoculture of a weed and they are, at present at least, supplied with a formidable array of broad spectrum chemical herbicides. Simple economics dictates that a single application of a herbicide will always be preferred to multiple applications. Thus, host-specific microbial face a considerable disadvantage.

The great strength of chemical herbicides i.e. that they give reliable weed control in a very wide range of edaphic and climatic conditions, thus opening up worldwide markets. This presents a difficult challenge for using organisms which have evolved to have quite specific requirements for successful infection. There are many examples of the lack of the required reliability of microbial herbicide efficacy. This is, indeed, why most companies will not become involved in development research until efficacy has been proven in field treatments. For example, Table 3 illustrates the problem of poor field performance of *Alternaria cassiae*, with efficacy ranging from zero to complete control of *C. obtusifolia*. The failure in South Carolina in 1982 was attributed to very dry weather and the development of a second flush of sicklepod seedlings. Similarly, inoculum desiccation was blamed for poor performance at other sites in other years. The same reason has been cited for poor performance in field trials by other mycoherbicides, notably CASST and BIOMAL, although in the second case the problem was overcome sufficiently well to allow introduction of the product to the market in 1992 to be contemplated.

8. FORMULATION TO IMPROVE FIELD RELIABILITY

The most serious constraint on efficacy arises from the lack of sufficient free-water to enable efficient spore germination and pre-infection growth of the microbial inoculum on the leaf surface. An obvious strategy to reduce or eliminate the problem is to inoculate the soil in which the weed target is growing. The soil inoculation allows the use of granular formulations which can be designed to promote germination and proliferation of the micro-organisms by the incorporation of suitable substitutes. Another considerable advantage is that there is a wealth of information available about the physiology, biology and ecology of soil-borne plant pathogenic micro-organisms.

Despite the fact that the first registered microbial herbicide, De Vine, is a soil-borne pathogen, *Phytophthora palmivora*, there has been relatively little research into soil borne pathogens as microbial herbicides. To date, the main genus studied in this context is *Fusarium*. The soil-borne pathogen *Rhizoctonia solani* Kuhn has been studied only as an agent against an aquatic weed species (*Eichhornia crassipes*).

Although having advantage over foliage-applied pathogens, soil inoculated microbial herbicides still suffer from disadvantages such as specificity. In addition, many of the pathogens used so far are closely related to serious crop pathogens. Thus, there may be former resistance to using them, stemming from the widely held concern that the applied organisms may mutate, or change in some other way, and so become virulent on the treated, or neighbouring crops. To overcome these constraints, considerable effort should be made to develop appropriate formulations. Much has been done already, but this has served to identify the even greater amount of research that is still needed, particularly as there is a growing awareness that many, if not all, of the pathogens being studied at present are likely to need individual formulations designed to meet their particular needs. The subject of formulations has been reviewed by many workers (8, 12, 31, 44).

There are many options available to overcome the principal constraint on reliability - moisture storage. These include formulation as invert emulsions (74), standard emulsions with vegetable oils, granules of alginate gels, inert solid carriers, or as colonized feeds. The liquid formulations, whether they are simple aqueous suspensions of spores or complex emulsions, may be variously mixed with surfactants, humectants, stickers etc., to suit particular applications. The progress towards meeting this objective as evidenced by the literature reviewed by Daigle and Connick (31) and Boyette *et al.* (14) is promising. Many incidental benefits of formulation are reported to reduce dew dependency and these may prove to be as or more important than protecting against desiccation. For example, formulation as an invert emulsion appears to reduce the inoculum threshold for pathogen infection of leaves (29). Thus, when formulated in this way only one spore of *Alternaria crassa* in 2 lit droplet effectively infects the target weeds (*Cassia obtusifolia* and *Datura straumarium*). In contrast, in aqueous suspension 1000 spores per droplet are required to ensure that every droplet produced a successful infection. According to Stowell (89), in order to ensure satisfactory economics, a potential microbial herbicide product should yield, in liquid fermentation, 2.7×10^{11} propagules per litre to control the weed. Obviously, then a 1000-fold increase in infectivity (E virulence) obtained by formulation has great economic significance. Other, simpler approaches include the addition of nutrients like sucrose to the formulation which help speed up germination and infection by the spores (95, 97, 107.)

One of the most promising formulation approaches for microbial herbicides, which offers many opportunities for practical exploitation, is perhaps a mixture with chemical herbicides. On the one hand, such mixtures can be used to provide control of a broad spectrum of weeds, using the microbial herbicide to control an important weed not controlled by the chemical. On the other hand, the known synergism between some chemical herbicides and their biological partners can be exploited usefully to component of the mixture when used separately. This synergism often allows very low doses of herbicide to be used effectively. For example, *Colletotrichum coccodes*, an effective microbial herbicide for velvet leaf (*Abutilon theophrasti*) is unreliable if environmental conditions are unfavourable, killing only infected leaves and not preventing continued growth of the weed. Mixing with the growth regulator, thidiazurom produces a synergistic interaction and weed mortality is enhanced (113, 114). Other similar interactions have been reported (28).

It can be concluded from above discussion that formulation of living agents demands a more rational approach which should exploit appropriate physiological and biochemical attributes of the microbial agents and this will only be achieved by intensive, focussed research.

9. GENETIC MANUPULATION OF MICROBIAL HERBICIDES

Inevitably, any consideration of the possibility of improving reliability of microbial herbicides includes reference to the need to increase pathogenicity. While this can be achieved, in some instances, by traditional methods, such as selection, induced mutation, or optimization of production and stability of the agent, increasingly the opportunities offered by genetic means are discussed such approaches include sexual and parasexual crossing, somatic hybridization, and the use of recombinant DNA, i.e. genetic engineering. The last offers unique opportunities, although public and political pressures may restrict the release of genetically engineered organisms into the environment.

The technology of transferring, and gaining expression of, specific genes in filamentous fungi has developed rapidly in recent years and gene transformation has been achieved by many different groups. The subject has been reviewed by Greaves (44) and Kistler (62).

The first stage in the consideration of how to improve the performance of a wild strain of plant pathogen, so that it can be developed as a commercial microbial herbicide, is to identify the factors that limit efficacy. Some of the characteristics, which could be targets for manipulation well identified by Greaves *et al.* (45) are discussed below.

a) Genes involved in the production of specialized structures essential for colonization and reproduction. The targets here include recognition of the host-plant surface, production of sexual and asexual spores, spores germination, production of appresoria and haustoria, and adherence of pathogens to host surface. All of these, if improved, would contribute significantly to reducing the necessary inoculum levels. However, they are all under multigenic control and thus, advances in the short term are most likely to come from strain selection rather than genetic manipulation.

   Pathogenicity has the greatest potential for enhancement in biological weed control by molecular improvement. Progress is being made in identifying the genes and the mechanisms responsible for numerous plant pathogens (15, 104, 113). The research in biological weed control with microorganisms presently is more involved with determining the mechanism of action and has not yet progressed fully to the manipulation of those genes for increased activity.

   A number of strategies for the improvement of biological weed control can be used to enhance pathogenecity (82). This can include alteration in virulence, modification of specificity by increasing host range, or enhancing aggressiveness. Also the genes responsible for detoxifying plant defences, producing enzymes that weaken the plant cell wall, or secondary metabolite production which exhibits plant growth can be altered. Virulence can be enhanced by alteration of regulatory genes, promoter regions, or by the insertion of gene material with greater virulence into a highly competitive organism (57). Pathogenicity to a variety of plant species has been shown to be independent of one another in many cases (60). Once the genes responsible for phytotoxic activity have been identified, insertion into the regulatory regions of another microorganism with differing host range, aggresiveness of competitive ability is possible. Gene insertion into promoter regions can overexpress or increase copy numbers. *gus* A genes have been used to alter promoter regions in and enhance the expression of genes (16). Another strategy in pathogenicity manipulation would be the addition of a gene of a known bioaction into a competitive receptor. Insertion of the gene responsible for a nonselective toxin, such as tabtoxin into a known pathogen may be one strategy to increase biological control activity (16).

b) The genes responsible for certain enzymes such as those involved in plant cell degradation may be altered to increase pathogenicity. Cutinase may play a role in the agressiveness of pathogens (31). Insertion of the cutinase genes results in greater infection (34). Mutants with reduced production of pectate lyase, a cell wall-degrading enzyme, caused less damage to plant cell wall than the wild type (73). Understanding the genetics of cutinase, pectate lyase, cellulases, proteases and other enzyme systems will further help the manipulation of these genes to enhance the pathogenicity of biological control agents. Superior strains may also be developed once mechanism of gene products are understood.

A possible constraint to successful infection by an introducted bioherbicide may be plant defense mechanisms. Phytoalexins produced by the plant may inhibit infection and subsequent suppression of the host by biocontrol agents (94). The genetic basis of this response and other resistance mechanisms needs further study; however no research in this area is known at this time. This could include resistance to these compounds or possible mechanisms which reduce the antagonistic effect or suppress production of phytoalexins. The importance of phytoalexin production in biological weed control systems has been noted (80).

c) Genes that regulate pathogen specificity; one of the most frequently cited 'disadvantages' of microbial herbicides is that they are too specific. Conversely, some of the most powerful plant pathogens have too broad a host range to be suitable as microbial herbicides. In order to minimize adverse effects to the environment, a narrow host range is recommended. However, from a marketing standpoint, a biological control agent that is too specific will not command the wide market need to attract industry support. All too often, host-specific pathogens lack virulence and many broad host range pathogens will attack many plant species including beneficial crops. However, it may be advantageous to begin with broader host range agents to ensure efficacy, and then genetically alter the agent for reduced host range (78). Yet in other cases, when a target host is a crop species, use of agents with broad host range would be devastating (36, 53).

Genetic technology can be used to alter host range and even increase plant production of certain toxins (74).

Research with a broad host range pathogen illustrates the use of molecular manipulations to reduce the host range of a mycoherbicide. *Sclerotinia sclerotium* causes diseases on both Canada thistle and spotted knapweed (*Centavrea maculosa* Lam). Mutants with altered virulence were used to limit host range of this disease (68). Auxotrophs, nonsclerotial farming, reduced virulence or altered host range. Pyrimidine auxotrophs have the potential to delimit the range of host activity (63, 78). Other examples in the literature demonstrate the use of genetically altering biocontrol agents, thus increasing potential of this technique (2, 63, 92).

In developing biological control systems, the pathogen-weed interaction is not limited and other biotic interactions can be involved as well. Narrow host range may be attained through a vector, such as an insect, which will deliver the agent to the host. In these cases host specificity may not be as critical as in other cases. The pathogen vector acts as a direct delivery system to the target host (58, 70).

d) Genes determining fitness for survival in the environment. Prior to colonization and/or infection of the host by a pathogen, biocontrol agent must survive and compete in the leaf, soil or aquatic environment with many other microorganisms. Even if a strain exhibits virulence, adaptation to the environment is necessary. Critical to the success of any biocontrol agent is its ability to withstand this initial exposure to biotic and abiotic stresses in the soil or aerial environment for a given time period so that it may establish itself in the environment. The complexity of competitive ability and the many factors that may influence survival allow for many molecular approaches to investigations and understanding of the survival and colonization.

In order for biological control to be successful, the biological control agent needs to survive. Because of limited resources, microbes compete with each other for food, nutrients and essential elements in the phyllosphere, soil, spermosphere, and rhizosphere (6, 39, 100). Environment stresses are many and include high temperature, pesticides, soil moisture, dew period, soil texture and temperature also have been shown to be important in influencing survival and colonization (1, 33, 38, 49). Those factors influencing spermosphere and root colonization are complex and many traits may be involved, therefore increasing the difficulty of study survival ability relies on saprophytic growth as well as colonization of host and growth on root or leaf (1).

The biotic stresses that an introduced microorganisms will face are competition. Increased competitiveness and survival may be determined by traits such as production of antibiotics, tolerance to antibiotics, motility, resistance to starvation, or survival structures (35, 67). The ability to utilize a wide variety of substrates will increase a microorganism's resiliency. Adaptation to adverse environments may involve increased antibiotic or siderophore production which have been shown to be important in other systems.

Survival structures that widen the dispersal of the biological control agents may influence efficacy (68). A nonsclerotial mutant of *Sclerotinia sclerotium* was used to study sclerotial development and survival of this fungus in the control of Canada thistle (*Cirsium arvense* scop). Lack of ascospores limits survival in adverse environmental conditions, which will reduce speed of the introduced organisms, even though pathogenicity is to be retained.

Survival also relies on the ability of an organism to withstand stresses imposed by crop production. Resistance to benomyl, a common fungicide, may increase survival of the biological control agent. *Colletotrichum gloeosporiodes*, the causal agent of anthracnose, and improve the use of the microherbicide Collego. The gene responsible for benomyl 'resistance' is B-tubulin (B Tub) an encoding gene, tub I gene. Fungicide resistant strains were obtained for tracking or determining phylogenetic relationships among *Colletotrichum* species (67).

The general public are, to some degree, conditioned by the media to perceive that biological control may pose a threat to the environment. Headline phrases, such as 'Biological Warfare Against Pests' are used frequently and serve to perpetuate the perceived threat. The addition of 'Genetically Engineered' to such media type is usually the final straw in breaking the camel's back. Such exaggerated concern does nothing to ease the strictness of regulatory control of the release of genetically manipulated organisms which clear constrains development of genetically manipulated microbial herbicides.

More recent studies tend to lead to the conclusion that genetically modified organisms offer no more risks than unmodified organisms. The report by Kelman *et al.* (55) is typical of many other studies in concluding that there is no evidence of unique risks in the use of recombinant techniques or gene transfer between unrelated organisms, that the risks associated with organisms release into the environment are the same for both genetically modified and unmodified organisms, and that the assessment of these latter risks should be based on the nature of the organism and the receiving environment, not on the method by which the organism was modified.

Assuming that these conclusions can be widely formulated and accepted, the genetic manipulation of plant pathogenic microorganisms selected as potential microbial herbicides offers significant potential for improvement of effective, reliable and environmentally safe products.

10. CONCLUSIONS

The broad scope of this chapter has necessitated a selective consideration of factors, that are of importance in the future development of reliable microbial herbicides. Research in the past two decades has demonstrated clearly that it is relatively easy to isolate, identify and characterize large numbers of plant pathogenic micro-organisms which offer real potential as microbial herbicides. Every plant species studied so far is susceptible to control in this way. Equally, however, the same research,programmes have demonstrated only too clearly, that it is very difficult to translate successful laboratory and glasshouse experiments into effective, reliable weed control in the field. It is essential, therefore, that researchers learn from the mistakes of the past and resist the temptation to add to the long and impressive list of laboratory winners that will never enter the commercial development race. Market identification and thus target selection must be much more controlled in future and perhaps, local needs made subservient to the greater needs of the science as a whole. After all, if commercial successes are not forthcoming, the strategy will lose credibility and funding sources, which are already limited, may dry up altogether. In this context, it is essential that there is a greater cohesion between public and private sector and more cooperative or club research should be carried out involving both sectors.

Another area of research, which needs greater funding and research activity is the genetic manipulation of biological control agents, which allow us to alter the ecological balance to favour the pathogen and suppression of the weed. Molecular investigations are imperative in the development of microorganisms for biological control to reduce synthetic chemical pesticide use.

11. REFERENCES

1. Ahmed, J.S. and Baker, R. 1987. Competitive saprophytic ability and cellulolytic activity of rhizosphere- competent mutants of *Trichoderma harzianum. Phytopath.* **77**: 358.

2. Altman, J., Neate S., and Rovira A.P. 1990. Herbicide-pathogen interactions and mycoherbicides as alternative strategies for weed control. In : *Microbes and Microbial products and Herbicides* (ed. R.E. Hoagland), ACS Symposium Series 429, American Chemical Society, Washington D.C., p. 240.

3. Anderson, D.M. and Mills, D. 1985. The use of transposon mutagenesis in the isolation of nutritional and virulence nutrients in two pathovars of *Pseudomonas syringae. Phytopath.* **75**: 104.

4. Auld, B. 1986. Potential for mycoherbicides in Australia, Workshop Proceedings, Agricultural Research and Veterinary Centre, Orange, NSW, Australia.

5. Auld, B.A. 1991. Economic aspects of biological weed control with plant pathogens. In: *Microbiol Control of Weeds* (ed. D.O. TeBeest), Chapman and Hall, New York, p. 262.

6. Auld, B.A., Say, M.M. and Millar, G.D. 1990. Influence of potential stress factors on anthracnose development on *Xanthium spinosum. J. Appl. Ecol.* **27**: 513.

7. Baker, R. 1981. Eradication of plant pathogens by adding organic amendments to soil. In : *Handbook of Pest Management in Agriculture*, (ed. D. Pimental), CRC Press, Boca Ratan, FL, 137.

8. Bannon, J.S. 1988. CASSTM herbicide (*Alternaria cassiae*): A case history of a mycoherbicide *Am. J. Alt. Agric.* **3**: 73.

9. Bowers, R.C. 1982. Commercialization of microbial biological control agents. In : *Biological Control of Weeds with Plant Pathogens* (ed. R. Charudattan and H.L. Walker), Wiley, New York, p. 157.

10. Bowers, R.C. 1986. Commercialisation of collego - an industralist's view. *Weed Sci.* **34(Suppl.)**: 24.

11. Boyette, C.D. 1988. Biocontrol of three leguminous Weed Species with *Alternaria cassiae*, Weed Technol. 2: 414.

12. Boyette, C.D., Quimby, P.C., Jr., Connick, W.J. Jr., Dougle, D.J. and Fulgham, F.E. 1991. Progress in the production, formulation and application of mycoherbicides. In : *Microbiol Control of Weeds* (ed. D.O. Tebeest), Chapman and Hall, New York, p. 209.

13. Boyette, C.D. Templeton, G.E. and Oliver, L.R. 1984. Iady Texas gourd (*Cucurbita texana*) control with *Fusarium solani* sp. *cucurbitae. Weed Sci.* **32**: 649.

14. Boyette, C.D., Templeton, G.E. and Smith R.J. Jr. 1979. Control of winged waterprimrose (*Jussiaea decurrens*) and northern jointretch (*Aeschynomene virginica*) with fungal pathogens. *Weed Sci.* **27**: 497.

15. Bridges, 1992 Mycoherbicides. In : *Patent Information Monograph 3(c)*, (ed. T.H. Lemon and M.J.R. Blackman). The Patent Office, London, p. 127.

16. Broglie, B. and Roby, C.H. 1988. Model suicide vector for containment of genetically engineered microorganisms. *Appl. Environ. Microbiol.* **54**: 2472.

17. Bronson, C.R. 1991. The genetics of phytotoxin production by plant pathogenic fungi. *Experientia*, **47**: 771.

18. Brooker, N.L., Mischke, C.F., Mischke, S. and Lydon, J. 1993. Transformation of *Colletotrichum* with the acetylation gene (bar) from *Streptomyces*, *Fungal Genet. Newsl.*, **40A** : 36.

19. Burnett, H.C., Tucker, D.P.H. and Ridings, W.H. 1974. *Phytophthora* root and stem rot of milk weed to vine. *Plant Dis. Rep.*, **58**: 335.

20. Carter, A.D., Hollis, J.M., Thompson, T.R., Uakes, D.B. and Binney, R. 1991. Pesticide contamination of water sources : Current policies for protection and a multidisciplinary proposal to aid future planning. In : *Brighton Crop Protection Conference - Weeds*, England, Nov. 1991, **Vol. 2**: 491.

21. Cowder, J.D., Gotleib, A.R., Stowell, L. and Watson, A.K. 1987. Herbicidal compositions comprising microbial herbicides and chemical herbicides or plant growth regulators. European Patent Office Application, No. EP0207 753 AI.

22. Chan, S.S. and Walsted, J.D. 1987. Correlations between overtopping vegetation and development of douglas sir in the Oregan Coast Range. *West. J. Appl. For.* **2**: 117.

23. Charudattan, R. 1984. Role of *Cercospora rodmanii* and other pathogens in the biological and integrated controls of water hyacinth. In: *Proceedings International Conference on Water Hyacinth*, UN Env. Prog. Nairobi, p. 834.

24. Charudattan, R. 1988. Inundative control of weeds with indigenous fungal pathogens. In: *Fungi in Biological Control Systems*, (ed. M.N. Burge), Manchester university Press, Manchester, p. 88.

25. Charudattan, R. 1990. Pathogens and potential for weed control. In : *Microbes and Microbial Products as Herbicides*, (ed. R.E. Hoagland), ACS Symposium Series 439, American Chemical Society, Washington DC, p. 132.

26. Charudattan, R. 1991. The mycoherbicide approach with plant pathogens. In: *Microbial Control of Weeds*, (ed. D.O. TeBeest), Chapman and Hall, New York, p. 24.

27. Charudattan, R. and Deloach, C.J. Jr. 1980. Management of pathogens and insects for weed control in agroecosystems. In: *Weed Management in Agroecosystems: Ecological Approaches*, (ed. M.A. Altieri and M. Liebman), CRC Press, Inc., Boca Raton, p. 246.

28. Charudattan, R., Walker, H.L., Boyette, C.D., Ridings, W.H., TeBeest, D.O., Van Dytce, C.G., and Worsham, A.D. 1986. Evaluation of *Alternaria cassiae* as a mycoherbicides for sicklepod (*Cassia obtusifolia*) in regional field test. *Soutern Coop. Ser. Bulletin* 377, Alboma Agricultural Experimental Station, Auburn University, Albama.

29. Connick, W.J., Daigle, D.J. and Quimby, P.C., 1991. An improved invert emulsion with high water retention for mycoherbicide delivery. *Weed Technol.* **5**: 442.

30. Crandall, B.S. and Baker W.L. 1950. The wilt disease of American persimmon caused by *Cephalosporium diospyri*. *Phytopath.* **40**: 307.

31. Daigal, D.J. and W.J., Connick, Jr. 1990. Formulation and application technology for microbial weed control. In : *Microbes and Microbiol Products as Herbicides*, (ed. R.E. Hoagland), ACS Symposium Series 439. American Chemical Society, Washington DC, p. 488.

32. Daniel, J.T., Templeton, G.E., Smith, R.J. and Fox, W.T. 1973. Biological control of northern jointvetch in rice with an endemic fungal disease. *Weed Sci.* **21**: 303.

33. DeJong, M.D. 1988. Risk to fruit trees and native trees due to control of Black Cherry (*Prunus serotina*) by silverleaf fungus (*Chondrostereum purpureum*), Ph.D. Thesis, L and Bouwuniversiteit te Wageningen, The Netherlands.

34. Department of Agriculture, 1988. Economic Indicators of the Farm Sector : Costs of Production. ECIF6-1, Washington, D.C., 128.

35. Dickman, M.B., Patil, S.S. and Kolattukudy, P.E. 1982. Purification, characterization and role in infection of an extracellular cutinolytic enzyme from *Colletotrichum gloeosporiodes* Penz. on *Carica papaya* L. *Physiol. Plant Pathol.* **20**: 333.

36. Dinoor, A. and Erhed, N. 1984. The role and importance of pathogens in natural plant communities. *Ann. Rev. Phytopath.* **22**: 443.

37. Elad, Y. and Chet, I. 1987. Possible role of competition for nutrients in biocontrol of *Pythium* damping-off by bacteria. *Phytopath.* **77**: 190.

38. Ettinger, W.F., Thukral, S.K. and Kolattuteudy, P.E. 1987. Structure of cutinase gene. cDNA, and the derived amino acid sequence from phytopathogenic fungi. *Biochemistry* **26**: 7883.

39. Fravel, D.R. 1988. Role of antibiosis in the biocontrol of plant diseases. *Ann. Rev. Phytopathol.* **26**: 75.

40. Fredrickson, J.K. and Elliot, L.F. 1985. Colonization of winter wheat roots by inhibitary rhizobacteria. *Soil. Sci. Soc. Amer. J.* **49**: 1172.

41. Garber, R.C., Fry, W.E. and Yoder, O.C. 1983. Conditional field epidemics on plants: a resource for research in population biology. *Ecology* **64**: 1653.

42. Garrett, S.D. 1970. *Pathogenic Root - Infecting Fungi.* Cambridge University Press, London.

43. Gianeesi, L.P. and Puffer, C. 1991. Herbicide use in the United States : National Summary Report, Resources for the Future, Washington, USA.

44. Greaves, 1993. Formulation of microbiol herbicides to improve performance in the field. In: *Proceedings 8th EWR & Symposium.* Quantitative Approaches in Weed and Herbicide Research and their Practical Application, Braunschweig p. 219.

45. Greaves, M.P., Bailey, J.A. and Hargreaves, J.A. 1939. Mycoherbicides, opportunities for genetic manipulation. *Pestic. Sci.* **26**: 93.

46. Griffith, C.A. 1970. 'Persimmon wilt research', Annual Report 1969-1970, Noble Foundation Agricultural Division, Ardmore, Oklahama.

47. Hasan, S. and Ayers, P.G. 1990. The control of weeds through fungi: principles and prospects. *New Phytol.*, **115**: 201.

48. Hoagland, R.E. 1990. Microbes and microbiol products as herbicides: an overview. In : *Microbes and Microbiol Products as Herbicides*, (ed. R.E. Hoagland), ACS Symposium Series 439, American Chemical Society, Washington DC, p. 2.

49. Holt, J.S. and LeBaron, H.M. 1990. Significance and distribution of herbicide resistance. *Weed Technol.* **4**: 141.

50. Howell, C.R. 1982. Effect of *Gliocladium virens* on *Pythium ultimum, Rhizoctonia solani* and damping off of cotton seedlings. *Phytopath.* **72**: 496.

51. Howell, C.R. and Stipanovic R.D. 1984. Phytotoxicity to crop plants and herbicidal effects on weeds of viridiol produced by *Gliocladium virens. Phytopath.* **74**: 1346.

52. Johnson, B.N., Kennedy, A.C. and Ogg, A.G. Jr. 1993. Suppression of downy brome growth by a rhizobacterium in controlled environments, *Soil. Sci. Soc. Am. J.* **57**: 73.

53. Jones, R.W. and Hancock, J.G. 1990. Soilborne fungi for biological control of weeds. In : *Microbes and Microbial Products as Herbicides*), (ed. R.E. Hoagland), ACS Symposium Series 439. American Chemical Society, Washington DC, p. 276.

54. Jones, R.W., Lanini, W.T. and Hancock, J.G. 1988. Plant growth response to the phytotoxin viridol produced by the fungus *Gliocladium virens, Weed Sci.* **36**: 683.

55. Kelman, A., Anderson, W., Falkow, S., Federoff, N.V. and Levin, S. 1987. Introduction of recombinant DNA-engineered organisms in the environment key issues. National Academy Press, USA.

56. Kennedy, A.C. 1996. Molecular biology of bacteria and fungi for biological control of weeds. In: *Molecular Biology of Biological Control of Pests and Diseases of Plants*, (eds. M. Gunasekaran and D.J. Webes). CRC Press, Inc. pp. 155-172

57. Kennedy, A.C., Bolton, H. Jr., Stroo, H.F. and Elliott, L.F. 1992. The competitive abilities of the Tn5Tox - Mutants of a rhizobacterium inhibitory to wheat growth. *Plant Soil.* **144**: 143.

58. Kennedy, A.C., Elliott, L.F., Young, F.L. and Douglas, C.L. 1991. Rhizobacteria Suppressive to the weeds downy brome. *Soil. Sci. Soc. Amer. J.* **55**: 722.

59. Kenny, D.S. 1986. Devine - the way it was developed - an industralist's view. *Weed Sci.* **34 (Suppl.)**: 15.

60. Khodayari, K. and Smith, R.J. Jr. 1988. A mycoherbicide integrated with fungicides in rice, *Oryza sativa. Weeds Technol.* **2** : 282.

61. Kirkpatrick, T.L., Templeton, G.E. and Oliver, L.R. 1984. Texas gourd (*Cucurbita texana*) control with *Fusarium solani* f. sp. *cucurbitae. Weed Sci.* **32**: 649.

62. Kistler, C.H. 1991. Genetic manipulation of plant pathogenic fungi. In : *Microbial Control of Weeds*, (ed. D.O. TeBeest), Chapman and Hall, New Hall, New York, 152.

63. Kremer, R.J. and Spencer, N.R. 1989. Impact of a seed - feeding insect and microorganisms on velvet leaf *Abutilon theophrasti* seed viability. *Weed Sci.* **37**: 211.

64. Lacey, C.A., Lacey, J.R., Chicoine, T.K., Fay, P.K. and French, R.A. 1986. Controlling knapweed on Mantana rangeland. Mant. Stat Univ., Coop Ext. Serv. Circ. No. 311.

65. Lindgren, P.B., Pano poulos, N.J., Staskawicz, B.J. and Dahlbeck, D. 1988. Genes required for pathogenicity and hyper sensitivity are conserved and interchangeable among pathovars of *Pseudomonas syringae*. *Mol. Gen. Genet*. **211**: 499.

66. McRae, C.F. and Auld, B.A. 1988. The influence of environmental factors on anthracnose of *Xanthium spinosum*. *Phytopath*. **78**: 1182.

67. McRae, C.F., Ridings, H.I. and Auld, B.A. 1988. Anthracnose of *Xanthium spinosum* - quantitative disease assessment and analysis. *Anstral. Plant Path*. **17**: 11.

68. Miller, R.V., Ford, E.J., Zidack, N.K. and Sands, D.C. 1989. A pyrimidine auxotroph of *Sclerotinia sclerotiorum* for use in biological weed control. *J. Gen. Microbiol*. **135**: 2085.

69. Mortensen, K. 1986. Biological control of weeds with plant pathogens. *Can. J. Plant Pathol*. **8**: 229.

70. National Research Council, 1989. Alternative Agricultural, National Academic Press, Washinton, D.C. 448.

71. Ohr, H.D. 1974. Plant disease impacts on weeds in the natural ecosystem. *Proc. Am. Phytopathol. Soc*. **1**: 181.

72. Panaccione, D.G., Mc Kiernan, M. and Hanau, R.M. 1988. *Colletotrichum gramminicola* transformed with homologous and heterologous benomyl-resistant genes retains expected pathogenicity to corn. *Mol. Plant-Microbe Interact*. **1**: 113.

73. Phatak, S.C, Sumner, D.R., Wells, H.D., Bell, D.K. and Glaze, N.C. 1983. Biological control of yellow nutsedge with the indigenous rust fungus *Puccinia conaliculata*. *Science* **219** : 1446.

74. Phatak, S.C., Callaway, M.B. and Vavrina, C.S. 1987. Biological control and its integration in weed management systems for purple and yellow nutsedge (*Cyperus rotunaus* and *C. esculentus*), *Weed Technol*. **1**: 84.

75. Purcell, A.H. 1982. *Evaluation of the Insect Vector Relationship in Phytopathogenic Prokaryotes*. Vol. I (eds. M.S. Mount and G.H. Lacy). Academic Press, New York, 121.

76. Quimby, P.C Jr. 1982. Impact of diseases on plant populations. In: *Biological Control of Weeds with Plant Pathogens*, (eds. R. Charudattan and H.L. Walkee), Wiley Press, New York, 47.

77. Quimby, P.C. Jr., Falgham, F.E., Boyetle, C.D. and Connick, W.J. Jr. 1988. An invert emulsion replaces dew on biocontrol of sicklepod - a preliminary study. In : *Pesticide Formulation and Application Systems* (eds. D.A. Houde and G.B. Beestman), ASTM-STP980, American Society for Testing and Materials, Philiadelphia, Vol. 8, p. 264.

78. Reid, J.L. and Collmer, A. 1988. Construction and characterization of an *Erwinia chrysanthemi* mutant with directed deletions in all of the pectate lyase structural genes. *Mol. Plant Microbe Interact*. **1**: 32.

79. Riddle, D.L. and Georgi, L.L. 1990. Advances in research on *Caenorhabditis elegans*: application to plant parasitic nematodes. *Annu. Rev. Phytopathol*. **28**: 247.

80. Ridings, W.H. 1986. Biological control of stranglevine in citrus - a researcher's view. *Weed Sci*. **34 (Suppl.)**: 31.

81. Ridings, W.H., Mitchell, D.J., Schoulties, C.L. and El-Gholl, N.E. 1976. Biological control of milkweed vine in Florida citrus groves with a pathotype of *Phytophthora citrophthora*. In: *Proceedings IV International Symposium on Biological Control of Weeds*, p. 224.

82. Roberts, D.P. and Fravel, D.R. 1995. Strategies and techniques for improving biocontrol soilborne plant pathogens, ACS Symposium Series, Washington, D.C., 323.

83. Sands, D.C., Ford, E.J. and Miller, R.V. 1990. Genetic manipulation of broad host-range fungi for biological control of weeds. *Weed Technol.* **4**: 471.

84. Scheepens, P.C. and Hoogerbrugge, A. 1988. Control of *Prunus serotina* in forests with the endemic fungus *Chondrostereum purpureum*. In: *Abstracts VII International Symposium on Biological Control of Weeds*, Biological Control of Weeds Laboratory, USPA - ARS, Rome.

85. Shason, A. and Gressel, J. 1991. Elicitation of a flavonoid phytoalexin in accumulation. In : *Cassia obtusifolia* by a mycoherbicide : estimation by AICIZ-spectrofluorimetry. *Pestic. Biochem. Physiol.* **41**: 1442.

86. Smith, R.J. Jr. 1982. Integration of microbial herbicides with existing pest management programes. In: *Biological Control of Weeds with Plant Pathogens*, (eds. R. Charudattan and H.L. Walfeer), Wiley, New York, p. 189.

87. Smith, R.J. Jr. 1986. Biological control of nothern jointretch in rice and soybeans - a researcher's view. *Weed Sci.* **34 (Suppl.)**: 17.

88. Stougard, R.N., Shea, P.J. and Martin, A.J. 1990. Effect of soil type and pH on adsorption, mobility and efficacy of imazaquin and imazethapyr. *Weed Sci.* **38**: 67.

89. Stowell, L.J. 1991. Submerged fermentation of biological herbicides. In : *Microbial Control of Weeds*, (ed. D.O. TeBeest), Chapman and Hall, New York, p. 225.

90. TeBeest, D.O. 1982. Survival of *Colbetotrichum gloeosporioides* f. sp. *aeschynomene* in rice irrigation water and soil. *Plant. Dis.* **66**: 469.

91. TeBeest, D.O. 1988. Additions to the host range of *Colletotrichum gloeosporioides* f. sp. *aeschynomene*, *Plant. Dis.* **72**: 16.

92. TeBeest, D.O. and Templeton, G.E. 1985. Mycoherbicides : Progress in the biological control of weeds, *Plant Dis.* **69**: 6.

93. Templeton, G.E. 1982. Status of weed control with plant pathogens. In: *Biological Control of Weeds with Plant Pathogens*, (eds. R. Charudattan and H.I. Walker), Wiley Press, New York, 29.

94. Templeton, G.E., TeBeest, D.O. and Smith, R.J. Jr. 1984. Biological weed control in rice with a strain of *Colletotrichum gloeosporiodes* (Pen Z.) Sace. used as a mycoherbicide, *Crop Prot.* **3**: 409.

95. Templeton, G.E., Smith, R.J. Jr. and TeBeest, D.O. 1986. Progress and potential of weed control with Mycoherbicides. *Rev. Weed Sci.* **2**:1.

96. Tinline, R.D. 1963. *Cochliobolus sativus* VII. Nutritional control of the pathogenicity of some auxotrophs to wheat seedlings. *Can. J. Bot.* **41**: 489.

97. VanEtlen, H.D., Mathews, D.E. and Mackintosh, S.F. 1987. Adaptation of pathogenic fungi to toxic chemical barriers in plants : the psiatin demethylase of *Nectria harmatoccocca* as an example, In : *Molecular Strategies for Crop Protection*, UCLA Symposium on Molecular and Cellular Biology, A.R. Liss, New York, **48**: 59.

98. Walker, H.L. 1981. Granules formulations of *Alternaria macrospora* for control of spurred anoda (*Anoda cristata*), *Weed Sci.* **29**: 342.

99. Walker, H.L. 1982. A seedling blight of sicklepod caused by *Alternaria cassiae*. *Plant Dis.* **66**: 426.

100. Walker, H.L. and Boyette, C.D. 1985. Biological control of sicklepod *Cessia obtusifolia* in Soybeans (*Glycine max.)* with *Alternaria cassiae. Weed Sci.* **33**: 212.

101. Walker, H.L. and Riteef, J.A. 1982. Evaluation of *Alternaria casiae* for the biocontrol of sicklepod (*Cassia obtusifolia*). *Weed Sci.* **10**: 651.

102. Watson, A.K. 1991. The classicial approach with plant pathogens. In: *Microbiol Control of Weeds*, (ed. D.O. TeBeest), Chapman and Hall, New York, p. 3.

103. Weidemann G.J. 1988. Effects of nutritional amendments on conidial production of *Fusarium solani* f. sp. *Cucurbitae* on sodium alginate granules and on control of Texas gourd, *Plant Dis.* **72**: 757.

104. Weidemann, G.J. and Templeton, G.E. 1988a. Control of Texas gourd, *Cucurbita texana* with *Fusarium solani* f. sp. *cucurbitae. Weed Technol.* **2**: 271.

105. Weidemann, G.J. and Templeton, G.E. 1988b. Efficacy and soil persistence of *Fusarium solani* f. sp. *cucurbitae* for control of Texas gourd (*Cucurbita texana*). *Plant Dis.* **72**: 26.

106. Weller, D.M. 1986. Biological control of soil borne plant pathogens in the rhizosphere with bacteria. *Ann. Rev. Phytopathol.* **26**: 379.

107. Willis, D.K., Bartar, T.M. and Kinscherf, T.G. 1991. Genetics of toxin production and resistance in phytopathogenic bacteria. *Experientia* **47**: 766.

16 Microwave Treatment and other New Methods for Eradicating the Dry Rot Fungus (*Serpula lacrymans*) Wulf. : Fr. (Schroet.)

Jorgen Bech-Andersen and Steen Andrew Elborne

CONTENTS

ABSTRACT 280

1. INTRODUCTION 280
   - 1.1 The Biblical Solution 280
   - 1.2 Bewitched Houses 280
2. DETECTING DRY ROT ATTACKS 281
3. TREATMENT OF MASONRY 281
   - 3.1 Microwaves 282
   - 3.2 Fungicides 282
   - 3.3 Wearing both Belt and Braces 284
4. REFERENCES 285

Advances in Microbial Biotechnology
J.P. Tewari, T.N. Lakhanpal, Jagjit Singh, Rajni Gupta & B.P. Chamola (eds.)
APH Publishing Corporation, New Delhi - 110 002, India.

ABSTRACT

In the old days dry rot was treated as something supernatural. Later remedial treatments with kitchen salt was used to prevent dry rot. The traditional repair methods after dry rot were rather extensive, with much damage to the buildings as a consequence. Today methods have been developed to lessen both extent and cost of repair. One method is to eradicate dry rot mycelium in timber with microwaves. Another method is to impregnate either timber or masonry with a boron diffusion compound, which is less toxic than the remedies previously used.

1. INTRODUCTION

Microwaves are actually radiowaves sent from an antenna. Generally speaking this means that twenty minutes with news on the radio is enough to kill the dry rot fungus. This is what is utilized during repair work. The damage caused by the true dry rot fungus and associated costs for repair work is not a new phenomenon. The majority of the treatments carried out to repair dry rot damage are often too destructive. This article deals with alternate approaches. The detailed account of alternate approaches are also given in 'Building Mycology, Management of decay and health in buildings' by Sigh (11).

1.1 The Biblical Solution

In the Bible in Leviticus eradication of the true dry rot fungus is described in detail. Leprosy of houses as it was called had to be reported to a clergyman and the house had to be emptied before he came to inspect it. If after the inspection the patches of infection continued to grow mortar and stones had to be removed and replaced. New patches of infection would mean that the house was impure and had to be demolished down. The infected stones, mortar and timber had to be carried outside the town and placed in an impure site. Those people who had slept in the house had to wash their clothes. On the other hand had the patches not grown after a week you were to take to birds, cedarwood, red yam and twigs of Isop and do the following with these:

One bird should be slaughtered and its blood sprinkled over the yarn, twigs, the other bird and the house. The second bird should be let loose to expiate the sins of the house when flying over the fields.

1.2 Bewitched Houses

In Denmark dry rot has caused problems in houses for a long time, especially where wooden joints are placed in contact with moist soil or foundations. In these cases it was recommended to spread common cooking salt around new timber, but rarely with any success (1). A house was thought of as bewitched when continuously you had to fight something mysterious that broke down the wood work which therefore had to be replaced. Fungi were not known to be implicated at that time. As a left over from that time it has been common practice to pull down all wood and masonry after dry rot attacks. On facades of old buildings one can still see scares in the form of several squaremetres of pale bricks where repair has taken place. The cost of repair was enormous.

2. DETECTING DRY ROT ATTACKS

Newly started housing co-operatives can often be completely knocked out financially by a dry rot attack. Therefore new as well as old means of detecting and repairing attacks of dry rot must be taken into account. Three conditions must be met when repairing after dry rot attack:

(i) The fungal attack must not be able to break out again.

(ii) The structural strength of the construction must be re-established and sources of moisture eliminated.

(iii) Repair methods must be economical.

The dry rot fungus can spread from the point of attack following windows, posts and notch boards through the entire house. Attacks are therefore difficult to delimit completely. If some infected wood is left in the building the fungus will continue to grow and in stead of one large attack you will find several small attacks at the margin of the original attack. Detection and delimitation of an attack is therefore very important (6).

3. TREATMENT OF MASONRY

The dry rot fungus is able to penetrate masonry and from there reinfect newly inserted timber. In stead of tearing down all masonry you can use the following method. Remove the plaster and scrape or burn of visible fungal mycelium, then scratch out the joints and apply a fungicide. After pointing and plastering the wall with new mortar a barrier has been created which is toxic to the fungus and which hinders it from reaching new timber. In this manner the fungus is starved to death inside the wall. Previously the scratched-out joints were pointed with mortar mixed with fungicide. This meant that bricklayers had to wear safety equipment all through the process.

Today the process has been changed so that the scratched-out joints are first treated with fungicide, then pointed with pure mortar, then a second round with fungicide, ending with plastering of the entire wall. This method ensures an effective two-layered fungicide barrier behind a layer of pure plaster without the indoor climate being affected.

Scratching of joints with modern fastrunning tools produce more carcinogenic dust particles than older methods. This means that masks must be worn and scratching-out must be limited as much as possible. Some experiments have indicated that a less destructive method can be applied. In this method the plaster is beaten down and fungal mycelium is swept of the wall. The wall is treated with fungicide and rendered, then one more round of fungicide treatment and rendering, ending with plastering.

Today scratching-out joints to a depth of 3 centimetres is only employed where timber is in direct contact with infected masonry, e.g. where joints rest on or run along masonry, as well as in embrasures.

The method means lower cost and less exposure to poisonous dust for the bricklayers who have to wear masks for a shorter period than before.

3.1 Microwaves

Previously all infected timber had to be removed. Today new options for sparing infected wood with sufficient residual strength are found allowing to keep up to half of the infected joists (Fig. 1).

Since 1985 civil engineer Claus Andersen and the senior author have conducted experiments with eradication of dry rot in both timber and masonry by means of microwaves. Small glass tubes with live mycelium of dry rot were inserted into either timber or masonry at different depths and treated with microwaves for varying periods of time. A thermometer built into the constructions showed in all cases that the fungus died when the temperature reached about 50^{0}C (23).

In practice today when the method is used a thermometer is inserted into the construction as far as possible from the microwave source. Treatment continues until the temperature reaches 50^{0}C. When you are sure that the lethal temperature has been reached it is unnecessary to check whether the fungus is still alive. The methods for showing the latter are both expensive and uncertain.

The effect of common household microwave ovens are well-known, e.g. allowing to heat a frozen spring roll in three minutes. From an antenna the oven emits radio waves with a frequency of 2450 MHz. Radio Denmark uses a frequency of 100 MHz so in both cases it is radiowaves. Water molecules are polarized with a north and a south pole. When water is influenced by radiowaves which change north and south pole 2450 million times per second the water molecules will rotate accordingly. The friction from this action will heat the water molecules.

When treating wooden joists which have full residual strength but are superficially decayed the microwaves will pass through the wood but will influence the water molecules in the dry rot mycelium. The water when heated to 50^{0}C will coagulate the proteins and thus kill the fungus (10) (Fig. 2).

Timber treated with microwaves will not burn as can happen with infrared treatment. The maximum temperature will be 100^{0}C, the boiling point of water. In practical use the microwave device is a box measuring about 50 by 40 centimetres which is placed over the exposed joist ends. The two antennae inside the device emit radiowaves which will influence an area of 50 by 40 centimetres. As mentioned previously the timber is heated until a temperature of 50^{0}C is reached at the most distant point of the joist (5, 8, 9).

To prevent dry rot from re-entering the sterilised wood it is impregnated with a boron diffusion fungicide both by surface application and by filling drill holes. A lethal level of 6 grams boric acid per litre wood must be reached. Adjacent masonry is treated as previously described to keen the fungus behind a barrier away from the wood. After the microwave treatment a metal sign is placed on the treated timber stating the temperature reached and by whom it was carried out. At present about 600 microwave treatments with subsequent diffusion impregnations have been carried out.

3.2 Fungicides

In Denmark new fungicides have been introduced and some of the old ones are no longer in use. Previously sodium pentachlor phenolate (NaPCP) was used for example to mix in mortar.

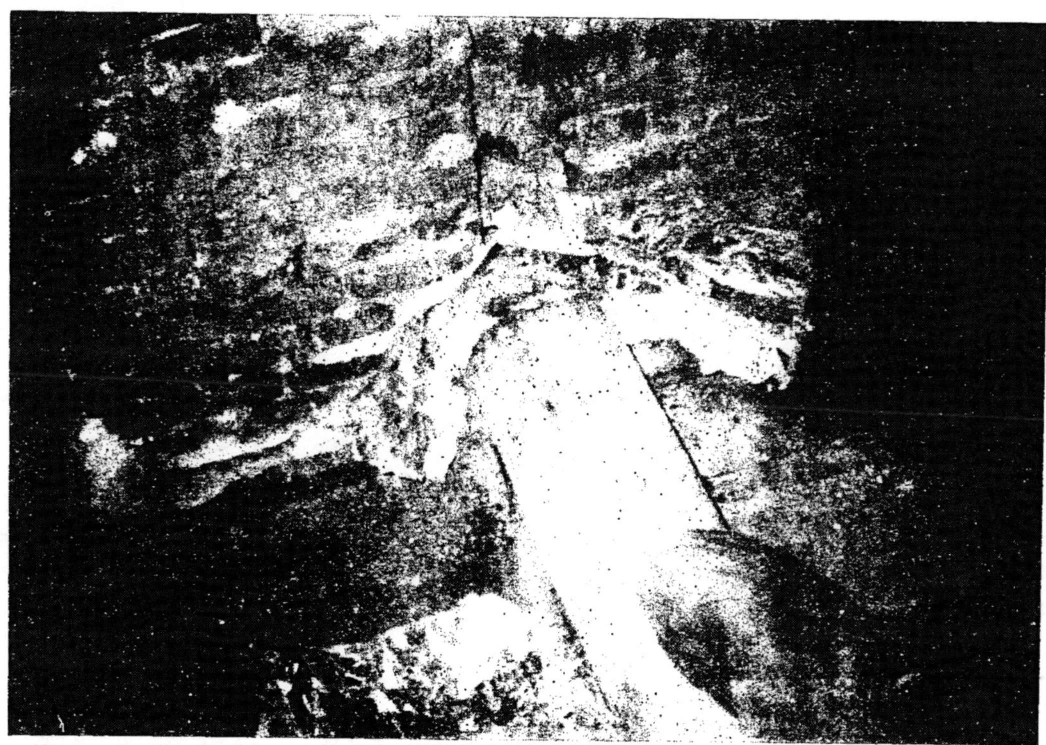

Fig. 1 : In preparation for treatment with microwaves an aluminium foil is placed behind the timber to reflect the microwaves

Fig. 2 : The treatment is complete when the temperature reaches 50°C. Ten minutes at 37°C

PCP however is poisonous and more than half of it would evaporate within, the first year after treatment thus harming the indoor climate.

PCP was replaced by organic tin compounds and boron diffusion agents. Especially the boron compounds are promising since they are less toxic than kitchen salt towards humans and animals, they are odourless and can penetrate timber completely by so-called diffusion. This means that both sapwood and heartwood can be impregnated by boric acid. No other remedy can impregnate heartwood.

Pressure impregnated wood used for replacement must be approved according to Danish Standard DS 2122 class A and must be impregnated at the cut ends by filling drill holes with a boron diffusion agent, thus impregnating the heartwood.

3.3 Wearing both Belt and Braces

In fighting the true dry rot fungus you have to wear both belt and braces, so you use both microwave treatment and boron diffusion impregnation just in case. Repair after dry rot infection is so costly that every means must be taken to reduce the cost. The method suggested by the senior author in the papers called BYG-ERFA from 1979 have to be modified to reduce cost of repair (4).

All infected timber no longer has to be replaced. As long as it has full residual strength microwave treatment and boron diffusion impregnation can be applied in stead.

All joints of masonry no longer have to be scratched out to a depth of three centimetres and then treated with fungicide. Scratching-out is only used when masonry is in contact with wood. In masonry without wood contact the plaster is beaten down and fungal mycelium is brushed away. The wall is then treated twice with fungicide and rendering. An other cost saving method is to repair floor partitions from above and at the same time not disturb the persons in the flat below. Infected joists can either have microwave treatment or be cut up and replaced from above without destroying the ceiling and stucco in the flat below. A shore is used to support the joists while they are being fully or partly replaced. Likewise laths are used to support the casting boards. A pressure impregnated wall plate is inserted or the battlement is walled up with bricks. New timber is spliced on and finally treated with a boron diffusion agent.

The different methods mentioned above will save between one and two thirds of the normal cost, while at the same time the residents are disturbed as little as possible (7).

The following points are now fulfilled :

(i) The strength of the timber is re-established

(ii) The fungus is prevented from re-infecting the construction

(iii) Costs are low

(iv) The building can be insured against fungal decay

Today it is not necessary to call for the clergyman or to slaughter any birds to eradicate dry rot. However one might serve roasted doves from the cornices now that a microwave oven is at hand.

4. REFERENCES

1. Aagaard, K. 1815. *Beskrivelse over Torning Lehn. Et Bidrag til Kundskab om Heilugdommet Slesvig.* Gyldendal. Kjobenhavn.

2. Andersen, C. 1989. Patentansogning S 745-89 (application for patent).

3. Andersen, C. 1991. Kvalitetssikring på mikrobolgebehandling. (folder)

4. Bech-Andersen, J. 1979. AEgte Hussvamp og indvendige reparationer. *BYG-ERFA erfaringsblad 790215.*

5. Bech-Andersen, J. 1987. Behandling af AEgte Hussvamp. Metode 2, reg. 450-11 11 3. Opfinderkontoret. Dansk Teknologisk Institut.

6. Bech-Andersen, J. 1989. Angreb af aegte hussvamp og andre traenedbrydende svampe i huse. *Andels- og ejerlejlighedsforeningernes telefon- og servicehdndbog 11.* årgang, 1989-90: 21-24.

7. Bech-Andersen, J. 1990. Anvendelse af mikrobolger og nye metoder gor reparation af AEgte Hussvamp billigere. *Andels- og Ejerlejlighedsforeningernes telefon- og servicehdändbog* 12. årgang, 1990-91: 22-24.

8. Bech-Andersen, J. 1993. Eradicating the true dry rot fungus with microwaves. Document no. IRGWP 93-10001: 32-34.

9. Bech-Andersen, J. and Andersen, C. 1992. Theoretical and practical experiments with eradication of the dry rot fungus by means of microwaves. Document no. IRGWP 1577-92.

10. Miric, M. and Willeitner, H. 1984. Lethal temperature for some wood destroying fungi with respect to eradication by heat treatment. Document no. IRGWP 1229.

11. Singh, J. 1994. *Building Mycology: Management of Decay and Health in Buildings. E.* & F.N. Spon, London, 326 pp.

17 *Serpula lacrymans,* the Dry Rot Fungus: A Much Misunderstood Organism

John W. Palfreyman, Gordon A. Low, Christopher Westwood and Nia A. White

CONTENTS

	ABSTRACT	288
1.	INTRODUCTION	288
2.	NATURAL SOURCE OF THE DRY ROT FUNGUS	291
3.	BIOLOGICAL CONTROL OF DRY ROT	294
4.	ENVIRONMENTAL CONTROL OF DRY ROT	297
5.	THE INCREASED OF *S.LACRYMANS* TO INCREASED TEMPERATURES	300
6.	THE RELATIONSHIP BETWEEN WOODY AND NON-WOODY RESOURCES	301
7.	CONCLUSION AND FUTURE WORK	302
8.	REFRENCES	305

Advances in Microbial Biotechnology
J.P. Tewari, T.N. Lakhanpal, Jagjit Singh, Rajni Gupta & B.P. Chamola (eds.)
APH Publishing Corporation, New Delhi - 110 002, India.

ABSTRACT

The 'true dry rot fungus', *Serpula lacrymans,* is the most significant fungal cause of damage to timber in the built environment in many temperate regions of the world. Though control methods to deal with the fungus have been developed in these regions, many aspects of the organism remain under researched and quite why it is so damaging to the built environment remains obscure. However, recent studies have thrown some light on a range of aspects of the organism, for example its envirorunental niche and sensitivity, reaction to potential biocontrol agents, utilisation of and dependence on nonwoody materials and nitrogen metabolism have all revealed information which may help in the development of new control strategies. In addition molecular analyses of the organism are underway and in recent years protein, antigen and nucleic acid systems have all been applied to the detection, identification and understanding of the organism.

In this review a number of aspects of the fungus that are under active research in the author's laboratory will be discussed and thier significance in relation to none elsewhere evaluated.

1. INTRODUCTION

The dry rot fungus *Serpula lacrymans* (Figs. 1-3) is responsible for many millions of dollars (US) of damage to timber within the built environment each year most significantly in northern, temperate regions of Europe and Japan. Other areas where the fungus causes damage include parts of Australia, northern India and temperate areas of the former Soviet Union though the fungus appears largely absent from the USA, at least in the built environment. Much of the early work on *S.lacrymans* was undertaken by Richard Falck in Germany in the early years of the 20th Century. It was Falck who, during his work on 'Hausschwamm', identified various basic features of the organism; for example that there appear to be two 'strains' - one from the built environment, one from the forest (now called *S.lacrymans* and *S. himantioides* respectively), that *S.lacrymans* is peculiarly sensitive to temperature, that osmotic pressure in the cell sap of *S.lacrymans*, and other fungi, drives the elongation of the hyphal tip and that the fungus could be prevented by 'respiratory poisons' (19). Much of Falck's pioneering work was published in his paper *Die Meruliusfaule des Bauholzes* (14) and, though directed mainly towards *S.lacrymans*, has had a significant impact in many areas of timber preservation. A resume of Falck's findings was published by Hornung and Jennings (18).

For many years after Falck's pioneering research on dry rot, little progress was made however the large numbers of damaged buildings in Europe after 1945, and a subsequent increase in reported outbreaks of dry rot stimulated a renewed interest in *S.lacrymans*, typified by the studies of Cartwright and Findlay in the UK (10) through to the physiological, biochemical and model building studies reported by Jennings and co-workers in Liverpool (UK) (11, 20). Further studies by Bravery in the 1980's demonstrated the link between woody and non-woody resources and the growth of *S.lacrymans* (8) and this theme was taken up by Bech-Andersen in

Fig.1: A basidiocarp of *S.lacrymans*; such structures may be up to 1m in diameter and a basidiocarp of this size is capable of producing spores at a rate of 8 × 10^7/min (Falck 1912)

Fig.2: The destructive effects of *S.lacrymans*; if left unchecked the dry rot fungus is capable of causing extensive damage to woodwork in buildings. Though lignin is left largely undegraded by the organism the removal of cellulose from the wood leaves it with very little structural strength

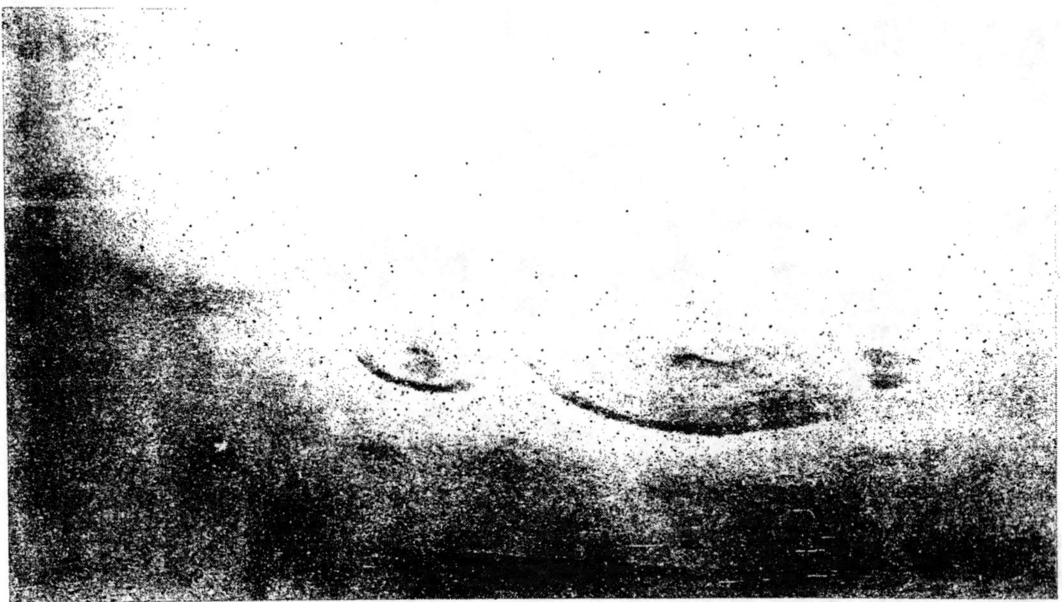

Fig.3: The mycelia of *S.lacrymans*; though ostensibly a wood decay fungus *S.lacrymans* is capable of penetrating non-woody materials and can appear, as shown in this diagram, as a mycelial flush on a plaster surface

the late 1980's and early part of the current decade (5). The development of numerous new techniques in molecular biology in the last decade has opened up many new avenues of research on dry rot; some of these will be discussed in this current paper. In addition new pressures have developed on professionals involved in the control of dry rot, most notably the consumers desire to find 'natural' and/or 'environmentally-friendly' solutions to biological problems such as dry rot, and a movement away from what are perceived as general toxicants which have been used in its control in the past.

Thus new foci for research into dry rot have become apparent, and a new perspective about the organism is developing. For many years *S.lacrymans* has been thought to be a difficult fungus to eradicate in buildings, with the financial implications of repairing problems being oppressive. However, new research, and new views of previous studies, have indicated that *S. lacrymans*, is a very sensitive organism and that its ability to cause damage to buildings is intimately related to maintenance problems in the built environment, inappropriate design and/ or usage of buildings, as well as the types of materials present. By understanding more about the physiology of the organism, in particular its dependence upon sources of moisture and non-woody resources, it seems likely that new control measures for the problem will be found. However, until the fear engendered in the general public by the possible consequences of dry rot is lessened, the application of new processes will be limited. In this review studies on a) the natural environment of *S.lacrymans*; b) potential biocontrol systems; c) environmental sensitivity of *S.lacrymans* both from a biochemical perspective and from that of the built environment; and d) the relationship between woody and non-woody resources will be discussed.

Finally it should be noted that the name of the organism forming the subject of this chapter has changed over the years. Von Wulfen (46) introduced the name *Serpula lacrymans*, whereas Falck (14) called the organism *Merulius lacrymans*. More recently the original, correct name *Serpula lacrymans* has been universally accepted (31). The name of the fungus is particularly apt as it is derived from *serpere* (Latin, to wind or to creep) and *lacrimans* (Latin, weeping) note that the original spelling of the epithet was *lacrymans* (46) and though this is incorrect 'Article 73.1 of the International Code of Botanical Nomenclature, 1988 states 'the original spelling of a name or epithet is to be retained" (31). For further background reading about dry rot *S.lacrymans* the authors would recommend the reviews contained in Jennings and bravery (20), Hennenbert *et al.* (17), Singh (38) and Palfreyman *et al.* (29). In addition, for reference to dry rot research prior to 1988, the comprehensive bibliography produced by Sehann and Hegarty (35) should be consulted.

2. NATURAL SOURCE OF THE DRY ROT FUNGUS

Despite its widespread occurrence in many temperate regions of the world reported findings of *S. lacrymans* in the wild are still rare. Currently there are only two natural locations where occurrence of the species has been substantiated and a third where occurrence remains unsubstantiated. The earliest recording of wild finds of *S.lacrymans* was made by Bagchee (1), who reported finding the dry rot fungus in Narkanda in Himachal Pradesh, India. In the late 1940's and early 1950's reports of *S.lacrymans* in this area were not uncommon, though from 1954 until 1993 there were no further reports; this absence may be attributed to changes in the mycological interests at the Forest Research Institute, Dehra Dun, India (5).A second report of a natural find, this time in California was published by Cooke in 1957 (12). Both the Indian and US habitats are in areas with relatively stable temperatures, high humidity and dense forests. In neither case is *S.lacrymans* a particularly conspicuous organism, its spread is presumably controlled by a range of environmental and biotic factors. This latter observation starkly contrasts with the often spectacular predominance of *S.lacrymans* in a typical domestic outbreak of dry rot.

A third report of a wild find of *S.lacrymans*, from the Czech Republic (22), remains unsubstantiated and it seems probable that the organism found was in fact the closely related species *S.himantioides*. For an organism which is so common in buildings these very restricted reports seem surprising. In recent years the Himalayn and US habitats of *S.lacrymans* have been revisited by a number of expeditions (5, 39, 40). *S.lacrymans* has been rediscovered at these sites (Fig. 4) and molecular techniques, notably SDS-Polyacrylamide Gel Electrophoresis, have been used to confirm the morphological identifications (Fig. 5). Observers on these recent expeditions have frequently noted the close relationship between wood decayed by *S.lacrymans* and cords and mycelial fans spreading between the wood and the soil (39, 40, 44). Actual habitats have been described by a number of recent authors; one reported as follows: 'Three *S.lacrymans* fruit bodies with strands and fan-shaped mycelium were found in woodland sites in the Narkanda region of the Western Himalayas at between 2800 and 3100 m above sea level. One basidiome was in contact with soil on the heart-rotted hollow of, *Picea smithiana*; a second on the underside of a fallen, burnt and well decayed stump of *Abies pindrow*; the third sample was located in partial contact with the soil, on the underside of a fallen, well decayed stump of *Taxus baccata*. In each case, the fruit bodies were north or north east facing, situated in relative shade or darkness, at a local air temperature of 10-14°C and r.h. 60-80%, adjacent to clay-loam soil (pH 7 approximately) with wood moisture content in the range 27-55%. The associated basidiomycetes were *Heterobasidion insular, Armillarea mellea* and *S.himantioides* (44). The similarities between the microclimate in this area of India and that found in badly maintained, damp buildings in the UK, northern Europe and beyond is striking.

Fig. 4 : Map of India showing the site of some of the recent finds of 'wild' *S.lacrymans*

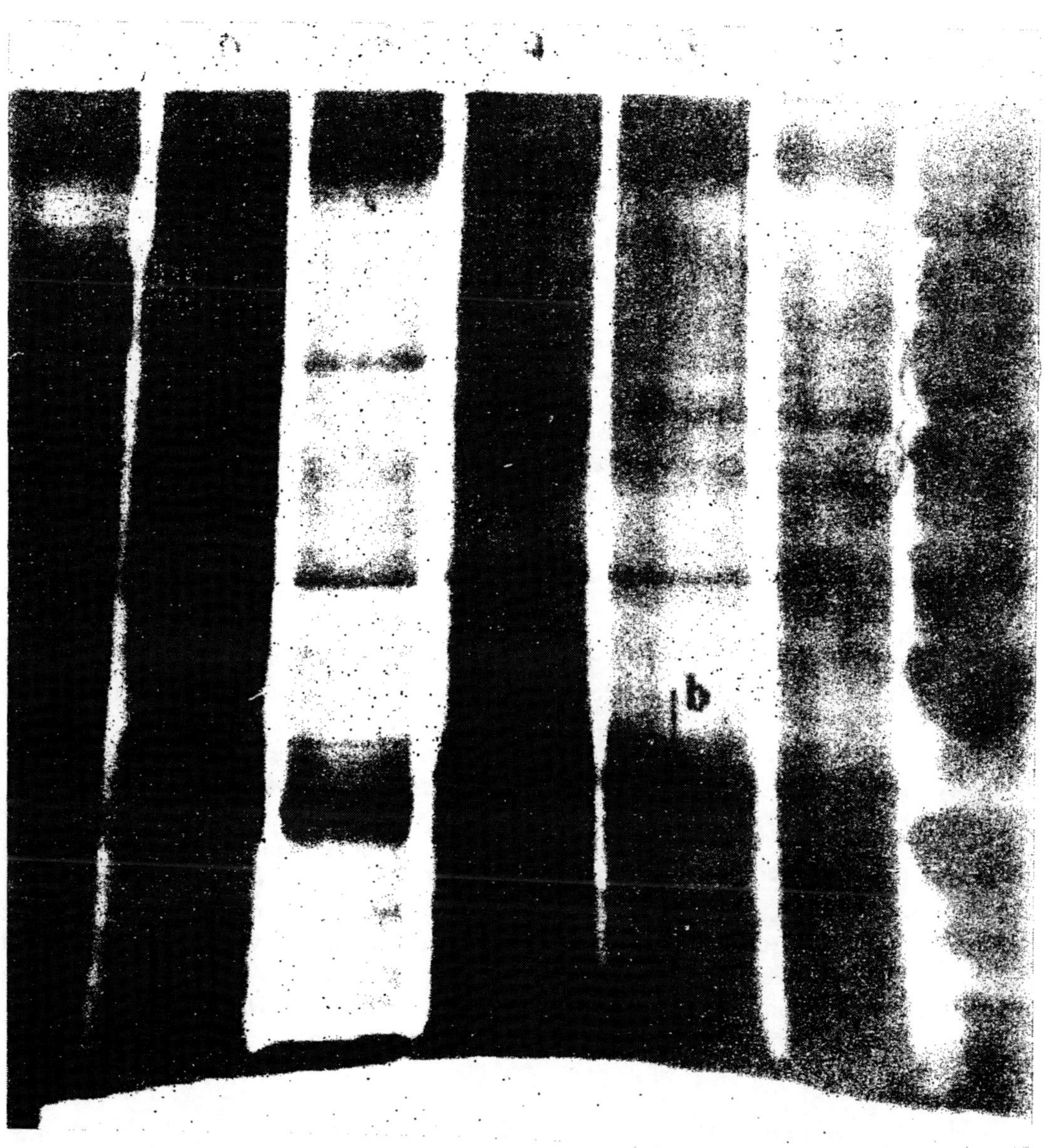

Fig.5: SDS-PAGE of isolates of wild *S.lacrymans* demonstrating the similarity between these isolates and those of building isolates from Europe (tracks 2 and 3 and 4 respectively). Differences between *S.lacrymans* and *S.himantioides* can also be seen (tracks 5 and 6)

White *et al.* (44) also report the first modern isolation of *S.lacrymans* from the wild, along with a comparison of the microclimate tolerances of the wild isolate with those of domestic isolates. Interestingly the wild and building isolates behaved as two separate cohorts, the wild isolates being less affected by extremes of temperature than those from the built environment and having markedly higher growth rates (as measured by linear extension rates on malt extract agar). In other ways, for example in response to osmotic stress, overall growth rates and the ability to alter substrate pH, there was little difference between the wild and building isolates. There was also very little difference in their overall ability to decay wood samples despite the fact that some of the building isolates had been first isolated over 50 years ago and not 'passed in wood' for at least 5 years (25). Unlike many other wood decay fungi, *S.lacrymans* does not appear to 'lose' its ability to degrade timber by repeated subculture on agar. Isolation of cultures from the US site on Mount Shasta in northern California has not yet been confirmed, but once it is, phylogenetic studies using PCR, as reported by Theodore *et al.* (41) on building isolates, will be undertaken.

Continued molecular research comparing natural and building isolates will hopefully resolve/elucidate the origins of *S.lacrymans*; are building isolates descendants of material exported in the past from areas such as the Indian Himalayas and the Californian Rocky Mountains or has the organism retreated, perhaps due to environmental changes, from a common dispersion in the world to a very specific niche.

3. BIOLOGICAL CONTROL OF DRY ROT

Biological control (biocontrol) can be defined as 'the suppression of a pest by means of the introduction, propagation and dissemination of the predators, parasites and diseases which attack it (9). Though the use of such systems is widespread in areas of agriculture, the application of biocontrol to timber preservation is more limited (9) and their use to combat *S.lacrymans* has been reported only in studies performed in Japan (13) and in the UK (29, 32, 34). Early laboratory studies demonstrated that a range of different species of the soil fungus *Trichoderma* were able to inhibit the growth of, and ultimately kill, *S.lacrymans* on a range of different media (32). Significantly variations in the levels of nitrogen and iron in the basal medium used in these studies had large effects on the outcomes of interactions, generally the more stringent that nutrient conditions became the less likely was effective killing of *S.lacrymans* by *Trichoderma*, though certain isolates of the latter organism were highly effective in all media tested. As might be expected, a full range of results from *Trichoderma* killing of *S.lacrymans* through stalemate reactions and *S.lacrymans* killing of *Trichoderma* was also found (Fig 6). Too date all *S.lacrymans* isolates tested have responded similarly to specific *Trichoderma* isolates.

Moving from agar to wood substrates demonstrated that preinoculation of the latter substrate with various *Trichoderma* spp. could afford effective protection from attack by *S.lacrymans*. However, when wood was precolonised with the decay fungus, *Trichoderma* spp. were unable to prevent further decay, as evidenced by continued weight loss. In related experiments Palfreyman *et.al.* (28) demonstrated that timber already partially decayed by *S.lacrymans* could be further degraded by *Trichoderma* spp. which by themselves are unable to degrade wood. Presumably the initial degradative action of *S.lacrymans* makes the cellulosic polymer, which is normally inaccessible to the *Trichoderma*, assimalable by this mould. However, *S.lacrymans* could never be re-isolated from blocks whether preinfected with this organism or *Trichoderma* even in the presence of selective media. This result is consistent with the hypothesis that even if decay is continuing it is not caused by the basidiomycete.

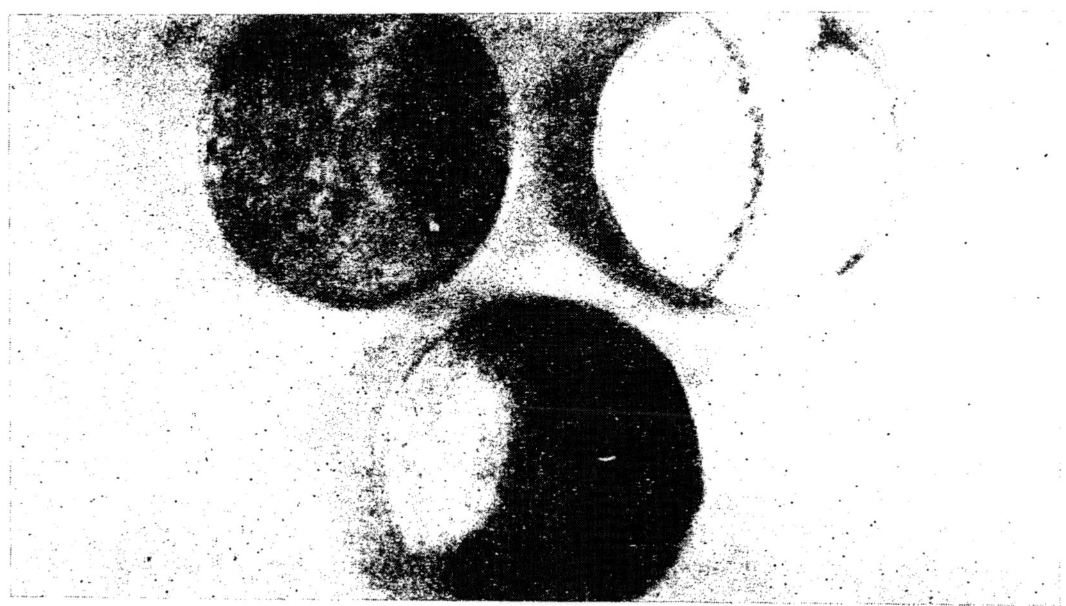

Fig.6: Reactions between *S.lacrymans* and *Trichoderma* spp.; a) *S.lacrymans* overcomes *Trichoderma longibrachiatum*, b) deadlock between *S.lacrymans* and *Trichoderma viride* and c) *S. lacrymans* is overcome by *Trichoderma harzianum*

More recent studies on natural substrates, i.e. wood, in a specifically designed microcosm (34) demonstrated that growth of *S.lacrymans* along wood sticks could be prevented by *Trichoderma* spp. However, the former organism was not necessarily killed by the latter and may be able to adopt survival strategies. Similarly in workshop sized model systems the interaction between *S.lacrymans* and control organisms is more ambiguous and, the *Trichoderma* spp. tested were unable to kill the dry rot fungus once it was well established within the woody substrate. *Trichoderma* merely slowed the growth of the dry rot fungus (Low *et al.* unpublished observations).

There are a number of possible explanations for the discrepancy between the laboratory and 'workshop' experiments. For example, the former experiments were carried out in a sterile environment and at a fixed temperature, whereas in the latter experiments sterility was not maintained and temperatures varied. Further more, the workshop models were more representative of natural infections since genuine building materials were used. Compared to those in small-scale microcosms, fungal colonies were less restrained spatially and temporally; each colony therefore had the potential to exhibit more of its developmental capabilities, especially with regard to the regulation, maturation and aging of the colony and to the allocation of its resources. Finally the workshop models were designed specifically to favour the dry rot fungus, whereas the *Trichoderma* used in the experiments was isolated from a soil sample in the UK and may not necessarily be adapted to growing in the built environment. A possible strategy for improving the performance of the control agent is to look for natural competitors in the natural environment of the decay organism, i.e. in northern India or in the USA, such studies are currently underway.

Biological control studies have revealed other interesting features of *S.lacrymans*. Specifically the organism can be induced to produce a range of, presumably, defensive enzymes once under attack by *Trichoderma*. Thus, Score *et al.* (33) demonstrated production of various extracellular phenoloxidases during interactions, most notably the production of laccase, an enzyme normally associated with white rot fungi. Various *Trichoderma* spp. could also be induced to produce laccase, again not an enzyme normally produced by these organisms. Additionally melanisation of *S.lacrymans* mycelium seems to be an early response of the basidiomycete to attack by potential competitors and, during interactions in microcosms (34), cord formation by *S.lacrymans* was a highly visible feature of its defensive response.

Effective protection of wood by *Trichoderma* crucially depends on effective colonisation by this antagonist (9), which itself depends upon nutrient resources and the presence of other, competing, organisms. In an attempt to understand, and eventually model, the interaction between *Trichoderma* and *S.lacrymans* in a range of substrates, a tessellated agar tile system for monitoring multiple interactions is under test (Fig 7). In this system, complex grids of up to 36 inoculated agar tiles can be established, with results to date demonstrating the effect that the spatial arrangement of interacting species has on the final interaction outcome (compare the 3 × 3 grids inoculated with *Trichoderma* either in the centre or in a corner tile). Other studies using this system (45) have revealed that stochastic processes are important in the development of final interaction patterns, particularly when more than 2 species are present in the initial inoculation. Data from these tessellations can be incorporated into mathematical or computer models and a cellular automation model has been shown to be able to predict outcomes of complex interactions under some circumstances (7).

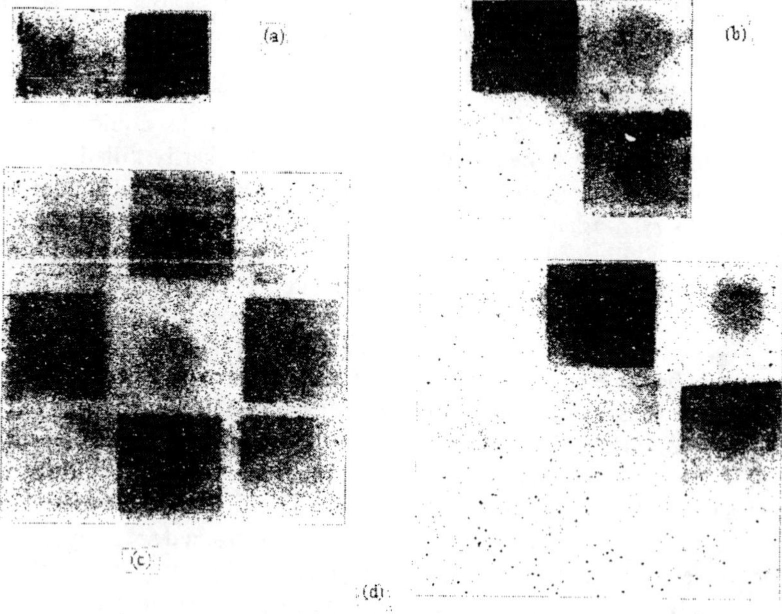

Fig.7: Tessellations of S. *lacrymans* and *Trichoderma* spp.; small 1 cm square tiles of malt extract agar (45) were inoculated with *S.lacrymans* on day 1, and once colonisation was established (normally after 4 days) other tiles were inoculated with *Trichoderma* spp. In the 2x1, 2x2 and 3x3 arrangements 1, 3 or 8 tiles were inoculated with *S.lacrymans* respectively. In each arrangement one tile was inoculated with *Trichoderma* T25 (32) - tile 1, 2, 5 or 3 (a) to d) respectively where the top left hand tile is number 1). Therefore in one set of 3x3 arrangements the central tile was inoculated with -*Trichoderma*, in one set it was the central tile. Killing of *S.lacrymans* was associated with discolouration of the tiles

4. ENVIRONMENTAL CONTROL OF DRY ROT

Treatment of dry rot in the UK, and beyond, normally involves the removal of fairly large quantities of both infected and un-infected timber together with introduction of preservative treated timber and, possibly, irrigation of walls with fungicides. An understanding of the biology of *S.lacrymans* suggests that many of these procedures are unnecessary as, if it is denied an adequate moisture source, the fungus in unable to grow (Though it produces water through its normal metabolic processes this alone are insufficient to support the processes of growth and decay). The most important feature of a dry rot treatment should involve the removal of moisture sources and the drying of infected timber. Despite this fact, dry rot repairs tend to be excessively expensive and there is a widespread fear of the effects of the organism. Use of natural drying processes to treat dry rot remedially are now being promoted by a number of companies under the description of 'Environmental Control'; recent studies have been directed toward defining the limits of such control methods (26). These studies have involved the development of various types of model from simple microcosms to complex mock-ups of parts of historic buildings. (Figs 8-10).

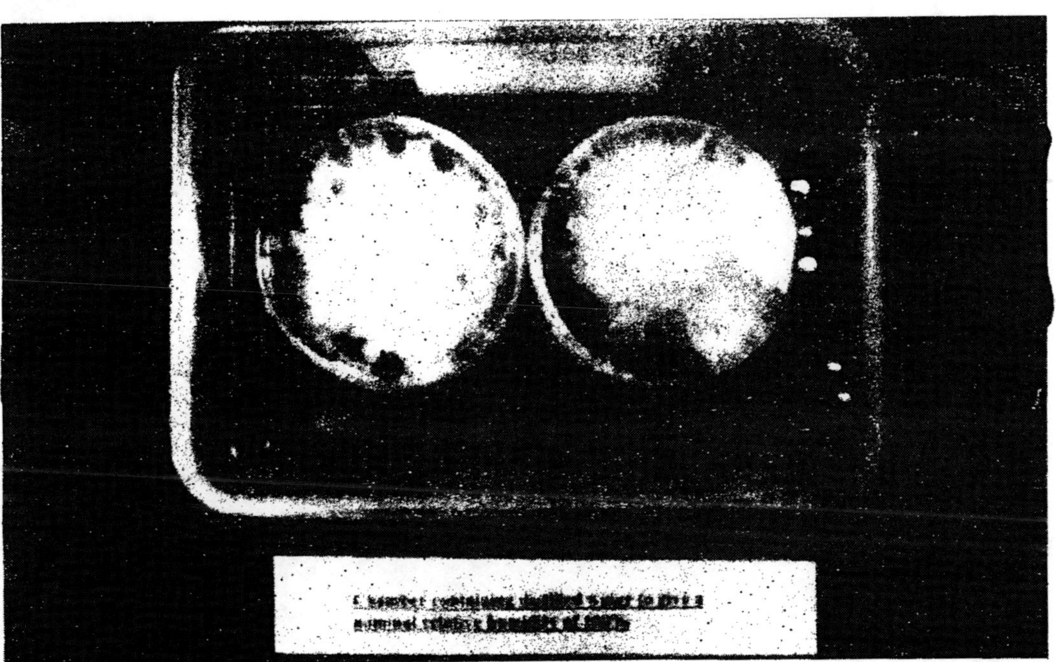

Fig.8: Microcosm experiment to determine the effects of air flow on the growth and decay capacity of *S.lacrymans*

The microcosms have been used to define the minimal humidity requirements of the dry rot fungus as well as investigating the effects of air movements on growth and decay. Using salt solutions to generate designed relative humidity (r.h.) values it has been confirmed that growth of the fungus declined abruptly on reducing r.h. from 100% to 93% but survival of the fungus was not lost until an r.h. regime of 76% was applied (25, 26). In airflow experiments flow rates of 1.5 l/min through the microcosms was sufficient to affect the growth of the dry rot fungus and at 2.5 l/min and above evidence of fungal growth on the surface of infected wood blocks was rapidly lost. However, at rates of 1.5 l/min occasionally marked trophic effects were seen on fungal growth which changed from multidirectional to unidirectional, away from the direction of the air flow, as if the fungus was trying to 'escape' from the stress. Such effects

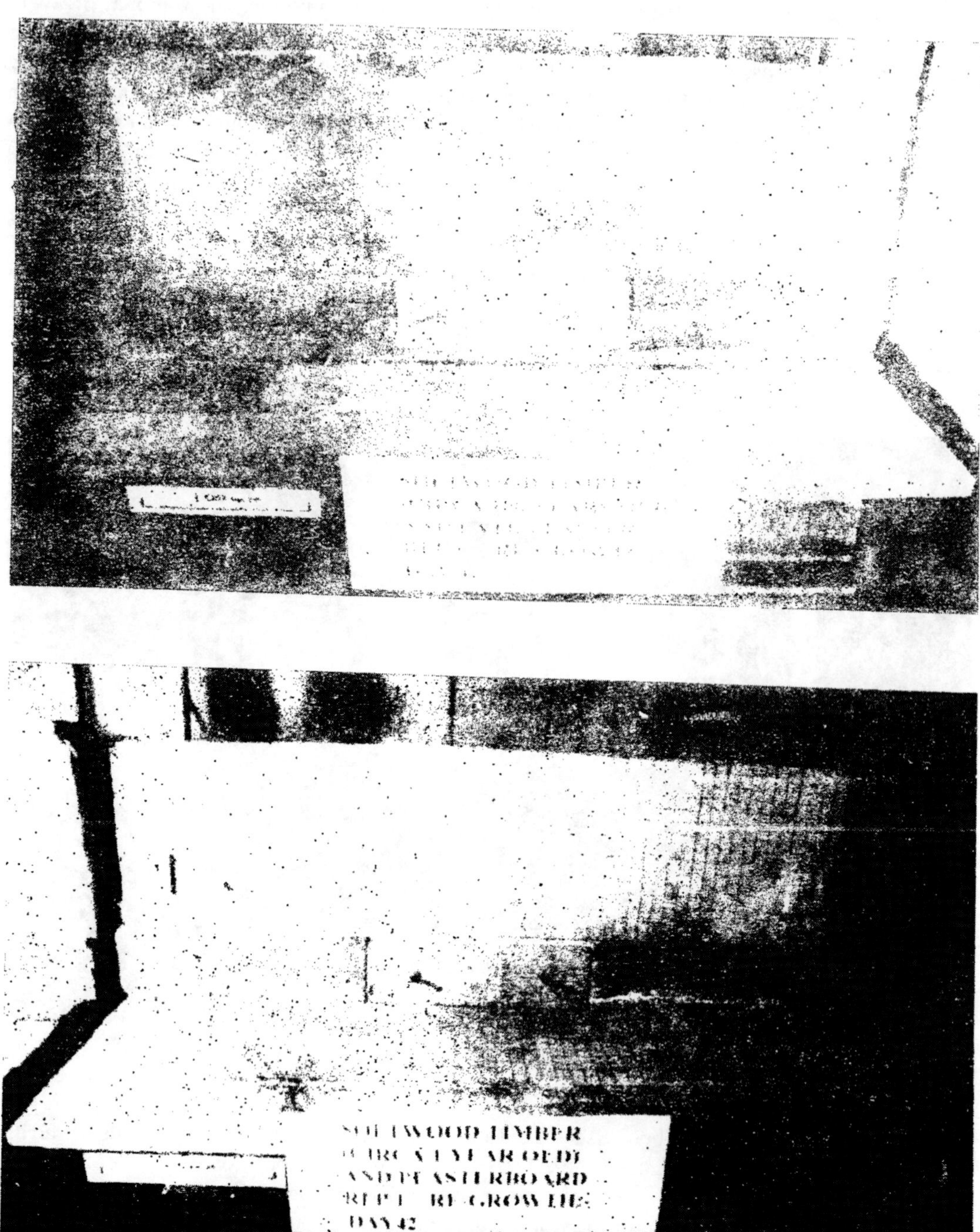

Fig.9: Workshop model; left hand model was constructed with old materials including lath and plaster, the right hand model was constructed with new materials including.plaster board

Fig.10: Full scale model; model of the wall, floor junction of an historic building showing multiple moisture sensors in place. The model is contained within the Dry Rot Research Room at the University of Abertay Dundee - a joint project with Historic Scotland

could have significance if dry rot is to be controlled simply by air movements or insufficiently stringent treatments could possibly stimulate the linear spread of *S.lacrymans*. As with the humidity experiments, regimes that killed the fungus were generally more harsh than those that merely inhibited its growth and decay capacities.

Despite their controllability the microcosm experiments bear little resemblance to field studies, therefore in an effort to make the work more representative of real life situations the models shown in Figs. 9 and 10 have been used. The model shown in Fig. 9 is one of a set of four, two of which were constructed from old building materials (including lath and plaster) and two of more modem materials, all placed in aquarium tanks. Both were moistened naturally, inoculated with the dry rot fungus in the simulated skirting board, sealed with plastic covers to ensure high humidity and incubated at a temperature of around 20°C. Colonisation of the horizontal woody material was rapid and followed by coverage of the vertical plaster surface. In the 'old' model, fungal mycelium rapidly penetrated the lath and plaster, with growth being particularly lush on the wood/ plaster interface of this material. Within 6 to 8 weeks cords of *Serpula lacrymans* had covered the surface of the bricks supporting the model and contact had been made with the water reservoir beneath the models.

After 8 weeks, treatment of the models was initiated. The plastic covers were removed and fans used to dry the interiors of the tanks, however the water reservoirs were retained. Within 24 h effects on the *S.lacrymans* mycelia were apparent and after 7 days there were no signs of active *S.lacrymans*. After a further week when to all appearances the *S.lacrymans* infection had been destroyed, the models were re-sealed and, within a few days, healthy mycelial growth re-emerged from the wood and plaster-work. A second treatment followed, but this time the water reservoir was removed from one pair of models. This time fungal re-growth occurred only in the systems which contained free water. These results indicate the necessity, if environmental control is to be used, of ensuring that no free water is available to allow reactivation of otherwise quiescent *S.lacrymans* (26). The results also indicate: i) why the dry rot fungus has developed a fearsome reputation amongst property owners and ii) that it can be successfully controlled environmentally if all free water is removed from its environment. Results from full scale models (Fig. 10) will be used to confirm, or otherwise, these observations and to optimise the use of the treatments in practice.

Interestingly, fungal contamination of models rarely affected the growth of the dry rot fungus, as also the situation in infected buildings. However, occasionally colonies of probable *Trichoderma* spp. were seen to inhibit growth of *S.lacrymans* within the models. This finding would suggest that a type of integrated control, utilising both environmental and biological antagonism, might be appropriate for the remedial treatment of dry rot. The models described here are currently being used to test this hypothesis.

5. THE RESPONSE OF *SERPULA LACRYMANS* TO INCREASED TEMPERATURE

The sensitivity of *S.lacrymans* to temperatures only marginally above its optimum for growth has been known since the studies of Falck (14); indeed Falck used this sensitivity to distinguish *S.lacrymans* from *S.himantioides*. A temperature range of around 3-26°C is normally considered necessary for growth of *S. lacrymans* (36), though temperatures below 3°C do not necessarily kill the fungus whereas temperatures above around 34°C are definitely lethal. The actual combination of time and temperature required to kill different isolates may vary (43). The relatively high sensitivity of *S.lacrymans* to supraoptimal temperatures, as compared with *S.himantioides*, has been postulated as the reason for the rarity of the former organism in the wild, and specifically its absence outside the built environment in northern Europe (10).

The relative lack of thermotolerance that *S.lacrymans* exhibits has prompted the development of heat treatment systems for killing dry rot in buildings (21), notably in Denmark where the necessity to develop chemical free treatment systems is particularly marked. Such heat treatment will only destroy the mycelial form of *S.lacrymans* since the basidiospores are considerably more resistant to elevated temperatures and can survive, for example for over one hour even at 90°C (16). Thus heat treatment of buildings is simply a way of temporarily inhibiting the growth of *S.lacrymans* and must be complemented with effective methods to inhibit moisture accumulation in wood if decay is to be prevented. However, this is true of any treatment methodology which does not involve rendering wood toxic.

The biochemical results of elevated temperatures were discussed by Languad and Goksoyr (24). Though respiration, in terms of oxygen consumption, was relatively unaffected by increasing temperatures, nucleic acid breakdown was apparent at temperatures above 29°C. The use of electron microscopy revealed that at slightly higher temperatures disruption of nuclei occured and after 1 hour at 37.5°C no nuclei could be observed (23). Whilst these studies indicate that cell death is accompanied by nucleic acid breakdown, they do not explain why this breakdown occurs at relatively low temperatures, which are tolerated by many other higher fungi. In an attempt to study the mechanistic reasons for temperature sensitivity White *et.al.* (43), and Sienkiewicz *et al.* (37) studied the heat shock response of both *S.lacrymans* and *S.himantioides*.

Thermotolerance, as evidenced by the heat shock response, has been documented for many years in a wide range of different species (15), as is normally defined as the transient ability of cells to survive normally lethal heat treatments. Classically it can be induced by incubating cells at sub-lethal temperatures for specific periods of time prior to incubating them at lethal temperatures. During the sub-lethal, or training, regime specific 'heat shock' proteins are produced which afford subsequent protection against the lethal temperature. Whilst originally discovered in response to temperature manipulations heat shock proteins are now known to be part of the cells general defence against environmental assault and, interestingly, heat shock proteins and their derivatives are thought to be involved in cell differentiation (6).

By preincubating *S.lacrymans* for specific times at sublethal temperatures it was indeed possible to generate a transient resistance to elevated temperatures (43). However, significant differences were found between the heat shock responses of *S.lacrymans* and *S.himantioides*. Specifically a very restricted training window was available for the former organism whereas a wider set of training regimes were consistent with survival of *S.himantioides* at elevated temperatures. Thus, prolonged incubation of *S.lacrymans* at sublethal temperatures resulted in a return to normal heat sensitivity, whereas similar incubation of *S.himantioides* merely improved the heat shock response. Other responses found during heat treatment of *S.lacrymans* included appressed and transparent colony formation linked to yellowing (a notable sign of stress of *S.lacrymans* which is also apparent early on during interactions with potential biocontrol agents). A further difference in the heat shock response of *S.lacrymans* and *S.himantioides* relates to the apparent breakdown of hsp 60 (a classic marker protein of the response) in the former organism (37). The significance of this is not understood at present.

These results indicate that the heat shock response of *S.lacrymans* is different from that of the closely related *S.himantioides*, however they do not explain fully the relatively increased heat sensitivity of the former organism. Currently we are investigating other putative defensive mechanisms such as the intracellular production of trehalose, glycerol and other protectants. Studies are also directed towards monitoring the enzymes involved in metabolism of trehalose during stress responses.

6. THE RELATIONSHIP BETWEEN WOODY AND NON-WOODY RESOURCES

The appearance in the wild of *S.lacrymans* at the interface of woody resources and the soil is very consistent with field observations of the organism in the built environment. Indeed the fungus is rarely found in buildings made only of wood - one of many anecdotal observations from the field which is not fully understood at a mechanistic level. For example, *S.lacrymans* is also often found in association with metal wall fixings, perhaps because it is mobilising iron from such fittings for use in the wood decay process.

The link between wood and calcium rich resources such as plaster and concrete in the promotion of the degradative effects of *S.lacrymans* has been discussed by many authors (3, 4, 27). It has been proposed that oxalic acid is produced by *S.lacrymans* for the initial attack of woody components, by allowing the secreted enzymes of the organism to reach the cellulosic polymer come into contact with their target molecule cellulose. The oxalic acid is ultimately neutralised by calcium present in the organisms environment. In the absence of calcium resources *S.lacrymans* acidifies its own environment until growth is inhibited. According to this theory, *S.lacrymans* is unable to modulate oxalic acid production, perhaps because this organic acid is essentially a waste product which is fortuitously used by the organism in the decay process, or because the organism has evolved a physiology sensitive to external acid environments. Calcium prevents excessive acidification of the substrate as calcium oxalate is highly insoluble and essentially causes neutralisation of the oxalic acid. In experiments where *S.lacrymans* was incubated with rock wool, EDaX analyses demonstrated that calcium was removed from the wool fibres and could be detected in the hyphae of the fungus (2). Experimental evidence indicating a role for calcium in the decay process was reported by Palfreyman *et al.* (30), who showed that in the absence of calcium, i.e. when grown on a calcium-free growth medium, *S.lacrymans* was unable to degrade small wood blocks. Other divalent ions, notably magnesium and strontium could mimic the effect of calcium to a limited extent, consistent with the relative solubilities of calcium, magnesium and strontium oxalates.

Surface pH measurements of colonies of *S.lacrymans* growing on a variety of different media indicated that the growth front of the organism was considerably more acidic than older regions of the mycelium, a finding consistent with a buffering effect by the divalent ions. In contrast, *Coniophora puteana*, another wood decay basidiomycete but one which does not appear to require divalent ions to support its decay, also acidified the medium on which it was growing, yet there was no evidence for divalent ion buffering of medium pH changes (29). However, *C.puteana* was unable to degrade wood on the minimum medium used for testing the effects of calcium on *S.lacrymans*, indicating other differences between the decay mechanisms of the two organisms. It is not known at present which organism, *S.lacrymans* or *C.puteana*, is the more typical of the brown rot decay fungi (essentially decay fungi which degrade cellulose of timber resources but not the lignin) though all such organisms are able to decay microcrystalline cellulose and will produce small size chemicals to initially open up the matrix for decay by larger enzymes.

Production of oxalic acid, and hence crystals of oxalate salts, is a relatively common finding amongst microorganisms. Recent studies have demonstrated that, when grown in the presence of various types of sandstone, hyphae of *S.lacrymans* become encrusted with a coating of what is probably calcium oxalate, as well as precipitating large quantities of calcium oxalate crystals (Fig. 11, 12). It seems likely that the organism may also erode the surface of stone work on which it is growing; preliminary electron microscopy evidence apparently supports this hypothesis though longer term experiments, currently underway, are required for confirmation or otherwise (Low, Palfreyman and Young, unpublished observations). In related studies, the significant effect of different building materials on wood decay has been demonstrated (unpublished observations). In these, and the model studies discussed under 'Environmental Control', the most effective building materials in promoting decay are 'lath and plaster' and certain sandstone materials particularly prevalent in older buildings. Lath and plaster consists of a layer of plaster (essentially calcium carbonate) keyed into thin strips of wood built onto a wooden frame. This traditional building material contains exactly the correct complement of components for promoting decay and, if consistenly damp, must be considered a major target for the dry rot fungus.

7. CONCLUSION AND FUTURE WORK

Despite many years of study, the dry rot fungus, *S.lacrymans*, is still a major problem in the built environment in many temperate areas of the world. However, the popular myth that has developed in many countries of Europe, notably the UK, about the organism being highly resistant to remedial treatment is incorrect and has probably developed because of the inability of many companies and. home owners to couple remedial treatment of wood to general repairs to buildings, along with stringent building maintenance.

S.lacrymans exhibits a number of interesting features compared with other wood decay fungi, most notably being its ability to cause total destruction of wood in the built environment. This is in marked contrast to organisms such as *C. puteana* which, whilst they are a problem in buildings, generally cause only localised problems and are considered easy to control. Two biological reasons account for the destructive nature of *S.lacrymans*. First the built environment

Fig.11: Hyphae of *S.lacrymans* growing across the surface of sandstone; encrustation of hyphae with calcium oxalate is clearly seen

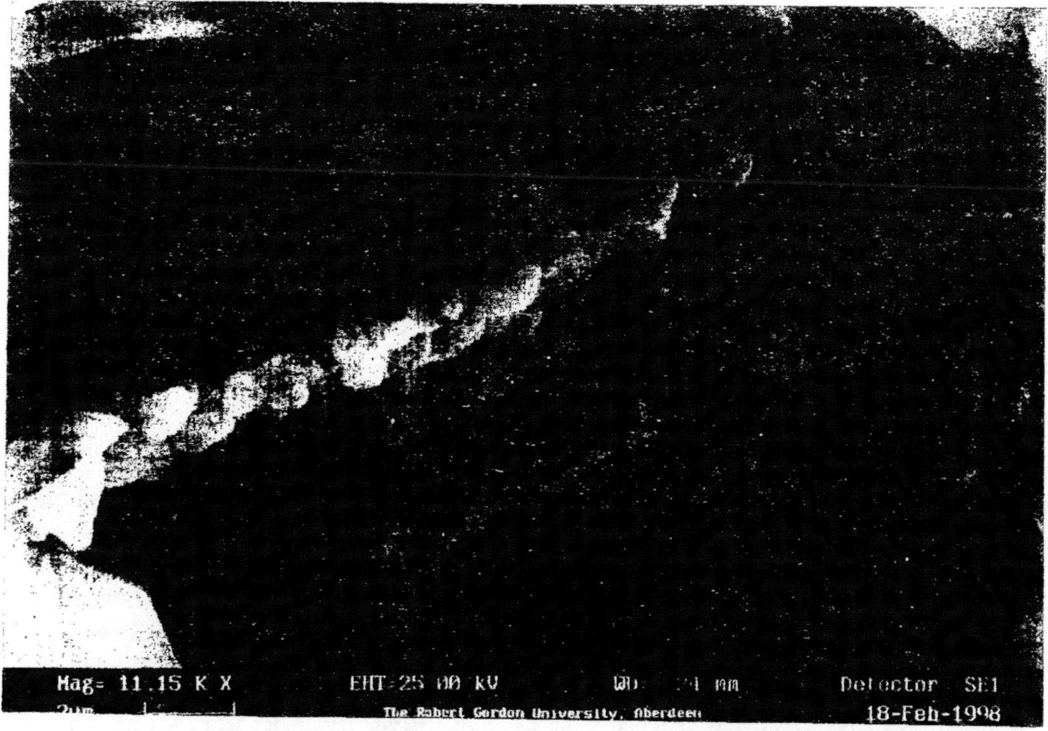

Fig.12: Crystals of calcium oxalate surrounding an individual hypha of *S.lacrymans*

mimics the natural environment of the fungus, in that it offers wood as a nutrient source, non-woody materials to neutralise oxalic acid production and maintain organism viability, relatively stable temperatures and, in the absence of correct maintenance or in the presence of inadequate design or use, a source of free moisture. Second the built environment is unlikely to contain any natural competitors of the dry rot fungus.

Future research on *S.lacrymans*, with the aim of improving its control, should concentrate on understanding the decay process of dry rot and in particular any differences between it, and decay caused by other types of higher fungi. Continued study of the natural environment of *S.lacrymans* is also likely to reveal interesting and useful information on the organism itself, as well as offering the possibility of identifying those parameters, chemical, physical or biological, which restrict the growth of the fungus in the natural environment.

In 1989 Thornton (42) remarked correctly that *S.lacrymans* was a relatively under-researched organism. In addition it is not utilised as a test organism in any of the European standard tests for new wood preservatives - for understandable reasons, these preservatives are primarily designed to combat decay fungi in the outdoor environment. However, for an organism which causes so much destruction in the built environment, this lack of research and absence of standard applied tests is surprising, and even today there are only a very limited number of research groups in the world whose primary interest is *S.lacrymans*. The authors of this article find that fact surprising, not just for economic reasons, but because *S.lacrymans* is an unusual organism, the academic study of which is likely to be rewarding in ways unpredictable at present.

In conclusion *S.lacrymans*, despite its unusual characteristics, is a relatively easy organism to control, simply because of its need for water. We believe that the almost mythical reputation that the organism has gained is totally unjustified and misguided. Indeed we fear that, should the organism become totally removed from the built environment - as many would hope, *S.lacrymans* could well become an endangered species and might even need protection in its natural habitat. The relationship between the isolates found in the natural environment in India and/or the USA and those isolates from buildings in Europe remains a mystery at present, though the opening up of global trade in the 17th and 18th century might have precipitated the migration of *S.lacrymans*. Alternatively, as a prolific sporulator, *S.lacrymans* could have disseminated itself around the world as spores. A third possibility, namely that the organism was once common in the forests of Europe yet in time became restricted to the built environment, perhaps due to climatic changes, cannot be ignored. It is possible that the type of phylogenetic study started by Theodore *et al.* (41) in Australia, and now underway in our own laboratories, may shed light on the unusual geographical distribution of the species.

ACKNOWLEDGEMENTS

Studies in the Dry Rot Research group at the University of Abertay Dundee have been supported by a number of organisations including the Arthur Quarmby Partnership and the UK DTI/EUREKA/EUROENVIRON programme (studies on biocontrol), Historic Scotland (studies on environmental control) the Karl Meyer Foundation (initial studies on ethnobotany for the development of new wood preservatives and support of the Shimla (1997) conferencefand the university itself (heat shock studies). Thanks are also due to Dr. Jagjit Singh (Oscar Faber Applied Research) for his continuing interest in the work of our group and for inviting members of the group on his expeditions to the Himalayas, and to Dr. Maureen Young of the Masonry Conservation Research Group at the Robert Gordon University, Aberdeen, for collaboration on the effects of *S.lacrymans* on sandstone (notably Figs. 11 and 12). Various students have been involved in the work reported here including Natasha Sienkiewicz, Beth Laing and Susan Harper who contributed to the studies on heat shock proteins, divalent ions requirements and tessellated agar tile experiments respectively. Other colleagues at the University of Abertay Dundee who have made contributions to the work reported here include Dr. Eldridge Buultjens (heat shock studies) and Dr Alan Bruce (biological control).

8. REFERENCES

1. Bagchee, K. 1954. *Merulius lacrymans* (Wulf) Fr. in India. *Sydowia* **8** : 191-202.

2. Bech-Andersen, J. 1987. Production function and neutralisation of oxalic acid produced by the dry rot fungus and other brown rot fungi. International Research Group on Wood Preservation Document No. IRG/WP/1330.

3. Bech-Andersen,J. 1991. The dry rot fungus in houses. Intemational Research Group on Wood Preservation Document No. IRG/WP/2389.

4. Bech-Andersen J. 1993. The dry rot fungus and other fungi in houses. Part 2. International Research Group on Wood Preservation Document No. IRG/WP/93 -I 0001

5. Bech-Andersen, J. 1995. The dry rot fungus and other fungi in houses. Hussvamp Laboratoriet Aps. Denmark.

6. Bond, U. and Schlesinger, M.J. (1987). Heat-shock proteins and development. *Adv. Genet.* **24** : 1-29.

7. Bown, J., Samson, W.B., Crawford, J.W., Staines, H.J., Palfreyman, J.W., Ritz, K., Sturrock, C.J. and White, N.A. 1998. Modelling the spatio-temporal dynamics of fungal communities using a stochastic cellular automaton (in press).

8. Bravery, A.F. and Grant, C. 1985. Studies on the growth of *Serpula lacrymans* (Schumacher ex Fr.) Gray. *Material und Organismen* **20**: 171-191.

9. Bruce, A. 1997. Biological control of wood decay. In : *Forest Products Biotechnology* (eds., A., Bruce, and J.W. Palfreyman), Taylor and Francis, London, pp. 251-266.

10. Cartwright, K.St.G. and Findlay, W.P.K. 1958. *The Decay of Timber and its Prevention.* HMSO, London.

11. Coggins, C.R. 1977. Aspects of the growth of *Serpula lacrymans* the dry rot fungus. PhD Thesis, University of Liverpool.

12. Cooke, W.B. 1957. The genera *Serpula* and *Merulipora. Mycologia* **49**: 197-225.

13. Doi, S. and Yamada, A. 1992. Antagonistic effects of three isolates of *Trichoderma* spp. against *Serpula lacrymans* (Fr.) Gray in laboratory soil inoculation tests. *Soc. Antibact. Antifungal Agents, Jap.* **20**: 345-349.

14. Falck, R. 1912. Die Merliusfaule des Bauholzes Hausschwammforschungen, Vol 6. (pp. 1-422). Dry rot in timber. Translated as an extended abstract by U. Hornung and D.H. Jennings (1980) Bull. Brit. Mycol. Soc. **14**: 119-130.

15. Hahn, G.M. and Li, G.C. 1982. Thermotolerance and beat shock proteins in mammalian cells. *Radiation Research* **92**: 452-457

16. Hegarty, B. and Schmitt, U. 1988. Basidiospore structure and germination of *Serpula lacrymans* and *Coniophora puteana.* International Research Group on Wood Preservation Document No. IRG/WP/1340.

17. Hennebert, G.L., Boulager, P. and Balon, F. 1990. La Merule. Editions Ciao, Brussels.

18. Hornung, U. and Jennings, D.H. 1980. Dry for in Timber. *Bull. Brit. Mycol. Soc.*14: 119-130.

19. Huttennan, A. 1991. Richard Falk, his life and work. In : *Serpula lacrymans: Fundamental Biology and Control Strategies* (eds. D.H. Jennings and A.F. Bravery) John Wiley and Sons, London. pp. 193-206.

20. Jennings, D.H. and Bravery, A.F. 1991. *Serpula lacrymans*: *Fundamental Biology and Control Strategies* (eds. D.H. Jennings and A.F. Bravery) John Wiley and Sons, London.

21. Koch, A.P., Kjerulf-Jensen, C. and Madsen, B. 1989. New experiences with dry rot in Danish. houses, heat treatment and viability tests. International Research Group on Wood Preservation, Document No. IRG/WP/1423.

22. Kotlaba, F. 1992. Finds of *Serpula lacrymans* in nature. *Ceska Mykology* **46**: 99-104.

23. Languad, F. 1972. The effects of supraoptimal temperatures on the fine structure of *Merulius lacrymans* (Jacq.) *FR. J. Gen. Microbiol* **70**: 157-159.

24. Languad, F. and Goksoyr, J. 1967. Effects of supraoptimal temperatures on *Merulius lacrymans. Physiol. Plantarum* **20**: 702-712.

25. Low, G.A., Palfreyman, J.W. , White, N.A., Staines, H.J. and Bruce, A. 1997. Preliminary studies to assess the effects of aeration and lowered humidity on the decay capacity, growth and survival of the dry rot fungus *Serpula lacrymans.* IRG Wood Preservation, Document No. IRG/WP/ 97-10208.

26. Low, G.A., Palfreyman, J.W., White, N.A. and Sinclair, D.C.R. 1998. Development of model systems for investigation of the dry rot fungus *Serpula lacrymans*: use for the analysis of the environmental sensitivity of the organism. *Holzforschung* (in press).

27. Paajanen, L.M. and Ritschkoff, A.C. 1991. Effects of mineral wools on growth and decay of *Serpula lacrymans* and some other brown rot fungi. International Research Group on Wood Preservation Document No. IRG/WP/1481.

28. Palfreyman, JW., White, N.A., Score, A.J., Phillips, E.M. and Buultjens, T.E.J. 1994. The effects of environmental conditions on the growth and decay capacity of the dry rot fungus *Serpula lacrymans.* Proceedings 1st Int. Conf. on the Development of Wood Science/Technology and Forestry. Missenden Abbey, UK.

29. Palfreyman, J.W., White, N.A., Buultjens, T.E.J. and Glancy, H. 1995. The impact of current research on the treatment of infestations by the dry rot organism *Serpula lacrymans. Internat. Biodeteri. Biodegrad.* **35**: 369-395.

30. Palfreyman, J.W., Phillips, E. and Staines, H. 1996. The effect of calcium ion concentration on the growth and decay capacity of *Serpula lacrymans* and *Coniophora puteana. Holzforschung* **50** : 3-8.

31. Pegler, D.N. 1991. Taxonomy, identification and recognition of *S.lacrymans*. In *: Serpula lacrymans*: *Fundamental Biology* and *Control Strategies*, (eds. D.H. Jennings and A.F. Bravery) John Wiley and Sons, London. pp. 1-7.

32. Score, A. and Palfreyman, J.W. 1994. Biological control of the dry rot fungus *Serpula lacrymans* and *Trichoderma* spp. Effects of media constraints on interaction and growth rates. *Internat. Biodeter. Biodegrad.* **33**: 115-128.

33. Score, A.J., Palfreyman, J.W. and White, N.A. 1997. Extracellular phenoloxidase and peroxidase enzyme production during interspecific fungal interactions. *Internat. Biodet. Biodegrad.* **39**: 225-233.

34. Score, A.J., Bruce, A. and Palfreyman, J.W. 1998. The biological control of *Serpula lacrymans* by *Trichoderma* species, Use in the treatment of dry rot infected wood. *Holzforschung* (in press).

35. Seehann, G. and Hegarty, B. 1988. A bibliography of the dry rot fungus, *Serpula lacrymans.* International Research Group on word Preservation, Document No. IRG/WP/1337.

36. Segmuller, J. and Walchli, O. 1981. *Serpula lacrymans* (Schumacher ex Fr.) S.F. Gray. In : *Some Wood Destroying Basidiomycetes,* (ed. R. Cockcroft) Vol. 1 of a collection of monographs, International Research Group on Wood Preservation, Boroko, Papua, New Guinea. pp. 141-159.

37. Sienkiewicz, N., Buultjens, T.E.J., White, N.A. and Palfreyman, J.W. 1997. *Serpula lacrymans* and the heat-shock response. *Internat. Biodeter. Biodegrad.* **39**: 217-224.

38. Singh, J. 1994. *Building Mycology.* Spon, London.

39. Singh, J., Bech-Andersen, J., Elbourne, S.A., Singh, S., Walter, B. and Goldie, F. 1993. The search for wild dry rot fungus (*Serpula lacrymans*) in the Himalayas. Part 1, expedition and discovery. *Mycologist* **71** : 124-131.

40. Singh, J., Bech-Andersen, J., Elbourne, S.A., Singh, S., Walter, B., Goldie, F., Palfreyman, J.W. and Koch, A.P. 1994. Origin and spread of dry rot in Europe. A comparative physiology, morphology, genetics and ecology. Proc. I st Int. Conf on the Development of Wood Science/ Technology and Forestry. Missenden Abbey, U.K.

41. Theodore, M.L., Stevenson, T.W., Johnson, G.C., Thornton, J.D. and Lawrie, A.C. 1995. Comparison of *Serpula lacrymans* isolates using RAPD-PCR. *Mycological Research* **99**: 447-450.

42. Thornton, J.D. 1989. Why don't more people work with *Serpula lacrymans* nowadays? A discussion of some of the different approaches to experimenting with this unique fungus. International Research Group on Wood Preservation Document No: IRG/WP/1383.

43. White, N.A., Buultjens, T.E.J. and Palfreyman, J.W. 1995. Induction of transient thermotolerance in *Serpula lacrymans* and *Serpula himantioides* following exposure to supraoptimal (sublethal) temperatures. *Mycological Research* **99**: 1055-1058.

44. White, N.A., Low, G., Singh, J., Staines H. and Palfreyman, J.W. 1997. Isolation and study of wild *Serpula lacrymans* from the Himalayan forest. *Mycological Research* **101**: 580-584.

45. White, N.A., Sturrock, C.J., Ritz, K., Samson, W.B., Bown, J., Staines, H.J., Palfreyman, J.W. and Crawford, J. 1998. Interspecific fungal interactions in spatially heterogeneous systems. *FEMS Microbiology Ecology* (in press).

46. Wulfen, F.X. von 1981. In : *Misicellanea Austriaca ad Botanicam, Chemiam, et Historiam Naturalem Spectantia.* Vol. 2, (ed., N.J. Jacquin), Vindobonae, pp. 98-113.

18 Microbial Technology in the Biocontrol of Post Harvest Diseases

N.K. Mehrotra and Neeta Sharma

CONTENTS

ABSTRACT 310

1. INTRODUCTION 310
2. HAZARDS FROM SYNTHETIC PESTICIDES 310
3. ALTERNATIVES TO CHEMICAL FUNGICIDES 312
4. PRINCIPAL BIOLOGICAL CONTROL METHODS 312
5. EFFECT OF PRESENT POST-HARVEST MANAGEMENT PRACTICES IN EPIPHYTIC MICROFLORA 313
6. MICRO-ECOLOGY OF THE SYSTEM 314
7. STRATEGIES FOR BIOLOGICAL CONTROL 315
8. MECHANISM OF BIOLOGICAL CONTROL 316
9. MANAGEMENT ACTIVITIES 317
10. BIOLOGICAL CONTROL AGENTS FOR POST HARVEST PATHOGEN 318
11. PRESENT STATUS OF THE BIOLOGICAL CONTROL OF POST-HARVEST DISEASES 319
12. OBSTACLES AND RECOMMENDATIONS 324
13. REFERENCES 326

Advances in Microbial Biotechnology
J.P. Tewari, T.N. Lakhanpal, Jagjit Singh, Rajni Gupta & B.P. Chamola (eds.)
APH Publishing Corporation, New Delhi - 110 002, India.

ABSTRACT

The use of "microbial technology" as an alternative to the synthetic pesticides has its roots in the total human health and environmental safety issues. The phenomenon of pest resurgence, pesticide resistance, secondary pest outbreak are mainly among the several other factors involved in providing motivation for the development of biological control of post-harvest diseases. Biological control is accomplished either by conservation or by augmentation. The habitat characteristics of the host pathogen and environment should be considered before utilizing biocontrol procedures. r-strategists species of pathogen can grow quickly and colonize new resources rapidly, with minimal nutrient and environmental restriction. Thus, these are better biocontrol agents than K-strategists. Antibiosis competition are the main process by which bio agents act on the intact surface as well as on the wound size. The ultimate potential of antagonistic microorganisms will depend on its effectiveness and its compatibility with current and holding practices.

1. INTRODUCTION

The use of "microbial technology" in the field of biological control is approaching a turning point in history. During the past decade, there has been a series of economic factors and scientific advances which has made the commercialization of biological control agents more attractive to the pesticide industry. The advent of biotechnology has provided an array of new tools to create added value and reduced costs of the biological control products. Accordingly, biological control is currently receiving a great deal of commercial attention as an alternative to the chemical pesticides, in the area of "Market-Pathology". The losses incurred by microorganisms is quite heavy (Table 1).

The promotion of these "living pesticides/fungicides" as an alternative to the chemicals has its roots in the total human health and environmental safety issues. The toxicological problems, real and potential, which can be associated with chemical pesticides, have led to the public fear of chemicals and extensive safety testing for their registration; there are, however, several other compelling incentives to develop the alternate biological pesticides.

2. HAZARDS FROM SYNTHETIC PESTICIDES

2.1 Pest Resurgence

Routine use of broad spectrum chemicals to control various pests and pathogens, was sometimes followed by quick return of such harmful microbes at menacing levels. This phenomenon is termed as pest resurgence. The rebound of pathogens in such situations is due to the more severe and prolonged damage done by the chemicals to the natural enemies of pests/pathogens than to the incitant itself. Once the residues of the pesticides have dissipated, the pathogens which have survived quickly begin to reproduce and exhibit increased survival rates.

Table 1: Crop losses in developing countries estimated by the US National Research Council (2)

Crop	Estimated loss of total crop (%)
Apples	14
Avocados	43
Bananas	20-80
Cabbage	37
Carrots	44
Cassava	10-25
Cauliflower	49
Citrus	20-95
Grapes	29
Lettuce	62
Papayas	40-100
Plantain	35-100
Potatoes	5-40
Onions	16-35
Raisins	20-95
Stone fruits	28
Sweet potatoes	35-95
Tomatoes	5-50
Yams	10-60

2.1.1 Secondary pest outbreak

Another phenomenon associated with the use of broad spectrum fungicides has been the outbreak of such species which are not usually pathogenic. These species termed as "secondary pests" become aggressive due to the destruction of natural enemies which mold these species at innocuous levels in the absence of pesticides/fungicides. The elimination of target pathogens along with the beneficial organisms leads to the better survival of secondary or pesticide created pests.

2.1.2 Pesticide resistance

Resistance to pesticide develops in a population when certain individuals possess genes which allow them to better avoid or survive contact with the chemicals. Treating such population with a pesticide confers differentially greater fitness to such tolerant individuals, and the frequency of the resistant genotype incitant when the tolerant individuals reproduce. Pesticide resistance may develop in species in which the surviving individuals remain together.

2.1.3 Other factors

Besides these, several other factors are also involved in providing motivation and incentives for the development of alternative pesticides. In 1988, the U.S. Congress mandated that all the pesticides registered under less comprehensive older data requirements were required to undergo a process using new guidelines. As a result of this, approximately 32,000 of 45,000 active product registrations were cancelled. Extensive safety testing is required for the registration of new pesticides. It is estimated that the development and data requirement costs for the newer chemicals ranges from 40-50 million dollars/product; however, the safety data requirements (SDR) for the registration of the biological pesticides are less extensive than for the chemicals. Accordingly, the registration cost estimate for the biological control agents ranges from 0.5-1 million dollars.

With an increasing interest displayed by workers throughout the world in the possibility of developing practical workable procedures for the utilisation of microorganisms to suppress plant pathogens, it is quite proper that a review to cover the recent advances in the use of micro-organisms native or induced, to control the post-harvest diseases of fruits and vegetables becomes necessary and our this intention is also being presently attempted.

Here groups of organisms which are antagonistic to pathogens of perishables are touched briefly. These organisms vary both in their innate ability to suppress the pathogens and in their ability to thrive and compete in different environments; consequently, the selection of an organism or organisms for any particular biological control program will be a compromise among such parameters and ability.

3. ALTERNATIVES TO CHEMICAL FUNGICIDES

These developments have lessened our ability to control post-harvest diseases and have stimulated exploration for alternative technologies. A considerable effort has been made over the past decade to determine the potential of biological control as an alternate to the chemical control of post-harvest diseases.

We have few commercial successes to point towards the biological control of plant diseases. However, the post-harvest environment provides a more favourable milieu for the application of biological control agents than in the fields. While temperature and relative humidity are more stable and controllable, exposed surfaces being delimited in harvested commodities allow more effective targeting and application of biological control organisms.

Another short coming of biological control agents has been their limited range of effectiveness. Because of this limitation, it is often economically not feasible to develop such agents commercially, but rather surprisingly the antagonists discovered for the biological control of post-harvest diseases have a wider range of activity.

4. PRINCIPAL BIOLOGICAL CONTROL METHODS

Biological control involves the use of natural enemies to suppress pest/pathogen population densities to levels lower than would other wise be. Organisms for the biological control of post-harvest diseases can be used in various ways, but maximum attention has been given to their conservation and augmentation in a particular environment rather than to their importation and addition of new species, as is often done for insect or weed control. The choice of these approaches is, in part, because of the presence of a diverse set of microbes already associated with plants, which provide ample opportunity for the development of resident species as the competitors or antagonists to the pathogenic organisms. Two major methods exist for the utilisation of such natural enemies, i.e., conservation and augmentation.

4.1 Conservation

Conservation as a form of biological control is the study and application of the extent to which natural enemies are able to realise their potential to suppress the pathogens. Fundamental to the biological control through conservation is the assumption that natural enemy species already exist locally and they have the potential to effectively suppress the pest, if given an opportunity.

Biological control of the phyto-pathogens through conservation is accomplished either by preserving the existing microbes which attack or compete with the pathogens or by enhancing the conditions for their survival and reproduction at the expense of the pathogenic organisms. Conservation is applicable in situations where microorganisms important in limiting disease-causing organisms already exit in the "microcosms". They may be conserved by avoiding the practices which negatively affect them.

Negative influence that harm natural enemies are the chemical pesticides having broad spectrum and long residual action (11). Positive form of conservation includes efforts to enhance the requisites which are needed by the natural enemies to flourish in the system, and which may involve creating or maintaining physical refuges, and moderating physical conditions.

4.2 Augmentation

When natural enemies are missing or late to arrive or simply too scarce to provide control then the method of augmentation is employed. Biological control through augmentation is based on mass-culturing antagonistic species and adding them to the system. The natural enemy is already present in the system, but at lower numbers or in locations different from that are desired. The purpose of augmentation is to increase the numbers or modify the distribution of the antagonist in the system. The activity of augmenting microbial agents is also referred as "introduction".

Augmentation of antagonists has two approaches. First is the direct augmentation at potential infection sites with organisms antagonistic or parasitic to the pathogens, thus being directly responsible for the disease suppression. The second approach is to inoculate the plants with non-pathogenic organisms which prompts up general plant defense mechanisms against infection by the pathogens (induced resistances).

Inoculative releases consist of small amounts of the inoculum (antagonist) which will increase and limit the pathogen population. However, on the other hand, in the inundative releases, large amount of initial inoculum (antagonist) is applied, with the expectation that the control will result directly from this large population.

5. EFFECT OF PRE-AND POST-HARVEST MANAGEMENT PRACTICES ON EPIPHYTIC MICROFLORA

Microflora on the surfaces of fruits and vegetables is undoubtedly affected by pre-harvest, harvest (transition) and post-harvest activities. However, environmental conditions in the field (pre-harvest) are different from conditions in storage (post-harvest).

Primarily in the fields, the environment is less controllable and cultural practice vary. Microflora develop in response to these factors over the growing season and area in some "state" by harvest time. The impact of this state to the development of storage diseases is usually known. This state includes both non-pathogenic and pathogenic microbes. It may also include latent infections. At harvest (the transition period), fruits vegetables are subjected to handling and processing. There may be bruising and wounding, washing and dipping, and heating and cooling during the transition to storage. This harvesting and processing transition has considerable impact on the epiphytic microflora and their physical-chemical environment; it provides an opportunity to intentionally modify the microflora and their environment in numerous ways prior to storage. Once the commodity moves to storage, post-harvest environmental conditions which drive the growth or development of microflora are regulated and are often fairly constant. This provides additional opportunities to manage (44). Pesticides used, fertilizers applied and also the nutritional makeup of the commodities may selectively promote certain microorganisms.

The natural epiphyllous microflora which includes fluorescent pseudomonads and lactic acid-type bacteria are altered, quantitatively and qualitatively when standard pesticide applications are made. Fertilizers and nutrients applied to plant surfaces as well as soil conditions may affect epiphytic microbial populations.

Post-harvest treatments profoundly affect the epiphytic microbial populations. Their direct effect on pathogens is well documented but, how these applications act upon other accompanying micro-organisms, is still not very clear. The dumping of commodities into biocidal solutions and spraying them with waxes, anti-oxidants and fungicides could also eliminate a number of beneficial micro-organisms, besides the pathogenic ones. Dumptanks are important source of pathogens for causing infection in perishables.

Storage conditions affect the development of post-harvest pathogens, antagonists and other microorganisms on plant surfaces. It was observed that *Botrytis cinerea* Pers caused severe damage in ice bank coolers to carrots, but was less of a problem under cold room storage conditions (44).

6. MICRO ECOLOGY OF THE SYSTEM

To understand the principles that apply to the biological control of plant pathogens, we must consider the habitat characteristics of the system at the level of the pathogens and the agents used to control them.

Various forms of competition in such environments are important in the ability of any particular organism to increase in numbers and consequently reduce the numbers or activity of the other organisms, including plant pathogens (1). There may be microbial competition during the initial establishment on a fresh resource that was not previously colonised by the micro-organisms, or that may be after initial establishment, to secure enough of the limited resources present to permit their survival and eventual reproduction.

6.1 Types of Antagonist

Microorganisms show many traits, which may characterize them particularly either at the colonisation phase or at the subsequent phases of competition. Ruleral species (r-Strategists) have a high reproductive capacity. In contrast to these r-Strategists, species found in more stable situations face competition for the space and limited resources. These organisms, termed as K-Strategists, become more dominant as the community matures and become more crowded.

Plant pathogens are spread through this r-K range of characteristics. These are opportunistic pathogens that are able to attack the young, weakened, or predisposed plants, but may be poor competitors *(Botrytis, Pythium).* There are pathogens that can produce antibiotics (such as *Penicillium)* that can inhibit competitors. It is important to understand the ecology of the target organism before one can effectively choose as to consider, what biocontrol strategy might be the most effective.

Similarly, the properties of an effective biological control agent will depend on the setting in which it is intended to function. A frequent need, therefore, is of a control agent that has the characteristics of an r-Strategist (5), which can grow quickly and colonise new resources rapidly, with minimal nutrient and environmental restrictions, function well in disturbed environments and should be able to survive on the plant near to the pathogen inoculum or the source or site of infection. Thus r-Strategist biocontrol agents can be considered as an approximate equivalent of the protectant fungicides, being in the infection court before the infection cycle of the pathogen starts. However, at sites where pathogen has already invaded the host tissue, a more competitive species will be required.

7. STRATEGIES FOR BIOLOGICAL CONTROL

Different biological control strategies are identified depending upon the life cycle of the pathogen involved (16). The efficacy of any biocontrol agent is mainly determined by the intensity of the interaction as well as by the duration of the interaction in any of the following strategies: i) microbial protection of the host against infection; ii) microbial interference with colonisation and sporulation of a pathogen; and iii) microbial interference with the surviving structures of a pathogen.

7.1 Prevention of Infection

The success in terms of the microbial prevention of necrotrophic infection process is based on the timely inoculation of the host with the antagonistic microorganisms. Properly timed inoculation with an antagonist is relatively easy to accomplish in a bioassay. Therefore, it is not surprising that the best examples of microbial protection against infections are those where man-made wounds are involved, such as post-harvest fruit decay caused by wounds that occur to be infected during harvesting and packaging procedures.

Laboratory studies on antagonistic microorganisms that are able to prevent rot in stored stone, pome, citrus, grapes and mango fruits (26, 27, 35, 36) have generated a number of interesting biological control agents that have shown efficacy on a semi-commercial scale (16). Researches under packing house conditions have demonstrated that effective and reliable control of post-harvest diseases could be achieved by applying antagonistic yeasts in combination with thiabendazole used at 10% of the normal dose (13).

Another example of man made wounds are the bruises made during the picking operations on cucumber and tomato which represent an important entry point for the necrotrophic grey mould fungus *Botrytis cinerea* Persoon Fries. The selection of antagonistic microogranisms during bioassays was based on the use of cucumber stem pieces which resulted in the discovery of promising yeast like biological control agents that are equally effective as Trichodex or standard fungicide under green house conditions. The high rate of success obtained in the microbial wound protection can probably be explained by the similarity between the conditions prevailing during the bioassays and those in practices.

The efficacy of Trichodex in vineyards has been demonstrated in field experiments (28), mainly under warmer climatic conditions. With a view to optimise reliability, the use of Trichodex has to be included in the frame of a more general IPM programme, thus achieving better levels of control of *B. cinerea* in various vegetables and fruit crops and allowing for a significant reduction in the inputs of chemicals and, as a consequence, in the development of fungicide-resistance which has already been well documented for this pathogen.

Many different mycoparasitic fungi have been described which may interfere with the sporulation capacity and consequently suppress the disease. The mycoparasitic species *Ampelomyces quisqualis* is now available in the USA under the trade name of "AQ 10" for use against powdery mildew of grapes (19). *Sporothrix flocculosa* Traquair is used against powdery mildew diseases (3). It was found that the relative humidity above 80% favoured biological control almost as effectively as the standard chemical fungicides.

8. MECHANISMS OF BIOLOGICAL CONTROL

There are several different ways in which a microbial biological control agent can operate against a targeted pathogen (14).

8.1 Antibiosis

Biological control agents, that owe their effectiveness to the elaboration of the fungicidal substances in the wound sites, are subject to exactly the same problems as of the synthetic chemical fungicides and possibly a few more that are unique to the living microorganisms. Published properties of such compounds should pose concern for their use on foods, even if the biocontrol formulations contained only the washed microbial cells which generate the antibiotics only in the colonized wounds. The application of living cells of *P. cepacia* and related pseudomonads to fresh fruits and vegetables would require a detailed evaluation of their pathogenicity to the immunosuppressed humans.

8.1.1. Competition

Recent investigations on the biological control of post-harvest diseases have emphasized the use of microorganisms that are not antagonistic to the pathogen *in vitro,* but which appear to control the disease by heavily colonizing the wound and competing with the pathogen for nutrients and space. Some evidence indicates that these microorganisms may elicit defense mechanisms of the host as well. The yeast *Aureobasidium* and *Debaryomyces* have been investigated in this connection. The application of *Debaryomyces* and related yeasts is encouraged by the fact that these microorganisms are frequently associated with foods and might be more acceptable to consumers than exotic bacteria that could contaminate treated produce with xenobiotics. Chalutz and Wilson (9) found that the treatment of grape fruit eradicative action against *Penicillium digitatum,* probably reflected the time lag required for *Debaryomyces* to become effective disease control agent only if applied simultaneously with the plant pathogen; however, the treatment failed when the yeast was applied seven hours after the inoculation of the fruit with *P. digitatum* spores.

8.1.2 Wound site

The picture is now emerging that such biocontrol agents will be the most useful against wound pathogens, since these sites are most readily colonized by the selected bacterial and fungal antagonists. Such biological agents must increase significantly in cell mass before they inhibit the pathogen, it is unlikely that the washed cell formulations can be depended upon for the eradication of the established wound infections or quiescent infection on peaches, tropical fruits or citrus. Chalutz and Wilson (9) noted that *Debaryomyces* was more resistant to thiabendazole and imazalil than *P. digitatum* suggesting that a chemical fungicide might be added to the biocontrol formulation to increase its eradicative action and to reduce the selection pressure for the emergence of pathogens biotypes resistant to either the biocontrol agent or the chemical fungicide.

9. MANAGEMENT ACTIVITIES

The epiphytic microflora live on surfaces. If one wants to manage such microflora to control the post-harvest diseases, some management activities which can have an impact on or alter the microflora have been identified (Table-2). The activities are selected because they are broad and inclusive, covering a range of sub-activities and listed according to-the time when they ought to be applied. Such activities are readily recognized as usable inputs or actions. They can be selected and implemented in many instances without a thorough knowledge of their total impact on future occurrences such as the development of post-harvest disease. Presently, our primary interest is focused on "select and apply biological and/or chemical agents". Such a management activity, perhaps more than the others identified, requires a considerable research effort. It should be emphasized that successful disease control depends on understanding of all the listed management activities and using them concertedly. There is, thus a need for research to support and develop such management activities. The knowledge, thus generated could be used to integrate data and research activities to develop computer based models for post-harvest disease management.

Table 2 : Management activities for biological control of post-harvest diseases

Pre-harvest activities

Select :

- Varieties
- Cultural practices - seed treatment, tillage, pruning
- Chemicals - pesticides, fertilizers
- Biological agents

Harvesting/Processing activities

Select :

- Harvest time, environmental conditions
- Harvesting, handling, culling methods
- Processing methods
- Chemical agents
- Biological agents

Post-harvest activities

Select :

- Environment-temperature, humidity, atmospheric composition
- Time in storage
- Chemical agents
- Biological agents

9.1 Research Activities Required to Develop Management Activities for Biological Control of Post-harvest Diseases

Most researches on the biological control of diseases has centred on the rapid procedure (bioassays) for selecting and thus introducing bio-control agents. This is usually a selection of one antagonist to control one pathogen referred to as the "silver bullet" approach (39). The result is growing recognition that biological control is dependent upon a thorough knowledge of eco-systems and their management.

Research activities which are required to generate knowledge upon which to develop management activities are listed in order as under :

(i) Select the commodity

(ii) Select the disease target on the commodity site.

(iii) Study the microbial ecology of the infected site.

(iv) Determine the make up of the microfloral populations.

(v) Study the impact of environment of the microflora.

(vi) Study the impact of pre-harvest, harvest and post-harvest activities on the disease development.

(vii) Study the impact of physical, chemical and biological inputs on the disease development.

(viii) Construct models to guide the management activity.

10. BIOLOGICAL CONTROL AGENTS FOR POST-HARVEST PATHOGENS

Biological control of pathogens causing post-harvest diseases of perishables is fundamentally a matter of ecological management of community of organisms, as in all biological control. In the case of plant pathogens, however, there are two distinctions. First, the ecological management occurs at the microbial level, typically in the microcosms of the ecosystem such as at the leaf, root and fruit surfaces (1). Second, the biological control agents include competitors, as well as the parasites (Table-3)

Diseases of the fruits are caused by a wide variety of pathogens. Because of this diversity, the antagonist species which negatively affect plant pathogens and the mechanisms by which they accomplish their beneficial action are also quite different.

The fruit diseases manageable through biological control include diseases in plant fruits and their post-harvest diseases. One of the first systems developed was against *Botrytis cinerea* Persoon Fries in vineyards, where sprays with spore suspension of *Trichoderma harzianum* Rifai were found effective in suppressing the disease incidence. Several organisms, including *Gliocladium roseum, Penicillium* sp., *Trichoderma viride, Colletotrichum gleosporioides, Chaetomium globosum* were as effective as fungicides in suppresing *B. cinerea* on strawberies, and *Ulocladium chartarum* on apples (26, 29, 34, 40).

Post-harvest diseases which can be responsible for 10-15% loss of the produce (22) have received considerable attention. Numerous reports deal with the suppression of post-harvest diseases in fruit crops (5, 27, 43) by such organisms as species of *Bacillus, Debaryomyces, Penicillium, Pseudomonas* and *Trichoderma;* competitive *Cladosporium herbarum* (Persoon Fries) Link gives control as good as the commercial fungicides.

Table 3 : Control of the post-harvest diseases through epiphytic microflora

Host	Disease	Antagonist	Reference
Rutaceae			
Citrus (Orange, lemon grape fruits)	Green, blue and sour rot	*Pichia guilliermondii*	(12)
		Debaryomyces hansenii	(26, 27)
Rosaceae			
Fragaria ananassa (Strawberry)	Rhizopus rot	*Trichoderma*	(41)
Malus pamila (apple)	Blue, gray and Rhizopus rot	*Pseudomonas cepacia*	(20, 21)
		Cryptococcus spp.	
	Botrytis rot	*Debaryomyces hansenii*	(43)
	Penicillium rot		(38)
		Bacillus subtilis	
Prunus armentane (apricot)	Brown rot		(30)
		Bacillus subtilis	
P. avnan (cherry)	Alternaria and brown rot		(42)
		Bacillus subtilis	
P.domestica (plum)	Brown rot		(45)
		Bacillus subtilis	
P. persicae (peaches)	Brown rot		(30)
Solanaceae			
Lycoperricon esculentum (tomato)	Alternaria rot	*Debaryomyces* sp.	(36)
	Botrytis, Rhizopus rot	*Pichia gulliermondii*	(7)
Vitaceae			
Vitis vinifera (grapes)	Rhizopus and Botrytis rot	*Hansemaspora uvarum*	(4)
		D. hansenii	(37)

11. PRESENT STATUS OF THE BIOLOGICAL CONTROL OF POST-HARVEST DISEASES

In the past ten years, a large body of information on the use of antagonist, both naturally occurring native microflora and artificially induced to control post-harvest diseases clearly demonstrates the potential of this emerging technology (Table-3).

Recent post-harvest pathology research has established the suitability of using certain naturally occurring epiphytic yeasts of the genus *Cryptococcus* which are effective in preventing the post-harvest disease development in apples, pears and cherries (33). This psychrophilic, microaerophilic and medically-unimportant yeast genus rapidly colonises and survives in wounds under a variety of temperature and atmospheric conditions.

Pichia guilliermondii inhibited green, blue and sour rot of citrus fruits and *Rhodotorula mucilaginosa* and *Hanseniaspora uvarum* effectively inhibited sour rot of oranges (13). A water suspension of the yeast antagonist, when applied to wounds on the surface of fruits prior to the inoculation with spore suspension of the pathogens (*Alternaria alternata, Botrytis cinerea* and *Rhizopus stolonifer*), reduced disease by 90% (9). *Debaryomyces hansenii* effectively reduced the incidence of post-harvest diseases of citrus and mango and tomato (27, 35). Efficacy of antagonist was affected by the yeast concentration in the wound as well as by the number of spores of the pathogen used for inoculation (Fig. 1). The antagonist grew readily on wound leachate (Fig. 2) and rapidly multiplied at the wound sites of injured fruit suggesting that the antagonist may retain its efficacy also under natural conditions. The mode of action of this yeast in antagonising fruit pathogens is not through the production of antibiotics, but by competition for the nutrients at the wound site.

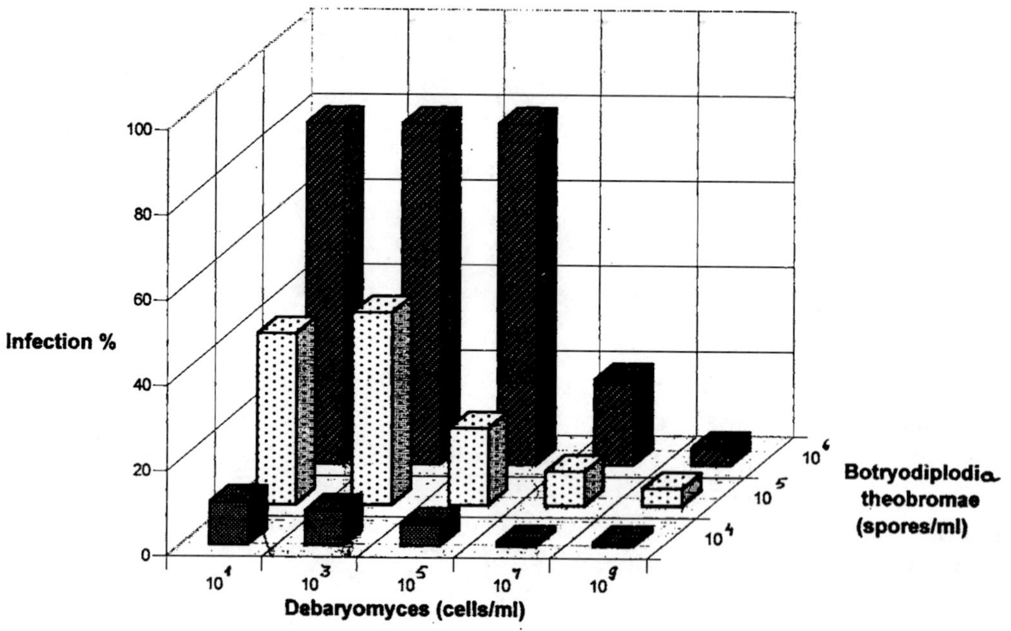

Fig.1: The relationship between concentration of the pathogen spore suspension and the antagonists cells in the inhibition of *Botryodiplodia theobromae* decay of mango by *Debaryomyces* sp.

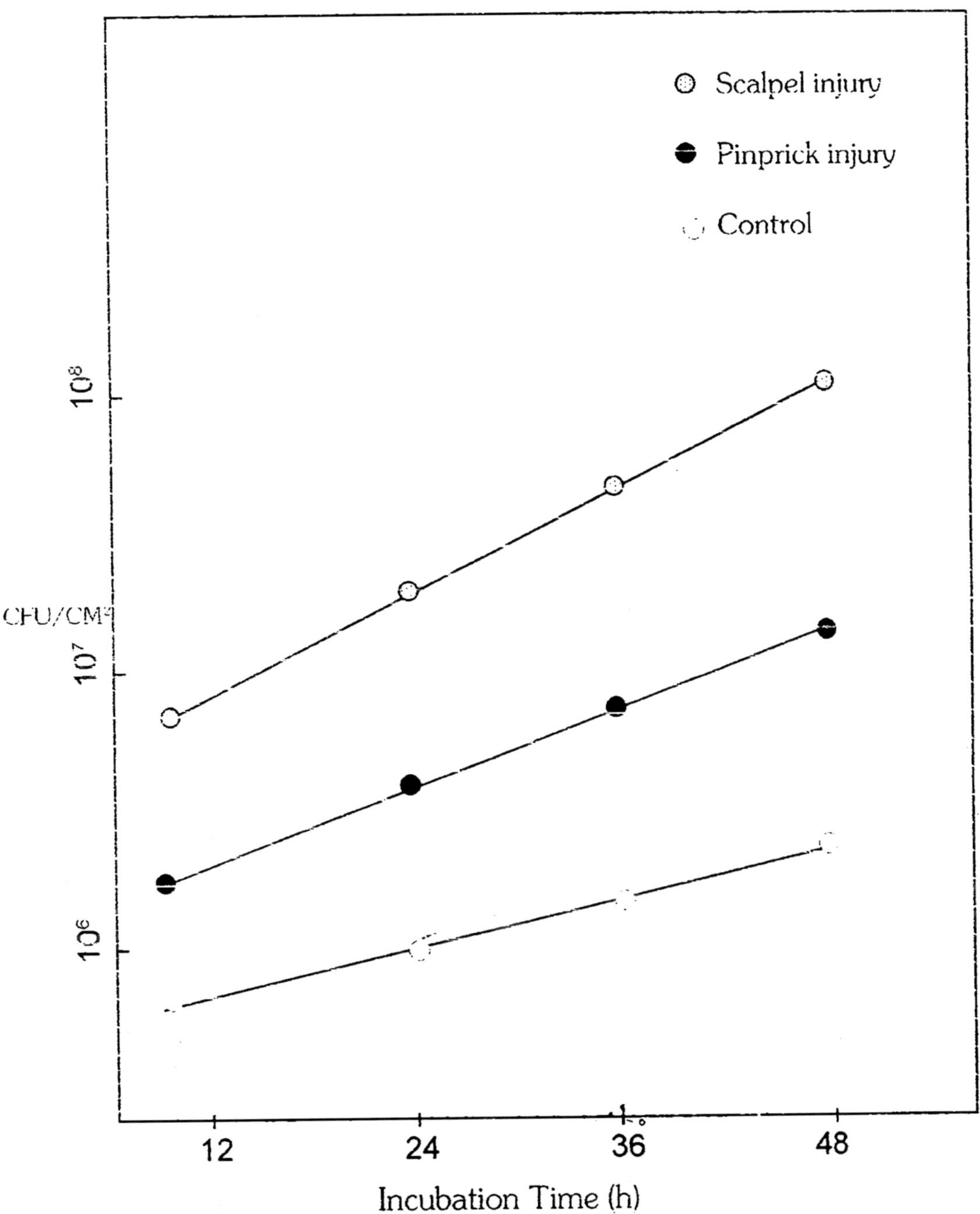

Fig. 2 : Growth of *Debaryomyces hansenii* on mango fruit leachate.

11.1 Enhancement of Biological Control by Using Calcium Salts

Biocontrol ability of *Debaryomyces hansenii, Candida* sp., *Klocekera apiculata* improved in the presence of calcium salts ($CaCl_2$ and $CaCO_3$) suggesting that Ca^{++} played an important role (Fig. 3, 4).

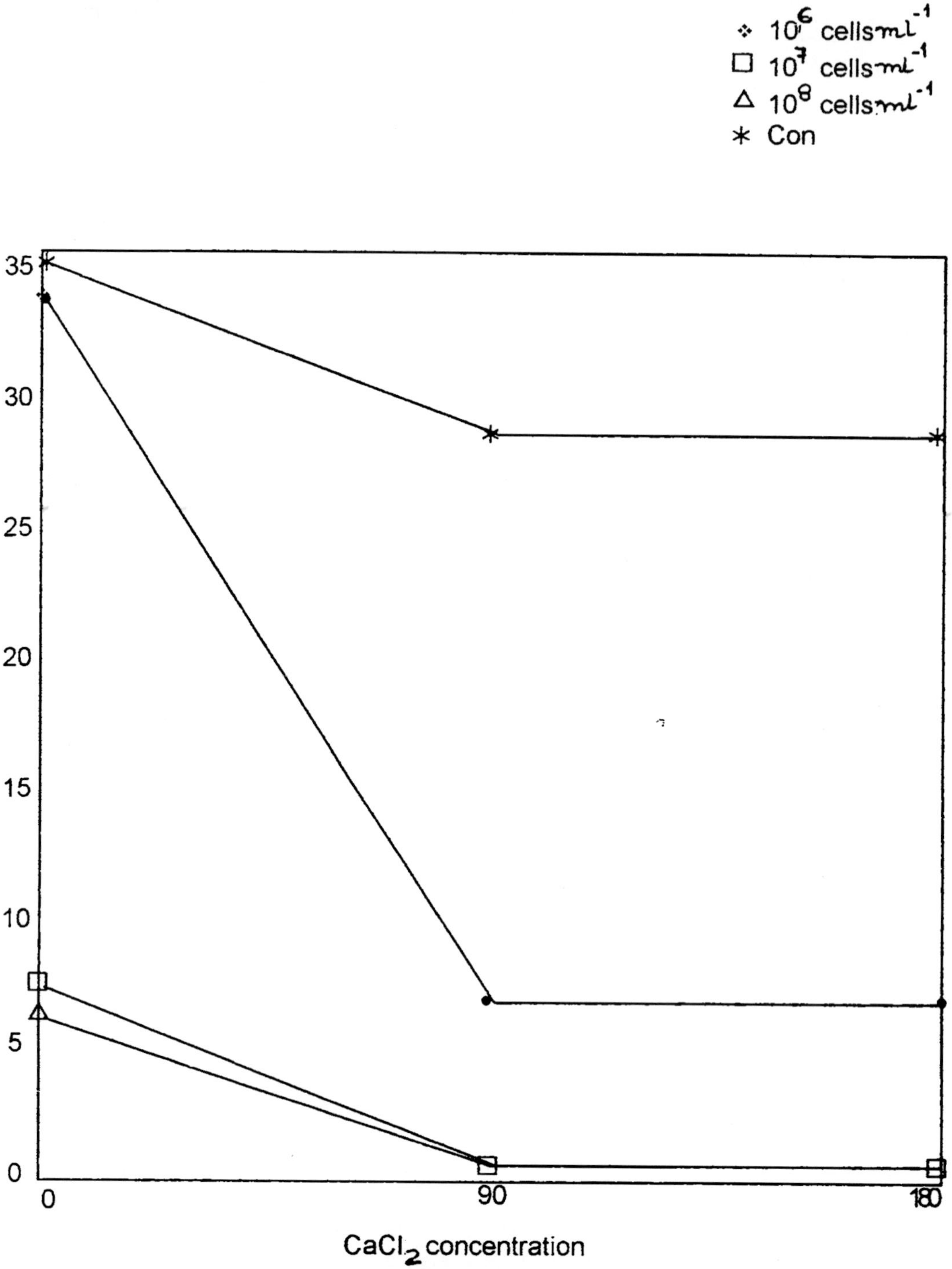

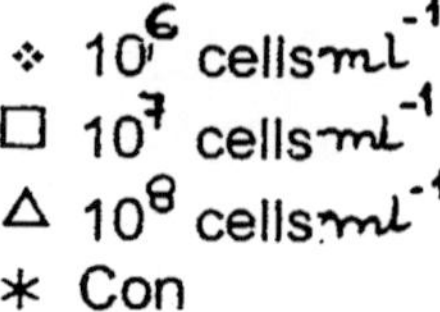

Fig.3: Effect of 90 or 180 m*M* $CaCl_2$ on biocontrol activity of *D. hansenii* against *B. cinerea*.

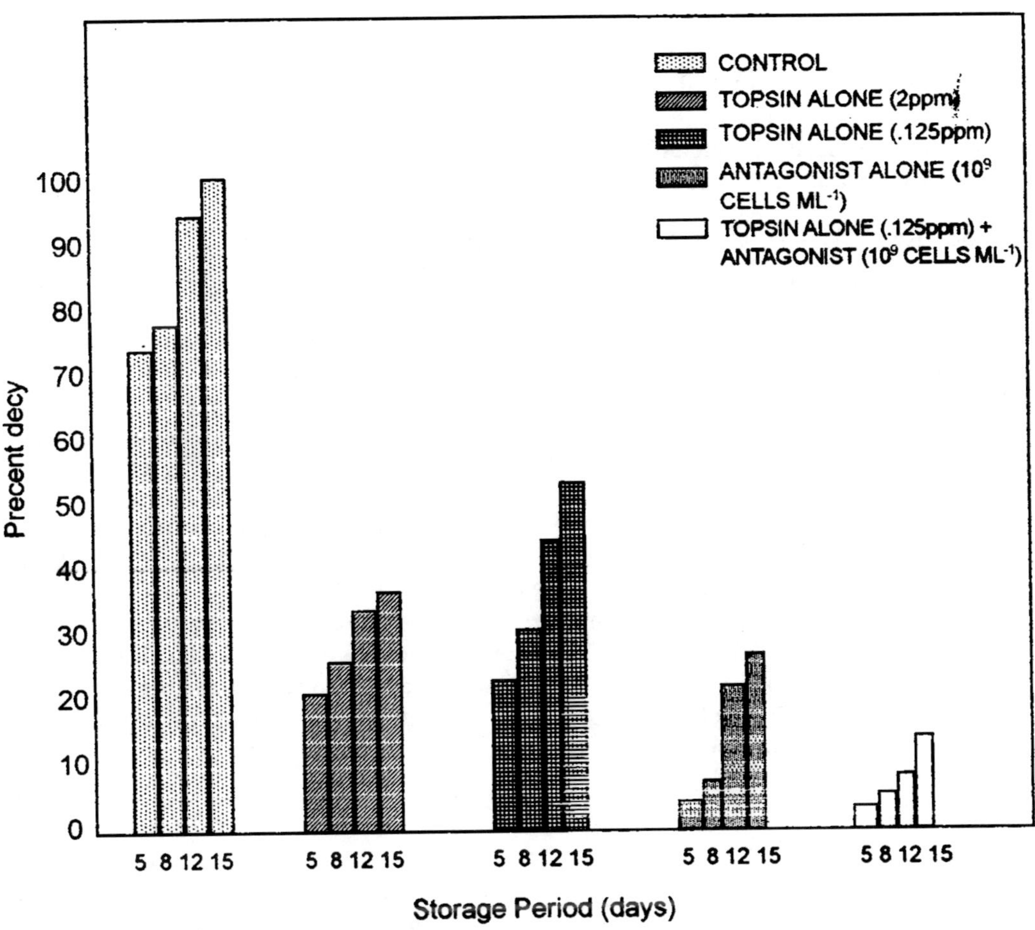

Fig.4: Compatibility of *Debaryomyces hansenii*, with Topsin for the control of post-harvest mold rots.

The control was facilitated by improving the colonisation ability of the yeast in wound site. This does not seem to be the case since the population of yeast, followed over a period of 96 hr after application, remained almost the same (36). Another possibility is that $CaCl_2$ could inhibit the growth of the pathogen in the wound site. The effect of $CaCl_2$ was found to be dependent on the concentration of reducing sugars, which confirms nutrient competition as the mode of action for the yeast strain (25).

11.1.1 Application methods

Most commonly practised methods for the application of bioagents are

(i) dipping of fruits in antagonist suspensions;

(ii) spraying the antagonist through standard nozzle; and

(iii) drenching of fruit with the antagonist suspension.

These methods proved to be effective but there exists some major technical problems in the case of dipping: one is the absence of dipping tanks as part of the inline systems in packing houses and the other is that presumably due to the recycling of the solution in the same tank for large quantities of fruit without replacement, there is an accumulation of the pathogen inoculum, which may decrease the potential of the antagonist.

12. OBSTACLES AND RECOMMENDATIONS

12.1 Economic Impetus for Biological Control

Annual pesticide revenues world wide are US $24-25 billion (23). Globally, about US $2.6 billion is invested each year in the research and development for chemical pesticides. Each pesticide requires at least 8-10 years and $15-30 million to develop and register, for an expected return of at least $50 million per year. Since 1970, a steady decline has occurred in the number of new pesticides and their profitability (6).

The current world wide market for the biological control agents is about $60 million per year, less than 0.5% of the pesticide sales (32). Knowledge of natural enemies of the pathogens takes about few years and $1-2 million to bring a biopesticide to the market in the U.S. Moreover, the biological control agents for niche markets may have more of the market share, increased relative profitability and a longer useful lifetime than the chemical pesticides. Immediate and rapid progress in adopting biological control can be made by recognising and better mobilising the existing research and development establishment. This increase from $100 to $200 million/ year would still be a better investment as compared to the $2.6 billion spent on discovering new pesticides.

12.1.1 Competitive cost

Augmentative biological control is slowly replacing chemical pesticides for controlling pests in the glasshouses, orchard crops, forests and fields (31), and at urban and industrial sites.

Classical biological control has documented the benefit cost ratios exceeding 30:1 (18) whereas the chemical pesticides produce the ratio less than 5:1. The cost of producing bio agents can be reduced by developing commercial diets, improving strains and increasing production levels to rationalise the economics of the scale.

12.1.2 Demonstrated efficacy

Biological control must solve the pest/pathogen problem and do it consistently under all field conditions (17). Use of *Bacillus thuringiensis* controls pest predictability, usually with one application, as do the chemical pesticides.

Reliable production and distribution of natural enemies necessitates scale-up and pilot testing research findings in agricultural and urban contexts. Large scale production of antagonist is not simply a multiplication of small scale tests but an organisation of optimised rearing operations. Evaluation of the impact of the agents on the pathogen populations and economic benefits should be major components of such pilot tests. Application strategies must preserve the quality of the natural enemies, whether natural or induced (46).

12.1.3 Targeted interdisciplinary research

Research is needed on artificial growth media for antagonist, mass production procedures, strain development using both genetic selection and molecular techniques, application methods and strategies and both impact and economic evaluation (8). Coordination and cooperative fundings are needed to focus the augmentation research on a few biological control projects that are likely to solve the specific pest problems within three to five years.

12.1.4 Technology development not implementation

Technology development could be accelerated by establishing more government/industry partnership (42), building municipal and grower's cooperatives and providing greater assistance with permits and registrations of the commercial products (10). Governments transfer research to the private sector through agreements, training and contracts for the products (15).

12.1.5 Regulation of biological control

Regulations of the importation and release into the environment of natural enemies is becoming increasingly complex and expensive world wide, however, very few of such organisms present a potential risk and suitable regulatory procedure have been proposed to assure their safety (24). Generally, the concerns are ownership of germplasm, endangered species, non-indigenous organisms, purity of culture, and efficacy of the product. Non-target effects are the only risks of using natural enemies and generally these are minimal. For this, simple, efficient regulatory procedures for importation, interstate movement and release are needed.

12.1.6 Compatability of biocontrol agents with present processing technology

Many biological and physical factors must be considered when selecting or developing biocontrol agents. The tremendous diversity of crop-pathogen-post-harvest environment combinations presents a challenge to the commercialisation of biological control. Biocontrol agent must be resistant to chemicals used for control, compatible with commercial handling system including dump tanks, drenches, and line spray applicators. In addition, biocontrol must be effective in wide range of temperature and storage atmospheres. With proper planning and innovative research, these challenges can be met.

A basic understanding of the ecological succession of antagonistic and other micro-organisms on the surfaces of commodities and wounds is crucial for the development of any viable disease management program. Such information would facilitate the development of post-harvest strategies for post-harvest disease control and the optimisation of the potential of antagonistic micro-organisms.

The ultimate potential of epiphytic microflora will depend on its effectiveness and its compatability with current handling and storage practices. However, much more research is required to determine the effects of various post-harvest practices on the population dynamics and the biological activity of the antagonists. It is equally important to know as to how the interaction between antagonist-pathogen-host and the environment could be manipulated to favour the antagonistic activity of the biological control agents.

13. REFERENCES

1. Andrews, J.H. 1992. Biological control in the phyllosphere. *Ann. Rev. Phytopath.* **6**: 603-635.

2. Anon, 1978. Report of the Steering Committee for study of Post-harvest food losses in developing countries. National Research Council, NSF, Washington DC. p. 206.

3. Belanger, R.R., Labbe, C. and Jarvis, WR. 1994. Commercial scale control of rose powdery mildew with a fungal antagonist. *Plant Dis.* **78**: 420-424.

4. Ben Arie, R., Droby, S., Cohen, L., Weiss, B. and Wilson C.L. 1988. Yeasts as biocontrol agents of post harvest diseases of fruits. *Phytoparasitica* **16**: 69.

5. Campbell, R. 1989. *Biological Control of Microbial Plant Pathogens,* Cambridge University Press, Cambridge, U.K.

6. Carlson, G.A. 1993. Economics of biological control. In: *Biological Control: Developing Strategies for 90's* (ed. R.C. McDonald), North Carolina Deptt. Agric. Releign, North Carolina, USA. 79pp.

7. Chalutz, E., Droby, S., Cohe, L., Weiss, B., Barkai Golan, R., Daus, A., Fuchs, Y. and Wilson, C.L. 1989. Biological Control of Botrytis, Rhizopus and Alternaria Rots of Tomato Fruit by *Pichia guilliermondii. Biological Control of Postharvest Diseases of Fruits and Vegetables,* Workship Proceedings Sept. 12-14.

8. Cerutti, F. and Bigler, F. 1995. Quality assesment of *Trichogramma brassicae* in the laboratory. *Entomol. Exp. Appl.* **75**: 19-25.

9. Chalutz, E. and Wilson, C.L. 1990. Post-harvest biocontrol of green and blue mold and sour rot of citrus fruit by *Debaryomyces hansenji. Plant Dis.* **74**: 134-137.

10. Cook,R.J. 1993. The role of biological control in pest management in the 21st century. In: *Pest Management: Biologically Based Technologies* (eds. R.D. Lumsden and J.L. Vaughn), American Chemical Society, Washington D.C. USA, 435PP.

11. Croft, B. A. 1990. *Arthropod Biological Control Agents and Pesticides,* John Wiley and Sons, New York.

12. Droby, S., Chalutz, E., Wilson, C.L. and Wisniewski, M. 1989. Characterisation of the biocontrol activity of *Debaryomyces hansenii* in the control of *Penicillium digitatum* of grape fruit. *Can. J. Microbiol.* **35**: 794-800.

13. Droby, S., Chalutz, E., Cohen, L., Weiss, B., Wilson, C.L. 1990. Biological control of post-harvest diseases of citrus. In: *Biological Control of Post-harvest Diseases of Fruits and Vegetables,* Workshop Proceedings, West Virginia.

14. Elad, Y. 1986. Mechanisms of interactions between rhizosphere microorganisms and soil borne plant pathogens. In: *Microbial Communities in Soil. Elsevier* (eds. V.A. Jensen, A. Kojoller and L.H. Sorensen), New York. pp. 49-60

15. Fitzner, M.S. 1993. The role of education in the transfer of biological control technologies. In: *Pest Management : Biologically Based Technologies* (eds. R.D. Lumsden and J.L. Vanghn), American Chemical Society Washington, D.C. USA, pp. 382-287.

16. Fokkema, N.J. 1993. Opportunities and problems of foliar pathogens with microorganisms. *Pestic Sci.* **37** : 411-416.

17. Gross, R. 1993. Insect behavioral and morphological defenses against parasitoids. *Ann. Rev. Entomol.* **38**: 251-273.

18. Habeck, M.H., Love joy, S.B. and Lee, J.G. 1993. When does investing is classical biological control research make economic sense? *Florida Entomol.* **76** : 96-1 0 1.

19. Hofstein, R. and Fridlender, B. 1994. Development of production, formulation and delivery systems. In: *Pest and Diseases,* Brighton crop protection conference BCPC, Farham, 1273-1280.

20. Janisiewicz, W.J. and Roitman, J. 1988. Biological Control of blue mold on apple and pear with *Pseudomonas cepacia. Phytopathology* **78**: 1697-1700.

21. Janisiewicz, W.J., Usall. J. and Bois B. 1991. Control of post harvest diseases of fruits with biocontrol agents. *Food and Fertilizer Tech. Ctr. Tech. Bull.* **125.**

22. Jeffries, P. and Jeger, M.J. 1990. The biological control of post-harvest diseases of fruit. *Biocontrol News and Information* **11** : 333 -336.

23. Klassan, W. 1993. Pest management and biologically based technologies: a lock to the future. In: *Pest Management: Biologically based Technologies* (eds. R.D. Lumsden and J.L. Vaughn), American Chemical Society, Washington, D.C. USA. 435pp.

24. Leppla, N. C. 1996. Environmentally friendly methods for reducing the damage caused by exotic weeds in natural habitats: conflicts of interest, safegaurds and national policy. *Castanea* **61**: 214-255.

25. McLaughhin, R.J. 1989. A review and current status of research on enhancement of biological control of post-harvest diseases by use of calcium salts with salts. In : *Biological Control of Post-haivest Diseases of Fruits and Vegetables*, Workshop Proceedings, West Virginia.

26. Mehrotra, N.K., Sharma, N., Ghosh (Nayek), R. and Nigam, M. 1996. Biological control of green and blue mold disease of citrus fruit by yeast. *Indian Phytopatho.* **49(4)**: 65-69.

27. Mehrotra, N.K.,Sharma, N., Nigam, M. and Ghosh (Nayek), R. 1998. Biological control of sour-rot of citrus fruits by yeast. *Proc. Nat. Acad. Sci. India* **68(B) II** : 133-139.

28. O'Neill, T.M., Elad, Y, Shitienberg, D. and Cohen, A. 1996. Control of grapevine grey mould with *Trichoderma harzianum. Phytopathology* **85**: 393-401.

29. Peng, G. and Sutton, J, C. 1991. Evaluation of microorganisms for biocontrol of *Borlytis cinerea* in strawberry. *Can. Jour. Pl. Path.* **13**: 247-257.

30. Pusey, P.L. and Wilson, C.L. 1984. Post harvest biological control of stone fruit brown rot by *Bacillus subtilis Plant Disease* **70** : 587-590.

31. Ravensberg, W.J. 1992. Biological control of pests: current trends and future prospects. In : *Pest and Diseases,* Brighton Crop Protection Conference, Brighton UK, 591-600.

32. Ridgway, R.L., Hoffman, M.P., Inscoe, M.N. and Glenister, C.S. (eds.) 1996. *Mass-reared Natural Enemies: Application, Regulation, and Needs.* Entomol. Soc. Amer, Thomas Say Publication in Entomology-ESA, Lanhan, Maryland, USA.

33. Roberts, R.G. 1989. Characterization of post-harvest biological control of deciduous fruit diseases by *Crytococcus* spp. In : *Biological Control of Post-harvest Diseases of Fruits and Vegetables*, Workshop Proceedings. West Virginia.

34. Sharma, N. 1989. Fruit rot diseases of grapes and amda - a new record. FAO Plant Bulletin, pp. 24-25.

35. Sharma, N., Nigam, M. and Ghosh, (Nayek) R. 1995. Biocontrol of black mould rot of mango with *Debaryomyces hansenii. Indian J. Pl. Path.* **13**: 51- 55.

36. Sharma, N., Nigam, M. and Ghosh (Nayek), R. 1996. Biological control of grey mold of tomato with antagonistic yeast. *Proceedings of International Conference on Plants and Environmental pollution,* Lucknow pp. 43-46.

37. Sharma, N., Nigam, M. and Ghosh (Nayek), R. 1997a. Post-harvest biocontrol of *Pencillium* rot of table grapes. *J. Biol. Control* **11**: 53-56.

38. Sharma, N., Ghosh (Nayek), R. and Nigam, M. 1997b. *Debaryomyces hansenii* - An effective biocontrol agent against fruit rot of apples in the proccedings of 49th Annual meeting of the Phytopathological society, Calcutta pp. 45. (Abst.)

39. Spurr, H.W. 1990. The phylloplane. In: *New Directions in Biological Control Alternative for Suppressing Agricultural Pests and Diseases* (eds. R. Baker and P. Dunn), UCLA Symposium on Molecular and Cellular Biology. New Series Vol. 112, A.R. List Inc., New York. pp. 271-278.

40. Sutton, J.C. and Peng, G. 1993a. Manipulation and vectoring of biocontrol organisms to manage foliage and fruit diseases in cropping systems. *Ann. Rev. Phytopath.* **83**: 615-621.

41. Tronsmo A. and Denis C. 1977. The use of *Trichoderma* species to control strawberry fruit rots. *Neth J. Pl. Path* **83 (Suppl. 1)** : 449.

42. Utkhede, R.S. and Sholberg, P.I. 1986. *In vitro* inhibition of plant pathogens by *Bacillus* subtilis and *Enterobactier acrogenes* and vivo control of two postharvest Cherry disease. *Can. J. Microbid.* **32**: 963-967.

43. Walters, T.L. 1996. Report to USDA: Science and Education Evaluation Study to increase the Commercialisation and Private Sector Co-Funding for Biological Pest Control Technology. American competitiveness Enterprise, Inc. Altadena, California, USA, 134p.

44. Wilson, C.L. and Wisniewski, M.E. 1989. Biological control of post-harvest diseases of fruits and vegetables - An emerging technology. *Ann. Rev. Phytopath.* **27**: 425-441.

45. Wilson, C.L. and Pusey, P.L. 1985. Potential for the biocontrol of post harvest diseases. Plant disease **69**: 375-378.

46. Zalom, F.G. and Fry, W.E. 1992. Food, crop pests and the environment, the need and potential for biologically intensive integrated pest management. APS Press. St. Paul, Minnesota, USA.

19 Microbes in Biocontrol of Heavy Metal Pollution

S.K. Katiyar and R. Katiyar

CONTENTS

ABSTRACT ... 330

1. INTRODUCTION ... 330
2. METAL-MICROBE INTERACTIONS ... 333
3. INDUSTRIAL APPLICATION OF MICROBIAL METAL ACCUMULATION ... 333
   - 3.1 Living Organisms ... 333
   - 3.2 Immobilized Cell System ... 334
   - 3.3 Immobilized Living Systems ... 334
   - 3.4 Immobilized Non-living Biomass ... 335
   - 3.5 Immobilized Metal-binding Components ... 336
   - 3.6 Growth-decoupled Enzymic Metal Removal ... 337
   - 3.7 Microbial Metal Transformation ... 337
   - 3.8 Metal Removal by Derived, Induced or Excreted Microbial Products ... 337
   - 3.9 Metallothioneins, Phytochelatins, and other Metal-binding Proteins ... 338
   - 3.10 Microbial Extracellular Polymers ... 338
   - 3.11 Siderophores ... 339
   - 3.12 Metal Removal and Compound Derived from Fungal Biomass ... 340
4. METAL RECOVERY ... 341
5. INDUSTRIAL AND ECONOMIC ASPECTS ... 341
6. CONCLUSION ... 344
7. REFERENCES ... 344

Advances in Microbial Biotechnology
J.P. Tewari, T.N. Lakhanpal, Jagjit Singh, Rajni Gupta & B.P. Chamola (eds.)
APH Publishing Corporation, New Delhi - 110 002, India.

ABSTRACT

Microorganism can accumulate heavy metals and radionuclide from their external environment. The amount accumulated can be large and variety of physical, chemical and biological mechanism may be involved including absorption, precipitation, complexation and transport. Living and dead cells, as well as products excreted by or derived from microbial cells, like cell wall constituents, pigments, polysaccharides, metal binding proteins and siderophores are capable of metal removal.

Bio-accumulation patterns of metals in different organism (algae, fungi and bacteria) can be utilized as effective indicators of environmental metal contamination. Resistance to metal ions has been observed in several microorganisms. Micro-algae, fungi and bacteria have been used to remove heavy metals from aqueous systems since they have high capacity to accumulate dissolved metals. At cellular and molecular level, toxicity of heavy metal is not only related to their interaction with thiol groups of proteins but also to their inophoretic properties and their ability to generate free radicals. Suspended and immobelized cultures of the microbes have shown that their cell responses depends mainly on the metal (type concentration and activity) on the species used. Immobilisation process may increase tolerance of heavy metals.

1. INTRODUCTION

Microorganisms can accumulate heavy metals and radionuclides form their external environment (9, 22, 23, 24, 27, 48, 49, 83, 97). The amount accumulated can be large and a variety of physical, chemical and biological mechanism may be involved including, absorption, precipitation, complexation and transport. Living and dead cells, as well as products, exerted by microbial cells e.g., cell wall constituents, pigments, polysaccharides, metal-binding proteins and siderophores, are capable of metal/radionuclide removal form solution as well as related elements or compounds including metalloids, organo-metalloids (29, 86) and metal containing particulates (85, 98). The removal of radionuclides, metal or metalloid species, compounds and particulates from solution by biological material, particularly by non-directed physico-chemical interactions, is now frequently termed "biosorption;" (83, 97). Although virtually all biological material has biosorptive properties (41, 60), most worked to date has been directed to microbial systems.

Biosorption and related phenomena, are of importance because the removal of potentially toxic and valuable metals and radionuclides form aqueous effluent can result in detoxification and therefore safe environmental discharge (1, 23, 59, 57). In addition to this appropriate treatment of loaded biomass can enable recovery of valuable elements for recycling or further containment (10, 93) Fig. 1. It is interesting to note that the maximum usefulness of organisms in metal accumulation have probably not been demonstrated, although further work on the accumulation of toxic metal ions was recommended. Considerable progress has been achieved since then in the study of metal-microbe interactions, particularly those relevant to the control of environmental pollution. This chapter emphasises the theoretical and practical evidence for microbial biotechnology as a feasible alternative to existing treatment methods for removal and recovery of heavy metals and radionuclides form liquid waste.

Table 1 : Some examples of microbial heavy metal and radionuclide accumulation

Organism	Element	Uptake (% dry weight)
Bacteria	Uranium	2-4
Streptomyces sp.	Uranium	30
S. viridrochromogenes	Uranium	30
Thiobacillus ferrooxidans	Silver	25
Bacillus cercus	Cadmium	4-9
Zogloea sp.	Cobalt	25
	Copper	34
	Nickel	13
	Uranium	44
Psedomonas aeruginosa	Uranium	15
Mixed culture	Copper	30
Mixed culture	Silver	32
Bacillus sp.	Lead	60.1
	Copper	15.2
	Zinc	13.7
	Cadmium	21.4
	Silver	8.6
Algae		
Chlorella vulgaris	Gold	10
Chlorella regularis	Uranium	15
Fungi		
Phoma sp.	Silver	2
Penicillium sp.	Uramium	8-17
Rhizopus arrhizus	Copper	1.6
	Lead	10.4
	Silver	5.4
	Mercury	5.8
Aspergillus niger	Mercury	5.8
Yeasts		
Saccharomyces cerevisiae	Uranium	10-15
	Thorium	12
	Zinc	0.5
Yeasts (14 Strains)	Silver	0.05-1

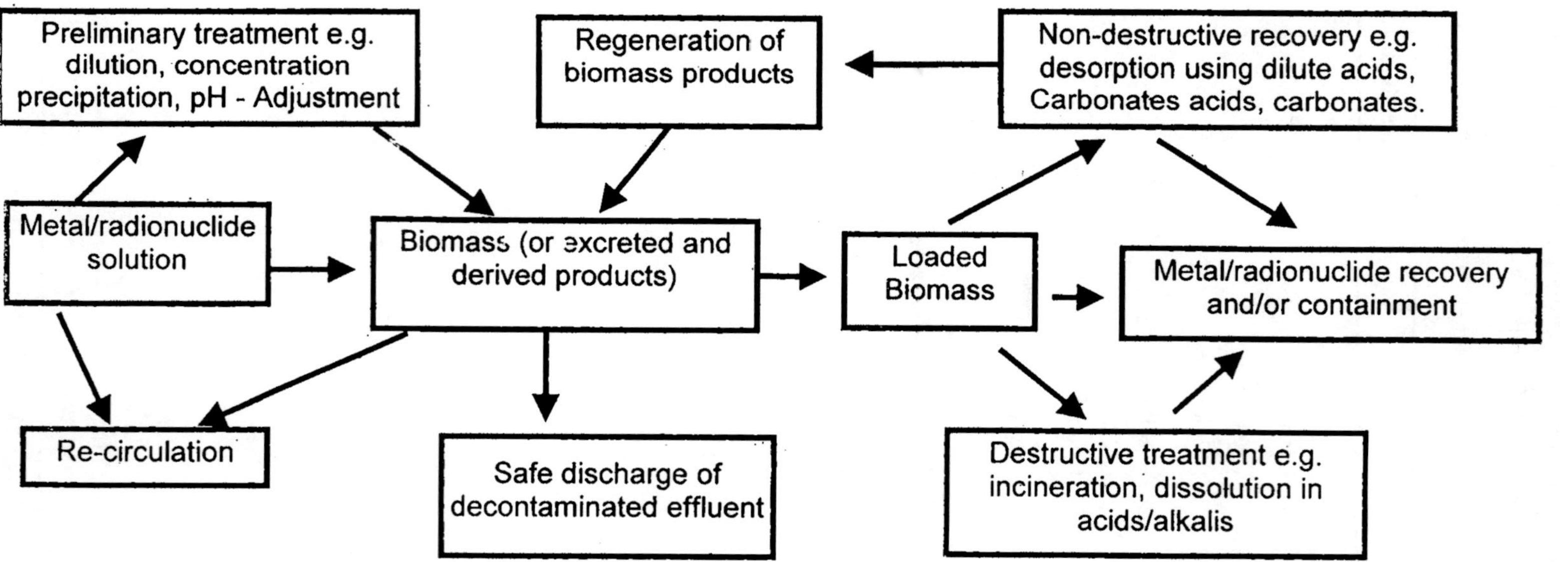

Fig.1: Generalized scheme of removal of heavy metals radionuclides, and related substances, from solutions by microbial biomass. The bisorbent may be living or dead, freely suspended or immobolized within or on inert matrices or as pellets, biofilms or aggregatese or derived or extracted products. The choice of destrictive or non-destrictive recovery may depend on the element and mechanism of removal

2. METAL-MICROBE INTERACTIONS

Metals are involved in all aspects of microbial growth, metabolism and differentiation. Essential metals, e.g. K, Ca, Mg, Cu, Zn, Fe, Co, Mn, and those with no essential biological function, e.g. Cs, Cd, Pb, Al, Sn, Hg, can be accumulated by microorganism by non-specific mechanism of sequestration or transport. Many of the so called heavy metals are toxic at high concentrations and a variety of resistance mechanisms have been documented in all microbial groups. (48, 80). Microorganism have encountered toxic metals in the environment throughout their, evolutionary history although it is now mainly a result of industrial activities that ecosystems are subject to contamination by heavy metals, organo-metals and radionucleides (2).

The ability to grow in the presence of high metal concentrations is found in a range of organisms and may be a result of intrinsic or induced features including specific mechanisms of resistance, and/or environmental factors, that may reduce or eliminate toxicity, e.g. pH, Eh., inorganic anions and cations, particulate are soluble organic matter, clay minerals and salinity (27). Precipitation, complexation, and crystallization of heavy metal and radionuclide species exterior to cells can result in detoxification and many examples of microbial metal deposition are of great significance in biogeochemical cycles, e.g. microfossil and mineral formation (7, 61), iron and manganese depsition (23). Some of these processes depends on the ability of microorganism to effect chemical transformations of heavy metals and their compounds by e.g. oxidation, reduction, methylation and demethylations (12, 15, 55, 87). Transport mechanism for entry of metal species into microbial cells have received considerable attention at physiological, biochemical, and genetical levels and a wealth of detailed information is available. (46, 84). Once inside cells, metal ions may be compartmentalized and/or converted to less toxic forms by e.g., precipitation or sequestration by metal binding proteins (104).

3. INDUSTRIAL APPLICATION OF MICROBIAL METAL ACCUMULATION

3.1 Living Organisms

Because toxicity, inactivation of living cells may be a problem though it may be feasible to separate microbial growth form the metal-confacting phase or to use- resistant strains. However, some, metal-resistant strains exhibit- reduced accumulation and effect conversion of a given metal species in to another of altered volatility and toxicity by e.g. oxidation, reduction or methylation. If internalized, metal ions may be sequestered by metal-binding protiens or localized with in certain organelles. Furthermore, the disposal of large quantities of loaded biomass, which contains a high proprotion of water, may be, problematical (11). Most living cell systems exploited to date have been used for decontamination of effluents containing metals at concentrations below toxic levels and may employ a mixture of microogranism as well as higher plants (50). Algal and cyanobacterial blooms, encouraged by addition of sewage effluent, reduced levels of Cu, Cd, Zn, Hg and Fe in mining effluents (45). It is observed that effluent containing Pb, Cu, Zn, Mn, Ni, Fe and Cd to pass through engineered channels containing cyanobecteria, algae and higher plants including *Potamogeton* and *Thpha*, Metals are removed by entrapment of particulate forms and general biosoprtion with an efficiency >79% (11). In these examples, decomposition of settled dead biomass in segments can result in H2S production by sulphate reducing bacteria, e.g. *Desulphovirbro* and *Desulphomacuatlum* sp., and metal precipation as sulphides e.g. ZnS, CdS, Cus and FeS. This process is virtually irreversible of reducing conditions are maintained and resolubilization of metal does not occur after biomass decay (3, 11). Sulphate-reducing bacteria have been used in a heavy metal and sulphate treatment plant that uses ethanol as the growth substrate; a wide range of heavy metals are removed by sulphide perception allowing discharge of decontaminated effluent.

Naturally occurring or artificial bogs appear to have potential for treatment of acid mine drainage and these rely on metal interactions with a complex array of microorganism and plants including mosses and sedges. Mn removal efficiencies up to 90 % have been recorded. Removal mechanism are varried and range form - cation exchange by wall constituents of mosses to precipitation by sulphide released from sulphate-reducing bacteria (under anaerobic conditions) or bacterial oxidation and precipitation (11). The most frequently used living cell system for metal removal from dischanged effluents is sewage treatment. In most treatments a primary sedementation step can remove upto 40-60% of total metals present : the remainder may pass in to biological treatment system which is usually an activated sludge or trickling filter system. The efficiencies of both for metal removal being similar. It appears that the metal are largely bound by the extra cellular polymers produced by the microorganisms present, particularly bacteria (31, 54). Many other batch and continuous processes employing living cells have been described, some using microbial consortia and efficient removal of e.g. Ag and other metals, has been demonstrated.

3.2 Immobilized Cell Systems

Freely suspended microbial biomass had disadvantages that include small particle size and low mechanical strength while a similar density to effluent may complicate biomass separation. Immobilized biomass particles appear of a greater potential in packed or fludized bed reactors with benefits including control of particle size, better capability of regeneration, easy separation of bio-mass and effluent and recirculation, high-bio-mass loading and minimal clogging under continuous flow (101). Further such systems may be mathematically defined (56). Derived products, as well as living or dead cells can be used in immbolised forms including bio-film and many options have been explored for heavy metal removal and recovery.

3.3 Immobilized Living Systems

Heavy metal toxicity and other extreme properties of waste effluents may limit the use of living cell systems. However, living cell systems may allow a greater potential for a long term continuos process, without the need for periodic desorption if there is continual biomass replenishment (59). Furthermore, metal removal mechanism only expressed by living cells may be exploited.

The method most often applied to metal removal is the use of cells immobilized as a biofilm or inert support. Ideal supports have a large surface area but are sufficiently porous to enable high flow rates and minimal logging. Materials include those with plane and smooth surfaces (glass, metal sheets and plates, plastics), uneven surfaces (wood shaving, clays, sand, crushed rock, coke etc. and porous materials (foams, sponges) and many of these systems have been used in a variety of bio-reactor configuration including rotating biological contactors fixed bed reactors trickle filters and fludized beds and air-lift bioreactors (42, 43). Living cell biofilms may provide an additional capacity for removal of other pollutants including hydrocrabons, pesticides and nitrates. *Pseudomonas aeruginosa* was immobilized on particles of poly vinyl chloride (PVC) and polypropylene webs and used in batch and column reactors for simultaneous denitrification and heavy metal removal form contaminated wasted water (40). Another system used a mixed bacterial culture, mainly - *Pseudomonas* sp., immobilized as a film on anthracite particles for denitrification and uranium removal. Particles with excess cell mass, which accumulated at the upper boundary of the fludized bed were continuously removed, passed through a vibrating screen and returned to the fluidized bed to maintain optimum bio-film thickness. Excess biomass arising from particle passage through the vibrating screen was used as uranium biosorbent in a separate stirred - tank reactor. An alternative tank was used with biosorbent particles circulating counter to uranium flow. Uranium in solution was reduced from 25 to 0.5 gm-3 in a mean liquid residence time of 8 minutes (83). A large scale commercial process (>5x106 gallons per day), which treats effluents from gold mining and milling uses rotating disc bio-film contacting units for simultaneous degradation of cyanide, thiocyanide and ammonia. Heavy metals are removed by biosorption to the microbial biofilm. Metal loaded biomass is periodically recovered for controlled disposal. In comparision to non-living cells, rather less attention has been paid to living cell immobilization by entrapment with in gels or other matrices, polyacrylamide gel immobilized Streptomyces species particle has good mechanical stability and removed uranium, copper and cobalt with some degree of selectivity. The order of efficiency being $UO_2^{2+} >> Cu^{2+} > Co^{2+}$. Desorption was achieved using 0.1 M NaCO3 and particles were not affected by up to five biosorption-desorption cycles (68). Polyacrylamide-immobilized *Citrobacter* sp. removed uranium, cadmium, copper and lead from solution containing glycerol-2-phosphate with high efficiency. Regenerated biomass was capable of functioning for extended periods over a temperature range of 2-450°C (57).

3.4 Immobilized Non-living Biomass

Almost all microbial groups have been used successfully in immobilized forms (18, 36, 52, 100). A variety of methods are available for whole cell immobilization, each of which may have an influence on metal removal efficiency. Entrapment of cells with in alginates, polyacrylamide and silica gels may be highly efficient in small scale systems, though diffusion limitation may be a problem (56). However, small scale systems may be adequate for low volume waste containing valuable elements, e.g. Au. In general immobilized biomass should have particle size similar to that of other commercial absorbents (0.5 - 1.5 mm) and posses particle strength, high porosity, hydrophilicity and chemical resistance (9). Ideally, particles should contain a maximum amount of biomass with minimal amount of required binding agents. Microbial cell can be immobilized using a variety of support materials including agar, cellulose, alginate, cross linked ethyl acrylate, ethylene glycol dimethyl acrylate, polyacrylamide, toluene di-isocyanate, glutaraldehyde (cross linking reagent) and silica gel (9). The biomass may be used in its natural state or modified by physical or chemical treatments to improve biosorption efficiency. Alkali treatment of *Bacillus subtilis* resulted in an increase in silver and copper loading capacity from 11.4 to 86.7 mg g-1 and 9.2 to 79.2 mg g-1 respecitvely (9).

Waste biomass of *Bacillus* sp. from industrial fermentations has been used in granular form for metal biosorption. Sieved concentrated biomass is mixed with granulating cross linking agents and extruded to give the desired final size : drying provides a product of indefinite shelf life. An alternative procedure is grinding of dried stablized biomass to irregular granules to the desired dimensions (9). Bacteria can be cross linked using glutaraldehyde; fungal biomass has been treated with xylene and then corss linked using formaldehyde, formaldehyde - resorcinol or formaldehyde urea solutions or polyvinyl acetate (9). The granulated *Bacillus* preparation is non-selective and can remove many heavy metals from solution independent of differing initial concentrations e.g. Cd, Cr, Cu, Hg, Ni, Pb, U and Zn but does not bind Ca, Na, K, or Mg, single or mixed metals are generally loaded to >10% of the dry weight giving a removal efficiency of >99% and effluents with total metal concentrations around 10 - 50 ppb.

Biosorption of Uranium and other elements, by - immobilized *Rhiozopus arrhizus* biomass has been rigorously examined (95). The immobilized particles (0.7 - 1.3 mm diameter) contained 12 - 23% added polymer with improved uranium removal at lower polymer contents and lower particle diameter (95). Complete Uranium removal was possible from dilute Uranium ore bioleaching solutions (<300 mg l g-1) with elute concentrations after desorption >500 mg l g-

1. The particles maintained full loading capacity (~ 50 mg (g-1) over multiple biosorption - disorption cycles.

Immobilized non-viable algal preparations have also received detailed attention (4). Fluidized beds of alginate and polyacrylamide - immobilized agar, e.g. *Chlorella vulgaris* and *Spirulina platensis* have been used to remove variety of metals, including Cu^{2+}, Pb^{2+}, Zn^{2+} and Au^{3+} from mixtures of several schemes for selective recovery have been devised (36). Alginate and polyacrylamide provide good resistance to hydrostatic pressure and mechanical degradation (67) though it is thought that polyacrylamide is not strong enough for commercial application (4). A commercial immobilized algal preparation is alga SORBTM which contains algal biomass immobilized in silica matrix and used in batch or column systems. Columns are slurry - packed with immobilised algal particles, 40 to 100- mesh size and used for metal biosorption, selective metal recovery is by treatment with appropriate reagents after which the regenerated bio-mass retains approximately 90% of original metal uptake efficiency even after >18 months of regular use. Alga SORBTM has been successfully used for removal of Ag^{+}, Al^{3+}, Au^{+}, Co^{2+}, Cr^{3+}, Cr^{6+}, Hg^{2+}, Ni^{2+}, Pb^{2+}, Pd^{2+}, Pt^{2+}, U^{6+} and Zn form contaminated effluents and process streams (4).

3.5 Immobilized Metal-binding Compounds

Procedures for immobilization or low molecular weight organic molecules, as well as whole cells and polymers, are similar to those used for enzyme immobilization. Of the methods available absorption, entrapment and cross linking are more suitable for whole cells or cellular components (4), whereas immbolization of lower molecular weight metal - binding compounds requires an association with a carrier material, usually by means of covalent attachment, to provide the size and integrity appropriate to practical use. Carrier materials have functional groups to which the biosorbent molecules are covalently attached directly by means of an intermediate linkage or spacer group (38). A range of carriers are available for immobilization of metal binding legands ranging from natural and synthetic organic substances like dextran, cellulose, agar, alginate carbon, polyacrylamide, nylon, polypropylene and inorganic substances, like controlled pore glass silica gel, natural and synthetic zeolites (aluminosilicates). Selection of these substances will depends on factors such as surface area, effective pore size, volume, particle size, mechanical strength, density, swellability, ease of preparation and cost (38). Controlled pore glass carriers with pore diameter of either 22 to 24 or 140 to 150 A° (with specific surface areas ranging from 826 to 211 m2 ml-1 respectively) have been found suited to variety of immobilization procedures with different legands (38, 39).

Most carriers require active ion to introduce functional groups to which covalent ligation of bridging agents or metal binding legands can be achieved. Polysaccharide carriers can be activated with cyanogen bromide which yield a carrier enriched with imidocarbonate groups to which metal binding compounds containing primary amino groups can be directly coupled. Organosilane, reagents particularly those containing amino groups, are suitable for functionalization of controlled pore glasses and zeolites, which may then be further funtionalized by glutaraldehyde which extends the linkage chain and introduces aldehyde groups for direct linkage of amine groups to the chain (38).

A number of metal binding molecules have been immobilized to functionalized carriers (16, 17) including the siderophore desferrioxamine and cysteine which both possesses primary amine groups not essential for complete metal binding but useful for ligation to carrier (38). In some cases, post immbolization technique may be necessary for the removal of prebound metal and reduction of C-N groups (38). Amino silanated glutaraldehyde activated controlled pore glass carriers with coupled cystine exhibited practical mercury-binding capacities of 783 and 6992 mg Hg Kg-1 for carriers of pore size 22 and 140 A0 respectively (38, 39).

3.6 Growth - Decoupled Enzymic Metal Removal

Metal accumulation by growth-decoupled "resting cells" of a *Citrobactor*. sp. is catalysed by a surface-located acid type phosphatese enzyme. Which releases, $HP0_4^{2-}$ from a supplied substrate e.g. glycerol-2 phosphate and precipitates divalent cation (M^{+2}) as $MHPO_4$ at cell surfaces (56, 59, 60). The enzyme has wide pH optimum (pH -5-9) and is markedly resistant to inhibition by heavy metals such as Co, Ni, and Cd. Phosphatese activity was increased by growth in carbon limited continuous culture enabling convenient production of reactive cells by pre-growth in metal-free conditons. The *Citrobactor* process has been mathamatically defined in terms of enzymic parameters which represents an advance over several other empirical methods used to date (56). The process is non-specific and depends on insolubility of the particular metal phosphate involved; Cd^{2+} Pb^{2+}, Cu^{2+} and UO_2^{2+} can be precipitated singly or in combination. With time metal phosphate formation may limit further metal accumulation because of column blockage (59). The need for phosphate donar, e.g. glycerol 2- phosphate, involves an economic consideration although the possibility of using a cheap or industrial waste product such as tributyle phosphate which is a by-product of nuclear fuel processing, may make this process attractive.

3.7 Microbial Metal Transformations

Microorganism can transform heavy metals and metalloid species by e.g. oxidation, reduction, methylations and demethylation (26, 103). Continous cultures of Hg^{2+} resistant bacteria, which can reduce Hg^{2+} to Hg^0 with mercuric reductase, volatilized Hg^0 from contaminiated sewage at a rate of 2.5 mg $l^{-1}h^{-1}$ (98% removal). Free and immobilized *Chlorella emersonii* cells were also capble of singificant mercury volatilization form solution (102). Organomercurials may be detoxified by organomercurial lyase, the resulting Hg^{2+} then being reduced to Hg^0 by mercuric reductase (12). Many bacteria, algae, fungi and yeasts can reduce Au^{3+} to elemental Au° (30) and Ag^+ to elemental Ag^0 which deposits on glass surface and in and around growing colonies (51). Microbial transformations of Arsenic and Chromium, species are also assoicated with a decrease in toxicity and may have relevance to waste water treatment (103). Treatment of Aresenic loaded sewage with arsenite oxidase-producing bacteria (which oxidase As^{3+} to As^{5+}) can improve certain arsenic removal methods since arsenate, As^{5+}, is more easily precipitated from waste water by Fe^{3+} than arsenite, As^{3+} (103). Chromate (Cr O42-) reducing bacteria, e.g. *Enterobacter cloacae*, are resistant to high levels of chromate (10mM) and can anaerobically reduce CrO_4^{2-} to Cr^{3+} which is precipitated. Both a fed-batch process and dialysis cultures were effective methods for CrO_4^{2-} detoxification. Biomethylated metal derivatives are often volatile and may be lost form a given system. (89). Though amount liberated are usually low and recovery is unlikely to be economical. Microbial dealkylation of organeometallic compounds, which may constitute detoxification mechanism, can eventually result in the formation of ionic species which could possibly be removed using a biosorptive process (25, 26, 59).

3.8 Metal Removal by Derived, Induced or Excreted Microbial Products

Virtually all biological material has high affinity for heavy metals and radionuclides (59). Some biomolecules may be induced by the presence of certain metals, e.g. metallothioneins and other metal-binding proteins, or by deprivations of an essential metal ion e.g. Fe^{3+} leading to sidero production (69). Other molecules with significant metal binding abilities may be over produced as a result of exposure to sublethal metal concentrations and interference with "normal" metabolism e.g. fungal melanins (28). However, the majority of metal binding biomolecules found in or excreted by micro-organisms are synthesised as a result of "noraml" growth and are important structural compounds. Metal binding by such compound may be fortuitous and relative efficiencies may largely depend on the metal species and the chemical -nature and reactivity of the metal binding legands present. (84, 97). It should be noted that there is wide diversity in the chemical composition of microbial cells and that microbial species can be manipulated by cultural and genetic procedures (60, 61).

3.9 Metallothioneins, Phytochelatins, and Other Metal-binding Proteins

Metal binding portiens have been recorded in all microbial groups examined e.g. cyanobacteria (71), bacteria (74). microalgae (77) and filamentous fungi (65). Although most detailed work has been carried out with yeasts (13, 104). Metallothioneins are small cystein-rich polypeptides that can bind essential metals, like Cd, Cu resistance in *Saccharomyces cervisiae* is mediated by the induction of a 6573 dalton cystine rich protein copper metallothionein (Cu-MT) (20). It has been suggested that yeast Cu-MT (and analogous proteins) may be of potenital in metal recovery since it can bind other metals besides Cu, i.e., Cd, Zn, Ag, Co, and Au, although these metals do not generally induce MT synthesis (13). It has been described that synthesis an inducible Cd binding protien in *S. cervisiae* which was cysteine rich (18 mol%) and high in Cd content [63μ g Cd (mg protein)-1] and showed a high similarity in amino acid composition with Cu-MT (Inouhe et al, 1989) the cup I gene has been cloned into *Escherichia coli* with resulting expression of a functional Cu, Cd, and Zn binding protein product and increased ability for metal accumulation by the bacterium (6). Two other approaches for exploitation of yeast Metallothionein have been suggested. The first is the engineering of yeast strains with constitutive and expression of MT genes which may then accumulate elevated amounts of metals that do not induce yeast MT transcription, eg. Gold, while the second relies on extracellular relase of Metallothionein by attaching the appropriate secretory signal sequence to MT genes (13). An ultimate aim may be production of different Metallothioneins-specific for different metals.

Another group of metal-binding molecules are short, cystine-containing (γ-glutamyl peptides, commonly termed phytochelatins (73, 104). The γ-glutamyl peptides are involved in heavy metal detoxification in algae and plants (32, 77) as well as certain fungi and yeasts (104). Metal ions are bound within a metal-thiolate cluster composed of an oligomer of peptides. *Saccharomyces cervisiae* does not synthesise phytochelatins but *Schizosaccharomyces pombe* can make at least seven different homologus γ-glutamyl peptides in response to metal exposure. A variety of metals including Cd, Cu, Pb, Zn, and Ag may be effective although most peptide synthesis with Cd (37). On exposure to Cu^{2+}, a Cu^{2+} γ-glutamyl peptide complex containing no labile Sulphor is synthesed in *S. pombe* with the metal bound as Cu (I) (75).

Cd induced γ-glutamyl peptides contained lavile S a major production which may arise from Cd stimulated production of S^{2-} in e.g., *S. pombe* and *Candida glabrata* (62, 75, 100). S2-lavels of 0.1 to 1.5 mol. (mol peptide)-1 have been recorded (104) and this enhances both the stability and metal content of the peptide complex.

3.10 Microbial Extracellular Polymers

Microbial exopolymers that have received most attention in relation to metal binding are those which form capsules or slime layers. Most of these are composed of polysacharides which may be associated with protein and certain other excretory products. The metal binding properties of many extra cellular polymers have been documented for a variety of microbial species, in pure and mixed cultures (31, 78, 82). In bacteria the wide range of polymers that can be synthesized fit broadly into major polysaccharide, protein and nucleic acid groups but with a wide occurrence of hybird polymers e.g. glycoproteins and lipopolysaccharides (60). Potential interactions of metals with these have been reviewed (31, 41, 60). Metal binding to polymeric material may involve simple exchange, with cationic groups involving carboxyl, organic phosphate, organic sulphate and phenolic hydroxyl complex formation. Additionally, there may be deposition of the metal in a chemically altered form (61).

In activate sludge, the bacterium *Zooloea ramigera* is important in flocculation because of extensive polysaccharides production. This has metal binding properties and a continuous process for metal accumulation using pre grown *Z. ramigera* removed approximately 3 m mol (g dry weight)-1 copper at a biomass concentration of <1 dry weight-1 (70). This was only 40 % of the value recorded for a batch process a probable reason being the formation of flocs with a high proportion of inaccessible copper binding silt (60, 70), other bacterial polymers have received attention , e.g. polysaccharide form capsulate *Klebsiella aerogenes* (Enterobacter) and *Pseudomanas* sp. (81) and *Arthrobacter viscosus* which comprise glucose, galactose and mannuronic acid (82). Extra cellular polysaccharides production is characteristic of many algae though the capacity for metal accumulation appears rather specific and not all poloysaccharides exhibit metal binding, a correlation existing between high anionic charge and metal complexing capacity (47). It appears that carboxylic acid groups (arrising from uronic acid) play a major role in several other bacterial and algal polymers (47, 63). Once inside cells, metal ions may be compartmertaligsic and are converted to less toxic froms by by, e.g. precipitiation or sequestrastion by metal binding proteins (104).

The application of an emulsifying agent, emulsan, and its deirvatives from bacteria such as *Acinetobacter*, *Pseudomonas* and *Acinitobacter* sp. For uranium - removal has been described. Emulsan has a polysaccharide backbone comprising of three amino sugars, D.-glucosamine, D.-galactosamine uronic acid and an unidentified haxosamine, with linked fatty acids (59). Metal binding ability is related to divalent cations for full emulsifying activity above pH 6 and this substance is capable of high uranium binding. If emulsion was sonicated and dispersed in water, the emulsano-sol product bound more than 800 mg uranium per gram (106).

3.11 Siderophores

Siderophores are low molecular weight Fe^{3+} co-ordination compounds that are excreted under iron-limiting conditions by many iron dependent micro-organism, particularly bacteria and fungi, to enable accumulation of iron from the enrironment. The maximum concentration of Fe^{3+} in solution at biological pH value is around 10-18 m which is too low to support microbial growth (69). Siderophores fit into two general classes. The phenolate/catecholate type and the hydroxamate type. Two prototype structures of catecholate and hydroxamate groups are exhibited by enterochelin (enterobactin) and ferrichrome respectively (8) from which many siderophore structures can be extrapolated (69, 59). The synthesis of both classes is dependent on the iron states of the medium with maximal production occurring under iron limiting conditions. Although virtually specific for Fe^{3+}, Siderophores and analogous compounds can combine with certain other metals, e.g., Ga(III), Scandium (Sc), Indium (In), Nickel, Uranium and Thorium etc. (72). Plutonium Pu (IV) can also be complexed by siderophores and a potential for actinide treatment has been suggested particularly as Pu^4 and Fe^3 and Pu^6 and Th^4 exhibit similar chemical properties (8). Cyanobacteria and algae may also produce siderophores, and analogues compounds, under iron deprivation which are capable of complexing e.g. copper.

Although most applied speculation has been directed towards Ga (III0 sequestration, there seems to be little progress in siderophore use from metal removal from industrial effluents. However, desferrioxamine (Desferal™), produced by mutant strain of *Streptomyces pilosus* are used in iron chelation therapy, has been immobilized to various functionalized carrier materials.

3.12 Metal Removal and Compound Derived from Fungal Biomass

Fungi and yeast have received considerable attention in connection with metal biosorption particularly because waste fungal biomass arises as a by product from several industrial fermentations (93, 97). Many fungi have high chitin content in cell wall and this polymer of N-acetyl glucosamine is an effective metal and radionuclide biosorbent (59). Actinide accumulation in intact biomass appears mainly to comparise metabolism-independent biosorption (88). The main site of uptake being the cell wall (97). Although, permeabilization of cells with carbonates or detergents can increase uptake (28). In *Saccharomyces cerevisiae*, uranium was deposited as a layer of needle like fibrils on cell walls, reaching upto 50% of the dry weight of individual cells. Similar precipitation has been observed for thorium. In *Rhizopus arrhizus*, rapid co-ordination of uranium to the amine nitrogen of chitin and simultaneous absorption in the cell wall chitin structure are followed by slower precipitation of uranyl hydroxide (96). Accumulation of hydrolysis products continues untill a final equilibrium is reached. (91). Chitosan and other chitin drivatives also have a significant biosorption capability (99). Insoluble chitosan - glucan complexes are glucans possessing amino or sugar acid group from *Aspergillus niger* exhibit biosorptive properties and effect efficient removal of transition metal ions from solution (66). Synthesized Chitosan derivatives, e.g., N-[2-(1.2 dihydroxyethyl) tetrahydrofuryl] chitosan (NDTC), are capable of uranium removal from brines at an order of magnitude greater than intact native biomass of R. arrhizus (66). After hydroxide treatment of variety of fungi to expose chitin/chitosan, and other metal binding legands, wet-led papers which incorporated treated mycilum were found to be efficient metals removal agents in laboratory trails (99).

Melanins are fungal pigments which enhance the survival capabilities of many species in response to environmetal stress (5). Fungal melanins are located in the exterior to cell walls where they may appear as electron-dense deposits and granules. Granules may be released into the external medium and be termed "extra cellular melanin" although this is generally of identical composition as wall associated melanin. Extra cellular melanins are more correctly defined as melanins synthesized-exterior to cells by secretion of phenol oxidase which oxidize exterioxidize phenolics or secretion of phenols that are subsequently oxidized (5).

Fungal phenolic polymers are melanins contain phenolic units, peptides, carbohydrates, aliphatic hydrocarbons and fatty acids and therefore possess many potential metal-binding sites (79, 80). Oxygen containing groups in these substances, including carboxylic, phenolic and alcholic hydroxyl, carbonyl, and methoxyl groups may be particularly important in metal binding.

A variety of melanin types occur in fungi. In Ascomycetes and Deuteromycetes, wall melenins are generally synthesized by the pentapeptide pathway via. 1,8 dihydrooxynapthalene (DHN) as the immediate percusor. Other extracellular dark pigments may be referred as 'heterogenous' melanins (5). A variety of heavy metal can induce or accelerate melanin production in fungi and melanized cell forms e.g., chalmydospores, can have high metal uptake capacities with virtually all the metal being located in the cell wall. Melanin from *Aureobasidium pullulans* can bind significant amount of metal like Cu^{2+}, and Fe^{3+} as well as organometallic compounds e.g., tributylin chloride (29).

4. METAL RECOVERY

Successful biotechnological exploitation of microbial metal accumulation may depend on the ease of metal recovery and biosorbent regeneration for use in multiple biosorption - desorption cycles (92, 93, 97). The mechanism used for metal recovery from loaded biomass depend on the case of removal from the biomass and this can depend on the element involved and the mechanism of accumulation. Metabolism-independent biosorption is frequently reversible by non-destructive, methods and may often be considered analogous to an ion exchange process. Metabolism-dependent accumulation and intracellular compartmention or sequestration with in organelles or by binding to induced proteins etc. is often irreversible requiring destructive recovery. The latter may be achieved by incineration or dissolution in strong acids or alkalis. The choice of metal recovery method used can depend on its commercial value. If a cheap and plentiful supply of waste biomass is used to recover valuable metals then the economics of destruction may be satisfactory. Most work has concentrated on non-destructive desorption which, for maximum benefit, should be highly efficient, economical and result in minimal damage to the biosorbent (19). Dilute mineral acids for removal of heavy metals although at higher concentrations (.>1 m) or with prolonged exposure, these may be damage to the biomass (92, 94). Sulphate ion in particular can cause irreversible alteration to cell wall biosorption sites in fungi which reduced the reuse potential of the bionmass. Between pH 5-7, a variety of metal cations including Cu^{2+}, Cr^{3+}, Ni^{2+}, Pb^{2+}, Zn^{2+}, Cd^{2+}, and Co^{2+} were strongly bound by algal biomass and there could be quantitatively released from the biomass using legand pH^2. By increasing the pH to higher values some anionic metal species e.g. SeO_4^{2-}, CrO_4^{2-} and MoO_4^{2-}, were removed after being bound at pH values around 2. However, Au^{3+}, Ag^+ and Hg^{2+} remained strongly bound at pH 2 and the addition of legands that formed stable complexes with metal ion was necessary to elute such metal ion that showed little pH dependence in biosoprtion. Mercapotethanol added at > pH 9 reversed Ag^+ biosorptoon by *Chlorella pyrenoidosa*, *Chlorella ulgaris* and *Spirulina platensis* although if high amounts of Ag^+ were taken up by the biomass, a reduction to pH 2 was effective, for removal such treatments revealed the existence of pH- independent primary strong binding sites e.g. sulphydryl groups, and pH dependent weaker secondary binding sites like carboxylate groups, to which Ag^+ may bind once the primary binding sites were saturated (35). It may be possible to apply selective desoprtion of chosen elements form a biosorbent loaded with a number of different elements with an appropriate choice of eluant. This has been demonstrated with polyacrylamide - immobilized biomass of *Chlorella vulgaris* loaded by exposure to a 0.1 M mixture of $aucl_4^{4-}$, Cu^{2+} , Zn^{2+} and Hg^2 at pH 6. At pH 2, Zn^{2+} and Cu^{2+} were desorbed from the biomass where as the other metals remained strongly bound Hg (II) and Au (III) were selectively removed using mercaptoethanol, at pH 2 and 5 respectively (4).

5. INDUSTRIAL AND ECONOMIC ASPECTS

In simple term biosorption is concidered to be a solid-liquid contact process comprising a metal/radionculcide uptake phase and a desorption phase. Contact between the contaminated solution and the biosorbent can be achieved in a variety of batch; semi-continuous or continuos flow reactors. These may be used in a series or in parallel , the latter increasing the operating capacity and facilitating biosorbent - regeneration or disposal. Simplicity of use, compactenss and cost efficiency are obvious key points (10). Separatioin of metal-saturated biosorbent for regeneration and metal recovery is usually an integral part of any process . For stirred batch and continuos flow reactors, separation of loaded biosorbent like setting, flotation, filtration or centrifugation, is possible with the biosorbent being-regenerated, ashed or disposed of (97). Ideally any process should produce low volume containable waste or elute. Several biosorbent systems described are based on packed - or fixed bed reactor system with effluent up flow or down flow (Fig. 1). Although the latter may be sensitive to suspended solids in the enfluent which may eventually clog the columns. A pulsating-bed contactor allows continuous operation by intermittent withdrawal of metal saturated biosorbent, for separate regeneration, and replacement by fresh biosorbent (97). Other methods include fludized - or expanded bed - systems (Fig. 2) which are well established in other industrial applications. These have effluent upflow with metal-saturated biosorbent being removed from the column base for regeneration (10). Several other configurations are possible including air-lift reactors which may be used in series to cope with large flow volume and provide adequate contact times (10, 14, 93, 101, 105) (Fig. 2). Naturally pelleted biomass from several fungal species removed thorium form solution in 1 M $HN0_3$, when used in air-lift reactor, and exhibited a well-defined breakthrough point after biosorbent saturation (28, 101). Separation of immobilized or particulate biosorbent from solution is relatively easy in contrast to non-immobilized biomass. However, in some cases freely suspended metal accumulating microbes may be separated by high-gradient magnetic separation (HGMS), a technique developed for extraction of weakly magnetic collides (10).

Industrial application of biosorption depends on such factors as loading capacities, efficiencies and selectivity, ease of metal recovery and an equivalenece at least to traditional physical and chemical treatments in performance (9, 10, 84, 93). In comparison with existing treatment methods, several biosorptive process offer advantages including high efficiency at low metal concentrations, low affinities for Ca^{2+} and Mg^{2+}, operation over broad pH and temperature ranges (particularly with dead biomass or derived products) in multiple biosorption -desoroption cycles, compatibility with conventional ion exchange columns and reactors , and low capital and running costs (97). Furthermore, microbial biomass may be supplied as a fermentation by-product or specifically grown using cheap substrates. Many biosorbent appear competitive in cost with ion exchange resins while frequently exhibiting higher efficiencies (4). The major disadvantages of traditional treatment methods include difficulties with suspended solids and land costs for evaporation lagooning; the requirement for low metal concentrations for membrane processes which are usually expensive and have a short-working life; high cost of chemicals for precipitation and clarification techniques and difficulties in disposing of metal hydroxide-contaminated sludges; loss of adsorbent efficiency with prolonged use and regeneration for activated carbon; interference by- Ca^{2+} and Mg^{2+}, chemical degradation and fouling, cost and sensitivity to thermal and osmotic shock for ion exchange resins (97). It has been suggested that biosorptive treatment and should exhibit >99% removals with metal loadings > 150 mg (g biomass)-1 for economic success although if decontamination of effluent for environmental discharge is the eventual aim, then such high biomass loadings may not be necessary. Most economic analyses have been made with the recovery of precious metals in mind. It should be stressed that the use of % removal as a term to describe the efficiency of a particular biosorption processes may be meaningless if the residual concentration is still above permissible limits; waste treatment processes may have to achieve discharge concentration of < 0.05 mgl-1 (10). Some processes e.g. evaporation and reverse osmosis, are economically feasible if metal concentrations are high. However it should be stressed that biosoprtive treatments need not necessarily replace existing methodologies but may be used as polishing systems for processes that are not completely efficient.

For commercial use, immobilized or pelleted preparations appear best with recovery utilizing a cheap desrobing agent which may be recycled (11, 93) several published commercial processes are competitive in terms of cost and operational characteristics with traditional treatments, and capable of handling effluent volume > 100 m^3 day^{-1}. However, significantly higher flow rates are not unusual in many industries, including those concerned with the

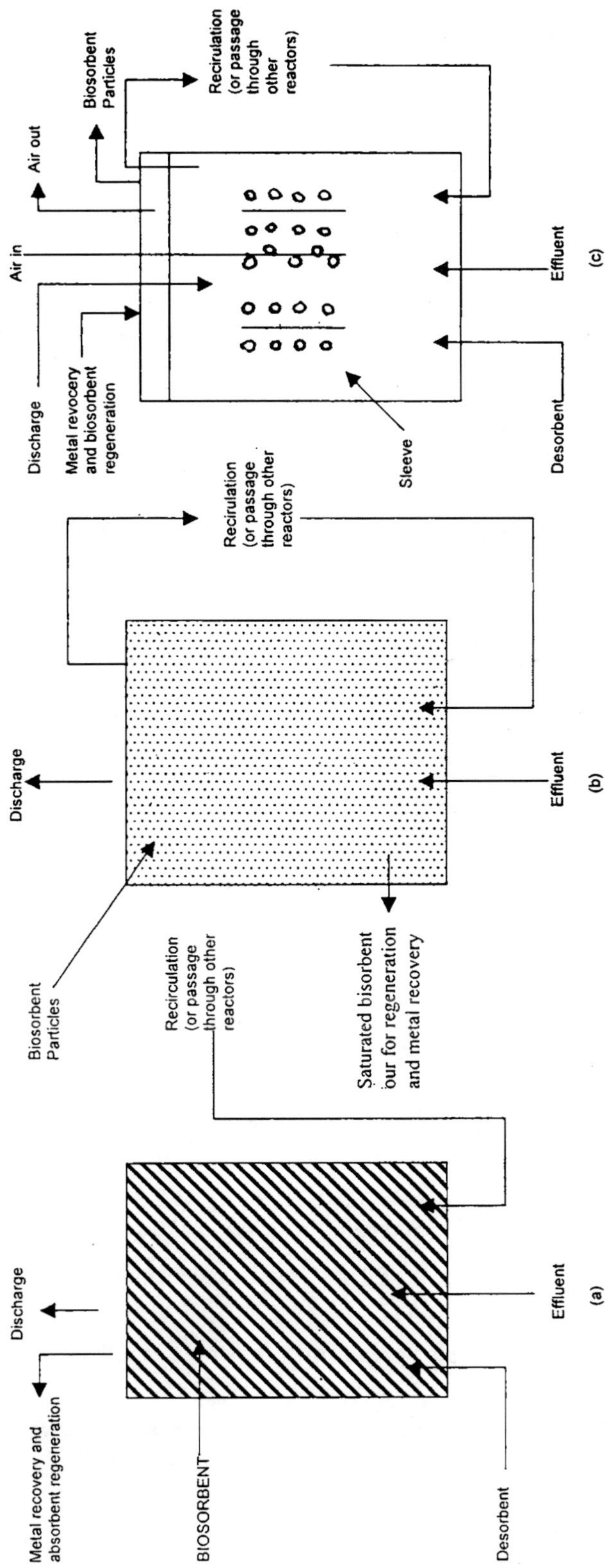

Fig.2: Simplified diagrame of (a) as packed bed, (b) a fluidized bed and (c), an airlift bioreactor for the removal of heavy metals and radionuclides by particulate or immobilized microbial biomass

provision of nuclear power (21) and these would require large amounts of biosorbent which may adversely influence operation and cost (59).

6. CONCLUSION

There is great awareness of the potential dangers of environmental pollution by heavy metal compounds as well as those radionuclides which arise in waste waters from nuclear power industry (21). The removal of these pollutant form contaminated solutions by living or dead in microbial biomass, and derived or excreted products can provide an economically feasible and technically efficient means for element recovery and environmental protection (53). Living cells have possibilities in some applications particularly where growth is allowed under metal contacting conditions or if metabolic products, e.g. H_2S, are involved in metal precipitation (3, 9). However, metal binding proteins, particulate metal accumulation, extra-cellular precipitation and complex formation by living cells have received less attention (48). For microbial exopolymers there is a current need to identify polymers selective toward metal like Cu, Cd, Co, Zn, Cr, and As with little reactivity towards, Na, K, Ca, or Mg (31). Many biological metal removing systems in commercial production use dead biomass, frequently immobilized in particulate form in processes analogous to those already used in ion exchange and activated carbon absorption and these appear competitive in capital and operational costs with existing treatment methods (93, 97) further development of this area.

7. REFERENCES

1. Ashley, N.V. and Roach, D.J.W. 1990. Review of biotechnology applications to nuclear waste treatment. *J. Chemi. Tech. Biotech.* **49** : 381-394.

2. Babich, H and Stozky, G. 1985. Heavy metal toxicity to microbe-mediated ecologic processes a review and potential application to regulatory policies. *Environ. Res.* **36** : 111-137.

3. Barnes, L.J., Janssen, F.J., Sherren, J., Versteegh, J.H., Koch, R.O. and Scheere, P.J.H. 1991. A new process for the removal of sulphate and heavy metals from contaminated water extracted by a geohydrological control system. In : *Research and Technology Symposium and Recruitment Fair Ist Chemical Environment Research Events*, Insitution of Chemical Engineers, Rugby pp. 33-36.

4. Bedell, G.W. and Darnall, D.W. 1990. Immobilization of nonviable, biosorbent algal biomass for the recovery of metal ions. In : *Biosorption of Heavy Metals* (ed. B. Volesky) CRC Press, Boca Raton, pp. 313-326,

5. Bell, A.A. and Wheeler, M.H. 1986. Biosynthesis and functions of fungal melanin's. *Ann. Rev. Phytopath.*, **24** : 411-451.

6. Berka, T., Shatzman, A., Zimmerman, J., Strickler, J. and Rosenberg, M. 1988. Efficient expression of the yeast metallothionein gene in *Escherichia coil. J. Bacteriol.* **170** : 21-26.

7. Beveridge, T.J. 1989. Role of cellular design in bacterial metal accumulation and mineralizatioin. *Ann. Rev. Microbiol.* **43** : 147-171.

8. Birch, L. and Bachofen, R. 1990. Complexing agents form microorganism. *Experentia* **46** : 827-834.

9. Brierley, C.L. 1990. Metal immobilization using bacteria. In : *Microbial Mineral Recovery* (eds. H.L. Ehrlich and C.L. Brierley) McGraw Hill Publishing Company, New York, pp. 303-323.

10. Brierley, J.A. 1990. Production and application of a *Bacillus*-based product for use in metals biosorption. In : *Biosoprtion of Heavy Metals* (ed. B. Volesky) CRC Press, Boca Raton, pp. 305-511.

11. Brierley, C.L., Brierley, J.A. and Davidson, M.S. 1989. Applied microbial processes for metals recovery and removal from wastewater. In : *Metal Ions and Bacteria* (eds. T.J. Beveridge and R.J. Doyle) John Wiley and Sons, New York, pp. 359-82.

12. Brown, N.L., Lund, P.A. and Ni Bhriain, N. 1989. Mercury resistance in bacteria. In : *Genetics of Bacterial Diversity* (eds. D.A. Hopwood and K.F. Chater) Academic Press, London, pp. 175-195.

13. Butt, T.R. and Ecker, D.J. 1987. Yeast metallothionein and applications in biotechnology. *Microbiolo. Rev.* **51** : 351-364.

14. Campbell, R. and Martin, M.H. 1990. Continouous flow fermentation to purify waste water by the removal of cadmium. *Water, Air, and Soil Pollution* **50** : 307-408.

15. Cooney, J. J. 1988. Microbial transformations of tin and tin compounds. *Indus. Microbio.* **3** : 195-204.

16. De Voe, I.W. and Holbein, B.E. 1985. Insolubel chelating compositions. US Patent 4530963.

17. De Voe, I.W. and Holbein, B.E. 1986. Insolubel chelating compositions. US Patent 4585 559.

18. Biomass for heavy metal radionuclide recovery. *Inds. Microbil.* **7**: 97-104.

19. Faison, B.D., Cancel, C.A., Lewis, S.N. and Adler, H.I. 1990. Binding of dissolved strontium by *Micrococcus luteus. App. Environ. Microbio.* **56**: 3649-3656.

20. Fogel, S., Welch, J.W. and Maloney, D.H. 1988. The molecular genetics of copper resistance in *Saccharomyuces cervisiae* - A paradigm for non-conventional yeast. *Basic Microbio.* **28** : 147-160.

21. Francies, A.J. 1990. Microbial dissolution and stabilization of toxic metals and radioinuclides in mix wastes. *Experientia.* **46** : 840-851

22. Gadd, G.M. 1986. The uptake of heavy metals by fungi and yeast; the chemistry and physiology of the process and applications for biotechnology. In : *Immobilizations of Ions by Biosorption* (eds. H. Eccles and S. Hunt) Chichester : Elis Horwood, pp. 135-147.

23. Gadd, G.M. 1988. Accumulation of metals by micro-organism and algae. In : *Biotechnology- A Comprehensive Treatise.* Vol. **6b**. *Special Microbial Processes*, (ed. H-J, Rehm) VCH VERlagsgesellschasft, Weinheim, pp. 401-433.

24. Gadd, G.M. 1990a. Heavy metal accumulation by bacteria and other micro-organisms. *Experentia.* **46** : 843-846.

25. Gadd, G.M. 1990b. Biosoprtion, Chemistry and Industry, Patent No.134216.

26. Gadd, G.M. 1992. Molecular biology and biotechnology of microbial interactions with organic and inorganic heavy metal compounds. In : *Molecular Biology and Biotechnology of Extremophiles* (eds. R.A. Herbert and R.J. Sharp) pp. 225-237.

27. Gadd, G.M. and Grifiths, A.J. 1978. Micro-organisms and heavy metal toxicity. *Microbial Ecology* **4**: 303-317.

28. Gadd, G.M. and White, C. 1988. The removal of thorium from simulated acid process streams by fungal biomass. *Biotech. Bioengineering* **33** : 592-597.

29. Gadd, G.M. Gray, D.J. and Newby , P.J. 1990. Role of melanin in fungal biosorption of tributyin chloride. *Appl. Microbio. Biotech.* **34** : 116-121.

30. Gee, A.R. and Dudeney, A.W.L. 1988. Adsorption and crystalization of gold at biological surfaces. In : *Biohydrometallugry*, (eds. P.R. Norris and D.P. Kelly) Science and Technology Letters, Kew, pp. 437-452.

31. Geesey, G and Jang, L. 1990. Extracellular polymers for metal binding. In : *Microbial Mineral Recovery* (eds. H.L. Ehrlich and C.L. Brierley) McGraw-Hill Publishing company, New York, pp. 223-247.

32. Gekesler, W., Grill, E., Winnacker, E.L. and Zenk, M.H. 1988. Algae sequester heavy metals via synthesis of phytochelatin complexes. *Archives of Microbiology* **150**: 197-202.

33. Ghirose, W.C. 1984. Biology of iron and maganese-depositing bacteria. *Ann. Rev. Microbio.* **37**: 515-550.

34. Goddard, P.A. and Bull, A.T. 1989. Accumulation of silver by growing and non-growing populations of *Citrobacter intermedius* B6. *App. Microbiol. Biotech.* **31**: 314-319.

35. Greane, B. and Darnall, D.W. 1990. Microbial oxygenic photoautotrophs (cyanobacteria and algae) for metal-ion binding. In : *Microbial Mineral Recovery* (eds., H.I. Ehrlich and C.L. Brierly) McGraw-Hill Publishing company, New York, pp. 277-302.

36. Harris, P.A. and Ramelow, G.J. 1990. Binding of metal ions by particulate biomass derived from *Chlorella vulgaris* and *Scendesmus quadricauda. Environ. Sci. Technol.* **24** : 220-228.

37. Hayashi, Y., Nakagawa, C.W., Uyakul, D., Imai, K., Isobe, M and Goto, T. 1988. The change of cystine components in Cd-binding peptides form the fission yeast during their induction by cadmium. *Biochem. and Cell Bio.* **66**: 288-295.

38. Holbein, B.E. 1990. Immobilization of metal-binding compounds. In : *Biosorption of Heavy Metals* (ed. B. Volesky) CRC Press, Boca Raton, pp. 327-338.

39. Holebin, B.E., Brener, D., Greer, C. and Brown, E.N.C. 1987. Insoluble compositions for removing mercury form a liquid medium. US patent 4654322.

40. Hollo, J., Toth, J., Tengerdy, R. P. and Johnson, J.E. 1979. Denitrification and removal of heavy metals form waste water by immobilized microorganisms. In : *Immobilized Microbial Cells*, (ed. K. Venkatsubranmanian) American Chemical Society, Washington, pp. 73-86.

41. Hunt, S. 1986. Diversity of biopolymer structure and its potential for ion binding applications. In: *Immobilisation of Ions by Bio-sorption*, (eds. H. Eccles and S. Hunt) pp. 15-46.

42. Huang, C.P. and Morehart, A.L. 1990. The removal of Cu(II) form dilute aqueous solutions by *Saccharomyces cervisiea. Water Research.* **24**: 433-439.

43. Hutchins, S.R., Davidson, M.S., Brierley, J. and Brierley, C.L. 1986. Microoganisms in reclamation of metals. *Ann. Rev. Microbio.* **40**: 311-336.

44. Inouhe, M., Hiyama, M., Tohoyama, H., Johao, M. and Murayama, T. 1989. Cadmium-binding protein in a cadmium-resistant strain of *Saccharomyces cervisiae. Biochimica et Biophysica Acta* **993**: 515-255.

45. Jackson, T.A. 1978. The biochemistry of heavy metal in polluted lakes and stream at Flin Flon, Canada, and a proposed method for limiting heavy-metal pollution of natural rivers. *Environmental Geology* **2**: 173-189.

46. Jones, R.P. and Gadd, G.M. 1990. Ionic nutrition of yeast-physiological mechanism involved and applications for biotechnology. *Enzyme Microbial Technology* **12**: 402-418.

47. Kaplan, D., Christiaen, D. and Arad, S.M. 1987. Chlating properties of extra-cellular polysaccharides from Chlorella spp. *Appl. Environ. Microbio.* **53**: 2953-2956.

48. Katiyar, S.K. 1997. Effects of Zinc on freshwater microbial communities of river Yamuna around Delhi, *Proc. Nat. Acad. Sci. India* **(B) 1**: 67-72.

49. Katiyar, S.K. and Katiyar, R. 1998. Effects of different concentrations of Zn on three native phytoplankton of river Yamuna (In Press).

50. Kauffman, J.W., Laughlin, W.C. and Baldwin, R.A. 1986. Microbiological treatment of uranium mine waters. *Environ. Sci. Techno.* **20**: 243-248.

51. Kierans, M., Staines, A.M. , Bennett, H. and Gadd, G.M. 1991. Silver tolerance and accumulation in yeast. *Biology of Metals* **4**: 100-106.

52. Kuhn, S. and Pfister, R.M. 1990. Accumulation of cadmium by immobilized *Zoolgoea ramigera* 115. *Journal of Industrial Microbiology* **6**: 123-8

53. Kujacak, N. 1990. Feasibility of biosorbent application. In : *Biosorptioin of Heavy metals* (ed. B. Volesky) CRC Press, Boca Raton, pp. 371-378.

54. Lester. J.N., Sterritt, R.M., Rudd. T. and Brown, M.J. 1984. Assessment of the role of the bacterial extra-cellular polymers in controlling metal removal in biological waste water treatment. In : *Micro-biological methods for Environmental Biotechnology*, (eds. J.M. Grainger and J.M. Lynch) Academic Press, London, pp. 197-217.

55. Lovely, D.R., Phillips, E.J.P., Gorby, Y.A. and Landa, E.R. 1991. Microbial reduction of uranium. *Nature*, **350**: 413-416.

56. Macaskie, L.E. 1990. An immobilized cell bio-process for the removal of heavy metals form aqueous flows. *Chem. Techno. Biotechno.* **49**: 357 -379

57. Macaskie, L.E. 1991. The application of biotechnology to the treatment of wastes produced from the nuclear fuel cycle; bio-degradation and bio-accumulation as a means of treating radioinuclide containing streams. *CRC Criti. Rev. Biotechno.* **II**: 41-112.

58. Macaskie, L.E. and Dean, A.C.R. 1987. Use of immobilized biofilm of Critrobacter sp. for removal of uranium and lead from aqueous flows. *Enzyme and Microbial Technology* **9**: 2-4.

59. Macaskie, L.E. and Dean. A.C.R. 1989. Microbial metabolism, desoulbilizatioin, and deposition of heavy metals; metal uptake by immobilized cells and application to the treatment of liquid wastes. In : *Biological Waste Treatment* (ed. A. Mizrahi) Alan R. Lss Inc., New York, pp. 150-201.

60. Macaskie, L.E. and Dean, A.C.R. 1990 : Metal sequestering biochemicals. In : *Biosorption and Heavy Metals*, (ed. B. Volseky) CRC Press, Boca Raton, pp. 199-248.

61. McLean, R.J.C and Beveridge, T.J. 1990. Metal-binding capacity of bacterial surfaces and their ability to form mineralized aggregates. In : *Microbial Mineral Recovery* (eds. H.L. Ehrlich and C.L. Brierly) McGraw-Hill publishing Company, New York, pp. 185-222

62. Mehra, R.K., Tarbet, E.B., Gray, W.R. and Winge, D.R. 1988. Metal-specific synthesis of two metallothioneins and γ-glutamyl peptides in *Candida glarbuta*. *Proceedings of the national Academy of Sciences*, USA. **85**: 8815-8819.

63. Menzini, G., Cesaro, A., Delben, F., Paoletti, S. and Reisenhofer, E. 1984. Copper^{++} binding by natural ionic polysaccharides, Part I: Potentionmetric and spectroscopic data. *Bio-electro-chemistry and Bioenergetics*. **12**: 443-454.

64. Mullen, M.D., Wolf, D.C., Ferris, F.G., Beveridge, T.J., Fleamming C.A. and Bailey, G.W. 1989. Bacterial sorption of heavy metals. *Applied and Environmental Microbiology* **55**: 3143-3149

65. Munger, K., Germann, N.A. and Lerch, K. 1985. Isolation and structural organization of the *Neurospora crassa* copper metallothionein gene. *EMBIO Journal* **4**: 1459-1462.

66. Muzzarelli, R.A.A., Bregani, F. and Sigon , F. 1986. Chelating abilities of amino acid gulcans and sugar acid gulcans derived form Chittosan. In Immobilisation of *Ions by Bio-sorption* (eds. H. Eccles and S. Hunt), Ellis Horwood, Chichester, pp. 173-82.

67. Nakijima, A., Horikoishi, T. and Sakaguchi, T. 1982. Recovery of uranium by immobilized micro-organism. *Euro. J. Appl. Micriobio. Biotechno.* **16**: 88-91.

68. Nakajima, A. and Sakaguchi, T. 1986. Selective accumulation of metals by micro-organisms. *App. Microbio. Biotechno.* **24**: 59-64.

69. Neilands, J.B. 1989. Siderophore systems of bacteria and fungi. In : Metal Ions and Bacteria, (eds. T.J. Beveridge and R.J. Doyle) John Wiley and Sons, New York, pp. 141-163.

70. Norberge, A. and Rydin. S. 1984. Development of a continuos process for metal accumulation by *Zogloea ramigera. Biotech. Bioengin.* **26**: 265-268.

71. Olfason, R.W. 1984. Prokaryotic metallothionein. *Interna. J. Pep. Res.* **24**: 303-308.

72. Premuzic, E.T., Francis, A.J., Lin, M. and Schbuert, J., 1985. Induced formation of chelating agents by *Pseduomanas aeruginosa* grown in the presence of thorium and uranium. *App. Environ. Microbio.* **14**: 759-768

73. Rauser, W.E. 1990. Phytochelatins. *Ann. Rev. Biochem.* **59**: 61-86.

74. Rayner, M.H. and Sadler, P.J. 1989. Cadmium accumulation and resistance mechanism in bacteria. In : *Metal-Microbe Interactions* (eds. R. K. Poole and G.M. Gadd) IRL Press, Oxford, pp. 39-47.

75. Reese, R.N., Mehra, R.K. and Tarbet, E.B. 1988. Studies on the γ-glutamyl Cu-binding peptide form *Schizosaccharomyces pombe. J. Biol. Chem.* **263**: 4186-4192.

76. Reese, R.N., and Winge, D.R. 1988. Sulfide stabilizatiion of the cadmium-γ-glutamyl peptide form *Schizosaccharomyces pombe. J. Biol. Chem.* **263**: 12832-51283.

77. Robinson, N.J. 1989. Algal metallothioneins: Secondary metabolites and proteins. *J. App. Phyco.* **1**: 5-18.

78. Rudd, T., Steritt, R.M. and Lester, J.N. 1984. Formation and conditional stability constants of complexes formed between heavy metals and bacterial extra-cellular polymers. *Water Research,* **18**: 379-384.

79. Sakaguchi, T. and Nakajima, A. 1982. Recovery of uranium by chitin phosphate and chitosan phosphaste. In : *Chitin and Chitosan* (eds. S. Mirano and Tokura) Japanese Society of Chitin and Chitosan, Tottori, pp. 177-182.

80. Sakaguchi, T. and Nakajima, A. 1987. Accumulation of uranium by biopigments. *J. Chem. Techno. Biotech.* **40**:133-141.

81. Scott, J.A. and Palmer, S.J. 1990. Sites of cadmium uptake in bacteria used for biosorption. *App. Microbio. Biotech.* **55**: 1153-1156.

82. Scott, J.A., Sage, G.K. and Palmer S.J. 1988. Metal immobilisation by microbial capsular coatings. *Bio-recovery.* **1**: 51-8.

83. Shumate, S.E. and Strandberge, G.W. 1985. Accumulation of metals by microbial cells. In : *Comprehensive Biotechnology* (ed. M. Moo-Young), C.N. Robinosn and J.A. Howell. Pergamon Press, New York, pp. 235-247

84. Silver, S.; Laddasga. R.A. and Misra, T.K. 1989. Plasmid determined resistance to metal ions. In: *Metal-Microbe Interactions* (eds. R.K. Poole and G.M. Gadd) Oxford: IRL Press, pp. 49-63.

85. Singleton, L., Wainright, M. and Edyvean R.G.J. 1990. Some factors influencing the absorption of particulate by-fungal mycelium. *Bio-recovery* **1**: 271-89.

86. Thayer, J.S. 1984. *Organometallic Compounds and Living Organisms.* Academic press, New York.

87. Thayer, J.S. 1989. Methylation: its role in the environmental mobility of heavy elements. *App. Organomet. Chem.* **3**: 123-128.

88. Tobin, J.M., Copper, D.G. and Nufeld, R.J. 1987. Influence of anions on metal adsoprtion by *Rhizopus arrhizus* biomass. *Biotechn. Bioengin.* **30**: 882-886.

89. Trevors, J.T. 1986. Mercury methyaltion by bacteria. *Jour. Basic Microbio.* **26**: 499-504.

90. Trevors, J.T., Oddie, K.M. and Belliveau, B.H. 1985. Metal resistance in bacteria. *FEMS Microbiology reviews* **32**: 39-54.

91. Tsezos, M. 1983. The role of chitin in uranium adsorption by *Rhizous arrizhus. Biotech. Bioengin* **22**: 2025-2040.

92. Tsezos, M. 1984. Recovery of uranium form biological absorbent-desorption equilibrium. *Biotech. Bioengin* **26**: 973-81.

93. Tsezos , M. 1990. Engineering aspects of metal binding by biomass. In :*Microbial Mineral Recovery* (eds. H.L. Ehrlich and C.I. Brierley), pp. 323-339.

94. Tsezos, M., Baird, M.H.I.and Shemilt, L.W. 1987. The elution of radium absorbed by microbial biomass. *Chem. Engin. Journ.* **34b**: 57-64.

95. Tsezos, M. and Deutschemann, A.A. 1990: An investigation of engineering parameters for the use of immobilized biomass particles in biosoprtion. *J. Chemi. Techn. Bioengin.* **48**: 29-39.

96. Tsezos, M. and Volesky, B. 1982. The mechanism of uranium biosoprtion by *Rhizopus arrhizus. Biotech. Bioengin.* **24**: 385-401.

97. Volesky, B. (ed) 1990. *Biosorption of Heavy Metals*, CRC Press, Boca Raton.

98. Wainwright, M. and Grayston, S.J. 1989. Accumulation and oxidation of metal sulphides by fungi. In : *Metal-microbe Interactions* (eds. R.K. Poola and G.M. Gadd) IRL Press, Oxford, pp. 119-130.

99. Wales, D.S. and Gadd Sagar, B.F. 1990. Recovery of metal ions by microfungal filters. *J. Chem. Techno. Biotechno.* **9**: 345-355.

100. Watson, J.S., Scott, C.D. and Faison, B.D. 1990. Evaluation of a cell-bioplymer sorbent for uptake of strontium form dilute solution. *Amercian Chmeical Society Symposium Series* **422**: 173-186

101. White, C. and Gadd, G.M. 1990. Biosorption of radioinuclides by fungal biomass. *Journal of Chemi. Techno. Biotechn.* **49** : 331-343

102. Wilkinson, S.C., Goulding, K.H. and Robinson, P.K. 1989. Mercury accumulation and volatilizatiion in immobilized system. *Biotechnology Letters* **11**:1.

103. Willaims, J.W. and Silver, S. 1984. Bacterial resistance and detoxification of heavy metals. *Enzyme and Microbial Technology* **6**: 530-537.

104. Winge, D.R., Reese, R.N., Mehra, R.K., Tarbet, E.B., Hughes, A.K. and Dameron, C.T. 1989. Structural aspects of metal (γ-glutamyl peptides). In : *Metal Ions Homeostais; Molecular Biology and Chemistry* (eds. D.H. Hamer and D.R. Winge) Alan R. Likss Inc., New York, pp. 301-311.

105. Yakubu, N.A. and Dudeney, A.W.L. 1986. Biosorption of uranium with *Aspergillus niger*. In : *Immobilisations of Ions in Bio-sorptioon* (eds. H. Eccles and S. Hunt) IRL press, pp. 183-200.

106. Zosim, Z., Gutnick, D. and Rosenberg, E. 1983. Uranium binding by emulsan and emulsanoskols. *Biotechno. Bioengin.* **25**: 1725-1735.

20 Root Exudation and its Implication on Rhizosphere Mycoflora

Rupam Kapoor

CONTENTS

ABSTRACT 352

1. INTRODUCTION 352
2. ROOT EXUDATION 352
   2.1 Composition of Root Exudates 353
   2.2 Site of Exudation 353
   2.3 Factors Affecting Root Exudation 353
   2.4 Rhizosphere Microflora 353
3. INFLUENCE OF ROOT EXUDATION OF RHIZOSPHERE MYCOFLORA 355
4. INTERRELATIONSHIP BETWEEN MYCORRHIZOSPHERE EFFECT AND ROOT EXUDATION 356
5. CONCLUSIONS 357
6. REFERENCES 357

Advances in Microbial Biotechnology
J.P. Tewari, T.N. Lakhanpal, Jagjit Singh, Rajni Gupta & B.P. Chamola (eds.)
APH Publishing Corporation, New Delhi - 110 002, India.

ABSTRACT

The loss of low-molecular weight soluble carbon compounds from intact roots into soil is a widespread phenomenon. It has been estimated that large part of carbon fixed by the plant is lost across the root-soil interface depending upon plant species, its genotype, age, stage of development and many abiotic factors. The quantity and quality of root exudates directly affect the soil microflora, including soil-borne pathogens and microbes for symbiotic association. Resting spores and sclerotia lying in the grip of soil fungistasis, may be induced to germinate when root exudation provides nutrients necessary to overcome the inhibitory factors. Certain root exudates affect the nutrient dynamics in the rhizosphere and are directly involved in mobilization of sparingly soluble minerals. Allelopathic effects and mycorrhizosphere effect in the soil may be, in part due to exudates.

1. INTRODUCTION

Plant growth and development are controlled largely by the soil environment in the root region, an environment which the plant itself helps to create and where microbial activity constitutes a major influencing force. Availability of nutrients in the rhizosphere is controlled by the combined effects of soil properties, plant characteristics and the interactions of plant roots with microorganisms and the surrounding soil (13). While there is no doubt that microorganisms play an important role in plant nutrition (23, 76), qualitative estimates of beneficial and/or harmful interactions between plants and microorganisms are largely speculative (23).

Different plant species as well as genotypes within a species, differently influence the quantitative and qualitative composition of microbial population in the rhizosphere (41, 49, 81). It was hypothesized that observed differences were due to quantitative and qualitative difference in root exudation among species or genotypes. Therefore understanding the relation of exudate-induced microbial activity in the rhizosphere to plant health and vigor is essential for the development of better crop production and system.

2. ROOT EXUDATION

The loss of carbon compounds from intact root into soil is a widespread phenomenon. It has been estimated that 1-40% of the total carbon fixed by the plant is lost across the root-soil interface indicating its importance as a major plant flux (50). Root exudates are metabolites released or leaked passively across the membrane from intact cells along a concentration gradient.

Root exudates differs from secretion -lysates, gases and mucilages. These differ in various ways such as mode of release, physical (solubility) and chemical characteristics, molecular weight, etc.

Water-soluble exudates comprise of low molecular weight substances which are lost passively without the involvement of metabolic activity. Exudation of a root proceeds along a concentration gradient (13).

Secretions are higher molecular weight substances, which depend on metabolic processes for their release. They can take place against electrochemical potential and chemical potential gradient (35).

Lysates are gelatinous materials at the surface of roots including bacterial cells, metabolic products, colloidal material, natural and modified mucilages and organic matter. Another term rhizodeposition is used to describe carbon loss from root which generates rhizosphere effect (50).

Gases are a part of low molecular weight exudates e.g. ethylene, carbon dioxide, hydrogen cyanide and sulphur dioxide (67).

Mucilage - covers the roots of many plants and is composed mainly of polysaccharides and polygalactonic acids of high molecular weight (80). When formed under non-axenic conditions, this mucilage is a mixture of plant and microbial origin and is termed as mucigel.

2.1 Composition of Root exudate

To know the effects of root exudates in the rhizosphere the requirement is to know the nature of exudation i.e., which all compounds can possibly be exuded by the root. Over 200 carbon compounds have been identified as root exudates (45) of which amino acids (26, 67, 68), sugars (18, 61, 78), organic acids (76, 18); nucleotides (29, 22); fatty acids; growth substances (30); enzymes (46, 20) and vitamins (65) are the major components of exudates. Other than these, few other compounds were also reported such as carbon and nitrogen (11); flavones (60); phytosiderophores, phenolics (54), etc.

2.2 Site of Exudation

Site of exudation is important as rhizosphere mycoflora is present around this root zone and varies with the quantity and quality of root exudate. Root caps and tips are site of active exudation, they release mucilaginous material and sloughed off cells (61, 26). Loss of sugars occur along the whole of the length of root, whereas loss of amino acids and organic acids mainly occur at the apical region of the root (Fig. 1) (40). Although root cap and tips are sites of active exudation, the main axis of root mostly releases the soluble and diffusible materials and some mucigel. Such losses occur all along the root. As all these releases are of passive nature, it can therefore be deduced that efflux is both temporally and spatially related.

The root exudates changes quantitatively and chemically, depending upon the stage from which it they have been collected, like germinating seeds or developing seedlings and not just different sites on roots e.g. spectrum of reducing sugars in bean seeds and seedling root exudates differed greatly (77).

2.3 Factors Affecting Root exudation

Several abiotic and biotic factors affect both quantity and quality of root exudation. The abiotic factors include pH, temperature, light intensity and stress conditions. The biotic factors include species, age and stage of development of plant. Besides these factors root exudation is also influenced by rhizosphere mycoflora.

2.4 Rhizosphere Microflora

Around the root, different types of microorganism are present. These are of diverse type in all respects including their behavioural pattern with roots. Amongst them, the ones which would affect the exudation are parasitic or symbiotic, as they would in one way or the other interact directly with the root.

Rovira and Davey (64) gave four ways in which microorganism may affect the nature of exudation:

(i) by affecting permeability of ıoot cells

(ii) affecting metabolism of roots

(iii) absorption of certain exuded compounds

(iv) altering nutrient status of the plant

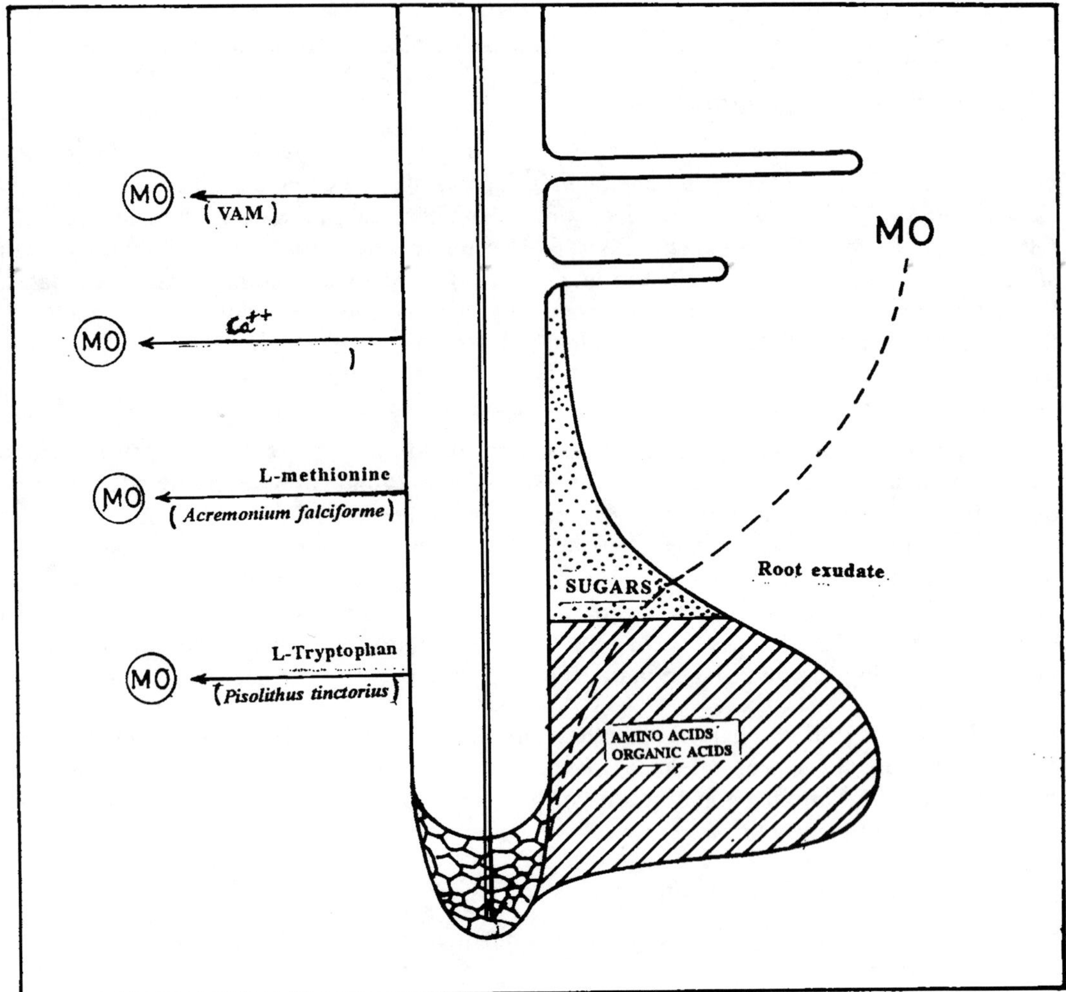

Fig.1: Schematic presentation of spatial separation of root exudates and microbial activity in the rhizosphere and possible role of low molecular weight root exudates on microorganisms (Mo) in the rhizosphere soil.

Pathogenic organism negatively affect the root and plant suffers comparatively more when infected by root diseases. Root exudation is significantly altered when root is associated with pathogens. Different plants behave differently, in the extent and quality of change in exudation when infected by a pathogen but in general, exudation is increased in their presence (52, 66). It was observed that wheat plants cultivated in the presence of *Pseudomonas putida* released double amount of exudates compared with control variants (58).

Dassi *et al.* (24) reported presence of chitinase and β-1, 3 glucanase isoforms in root exudates, during pathogenic and symbiotic associations with pea roots. Burseton (14) observed that root exudates of pea plants infected with *Fusarium solani* inhibit the germination of spores of *F. oxysporum* whereas root exudates from normal plants did not.

Both free living and symbiotic nitrogen fixers change root exudate composition. Lee and Gaskin (48) observed changes in flavonoid composition of root exudates in *Trifolium subterraneum* by the presence of *Rhizobium*. Kipe-Nolt *et al.* (42) observed variation in amounts of soluble carbon exuded from root of *Sorghum bicolor* in presence of diazotrophic bacteria. There are several reports on changes in root exudation on mycorrhization (7).

3. INFLUENCE OF ROOT EXUDATION ON RHIZOSPHERE MYCOFLORA

In order for fungus-root interaction to be initiated, dormant propagules of soil-borne pathogens must germinate. Soluble and volatile exudates from germinating seeds and developing roots are the primary stimuli for such responses through nutrient input and will stimulate propagule germination of soil fungi under certain conditions (6, 21). Propagules of nearly all common genera of soil-borne pathogens respond to seed or root exudates. However less is known about the specific exudate molecules that elicite these responses.

The particular compound in root exudates involved in stimulation of germination of soil-borne fungal propagule, could not be concluded in many studies. However, an evident stimulation of germination of many plants pathogenic fungi is reported in response to root exudates. Some of the significant fungi are, *Alternaria alternata* (36), A. *solani; Colletotrichum capsici, Curvularia lunata* (19), *Fusarium oxysporum* (71, 72); *F. solani* (39); *Verticillium dhahliae* (17), etc.

The specific soluble molecules involved in stimulating the soil-borne fungal propagules are largely unknown, but primarily sugars and amino acids are known to involve germination of *Verticillium albo-atrum, Phytophthora cinnamoni* and *P. megasperma* (28). Similarly germination of conidium of *Ceratosystis fimbricata* is stimulated by Ca^{++} in root exudates of *Ipomoea batatas* (Fig. 1) (44).

Mechanism involved in breaking dormancy of soil-borne pathogens by root exudates is yet to be found out. Many authors argue to the nutritive theory which states that many fungi require certain nutritive substance present in root exudates for germination (22, 31, 46). Conversely, some authors (34, 39) concluded that fungal toxins rather than the nutrients are the limiting factor for fungal growth in soil. Gowda *et al.* (32) visualised presence of a germination triggering factor in root exudates rather than nutrient factor responsible for germination of oospores.

The role of plant root exudates in attracting the spores of different fungi, i.e., chemotaxis, is summarized by many workers (4, 9). Chemotaxis is particularly important in case of species of *Phytophthora, Pythium, Aphanomyces* and *Cochliobolus* where in preparation phase of zoospores, after germination accumulate around the root in chemotactic response of root exudates.

Rhizosphere microflora or root exudate determines the resistance or susceptibility of the host. If given adequate inoculum of pathogen and the right type of root exudate then, the germination of pathogen propagule is higher probable. Even the plant which is susceptible genetically may escape infection if environment around the root is not optimal for pathogen germination. In addition to sugar, amino acids and organic acids, root exudates also contain a variety of toxic compounds viz., saponin, sapolatin, glyceride, HCN, etc. which influence activity of a pathogen on root surface (63).

The release of low molecular weight root exudates is of particular importance in relation to the nutrient dynamics in the rhizosphere. Some of these exudated solutes are in mobilization of sparingly soluble mineral nutrients, others more indirectly via affecting the microbial activity in the rhizosphere. For example tryptophan may serve as source for auxins production by ectomycorrhizal fungi such as *Pisolithus tinctorius* and methionine as source for ethylene production by soil fungi like *Acremonium falciformi.* Increased concentration of these phytohormones in the rhizospheres may strongly affect growth, morphology and physiology of the roots and, thus, indirectly acquisition of mineral nutrition (41, 61). Also, certain flavonoids in root exudates, such as luteolin and biochanin, many respectively act as 'signals' for rhizobium and as a stimulatory factor for hyphal growth of VAM fungi.

Manipulation of root-exudate stimulant of plant pathogens is a promising approach by which it is possible to interfere with pathogenic activity and achieve biological control (55).

4. INTERRELATIONSHIP BETWEEN MYCORRHIZOSPHERE EFFECT AND ROOT EXUDATION

With establishment of VA mycorrhizae there is accompanied slight change in root morphology (10, 27, 37, 38), but root physiology changes significantly, as does the physiology of the rest of the plant. For example, when plants become mycorrhizal there are changes in concentrations of growth-regulating compounds such as auxins, cytokinins and gibberellins (1, 8), photosynthetic rates increase, and the partitioning of photosynthate to shoots and roots changes (43, 73). The nutritional status of the host tissues changes in response to altered uptake of minerals from the soil and this in turn can modify structural and biochemical aspects of root cells that can alter membrane permeability and thus the quantity and quality of root exudation.

External hyphae of VAM fungi provide a physical or nutritional substrate for bacteria. Analysis of rhizosphere soil of VAM (*Glomus fasciculatum*) and non-VAM tomato plants (*Locypersicon esculentum* Mill.) by Bagyaraj and Menge (5) showed that greater population of bacteria and actinomycetes occured in the mycorrhizosphere, compared to the non-inoculated control. McAllister *et al.* (51) while studying the interaction between saprophytic fungi (*Aspergillus niger, Trichoderma koningii, Fusarium solani*) and *Glomus mosseae* observed that *G. mosseae* decreases the fungal population through its effect on plant. Modification in root exudates due to the VAM fungus has been proposed as an explanation for such findings (59, 67). Ames *et al.* (2) demonstrated mycorrhizosphere effects by adding culturally distinct bacteria to non-inoculated and *Glomus mosseae* inoculated pot-grown plants, and monitoring their population densities in the nonrhizosphere and rhizosphere soil. No population differences were observed in non-rhizosphere soil, but differential population shifts of introduced bacteria occurred between the VAM and non-VAM plants even though colonization in the VAM treatments was only 5.5%. Bansal and Mukerji (7) showed alterations and decrease in the total number of fungal colonies in the rhizosphere and rhizoplane of VAM inoculated *Lucaena leucocephala.* They observed that *G. macrocarpum* had a tendency towards the suppression of pathogenic fungi and stimulation of saprophytes. All these studies demonstrated that VAM fungi influenced the microbial populations in the mycorrhizosphere (41) and many of these microbial shifts could influence the growth and health of plants.

Many workers have correlated (7, 51) mycorrhizosphere effect with changes in root exudates. Colonization of VAM fungi is known to reduce the leakage of metabolites from roots of *Pinus radiata, Zea mays, Citrus aurantium* and *Sorghum* sp. (12, 33, 47, 59, 67, 68). In citrus seedlings inoculated with VAM fungi, exudation of reducing sugars and amino acids was lower in VAM than in non-VAM plants (25). Higher rates of root exudation (amino acids, sugars and phenolics) may also be the result of Zn deficiency and role of VAM in Zn acquisition is well documented (15).

In plants which typically form VAM, the quantitative exudation of metabolites as compared to non-mycorrhizal plant species is significantly higher (68). Even the extent of VAM colonization and proportion of root colonized is also correlated with exudation of carbohydrates from roots. Reduction in leakage of metabolites upon mycorrhizal formation in phosphorus deficient soils is also interpreted as conservation of photosynthates. The different species of plants vary in their ability to conserve photosynthates by reducing root exudation.

The nutritional status of the host tissue change in response to mycorrhization, this in turn can modify structural and biochemical aspects of root cells that can alter membrane permeability and thus the quantity and quality of root exudation (59). It has also been proposed that VAM fungi modify root exudation, thereby decreasing Mn acquisition by VAM plants either by decreasing the number of Mn-reducing bacteria (57) or by increasing the number of Mn-oxidising bacteria in the rhizosphere (3).

5. CONCLUSIONS

The rhizosphere is a dynamic microbial niche. The success of rhizosphere soil microbial technologies will depend on isolating and understanding the mechanism by which microorganisms influence plant growth as well as a basic understanding of the traits that constitute a competitive rhizosphere colonizer. Improved nutrition, status of the host by Vesicular-arbuscular mycorrhizal fungus has quantitative and qualitative effects on root exudates thereby altering rhizosphere microflora and population densities which might inhibit the pathogen resulting in increased tolerance of the plant towards the pathogen.

Integrated multi-disciplinary research is needed to understand the complex interactions of biotic, chemical and physical processes that interact to define the environment at the root surface and to define the exact mechanism underlying the changes in root exudation.

6. REFERENCES

1. Allen, M.F., Moore, T.S. and Christenen, M. 1980. Phytochrome changes in *Boutelou gracilis* infected by vescicular-arbuscular mycorrhizae. I Cytokinin increase in the host plant. *Can. J. Bot.* **58**: 371-374.

2. Ames, R.N., Reid, C.P.P. and Ingham, E.R. 1984. Rhizosphere bacterial populations responses to root colonization by a vesicular arbuscular mycorrhizal fungus. *New Phytologist* **96**: 555-563.

3. Arines, J., Porto, M.E. and Vilarino, A. 1992. Effect of Manganese on vesicular-arbuscular mycorrhizal development on red clover plant and on soil Mn-oxidising bacteria. *Mycorrhiza* **1**: 101-131.

4. Arshad, M. and Frankenberger, Jr. W.T. 1991. Microbial production of plant hormones. *Plant and Soil* **133**: 1-8.

5. Bagyaraj, D.J. and Menge, J.A. 1978. Interactions between a VA mycorrhizae and *Azotobacter* and their effects on rhizosphere microflora and plant growth. *New Phytologist* **80**: 567-573.

6. Baker, K.F. and Cook, R.J. 1982. Biological control of plant pathogens. *Annals of Phytopathological Society*, St. Paul Hin.

7. Bansal, M. and Mukerji, K.G. 1994. Positive correlation betweem VAM induced changes in root exudation and mycorrhizosphere mycoflora. *Mycorrhiza* **5**: 39-49.

8. Barea, J.M. 1986. Importance of hormones and root exudates in mycorrhizal phenomena. In: *Physiological and Genetical Aspects of Mycorrhizae* (eds. V. Gianinazzi-Pearson and S. Gianinazzi) INRA, Paris pp. 177-187.

9. Baur, W.D. and Caetano-Annolies, G. 1990. Chemotaxis, induced gene expression and competitiveness in the rhizosphere. *Plant and Soil* **129**: 45-52.

10. Berta, G., Fusconi, A. Trolta, A. and Scannerini, S. 1990. Morphogenetic modifications induced by the mycorrhizal fungus, *Glomus* strain E-3 on the root system of *Allium porrum* L. *New Phytologist* **114**: 207-215.

11. Biondini, M., Klein, D.A. and Redeute, E.F. 1988. Carbon and nitrogen losses through root exudation by *Agropyron cristalum, A. smithitic* and *Boutiloua gracilis. Soil Biology and Biochemistry* **20**: 477-482.

12. Bowen, G.D. 1969. Nutrient status effects on loss of amides and amino acids from pine root. *Plant and Soil* **30**: 139-142.

13. Bowen, G.D. and Rovira, A.D. 1992. The hidden half of the hidden half. In: *Roots*: *The Hidden Half* (eds. Y. Warsel, A Eshel and U. Kaf Kafe) Marcel Dekker Inc. New York USA, pp. 641-669.

14. Burston, E.W. 1957. Some effects of pea root exudates on physiologic races of *Fusarium oxysporum* f. sp. *pisi* (Lin.) Synder and Hansen, *Transactions of British Mycological Society* **40**: 145-154.

15. Cakmak, I. and Marschner, H. 1988. Increase in membrane permeability and exudation in roots of zinc deficient plants. *Journal of Plant Physiology* **132**: 356-361.

16. Catska, V. 1980. Effect of volatile and gaseous metabolites of germinating pea seeds on micro mycetes. *Folia Microbiology* **25**: 174-176.

17. Catska V., 1994. Interrelationship between vesicular arbuscular mycorrhiza and rhizosphere microflora on apple replant disease. *Biology of Plants* **36**: 99-104.

18. Chang, Ho Y. 1970. The effect of pea root exudate on the germination of *Pythium aphanidermatum* zoospore cysts. *Can. J. Bot.* **48**: 1501-1514.

19. Chaturvedi, S.N., Siradhana, B.S. and Muralla, R.N. 1974. The influence of cumin seed exudates on fungal spore germination. *Plant and Soil* **39**: 49-56.

20. Chhonkar, P.K. and Pal, S. 1987. Locus of Urease (Carbamido-amindo hydrolase) production in wheat soil ecosystem; an *in vitro* study. *Journal of Indian Society of Soil Science* **35**: 400-403.

21. Cook, R.J. and Baker, K.F. 1983. The nature and practice of biological control of plant pathogens. *American Phytopathological Society*, St. Paul. Minn. pp. 539.

22. Curl, E.A. and Truelove, B. 1986. *The rhizosphere.* Springer-Verlag, Berlin, Heidelberg, New York/Tokyo.

23. Darrah, P.R. 1993. The rhizosphere and plant nutrition: a quantitative approach. *Plant and Soil* **155/156**: 1-20.

24. Dassi, B., Dum as-Gaudot, E., Asselin, A., Richard, C. and Gianinazzi, S. 1996. Chitinase and β-1,3-gluconase isoforms expressed in pea roots inoculated with arbuscular mycorrhizal or pathogenic fungi. *European Journal of Plant Pathology* **102**: 105-108.

25. Dixon, R.K., Garrett, H.E. and Cox, G.S. 1988. Carbohydrate relationships of *Citrus jambhiri* inoculated with *Glomus fasciculatum. Journal of American Society of Horticultural Science* **113**: 239-242.

26. Egeraat, A.V. 1975. Exudation of ninhydrin positive compounds by pea seedling roots: A study of sites of exudation and of the composition of exudates. *Plant and Soil* **42**: 37-47.

27. Fitter, A.H. 1985. Functional significance of root morphology and root system architecture In: *Ecological Interactions in Soil* (eds. A.H. Fitter, D. Atkinson and D.J. Read), Blackwell Scientific Publications, Oxford, pp. 87-106.

28. Forster, M., Ribiero, O.K. and Erwin, D.C. 1983. Factors affecting oospore germination of *Phytophthora megaspermea* f. sp. *medicaginis. Phytopathology* **73**: 442-448.

29. Fries, N. and Forsman, B. 1951. Quantitative determination of certain nucleic acid in pea root exudate. *Physiologia Plantarum* **4**: 410-420.

30. Gogala, N. 1991. Regulation of mycorrhizal infection by hormonal factors produced by hosts and fungi. *Experientia* **47**: 331-339.

31. Good, M.H. and Spanis, W. 1956. Some factors affecting the germination of spores of *Fomes igniarius* var. *populinus* (Neuman) Campbell and significance of these factors in infections. *Can. J. Bot.* **36**: 421-437.

32. Gowda, P.S., Bhavanishankara and Shankara, B.S. 1986. Germination of oospores of *Peronosclerospora sorghi. Transactions of British Mycological Society* **87**: 653-655.

33. Graham, J.H., Leonard, L.T. and Menge, J.A. 1981. Membrane mediated decrease in root exudation responsible for phosphorus inhibition of vesicular arbuscular mycorrhizae formation. *New Phytologist* **91**: 683-690.

34. Griffin, G.J. and Roth, D.A. 1979. Nutritional aspects of soil mycostasis. In: *Soil Borne Plant Pathogens* (eds. B. Schippers and W. Cams) Academic Press, London. New York, pp. 79-96.

35. Hale, M.G., Moore, L.D. and Griffin, G.J. 1978. Root exudates and exudation. In: *Interations between Non-pathogenic Soil Microorganims and Plants* (eds. Y.R. Dommergues and S.V.B. Krupa) Elsvier, Amsterdam, pp. 163-203.

36. Harman, G.E., Nedrow, B. and Nash, G. 1978. Stimulation of fungal spore germination by volatiles from aged seeds. *Can. J. Bot.* **56**: 2124-2127.

37. Hetrick, B.A.D., Wilson, G.W.T. and Schwab, A.P. 1994. Mycorrhizal activity in warm and cool-season grasses: variation in nutrient uptake strategies. *Can. J. Bot.* **72**: 1002-1008.

38. Hooker, J.E., Munro, H. and Atkinson, D. 1992. Vesicular-arbuscular mycorrhizal fungi induced alteration on poplar root system morphology. *Plant and Soil* **145**: 207-214.

39. Jakson, R.M. 1957. Fungistasis, as a factor in rhizosphere phenomenon. *Nature* **180**: 96-97.

40. Jones, D.L. and Darrah, P.R. 1994. Amino acid influx and efflux at the soil-root interface of *Zea mays* L. and its implication in the rhizosphere. *Plant and Soil* **163**: 1-12.

41. Khanna, R., Chandra, S. and Khanna, K.K. 1993.. Rhizosphere mycoflora of triticale. *Biological Memoirs* **19**: 111-121.

42. Kipe-Nolt, J.A., Avalakki, U.K. and Dart, P.J. 1985. Root exudation of Sorghum and utilization of exudates by Nitrogen-fixing bacteria. *Soil Biology and Biochemistry* **17**: 859-863.

43. Koch, K.E. and Johnson, C.R. 1984. Photosynthate partitioning in split-root citrus seedlings with mycorrhizal and non-mycorrhizal root systems. *Plant Physiology* **75**: 26-30.

44. Kojima, M. and Uritani, L. 1978. Studies on factors in sweet potato root which induce spore germination of *Ceratocystis fimbriata. Plant Cell Physiology* **19**: 91-97.

45. Kroffczyk, I., Trolldeiner, G. and Beringer, H. 1984. Soluble root exudates of maize. Influence of potassium supply and rhizosphere microorganism. *Soil Biology and Biochemistry* **16**: 315-322.

46. Krasilinikov, N.A. 1952. Elimination of enzymes by roots of higher plant. *Dokl. Acad. Nacck. SSSR*. **87**: 309-312.

47. Laheurte, F. and Berthelin, J. 1988. Effect of phosphate solubilising bacteria on growth and root exudation over four levels of labile P. *Plant and Soil* **105**· 11-17.

48. Lee, K.J. and Gaskin, M.H. 1982. Increased root exudation of "C" compounds by sorghum seedlings inoculated with nitrogen fixing bacteria. *Plant and Soil* **69**: 391-399.

49. Lemanceau, P., Corberand, T., Gardan, L., Latour, X., Laguerre, G., Boeufgras, J.M. and Alabouvette, C. 1995. Effect of two plant species *(Linum usitatissitntim* L.) and tomato *(Lycopersicon esculentum* Mill) on the diversity of soil borne populations of flourescent pseudomonads. *Applied Environmental Microbiology* **61**: 1004-1012.

50. Lynch, J.M. and Whipps, J.M. 1990. Substrate flow in the rhizosphere. *Plant and Soil* **129**: 1-10.

51. McAllister, C.B., Garcia-Romera, I., Martin, J., Crodias, A. and Ocampo, J.A. 1995. Interaction between *Aspergillus niger* van Teigh and *Glomus mosseae* (Nicol & Gerd.) Gerd & Trappe. *New Phytologist* **129**: 309-316.

52. Meharg, A.A. and Killham, K. 1995. Loss of exudates from the roots of perennial rye grass inoculated with a range of microorganisms. *Plant and Soil* **170**: 345-349.

53. Morandi, D. 1996. Occurrence of phytoalexins and phenolic compounds in endomycorrhizal interactions and their potential role in biological control. *Plant and Soil* **185**: 241-251.

54. Nagahashi, G., Douds, Jr. D.D. and Abney, G.D. 1996. Phosphorus amendment inhibits hyphal branching of the VAM fungus *Gigaspora margarita* directly and indirectly through its effect on root exudation. *Mycorrhiza* **6**: 403-408.

55. Nelson, E.S. 1990. Exudate molecules initiated fungal responses to seeds and roots. *Plant and Soil* **129**: 61-73.

56. Papavisas, G.C. and Lumsden, K.D. 1980. Biological control of soil-borne fungal propagules. *Annual Review of Phytopathology* **18**: 389-393.

57. Posta, K., Marschner, H. and Romheld, V. 1994. Manganese reduction in the rhizosphere of mycorrhizal and non-mycorrhizal maize. *Mycorrhiza* **5**: 119-124.

58. Prinkyl, Z and Vancura, V. 1980. Root exudates of plant VI. Wheat Root exudation as dependent on growth, concentration, gradient of exudates and the presence of Bacteria. *Plant and Soil* **57**: 69-83.

59. Ratnayake, M., Leonard, L.T. and Menge, J.A. 1978. Root exudation in relation to supply of phosphorus and its possible relevance to mycorrhizal formation. *New Phytologist* **81**: 543-552.

60. Redmond, J.W., Battey, M., Djordjevie, M.A., Innes, R.W., Knempel, P.L. and Rolfe, B.G. 1986. Flavones induce expression of nodulation genes in *Rhizobium. Nature* **323**: 631-635.

61. Rengel Z. 1997. Root exudation and microflora populations in rhizosphere of crop genotypes differing in tolerance to micronutrient deficiency. *Plant and Soil* **196**: 255-260.

62. Rovira, A.D. 1956. Plant root excretions in relation to the rhizosphere effect. I - The nature of root exudates from oats and peas. *Plant and Soil* **7**: 178-194.

63. Rovira, A.D. 1969. Plant root exudates. *Botanical Review* **35**: 35-37.

64. Rovira, A.D. and Davey, C.B. 1974. Biology of the rhizosphere. In: *The Plant Root and its Environment* (ed. E.D. Carson), University Press of Virginia, Virginia, pp. 153-204.

65. Rovira, A.D. and Harris, J.R. 1961. Plant root excretion in relation to the rhizosphere effect. V. The exudation of the B-group vitamins. *Plant and Soil* **15**: 199-214.

66. Rovira, A.D., Bowen, G.D. and Foster, R.C. 1983. The significance of rhizosphere microflora and mycorrhiza in plant nutrition. In: *Encyclopedia of Plant Physiology* (eds. A. Lauchi and R.L. Bicheski) New Series, Vol. 15A, New York, pp. 61-68.

67. Schwab, S.M., Menge, J.A. and Leonard, R.T. 1983a. Quantitative and qualitative effects of phosphorus on extracts and exudates of sudangrass in relation to vesicular-arbuscular mycorrhiza formations. *Plant Physiology* **73** : 761-765.

68. Schwab, S.M., Menge, J.A. and Leonard, R.T. 1983b. Comparison of stages of vesicular-arbuscular mycorrhiza formation on sudan grass at two levels of phosphorus nutrition. *American Journal of Botany* **70** : 1225-1232.

69. Schwab, S.M. Leonard, R.T. and Menge, J.A. 1984. Quantitative and qualitative comparison of root exudates of mycorrhizal and non-mycorrhizal plant species. *Can. J. Bot.* **62**: 1227-1231.

70. Simon, E.W. 1974. Phospholipid and plant membrane permeability. *New Phytologist* **73**: 377.

71. Sivan, A. and Chet, I. 1989. The possible role of competition between *Trichoderma harzianum* and *Fusarium oxysporum* on rhizosphere colonization. *Phytopathology* **79**: 198-203.

72. Smith, W.H. 1977. Root exudates of seedlings and mature sugar maple. *Phytopathology* **60**: 701-703.

73. Snellgrove, R.C., Splittstrosser, W.E., Stribley, D.P. and Tinker, P.B. 1982. The distribution of carbon and the demand of the fungal symbiont in leek plants with vesicular arbuscular mycorrhizas. *New Phytologist* **92** : 75-87.

74. Thompson, K.K. and Hale, M.G. 1983. Effects of Kinetin in rooting medium on root exudation of free fatty acids and sterols from roots of *Arachis hypogea* L. Argentine under axenic conditions. *Soil Biology and Biochemistry* **15**: 125-126.

75. Tinker, P.B. 1984. The role of microorganism in mediating and facilitating the uptake of plant nutrients from soil. *Plant and Soil* **76**: 77-91.

76. Vancura, V. 1964. Root exudates of plants. I. Analysis of root exudates of barley and wheat in their initial phases of growth. *Plant and Soil* **21** : 231-248.

77. Vancura, V. and Hauzlikova, A. 1972. Root exudates of plants. IV. Differences in chemical composition of seed and seedling exudates. *Plant and Soil* **36**: 271-282.

78. Vejsadova, H., Hrsdora, H., Prikryl, S. and Vancura, V. 1989. Effect of different phosphorus and nitrogen levels on development of VA mycorrhiza, rhizobial activity and soybean growth. *Agriculture, Ecosystem, Environment* **29**: 429-434.

79. Watt, M., Van Des Weele, C.M., Mccully, H.E. and Canny, M.J. 1996. Effect of local variations in soil moisture on hydrophobic deposits and dye diffusion in corn roots. *Bot. Acta* **109**: 492-501.

80. Wiche, W. and Hofflich, G. 1995. Survival of plant growth promoting rhizosphere bacteria in the rhizosphere of different crops and migration to non-inoculated plants under field conditions in north-east Germany. *Microbial Research* **150**: 201-206.

21 Microbial Ecology of Plant Surfaces Living and Conserved Materials

Anjula Pandey

CONTENTS

ABSTRACT 364

1. INTRODUCTION 364
2. ECOLOGY OF MICROFLORA 364
   2.1 Microflora Associated with Seed Surfaces 366
   2.2 Terrestrial Subterranean Plant Organ Surfaces and Associated Microflora 366
   2.3 Microflora on Terrestrial Aerial Plant Surfaces 369
   2.4 Microbial Colonization on Dried or Preserved Materials 373
3. CONCLUSIONS 374
4. REFERENCES 375

Advances in Microbial Biotechnology
J.P. Tewari, T.N. Lakhanpal, Jagjit Singh, Rajni Gupta & B.P. Chamola (eds.)
APH Publishing Corporation, New Delhi - 110 002, India.

ABSTRACT

The plant surfaces provide distinct habitats for various microbes that exist at the interface of atmospheric and subterranean regions. The microflora associated with different plant organs are influenced by various physical, physiological and genetic factors which are responsible for bringing about the qualitative and quantitative variations.

The present paper includes different aspects of the microflora on dormant, subterranean and aerial terrestrial plant organ surfaces and materials of plant origin in conserved form. The understanding of their role as symbionts, antagonists, parasites, saprophytes or biocontrol agents would help in better reviewing of the complex phenomenon of plant-environment-dynamics and thus its manipulation. The success of different microbial techniques surely depends on understanding the mechanisms by which plants are influenced by microbial interactions.

The role of different causal factors responsible for the process of biodeterioration is important and thus, its management. By manipulating the storage conditions, this process could be reduced to an appreciable levels.

1. INTRODUCTION

The soil is a living system with dynamic physico-chemical, chemical and biological properties. It supports growth of various biological systems including plants, and also serves as a store-well for organic residues of plants and animal wastes in the form of disintegrated and decomposed material. In this process of decomposition and disintegration, the nutrients from soil are available to plants by means of their intimate relationships with the root system.

The microbial community may be viewed as a composite of small interconnected dynamic systems within the soil or above terrestrial regimes. The type of microflora associated with surfaces of different plant organs is affected not only by the general climate within which the plant is growing but also the gross morphology of the plant, the fine structures of its surfaces and stage of development. These factors must be added to the nature of biochemical compounds, principal availability of nutrients and presence or absence of fungistatic compounds, in addition to the pre-determined composition and physiology of the host organs and its microflora (both of which vary with environmental conditions and stage of host). The living plant surfaces of terrestrial plants, have aerial and subterranean organs, which chiefly differ in the physico-chemical and biotic factors, and are directly or indirectly interconnected (Fig. 1). The dormant, active and conserved plant materials harbour different microorganisms on their surfaces.

2. ECOLOGY OF MICROFLORA

For the purpose of discussion of the present topic, following major types of plant surfaces have been dealt with respect to ecology of microflora.

2.1 Microflora associated with seed surfaces

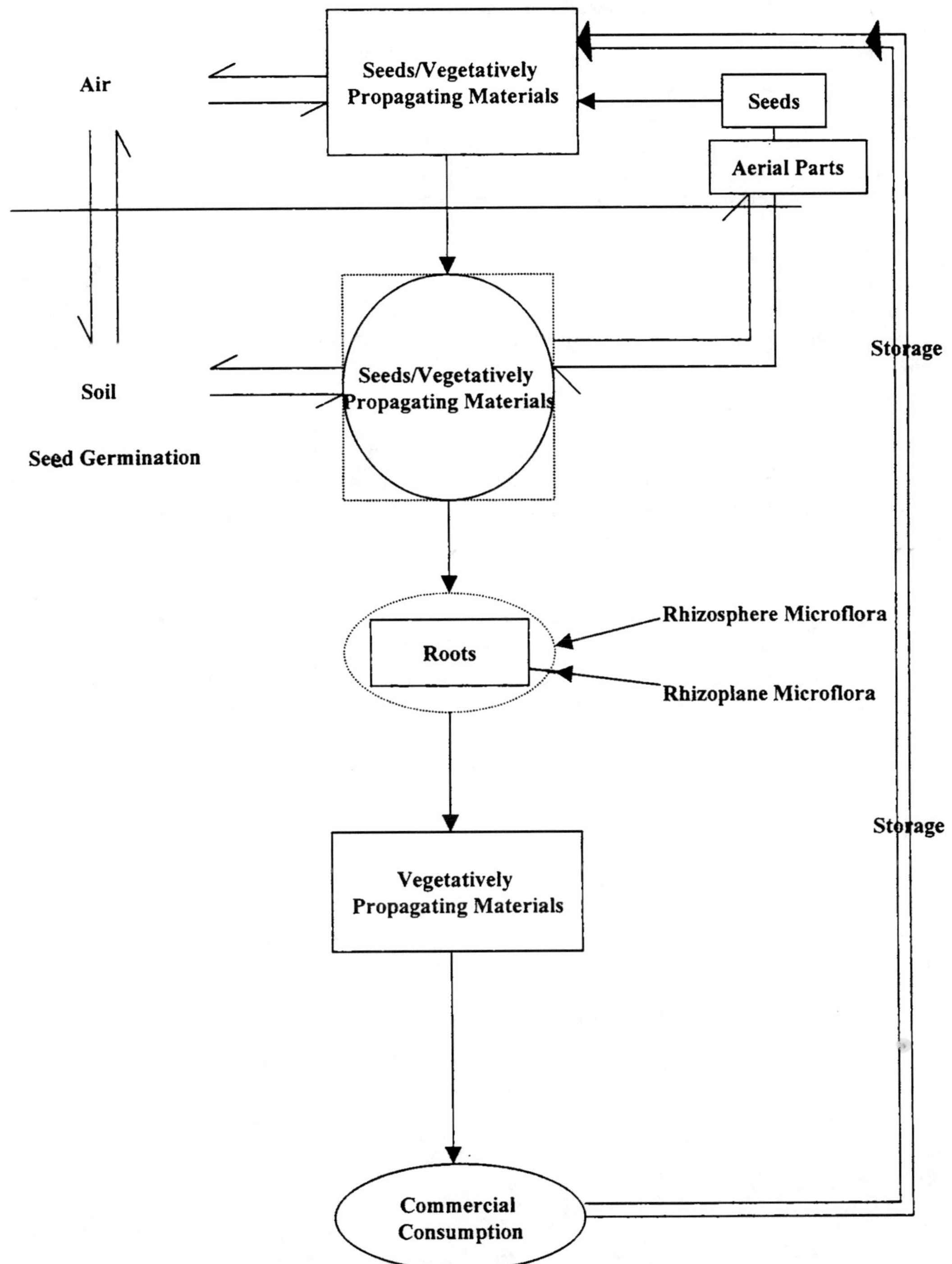

Fig. 1 : Schematic representations of origin and development of the Microflora on living plant system surface.

2.2 Terrestrial subterranean plant organ surfaces and associated microflora

2.3 Microflora on terrestrial aerial plant surfaces

2.4 Microbial colonization on dried or preserved materials

2.1 Microflora Associated with Seed Surfaces

Microbial succession follows a defined pattern in soil ecosystems during the plant growth from seed germination to maturity. The seeds of different plant species harbour seed-borne microflora, present on external and internal surfaces.

The seed may exude large amounts of organise compounds either through the micropile or broken surface (cracks) on the seed coat, that stimulate microbial activity at distance than comparable with those from roots (75, 100). Exudation from seeds, as in roots, stimulates the microbial activity that develops around the micropile and may compete with oxygen demand of the embryo. The resultant oxygen deficiency around the embryo induces more metabolite leakage and microbial growth (40, 62).

The region of influence around the seed is known as 'Spermosphere' and the phenomenon as 'Spermosphere effect'. When a seed carrying the natural and artificially pre-treated microorganism is sown in soil, certain microorganisms are selectively activated while others remain suppressed. The soil microorganisms compete with seed-borne microbes, for space and nutrients.

During germination, the emerging root starts sharing the spermosphere microflora and constitute the rhizosphere by selecting a few additional from soils. The nature of microorganisms moving from spermosphere to root depends on selective preference of the microbial elements, soil-plant species relationship and other factors. The epiphytic microflora of the above ground plant parts is broadly common to the seed microflora. This is supported by the fact that on seed treatment with fungicides, epiphytic microflora shows remarkable reduction. Pugh and Buckley (79) demonstrated that *Aureobasidium pullulans* isolated from the buds was common to the surface sterilised seeds, the hypocotyl and young roots of the seedlings of *Acer pseudoplatanus.*

In the seeds infected with pathogens, internally and externally, the pathogens compete with other saprobes for space and nutrients, and take over the lead to cause diseases in plants. In pretreated seeds (with chemicals or *Rhizobium),* the inoculum of useful microbes competes and does not allow the harmful microorganisms to proliferate and cause virulency in the crops.

2.2 Terrestrial Subterranean Plant Organ Surfaces and the Associated Microflora

The selective colonisation of the plant root surface by microorganisms was first reported by Hiltner (42) from Germany in 1904 and this phenomenon was known as 'Rhizosphere effect'. During the plant growth, there exists a dynamic equilibrium between root microorganism and those capable of competing with others and establishing, multiplying and attaining a dominant position.

Microbial ecology of the rhizosphere refers to the study of the interactions of microorganisms with each other and the environment surrounding the plant roots. The term rhizosphere is generally defined as the volume of soil that is adjacent to and influenced by the plant roots. The term comes from the Greek word for root (Rhizo or Rhiza) and includes both, the area of influence and the physical location around the root (sphere). The ecological niche created by roots, is a rich source of microbial nutrients. This defines two habitats, the root surface or Rhizoplane, and the Rhizosphere, a zone of enhanced nutrient status around the root resulting from the production of root debris and the emanation of diffusates.

The rhizosphere has been further subdivided into ectorhizosphere (outer rhizosphere) and the endorhizosphere (inner rhizosphere), where invasion or colonisation of root cortical cells by soil microorganisms occurs (26) called the mycorrhizosphere when there are mycorrhizal fungi associated with roots (59). The rhizosphere and the bulk soil merge one into the other without defining the boundary. The diametef of the rhizosphere approximates to the cylinder of soil in which root hairs explore and release exudates.

As a result of root growth, a new root area is generated which creates high nutrient conditions for brief duration. Subsequently because of the high nutrient status of rhizoplane and rhizosphere, their relatively equable physico-chemical environment and high density of soil-borne microorganisms, there is intense microbial competetion for availability of resources (13). Several works have reviewed on the microbial interactions with roots (5, 10, 12, 13, 20, 101, 108). Although the crop plants, in relation to pathogenis fungi in the rhizosphere have been studied greatly yet comparatively less work has been done on dynamics of saprophytes.

2.2.1 The Rhizosphere environment in microbial dynamics

The rhizosphere soil usually contain much higher fungal numbers than non-rhizosphere soil (38). The tropical soil with high rate of decomposition generally have more or less similar microflora in non-rhizosphere and rhizosphere soils, i.e. qualitative variations in microflora are less pronounced (70). The rhizosphere activity of the host may selectively favour the growth of some microorganisms that others (82, 103). The mycoflora of roots of white clover (*Trifolium repens*) had *Fusarium oxysporum, Cylindrocarpon destructans* and *Gliocladium roseum* and rye grass (*Lolium perenne*) from same pasture was having *Rhizoctonia, Trichoderma,* and *Penicillium* sp. together with *F. culmorum* which was virtually restricted to their location (108).

There exists an equilibrium between plant pathogens and the soil microflora. In case of disease occurrence, the pathogenic microflora breaks the barriers maintained between soil microorganisms in the rhizosphere and other factors and penetrates to cause infection.

The qualitative and quantitative variations of microorganisms are reported in the rhizosphere of diseased and resistant crop varieties from those of susceptible ones. There exists an antagonism between the pathogenic and saprophytic microflora of the resistant varieties. Using artificial amendments such as fertilizer application and organic amendments of soil and foliar chemical sprays, the dynamics between the two can be altered to cause favourable balance in the susceptible varieties. The herbicides are known to increase the exudation potential of roots (46).

The plant roots create a selective environment for microbial population (64). The nutrient exuded from the roots differ from plant species to species, at development stages and in different environments, which subsequently result in the qualitative and quantitative variations in microflora. These factors result in altered microbial population within the rhizosphere which differs from the microbial population in root free soil (42).

The diffusible exudates create gradation in the nutrition zone of influence of the roots from rhizoplane to rhizosphere. These principally consist of sugars such as oligosaccharides, amino acids, together with organic acids and miscellaneous organic compounds (45, 89). The zone behind the tip, extends to the root-hair zone is the area of maximum exudation (74, 88).

The differences between pH of leachates with and without plants could be an index of the chemical reactions occurring in the rhizosphere (95).

The root exudates have both stimulatory or inhibitory effects on different groups of microbes (87). Besides, microbes derive energy for growth from organic debris as sloughed-off root hairs, root cap tissue, cortical cells in young tissues (roots) and cortical fragments in the older roots (63). The young roots also have a layer of mucilage extending over the root hairs and root caps. The greatest exudation takes place at the root apex at the point of lateral root emergence. Exudates from these sites are largely non-diffusible and possibly derived from cells of the root cap. However, the nutritional effects of exudates extend for only a short distance, perhaps upto 12 mm, from the root surface due to their rapid utilisation by rhizoplane and rhizosphere microorganisms.

Among the abiotic factors, soil types, moisture contents, pH and soil compactness govern the exudation/diffusion process. Among the biotic factors, the quantity of organic matter such as roots debris, root growth can increase exudation (88); physical factor such as low temperature for short period, water stress promote exudation.

The rhizosphere colonisation is rapid in the older portions which are first to be colonised by microorganisms followed by region towards the root tip (102). The latter remains uncolonized, particularly in younger roots where the length of these fungus-free surface is greatest. The colonization is effected by great variety of propagules encountered by root as it moves through the soil, but small particles or organic debris are seen, particularly important inoculum source (12).

The virgin region of the root coincides with the zone of maximum exudates release so that a high enriched habitat, suitable for the development of rhizosphere selected fungi, is created.

The acidified and limed habitats show changes in the frequency of a particular fungal species not only due to shift of chemical soil factors but also by antagonistic interactions between the microfungi, thus reducing pathogenic fungi. The acidification has been known to promote the frequency of *Mycelium radices-atrovirens, Oidiodendron maius* but reduced *Ttichoderma viride* while liming increased *Sesquicillium candelabrum* and *Cylindrocarpon destructans* (82). In water-logged conditions, salt marshes and other stress borne habitats plants have been known to have preponderance of sterile mycelium in their rhizosphere (23, 24, 80, 81).

2.2.1 Microflora of the other underground organs

The microflora associated with bulbs, rhizomes, tubers and other underground organs, is supposed to be broadly similar to that of roots (39). This area is least worked out and needs special study parallel to that of roots. The nutritional status of the surrounding soil as result of exudation has been found to be responsible for qualitative and quantitative variations in microflora associated with these organs (71).

2.2.3 Rhizosphere microflora and pathogenesis

The role of rhizosphere microorganisms is largely associated with the occurrence of a number of diseases (102, 104). The metabolic activity of higher plants infected by microorganisms results in the formation of more amino acids, carbohydrates, etc. (110). The diseased plants show higher fungal population (quantitative and qualitative) due to root exudation in comparison to their control. The barley infected plants with *Ustilago hordei* showed higher values of total free amino acids, phenols and ortho-dihydroxy phenols in roots (3). Similar results have been observed in several ornamental plants infected with powdery mildew *Erysiphe cichoracearum* (77, 83).

Availability of nutrients in the rhizosphere is controlled by the combined effects of soil properties, plant characteristics and interaction of plant roots with microorganisms in the surrounding environment (13). Organic acids of plant root or microbe origin in the rhizosphere or by their mutual effects influences the rhizosphere activity that influences both equally or mutually (95). Thus different plant species as well as genotypes within a species are known for qualitative and quantitative variations in the rhizosphere microflora (50, 86, 109). This clearly explains the built-in disease resistance/tolerance in plant species towards different pathogens.

2.3 Microflora of Terrestrial Aerial Plant Surfaces

The living leaf acts as landing stage for spores and other propagules. The ecological studies have been aimed to characterise the microflora at stages during leaf development and to relate its compositions to the constantly changing capacity of the leaf to support microbial growth (9, 56, 64).

The leaves are colonised by various microorganisms from the stage of its development to senescence and death. Initially the leaf colonizers are host specific or restricted parasites, accompanied by primary saprophytes (18). These get their nutrition from the easily decomposable sugars exuded from the leaf surface, honey drew and faecal material from the leaf fauna, dead or dying parts of the leaves, and healthy tissue. On senescence, the phylloplane fungi begin to disintegrate the leaf tissue.

The leaf colonisers belong to two different types: casual inhabitants (represent airborne spores) and the resident inhabitants, which usually outnumber the 'casuals' and are more adapted to the phylloplane. However, Hudson (44) considered that, of the fungi that are almost ubiquitous, those which succeed in colonising expanding green leaves at an early stage should be regarded as the common primary saprophytes, and those which are more restricted in their host range, the restricted primary saprophytes.

The microbes growing or reproducing on leaves establish a dynamic equilibrium resulting into a micro and macro-environments (16, 17). Based on the type of equilibrium with leaf surface they act as symbionts or antagonistic against fungal parasites or as biocontrol agents. They may degrade plant surface waxes and cuticles, produce plant hormones, take part in decomposing plant material after senescence or influence the growth behaviour and root exudation of plants (53, 96).

2.3.1 Phylloplane environment and microbial succession

The phylloplane climate reflects general climate of microbes. Seasonal climate, as determined by regional climate zone has an important, broad and long-term influence on phylloplane surface. The climatic changes can bring about short-term alterations in phylloplane conditions. The age of plant, maturity level, etc. influence on phylloplane microflora. Qualitative and quantitative changes occur with in different plant genotypes, leaf morphology and leaf environment (19, 28, 30, 41).

The microbial succession on leaves has been the subject matter for investigation since 1955. As a result of investigations on the leaf surface microorganisms, Last (55) and Ruinen

(90) separately defined the "Phyllosphere" as environment on the surface of leaves. However, Kerling (49) defined the term phyllosphere for the zone actual on leaf surface.

The leaf surface due to its position on plant is exposed to different environmental fluctuations such as temperature, relative humidity, sunlight, etc. and pollution (7) that govern the stress on the microniche seasonal and diurnal fluctuations are also responsible for similar changes (29).

The microbial population changes with age, the least number of microbes are found on seedling leaves and increase as the plant grows reaching the maximum population only on yellowing at maturity. Among the most active fungi *Cladosporium herbarum* and *Ascochytula obiones* predominantly reported of green leaf surface. *Cephalosporium* sp., *Fusarium culmorum* and *Stemphylium bortryosum* follow the succession only at maturity of leaf towards senescence (22).

During microbial succession on *Beta* leaf, the spores of *Alternaria tennuis* and *Cladosporium herbarum* have been observed to germinate and penetrate the epidermis in mature leaves but the mycelial growth of *Epicoccum* remained superficial (49). The high population of *Alternaria alternata* and *Cladosporium cladosporioides* were also observed by Sharma and Mukerji (93, 94) on leaves of *Gossypium hirsutum* and *Sesamum orientale.*

The most frequent fungal colonizers on the living leaves at the time of leaf fall belong to species of *Alternaria, Aureobasidium, Botrytis, Cladosporium, Epicoccum, Phoma* and some host specific pathogens. Organic substances from the atmosphere, absorbed or deposited on the surface of the leaves fulfil a part of the nutritional needs. However, the rest of the nutrients are derived directly or indirectly from the leaves themselves. Among the fungi several of the primary colonizers are host specific or host-restricted weak parasites, which can utilise the living leaf tissue, often without causing visual disease symptoms (44, 76).

2.3.2. Microflora and leaf exudation

The process of leaching from plant surfaces is the removal of substances by the action of aqueous solution, such as rain, dew, moist or fog. Besides foliage, stem and branches are also susceptible to loss of valuable substances by leaching process. The phenomenon of leaching of carbohydrates in plants generally increases at flower/bud initiation stage, reaches maximum at full bloom stage and decreases again during senescence.

Leaf leachates contain components which can be utilised as nutrients by microorganisms. They may also inhibit their germination and growth and may be of considerable importance in control of the microbial population of plant surface (11, 36). Leaves like roots are bound by an environment which is rich in microbial propagules. In comparison to roots, the leaves are easier to study. They are generally uniform organ, whose, age and metabolic vigour may be estimated, at least crudely. The leaf exudates/leachates alongwith leaf surface make a niche for growth of microbial population (29).

2.3.3 Origin and development of the microflora on leaf surface

The newly expanded leaves are rarely virgin and commonly carry a small yeast mycoflora which originate from bud. *Aureobasidium pullulans* and *Sporobolomyces* have been isolated from internal surfaces of buds hybernating from one season to another. There are possibilities for continuity between the fungal flora of bud and leaf. The leaf morphological characters (trichomes, hairs) are responsible for trapping the fungal spores. As soon as leaf expands, fungal spores are recruited on it through water/wind splashes.

The phylloplane microflora is able to reproduce on green and healthy leaf. The young leaves harbour mainly yeast characteristically *Aureobasidium pullulans,* species of *Rhodotorula* and *Sporobolomyces.* Soon after its expansion conidia of filamentous fungi arrive on the phylloplane and may germinate and produce hyphal development on getting suitable conditions. The hyphal growth result in damage of waxy layers of the cuticle.

In initial stages of colonization, *Cladosporium herbarum* accompanied by a relatively few other Deuteromycotina, principally *Epicoccum pupurascens* and species of *Alternaria* and *Stemphylium.* After the leaf senescens, this mycoflora increases by a wide range of species that participate in the accelerating decay of host tissues.

The environmental fluctuations and others factors do have direct impact on development of fungi. High stress on leaf surface is responsible for low fungal biomass. The leaf morphology, anatomy (lower and upper parts) play an importance role (41). The upper leaf surface recruits greater number of fungi than do lower surfaces, subsequent development is commonly better on the latter surface being less exposed to pollutants, radiations, gets more protections from the abiotic factors (14, 60). Surface grooves not only tend to trap cells or conidia during their redistribution on the phylloplane by rain but may also coincide with sites of nutrient diffusion from the leaf.

2.3.4 Role of phyllosphere microflora in senescence

The fungal-host interaction studies have demonstrated that *Alternaria alternata, Cladosporium herbarum* and *C. cladosporioides* peneterated and developed intercellular hyphae in *Phaseolus* leaves (69). Species like *Alternatia* caused necrosis on living leaf tissue on penetration, *Cladosporium* and *Epicoccum* invaded only damaged or senescent leaves *Sporobolomyces* grew superficially only.

The colonization of phylloplane yeasts increase permeability of the leaf surface by breaking/dissolving cutin or pectin and thus damaging leaf cuticle. This result in high rate of nutrient secretion or leaching from the leaf.

In vitro studies have demonstrated the production of Indole Acetic Acid in *Aureobasidium pullulans* and *Cladosporium herbarum* (15, 105). Apparently these symptomless penetrations by facultative parasites would be unimportant in terms of disease incidence but significant in future decomposition process.

2.3.5 Microflora on the other aerial structures

The surface microflora associated with fruits and flowers reflects wide range of habitats than do leaves. Species of *Alternaria, Aureobasidium, Cladosporium, Epicoccum* and *Sporobolomyces* have been isolated from flowers of wheat, which are probably recruited from the atmosphere from the time of anthesis onwards. The larger flowers harbour species such as of *Aspergillus, Chaetomium, Fusarium, Penicillium* and *Rhizopus* which are not stress-tolerant (1). The most favourable microorganisms on the flowers surface is yeast which may establish outside or inside at the time of fruit formation (56). Subsequent development of other yeasts and filamentous fungi is dependent on different physical or physiological conditions.

The ecology of saprophytes on living stems, fruits and flowers are least worked out area. With respect to green stems, it might be expected that colonization would be parallel to that of the phylloplane, but there may be important differences between recruitment patterns for stem and leaves. The external stem morphology, age of the tissue and its other contents of cells are important factors.

The articulate stems are favourable niche for harbouring the fungi; woody stems harbour a variety of microflora (34, 51). Some fungi can degrade bark rich in substrates like suberized barrier tissues, high moisture contents and presence of fungistatic resins. Highest fungal activity occurs at nodes, lenticels and other irregular places where stress may be somewhat reduced (8, 68). The contamination of the lower parts of stems by rain splash from soil may play an important part in the establishment of microflora.

2.3.5.1 Buds as an ecological site for microorganisms

The tropical woody plants and some annuals are known to harbour a number of pathogenic and saprophytic microbes in their vegetative buds (21, 57, 67). The microflora associated with the premature leaf/buds. The gemisphere was defined by Leben (51), as a primary active site for the growth of non-pathogenic bacteria. Pathogenic types such as *Tilletia caries* (57) and *Venturia inaequalis* are known to stay in buds and act as source of inoculum. Certain fungi have been reported to be present from the unfolding of the leaves like *Aureobasidium pullulans* (79) and *Cladosporium herbarum* (47, 48).

Comparatively little has been investigated with respect to plant buds as the ecological site for non-pathogenic fungi in normal foliar buds of different plants. The bud appears to be most favourable area for the growth of microorganisms. They are totally dependent on other parts of the plant for nutrients. A few workers (21, 43, 47, 48, 57) have studied gemisphere microflora of a variety of plants. The dormant fruit bud contained approximately the same types of yeast found on the mature fruit indicating that the bud as an overwintering reservoir.

In a study on microfungi associated with buds of *Fagopyrum esculentum*, Singh et al. (98) observed certain frequent fungi in all phases of development *Alternaria alternata, Aspergillus flavus, A.niger, Candida albicans, Chaetomium globosum, Cladosporium herbarum, Nigrospora sphaerica, Penicillium frequentans, Mucor racemosus, Trichothecium roseum, Trichoderma viride*. Of these, *A.flavus, A.niger, Candida albicans* and *Cladosporium herbarum* were most dominant.

2.3.6 Phyllosphere-rhizosphere : Alternations and impact on microbial population

Role of leaf surface microorganism in biological control of plant pathogens and decomposition has been discussed by several workers (65). The fungi are known to disturb the natural equilibrium of microbes by selective destruction of flora on leaf surface (4, 54). The fungicides are detrimental for the growth of most microorganisms which generally result into a decrease in the microbial population of leaf surfaces (65). Leaf exudates increase with age; the amount of carbohydrates and amino acids may be responsible for the continued increase of microbial population of treated as well as control plants on leaf surface. The microflora may also increase gradually due to deposition from the air-spore and/or multiplication of microbes on the phyllosphere of older leaves (25).

Despite of the global concern for pollutants in recent years, there are only, a few studies on their specific effects on microorganisms in their natural environments (2, 61, 92). In the environment polluted with heavy metals, a decline in the microorganism number and species diversity and the development of metal resistant microbial population are encountered (6, 35).

Large variety of microbes, both pathogens and saprophytes, are found on the leaf surface. The stimulation of growth of the microorganisms varies widely, and is often selective. The alteration in the phyllosphere influences the plant growth which in turn affects the physiology of roots and thus an altered spectrum of chemical exudation causing change in the rhizosphere microflora. Thus, there exists a link between phyllosphere and rhizosphere microflora. Plant foliage/chemical spray may alter the leaf environment directly or influence on plant growth thus changing the micro-environment on the leaf surface as well as in roots by alteration in their micro-environment.

In recent years, much attention has been diverted towards the role of plant metabolism in controlling the rhizosphere mycoflora population by foliar application of antibiotics, growth regulators and fungicide chemicals (27, 84). Foliar application of organic substances, growth regulators hormones, plant nutrients or plant protecting chemicals, plant nutrients will result in altered exudation pattern of the root system and thus alteration in the rhizosphere microflora both quantitatively and qualitatively (85).

The concept of antibiosis has come to the significant level in the biological control of root borne diseases. An association has been found between the high incidence of *Trichoderma viride* in the rhizoplane of tomato varieties resistant to *Verticillium* wilt where as no *T. viride* could be isolated in the rhizosphere and rhizoplane of susceptible varieties.

The crop cultivation practices are known to alter the root environment by providing fertilizers and thus enhanced effect on rhizosphere microflora. Often such effects have been found to be helpful in suppressing the soil plant pathogens. The foliar spray of urea markedly changed the qualitative and quantitative increase in fungal population in rhizosphere of tea plants (106). Different hormones have been found to increase the rhizosphere microflora of legumes and oil yielding plants on foliar application (56, 97). Thus these techniques are very valuable tools in the hands of agricultural scientists to manipulate and influence on beneficial changes in plant rhizosphere. This field needs wide open stick for exploration.

2.4 Microbial Colonization on Dried or Preserved Plant Materials

Several fungi are known to occur on the surface of plants as weak parasites or invadors but act on decomposed leaves when fallen. Considerable amount of work has been done on the fungal colonization of dead leaves (94). The primary colonizers may be inside the tissue but as decomposition starts when facultative saprophytes are replaced by typical soil saprophytes such as *Pencillium* sp., *Trichoderma viride* and other (37).

The materials of plant origin harbour various fungi on their surface due to which they suffer a great loss due to deterioration by the storage moulds. In the preparation of foodstuffs for market, the use of mould inhibitors on edible materials and of relatively aseptic techniques in other preparations tend to reduce the incidence of microbial contamination.

Seeds of crop plants are consistent carriers of moulds on their surface and when stored improperly, they tend to deteriorate by the activity of many kinds of microorganisms. Untreated seeds are also carriers of spores or mycelium which can also infect the new crops. Among the other agricultural produces, timber is also subjected to colonization by a number of fungus-caused quality deterioration such as discolouration, loss of strength and disfiguring. The paper and wood industry face a great threat due to infection by fungi leading to quality deterioration of the products.

The natural objects of plant origin are bound to get colonized by microorganisms that bring several irreversible changes (66). Common articles such as clothing, food stuffs, housing materials, paper, wood, fibre and other commonly used products of industrial importance are deteriorated by the moulds (33). To combat such damage, the production technology and methods ought to be modified. The use of food preservatives in edible products against infection or deterioration is increasing day by day. The toxicity by microorganisms as a result of growth and accumulation of contaminants has drawn attention of the environmentalists and great concern for human health hazards.

The agricultural commodities under storage face a major problem of deterioration by the fungi under tropical conditions (72). Dried plant products, material of ethnological importance kept in musea, such as cereals, pulses, nuts, wooden objects, musical instruments, paintings, textile materials, historical collections and herbarium specimens are damaged by storage fungi. The fungal components mainly responsible for deterioration of these articles belong to species of *Aspergillus, Fusarium, Penicillium, Alternaria, Trichoderma* and *Cladosporium*. The favourable conditions for their development are relative humidity between 60-70% (66).

The woods, paper and other cellulose rich material under storage are also susceptible to microorganisms attack (31). The fibre, paper pulp or their derivatives such as library material has been seen to be spoiled by powdery and downy formations. The fruiting bodies, hyphal clusters or spores are frequently observed and may spread from one place to other via air (32).

Several storage fungi are known for their potential to cause mycosis or allergies in human beings. The staff working in inhouse environment of musea, herbaria, libraries, xylaria and other places, where plants or their products are housed, face the health hazards due to the negative impact of those fungi (33, 52, 99). The fine dust deposited on stored materials is the carrier of spores of different microorganisms (73). The environmental factors mainly responsible for the process of biodeterioration are temperatures (above 20^{0}C) and high relative humidity (above 65%).

Besides causing deterioration, of conserved materials, these plant surface fungi also play a positive role in recycling of organic materials, and help in maintaining the ecological balance of earth's natural resources.

3. CONCLUSIONS

In general the plant surfaces are colonized by facultative parasites when they are young and gradually at maturity common saprophytes like *Cladosporium* spp., *Aureobasidium pullulans* and other saprophytes replace them. On preservation or in storage common saprophytes already present on the surfaces continue and multiply on getting favourable conditions.

Thus, deterioration of plant material is a complex and prolonged process. This involves maximum decomposition and utilisation of cellulose. This is generally at a passive or neutral stage when plant is living but initiated after the material is dead or partially so and continue to do so till its death. The plant surface fungi play a significant role in this regard. The conditions favourable for the growth of microorganisms enhance the rate of biodeterioration under storage and thus play a negative role in conservation of biological objects.

The study of soil microbes-plant root interactions may help in better understanding of the complex phenomenon influencing growth of crops in environments with low availability of micronutrient. This knowledge, in an integrated manner with abiotic (soil chemistry, water, etc.) and biotic factors (plants, microorganisms, etc.) may help in understanding the dynamics of agricultural ecosystem.

4. REFERENCES

1. Abdalla, M. H. and El-Tayeb, N. Y. 1981. Preliminary survey of cotton flower mycoflora from Sudan. *Tran. Brit. Mycol. Soc.* **76** : 367-370.

2. Allen, G., Nickless, G., Wibberley, B. and Pic, K.A. 1974. Heavy metal particle characterization. *Nature* **252** : 571-572.

3. Ansari, A.R. and Prakash, D. 1986. Rhizosphere and rhizosplane mycoflora of barley infected with *Ustilago hordei. Indian Phytopath.* **39(3)** : 380 - 384.

4. Bainbridge, A. and Dickinson, C. H. 1972. Effect of fungicides on the microflora of potato leaves. *Trans. Brit. Mycol. Soc.* **59** : 31-41.

5. Bansal, M. and Mukerji, K. G. 1996. Root exudates in rhizosphere biology. In : *Concepts in Applied Microbiology and Biotechnology* (eds. K. G. Mukerji, V.P. Singh and S. Dwivedi). Aditya Books Pvt., New Delhi, pp. 97-119.

6. Bewley, R.J.F. 1980. Effect of heavy metal pollution on oak leaf microorganisms. *Appl. Environ. Microbiol.* **40** : 1053-1059.

7. Bewley, R.J.F. and Campbell, R. 1978. Scanning electron microscopy of oak leaves contaminated with heavy metals. *Tran. Brit. Mycol. Soc.* **71** : 508-51 1.

8. Bier, J.E. and Rawat, M.H. 1962. The relation of bark moisture to the development of canker diseases caused by facultative parasites. VII. Some effects of the saprophytes on the bark of poplar and wilow on the incidence of *Hypoxylon* canker. *Can. J. Bot.* **40** : 61-69.

9. Blakeman, J.P. (ed.). 1981. *Microbial Ecology of the Phylloplane.* Academic Press, London and New York.

10. Bolton, Jr. H. and Fredrickson, J. K. 1993. Microbial ecology of the rhizosphere. In : *Soil Microbial Ecology - Applications in Agricultural Environment Management* (ed. F. Blaine), New York, 646p.

11. Bora, L.C. and Chand, J.N. 1994. Leaf leachates and leaf surface colonizers of mung beans *(Vigna radiata* L.) cultivars at different growth stage. *Indian J. Mycol. Pl. Pathol.* **24(1)** : 6-10.

12. Bowen, G.D. and Rovira, A.D. 1976. Microbial colonization of plant roots. *Annual Review Phytopath.* **14** : 121-144.

13. Bowen, G.D. and Rovira, A.D. 1992. The rhizosphere in the hidden half of the hidden half. In : *Roots : The Hidden Half* (eds. Y. Waisel, A. Eshel and U. Kafkasfi) Marcel Dekker. Inc., New York, USA. pp. 641-669.

14. Breeze, E.M. and Dix, N.J. 1981. Seasonal analysis of the fungal community on *Acer platanoides. Trans. Brit. Mycol. Soc.* **77**: 321-328.

15. Buckley, N. and Pugh, G.J.F. 1971. Auxin production by phylloplane fungi. *Nature* **231** : 332.

16. Burrage, S. W. 1971. The microclimate at the leaf surface. In : *Ecology of Leaf Surface Microorganisms* (eds. T. F. Preece and C. H. Dickinson), Academic Press, London, pp. 91-101.

17. Burrage, S. W. 1976. Aerial microclimate around plant surfaces. In : *Microbiology of Aetial Plant Surfaces* (eds. C. H. Dickinson and T. F. Preece), Academic Press, London and New York, pp. 173-184.

18. Collins, M. A. 1976. Colonization of leaves by phylloplane saprophytes and their interactions in this environment. In : *Microbiology of Aerial Plant Surfaces* (eds. C. H. Dickinson and T. F. Preece), Academic Press, London, pp. 401-418.

19. Cowan, A. K., Botha, C. E. J., Hartley, B. J. and Cross, R. H. M. 1995. Ultrastructural changes in ageing leaves of a light grown achlorophyllous mutant of barley. *Physiologia Plantarum* **94** : 391-398.

20. Curl, E. A. and Truelove, B. 1986. *The Rhizosphere.* Springer-Verlag, New York, USA. 288p.

21. Davenport, R. R. 1970. Epiphylic Yeasts Associated with the Developing Grape Vine. M.Sc. Thesis, Univ. of Bristol, Bristol.

22. Dickinson, C. H. 1965. The microflora associated with *Halimione portulacoides* III. Fungi on green and moribund leaves. *Trans. Brit. Mycol. Soc.* **48** : 603-610.

23. Dickinson, C. H. and Pugh, G. J. F. 1965a. The mycoflora associated with *Halimionė portulacoides.* I. The establishment of the root surface flora of mature plants. *Trans. Brit. Mycol. Soc.* **48**: 381-390.

24. Dickinson, C. H. and Pugh, G. J. F. 1965b. The mycoflora associated with *Halimione portulacoides.* II. Root surface fungi of mature and excised plants. *Trans. Brit. Mycol. Soc.* **48** : 597-602.

25. Dickinson, C. H., Austin, B. and Goodfellow, M. 1975. Qualitative and quantitative studies on phylloplane bacteria from *Lolium perenne. J. Microbiol.* **91**: 157-166.

26. Dommergues, Y. R. 1978. The plant microorganism. In : *Interactions Between Nonpathogenic Soil Microorganisms and Plants,* (eds. Y.R. Dommergues and S.V. Krupa), Elsevier Scientific Publishing, New York, pp. 1-37.

27. Dublish, P. K. 1986. Foliar application of antibiotics, growth hormones and urea in relation to rhizosphere mycoflora of *Abelmoschus esculentus* and *Lageneria vulgaris. Indian Phytopath.* **39(2)** : 264-268.

28. Evans, J. R. 1995.Carbon fixation profiles do reflect light absorption profiles in leaves. *Aus. J. Plant Physiol.* **22** : 865-873.

29. Fiala, V., Glad, C., Martin, M., Jolivet, E. and Derridj, S. 1990. Occurrence of soluble carbohydrates on phylloplane of maize *(Zea mays* L.) variations in relation to leaf heterogeneity and position on the plant. *New Phytol.* **115** : 609-615.

30. Frossard, R. and Oertli, J. J. 1982. Growth and germination of fungal spores in guttation fluids of barley grown with different nitrogen sources. *Trans. Brit. Mycol. Soc.* **78** : 239-245.

31. Gallo, F. 1993. The biodeterioration of library materials. In : *Recent Advances in Biodeterioration and Biodegradation* (eds. K.L. Garg, N. Garg and K. G. Mukerji). Naya Prokash, Calcutta, pp. 89-143.

32. Garg, K.L., 1995. Use of Homoeopathic drugs as antifungal agent for protection of books and paper materials. In: *proc. 3rd International Conference on Biodeterioration of Cultural Properly.* 3, July 4-7, 1995. Bangkok, Thailand, 718 p.

33. Garg, K.L., Garg, N. and Mukerji, K.G. (eds.). 1993. *Recent Advances in Biodeterioration and Biodegradation*, Naya Prokash, Calcutta.

34. Garner, J.H.B. 1967. Some notes on the study of bark fungi. *Can. J. Bot.* **45** : 540-541.

35. Gingell, S.M., Campbell, R. and Martira, M. 1976. The effect of zinc, lead and cadmium pollution on leaf surface microflora. *Environ. Pollut.* **11** : 25-37.

36. Godfrey, B.E.S. and Clements, D. M. 1978. Effect of lilac leaf leachate on germination of *Alternaria alternata* and *Botrytis cinerea. Trans. Brit. Mycol. Soc.* **70**: 163-165.

37. Gray, T.R.G. and Williams, S. T. 1971. *Soil Microorganisms.* Oliver and Boyd, Edinburgh.

38. Greaves, M.P. and Webley, D.M. 1965. A study of the breakdown of organic phosphates by microorganisms from the root region of certain pasture grasses. *J. Appl. Bact.* **28** : 454-465.

39. Hale, M.G. and Graffin, G. J. 1976. The effect of mechanical injury on exudation from immature and mature peanut fruits under axenic conditions. *Soil Biology and Biochemistry* **8**: 225-227.

40. Harper, S. H. T. and Lynch, J. M. 1979. Effects of *Azotobacter chroococcum* on barley seed germination and seedling development. *Journal of General Microbiology* **112**: 45-51.

41. Helander, M.L. Ranta, H. and Neuvonen, S. 1993. Responses of phyllosphere micro-fungi to simulated sulphuric and nitric acid deposition. *Mycol. Res.* **97(5)** : 533-537.

42. Hiltner, L. 1904. Uber neuere erfahrungen and probleme auf dem gebiet der bodenbak teriologie und unter besonderer berucksichtigung der grundungung und brache. *Arb. Dtsch. Landwirt. Ges.* **98** : 59-78.

43. Hislop, E.C. and Cox, T. W. 1969. Effects of captan on the non-parasitic microflora of apple leaves. *Trans. Brit. Mycol.Soc.* **52** : 223-235.

44. Hudson, H.J. 1968. The ecology of fungi on plant remains above the soil. *New Phytol.* **67**: 837-874.

45. Jalali, B.L. and Suryanarayana, D. 1971. Shift in the carbohydrate spectrum of root exudates of wheat in relation to its root rot disease. *Plant and Soil* **34**: 261-267.

46. Kaiser, P. and Reber, H. 1970. Interactions entre la simazine- herbicide et les microorganises de la rhizosphere deu mais, meded Ryksfac. *Landb. Wet. Gent.* **35** : 689-705

47. Keener, P.D. 1950a. Mycoflora of buds I Results of cultures from non irradiated materials of certain woody plants. *Am. J. Bot.* **37** : 520-527.

48. Keener, P.D. 1950b. Mycoflora of bud.II. Results of histological studies of non-irradiated buds of certain woody plants. *Am. J. Bot.* **38** : 105-110.

49. Kerling, L.C.P. 1958. De microflora op let blad van *Beta vulgaris* L. *Tijdschr. PlZiekt.* **64** : 402-410.

50. Khanna, R., Chandra, S. and Khanna, K. K. 1993. Rhizosphere microflora of *Triticale. Biological Memoirs.* **19**: 111- 121.

51. Kohlmeyer, J. 1969. Ecological notes on fungi in mangrove forests. *Trans. Brit. Mycol. Soc.* **53**: 237-250.

52. Kowalik, R. 1977. Paper and parchment deteriorating fungi pathogenic to man *Wolfenbutteler for Schungen Herausgegeben Von der Herzog Augus Bibliothek.* **1** : 85-90.

53. Kranz. J. 1980. Hyperparasitsm of biotrophic fungi. In : *Microbial Ecology of the Phylloplane.* (ed. J.P. Blakeman). Acad. Press, London 502 p.

54. Kuthubutheen, A. J. and Pugh, G. J. F. 1978. Effect of fungicides on physiology of phylloplane fungi. *Trans. Brit. Mycol. Soc.* **71** : 261-269.

55. Last, F.T. 1955. Seasonal incidence of *Sporobolomyces* on cereal leaves. *Trans Brit. Mycol. Soc.* **38** : 221-239.

56. Last F.T. and Price, D. 1969. Yeasts associated with living plants and their environs. In: *The Yeasts, Biology of Yeasts* (eds. A.H. Rose and J.S. Harrison), Academic Press, London and New York, Vol. 1. pp. 183-218.

57. Leben, C. 1971. The bud in relation to the epiphytic microflora. In: *Ecology of Leaf Surface Microorganisms* (eds. T. F. Preece and C.H. Dickinson), Academic Press, London and New York. pp. 117-127.

58. Leelavathy, K.M. 1966. Studies on some aspects of the rhizosphere mycoflora of the important fodder grasses of U.P. Ph.D. thesis, Banaras Hindu University, Varanasi.

59. Linderman, R.G. 1988. Mycorrhizal interaction with the rhizosphhere microflora: The mycorrzhizosphere effect. *Phytopath.* **78** : 366 - 370.

60. Lindsey, B.I. and Pugh, G.J.F. 1976. Succession of microfungi on attached leaves of *Hippophae rhamnoides* L. *Trans. Brit. Mycol. Soc.* **67** : 61-67.

61. Little, P. and Martin, M.H. 1972. A survey of zinc, lead and cadmium in soil and natural vegetation around a smelting complex. *Environ. Pollut.* **3** : 241-254.

62. Lynch, J.M. and Pryn, S.J. 1977. Interaction between a soil fungus and barley seed. *Journal of General Microbiology* **103** : 193-196.

63. Lynch, J.M. and Whipps, J.M. 1990. Substrate flow in the rhizosphere plant and soil. 129: 1-10.

64. Marilley, Laurent, Vogt, G., Blanc, M. and Aragno, M. 1988. Bacterial diversity in the bulk soil and rhizosphere fractions of *Lolium perenne* and *Trifolium repens* as revealed by PCR restriction analysis of 16S rDNA. *Plant and Soil* **198** : 219-224.

65. Mishra, R.R. and Tiwari, R.P. 1979. Studies on phyllophere microflora. Effect of antibiotics and fungicides on leaf surface fungi and bacteria. *Acta Bot. Indica* **6** : 57-63.

66. Mukerji K.G., Garg, K.L. and Mishra A.K. 1995. Fungi in deterioration of museum objects. In : *Proc. 3rd Internatioinal Conference on Biodeterioration of Cultural Property.* 3. July 4-7, 1995, Bangkok, Thailand, 718p.

67. Naaz, Shagufta. 1992. Phylloplane Mycoflora of *Fagopyrum esculentum.* Ph.D. thesis, Meerut Univ., Meerut.

68. Nordström, U.M. 1974. Bark degradation by *Aspergillus fumigatus* - growth studies. *Canadian Journal of Microbioology* **20** : 299-308.

69. O'Donnell, J. and Dickinson, C.H. 1980. Pathogenicity of *Alternaria* and *Cladosporium'* isolates on *Phaseolus. Trans Brit. Mycol. Soc.* **74** : 335-342.

70. Odunfa, V.S.A. and Oso, B.A. 1979. Fungal populations in the rhizosphere and rhizoplane of cowpea. *Trans. Brit. Mycol. Soc.* **73** : 21-26.

71. Pandey, Anjula and Mukerji, K. G. 1997. Studies on Mycoflora Associated with underground plant parts of garlic *(Allium sativum L.).* In : Abstract *Proc. Microbes in Plant Improvement and Environment Protection.* University of Delhi, 23-24 December 1997. p. 243 (Abst).

72. Pandey, Anjula and Mukerji, K. G. 1998. Storage fungi in edible agricultural commodities. In: *Ethnomycology to Fungal Biotechnology : Exploiting Fungi from Natural Resources for Novel Products* (eds. Jagjit Singh and K. R. Aneja), Plenum Publishing Co., U.K. (In Press).

73. Parker, T.A. 1987. Integrated pest management for libraries. In : *Preservation of Library Materials.* London. IFLA P.A.C. Publ. **2** : 103 - 123

74. Pearson, R. and Parkinson, D. 1961. The sites of excretion of ninhydrin-positive substancs by broad bean seedlings. *Plant and Soil* **13** : 391-396.

75. Perry, D.A. 1973. Infection of seeds of *Pisum sativum* by *Pythium ultimum*. *Trans. Brit. Mycol. Soc.* **61** : 135-144.

76. Petrini, 0. and Carroll, G. 1981. Endophytic fungi in foliage of some Cupressaceae in Oregon. *Can. J. Bot.* **59** : 629-636.

77. Prakash, A., Akram, M. and Husain, S.I. 1994. Studies on rhizosphere mycoflora of some ornamental plants infected with powdery mildew *(Erysiphe cichoracearum* DC ex Merat.). *J. Phytol. Res.* **7(2)** : 135-138.

78. Preece, T.F. 1963. Micro-exploration and mapping of apple scab infection. *Trans. Brit. Mycol. Soc.* **46** : 523-525.

79. Pugh, G.J.E. and Buckley, N.G. 1971a. The leaf surface as a substrate for colonization by fungi. In: *Ecology of Leaf Surface Microorganisms* (eds., T.F. Preece and C.H. Dickinson), Academic Press, London, pp. 431-445.

80. Pugh, G.J.F. 1960. The fungal flora of tidal mud flats. In: *The Ecology of Soil Fungi* (eds. D. Parkinson and J. S. Waid), Liverpool University Press, Liverpool, pp. 202-208.

81. Pugh, G.J.F. 1967. Root colonization by fungi. In: *Progress in Soil Biology* (eds. O. Graff and J.E. Satchell), North Holland, Amsterdam, pp. 21-26.

82. Qian, X.M., Ashker, A. El., Kottke, I. and Oberwinkler, F. 1998. Studies of pathogenic and antagonistic microfungal populations and their potential interactions in the mycorrhizoplane of Norway Spruce *(Picea abies* (L.) Karst.) and beech (*Fagus sylvatica* L.) on acidified and limed plots. *Plant and Soil.* **199** : 111 - 116.

83. Rai, B. and Upadhyay, R. S. 1980. The populaion dynamics of *Fusarium udum* Butler in soil in relation to root region microflora of healthy and wilted plants of pigeonpea. In : *Proc. Nat Acad. Sci.* Golden Jub. Session (Biol. Sci.), 68 pp.

84. Ranga, Rao, V., Jayakar, M., Sharma, K. P. and Mukerji, K. G. 1972. Effect of foliar spray of morphactin on fungi in the root zone of *Capsicum annuum. Plant and Soil* **37** : 179-182.

85. Reddy, T. K. R. 1968. Foliar application of certain chemicals and antibiotics in relation to the rhizosphere microflora of rice *(Oryza sativa). Plant and Soil* **29** : 102-112.

86. Rengel, Z. 1997. Root exudation and microflora populations in rhizosphere of crop genotypes differing in tolerance to micronutrient deficiency. *Plant and Soil* **196** : 255-260.

87. Rovira, A.D. 1965. Interactions between plant roots and soil microorganism. *Ann. Rev. Microbiol.* 19 : 241-26.

88. Rovira, A.D. 1969a. Plant root exudates. *Botanical Review* **35**: 35-57.

89. Rovira, A.D. 1969b. Diffusion of carbon compounds away from wheat roots. *Australian Journal of Biological Sciences* **22** : 1287-1290.

90. Ruinen, J. 1956. Occurrence of *Beijerinckia* species in the phyllosphere. *Nature* **177** : 220-221.

91. Ruinen J. 1961. The Phyllosphere I. An ecologically neglected milieu. *Plant and Soil* **15** : 81 109.

92. Sadler, W.R. and Trudinger, P.A. 1967. The inhibition of microorganisms by heavy metals. *Mineral Deposits.* 2 : 158-168.

93. Sharma, K.R. and Mukerji, K.G. 1974. Incidence of pathogenic fungi on leaves. *Indian Phytopath.* **27** : 558-566.

94. Sharma, K. R. and Mukerji, K.G. 1976. Microbial ecology of *Sesamum ofientale* and *Gossypium hirsutum* L. In: *Microbiology of Aerial Plant Surfaces* (eds. C.H. Dickinson and T.F. Preece). Academic Press, London, pp.375-390.

95. Silber A., Ganmore Neumann, R. and Ben-Jaacov, J. 1998. Effects of nutrient addition on growth and rhizosphere pH of *Leucadendron* 'Safari Sunset'. *Plant and Soil* **199** : 205-211.

96. Singh, J.S. and Faull, J.L. 1985. Hyperparasitism and biological control of late blight fungus. In: *4th Int. Symp. Microbiology Phylloplane*, Wageningen, Holland.

97. Singh, P.V. 1966. Comparative Study of the Fungal Population in the Rhizosphere of Forbs Growing Under Material and Artificial Conditions. Ph.D. thesis, Banaras Hindu University, Varanasi.

98. Singh, P.N., Shagufta, Naaz and Sharma, Deepti. 1996. Microfungi associated with buds of *Fagopyrum esculentum* Moench. *Acta Botanica Indica* **24** : 57-59.

99. Staib, F. 1980. Deteriorating material as a possible source of fungi pathogenic to man. *Aspergillus fumigatus* as an example. In : *Biodeterioration Proceedings of the Fourth International Symposium* (eds. T. A. Oxley, D. Allsopp and G. Becker) Berlin Pitman Publ. London and Biodeterioration Society. pp. 341-343.

100. Stanghellini, M.E. and Hancock, J.G. 1971. Radial extent of the bean spermosphere and its relation to the behaviour of *Pythium ultimum. Phytopath.* **61** : 165-168.

101. Suvercha and Mukerji, K.G. 1990. Microbial ecology of rhizosphere and phylloplane. In : *Economic Plants and Microbes* (ed. R.P. Purkayastha). Today and Tomorrow's Printers and Publishers, New Delhi.

102. Taylor, G.S. and Parkinson, D. 1961. The growth of saprotrophic fungi on root surfaces. *Plant and Soil* **15**: 261-267.

103. Thomas, A. and Parkinson, D. 1967. The initiation of the rhizosphere mycoflora of dwarf bean plants. *Can. J. Microb.* **13**: 439-446.

104. Timonin, M.I. 1966. Rhizosphere effect of healthy and diseases Lodge pole pine seedlings. *Canad. J. Microbiol.* **12** : 531-537.

105. Valadon, L. R. G. and Lodge, E.1970. Auxins and other compounds of *Cladosporium herbarum. Trans Brit. Mycol. Soc.* **55** : 9-15.

106. Venkata Ram, C.S. 1960. Foliar applicatioin of nutrients and rhizosphere microflora of *Camellia sinensis. Nature* **187** : 621-622.

107. Wadhwani, K. and Mehrotra, N. 1982. Fungi associated with roots of smut infected plants of *Cynodon dactylon. Indian Phytopath.* **35** : 722-724.

108. Waid, J.S. 1974. Decomposition of roots. In: *Biology of Plant Litter Decomposition* (eds. C. H. Dickinson and G.J.F. Pugh), Academic Press, London and New York. pp. 175-211.

109. Wiehe, W. and Hoflich, G. 1995. Survival of plant growth promoting rhizosphere bacteria in the rhizosphere of different crops and migration to non-inoculated plants under field conditions in north-east Germany. *Microbiol. Res.* **150** : 201-206.

110. Wood, R.K.S. 1967. *Physiological Plant Pathology*. Blackwell Scientific Publications. Oxford and Edinburgh.

22 Management of Soil Borne Pathogens through Plants and their Products

M.N. Khare and H.K. Jharia

CONTENTS

ABSTRACT 382

1. INTRODUCTION 382
2. DISEASE CYCLE 382
3. MANAGEMENT 383
4. PLANT PRODUCTS 383
5. PERFORMANCE OF PLANT PRODUCTS 383
6. USES OF PLANTS 385
   - 6.1 Mulches 385
   - 6.2 Green Manuring 385
   - 6.3 Decoy Crops 386
   - 6.4 Inter and Mixed Cropping 386
7. CONCLUSIONS 386
8. REFERENCES 386

Advances in Microbial Biotechnology
J.P. Tewari, T.N. Lakhanpal, Jagjit Singh, Rajni Gupta & B.P. Chamola (eds.)
APH Publishing Corporation, New Delhi - 110 002, India.

ABSTRACT

Soil borne pathogens are responsible in causing seed rots, seedling rots, damping off, root rots, collar rot, stem rot, blights, pod and fruit rots. The pathogens survive in soil on dead organic material. Several methods are being used to control such pathogens and the diseases like cultural, chemical, physical and biological. Chemicals are hazardous to human and animal life. The biological control through microbes is also used. Attempts have been made to control soil borne pathogens through plants and their products which is quite safe. It should find a place in the integrated management system of soil borne pathogens and diseases.

1. INTRODUCTION

To meet the requirements of food grains to feed the population in the country which is increasing 1.8 per cent yearly it is essential to increase the crop yield per unit area. By the year 2002 nearly 225 million tonnes of food grains will be needed. Last year the production has crossed 200 million tonnes. All out efforts are required to check the losses caused by diseases and insect pests. The diseases not only reduce the quantity but also quality of the produce.

Soil plays an important role in harbouring the pathogens. Some pathogens are soil inhabitants as they can live in soil for a long time as saprophyte when host plants are not available for their parasitic phase. Some pathogens invade the soil and pass some period as a dormant phase in absence of a host. Their low competitive saprophytic ability reduces the possibility of their host to host spread through the soil. They infect new hosts through the contact of roots between the plants and from tissues of infected host plants.

A number of factors like soil texture, soil moisture, soil temperature, soil reaction, hosts's susceptibility or resistance, rhizosphere effect and other parameters including ecological relationship of microorganisms in soil, influence the initiation and severity of diseases caused by soil borne pathogens.

Among soil borne pathogens fungi have the major role followed by bacteria, actinomycetes and viruses. Nematodes form a separate group causing maladies in plants. Soil borne pathogens are responsible for causing seed rots, seedling rots, damping off, root rots, collar rot, stem rot, blights, pod and fruit rots

2. DISEASE CYCLE

Those pathogens which survive in soil on dead organic matter are always available in the soil and attack the germinating seeds causing preemergence losses, post emergence damages, root rot, collar rot and diseases of other plant parts due to direct infection or rain splashes and other agencies at different stages of crop growth. The infection may be systemic or local.

Seeds carry the pathogens internally or externally to new areas like *Tilletia indica* causing Karnal bunt of wheat, *Ascochyta* spp. causing blight in chickpea, lentil etc. They introduce the pathogens in pockets which later function as focii for further spread of the disease. In case of pathogens associated with seed coat and cotyledons on the soil surface and later on other suitable organic matter and get transmitted to plant parts through rain splashes. In case of covered smut of barley and smuts of sorghum the smut spores get dispersed in soil during harvest and are also transmitted as contaminant on the seed surface which result into systemic infection through soil. Such pathogens pass a short phase in soil. In case of ergot, the sclerotia reach the field with seed as inert matter, on germination, pass a saprophytic phase in soil and the sporidia attack the flowers through air.

3. MANAGEMENT

The basic idea in management of a soil borne disease is to reduce the inoculum potential to such a low level that the disease is not caused or the host develops resistance to the pathogen. Several methods like cultural, chemical, physical and biological are available. In biological method of control, the microorganisms antagonistic to the pathogens are introduced in soil and conditions are provided to increase their population so as to interfere in the ecological relationship of microorganisms in the rhizosphere as well as nonrhizosphere zone. One has to be cautious that the introduced microorganisms may not prove pathogenic later to plants or hazardous to humans and animals. Plant products and plants have also been used to check soil borne pathogens and diseases caused by them which are quite ecofriendly.

4. PLANT PRODUCTS

Plant parts and their products are used as soil treatment to have their influence on the pathogens present in soil and as seed treatment to control seed borne pathogens as well as the pathogens in the spermosphere and rhizosphere. In past seed treatment was being done with powder and ash of amla, sesame, mustard, barhati, vidanga, etc. before planting. Decoctions of plants were also used to control diseases (40, 41). Fewcett and Spencer (15) have reviewed the earlier work on antifungal natural products from plants. Khare and Shukla (23) have given a vivid account of utility of plants in crop disease control.

Extracts of different plants parts like seed, leaf, bark, stem, root, kernel, flower, pollengrain, bulb, rhizome, oilcakes are made in cold water, hot water and organic solvents. Leaf powder, resins, plants latex, oils and essential oils have also been used in plant disease control. *In vitro* assay of the material is done by filter paper disc method, poisoned food technique, growth in liquid media, cavity slide method, hanging drop method, etc. The *in vivo* testing is done by pot culture technique by soil and seed treatment as well as in field (23).

5. PERFORMANCE OF PLANT PRODUCTS

Dubey and Dwivedi (11) have reported inhibitory effect of extract of leaf and bulb of *Allium cepa* and *A. sativum* and fruit and bark extract of *Acacia arabica* on the growth and sclerotial viability of *Macrophomina phaseolina*. Leaf extracts of *Azadirachta indica, Lantana camera* and *Catharanthus roseus* checked the mycelial growth and spore germination of *Fusarium moniliforme* (29). Aqueous extract of *Ranunculus asiaticus* completely inhibited the growth and sporulation of *Alternaria solani*. Saxena and Mathela (42) reported that iridodial B-Monoenol acetate obtained from essential oil from *Nepata leucophylla* was most effective against *Sclerotium rolfsii* and actinidine from *N. clarkei* against *M. phaseolina*. Methanol extract of *Aegle marmelos* leaves at 150 ppm inhibited sclerotia formation in *S. rolfsii* (35). Sivasithambaram *et al.* (52) checked *Phytophthora cinnamomi* by adding fresh saw dust and composted tree bark and Wyk *et al.* (56) inhibited mycelial growth by antifungal substance of root bark of *Protea cyanaroides*. Rai *et al.* (36) observed that power density memory of water has synergistic effect on the inhibitory activity of ajoene, disturbing the enzymatic activity of spores of *Fusarium udum* and *F. lini*. Partially decomposed *Eucalyptus* leaf when added in soil reduced pre and post emergence damping off caused by *Pythium aphanidermatum* and *P. debaryanum* on tomato, brinjal and chilli (27).

Root extracts of clover and maize inhibited the growth of *Ophiobolus graminis* causing take all disease of cereals (9).

Bambawala *et al.* (3) evaluated 14 medicinal plants against cotton pathogens *in vitro* of which extracts from *Leucaena leucocephala* and *Casurina equisetifolia* inhibited spore germination of *Alternaria macrospora. Lawsonia alba* extract inhibited spore germination and mycelial growth of *Macrospora roridum.* Extracts of *L. alba, Datura metal* and *D. stramonium* gave good control of *Xanthomonas campestris* pv. *malvacearum.* In *Sarocladium oryzae* causal organism of sheath rot of paddy, the spore germination was checked by pepper seed extract and garlic bulb extract (21). Seed extract of Ajwain (1 : 20) completely checked the growth of *Rhizoctonia solani* causing sheath blight and in pot experiment the disease was reduced by 72.25 per cent (1). Leaf and oilcake extracts of *Thevatia peruviana, Prosopis juliflora* and *Eucalyptus globulus* resulted in maximum inhibition of sclerotial germination (14). Ethanolic leaf extracts of *Calotropis procera, A. indica* and *Datura stramonium* were effective against *R. solani* and *F. moniliforme* (30). Tewari and Dath (53) have reported inhibitory effect of garlic extract on *Corticium sasakii.* Leaf extract of *Lippia alba* gave complete inhibition of mycelial growth of *R. solani* (24) and essential oil of *Alpinia carinata* rhizomes at 300ppm checked the fungus when used as seed treatment (25). Benzene extract of *Piper nigrum* was highly fungitoxic to *R. solani* and *S. rolfsii* (5). In chickpea Singh *et al.* (48) reported efficacy of garlic leaf extract against *Fusarium oxysporum* f. sp. *ciceri* and *Sclerotium sclerotiorum* in terms of growth and sclerotial germination respectively. Extract treated seeds sown in infested soil produced wilt free plants. Singh *et al.* (49) observed induction of resistance in chickpea against *S. sclerotiorum* by the treatment of *A. marmelos* leaf extract. Ratnoo and Bhatnagar (39) reported control of gray stem blight by soil amendment with neem cake.

Garlic extract was inhibitory to *F. udum* and ajoene a product isolated from garlic extract checked the conidial germination of *F. oxysporum, F. semitectum* and *F. udum* (49). Prasad and Simlot (34) observed fruit extract of *Diospyros cordifolia* to inhibit germination of *F. oxysporum* f. sp. *udum*, f. sp. *cumini* and f. sp. *lycopersici*. The mycelial growth of *F. udum* was inhibited by leaf extract of *Callistemon lanceolatus* (32). The extract of *Vinca rosea* inhibited spore germination, sporulation and mycelial growth of *S. rolfsii* and *F. oxysporum* (31). Chauhan and Singh (6) and Singh and Chauhan (51) reported inhibitory effect of ajoene on the growth, sporangium formation, its germination and zoospore germination of *Phytophthora drechsleri* f. sp. *cajani* at 20ppm concentration. Singh and Singh (47) reported inhibition in growth of *F. udum* when infested soil was amended with ether extract of margosa cake.

In lentil, Singh *et al.* (46) evaluated extracts of *Ranunculus scleratus* at different growth stages, the leaves harvested just before flowering were most inhibitory to *F. oxysporum* f. sp. *lentis* followed by post flowering stage. The alkaline extract of pine bark has been reported very effective in checking the growth of *R.solani, S. rolfsii, F. oxysporum, Phytophthora parasitica* and *S. sclerotiorum*. The emergence of lentil was significantly increased by the addition of fresh or composted pine bark powder to soil infested with *R. solani* and *S. rolfsii*. In watermelon garlic clove juice effectively controlled *Fusarium* wilt (13).

Rhizome rot of ginger caused by *Pythium aphanidermatum* and *F. solani* was checked by soil amendment with cakes of neem, mustard, *Pongamia glabra* and *Callophyllum inophyllum*

(54). Neem cake application in soil highly reduced pre and post emergence rots (*R. solani*) in cotton root rot (*R. bataticola*) of soybean, wilt of cotton (*Ganoderma capsici*), root rot (*R. solani*) of fenugreek and betelvine rot (*Phytophthora capsici*) (37). Ratnoo and Bhatnagar (39) controlled ashy stem blight (*M. phaseolina*) by adding straw and oil-cakes. Cakes of groundnut and mustard at 2 per cent concentration of soil w/w reduced the pathogen in soil and wilt in tomato due to *F. oxysporum* f. sp. *lycopersici* (38). Dwivedi *et al.*(12) also observed similar effect when powdered neem cake, neem bark and neem seed were added in soil.

The mycelial growth of *R. solani* and *S. sclerotiorum* was completely inhibited at 7000 and 2000 ppm concentration of asafoetida in Czapeck's dox broth. Formation of sclerotia was completely checked in both the fungi at 1000 ppm (7). Latex of papaya checked the disease due to *Gloeosporium papayee* (22).

Banerjee *et al.* (4) reported inhibition of germination of sclerotia of *S. rolfsii, R. oryzae sativae* when soaked in 5 per cent emulsion of oil from *A. indica* and *Cymbopogon nardus*. Extracts of *Tinospora coniocordia* exhibited antifungal activity against *S. rolfsii* (33). Neem oil (10%) resulted in 100 per cent inhibition of *F. moniliforme* and *M. phaseolina* (55). Essential oil of *Ocimum canum* had most toxic effect and checked the germination of sclerotia of *M. phaseolina* (10). Singh and Singh (45) reported inhibition of growth and sclerotial formation of *R. solani* by the treatment with garlic oil.

Seed leachates of *Vigna radiata* and *V. mungo* retarded the germination of conidia of *F. moniliforme, F. semitectum, F. oxysporum, F. poae, Colletotrichum truncatum* and *Trichothecium roseum* upto 61.3 per cent (43). Hashmi *et al.* (19) controlled seed borne *F. moniliforme, F. oxysporum* and *F. semitectum* by treating seeds treated with aqueous extract of garlic (30g/100ml) water) for 12 hours controlled seed borne *Xanthomonas campestris* pv. *vesicatoria* and reduced disease severity (28).

6. USE OF PLANTS

Plants check pathogens and diseases caused by them when used as mulches, green manuring, decoy crops and inter and mixed cropping.

6.1 Mulches

Plant parts are incorporated in soil as organic matter as soil amendment. They control some pathogens and diseases by reducing their population directly or indirectly. Soil amendment with barley straw decreased *Verticillium* wilt of potato (20) and cotton root rot due to *M. phaseolina*; wheat straw decreased black scurf of potato caused by (18); cat straw reduced bean root rot due to *Thielaviopsis bassicola*, cruciferous crop refuge checked pea root rot due to *Aphanommyces euteiches* and sugarcane residues decreased banana wilt caused by *F. oxysporum* f. sp. *cubense* (44). Singh *et al.* (50) observed reduction in the incidence of Sclerotinia stem rot of chickpea when decomposed leaves of *A. marmelos* were added in soil 30 days before planting.

6.2 Green Manuring

Green manuring with rape, pea reduced take all disease of wheat (17); pea and *Melilotus officinalis* reduced root rot of cotton caused by *Phymatotrichum omnivorum* (8) and barley and oats controlled black scurf of potato caused by *R. solani*.

6.3 Decoy Crops

These are non host crops which activate the dormant propagules of fungi in absence of the actual host. *Tagetes minuta* is used as decoy crop for *Verticillium albo atrum* causing disease in olive (2); *Datura stramonium* for *Spongospora subterranea* in potato; Rye grass (*Rosea odorata*) for *Plasmodiophora brassicae* (16).

6.4 Inter and Mixed Cropping

In crop production programme inter and mixed cropping is recommended as the same pathogen does not infect both the crops. The root exudates of one crop may contain chemicals which are toxic to the soil borne pathogens infecting the other crop. When pigeonpea and sorghum are planted together, the root exudates of sorghum check *Fusarium udum* causing wilt in pigeonpea.

7. CONCLUSION

It is evident that plants and their products are quite useful in controlling soil borne plant pathogens and the diseases caused by them. Specific plants and their parts with higher quantity of active principles should be continuously searched. The selected plants can be further enriched with the desired active principle by molecular biological and biotechnological methods. Botanical pesticides appear to be of immense value in future for seed and soil treatment, being ecofriendly and safe. Further researches are needed in this direction. A coordinated approach among the scientists working at various locations on different diseases of crops will be of immense value.

8. REFERENCES

1. Ansari, M.M. 1995. Control of sheath blight of rice by plant extracts. *Indian Phytopath.* **48**: 268-270.
2. Baker, K.F. and Cook, R.J. 1974. *Biological Control of Plant Pahogens*. W.H. Freeman & Co. San Francisco. pp.433.
3. Bambawale, O.M., Mohan, P., Charabarty, M. and Chakrabarty, M. 1995. Efficacy of some medicinal plants against cotton pathogens. *Ad Pl. Sci.* **8**: 224-229
4. Banerjee, S., Bhattacharya, I. and Mukherjee, N. 1989. Sensitivity of three sclerotial rice pathogens to plant oils. *Int. Rice Res. Newsl.* **14(6)**: 23.
5. Chaudhuri, T. and Sen, C. 1983. Effects of some plant extracts on three sclerotia forming fungal pathogens. *Z. PflanzenkrPfanzenschutz.* **89**: 582-585.
6. Chauhan, V.B. and Singh, U.P. 1991. Effect of volatiles of some plant extracts on germination of zoospores of *Phytophthora drechsleri* f. sp. *cajani. Indian Phytopath.* **44**: 197-200.
7. Chaurasia, S.N.P. and Dayal, R. 1990. Inhibition of mycelial growth and sclerotium formation in *Rhizoctonia solani* Kuhn. and *Sclerotinia sclerotiorum* (Lib.) de Bary by asafoetida. *Trop. Sci.* **30**: 15-19.
8. Cook, R.J., Boosalis, M.G. and Doupnik, B. 1978. Influence of crop residues on plant diseases. In : *Crop Residue Management System*. Amer. Soc. Agron. Spec. Pub. **31**: 147-163.
9. Deffosse, L. 1966. Note on the inhibitory effect of root extracts of clover and maize on the growth of *Ophiobolus graminis* Sacc. *Parasitica* **22**: 17-19.
10. Dubey, R.C. 1991. Fungicidal effect of essential oils of three higher plants on sclerotia of *Macrophomina phaseolina. Indian Phytopath.* **44**: 241-243.

11. Dubey, R.C. and Dwivedi, R.S. 1991. Fungitoxic properties of some plant extracts against vegetative growth and sclerotial viability of *Macrophomina phaseolina. Indian Phytopath.* **44**: 411-413.

12. Dwivedi, S., Williams, P. and Sobita Devi. 1997. Effect of neem products on rhizoplane mycoflora of tomato plants. National Symposium on Recent Advances in Diagnosis and Management of Important Plant Diseases. Department of Plant Pathology, C.S.A. University of Agriculture and Technology, Kanpur. p.p. 98-99. (Abstr.).

13. El-Shami, M.A., Fadk, F.A., Tawfick, K.A., Sirry, A.R. and El-Zayat, M.M. 1986. An antifungal property of garlic clove juice compared with fungicidal treatment against *Fusarium* wilt of watermelon. *Egyptian J. Phytopath.* **17** : 55-62.

14. Ezhilan, J.G., Chandrasekar, V. and Kurucheve, V. 1994.Effect of six selected plant products and oil cakes on the sclerotial production and germination of *Rhizoctonia solani. Indian Phytopath.* **47** : 183-185.

15. Fawcett, C.H. and Spencer, D.M. 1970.Plant chemotherapy with natural products. *Ann Rev. Phytopath.* **8**: 403-418.

16. Garrett, S.D. 1970. *Pathogenic Root Infecting Fungi.* Cambridge Univ. Press, Cambridge, U.K. pp. 294.

17. Grossman, F. 1967. Grundung als Pflanzenschutzmabnahme. *Z. Pflanzenkrankh Pflanzschutz* **74**: 143-149.

18. Gudmested, N.C., Huguelet,J.E. and Zink, R.T. 1978. The effect of cultural practices and straw incorporation into the soil on *Rhizoctonia* disease of potato. *Plant Dis. Rep.* **62**: 985-989.

19. Hashmi, R.Y., Shahzad, S., Khanzada, A.K., Aslam, M., Kazmi, S.A.R. and Shazad, S. 1992. Studies on the seed mycoflora of lentil and its control. *Pakistan Scie. Indus. Res.* **35**: 345-347.

20. Huber, D.M. and Watson, R.D. 1970. Effect of organic amendment on soil-borne plant pathogens. *Phytopath.* **60**: 22-26.

21. Kanagarajan, R.S. 1974. Studies on certain aspects of seed borne fungi, IV. Seed borne fungi of some starchy seeds. *Acta. Bot. Indica* **2**: 129-135.

22. Khare, M.N. and Dhingra, O.D. 1974. Laboratory studies on preharvest rot of papaya fruits caused by *Gloeosporium papayae. Mysore J. Agric. Sci.* **8**: 115-120.

23. Khare, M.N. and Shukla, B.N. 1998. Utility of plants in crop disease control. *Vasundhara* **3**: 1-15.

24. Kishore, N., Dubey, N.K., Shrivastava, O.P. and Singh, S.K. 1983. Volatile fungitoxicity in some higher plants as evaluated against *Rhizoctonia solani* and some other fungi. *Indian Phytopath.* **36**: 724-726.

25. Kishore, N., Asthana, A. and Dubey, N.K. 1987. Antifungal activity of rhizome vapours of *Alphinia carinata* against *Rhizoctonia solani. Trans. Brit. Mycol. Soc.* **88**: 136-138.

26. Lewis, J.A. and Papavizas, G.C. 1975. Survival and multiplication of soil-borne plant pathogens as affected by plant tissue amendments. In; *Biology and Control of Soil-borne Plant Pathogens* (ed. G. Bruehl) American Phytopathological Society, pp. 84-89.

27. Madduleti, A. and Moses, G.J. 1995. Plant products and damping off disease management in vegetables. *Indian J. Mycol. Pl. Pathol.* **25**: 86.

28. Mangamma, P. and Sreeramulu, A. 1991. Garlic extract inhibitory to growth to *Xanthomonas campestris* pv. *vesicatoria. Indian Phytopath.* **44**: 372-373.

29. Meena, S.S. andMariappan, V.1993. Effect of plant products on seed borne mycoflora of sorghum. *Madras Agric. J.* **80**: 383-387.

30. Mishra, M., Tewari, S.N. and Mishra, M. 1990. Ethanolic extract toxicity of three botanicals against five fungal pathogens of rice. *Nat. Acad. Sci. Lett.* **13**: 409-412.

31. Narain, A. and Satapathy, J.N. 1977. Antifungal characteristics of *Vinca rosea* extracts. *Indian Phytopath.* **30**: 36-40.

32. Pandey, D.K., Chandra, H., Tripathi, N.N. and Dixit, S.N. 1983. Fungitoxicity of some higher plants with special reference to the synergistic activity amongst some volatile fungitoxicants. *Phytopath. Z.* **106**: 226-232.

33. Pariya, S. and Chakravarti, D.K. 1977. Antifungal activity of some Indian medicinal plant extracts on phytopathogenic fungi. *Phytopathologia Mediterranea* **16**: 33-34.

34. Prasad, B. and Simlot, M.M. 1983. Antifungal activity of fruit of temru (*Diospyros cordifolia*). *Sci. Cult.* **48**: 290-291.

35. Prithiviraj, B., Kishore, S., Ram, D. and Singh, U.P. 1996. Effect of methanol extract of *Aegle marmelos* leaves on *Sclerotium rolfsil. Inter Pharmacognosy* **34**: 148-150.

36. Rai, S., Singh, U.P., Mishra, G.D., Singh, S.P., Samerketu and Wagner, K.G. 1995. Synergistic effects of ajoene and the microwave power density memories of water on germination inhibition of fungal spores. *Medical and Biological Engineering & Computing* **33**: 313-316.

37. Rajan, F.S., Vedamuthu, P.G.B., Khader, M.D.A. and Jeyarajan, R. 1991. Management of root disease of fenugreek. *South Indian Horticul.* **39**: 221-223.

38. Raj, H. and Kapoor, I.J. 1996. Effect of oil cake amendment of soil on tomato wilt caused by *Fusarium oysporum* f. sp. *lycopersici. Indian Phytopath.* **49**: 355-361.

39. Ratnoo, R.S. and Bhatnagar, M.K. 1993. Effect of straw, oilcakes on ashy stem blight *Macrophomina phaseolina* (Tassi) Goid of cowpea. *Indian J. Mycol. Pl. Pathol.* **23**: 186-187.

40. Raychaudhuri, S.P. 1964. Agriculture in ancient India. ICAR, New Delhi, pp. 167.

41. Sadhale, Nalini, Mehra, K.L., Virmani, S.M. and Nene, Y.L. 1996. Surpala's Vrikshayurveda. Asian Agri-History Foundation, Secunderabad, pp.104.

42. Saxena, J. and Mathela, C.S. 1996. Antifungal activity of new compounds from *Nepata leucophylla* and *N. clarkei. App. Environ. Microbiol.* **62**: 702-704.

43. Saxena, R.M. and Gupta, J.S. 1982. Effect of seed leachates on spore germination of seed borne fungi of *Vigna radiata. Indian Phytopath.* **35**: 236-240.

44. Sequeira, L. 1962. Influence of organic amendments on survival of *Fusarium oxysporum* f. sp. *cubense* in the soil. *Phytopath.* **52**: 976-983.

45. Singh, H.B. and Singh, U.P. 1980. Inhibition of growth and sclerotial formation in *Rhizoctonia solani* by garlic oil. *Mycologia* **72**: 1022-1025.

46. Singh, J., Tripathi, N.N. and Singh, J. 1995. Effect of growth stages of *Ranunculus scleratus* L. on its fungitoxicity. *Indian Phytopath.* **48**: 363-364.

47. Singh,.N. and Singh, R.S. 1983. Inhibition of *Fusarium udum* pigeonpea wilt pathogen by ether distillate of margosa cake amended soil. *Indian J. Mycol. Pl. Pathol.* **13**: 329-330.

48. Singh, U.P., Pathak, K.K.,Khare, M.N. and Singh, R.B. 1979. Effect of leaf extract of garlic on *Fusarium oxysporum* f. sp. *ciceri, Sclerotinia sclerotiorum* and on gram seeds. *Mycologia* **71**: 556-564.

49. Singh, U.P., Ram, D. and Tewari, V.P. 1990. Induction of resistance in chickpea (*Cicer arietinum*) by *Aegle marmelos* leaves against *Sclerotinia sclerotiorum. Zeitschriftfur Pflanzenkrankheiten and Pflanzenschutz* **97**: 439-443.

50. Singh, U.P., Pandey, V.N., Wagner, K.G. and Singh, K.P. 1990. Antifungal activity of a ajoene, a constituent of garlic (*Allium sativum* L.). *Can. J. Bot.* **68**: 1354-1356.

51. Singh, U.P., and Chauhan, V. B. 1992. Effect of ajoene, a compound derived from garlic (*Allium sativum*) on *Phytophthora dreschsleri* f. sp. *cajani. Mycologia* **84**: 105-108.

52. Sivasithambaram, K., Smith, L.D.J. and Gross, O.M. 1981. Effect of potting media containing fresh saw dust and composted tree barks on *Phytophthora cinnamomi. Aust. Plant Pathol.* **10**: 20-21.

53. Tewari, S.N. and Dath, P.A. 1984. Effect of leaf extract media on some plants on the growth of three fungal pathogens of rice. *Indian Phytopath.* **37**: 458-461.

54. Thakore, B.B.L., Mathur, Sneh, Singh, R.B., Chakravarti, B.P. and Mathur, S. 1987. Soil amendment with oil cakes in finger field for rhizome rot control. *Korean J. Plant Prot.* **26**: 267-268.

55. Vir, D. and Sharma, R.K. 1985. Effect of fungicide XXXII. Evaluation of neem oil for control of plant pathogens. *Asian Farm Chemicals* **1 (7/8)**: 23-24.

56. Wyk, P.S. Van and Koeppen, B.H. 1974. P-hydroxybenzoycalleryanin: an antifungal substance in the root bark of *Protea cynaroides. South African J. Sci.* **70**: 121-122.

23 Interactions Between Soil Mycoflora and VA Mycorrhizal Fungi in Relation to Growth of *Vigna umbellata*

Vandana Joshi

CONTENTS

	ABSTRACT	392
1.	INTRODUCTION	392
2.	MICROBIAL SUCCESSION AND NUTRIENT AVAILABILITY	393
3.	SELECTIVE INFLUENCE OF VAM ON RHIZOSPHERE FUNGAL POPULATION	394
4.	MYCORRHIZOSPHERE BIOLOGY OF *VIGNA UMBELLATA*	396
5.	CONCLUSION	399
6.	REFERENCES	401

Advances in Microbial Biotechnology
J.P. Tewari, T.N. Lakhanpal, Jagjit Singh, Rajni Gupta & B.P. Chamola (eds.)
APH Publishing Corporation, New Delhi - 110 002, India.

ABSTRACT

Roots of living plants support growth of a complex of microbes which create a special habitat that influences growth and survival of the plant. Considerable amount of carbohydrates are released by the roots in the form of exudations and other root debris and dead cells are sloughed off during root growth. Exudation and other sloughed off cells, in turn supports a particular population of microflora and microfauna in the rhizosphere resulting in a specific microbial equilibrium. VAM fungi, being ubiquitous in occurrence, colonise the roots and cause changed exudation pattern of the roots and also alter the microbial equilibrium in the VA mycorrhizosphere. Therefore, the effect of colonisation by VAM fungus, *Glomus macrocarpum* was studied on the rhizosphere fungal population and plant growth of *Vigna umbellata* (rice bean). Seeds of rice bean plants were surface sterilized with chlorine water and sown in pots containing fumigated soil with two soil amendments : (i) + VAM (inoculated with *Glomus macrocarpum*), (ii) – VAM (uninoculated control). Observations in terms of rhizosphere, rhizoplane mycoflora, seed yield and root nodule numbers were estimated at the interval of one month and different stages of plant growth.

1. INTRODUCTION

It was generally held that organic materials were decomposed by bacteria, whereas fungi were thought to occur in soil only as dormant spores deposited by chance from the air. However, Waksman's (111), discovery of fungal mycelium in soil strongly suggested that fungi must have a role in decomposition. He further postulated the existence of two ecological groups of fungi in soil i.e., soil inhabitants and soil invaders. Reinking and Manns (91) applied these categories of fungi to plant pathogens, who studied the genus *Fusarium* in Central American soils. This later finding led Garret (44, 45, 46) to recognize two contrasting types of behaviour among root infecting fungi,

(i) the soil inhabiting parasities e.g. *Rhizoctonia* and *Pythium* species, for whom parasitism was incidental to a saprophytic existence in the soil; and

(ii) the root inhabiting parasites like *Fusarium* species, which were considered more highly specialised in that they were often host specific; their destructive effect on the host was delayed, and their existence in soil was transitory. The pathogenic propagules away from root remain ineffective due to the other microbes near roots. Only symbiotic microbes grow.

Harley (56) gave another concept for root infecting fungi; interference from saprophytic soil microorganism is generally a more powerful limiting factor, than is restriction by the physical conditions of the soil environment. Mechanical barrier to invasion results in a parallel evolution of infection habit amongst root infecting fungi parasitizing them.

New roots from well nourished plants are more susceptible than from poorly nourished plants. Seedlings of most plants owe their survival largely to escape from such diseases like damping off (*Pythium*), seedling blight (*Rhizoctonia*) and wilt (*Fusarium*). As the root senescence, the plant resistance is declined by various ways. Baker (12) reported that infection by root borne pathogen depends upon the type and cultivar of host i.e. host specialisation versus resistance.

Hiltner (58) defined the rhizosphere as the volume of soil that is adjacent to and around roots. The term comes from the Greek work for root (Rhizo or rhiza) and includes both the area of influence and the physical location around the root (sphere). Later on, Starkey (102) and Katznelson *et al.* (65) enumerated different ways in which microorganisms can influence plant growth and vice-versa. The rhizosphere has been subdivided and renamed by different workers. Perotti (84) subdivided the rhizosphere into 'endosphere' (corresponding to the rhizosphere of Hiltner) and 'histosphere' or surface of the roots. Clarke (29) suggested the term 'rhizoplane' as the zone of actual root influence and defined it as the external surface of plant roots with closely adhering soil particles. Timonin (106) recognized three zones near the root namely; rhizosphere, rhizoplane and histoplane (same as histosphere of Starkey). Curl and Truelove (30) described rhizosphere as a very thin soil zone influenced by root.

There are some microbes in the rhizosphere which symbiotically associate with roots of the plants and they form this association against all odds. This phenomenon as symbiosis is ecologically and agriculturally important providing organically bound nitrogen, phosphorus and other elements to the plant. Different zones of roots have been explored for fungal, bacterial and actinomycete flora by many workers (7, 14, 15, 16, 28, 37, 54, 59, 64, 87, 88,105, 110).

The influence of root on soil microorganisms starts immediately after seed germination, increases as the plant grows and reaches a maximum when plants have reached the peak of their vegetative growth (64). Also, the roots exert selective action on microorganisms, stimulating growth of some microorganisms while suppressing others. Simultaneous associative and antagonistic phenomenon between microorganisms are initiated (64). The microorganisms present in the soil affect transformation of organic and inorganic materials by way of production of Co_2, liberation of ammonia and minerals, consumption of O_2, denitrification and dissolution of phosphates (23, 65). They have also been found to synthesize vitamins, auxins and gibberllins in the rhizosphere (63, 64). Microbial synthesis of auxins in very small amounts stimulate growth but in higher amounts inhibit shoot elongation (6, 72).

The nutrients released by the roots vary at different stages of growth, due to which the microbial population also varies at every stage of plant (105). Roots support the growth of microbial population which in turn affect the growth and survival of the plant. A wide variety of microorganisms live in close vicinity of symbiotically associated plants i.e. mycorrhizae. Considerable amount of plant metabolites are released as root exudates in addition to the cell sloughage in the soil, which in turn affects the colonization and infection of plant root by mycorrhizal fungi, pathogens and by other microbes which leads to modification in each other's activity. Initially, these secretions contain complex substances. As senescence progresses, the root effect diminishes and microbial picture becomes complex (95). A few aspects of the microbial populations and ecology of rhizospheres of some crop plants have been studied by different workers (1, 2, 3, 14, 18, 19, 29, 40, 42, 43, 77, 85, 96, 97, 100, 105).

2. MICROBIAL SUCCESSION AND NUTRIENT AVAILABILITY

The development of a distinct microbial community is consistent with a successional pattern of microorganisms based on release of nutrients by roots and inter-microbial community competition for the available resources. The root exudate can have both stimulatory or inhibitory effect on different groups of microbes (95). Environmental fluctuations preclude succession of microbial communities in different habitats. The serial change of population is predictable if the succession is regular and stable (49). Initial random events may determine which microorganisms fill the niche of the ecosystem and direct the serial changes of succession. The pioneers are usually replaced by the better adapted population of microbes. This process continues until a stable or climax community is achieved (94).

Katznelson (64) pointed out that in rhizosphere with a diverse population, antagonism must occur quite frequently, Weindling *et al.* (112) and Garrett (44) also explained that antagonistic effect may involve production of inhibitory or antibiotic substances, competition for food and space, lysis or even parasitism. The microorganisms compete with each other in and around the rhizosphere soil for food and essential elements (13). Ranga Rao (87) observed that *Trichoderma viride* and *Streptomyces* sp. were found antagonistic to *M.phaseolina* and many species of *Aspergillus, Curvularia, Penicillium* and bacteria were found inhibiting the growth of *Rhizoctonia solani.*

Domsch (38) drew attention to the fact that some soil organisms, not regarded as pathogens, can retard plant growth while others stimulate it. This later response could be due to the beneficial organisms competing for substrates with the injurious ones. Both hypotheses - the production of growth substances and interactions between organisms are compatible with the wide variability in the effects of inoculation.

Availability of nutrients in the rhizosphere is controlled by the combined effects of soil properties, plant characteristics, and the interactions of plant roots with microorganisms and the surrounding soil (24). Organic and inorganic compounds released from roots can stimulate microbial activity in the rhizosphere because the rhizosphere microorganisms derive energy from root exudates and secretions, sloughed off cells and other root debris (71). Bacteria directly supported by root exudates may have a biomass of upto 36% of root dry weight in sand cultures (113). Different plant species (68, 114) as well as genotypes within a species (67, 107) differentially influence the quantitative and qualitative composition of microbial populations in the rhizosphere.

There is no doubt that microorganisms play an important role in plant nutrition (31, 108), quantitative estimates of the beneficial and harmful interactions between plants and microorganisms are largely speculative (31). Interactions are likely to be more important in deficient soils with low availability of particular nutrients. Mechanisms of tolerance of zinc and mangenese deficiency have been studied by many workers (52, 53, 60, 83, 92, 107).

3. SELECTIVE INFLUENCE OF VAM ON RHIZOSPHERE FUNGAL POPULATION

Rambelli (86) used the term 'mycorrhizosphere' to describe the soil surrounding and influenced by mycorrhizae. Portions of the rhizosphere can also be called the mycorrhizosphere when there are mycorrhizal fungi associated with the roots (70) or the actinorhizosphere or actinorhiza (109) when actinomycetes are associated with nodules on the root. The concept of mycorrhizosphere implies that mycorrhiza significantly influences the microflora of the rhizosphere by altering its root physiology and exudation pattern. As a consequence, the microbial community in a mycorrhizosphere differs quantitatively and qualitatively from that of rhizosphere of nonmycorrhizal plants.

Mycorrhizal fungi occur in nearly all soils on earth and form a symbiotic relationship with roots of most terrestrial plant, their morphology is the basis for grouping them into seven groups.

Mycorrhizal fungi also interface directly with the surrounding soil at the mantle (ectomycorrhizae only) and by means of hyphal, hyphal strands or rhizomorph that extend into the soil and increase the nutrient and water absorption potential of the root system and contribute greatly to the improvement of soil texture for better aeration and water percolation.

Microbial interactions in the mycorrhizosphere can be either direct interactions in the soil due to metabolic exchanges between organisms or indirect mediated by host plant via root exudates or within host plant. The physiological changes that have great impact on microorganisms in the surrounding soil are altered permeability of membrane and hence mycorrhiza formation leads to changed exudation pattern resulting in a new microbial equilibrium (4, 22, 55, 69, 90). Graham *et al.* (50) and Bansal and Mukerji (17) found significant reduction in reducing sugars and total amino acids in root exudates from mycorrhizal sudangrass *(Sorghum vulgare)* when compared with non-mycorrhizal controls. These changes coupled with chemical and physical impact of the fungal symbiont hyphae in the surrounding soil result in a very different potential in the rhizosphere.

The difference between the rhizosphere effect of soil surrounding non-mycorrhizal roots and the mycorrhizal roots is the presence of extra matrical hyphae of the mycorrhizal fungus, that extend out some distance from the host tissues into the soil (51). The hyphae produced by VA mycorrhiza are relatively sparse compared with profusion of hyphae and rhizosphere associated with ectomycorrhizae. Regardless of the type of mycorrhizae the extramatrical hyphae have profound effect on the soil microflora. Gilbert and Linderman (48) suggested the term 'mycosphere effect' to describe the enhanced microbial population of selected microorganisms near fungal structures in soil. The extramatrical byphae of VAM fungi exude substances that cause soil and organic fractions to aggregate (104).

The interactions in the mycorrhizosphere may involve a variety of bacteria and fungi which may influence plant growth. In most of the recent studies dual inoculation of bacteria, fungi and VAM fungi has been done in relation to plant growth (11, 20, 61). Meyer and Linderman (74) studied rhizosphere and rhizoplane microflora for different groups of bacteria as influenced by mycorrhiza. Dual inoculation with VA mycorrhizal fungi and rhizobacteria results in increased mycorrhizal colonization (10, 73). Bagyaraj and Menge (10) reported larger bacterial and actinomycete populations in the rhizosphere of mycorrhizal plants as compared to non-mycorrhizal tomato plants. The effects vary with time, the species of mycorrhizal fungus, the host and other environmental parameters. Newman *et al.* (81) used a multivariate statistical approach to study mycorrhizae and root - surface microorganisms on *Plantago.* Paulitz and Linderman (82) explained that the species composition of the rhizosphere microflora may be altered by the formation of mycorrhizae. This hypothesis was supported by Meyer and Linderman (73), who demonstrated difference in populations of taxonomic and functional groups of bacteria associated with VAM and non VAM plants. Populations of total bacteria were higher in the rhizoplane of mycorrhizal sweet corn and clover, compared to nonmycorrhizal plants. Ames *et al.* (5) demonstrated that microbial populations differ between VAM and non VAM plants. Populations of bacteria and actinomycetes were quantitatively and qualitatively analysed in pot cultures of different VA mycorrhizal fungi by Secilia and Bagyaraj (99). Urea hydrolyzers were present in all the VAM plants, but were absent in the non-mycorrhizal plants. This study confirmed that VAM fungi influence the populations of bacteria in green house pot cultures and that this effect varies with each VAM fungus.

The stimulation or inhibition of microbial activity by vesicular arbuscular mycorrhizal fungi has been postulated in recent years (18, 25, 70). The microbial activity this region is influenced by different organic substances released by the roots (8). Therefore, the mycorrhizosphere microbiota differ qualitatively as well as quantitatively from rhizosphere of non-mycorrhizal plants. Vesicular arbuscular mycorrhizal fungi are widespread in occurrence and due to their potential as biofertilizers have been investigated extensively (18, 19, 54, 78, 79).

4. MYCORRHIZOSPHERE BIOLOGY OF *VIGNA UMBELLATA*

In the present investigation, efforts have been made to study the effect of VA mycorrhiza on the rhizosphere, rhizoplane mycoflora and different growth parameters of an underutilized pulse crop rice bean [*Vigna umbellata* (Thunb.) Ohwi and Ohashi]. It is an important multipurpose crop grown by various ethnic tribal groups of north-eastern India and utilised for food, fooder, green manure, as a cover crop and soil enricher. This crop posseses immense potential, also, due to its high nutritional quality. *Vigna umbellata* is an annual, stem erect/ suberect or flexuose, tending to be viny, leaflets entire or lobed, inflorescence a raceme, flowers medium sized and yellow in colour. Pods are cylindrical, glabrous and 8-12 seeded.

Seeds of *Vigna umbellata,* cultivar RBL-1 were procured from National Bureau of Plant Genetic Resources, New Delhi. Seeds were surface sterilized with-0.1% mercuric chloride, washed with distilled water and sown in earthern pots filled with sterilized (by ushing 0.1% formalin solution) field soil of the garden of Botany Department, during kharif season of the year 1996. Five seeds were sown in each pot. The pots were divided in two series, one series inoculated with VAM fungus and other series without VAM fungus i.e. uninoculated control. Ten plants from each treatment were selected randomly for estimating microorganisms in the rhizosphere and rhizoplane mycoflora. Soil dilution and plate counts method (107) was employed for the qualitative and quantitative analysis of the mycoflora in the rhizosphere. Rhizoplane studies were conducted by serial root washing technique of Dickinson (36) using Czapek Dox agar medium (89). Both the qualitative and quantitative aspects of fungal population in the rhizosphere and rhizoplane of VA mycorrhizal inoculated and uninoculated plants were studied in relation to the age of plant at different stages of plant growth. The interaction between crop cultivar and VA mycorrhiza was studied by recording seed yield per plant and number of root nodules at regular intervals of growth period. Ten plants were harvested at an interval of 30 days upto a period of 90 days (when plants become fully mature) and studied for different growth parameters.

The earlier reports on rhizosphere mycoflora of *Abelmoschus esculentus* by Ranga Rao and Mukerji (88) showed consistent enhancement in the rhizosphere mycoflora from the seedling stage to maturity. Ranga Rao (87) observed that the rhizosphere population differed in total counts, in different cultivars of wheat and rice. Also the peak population levels in each cultivar reached at different stages of plant growth.

Earlier workers noted difference in the total counts of rhizosphere within the genera, species, cultivars and even in genetically related lines (8, 18, 19, 41, 63, 64, 66, 70, 80). The decrease in root exudation discourages the colonization of fungal population in the rhizosphere which depend on simple organic molecules released by the plant root. Specific changes in population of bacteria in the rhizosphere of mycorrhizal plants have been observed by Secilia and Bagyaraj (99). The level of decline in microbial population varied with host plant and VAM fungal species used which correlates with quantitative decrease in root exudation of sugars and amino acids (18, 19, 103).

The qualitative analysis of mycoflora in the rhizosphere vary in the treated and control plants. Qualitative analysis of mycoflora showed more number of saprophytic fungi in the rhizosphere of non-micorrhizal plants as compared to mycorrhizal plants. Different species of *Aspergillus (A. niger, A.nidulans, A.flavus* and *A.fumigatus)* and *Penicillium (P.citrinum* and *P.funiculosum)* dominated in the uninoculated control. Different fungal species viz. *Alternaria alternata, Gliocladium fimbriatum, Cladosporium herbarum, Acrophialophora nainianna, Trichoderma viride* and *Zygorhyncus homothallicus* were observed in the VAM inoculated plants. *P.citrinum* and *P.chrysogenum* were observed at 20 days growth stage in the rhizosphere. Percent frequency of occurrence of saprophytes was more in uninoculated soil as compared to mycorrhizosphere of inoculated plants. Pathogenic fungal species like *Fusarium moniliforme, F.oxysporum, F.solani* and *Macrophomina phaseolina* were observed only in uninoculated control.

In the rhizoplane of control soil, different species of fungi like *Aspergillus fischeri, A. flavus, Aureobasidium pullulans, Curvularia lunata* and *Rhizopus nigricans* predominated as compared to inoculated control. Pathogenic fungal species like *F.moniliforme, F.oxysporum, F.semitectum* and *Macrophomina phaseolina* were absent in inoculated soil and observed only in uninoculated control. Bali and Mukerji (15) reported increase in population of certain saprophytic fungi like *Trichoderma viride, Cladosporium cladosporioides* and *Gliocladium* sp. in both rhizoplane and rhizosphere of VAM inoculated roots of cotton plants in comparison to uninoculated ones. Summerbell (103) reported the inhibitory effects of *Trichoderma* species and other soil microfungi in formation of mycorrhiza *in vitro*. Gupta and Mukerji (54) studied the mycorrhizal allelopathy in the rhizosphere of *Trigonella foenum-graceum* and reported that mycorrhizae inhibit the growth of some fungi and enhance the growth of other fungi.

Therefore, it's clearly evident that inhibition of root borne pathogen was due to VAM inoculation. In recent years several reports have indicated that fungal symbiont exhibit increased resistance to several diseases (62, 98) or VAM fungi cause some beneficial effects on host plants which lead to reduction of disease symptom or pathogen population in soil (25, 26, 27,101). Induction of disease resistance mechanism by the mycorrhizal fungus in host plants is controversial.

Baby and Manibhushanrao (9) reported the potential of fungal antagonists *Trichoderma* spp., *Gliocladium virens* and vesicular arbuscular mycorrhizal fungi in controlling rice sheath blight caused by *Rhizoctonia solani* and their effect on plant growth. Dugassa *et al.* (39) reported that AM plants were highly susceptible against shoot pathogen *Oidium lini* but suffered less than non-AM plants in terms of shoot fresh weight, CO_2 assimilation and content of sucrose in shoot apex. This indicates that AM not only activates resistance mechanisms but also can induce tolerance against pathogens. AM plants are more resistant to some root pathogens but more susceptible to shoot pathogens and viruses (32, 33, 98). Localized morphological and biochemical alterations in AM roots were suggested to increase the resistance against wilt diseases in tomato and cucumber (34, 35). Furthermore, many details are known about the physiological and biochemical changes in plants due to symbiosis. But no clear evidence exists which of such alterations might be responsible for the mycorrhizal effect on pathogenesis.

In the present study, the mycorrhizal fungi reduced the total number of colonies per gram of soil in rhizosphere as well as rhizoplane of VAM inoculated plants. Peak population level of rhizosphere and rhizoplane mycoflora was observed at 30 days growth stage (Fig. 1, 2).

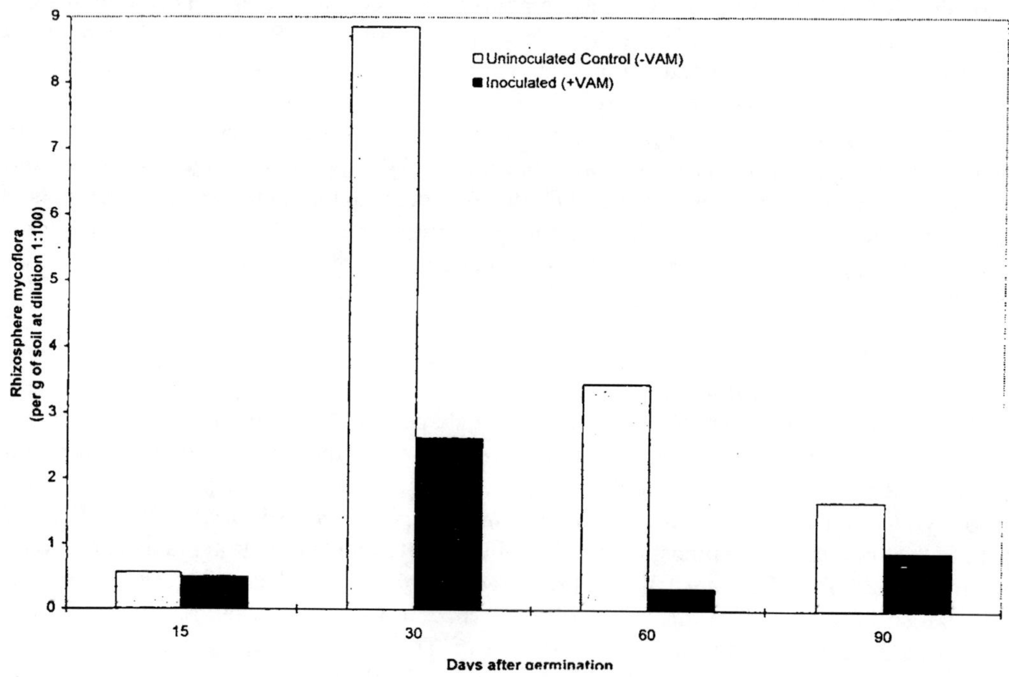

Fig. 1 : Rhizosphere mycoflora of *Vigna umbellata* (CV. RBL-1)

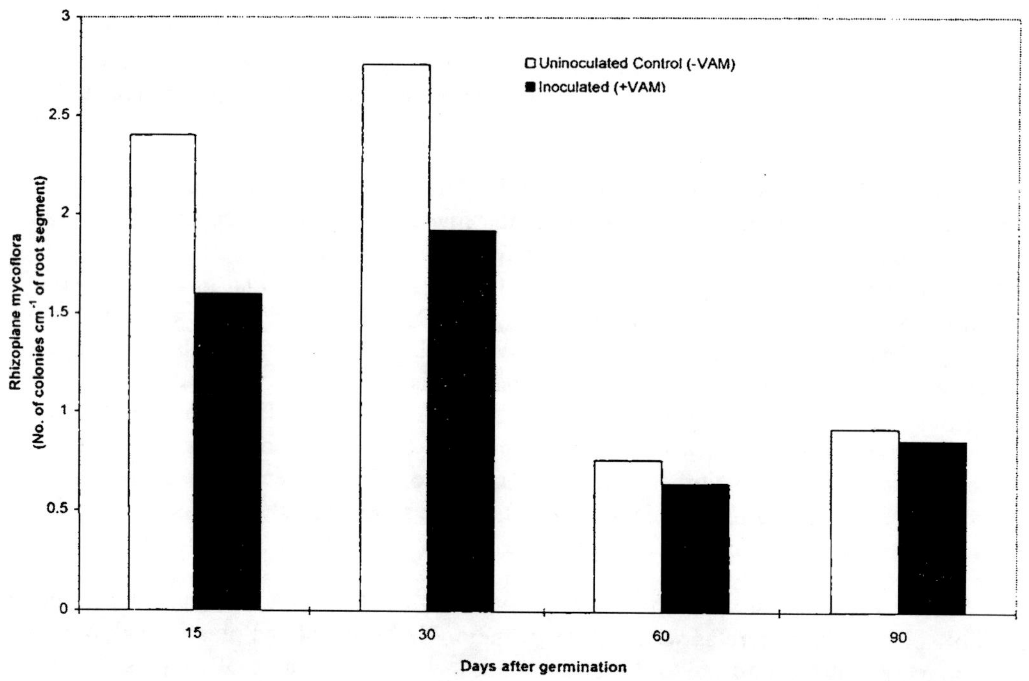

Fig. 2 : Rhizoplane mycoflora of *Vigna umbellata* (CV. RBL-1)

Effect of mycorrhiza on growth studies of different crop plants have been shown to cause an enhancement in various growth parameters by many workers. In the present study, effect of VAM inoculation on number of root nodules and seed yield per plant of CV.RBL-1 *Vigna umbellata* showed a stimulatory effect and an appreciable increase in root-nodule number (Fig. 3) dry matter and seed yield was observed. However, there was consistent increase in root nodule number in both VAM inoculated and uninoculated plants from earlier stages of plant growth till maturity. Barea *et al.*(21) and Harris *et al.* (57) have also shown in soybean, that nodulated root systems of mycorrhizal plants fixed more N_2 than nodulated root systems of non-mycorrhizal plants. In *Vigna umbellata* an increase of 53.8% in seed yield, 50.79% in number of root-nodules and 87.29% in shoot dry weight was observed. Higher percentage or root colonization was observed at all stages of growth in VAM inoculated plants of *Vigna umbellata* (Fig. 4).

Mycorrhizal association alters the physiology and morphology of roots and plants in general which in turn cause a new microbial equilibrium to be established in the rhizosphere of host plants. Linderman (70) observed that these changes presumably involve the same type of organisms as were involved in the rhizosphere before formation of mycorrhizae but quantitative changes occur within these types as a result of direct metabolic interaction with the mycorrhizal fungal hyphae or spores or the indirect effect mediated by the host. Present results support the hypothesis that a direct interaction between the hyphal network of VAM fungi and rhizosphere microorganisms accounts, at least in part, for the observed modifications in the microbial equilibrium.

To evaluate the influence of VAM fungi on disease incidence and development, different variables i.e. plant pathogen, symbiotic fungus and environmental conditions are to be considered simultaneously. As the VAM fungi could not be axenically cultured as yet the effect can be observed only in the host which is colonized. Several workers have studied the use of VAM fungi in biological control of root borne fungal pathogens. In general, mycorrhizal plants were found to suffer less damage and incidence of disease is decreased or development of pathogen is inhibited. However, increased severity of disease was also observed by some workers (47, 75, 76, 93, 98). Inoculations with VAM fungi are also known to provide resistance against several soil borne root pathogens and nematode pests. The interaction between the mycorrhizal fungi and plant pathogens - fungi, viruses and nematodes separately have been studied in detail by several workers.

5. CONCLUSION

VAM fungi substantially helped the rice bean plants and increased their biomass significantly. This activity finally resulted in healthier pods, seeds and high seed yield. In the mycorrhizosphere and rhizoplane of *Vigna umbellata,* CFU number showed a decreasing trend as compared to non-mycorrhizal plants. Non inoculated plants were found to show more number of pathogenic fungi implying thereby that VA mycorrhiza induced resistance in mycorrhizal plants.

VA mycorrhizae can prove to be an effective tool as a biofertilizer in the maintenance of sustainable agriculture and increased productivity in economically important plants. Mycorrhizae may be of significant importance in the managed ecosystems of agricultural lands in the design of integrated system of pest control and maintaining a microbial equilibrium of growth stimulation.

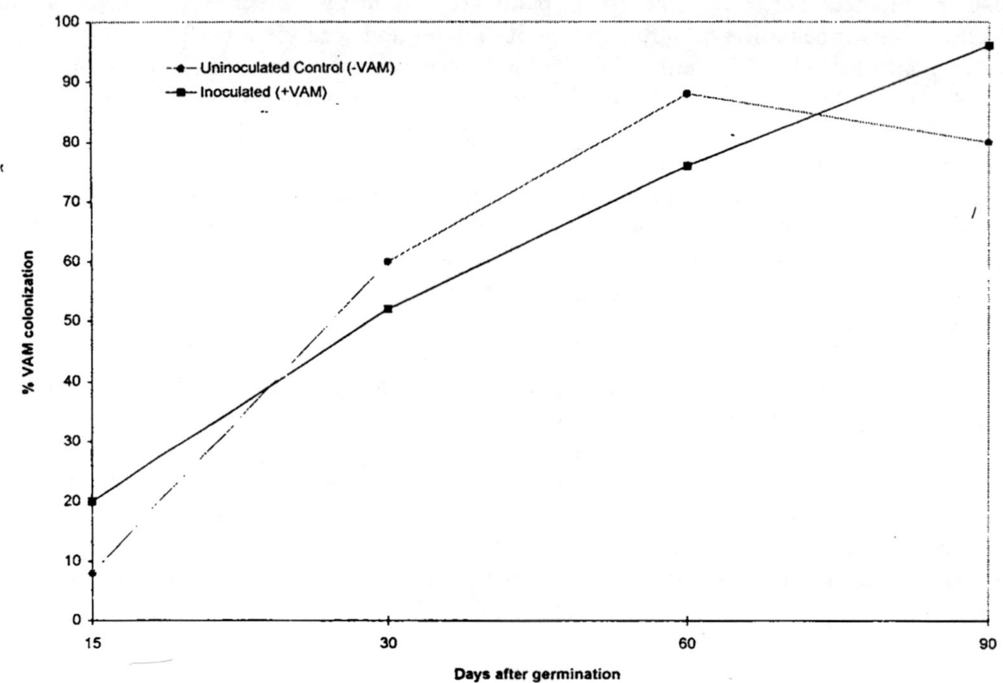

Fig. 3 : Effect of VAM on number of root nodules in *Vigna umbellata* (CV. RBL-1)

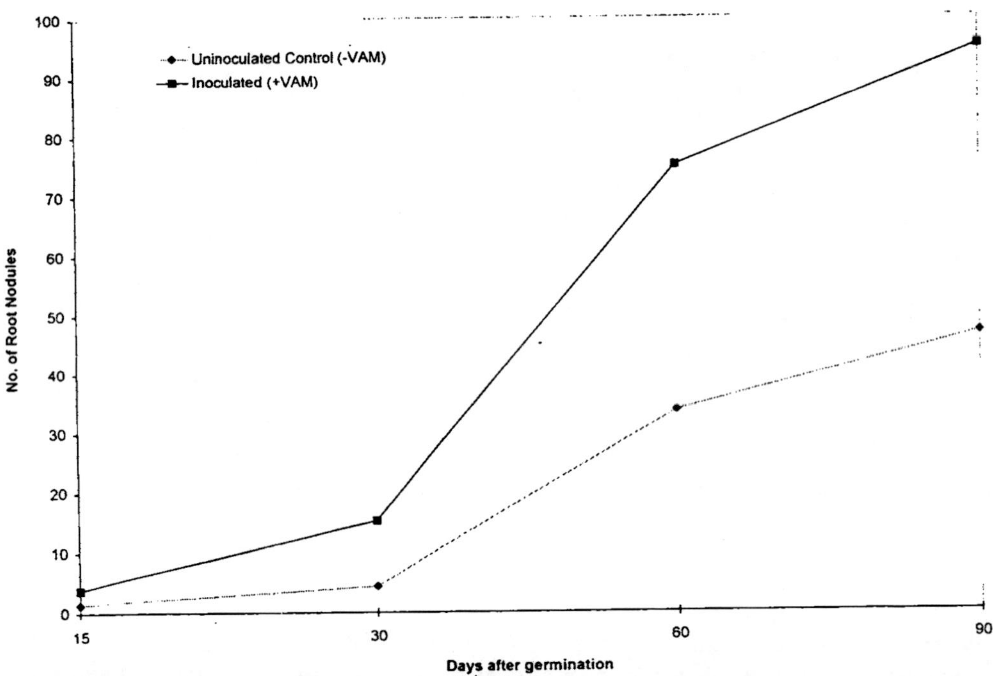

Fig.4 : Vesicular - arbucular mycorrhizal fungal colonization in roots of *Vigna umbellata* (CV.RBL-1)

6. REFERENCES

1. Agnihothrudu, V.1953. Soil conditions and root diseases. III. Rhizosphere microflora of important plants of India. *Proc. Indian Acad. Sci.* **37B** : 1-13.

2. Agnihothrudu, V. 1955. Incidence of fungistatic organisms in the rhizosphere of *Pigeon pea (Cajnus cajun)* in relation to resistance and susceptibility to wilt caused by *Fusarium udum* Butler. *Naturwissenschaften* **42** : 373.

3. Agnihothrudu, V. 1961. Rhizosphere microflora of tea *(Camellia sinensis* L.) in relation to root rot caused by *Ustulina zonata.* Lev. Sacc. *Soil Sci.* **91** : 133-137.

4. Alexander, M. 1977. *Introduction to Soil Microbiology.* John Wiley, New York.

5. Ames, R.N., Mihara, K.L., and Bethlenfalvay, G.J.1987. The establishment of microorganisms in vesicular - arbuscular mycorrhizal and control treatments. *Biol. Fertil. Soils* **3** : 217-223.

6. Audus, L.J.1959. *Plant Growth Substance,* 2nd ed. Leonard Hill(Books), London, pp. 5

7. Azcon-Aguilar, C. and Barea, J.M.1985. Effect of soil microorganisms on formation of vesicular-arbuscular mycorrhizas. *Trans. Br. Mycol.Soc.* **84** : 536-537.

8. Azcon, R., Aguilar, C. and Barea, J.M. 1992. Effects of interactions between different culture fractions of 'phosphobacteria' and *Rhizobium* on mycorrhizal infection, growth and nodulation of *Medicago sativa. Can. J. Microbiol.* **24** : 520-524.

9. Baby, U.I. and Manibhushanrao, K. 1996. Fungal antagonists and VA mycorrhizal-fungi for biocontrol of *Rhizocionia solani,* the rice sheath blight pathogen. In : *Current Trend in Life Sciences.* Vol. XXI : *Recent developments in Biocontrol of Plant Pathogens* (eds. K. Manibhushan Rao and A. Mahadevan) Today & Tomorrow's Printers and Publishers, New Delhi, India, pp. 1-9.

10. Bagyaraj, D.J. and Menge, J.A.1978. Interactions between a VA mycorrhiza and *Azotobacter* and their effects on rhizosphere microflora and plant growth. *New Phytol.* **80**: 567-573.

11. Bagyaraj, D.J. 1984. Biological interactions with VA mycorrhizal fungi. In : *VA Mycorrhiza.* (eds. C.L. Powells and D.J. Bagyaraj), C.R.C. Press, Boca Raton, Fl. pp. 131-153.

12. Baker, R. 1965. The dynamics of inoculum. In : *Ecology of Soil-Borne Plant Pathogens,* (eds. K.F. Baker and W.C. Synder), Univ. Calif. Press. Berkley, pp. 395-403.

13. Baker, R.1981. Eradication of plant pathogens by adding organic amendments to soil. In : *Handbook of Pest Management in Agriculture.* Vol. 2. (ed. D. Pimental), Chemical Rubber Company Press, Boca Raton, Florida, pp. 137-157.

14. Balandreau, J. and Knowles, R.1978. The rhizosphere. In : *Interactions Between Nonpathogenic Soil : Microorgcinisms and Plants,* (eds. Y.R. Dommergues and S.V. Krupa), Elsevier Scientific Publishing, New York, pp. 243-268.

15. Bali, M. and Mukerji, K.G. 1990. Interaction between soil microflora and VAM fungi in relation to growth of cotton. Proc. 7th NACOM, Jackson, Wyoming. USA, pp. 14.

16. Bali, M. and Mukerji, K.G. 1991. Interactions between VA mycorrhizal fungi and root microflora of jute. In : *Plant Roots and their Environment,* (eds. Mc Michael and H. Pearson), Elseviar, Amsterdom, pp. 396-401.

17. Bansal, M. and Mukerji, K.G.1994. Root exudates in rhizosphere biology. In : *Concepts in Applied Microbiology and Biotechnology,* (eds. K.G. Mukerji and V.P. Singh), Aditya Books, New Delhi, India, pp.98-120.

18. Bansal, M. and Mukerji, K.G. 1994a. Efficacy of root litter as a biofertilizer. *Biol. Fertel Soils,* **18** : 228-230.

19. Bansal, M. and Mukerji, K.G. 1994b. Positive correlation between VAM-induced changes in root exudation and mycorrhizosphere mycoflora. *Mycorrhiza* **5**:3 944.

20. Barea, J.M. and Azcon-Aguilar, C.1985. Mycorrhizas and their significance in modulating nitrogen-fixing plants. *Adv. Agron.* **36**:1-54.

21. Barea, J.M., Azcon-Aguilar, C. and Azcon, R. 1987. Vesicular - arbuscular mycorrhiza improve both symbiotic N fixation and N_2 uptake from soil as assessed with a ^{15}N technique under field conditions. *New Phytol.* **106** : 717-725.

22. Bowen, G.D. and Rovira, A.D. 1976. Microbiol colonization of plant roots. *Annu. Rev. Phytopathol.* **14** : 121-144.

23. Bowen, G.D. and Rovira, A.D. 1968. The influence of microorganisms on growth and metabolism of plant roots. In : *Root Growth,* (ed. W.J. Whittington), Butterworths, London, pp. 170-199.

24. Bowen G.D. and Rovira, A.D. 1992. The rhizosphere: the hidden half of the hidden half. In : *Roots : The Hidden Half,* (eds. Y.Waisel, A.Eshel and U.Kafkafi), Marcel Dekker. Inc., New York, USA. pp. 641-669.

25. Caron, M., Forton, J.A. and Richard, C. 1985. Influence of sustrate on the interaction of *Glomus intraradices* and *Fusarium oxysporum* f sp. *radicis-lycopersici* on tomatoes. *Pl. Soil* **87** : 233-239.

26. Caron, M., Forton, J.A. and Richard, C. 1986a. Effect of *Glomus intraradices* on infection by *Fusarium oxysporum* f. sp. *radicis lycopersici* on tomatoes over a 12-week period. *Can. J. Bot.* **64**: 552-556.

27. Caron, M., Forton, J.A. and Richard, C. 1986b. Effect of phosphorus concentration and *Glomus intraradices* on *Fusarium* crown and root rot of tomatoes. *Phytopath.* **76** : 942-946.

28. Catska, V., Macura, J. Vagnerova, K.1960. Rhizosphere microflora of wheat. III. Fungal flora of wheat rhizosphere. *Folia Microbial. Praha.* **5**: 320-330.

29. Clark, F.E.1949. Soil microorganisms and plant roots. *Adv. Agron.* **1** : 241-288.

30. Curl, E.A. and Truelove, B. 1986. *The rhizosphere,* (Advanced series in agriucltural sciences, Vol. 15) Springer, Berlin Heidelberg, New York.

31. Darrah, P.R.1993. The rhizosphere and plant nutrition : a quantitative approach. *Plant Soil* **155/156** : 1-20.

32. Dehne, H.W. 1977. Utersuchungen Liber den Einfluss der endotrophen Mycorrhiza aufdie Fusarium-Welke an Tomate und Gurke. Diso. Univ. Bonn. W. Germany. 150 pp.

33. Dehne, H.W. 1987. Zur Bedeutung der vesikuldr arbuskuldren (VA) Mykorrhiza firdie Pflanzengesundheit. Habil. Univ. Hannover, Germany.

34. Dehne, H.W. and Schönbeck, F. 1979. Untersuchungen Zum EinfluB der endotrophen Mykorrhiza auf Pflanzenkrai-ikheiten. II. Phenolstoffwechsel und Liginifizierung. *Phytopathol. Z.* **5**: 210-216.

35. Dehne, H.W., Schönbeck, F. and Baltruschat, H. 1978. Untersuchungen Zum EinfluB der endotrophen Mykorrhiza auf Pflanzenkrankheiten, III. Chitinase-Aktivitat und omithinzyklus. *Z. Pflkrankh-Pflschutz.* **85** : 666-678.

36. Dickinson C.H. 1971. Cultural studies of leaf saprophytes. In : *Ecology of Leaf Microorganisms* (eds. C.H. Dickinson and T.F. Press), Academic Press, London, pp. 292-324.

37. Dommergues, Y.R.1978. The plant-microorganism system. In : *Interactions Between Nonpcithogenic Soil Microorganisms and Plants,* (eds. Y.R. Dommergues and S.V. Krupa). Elsevier Scientific Publishing, New York, pp.1-37.

38. Domsch, K.H. 1969. Microbial stimulation and inhibition of plant growth. *Trans. Int. Cong. Soil Sci.* **3** : 455-463.

39. Dugassa, G.D., Alten, H. Von and Schbnbeck, F. 1996. Effects of arbuscular mycorrhiza (AM) on health of *Linum usitatissimum* L., infected by fungal pathogens. *Plant Soil.* **185** : 173-182.

40. Edward, J.C., Shrivastava, R.N. and Nair, Z. 1962. Microflora of soils and rhizospheres of various fields crops of the Allahabad Agricultural Institute Farm. *Allahabad Fmr.* **6** : 1-14.

41. Elkan, G.H. 1962. Comparison of rhizosphere microorganisms of genetically related modulating and non-nodulating soybean lines. *Can. J. Microbiol.* **8** : 79-87.

42. Elliott, L.F., Gilmour, G.M.,Lynch, J.M. and Tillenmore, D.1984. Bacterial colonization of plant roots. In *: Microbial- Plant Interactions,* (eds. R.L.Todd and J.E.Giddens) Soil Science Society of America, Madison, Wis., pp.1-16.

43. Foster, R.C. and Bowen, G.D. 1982. Plant surfaces and bacterial growth on the rhizosphere and rhizoplane. In : *Phytopathogenic Prokaryotes,* Vol. 1, (eds. M.S.Mount and G.H.Lazcy). Academic Press, New York, pp.159-185.

44. Garrett, S.D. 1956. *Biology of Root Infecting Fungi.* Cambridge Univ. Press, London.

45. Garrett, S.D.1959. Biology and ecology of root - disease fungi. In *: Plant Pathology: Problems and Progress.* 1908-1958, (eds. C.S. Holton, G.W.Fischer, R.W. Fulton, H. Hart and S.E.A Mc Callan). The Univ.Wis. Press, Madison. p.309-316.

46. Garrett, S.D.1970. Pathogenic Root-Infecting Fungi. Cambridge Univ. Press, Cambridge.

47. Gerdemann, J.W.1968. 'Vesicular-arbuscular mycorrhiza and plant growth. *A. Rev. Phytopathol.* **6**: 397-418.

48. Gillbert, S.D. and Linderman, R.G. 1971. Increased activity of soil microorganism near schlerotia of *Sclerotium rofusii* in soil. *Can. J. Microbio.* **17** : 557-562.

49. Golley, F.B.(ed.)1977. *Ecological Succession.* Benchmark papers in Ecology 15. Dowden Hutchinson and Stroudsburg, Pa.

50. Graham, J.H., Leonard, R.T. and Menge, J.A.1981. Membrane - mediated decrease in root exudation resposible for phosphorus inhibition of vesicular arbuscular mycorrhizae formation. *Plant Physiol.* **68** : 548-552.

51. Graham, J.H. and Menge, J.A.1982. Influence of vesicular arbuscular mycorrhizal and soil phosphorus on take-all disease of wheat. *Phytopath.* **72**: 95-98.

52. Graham, R.D.1988. Genotypic differences in tolerance to magnese deficiency. In: *Manganese in Soils and Plants,* (eds. R.D.Graham, R.J. Hannam and N.C. Uren). Kluwer Academic Publishers, Dordrecht, The Netherlands pp. 261-276.

53. Graham, R.D. and Rengel, Z.1993. Genotypic variation in zinc uptake and utilization by plants. In *: Zinc in Soils and Plants,* (ed. A.D. Robson). Kluwer Academic Publishers, Dordrecht, The Netherlands pp.107-118.

54. Gupta, Rajni and Mukerji, K.G. 1997. Mycorrhizal allelopathy in *Trigonella.foenum graceum,* In: *Souvenir and Abstracts; National Symposium on Microbes in Plant Improvement and Environmental Protection*; Dec. 23-24, 1997; Department of Botany, University of Delhi.

55. Hale, M.G. and Moore, L.D. 1979. Factors affecting root exudation. II. 1970-1978. *Adv. Agron.* **31** : 93-124.

56. Harley, J.L. 1969. *The Biology of Mycorrhiza.* Leonard Hill, London.

57. Harris, D., Pacorsky, R.S. and Paul, E.A. 1985, Carbon economy of soybean, *Rhizobium-Glomus* association. *New Phytol.* **101** : 427-440.

58. Hiltner, L. 1904. Uber nevere Erfahrungen and probleme auf dem Gebiet der Bodem bakteriologie und unter besomderer Berucksi chtigung der Grundung ung and Brache. *Arb. Dtsch. Landwirt. Ges.* **98**: 59-78.

59. Howard, A. 1943. *An Agricultural Treatment.* Oxford University Press, Geoffrey, Cumberlege, London, U.K.

60. Huang, C., Webb, M.J. and Graham, R.D. 1996. Pot size affects expression of Mn efficiency in barley. *Plant Soil* **178** : 205-208.

61. Jagpal, R. 1989. Ecophysiological effect of VAM on two legumes *Leucaena leucocephala* and *Sesbania sesban.* Ph.D. Thesis. Univ. of Delhi.

62. Jalali, B.L. and Jalali, I. 1991. Mycorrhiza in plant disease control. In *: Handbook of Applied Mycology,* Vol. II: *Soil and Plants* (eds. D.K., Arora, B, Rai, K.G., Mukerji and G.R. Knudson), Marcel Dekker, New York, pp.131-154.

63. Katznelson, H. 1961. Microorganisms in the rhizosphere. In : *Recent Advances in Botany.* Vol. 1. Univ. Tomoto Press, Toronto. pp. 610-614

64. Katznelson, H.1965. Nature and importance of the rhizosphere. In *Ecology of Soil-borne Plant Pathogens Prelude to Bilogical control,* (Eds. K.F. Baker and W.C. Snyder), Univ. Calif. Press, Los Angeles, Berkeley. pp. 187-209

65. Katznelson, H.,Lochhead, A.G. and Timonin, M.I. 1948. Soil microorganisms and the rhizosphere. *Bot. Rev.* **19** : 543-587

66. Katznelson, H.and Richardson, L.T.1948. Rhizosphere studies and associated microbiological phenomenon in relation to strawberry root rot. *Scient. Agric.* **28**: 293-308.

67 Khanna, R., Chandra, S. and Khanna, K.K. 1993, Rhizosphere microflora of triticale. *Biol. Mem.* **19** : 111-121.

68. Lemanceau, P.,Corberand, T.,Gardan, L., Latour, X., Laguerre, G., Boeufgras, J.M. and Alabouvette,C. 1995. Effect of two plant species. Flax (*Linum usitatissimum* L.) and tomato (*Lycopersicon esculentum* Mill.) on the diversity of soil borne populations of flourescent pseudomonads. *Appl. Environ. Microbiol.* **61** : 1004-1012.

69. Linderman, R.G. 1984. Microbial interactions, in the mycorrhizosphere. In : *Proc. 6th NACOM,* (ed. R. Molina). Oregon, U.S.A. pp. 117.

70. Linderman, R.G. 1988. Mycorrhizal interactions with the rhizosphere microflora. The mycorrhizosphere effect. *Phytopath.* **78** : 366-371.

71. Lynch, J.M. and Whipps, J.M. 1990. Substrate flow in the rhizosphere. *Plant Soil* **129** : 1-10.

72. Mc Manus, M.A.1960. Certain mitotic effects of Kinetin, G.A., I.A.A. and maleic hydrazide on the root of *Allium cepa. Nature,* (London) **185** : 44-45.

73. Meyer, J.R. and Linderman, R.G. 1986a. Response of subterranean clover to dual inoculation with vesicular arbuscular mycorrhizal fungi and plant growth promoting bacterium. *Pseudomonads putida. Soil Biol. Biochem.* **18**:185.

74. Meyer, J.R. and Linderman, R.G. 1986b. Selective influence on populations of rhizosphere or rhizoplane bacteria and actinomycetes by mycorrhiza formed by *Glomtys fasciculatum. Soil Biol. Biochem.* **18** : 191-196.

75. Mosse, B. 1963. Vesicular - arbuscular mycorrhiza : an extreme form of fungal adaptation, Symblotic associations. In : *Proceedings of 13th Symp. Soc. Gen. Microbiol* (eds. P.S. Nutman and B. Mosse), Cambridge Univ. Press. pp. 145-170.

76. Mosse, B., Hayman, D.S. and Arnold, D. J. 1973. Plant growth responses to vesicular arbuscular mycorrhiia. V. Phosphate uptake by three plant species from P defficient soils labelled with ^{32}p. *New Phytol.* **72** : 809-915.

77. Mukerji, K.G. 1966. Ecological studies on the microorganic population of usar soils. *Mycopath. Mycol. Appl.* **29** : 339-349.

78. Mukerji, K.G., Mandeep and Verma A. 1997. Mycorrhizosphere microorganisms screening and evaluation. In *: Mycorrhiza Manual (*ed. A.K.Verma), Springer, Geidellberg; Germany. pp. 85-97.

79. Mukerji, K.G. 1997. Establishment of *Albizia lebbek* in arid and semi-arid tropical forest lands. In : *Souvenir and Abstracts*; *National Symposium on Microbes in Plant Improvement and Environmental Protection;* Dec.23-24, 1997; Department of Botany, University of Delhi.

80. Neal, J.L., Atkinson, T.G. and Lavon, R.I. 1970. Changes in the rhizosphere microflora of spring wheat induced by disomic substitution of a chromosome. *Can. J. Microbio.* **16** : 153-158.

81. Newman, E.I., Heap, A.J. and Lawley, R.A.1981. Abundance of mycorrhizas and rootsurface microorganisms of *Plantago lanceolata* in relation to soil and vegetation-amultivariate approach. *New Phytol.* **89**:95-108.

82. Paulitz, T.C. and Linderman, R.G.1988. Interactions between VA mycorrhizal fungi and fluorescent pseudomonads in the rhizosphere. *Phytopath.* **78**:1967.

83. Pearson, N.J. and Rengel, Z.1977. Mechanisms of plant resistance to nutrient diffeciency stresses. In *: Mechanisms of Environmental Stress Resistance in Plants,* (eds. A.S. Basra and R.K. Basra). Harwood Academic Publishers, Amsterdam, pp. 213-240.

84. Perotti, R.1926. On the limits of biological inquiry in soil science. *Proc. Int. Soc.Soil Sci.* **2**:146-161.

85. Ramachandrareddy, T.K.1959. Rhizosphere microflora of pteridophytes. *Curr.Sci.* **28** : 113-114.

86. Rambelli, A.,1973. The rhizosphere of mycorrhiza. In: *Ectomycorrhizae,* (eds. G.L.Marks and T.T.Kozlowski), Academic Press, New York, pp. 299-343.

87. Rao Ranga, V. 1971. Studies on fungi in the root zone of cultivated plants and cytology of the *Ascus* in Chaetomiaceae. Ph.D. Thesis, Department of Botany, University of Delhi.

88. Rao Ranga, V. and Mukerji, K.G.1972. Fungal flora in the root zone of healthy and infected plants. *Ann. Ist. Pasteur* **122**: 81-90.

89. Raper, R.B. and Charles Thom. 1949. *A Manual of the Penicillia.* The Williams and Wilkins Company, Baltimore pp. 64-65.

90. Ratnayake, M., Leonard, R.T. and Menege, J.A. 1978. Root exudation in relation to supply of phosphorus and its possible relevance to mycorrhiza formation. *New Phytol.* **81** : 543-552.

91. Reinking, O.A. and Manns, M.M.1933. Parasitic and other Fusaria counted in tropical soils. *Z. Parasitenkd.* **6** : 23-75

92. Rengel, Z. and Graham, R.D.1996. Uptake of zinc from chelate-buffered nutrient solutions by wheat genotypes differing in Zn efficiency. *J.Exp. Bot.* **47**: 217-226.

93. Rhodes, L.H. 1980. The use of mycorrhizae in crop production systems. *Outlook Agriculture* **10** : 275.

94. Rice, E.L. and Pancholy, S.K.1972. Inhibition of nitrification by climax ecosystems. *Am. J. Bot.* **59** : 1033-1040.

95. Rovira, A.D.1965. Interactions between plant roots and soil microorganism. *Ann.Rev. Microbiol.* **19** : 241-266.

96. Rovira, A.D. 1979. Biology of soil-root interface. In *: The Soil Root Interface,* (eds. J.L. Harley and R.Scott Russell). Academic Press, New York, pp.145-160.

97. Sadasivan, T.S. 1960. The problem of rhizosphere microflora. *Proc. Natn. Inst. Sci. India* **20B**: 71-79.

98. Schonbeck, F. 1979. Endomycorrhiza in relation to plant diseases. In *: Soil Borne Plant Pathogens,* (eds. B. Schippers and W. Gams). Academic Press, London, U.K. pp. 272-280

99. Secilia, J. and Bagyaraj, D.J.1987. Bacteria and actinomycetes associated with pot cultures of vesicular - arbuscular mycorrhizas. *Can. J. Microbiol.* **33**: 1069-1073.

100. Sharma, K.R. and Mukerji, K.G. 1976. Microbial ecology of *Sesamum orientalo* L. and *Gossypium* plant surfaces, 1969. In *: Microbiology of Aerial Plant Surfaces,* (eds. C.H. Dickinson and T.F. Preece) Academic Press, London, pp. 81-89.

101. Sharma, M. and Mukerji, K.G. 1992. Mycorrhiza tool for biological control of plant diseases. In: *Recent Developments in Biocontrol of Plant Diseases,* (eds. K.G., Mukerji, J.P. Tewari, D.K. Arora and G. Saxena). Aditya Books, New Delhi. India, pp. 52-80.

102. Starkey, R.L. 1929. Some influences of higher plants upon the microorganisms in the soil. II. Influence of the stage of plant growth upon abundance of microorganisms. *Soil Sci.* **27** : 355-379.

103. Summerbell, R.C. 1987. The inhibitory effects of *Trichoderma* species and other soil microfungi on fomiation of mycorrhiza by *Laccaria bicolar in vitro. New Phytol.* **105** : 437-448.

104. Sutton, J.C. and Sheppard, B.R. 1976. Aggregation of sand-dune soil by endomycorrhizal fungi. *Can. J. Bot.* **54** : 326-333.

105. Suvercha and Mukerji, K.G. 1990. Microbial ecology of rhizosphere and phylloplane. In - *Economic Plants and Microbes,* ed. R.P. Purkayastha). Today & Tomorrow's Printers and Publishers, New Delhi - 110 005, India, pp. 155-162

106. Timonin, M.I., 1964. Interaction of higher plants and soil microorganism. Microbiology and Soil Fertility Proc. Biology Cholloquim, Oregon State University Press, Oregon, USA, pp. 135-157

107. Timonin, M.I. 1946. Microflora of the rhizosphere in relation to the deficiency disease of oats. *Soil Sci. Soc. Am. Proc.* **11** : 284-292.

108. Tinker, P.B. 1984. The role of microorganisms in mediating and facilitating the uptake of plant nutrients from soil. *Plant Soil,* **76** : 77-91.

109. Torrey, J.G. and Tjepkemena, J.D.1979. Symbiotic nitrogen fixation in actinomycete modulated plants. *Bot.Gaz.* **140 (Suppl.)** : Si-Sv.

110. Venkatraman, C.S. 1960. Foliar application of nutrients and rhizosphere microflora of *Camellia sinensis. Nature.* **187** : 621-622.

111. Walkman, S.A. 1922. A method of counting the number of fungi in the soil. *J. Bact.* **7**: 339-341.

112. Weindling, R., Katznelson, H.and Beale, H.P.1950, Antibiosis in relation to plant disease. *A. Rev. Microbiol.* **4** : 247-260.

113. Whipps, J.M. and Lynch, J.M.1983. Substrate flow and utilization in the rhizosphere of cereals [*Triticum aestivum* (Wheat), *Hordeum vulgare* (Barley)]. *New Phytol.* **95**: 605-623.

114. Wiehe, W. and Hoflich, G.1995. Survival of plant growth promoting rhizosphere bacteria in the rhizosphere of different crops and migration to non inoculated plants under field conditions in north-east Germany. *Microbiol. Res.* **150** : 201-206.

24 Arbuscular Mycorrhizal Technology for the Future

C. Manoharachary

CONTENTS

ABSTRACT 410

1. INTRODUCTION 410
2. INFECTION PROCESS 410
3. TAXONOMY 411
4. DISTRIBUTION 411
5. TECHNOLOGY APPLICATION 411
6. PHOSPHORUS UPTAKE 412
7. AM FUNGI AND LEAFY VEGETABLES 413
8. CONCLUSIONS 413
9. REFERENCES 413

Advances in Microbial Biotechnology
J.P. Tewari, T.N. Lakhanpal, Jagjit Singh, Rajni Gupta & B.P. Chamola (eds.)
APH Publishing Corporation, New Delhi - 110 002, India.

ABSTRACT

The mycorrhiza are one of the naturally existing component of dynamic microbial communities. The mycorrhiza make the plants resistant to diseases, stress and other besides helping in 'P' uptake, enhanced plant growth, biomass and yield. Majority of the plants are mycorrhiza dependent for their growth and survival. The important AM fungi are *Acaulospora, Gigaspora, Glomus, Scutellospora, Entrophospora* and *Sclerocystis* and are considered as the potential bioinoculants for commercial exploitation. The present paper, review, infection process, taxonomy, distribution, technology application, 'P' uptake and role of AM fungi in the establishment of plants.

1. INTRODUCTION

The beneficial role of fungal mycelia which form mantle on plant root was brought to the notice for the first time in 1885 by Frank. Since then our knowledge on Mycorrhizae and plant relationship has grown and many valuable informations have come to our notice. Presently the understanding is that the mycorrhizae promote plant growth, increase the yield, offer disease resistance and mobilise nutrients. These qualities have been found to be highly beneficial to the agricultural microbiologists and crop scientists etc. The mycorrhizal fungus derives nutrients i.e., carbohydrates from the plant roots and in turn helps the plant to draw phosphorus and other nutrients and also more water from the soil even under stress conditions. The two main groups of mycorrhizae are, Ectomycorrhizae and Endomycorrhizae. The best known ectomycorrhizal fungus is *Pisolithus* which has been commercially exploited and has come into wide use as biofertiliser in the Western countries and in the Asia-pacific region (20). These countries have richly improved their economy through the export of pine wood and other timbers. Mostly the ectomycorrhizae belong to Basidiomycotina and are known to promote the growth and establishment of forest tree seedlings. The three major morphological groups of endomycorrhizae are (1) Vesicular Arbuscular, (2) Orchidaceous, and (3) Ericaceous mycorrhizae. Of these the most important is vesicular arbuscular mycorrhizae (VAM). The wide spread occurrence of VAM in various plants and under different agro-climatic conditions covering a broad ecological range is well known.

2. INFECTION PROCESS

These are obligate symbionts in plant root and life cycle of VAM fungi has been partially understood. The VAM spores germinate in the soil and this has been not observed under laboratory conditions. The penetration of the root by VAM fungus is by appresorium and infection peg formation. Once AM fungus enters the host, the hyphae spread. In course of time, vesicles, arbuscules and hyphae are formed intercellularly and intracellularly in the root cortex. The extra matrical hyphae produces chlamydospores or sporocarps. These fungi belong to Zygomycotina. The important VAM fungi are *Acaulospora, Gigaspora, Glomus, Scutellospora, Entrophospora* and *Sclerocystis* and are also the potential bioinoculatns for commercial exploitation (3, 4).

VAM fungi are ubiquitous and worldwide in distribution and are found associated with soils deficient in phsophorus in association with roots of crop plants including vegetables. Enormous amount of work has been done on the taxonomy ecology and ultrastructure of these fungi in relation to uptake of nutrients and improvement in plant growth (10, 11, 13, 14, 17, 18, 19, 22, 25, 26, 29, 30, 31, 33, 34, 37, 38, 39, 41, 43, 46, 47, 48, 50).

3. TAXONOMY

VAM fungal taxonomy has received considerable importance and still forms most puzzling question to many of the scientist (59, 60, 81, 82, 100, 101, 102). Gibson *et al.* (28), Trappe (96) and others have worked on the affinities of VAM fungi. Berch (13), Hall (30), Morton and Benny (54), Schenck and Perez (82), Trappe (97) and Walker (100) have contributed significantly to the understanding of the taxonomy of AM fungi. The utilisation of ELISA techniques, immunoblotting procedures and gel electrophorosis have been found to help the seggregation of AM fungi (5, 53, 83).

4. DISTRIBUTION

The distribution and occurrence of AM fungi differ both qualitatively and quantitatively with the change in edaphic factors and the type of vegetation. The number of spores in soil is also influenced by the seasonal changes (17, 35, 59, 60, 74, 90).

VAM fungi are symbiotically present in the roots of most of the angiospermic and crop plants (91). The VAM fungi have been recognised as major contributors into soil aggregation process, waste land development, nutrient uptake and crop growth. Mosse (58) considered that VAM fungi have practical application in establishment of plants in soils under stress and lower nutrition status (1, 2, 3, 7, 79, 80, 85).

5. TECHNOLOGY APPLICATION

Mycorrhizal response is related to host genotype, Mosse (58) considered that establishment of the symbiotic association is important rather than its ultimate effect on plant growth (47). The study of plant response to mycorrhizal infection and 450 back crossed plants between high yielding low VAM responsive *Triticum spelta* and low yielding VAM responsive wild type, yielded plants ranging from well infected to non-infected. This suggests the possibility of breeding for mycorrhizal responsiveness. Field inoculation is important from the point of crop productivity. Spore inoculation is important from the point of crop productivity. Spore number can be increased by crop rotation, fertilizer application, liming, etc. The other method is to induce earlier infection in seedlings or cuttings by increasing soil infectivity, nutrient flow techniques, aeroponics and auxenic culturing have also been tried but with little success. It has also been suggested that use of expanded montmorillonit clay, clay aggregates or a substance marked under the name of terra green have proved very successful and supporting media for inoculum production in pot cultures, and mono auxenic culture of VAM fungi with Ri t-DNA transformed roots (34).

The transformed roots showed more growth. Auxenic culturing has been successful in ectomycorrhizal fungi. It took several decades to produce the inoculum of *Pisolithus* commercially (31). Whereas, auxinic culturing of VAM fungi will be more rewarding and exciting. However, all the above involves lot of money, labour and patience with little or no results. The commonly employed method is to generate AM inoculum through soil pot culture. In this method the spores of AM fungi will be introduced into soil and perlite mixture under aseptic conditions with the addition of nutrient solution under the crop cover of cenchurus, jowar, tomato, groundnut, maize and others (75, 76, 87, 92, 94, 98). In this age of biotechnology it should not be difficult to work out the requirements for rapid and mass multiplication of root system of selected host plants so as to use the inoculum for large scale field application.

The choice of inoculant and efficient AM fungus is important for crop production. Further it is better to use indigenous endophytes to mobilise unavailable soil phosphorus. The carbon supply from the host plant is essential to the formation and functioning of VA mycorrhizae and there is increasing evidence that fungal use of host carbon may lead to growth depression under conditions of ample nutrient availability. The revolutionary success of arbuscular mycorrhizal fungi reflects the unique combination of superior biotrophic mode of fungal carbon acquisition and the ability of living plant to absorb nutrients especially phosphorus from the fungus. Improved plant growth on response of carbon to mycorrhizae is closely linked to this bi-directional exchange of carbon and phosphorus between the symbionts and exchange which is strongly influenced by plant and fungal genotype and by the environment (4, 29, 55, 86).

6. PHSOPHORUS UPTAKE

Recent reviews on phosphorus and VAM fungi have discussed phosphorus metabolism (27, 67, 103). Phosphorus transport with emphasis on transfer across the symbiotic interface (87). The role of VAM fungi is well known in P uptake and plant growth (18) growth responses in relation to plant characteristics (31, 46) and research approaches to the study of transport of phosphorus and carbon in VAM fungi (41). In view of growth depressions and ample nutrient availability, the benefit of mycorrhizae in terms of increased nutrient supply to the plants will over-ride any negative impact of the fungal carbon drain on plant growth, like wise differences in symbiotic efficiency between arbuscular mycorrhizal fungi seemed to be linked much more to variation in fungal phosphorus, transport than to variation in quantities of host carbon used by the fungus. The possible mechanism for translocation and transfer of phosphorus to the host was reviewed by Jennings (42). Phosphorus is believed to be transported predominantly as polyphosphate down concentration gradient between hyphal tips and a sink. At the symbiotic interface, it is unknown whether translocation differ between VAM fungi. The phosphorus is tapped from beyond the depression zone and the non-soluble phosphorus is then circulated by other microbes following its entry into mycorrhizal hyphae and then translocated into the space of the symbiotic interface should receive special attention as there is the unusual step in the transport process.

Technology has brought a greater increase in world agricultural productivity in the last decades by improving productivity and by making the possible incorporation of areas until then considered marginal. These benefits have come from the increase in the quantity of grains, oil seeds and feed. However, urbanisation and changes in eating habits have caused a shift towards consumption of fruits and vegetables. This has been followed by growing commercialisation of ornamental plants. These products come from a laboratory and capital intensive activity, where chemical input play an essential role but also brings up lot of problems linked with degradation of natural environment and resource base. The US National Research Council (1989) records the role of arbuscular mycorrhizae in sustained agriculture. There is now an urgent need to define realistic objective and strategies to employ mycorrhizae in sustainable agriculture. On one hand it is necessary to characterise biodiversity of AM fungi and on the other hand the genetic determinants governing AM fungi plant interactions have to be identified in host plants. The European scientific network on AM (1994) stated that "selection system in the past have inadvertently affected for AM fungi that can tolerate high nutrient levels but which may not be most beneficial to plants under low input systems the symbiotic genes of both the partners must be identified in order to select plants and fungi for optimum symbiotic efficacy".

7. AM FUNGI AND LEAFY VEGETABLES

Leafy vegetables growth is the field, in which AM fungal biotechnology can be easily applied due to their seasonal growth and frequent use of controlled conditions etc. In comparison to the high production cost, the application of AM fungal inoculum will be easier for cultivation of leafy vegetables through soil root-bit inoculum application. As far as AM fungi concerned it is important to improve inoculum production technologies in order to produce abundant amounts of homogeneous fungal material for research and experimentation and application of arbuscular mycorrhizal technology in plant production (98).

Vegetable growing yields a much higher income than any other type of farming, a valid reason for the vegetable growing communities to make good living out of small savings. Vegetable crops can be produced in the same plot of land because of their short duration and seasonal growth. For any normal person, daily minimum requirement of vegetables is around 20%. The present consumption of vegetables per capita in this country is very inadequate. The production of vegetables is not sufficient to meet the required quantity. Therefore, it is essential that production of vegetables must be increased to 100%. There are nearly 20 leafy vegetables grown in this country which are grown all the year around.

They are easy to grow and are nutritious besides being rich in vitamins and minerals. All these crops are propagated from seed sown direct in the field where the crop is to stand. There are less number of varieties of leafy vegetables available in India.

It is important to mention here that the role of AM fungi in the production of leafy vegetables need much attention. Further AM fungal technology application as biofertilizer needs innovative approach.

8. CONCLUSIONS

AM fungi are the ubiquitous living organisms associated with 95% of plants excepting few plants. Nutritionally deficient soils support large population of such benefactors. Naturally occurring inoculum being very less the mycorrhizal dependent plants require artificial addition of AM fungal inoculum. The review of the published data indicated that indigenous species and strains of AM fungi have to be isolated, identified and characterized besides establishing their efficiency. Molecular approaches have only cleared certain doubts about their genotypic and phenotypic differences but did not help much in their commercialization as biofertilizer. The essential requirement being to understand the micro-environmental parameters responsible for the survival, germination, establishment and multiplication of AM fungi under natural conditions. Role of AM fungi in the growth and productivity of crop plants, viz. cereals, pulse crops, fibre crops, plantation crops, horticulture, oil seed crops and others has been worked out at length with maximum benefits only under limited and controlled environmental variables. However, leafy vegetables have been neglected even for preliminary survey of AM fungi. The present paper reviews a number of aspects of AM fungi.

9. REFERENCES

1. Abbott, L.K. 1982. Comparative anatomy of vesicular-arbuscular mycorrhizas formed on subterranean clover. *Aust. J. Bot.* **30**: 485-499.

2. Abbott, L.K. and Robson, A.D. 1984. The effect of root density, inoculum placement and infectivity of inoculum on the development of vesicular-arbuscular mycorrhizas. *New Phytol.* **97**: 285-299.

3. Abbott, L.K. and Robson, A.D. 1991. Factors influenceing the occurrence of VA mycorrhizal fungi. *Agriculture Eco. Envrion.* **35** : 121-150.

4. Abbott, L.K., Robson, A.D. and Gazey, C. 1992. Selection of inoculant vesicular-arbuscular mycorrhizal fungi. In : *Methods in Microbiology,* Vol. 24, *Techniques For the Study of Mycorrhiza,* (eds. J.R. Norris, D.J. Read and A.K. Varma), Academic Press, London. pp. 1-22.

5. Aldwell, F.E.B. and Hall, I.R. 1986. Monitering the spread of *Glomus mosseae* through soil infected with *Acaulospora laevis* using serological and morphological techniques. *Trans. Br. Mycol. Soc.,* 87 : 131-134.

6. Adholiya, A. and Sujan Singh, 1995. *Mycorrhizae; Biofertilizers for the Future*, Proceedings of the third National Conference on mycorrhizae. TERI, New Delhi, p. 548.

7. Ammani, K., Ventakeshwarlu, K. and Rao, A.S. 1985. Development of vesicular-arbuscular mycorrhizal fungi on an upland rice variety. *Current Science* **54 (21)** : 1120-1122.

8. Arriola, L., Niemira, B.A. and Safir, G.R. 1997. Border cells and arbuscular mycorrhizae in four Amaranthaceae species. *Phytopathology* **87**: 1240-1242.

9. Bagyaraj, D.J., Manjunath, A. and Patil, R.B. 1979. Interaction between a VA mycorrhiza and *Rhizobium* and their effects on Soybeen in the field. *New Phytol.* **82** : 141-145.

10. Barea, J.M., Escudero, J.L. and Azcon- Aguilar, C. 1980. Effects of introduced and indegenous VA mycorrhizal fungi on nodulation, growth and nutrition of *Medicago sativa* in phosphate fixing soils as affected by P-fertilizers. *Plant Soil* **54** : 283-296.

11. Baylis, G.T.S. 1972. Minimum levels of phosphorus for mycorrhizal plants. *Plant Soil* **36** : 233-234.

12. Becana, M., Paris, F.J., Sandalio, I.M: and del. Rio, I.A. 1989. Isoenzymes of superoxide dismutase in nodules of *Phaseolus vulgaris* L., *Pisum sativum* L., and *Vigna unguiculata* (L.), Walp. *Plant Physiol.* **90**: 1286-1292.

13. Berch, S.M., 1986. Endogonaceae : taxonoy, specificity, fossil record, phylogeny. *Front Appl. Microbiol.* **2**: 161-188.

14. Bethenfalvay, G.J., Ulrich, J.M. and Brown, M.S. 1985. Plant response to mycorrhizal fungi; host, endophyte, and soil effects. *Soil Sci. Soc. Am. J.* **49** : 1164-1168.

15. Bharat, Rai and Singh, D.B., 1980. Antagonistic activity of some leaf surface microfungi against *Alternaria brassicae* and *Drechslera graminea. Trans. Brit. Myco. Soc.* **75**: 363-369.

16. Bhargava, B.S. and Raghupathi, H.B. 1989. Chemical analysis of NPK. In : *Methods of Analysis of Soils, Plants, Waters and Fertilisers,* (ed. H.L.S. Tandon), pp. 58-60.

17. Black, R. and Tinker, P.B. 1979. The development of endomycorrhizal root systems. II. Effect of agronomic factors and soil conditions on the development of vesicular-arbuscular mycorrhizal infection in barley and on endophyte spore density. *New Phytol.* **83**: 401.

18. Bolan, N.S. 1991. A critical review on the role of mycorrhizal fungi in the uptake of phosphorus by plants. *Plant Soil* **134**: 189-207.

19. Bolan, N.S., Robson, A.D. and Barrow, N.J. 1987. Effects of phosphorus application and mycorrhizal inoculation on root characteristics of subterranean clover and ryegrass in relation to phosphorus uptake. *Plant Soil* **104** : 294-298.

20. Bowen, G.D. 1994. The ecology of Ectomyeorrhiza formation and functioning. Management of mycorrhizal in agriculture. *Horticul. Fores.* **6**: 61-67.

21. Daniels, B.A. and Trappe, J.M. 1980. Factors affecting spore germination of the vesicular-arbuscular mycorrhizal fungus *Glomus epigaeus. Mycolog.* **71**: 457-471.

22. Eissenstat, D.M., Graham, J.H., Syvertsen, J.P. and Drouillard ,D.L. 1993. Carbon economy of sour orange in relation to mycorrhizal colonization and phosphorus status. *Ann. Bot.* **71**: 1-10.

23. Fitter, A.H. 1985. Functioning of vesicular-arbuscular mycorrhizal under field conditions. *New Phytol.* **99**: 257-265.

24. Fridovich, L. 1986. Superoxide dismutases. In : *Advances in Enzymology and Realted Areas of Molecular Biology*, Vol. 58 (Ed. A. Meister), John Wiley & Sons, New York, NY, pp. 61-97

25. Gerdemann, J.W. 1961. A species of *Endogone* from corn causing vesicular-arbuscular mycorrhiza. *Mycologia* **53**: 254-261.

26. Gerdemann, J.W. and Nicolson, T.H. 1963. Spore of mycorrhizal *Endogone* species extracted from soil by wet-sieving and decanting. *Trans. Br. Mycol. Soc.* **46**: 235-246.

27. Gianinazzi-Pearson, V. and Gianinazzi, S. 1978. Enzymatic studies on the metabolism of vesicular-arbuscular mycorrhiza II. Soluble alkaline phosphatase specific to mycorrhizal infection in onion roots. *Physiol. Plant Pathol.* **12**: 45-53.

28. Gibson, J.I., Kimbrogh, J.W. and Benny, G.L. 1986. Ultra structural observation on Endogonaceae (Zygomycetes) II. Glaziellales ord, Nov. and Glaziellaceae fam. nov. New tax based upon light and electron microscopic observation of *Glaziella autowiaca. Mycologia* **78**: 941-954.

29. Gilmore, A.E. 1971. The influence of endotrophic mycorrhizae on the growth of peach seedlings. *J. Am. Soc. Hort. Sci.* **96**: 35-38.

30. Hall, I.R. 1984a. Taxonomy of VAM fungi. In : *Vesicular Arbuscular Mycorrhizae*, (eds. C.L. Powell and D.J. Bagyaraj), CRC Press, Florida, USA, pp. 57-94.

31. Hall, I.R. 1984b. Field trials assessing the effect of inoculating agricultural soils with endomycorrhizal fungi. *J. Agric. Sci.* **102**: 725-731.

32. Hanway and Heidal, 1952. Soil analysis : Methods used in Iowa state college soil testing laboratory, *Iowa Agronomy* **57**: 1-31.

33. Harley, J.L. 1989. The significance of mycorrhiza in ecological research, pp. 129-139.

34. Hayman, D.S. 1975. The occurence of mycorrhiza in crops as affected by soil fertility. In : *Endomycorrhizas*, (eds. F.E. Sanders, B. Mosse and P.B. Tinker), Academic Press, London, pp. 495-509.

35. Hayman,.D.S. 1983. The physiology of VAM symbiosis. *Can. J. Bot.* **61**: 944-963.

36. Hayman, D.S. 1987. VA mycorrhizas in field crop systems. In : *Ecophysiology*, CRC Press, Inc. Boca Raton, Florida, pp. 171-192.

37. Hayman, D.S. and Mosse , B. 1977. Plant Growth responses to vesicular Arbuscular Mycorrhizae. I. Growth of *Endogone* inoculated plants in phosphate deficient soils. *New Phytologist* **70**: 19-27.

38. Hetrick, B.A.D. and Bloom, J. 1983. Vesicular-arbuscular mycorrhizal fungi associated with native tall grass praire and cultivated winter wheat. *Can. J. Bot.* **61**: 2140-2146.

39. Hisch, P. 1984. Improved crop plant productivity through genetic manipulation of mycorrhizal fungi. *Chem. Ind.* **23**: 833.

40. Hoagland, D.R. and Arnon, D.I. 1950. The water culture methods for growing plants without soil. *California Agric. Exp. Sta. Circ.* p. 347.

41. Jackobson, I. 1995. Transport of phosphorus and carbon in VA mycorrhizaes. In : *Mycorrhiza,* (eds. A. Varma and B. Hock), Springle and Warlock, pp. 297-324.

42. Jennings, D.H. 1987. Translocations of solutes in fungi. *Biol. Rev.* **62**: 215-243.

43. Jenson, A. and Jackobson, I. 1980. The occurence of vesicular-arbuscular mycorrhiza in barley and wheat grown in some Danish soils with different fertilizers treatment. *Plant Soil* **55** : 403-414.

44. Janos, D.P. 1987. VA mycorrhizas in humid tropical ecosystems. In : *Ecophysiology of VA Mycorrhizal Plants,* (ed. G.R. Safir), CRC, Boca Raton, pp. 107-134.

45. Kehru, S.S., Kehry, H.K. and Sudhir Chander, 1995. A package of microbial inoculants for linseed. In : *Mycorrhizae : Biofertiliser for the Future.* Proc. of the Third National Conference on mycorrhizae (eds. Alok Adholeya and Sujan Singh), pp. 386-389.

46. Koide, R. 1991. Nutrient supply, nutrient demand and plant response to mycorrhizal infection. *New Phytol.* **117**: 365-386.

47. Krishna, K.R. and Bagyaraj, D.J. 1982. Influence of VA mycorrhiza on growth and nutrition of *Arachis hypogea. Legume Res.* **5 (1)** : 18-22.

48. Manoharachary, C. and Jagan Mohan Reddy, P. 1995. Role of vesicular-arbuscular mycorrhizal fungi in forestry. In : *Mycorrhizae :* Biofertilizers for the Future. Proceddings of the Third National Conferenc on mycorrhiza, (eds. Alok Adholeya and Sujan Singh) pp. 297-302.

49. Marklunk, S. and Marklund, G. 1974. Involvement of the superoxide anion radical in the autoxidation of pyrogallol and a convenient assay for superoxide dismutase. *Eus. J. Biochem.* **47**: 469-474.

50. Marscher, H. and Dell, B. 1994. Nutrient uptake in mycorrhizal symbiosis. *Plants Soil* **159**: 89-102.

51. Menge, J.A., 1983. Utilization of Vesicular-arbuscular mycorrhizal fungi in agriculture. *Can. J. Bot.* **61**: 1015-1024.

52. Morton, J.B. 1988. Taxonomy of VAM fungi: Classification nomenclature and identification. *Mycotaxon* **32**: 267-324.

53. Morton, J.B. 1990. Evolutionary relationships among arbuscular mycorrhizal fungi in the endogonaceae. *Mycologia* **82**: 192-207.

54. Morton, J.B. and Benny, G.L. 1990. Revised classification of arbuscular mycorrhizal fungi (Zygomycetes) : New order, Glomales, two new sub-orders: Glominae and Gigasporinae and two new families: Acaulosporaceae and Gigasporaceae, with an amendation of Glomaceae. *Mycotaxon* **37**: 471-491.

55. Mosse, B. 1972. The influence of soil type and *Endogone* strain on the growth of mycorrhizal plants in phosphate deficient soils. *Rev. Ecol. Biol. Sol.* **9**: '52-537.

56. Mosse, B. 1973. Advances in the study of vesicular-arbuscular mycorrhiza, IV. In soil given additional phosphate. *New Phytol.* **72**: 127-136.

57. Mosse, B. 1977. Plant growth responses to vesicular-arbuscular mycorrhiza. X. Response to *Stylosanthes* and Maize to inoculation in unsterile soils. *New Phytol.* **78**: 277-288.

58. Mosse, B. 1990. Future VA mycorrhiza research and prospects for a practical application. In : *Mycorrhizal Symbiosis and Plant Growth,* (eds. D.J. Bagyaraj, and A. Manjunath), Proceedings of Second National Conference on Mycorrhiza, Bangalore, India, pp. 5-10.

59. Mosse, B. and Bowen, G.D. 1986. A key to the recognition of some *Endogone* spore types. *Trans. Brit. Mycol. Soc.* **51**: 485-492.

60. Mukerji, K.G., Sabharwal, A., Kochar, B. and Ardey, J. 1984. Vesicular-arbuscular mycorrhizae: Concepts and advances. In : *Progress in Microbial Ecology,* (eds. K.G. Mukerji, V.P. Angihotri, and R.P. Singh), Print House (India), Lucknow, pp. 489-524.

61. Olsen, S.K., Cole, C.E., Watnabe, F.S. and Dean, L.A. 1954. Estimation of available phosphorus in soils by extraction with sodium bicarbonate. *U.S. Dept. of Agric. Circular* No. 939: 1-9.

62. Palma, J.M., Longa, A., del Rio, L.A. and Arines, J. 1993. Superoxide dismutase in vesicular-arbuscular mycorrhizal red clover plants. *Physiol. Plant.* **87**: 77-83.

63. Parvathi, K., Venkateswarlu, K. and Rao, A.S. 1984. Occurrence of VA mycorrhizas on different legumes in a laterite soil. *Current Science* **53**: 1254-1255.

64. Perrin, R. 1990. Interactions between mycorrhizae and diseases caused by soil-borne fungi. *Soil Manag.* **6**: 189-195.

65. Phillips, J.M. and Hayman, D.S. 1970. Improved procedures for clearing root and staining parasitic and vesicular-arbuscular mycorrhizal fungi for rapid assessment of infection. *Trans. Br. Mycol. Soc.* **55**: 158-161.

66. Pirozynski, K.A. and Malloch, D.W. 1975. The growth of land plants : a matter of mycotrophism. *Biosystems* **6**: 153-164.

67. Plenchette, C. 1982. Les endomycorrhizas a vesicules et arbuscules (VA) : unpotential a exploiter en agriculture. *Phytoprotection* (Qebec), **63**: 86.

68. Prabhakar, B. 1995. Taxonomy and ecoloyg of vesicular-arbuscular mycorrhial fungi associated with *Casurina equisetifolia* plantations. Ph.D. thesis Osmania University, Hyderabad, p. 102.

69. Rabinowitch, H.D. and Fridovich, L. 1983. Superoxide radicals, superoxide dismutases and oxygen toxicity in plants. *Photochem. Photobiol.* **37**: 679-690.

70. Raman, N. and Mohan Kumar, V. 1988. Techniques in mycorrhizal research. University of Madras, Madras p. 279.

71. Rao, A.S. and Parvathi, K. 1982. Development of VA mycorrhiza in groundnut and other hosts. *Plant Soil* **66**: 133-0137.

72. Raverkar, K.P., Tripathi, D. and Bhandari, A.R. 1994. Response of tomato to VAM inoculation at different levels of phosphorus. In : *Proceedings of the 34th Annual Conference of AMI,* Punjab Agri. Univ., Ludhiana, p. 149.

73. Reddy, C.D. and Venkaiah, B. 1984. Purification and characterization of Mn superoxide dismutases from mungbean. *Biochem. Int.* **8**: 707-714.

74. Red head, J.F. 1977. Endotrophic mycorrhizas in Nigeria species of the endogonacease and their distribution. *Trans. Brit. Mycol. Soc.* **69**: 275-280.

75. Robson, A.D., Abbott, L.K. and Malajczuk, N. 1994. Proceedings of an International Symposium on Management of mycorrhizas in agriculture, Horticulture and Forestry. Perth, WA, Australia Kluwer Academic Publishers, Vol. 159, No. 1, Dordrecht/Boston/London.

76. Rosendahl, S., Sem, R., Hopper, C.M. and Azcon Aguilar, C. 1989. Quantification of three vesicular-arbuscular mycorrhizal fungi *Glomus* spp. in roots of leek *Allium porrum* on the basis of activity of diagnostic enzymes after polyacrylamide gel electrophoresis. *Soil. Biol. Biochem.* 21: 519-522.

77. Safir, G.R. 1987. *Ecophysiology of VA Mycorrhizal Plants.* CRC Press, Florida, p. 234.

78. Sanders, F.E. and Tinker, P.B. 1971. Mechanism of absorption of phsophate from soil by *Endogone* mycorrhizas. *Nature* **233**: 278-279.

79. Sanders, F.E., Tinker, P.B., Bloack, R.L.B. and Palmerly, S. 1977. The development of endomycorrhizal root system. I. Spread of infection and growth promoting effects with four species of vesicular-arbuscular mycorrhizae. *New Phytol.* **78**: 257-268.

80. Schenck, N.C. 1982. *Methods and Principels of Mycorrhizal Research*-American Phytopathological Society, St. Paul, Minnesota, pp. 1-204.

81. Schenck, N.C. and Perez, Y. 1988. *Manual for the Identification of VA Mycorrhizal Fungi.* 2nd ed. INVAM, University of Florida, Gainesville.

82. Schenck, N.C. and Perez, Y. 1990. *Manual for the Identification of VA Mycorrhizal Fungi,* 3rd ed. Synergistic Pufbl., Gainesville, Florida.

83. Sen, R. and Harper, C.M. 1986. Characterization of vesicular-arbuscular mycorrhizal fungi (*Glomus* spp.). Selective enzyme staining following polyacrylamide gel electrophoresis. *Soil Biol. Biochem.* **18** : 29-34.

84. Sestak, Z., Catzky, J. and Jarvis, P.G. 1971. In : *Plant Photosynthetic Production Manual of Methods.* The Hague, p. 818.

85. Sieverding, E. and Toro, S. 1988. Influence of soil water regime on VA mycorrhiza. V. Performance of different VAM fungal species with Cassava. *J. Agron. Crop. Sci.* **161**: 322-332.

86. Smith, S.E. and Gianinazzi-Pearson, V. 1988. Physiological interactions between symbiont in vesicular-arbuscular mycorrhizal plants. *Ann. Rev. Plant Physiol. Mol. Biol.* **39**: 221-244.

87. Smith, S.E., Gianinazzi-Pearson, V., Koide, R. and Cairney, J.W.G. 1994. Nutrient transport in mycorrhizas : Structure, physiology and consequences for efficiency of the symbiosis. In : *Proc. Int. Symp. on Management of Mycorrhizas in Agriculture, Horticulture and Forestry.* (eds. A.D. Robson, L.K. Abbott and N. Malajczuk) Perth, W., Australia, *Plant Soil*, Special Issue, **159**: 103-114.

88. Stribley, D.P. and Snellgrove, R.C. 1984. Mycorrhizas and growth of transplated onions, *Rothamsted Rep.*, 1983, Part 1, p. 166.

89. Subbaiah, B.V. and Asija, G.L. 1956. A rapid procedure for determination of available nitrogen in soils. *Curr. Sci.* **25** : 259-260.

90. Sutton, J.C. 1973. Development of VAM in crop plants. *Can. Jour. Bot.* **59**: 2487-2489.

91. Tester, M., Smith, S.E. and Smith, F.A. 1987. The phenomenon of "non-mycorrhizal" plants. *Can. J. Bot.* **65**: 419-431.

92. Thompson, J.P. 1994. What is the potential for management of mycorrhizas in agriculture. In : *Management of Mycorrhizas in Agriculture, Horticulture and Forestry*, (eds. A.D. Robson, L.K. Abbottl and N. Malajczuk), Kluwer Academic Publishers, pp. 191-201.

93. Thronley, J.H.M. 1976. *Mathemetical Models in Plant Physiology.* Academic Press, London, p. 313.

94. Tilak, K.V.B.R. 1993. Associative effects of vesicular-arbuscular mycorrhizae with nitrogen fixers. *Proceedings of the Indian National Science Academy* **B59 (3 and 4)** : 325-332.

95. Toth, R. and Toth, D. 1982. Quantifying vesicular-arbuscular mycorrhizae using a morphometric technique. *Mycologia* **74**: 182-187.

96. Trappe, J.M. 1982. Synoptic keys to the genera and species of Zygomycetous mycorrhizal fungi. *Phytopathology* **72**: 1102-1108.

97. Trappe, J.M. 1988. Lessons from alpine fungi. *Mycologia* **80**: 1-10.

98. Verma, A. and Hock, B. 1995. In : *Mycorrhiza: Structure, Function, Molecular Biology and Biotechnology*, Springer-Verlag Publication, p. 747.

99. Vijayalakshmi, M. and Rao, A.S. 1987. Development of vesicular arbuscular fungi in sunflower. *Proc. Indian Nation. Sci. Acad.* **B53(3)** : 279-281.

100. Walker, C. 1983. Taxonomic concepts in the endogonacease, spore wall characteristics in species description. *Mycotaxon* **18**: 443-455.

101. Walker, C. 1986. Taxonomic concepts in the endogonaceae II. A fifth morphological wall type in endogonaceous spores. *Mycotaxon* **25**: 95-99.

102. Walker, C. 1987. Current concepts in the taxonomy of endogonaceae. In : *Mycorrhizae in the Next Decade: Practical Applications and Research Priorities. Proc. of 7th NACOM* (eds. D.M. Sylvia, L.L. Hung, and J.H. Graham), Florida, pp. 300-302.

103. Wani, S.P. and Lee, K.K. 1995. Effects of crop and soil management practices on mycorrhizal association with crops in the semi-arid tropics. In : *Mycorrhizae: Biofertilizers for the Future.* Proceedings of the Third National Conference on mycorrhiza (Eds. Alok Adholeya and Sujan Singh), TERI, New Delhi., pp. 284-290.

25 Vesicular Arbuscular Mycorrhizal Fungi Under Salinity and Drought Conditions

Bhoopander Giri and B.P. Chamola

CONTENTS

ABSTRACT 422

1. INTRODUCTION 422
2. VESICULAR ARBUSCULAR MYCORRHIZAL FUNGI 423
3. EFFECT OF SALINITY STRESS ON VAM FUNGI 423
   3.1 VAM Fungi Colonization 423
   3.2 VAM Fungal Spores Germination 423
4. ROLE OF VAM FUNGI IN STRESSED AND SALINE CONDITIONS 424
   4.1 Effect on Nutrient Uptake Under Stressed/Saline Conditions 424
   4.2 Effect on Plant Growth Under Stressed/Saline Conditions 425
   4.3 Effect on Plant Growth Under Drought Stress 425
   4.4 Effect on Hydraulic Conductivity 426
   4.5 Effect on Transpiration 426
5. CONCLUSION 426
6. REFERENCES 427

Advances in Microbial Biotechnology
J.P. Tewari, T.N. Lakhanpal, Jagjit Singh, Rajni Gupta & B.P. Chamola (eds.)
APH Publishing Corporation, New Delhi - 110 002, India.

ABSTRACT

Salinity and drought stress affect the establishment, growth, development and production of crops in billion of hactares of earth's land. These alters physiochemical and metabolic activities of host plants and it is a world wide problem. In arid and semi arid areas, concentration of salts and availability of water is enough to damage the growth of plants.

Vesicular arbuscular mycorrhizal fungi are ubiquitous soil microorganisms and play pivotal role under stressed condition. In arid and semi arid regions soil has low fertility and nutrient content. Mycorrhizal fungi improve nutrient uptake and can be utilize for the plant growth and development in the stressed conditions in various ways.

1. INTRODUCTION

Approximately one fifth of the world's land area is covered by arid and semi arid regions. The soils of arid and semi arid region are of low inherent fertility which deposed the development of drought stress, erosion, compaction, crusting, hard setting and high soil temperature. In arid regions drought stress has been considered to be a major limiting factor which alters metabolic pathways of plant led to the less survival of plant and biomass production. One third of the world's potentially arable land are suffering from poor supply of water which affects almost all physiological and biochemical parameters of the plant.

Salinity stress is a world wide phenomenon which leads to the poor development of crop. Several national and international laboratories/institutions are involved in the production of crops in saline soil which covers about one billion hectare of the world's land area. In India salinity is a major problem, nearly 7Mh of the land is not in use due to saline/stressed conditions. Plants growing in saline conditions mainly in arid and semi arid zones may suffer from high salt stress and temperature conditions. Absorption of constituent ions of saline habitat led to the detrimental accumulations of ions and decreased absorption of essential nutrients which result in the imbalanced nutrient level of plants. Plant productivity in stressed areas is often limited by lack of nutrients especially nitrogen. Salinity stress changes host physiology in terms of protein content, carbohydrates, proteases and nitrate reductases (15). In arid and semi arid areas, concentration of salts and availability of water is enough to damage the growth of plants. The specific ions such as Na and Cl are toxic at high levels of concentration (16).

During last two decades much attention has been paid towards the role of stresses on the growth and development of plants. Scientists have developed their confidence to protect the crop from salinity by various methods such as:

i. Breeding

ii. Cell and tissue cultue

iii. Inoculation with appropriate bacteria and other microorganisms

iv. Inoculation with appropriate strains of mycorrhizal fungi.

In the present review article, more emphasis has been given on the role of Vesicular Arbuscular Mycorrhizal fungi in stressed conditions.

2. VESICULAR ARBUSCULAR MYCORRHIZA (VAM)

Vesicular Arbuscular Mycorrhizal fungi makes symbiotic association with the root of higher plants, utilizes carbohydrates produced by plant, in turn plant is benefited by the increased uptake of mineral nutrients especially P and N. The fungus belongs to order Glomales or class Zygomycetes. The role of VAM fungi in saline and drought conditions is very important for increasing fertility. The potential of mycorrhizal inoculum can be utilized for the growth and development of plants in the stressed conditions through various ways such as:

i. Enhancing establishment and survival of transplanted seedlings in adverse conditions.

ii. Improving plant growth rate by increasing nutrient uptake.

iii. Improving ability of host plant to compete root/soil born plant pathogens.

iv. Boosting capacity of host plant against stresses.

v. Increasing efficiency of nutrient recycling.

vi. Stabilization and aggregation of the soil.

Mycorrhizal technology therefore, has assumed greater relevance in crop production in stressed conditions. They play significant role in the establishment and growth of plants/ seedlings.

3. EFFECT OF SALINITY STRESS ON VAM COLONIZATION

3.1 VAM Fungi Colonization

Juniper and Abbott (33) suggested that activity of VAM fungi may be affected by soil salinity. They found that environmental factors which affect the physiology of host plant are likely to affect their mycosymbionts also. It is well known that mycorrhizal associations are dependent on carbohydrate nutrition for the existence but photosynthetic activity of host plants also affects the carbohydrate status and root colonization in saline soils (32, 62). It has been found that effect of soil salinity differs in different plant species. The toxic effect of specific ions such as sodium, calcium and chloride, prevalent in saline soil affects VAM colonization (19) and altered enzyme status and absorption of several other macro and micro elements, therefore, their induction reduced photosynthesis, respiration and protein synthesis (16). Salinity of soil is also known to interfere with the uptake of cations such as calcium and magnesium and induced physiological drought which reduced VA mycorrhizal formation (10, 35).

3.2 Effect of Salinity on Germination of Spores

Besides lowering VA mycorrhiza formation, soil salinity also inhibits germination of VAM spores. Many workers have reported restricted germination of VAM spores in saline soils (19, 27, 33, 34, 35). Estaun (17) reported inhibitory effect of NaCl on germination of VAM fungal spores. He found similar effects when applied manitol and matric potential. *Gigaspora gigantea* and *Glomus epigaeus* showed reduced spore germination when spores were treated to polyethylene glycol (PEG) (37). Sylvia and Schenck (61) reported that soil matric potential and fungal contamination affects germination of spores of three *Glomus* spp.

Application of NaCl reduced hyphal growth of *Acaulospora trappei, Gigaspora decipiens* and *Scutellospora calospora* (33). However diameter of hyphae of *Scuttellospora calospora* was not affected by the increasing concentration of NaCl (34).

4. ROLE OF VAM FUNGI IN STRESSED AND SALINE CONDITIONS

4.1 Effect on Nutrient Uptake Under Stressed/Saline Conditions

Vesicular arbuscular mycorrhiza stimulates the growth of plants and their ability to survive under stressed conditions mainly where nutrient concentration is very low. VAM fungi enhance ability of plant to recover phosphorus from soils. Many workers found increased rate of phosphorus uptake in mycorrhizal plants (12, 25). There are some mechanism of mycorrhizal activities which increase concentration of mobile soil phosphorus and its transport to host plant root (52) these are :

i. Interaction between microphose bacteria (phosphate solublising bacteria) and VA mycorrhizal fungi.

ii. Production of enzymes like phosphatase by VA mycorrhizal fungi.

iii. Production of organic acids by VAM fungi, solubilizing immobile phosphate.

Poi *et al.* (49) found that microphose bacteria interact well with the mycorrhizal fungi in phosphorus deficient soils. Azcon *et al.* (6) postulated that VAM fungal hyphae stimulate phosphate ions and transport them to plants, led increased growth of plants. However, addition of NaCl and phosphorus fertilization reduced level of root colonization by VAM fungi (17). Less dry mass of mycorrhizal plants was also reported when soluble phosphate was added (59). In turn, increased acid phosphatase activity was found when plants were inoculated with VAM fungi, (14) Poss *et al.* (51) found increased onion growth in saline soil depending primarily on P-nutrition in presence of VAM fungus and main role of VAM in saline soil appeared to increase phosphorus accumulation where phosphorus level was very low.

VA mycorrhizal hyphae have the ability to extract nitrogen and transport it from soil to plant. VAM fungi contain some enzymes which influence nitrogen fixation rates by reducing stress imposed on plants (52). Nitrogen exists in many forms *viz.* free nitrogen, nitrate, nitrite ammonium ions and organic nitrogen. Ammonium is less mobile in soil. VAM fungal hyphae transport such immobile ammonium to plant roots. VAM fungi showed increased nitrate reductase activity which confers the importance to VAM in absorption of nitrate (29). Increased concentration of potassium in mycorrhizal tomato plants was reported. Bethlenfalvay (9) found that growth response of soybean to VAM inoculation was more related to improved rather than phosphorus nutrition of host plant.

Microelements are needed by the plant in very small quantities, but play very prominent role in the growth and development of plants. These elements includes, Cu, Zn, Mg, Mn, Ca, Co, Cd, etc. which are significantly correlated with developmental stages of the plants. Mycorrhizal fungi are known to accumulate greater amount of some microelements especially under stressed conditions (18, 56). The absorption of microelements is usually limited by the rate of diffusion. Uptake of these elements results in formation of depletion zones around actively growing plant roots (50). VAM fungus prevents plant from Mn toxicity and reduced amount of Mn in red clover (4), alfalfa (57) and maize (39).

Enhanced uptake of Zn in mycorrhizal *Zea mays* growing under calcareous soil was reported (32) which ranged from 16 to 25%. Similar results were also obtained in mycorrhizal

Linum usitatissimum plants by Wellings and Thompson (63). Translocation of Ca by external hyphae of VAM fungus to host plant has been shown (53).

4.2 Effect on Plant Growth Under Saline Conditions

Several workers have paid attention towards the role of VAM fungi in improved growth and nutrition of plants under different habitats. Several reports are available which are dealing with the possible role of VAM fungi to increase, yield and quantity losses of plant under salinity (28, 45, 50, 51). This may be due to increased uptake of phosphorus and other mineral elements which led to increased growth of plant. VAM fungi reduce salinity stress and enhance resistance against plant pathogens and reduce transplantation shock in the multipurpose tree species (20). VAM fungi have increased water absorption capacity of the plants, elevated rate of photosynthesis and improved growth rate under stressed conditions. Many workers (24, 54) pointed out that VAM inoculated plants increased the number of adventitious roots, their branching and also suggested that these modifications may be relevant to salt tolerance. Many workers also reported improved growth in mycorrhizal plants like *Parthenium argentatum* (40), *Allium cepa* and *Capsicum annum* (20) and *Arachis hypogaea* (24) under salinity/stressed conditions.

Pond *et al.* (50) reported distinct improved growth of *Lycopersicum esculantum* in saline soil inoculated with *Glomus fasciculatum*. Poss *et al.* (51) summerized that VA mycorrhizal inoculated plant increased growth over uninoculated control plants, under low phosphorus level. However, no response of VAM inoculation was reported when 0.8 and 1.6 nmol P^{k-1} were added. When guagule (*Parthenium argentatum*) plant was inoculated with VAM fungus *Glomus intraradices* showed increased growth in highly saline soil and decreased amount of arbuscules and vesicles in mycorrhizal roots (48).

Recently Gupta and Krishnamurthy (24) found that peanut plants were able to tolerate 1% NaCl and showed moderate resistance to salinity. In our experiments we have observed increased growth with more number of nodules and mycorrhizal cononization in VA mycorrhizae inoculated seedlings of *Sesbina aegyptiaca* and *Sesbania grandiflora* under salt stressed soil in comparison to uninoculated seedlings.

4.3 Effect on Plant Growth Under Drought Stress

In arid and semi-arid lands drought stress is considered a major limiting factor in crop yield which affects physiological and biological process in plants led to the altered metabolic pathways. Soil of arid and semi arid regions suffering from poor supply of water affects almost all physiological and biochemical parameters of plants like growth, cell wall synthesis, nitrate reductase activity, respiration and photosynthesis. It has also reported that VAM fungus improve growth and survival of plants in different habitats and has capacity to alter crop production. In these regions water is generally the major limiting factor for crop production. Nelson and Safir (44) found increased drought tolerance of onion plant was associated with improved phosphorus nutrition and water uptake (47). Osonubi (46) found that leaf area, shoot dry weight, xylem pressure and soil potentials were higher in mycorrhizal and non-mycorrhizal phosphorus fertilized plant but under stressed conditions root length was greater only in mycorrhizal plants. According to Subramanian and Charest (60), under drought conditions mycorrhizal maize plants accumulate organic solutes such as sugar and nitrogenous compounds which contribute to drought tolerance of the host plant. VAM fungal inoculation alters the mechanism that control plant water relation, leaf gas exchange (26), leaf expansion and nutrient status (36) and phytohormone production (21). Modifications in phytohormone production by mycorrhizal fungi influence plant development and function under drought conditions (2, 22).

Chlorophyll content was not altered either by water stress or the presence of mycorrhiza. Higher chlorophyll and leaf starch levels are observed in VAM inoculated rose plant under water stress. However, no correlation was found between carbohydrates and osmotic adjustment in mycorrhizal *Capsicum annum* whereas soluble proteins increased with mycorrhizal maize (5, 13).

4.4 Effect on Hydraulic Conductivity

Many workers have studied water transport in terms of hydraulic conductivity of root. Levy and Krikum (40) determined that VA mycorrhizal inoculation increased stomatal conductance but did not effect conductivity under well watered condition. It has also been reported that VAM association affects stomatal regulation. Root conductivity of P-red clover was measured by Hardie and Leyton (26). They found that higher transpiration rate and water demand of VA mycorrhizal plant were not a result of increased plant high, but higher root conductivity. Increased drought tolerance of onion plant was also associated with increased P-nutrition and hydraulic conductivity (43).

Graham and Syvertson (23) pointed out that mycorrhizal inoculation of citrus cultivars reduced root-shoot ratio whereas increased root hydraulic conductivity of mycorrhizal plant and was twice that of the non-mycorrhizal control plants.

Higher transpiration rates and phosphorus nutrition in mycorrhizal plant were associated to the increased conductivity of roots. Drought stressed root showed lower hydraulic conductivity in comparison to well watered plants (41). Higher transpiration rates and root density of VAM inoculated plants were responsible for rapid depletion of soil water leading to more severe stress condition during stress period.

4.5 Effect on Transpiration

Many reports have shown the influence of VAM fungi on transpiration rates and concluded that VAM fungi inoculated plants showed increased transpiration rates (26, 30, 41). Allen *et al.* (3) reported increased transpiration rates in *Boutelova gracilis* in which resistance to water transport was reduced by 50%. VAM inoculation in citrus seedlings increased root growth and transpiration rates and reduced water potential (23). Greater transpiration was also reported in red clover (26); and not in sour orange (23). Transpiration rates and leaf conductance were significantly greater in mycorrhizal plants (30, 36).

5. CONCLUSION

In view of increase establishment, growth, development and survival of plants in stressed soil, sustainable technologies need to be incorporated into agriculture and forestry systems. Application of VA mycorrhizal fungi is an important aspect among such approach (1, 42). Number of reports are available regarding influence of VA mycorrhiza on drought and salinity stress. It is well known, that VA mycorrhizal colonization improved water uptake in host plants and enhanced nutrient status of plants in the nutrient deficient soil. In arid regions where lack of water restricts the production of crops, VAM fungi alters hormonal and physiological mechanisms of host plants and control loss of water (22). VAM fungi also influence photosynthesis, photorespiration and photosynthetic pigments under drought conditions. VAM fungi can help to root in stressed soil by increasing uptake and mineral nutrients, change in root hydraulic conductivity and lack of toxic ions.

6. REFERENCES

1. Allen, M.F. and Bossalis, M.G. 1983. Effect of two species of VA mycorrhizal fungi on drought tolerance of winter wheat. *New Phytol.* **93**: 67-76.

2. Allen, M.F., Moore, T.S. and Christensen, M. 1980. Phytohormone changes in *Bouteloua gracilis* infected by vesicular arbuscular mycorrhizae. I. Cytokinin increase in the host plant *Can. J. Bot.* **58**: 371-374.

3. Allen, M.F., Smith, W.K., Moore T.S. and Christensen, M. 1981. Comparative water relations and photosynthesis of mycorrhizal and nonmycorrhizal *Boutelous gracilis* (H.B.K.) Lag ex Steud. *New Phytol.* **88**: 683-693.

4. Arines, J., Vilarino, A. and Sainz, M. 1989. Effect of vesicular arbuscular mycorrhizal fungi on Mn uptake by red clover. *Agri. Eco. Environ.* **29**: 1-4.

5. Auge, R.M., Schekel, K.A. and Wample, R.L. 1987. Leaf water and carbohydrate status of VA mycorrhizal rose exposed to drought stress. *Plant Soil* **99**: 291-302.

6. Azcon-Aguilar, C., Gianinazzi-Pearson, V., Fardaeu, J.C. and Gianinazzi, S. 1986. Effect of vesicular arbuscular mycorrhizal fungi and phosphate solubilising bacteria on growth and nutrition of soybean in a natural calcareous soil amended with ^{32}P-^{45}Ca tricalcium phosphate. *Plant Soil.* **96**: 3-15.

7. Berta, G., Fusconi, A., Trotta, A. and Scannerini, S. 1990. Morphogenetic modifications induced by the mycorrhizal fungus, *Glomus* strain E_3 in the root system of *Allium porrum* L. *New Phytol.* **114**: 207-215.

8. Bethlenfalvay, G.J., Browth, M.S., Mishra, K.L. and Stafford, E. 1987. *Glycine-Glomus-Rhizobium* symbiosis. *Plant Physiol.* **85**: 115-119.

9. Bethlenfalvay, G.J. and Franson, R.L. 1989. Manganese toxicity alleviated by mycorrhizae in soybean. *J.Plant. Nutri.* **12**: 952-972.

10. Bernstein, L. 1975. Effects of salinity and solicity on plant growth. *Annu. Rev. Phytopathol.* **13**: 295-312.

11. Bloss, H.E. 1980. Vesicular arbuscular mycorrhiza in guayule. *Mycologia,* **72**: 213-216.

12. Bowen, G.D. 1987. The biology and physiology of infection and its development. In : *Ecophysiology of VA Mycorrhizal Plants,* (ed. G.R. Safir), CRC Press, Boca Raton, pp. 27-57.

13. Charest, C., Dalpe, Y. and Brown, A. 1993. The effect of vesicular arbuscular mycorrhizae and chilling on two maize hybrids. *Mycorrhiza* **4**: 89-92.

14. Dodd, J.C., Burton, C.C., Burns, R.G. and Jeffries, P. 1987. Phosphate activity associated with the roots and the rhizosphere of plants infected with vesicular arbuscular mycorrhizal fungi. *New Phytol.* **107**: 163-172.

15. Durga prasad, K.M.R., Muthukumarsamy, M. and Paneerselvum, R. 1996. Change in protein metabolis induced by NaCl salinity in soybean seedlings. *Ind. J. Pl. Physiol.* **2**: 98-101.

16. Epstein, E. 1972. *Mineral Nutrition of Plants : Principles and Perspectives.* Wiley Industrial, New York.

17. Estaun, M.V. 1989. Effect of radium chloride and manitol on germination and hyphal growth of the vesicular arbuscular mycorrhizal fungus *Glomus mosseae. Agric. Ecosyst. Environ.* **29**: 123-129.

18. Faber, B.A., Zasoki, R.J., Burau, R.G. and Uriu, K. 1990. Zinc uptake by corn as affected by vesicular arbuscular mycorrhizae. *Plant Soil* **129**: 121-130.

19. Gildon, A. and Tinker, P.B. 1981. A heavy metal tolerant strain of mycorrhizal fungus. *Trans. Brit. Myco. Soc.* **77**: 648-649.

20. Giri, B. 1997. VAM colonization in MPTS. M.Phil. Thesis, Department of Botany, University of Delhi, Delhi.

21. Gogala, N. 1991. Regulation of mycorrhizal infection by hormonal factors produced by hosts and fungi. *Experimentia* **47**: 331-340.

22. Graham, J.H. and Menge, J.A. 1982. Influence of vesicular arbuscular mycorrhizae and soil phosphorus on take - all disease of wheat. *Phytopathology* **72**: 95-98.

23. Graham, J.H. and Syvertsen, J.P. 1984. Influence of vesicular-arbuscular mycorrhiza on the hydraulic conductivity of roots of two citrus root stocks. *New Phytol.* **97**: 277-284.

24. Gupta, R. and Krishnamurthy, K.V. 1996. Response of mycorrhizal and non-mycorrhizal *Arachis hypogaea* to NaCl and acid stress. *Mycorrhiza* **6**: 145-149.

25. Hale, K.A. and Sanders, F.E. 1982. Effect of benomyl on vesicular arbuscular mycorrhizal infection of red clover (*Trifolium pratense* L.) and consequences for phosphorus inflow. *J. Plant. Nutrit.* **5**: 1355-1367.

26. Hardic, K. and Leyton, L. 1981. The influence of vesicular arbuscular mycorrhiza on growth and water relations of red clover. *New Phytol.* **89**: 599-608.

27. Hirrel, M.C. 1981. The effect of sodium and chloride salts on the germination of *Gigospora margarita. Mycologia* **73**: 610-617.

28. Hirrel, M.C. and Gerdemann, J.W. 1980. Improved growth of onion and bell pepper in saline soils by two vesicular mycorrhizal fungi. *Soil Sci. Amer.* J. **44**: 654-655.

29. Ho., I. and Trappe, J.M. 1975. Nitrate reducting capacity of two vesicular arbuscular mycorrhizal fungi. *Mycologia* **67**: 886-888.

30. Huang, R.S., Smith W.K. and Yost, R.S. 1985. Influence of vesicular arbuscular mycorrhiza on growth, water relations and leaf orientation in *Leucaena leucocephala* (Lam.) de Wit. *New Phytol.* **99**: 229-243.

31. Jasper, D.A. 1994. Management of mycorrhizas in revegetation. In : *Management of Mycorrhizas in Agriculture, Horticulture and Forestry*, Kluwer, Academic Publishers, Dordrecht, Netherland. pp. 211-219.

32. Jasper, D.A., Robson, A.D. and Abott, L.K. 1979. Phosphorus and formation of vesicular arbuscular mycorrhizal. *Soil Biol. Biochem.* **11**: 501-505.

33. Juniper, S. and Abott, L.K. 1991. The effect of salinity on spore germination and hyphal extension of some VA mycorrhizal fungi. 3rd European symposium on mycorrhiza diversity of Sheffield (Abstr.) Sheffield, U.K.

34. Juniper, S. and Abott, L.K. 1992. The effect of change of soil salinity on growth of hyphae from spores of *Gigospora decipiens* and *Scutellospora calospora.* (Abstracts), Interntional Symposium on Management of Mycorrhizas in Agriculture, Horticulture and Forestry. University, of Western Australia.

35. Juniper, S. and Abott, L.K. 1993. Vesicular arbuscular mycorrhiza and soil salinity. *Mycorrhiza* **4**: 45-57.

36. Koide, R. 1985. The effect of VA mycorrhizal infection and phosphorus status on sunflower hydraulic and stomatal properties. *J. Exp. Botany*, **36**: 1087-1098.

37. Koske, R.E. 1981. *Gigaspora gigantea*: Observations on spore germination of a VA mycorrhizal fungus. *Mycologia* **73**: 300.

38. Kothari, S.K., Marchher, H. and Romheld, V. 1990. Direct and indirect effects of VA mycorrhizal hyphae in acquisition of of P and Zn by maize grown in a calcarious soil. *Plant Soil* **131**: 177-185.

39. Kothari, S.K., Marchner, H. and Romheld, V. 1991. Effect of a vesicular arbuscular mycorrhizal fungus and rhizosphere microorganisms on manzanese reduction in the rhizosphere and manganese concentrations in maize (*Zea mays* L.) *New Phytol.* **117**: 649-655.

40. Levy, Y. and Krikum, J. 1980. Effect of vesicular and arbuscular mycorrhiza on *Citrus jambhiri* water relations. *New Phytol.* **85**: 25-31.

41. Levy, Y., Syvertsen, J.P. and Nemec, S. 1983. Effect of drought stress and vesicular arbuscular mycorrhiza on citrus transpiration and hydraulic conductivity of roots. *New Phytol.* **93**: 61-66.

42. Linderman, R.G. and Bethlenfalvay, G.J. 1992. VA mycorrhiza and sustainable agriculture. *Soil Science Society of America Madison*, WI.

43. Nelson, C.E. and Safir, G.R. 1982a. The water relations of well-watered mycorrhizal and non-mycorrhizal onion plants. *J. Amer. Soc. Hort. Science.* 107: 281-288.

44. Nelson, C.E. and Safir, G.R. 1922b. Increased drought tolerance of mycorrhizal onion plants caused by improved phosphorus nutrition. *Planta* **154**: 407-413.

45. Ojala, J.C., Jarell, W.M., Menge, J.A. and Johnson, E.L.V. 1983. Influence of mycorrhizal fungi on the nutrition and yield of onion in saliene soil. *Agron. J.* **75**: 255-259.

46. Osonubi, O. 1994. Comparative effects of vesicular arbuscular mycorrhizal inoculation and phosphorus fertilization on growth and phosphorus uptake of maize (*Zea mays* L.) and sorhgum (*Sorghum biocolor* L.) plants under drought stressed conditions. *J. Plant Nutri.* **18**: 55-59.

47. Osonubi, O., Mulongoy, K., Awotoye, O.O., Atayese, M.O. and Okali, D.U.U. 1991. Effect of ectomycorrhizal and vesicular -arbuscular mycorrhizal fungi on drought tolerance of four leguminuous woody seedlings. *Plant Soil* **136**: 131-143.

48. Pfeiffer, C.M. and Bloss, H.E. 1988. Growth and nutrition of guagule (*Parthenium argentatum*) in a saline soil as influenced by vesicular arbuscular mycorrhiza and phosphorus fertilizer. *New Phytol.* **108**: 315-321.

49. Poi, S.C., Ghosh, G. and Kabi, M.C. 1989. Response of chickpea (*Cicer arietinum* L.) to combine inoculation with *Rhizobium*, phosphobacteria and mycorrhizal organisms. *Zentral. Fur. Microbiol.* **114**: 249-253.

50. Pond, E.C., Menge, T.A. and Jarrell, W.M. 1984. Improved growth of tomato in salinized soil by vesicular arbuscular mycorrhizal fungi collected from saline soils. *Mycologia* **76**: 74-84.

51. Poss, J.A., Pond, E., Menge, J.A. and Jorrell, W.M. 1985. Effect of salinity on mycorrhizal onion and tomato in soil with and without additional phosphate. *Plant Soil* **88**: 307-319.

52. Raman, N. and Mahadevan, A. 1996. Mycorrhizal research - a priority in Agriculture. In : *Concepts in Mycorrhizal Research* (ed. K.G. Mukerji), Kluwer Academic Publishers, Netherlands, pp. 41-75.

53. Rhodes, L.H. and Gerdemann, J.W. 1978. Translocation of calcium and phosphate by external hyphae of vesicular arbuscular mycorrhizae. *Soil Science* **126**: 125-126.

54. Rosendahl, C.N. and Rosendahl, S. 1990. The role of vesicular arbuscular mycorrhizal fungi in biological control or post emergence damping caused by *Pythium ultimum*. *Symbiosis* **9**: 363-366.

55. Rosendahl, C.N. and Rosendahl, S. 1991. Influence of vesicular arbuscular mycorrhizal fungi (*Glomus* spp.) on the response of cucumber (*Cucumis sativus* L.) to salt stress. *Environ. Exp. Botany* **31**: 313-318.

56. Smith, S.E. 1980. Mycorrhiza of autotrophic higher plants. *Biological Reviews* **55**: 475-510.

57. Srivastava, D. and Mukerji, K.G. 1995. Field response of mycorrhizal and non-mycorrhizal *Medicago sativa* var. local in the F_1 generation. *Mycorrhiza* **5**: 219-221.

58. Srivastava, D., Kapoor, R., Srivastava, S.K. and Mukerji, K.G. 1996. Vesicular arbuscular mycorrhiza : an overview. In : *Concepts in Mycorrhizal Research*, (ed. K.G. Mukerji), Kluwer Academic Publishers, Netherlands. pp. 1-39.

59. Stribley, D.P., Tinker, P.B. and Royner, J.H. 1980. Relation of internal phosphorus concentration and plant weight in plants infected by vesicular arbuscular mycorrhizas. *New Phytol.* **86**: 261-266.

60. Subramanian, S. and Charest, C. 1995. Influence of arbuscular mycorrhizae on the metabolism of maize under drought stress. *Mycorrhiza* **5**: 273-278.

61. Sylvia, D.M. and Schenck, N.C. 1983. Germination of chalmydospores of three *Glomus* species as affected by soil matric potential and fungal contamination. *Mycologia* **75**: 30-35.

62. Thomson, B.D., Robson, A.D. and Abott, L.K. 1990. Mycorrhiza formed by *Gigaspora calospora* and *Glomus fasciculatum* on subterranean clover in relation to soluble carbohydrate concentration in roots. *New Phytol.* **114**: 217-225.

63. Wellings, N.P. and Thomson, J.P. 1991. Effect of VAM and P-fertilizers rate on Zn fertilizers requirements on Linseed. In : *Proc. Second Asian Conference on Mycorrhiza*, (eds. I. Soerianegara, and Supriyanto) SEAMEO-BIOTROP, Indonesia, pp. 143-152.

26 Mineral Nutrition in VA Mycorrhizal Plants

P. Ramarao, C. Manoharachary and B. Bhadraiah

CONTENTS

ABSTRACT 432

1. INTRODUCTION 432
2. ARBUSCULAR MYCORRHIZAE AND SYMBIOTIC BENEFITS 432
   2.1 Phosphorus Uptake 433
   2.2 Nitrogen Uptake 436
   2.3 Other Nutrients 438
3. NUTRIENT UPTAKE IN FOREST TREE SPECIES 438
4. CONCLUSIONS 441
5. REFERENCES 441

Advances in Microbial Biotechnology
J.P. Tewari, T.N. Lakhanpal, Jagjit Singh, Rajni Gupta & B.P. Chamola (eds.)
APH Publishing Corporation, New Delhi - 110 002, India.

ABSTRACT

Vesicular arbuscular mycorrhizae form obligate symbiotic association with majority of plant roots and are of ubiquitous occurrence in nutritionally deficient soils. These fungi play a significant role in augmenting the availability of phosphorus, increased uptake of P and other nutrients besides increasing plant growth, biomass and productivity. Exhaustive information is available on 'P' uptake by mycorrhizal plants. Mycorrhizae play a critical role in nutrient cycling in ecosystem and, also modify plant root systems. Radio-isotopes were employed in elaborating and proper understanding of the 'P' uptake. Mycorrhizas enhance the fitness of most plant species but not in all environmental situations. Modelling systems were also developed to better understand the role of mycorrhizas in N and P immobilization and release. Soil type, soil phosphorus levels and availability and inherent mycorrhizal dependency of the host plants are of much significance in mycorrhizal response. The present review attempt symbiotic efficiency, uptake of phosphorus, nitrogen and other nutrients in VAM fungal inoculated plants over non-mycorrhizal plants. Further role of VAM fungi in the establishment of seedlings, plant growth, nutrient uptake and other related aspects studied in *Tectona grandis* and *Terminalia arjuna*, are also presented.

1. INTRODUCTION

Mycorrhizas are the obligate symbiotic associations between fungi and plant roots. Mycorrhizal fungi are known to utilize the humus to release and transfer nutrients from plants thus play an important role in nutrient cycling. In recent years, the role of VA mycorrhizae in the nutrition of plants especially in infertile soils has received much attention. It has been proved beyond doubt that the improvement in the growth of plants is due to the increased mineral uptake by mycorrhizal roots. Three possible explanations for the greater uptake of mineral nutrients by mycorrhizal plants compared to non-mycorrhizal plants are suggested: i) mycorrhizae may increase nutrient uptake by shortening the distance that nutrients must diffuse through the soil to the host plant; ii) there is difference in relationship between rate of nutrient uptake, absorption and nutrient concentration at the absorbing surface and iii) mycorrhizal hyphae may chemically modify the availability of nutrients for uptake by plants. Went and Stark (48) proposed that mycorrhizal fungi formed a large component of the soil fungi involved in rapid recycling of nutrients from decaying organic matter to plants. Successional habitats have a high degree of symbiosis, organically bound nutrients and low entropy, suggesting that mycorrhizae are important in the successional process and ecosystem development.

2. ARBUSCULAR MYCORRHIZAE AND SYMBIOTIC BENEFITS

Arbuscular mycorrhizal fungi the obligate symbionts play a significant role in augmenting plant nutrient availability in ecosystems deficient in nutrients. The research data has shown improved nutrient transfer to the plant tissue through the augmentation of the absorbing surface of roots by extending the fungal mycelium into non-rhizosphere soil. Gains in phosphate, nitrogen, water transfers and others are most commonly reported. Such benefits have been well documented in reviews made by Allen (2), Gupta (19), Mukerji *et al.* (32) and Tate (43).

Camel *et al.* (8) have reported that arbuscular mycorrhizal fungi may not only enhance soil-plant transfer nutrients, but they may also be instrumental in movement of nutrient between plants. Chiariello *et al.* (9) and Read *et al.* (37) have demonstrated such nutrient uptake through the use of $^{14}CO_2$ and ^{32}P also. Plant biomass productivity is associated with the capacity of mycorrhizal associations to reduce or prevent plant disease development (28).

Increased fungal biomass in mycorrhizal root tissue is attributed to augmented availability of photosynthates in the form of simple sugars for fungal energy and carbon needs. However, the above effect must coincide with the reduced competition among rhizosphere microorganisms. Some research data indicates that root exudates influence the spore germination of AM fungi. However, Bowen (6) concluded that root exudates had no effect on spore outgrowth and spores are sensitive to root exudates. Tate (43) has stated that there exists a potential interaction between rhizosphere microbes and AM fungi.

By their role in the acquisition of nutrient resources, mycorrhizae form critical component of nutrient cycling. As such Mosse (31) has suggested that mycorrhizae should be considered as plant-fungus-soil partners.

2.1 Phosphorus Uptake

Arbuscular mycorrhizae are the most wide spread symbionts associated with plant-root systems. Harley (21) stated that AM fungi are the most abundant and ubiquitous fungi occurring in soil of all the continents. Research on VAM fungi has completed more than 100 years, but many researchers have yet to understand the immense variety of types and roles that VAM play in both native and man-made environmental systems. VAM fungi play a critical role in nutrient cycling in ecosystem and also modify plant root systems. Mycorrhizal mycelium extends beyond the depletion zone surrounding the root and exploit micro-habitat beyond the nutrient depleted area where rootlets or root hairs cannot thrive. VAM fungal association with plant root appears to allow plants to acquire resource under time of acute stress as shown by cacti which grow and flourish best in arid and semi-arid environments having heavily mycorrhizal roots.

AM fungi can also increase the uptake of mineralized 'P' by occupying the microsites of active decomposition. Bacteria like *Escherichia, Bacillus* and *Pseudomonas* and several species of *Streptomyces, Aspergillus, Curvularia, Penicillium* and *Trichoderma* are commonly associated in dissolving phosphatic component such as rock phosphate, bone meal, etc. (45).

Many pot experiments including few field experiments have demonstrated that mycorrhizal plants are more efficient at taking up soil 'P' than non-mycorrhizal plants. Hyphae extend from internal infection out into the surrounding soil. Mycorrhizal roots effectively increase the absorbing area of the root. 'P' is taken up by this external mycelium and translocated as polyphosphate to the root (5). Bolan *et al.* (4) have shown that using iron hydroxide it is possible to reduce the chemical availability of 'P'. Interesting data has been collected by them using ^{32}P, they found that differences in absorbing 'P' and translocating 'P' are, intact, incapable of revealing differences in the pattern of absorption of 'P' from various soil fractions. The question of phosphorus utilization by mycorrhizal plants, therefore, needs urgent reinvestigation. Particularly in soils where much 'P' is in the form other than Pi (40). Hyphal transport of nutrient and 'P' in particular from soil to plant is the most fascinating aspect in plant-root and AM fungal symbiotic association and efficiency. The large 'P' efflux from the fungus into the space of the symbiotic interface must receive special attention. Length, spread and longevity of hyphae have greater impact on 'P' uptake. 'P' uptake by the VA mycorrhizal hyphae occurs against a steep concentration gradient (17). The 'P' uptake kinetics of AM fungi are similar to those of plant root but no kinetic parameters were determained for the external AM fungal mycelium. Further the 'P' uptake is controlled by the ability of the fungus to translocate and transfer 'P' to the host. Barber (3) has concluded that 'P' concentration is higher at the surface of the hyphae than at the surface of the root and it can not be excluded that 'P' uptake by a fungus is influenced by its transport kinetic parameters. Organic and inorganic substances released from roots result in the formation of the rhizosphere where biological activity and availability of nutrients differ from bulksoil conditions. Parallel processes might occur at the hyphal surface and give rise to a hyphosphere. Inorganic 'P' increase in rhizosphere also affect pH changes VA mycorrhizal fungi colonize the roots and after infection increased levels of phosphotases were noted.

Harley and Smith (20) have reviewed mechanisms responsible for the hyphal translocation and measured rates of translocation phosphorus and it is believed to be transported predominantly as polyphosphate down a concentration gradient between the hyphal tips and a sink at the symbiotic interface. The 'P' transfer across the symbiotic interface is believed to follow a pattern similar to the oppositely directed C transfer; passive transport from fungus into interface is followed by active uptake by the plant cell.

'P' loss from the fungus into the periarbuscular space appears to be the most unusual step in the hyphal 'P' transport from soil to plant and could be an important determinant for the 'P' transport efficiency of VA mycorrhizal fungi (42). Alkaline phosphatase is a possible marker for symbiotic efficiency in terms of 'P' transport as its activity increases in the early stages of root colonization. Bacterial immobilization, water uptake and other play significant role in the hyphal 'P' transport and P uptake by roots. The available data suggests that the present understanding of mutual effects on 'P' uptake by roots and VA mycorrhizal fungi is preliminary. It is also suggested that the fungal contribution to 'P' uptake of mycorrhizas cannot be estimated from comparisons with non-mycorrhizal plants and the 'P' transport efficiency of VA mycorrhizal fungi can not be evaluated from measurements of plant 'P' contents.

In humic and forest soils, much of the phosphate (upto 80% or more) present within the rooting depth of plants may be in the form of phytate, (inositol phosphates) which, are insoluble, but may be brought into solution by phosphatases positioned on the surface area of roots or fungal hyphae.

Slow diffusion of phosphate ions in the soil solution contrasted with rapid absorption of phosphate by roots and other absorbing organs result in the development of depletion zones around them. Uptake may be limited by the rate of diffusive movement of ions into these depletion zones, rather than by the rate of the transport across living membranes into the root or mycorrhizae.

Increased inflow of phosphorus in mycorrhizal roots was in average about 3-4 times greater than into uninfected roots. This is due to increased efficiency uptake and the existence and continued growth of extra matrical mycelium into soil. This hyphal system can be envisaged as extending beyond the P depletion zone around the root and exploiting a greater and less depleted volume of soil. There is no doubt that uptake would be maintained as long as the hyphae continued to colonize undepleted soils. It is possible to calculate the apparent contribution of the fungus to the total uptake process and the flux which might be expected to occur through mycorrhizal entry points. Once the fungus has become established, the rate of P uptake by hyphae is related to the length of the external mycelium (18).

The increased rate of P uptake by mycorrhizal roots is due to improved exploitation of a given volume of soil, coupled with more rapid translocation of P through hyphae to roots than diffusion of P through soil to the root surface.

Apart from the physical extention of the root system, there are two other mechanisms which would result in increased efficiency of uptake by infected roots; these are: i) the possibility that mycorrhizal hyphae are able to absorb P effectively from a lower concentration in the soil solution than the associated roots; and ii) infected plants can exploit sources of soil P not available to uninfected ones (e.g., rock phosphate fertilizer or fixed inorganic or organic soil P). The importance of mycorrhizal infections to a particular species or variety of a host plant may depend upon the concentration of P in the soil, the relative affinities of the root and fungal system for P and also upon the P requirement of the host. Most studies of VAM have concentrated upon P nutrition of plants in legumes and citrus.

Mycorrhizal roots can exploit the sources of P in soil. It can be seen that the growth of mycorrhizal maize responded to the application of rock phosphate or tricalcium phosphate where as these fertilizers had no effect on the growth of non-mycorrhizal plants. In contrast, both mycorrhizal and non-mycorrhizal plants responded to monocalcium and super phosphate with no significant difference between them. Similar results have been obtained for insoluble phosphorus fertilization of a variety of host plants, usually in low pH soils.

Mycorrhizal plants were clearly more effective at extracting P from the fertilizer. The mechanism underlying the increased uptake might depend upon hyphal exploitation of the soil volume or lower km of mycorrhizal roots and also synergistic action between mycorrhizae and phosphate solubilizing bacteria. Bacteria associates with mycorrhizal fungus solublizes the P and incorporates to the plant.

It seems likely that synthesis of polyphosphates prevents excess accumulation of inorganic phosphorus in fungal cells when external supply is plentiful and uptake is rapid. In VA mycorrhizae formation of polyphosphates might account for the increased P concentration and translocation from the soil to the roots of host plant.

Histochemical and EM studies have shown the arbuscule to be a specialized haustorial organ capable of tremendous metabolic activity and well adopted as a site of nutrient exchange between the fungus and host (11). Alkaline phosphatases may be involved in active transport and transfer processes in VAM fungi and also involved in penetration of hyphae into the host cells (12). Mycorrhizal hyphae, by growing into the soil matrix, can gain access to bulk soil phosphorus behind the depletion zones created by plant roots. The depletion zone results in a change in the overall dispersion pattern of P at a site.

Mycorrhizal fungi can also increase the uptake of mineralized P by occupying the microsites of active decomposition and being involved in the degradation of litter. Herrera *et al.* (23) found mycorrhizal hyphae on decomposed leaf and rapid transport of ^{32}P form to the host plant.

Mycorrhizal fungi, by immobilizing nitrogen and phosphorus reduce the availability of these nutrients, so that nitrogen and phosphorus limit the growth of saprophytic microbes and there by retards decomposition (16). The implications of this activity is interesting, especially where levels of N and P are low and mycorrhizal activity is high (10, 22).

Mycorrhizal fungi also determine 'P' allocation among plants interconnected by a mycelial network in a soil matrix. Finlay and Read (14) found that P could be taken up by a single hypha and transported throughout mycelial network in several plants. The relative allocation of P with in the mycelium and between plants may depend upon the photosynthetic activity of the host and the ability of the plants to compete for mycorrhizae.

Caldwell *et al.* (7) noted that plants with a greater rooting density and mycorrhizal hyphal density gained more phosphorous from the interspace than neighbouring plants. This could result in the redistribution of P in the soil, creating habitats where resources are distributed in discrete patches (1).

Many workers have suggested that mycorrhizal roots mostly increase the surface area for P uptake. However, three mechanisms of mycorrhizal activity that contribute to weathering of soil P and transport to host plant. These are: i) the interaction of mycorrhizal fungi and P solubilizing bacteria; ii) the production of phosphatases by the mycorrhizal fungus and iii) the production of organic acids by mycorrhizal hyphae that mineralize P.

Recent studies have suggested that mycorrhizal hyphae have the capacity to alter the weathering rates of soil P and increase in total P cycling in ecosystems. In amended soils with iron hydroxide with ^{32}P labelling, Bolan *et al.* (4) found that mycorrhizal fungi had access to the introduced P source and it was not available to non-mycorrhizal plants. Duce (13) noted that the bound inorganic P pool was primarily decreased with the increased plant growth associated with VA mycorrhizae and also reported no significant increase in rhizosphere phosphates activity.

The mycorrhizae could enhance the availability of soil P by weathering P from the clay matrix and maintaining the solution and also P by binding Ca with the secreted oxalates. These oxalates would be degraded by actinomycetes that would enhance soil CO_2 and weathering (29). Mycorrhizae can provide a large proportion of the P taken up by the plant over a growing season and substantial fraction of biologically active P should be in the fungal mycelium (46, 47).

The mycorrhizae immobilize large quantities of nutrients important for decomposition and plant growth. Mycorrhizal fungi transport these nutrients to the host. They also appear to be localized where they can rapidly take up mineralized nutrients resulting in enhanced biotic regulations of nutrient cycling. Phosphorus toxicity has been explained for certain growth depressions in mycorrhizal plants. But the characteristic symptoms of P toxicity have not been observed by inoculation with mycorrhizal fungi.

2.2 Nitrogen Uptake

Mycorrhizae are important in the uptake of nitrogen (N) and transport organic N from substrate to the host plant. From several studies, the role of mycorrhizea in N cycling has occupied a central focus in mycorrhizal research efforts. Mycorrhizae can affect N-cycling via several means. Mycorrhizal hyphae have the capacity to extract N and transport it from soil to the host plant because enhanced absorptive surface area. It's absorption and accumulation involve reduction mediated by nitrate and nitrite reductase. Mycorrhizal interactions with the synthesis of nitrate reductase may be related to the improved availability of P to VA mycorrhizal plants. Information on nitrate assimilation by VAM fungi is scanty. Ho and Trappe (25) found that nitrate was reduced to nitrite by isolated spores of *Glomus mosseae* and *G. macrocarpum.*

In addition, mycorrhizal fungi contain enzymes that breakdown organic nitrogen and contain nitrogen reductase for altering forms of nitrogen in soil. The range of estimates of N immobilized by mycorrhizae is extreme. Vogt *et al.* (46) estimated that the fungal component of the mycorrhizae contributed about 27 kg per ha per year (27% of N turnover). Fogel and Hunt (15) estimated that the return of N in the soil by mycorrhizae comprised about 83-87% of the tree return which include the extramatrical hyphae, at large and rapidly cycling biomass component.

A solid quantitative modelling effort is needed to understand the role of mycorrhizae in nitrogen immobilization and release. As mycorrhizae can increase the water flow through plant there is a potential for increased nitrate migration to roots. Due to low nitrate reductase activity, mycorrhizae show less importance in absorption of nitrate (33). Mycorrhizae might have an NAD-dependent enzyme which might contribute to nitrate reductase activity. This has to be ascertained through reliable techniques by which it is possible to produce a large amount of harvestable extramatrical mycelium. The main site of nitrate reductase always is the host and that nitrate may be absorbed by the fungal hyphae and translocated to its cells.

Mycorrhizal fungi have the ability to utilize and transport organic nitrogen particularly as amino acids to the host. The ability of ericoid mycorrhizae to utilize organic nitrogen is specially important. These associations are found predominantly in highly acidic organic soils in which nitrogen uptake by plants is virtually non existent without mycorrhizae (38).

A large fraction of available nitrogen in the soil is ammonium which is transported by VA mycorrhizae or ectomycorrhizae (30) when the nitrogen is distributed in descrete patches. Direct investigation of ammonium uptake by VA mycorrhizae have been few. It was reported that loss of ammonium, nitrate and nitrite from soil by leaching with water was retarded when plants were inoculated with *Glomus mosseae.* Mycorrhizal infections certainly stimulated growth of the ammonium fed plants more than that of the nitrate fed plants. This could have been either because the infection increased ammonium uptake directly or because the fungal hyphae compensated for short root length and maintained the uptake of other nutrients, particularly of P.

Termite nests represent large pools of nitrogen in many ecosystems and the ability of plants to extract nitrogen via mycorrhizal hyphae could be an important pathway for nitrogen incorporation into actively cycling fractions (41).

Mycorrhizae can fix atmospheric nitrogen. Mycorrhizal associations can enhance nitrogen gain in ecosystems by increasing the nitrogen fixation rates of plant bacterial nitrogen-fixing associations (24). The nitrogen-fixation rates have been observed in legumes, actinorrhizal roots and free living associations. Increase in nitrogen fixation rates by mycorrhizal associations is variable and dependent upon P concentration, photosynthetic rates and hormones (35). By increasing plant growth, mycorrhizae also increase the nitrogen requirement of plant and where nitrogen is not deposited and it may not increase nitrogen-fixation in the local soil. However, the nitrogen requirements apparently were not met by fixation alone as there was still a net depletion in soil N. Thus, the spatial and temporal characteristics of mycorrhizae and nitrogen-fixing activity need to be understood when the ability of mycorrhizae and nitrogen fixation to increase soil nitrogen can be validated.

Recently several investigators demonstrated that nitrogen-fixed in association with one plant can be transported to an adjacent, non-fixing plant via the mycorrhizal fungal mycelium.

In crop systems, nitrogen-fixed by soybeans was transported to maize via the VA mycorrhizae and significantly increased the growth and nitrogen status of the maize plants (44). Similar responses have been hypothesized between actinorhizal and adjacent plants via connecting ectomycorrhizae. These interactions may be of critical importance to nitrogen distribution in any ecosystem.

2.3 Other Nutrients

Clear evidence for mycorrhizal involvement in the uptake of nutrients other than phosphate is scanty. Many of the analyses of the elemental composition of mycorrhizal plants revealed no clear picture of uptake. Two micro nutrients like zinc and copper have been shown to be consistently at higher concentration in mycorrhizal plants. Interactions between phosphate fertilization and deficiencies of trace elements particularly copper and zinc are well known in several species of potentially mycorrhizal plants. Increased uptake of minor elements such as copper and zinc and P may be involved in improved nodulation and nitrogen-fixation of mycorrhizal legumes.

VAM fungi increase the uptake of any nutrient that move to plant root by diffusion. Mycorrhizal roots can absorb zinc and sulphur from solution and this may reflect differences in the relationship between the surface area and weight of roots.

Elemental concentration in mycorrhizal plants can decrease with increasing levels of supplied P so that concentration of Zn, Cu and S converge in mycorrhizal plants as soil P increases. Thus convergence has been attributed to the decrease in mycorrhizal infection which occur at high P levels but P was not limiting factor. Direct uptake of zinc has been observed in VAM plants, although experiments on radioactive labelling of the zinc pools in soil suggest that VAM have access only to the more soluble zinc fractions.

It has been demonstrated that hyphal translocation of zinc from soil to the plant is influenced by phosphorus nutrition, at high P levels plants were unable to obtain zinc because of the suppression of mycorrhizal infection and the elements of hyphal translocation. Enhanced uptake of sulphur by VAM and hyphal translocation with the increased absorption in improved P nutrition (39).

X-ray microanalysis has shown that calcium to be a strong secondary constituent of polyphosphate granules and balancing in P transfer to the host because its ability to stimulate alkaline phosphatase activity.

There is insufficient data about differential effects of other changes in soil eg., phosphate level, pH, nitrogenase fertilizers, temperature, placement of inoculum in the effectiveness of different hosts and endophytes combinations. Mathematical modelling of an infection might well provide useful information in this sphere as the modes calculating rates of infections and fungal growth under different conditions are not available. The mycorrhizae immobilize large quantities of nutrients important for decomposition and plant growth. Mycorrhizal fungi transport these nutrients to the host plant. They also appear to be localized where they can rapidly take up mineralized nutrients, thereby, increasing the biotic regulation of nutrient cycling.

3. NUTRIENT UPTAKE IN FOREST TREE SPECIES

In the light of above review, it has been observed in our laboratory that *Glomus caledonium* and *G. fasciculatum* were found to be most dominant and successfully colonising VAM fungi in the roots of *Tectona grandis* L. (Verbenaceae) and *Terminalia arjuna* (Roxb.)

Wight and Arn. (Combretaceae). Therefore, these two VAM fungi were multiplied on *Cenchrus ciliaris* and their effect on growth and mineral uptake in *Tectona* and *Terminalia* were assessed (36).

Glomus caledonium and *G. fasciculatum* isolated from forest soils of Nirmal (Adilabad district, A.P.) were selected. The experiments were conducted in fresh polythene bags (each bag containing 4.95 kg soil) collected from Mulugu forests. Soil inoculum (80-100 spores/100 g of soil) was thoroughly mixed with 1.5 kg of sterilized soil filled in polythene bags of 4.5 kg of soil capacity. The treatments consisted of an addition of two VAM fungi *Glomus caledonium* and *G. fasciculatum.* Control plants were not inoculated with any test fungus. Moisture levels were maintained and Hoagland nutrient solution without phosphorus mixture was added. The levels of N, P, K, Cu, Zn, Fe and Mn of shoot and root tissues were estimated both in mycorrhizal and non-mycorrhial plant grown under glass house conditions. Phosphorus was estimated by Olsen's method (27), nitrogen was estimated by Kjeldahl method (34); Potassium (K) by ammonium acetate method and Cu, Zn, Fe and Mn were estimated by using atomic absorption spectrophotometer (AAS) method (26).

It is clear from Tables 1,2,3 and 4 that infection with mycorrhizal fungi enhanced the growth, root/shoot dry weight and mineral uptake in general with slight variations. Of the two VAM fungi *G. fasciculatum* proved better than *G. caledonium* in increasing the plant height, root/shoot dry weight in seedlings of *Tectona* and *Terminalia* (Table 1, 2). It is seen from Table 3 that the nitrogen uptake was more in test plants infected with *G. fasciculatum* than *G. caledonium.*

However, the uptake of phosphorus by *G. fasciculatum* was more with regard to *Terminalia arjuna* and in the shoots of *Tectona.* It is of interest to note that *G. caledonium* treatment resulted in more uptake of phosphorus and potassium in the roots of both the test plants compared to *G. fasciculatum* treated plants.

Table-1 Effect of VAM fungi on plant height (in cm) in two forest tree

Treatment	*Tectona grandis* (age in days)			*Terminalia arjuna* (age in days)		
	30	60	90	30	60	90
Control	30	35	39	31	34	40
Glomus caledonium	51	51	60	30	52	59
Glomus fasciculatum	52	60	70	53	62	72

Table 2: Effect of VAM fungi or root/shoot dry wt. (g/pl^{-1}) of two forest tree seedlings

	Tectona grandis						*Terminalia arjuna*					
	Root (age in days)			Shoot (age in days)			Root (age in days)			Shoot (age in days)		
	30	60	90	30	60	90	30	60	90	30	60	90
Control	0.4	0.5	0.7	0.5	0.9	1.4	0.8	1.2	2.0	2.1	2.8	3.5
G. caledonium	1.0	1.7	2.4	3.9	4.8	5.2	1.8	2.0	2.6	3.8	4.8	5.5
G. fasciculatum	2.0	2.2	2.9	5.0	5.8	6.6	1.8	2.2	2.7	3.0	5.0	6.0

Table 3 : Effect of VAM fungi on N, P (g/pt^{-1}) and K in 90 days old seedlings of *T. grandis* and *T. arjuna*

Tree species	*Tectona grandis*		*Terminalia arjuna*	
Nutrient/treatments	Root	Shoot	Root	Shoot
Nitrogen				
Control	3.0	9.0	6.0	12.0
Glomus caledonium	3.8	13.0	6.8	18.7
Glomus fasciculatum	4.0	15.0	7.7	19.0
Phosphorus				
Control	4.3	16.0	4.0	18.0
Glomus caledonium	6.6	28.0	4.5	22.0
Glomus fasciculatum	5.4	29.0	5.8	28.0
Potassium				
Control	1.1	4.8	1.0	3.5
Glomus caledonium	1.9	5.3	2.2	3.8
Glomus fasciculatum	1.0	5.7	1.2	4.6

Table 4: Effect of VAM fungi on micronutrient uptake(mg/g/dry wt.) in 90 days old seedlings of *T. grandis* and *T. arjuna*

Tree species	*Tectona grandis*		*Terminalia arjuna*	
micronutrients	Root	Shoot	Root	Shoot
Zn				
Control	0.57	1.24	0.62	1.4
Glomus caledonium	0.38	1.44.	0.90	1.5
Glomus fasciculatum	1.30	1.70	1.20	1.8
Cu				
Control	0.6	0.72	0.62	0.82
Glomus caledonium	0.9	1.20	0.80	1.00
Glomus fasciculatum	1.00	1.30	1.00	1.20
Fe				
Control	172.7	247.7	160.0	260.0
Glomus caledonium	190.0	260.0	198.0	258.0
Glomus fasciculatum	201.0	268.0	208.0	256.0
Mn				
Control	3.5	3.9	3.2	3.8
Glomus caledonium	3.6	4.1	3.2	3.8
Glomus fasciculatum	3.8	4.2	3.6	4.5

From Table 4 it is evident that the two VAM fungi enhanced the Zn, Cu, Fe and Mn contents in roots and shoots of *Tectona* and *Terminalia* in 90 day old seedlings excepting in 90 day old shoots of *Terminalia* which showed a negligible decreasing trend in Fe by both the VAM fungi.

The VAM inoculated plants were found to be more resistant to diseases. Seedling establishment and survival rates of *Tectona* and *Terminalia* were high after treatment with VA mycorrhizae. Further field experimentation using VAM fungi as bioinoculants needs to be conducted under naturally existing conditions of a forest ecosystem.

4. CONCLUSIONS

(i) Proper understanding of VA mycorrhizal fungi and host plant root infection process with reference to molecular approach is essential.

(ii) In view of controversial data reported in literature about 'P' uptake, the question of 'P' utilization by mycorrhizal plants needs reinvestigation particularly in soils where 'P' is in forms other than Pi.

(iii) VA mycorrhizal plant took up more toxic metals when exposed to acid rain, hence needs further investigation.

(iv) Increased uptake of nitrogen and potassium is the consequence of improved supply of 'P'.

(v) Role of VA mycorrhizal fungi has been established with reference to the uptake of N, P and K in *Tectona* and *Terminalia.*

(vi) Mycorrhizal dependency of *Tectona* and *Terminalia* has been established.

(vii) Data has to be strengthened with reference to commercialization and field application of VAM fungi and in particular the biochemistry of VA mycorrhizal colonized plants.

(viii) Mineral nutrition needs to be studied in VA mycorrhizal plants under different ecological variables, agroclimatic conditions, different crops, forest plants and soil conditions.

ACKNOWLEDGEMENTS

PR and CM express their grateful thanks to the DST, New Delhi, for financial support of the project entitled, "VA mycorrhizae from *Tectona* and *Terminalia*". We are also thankful to the UGC for granting funds under SAP.

5. REFERENCES

1. Allen, E.B. and Allen, E.F. 1990. The mediation of competition by mycorrhizae in successional and Patchy environments. In : *Perspectives on Plant Competition* (eds. J. Gaced and G.D. Tilman). Academic Press, New York. pp. 367-389

2. Allen, M.F. 1991. *The Ecology of Mycorrhizae.* Cambridge Univ. Press. N.Y, p. 184.

3. Barber, S.A. 1984. *Soil Nutrient Bioavailability.* John Wiley. N.Y, p. 398.

4. Bolan, N.S., Robson, A.D., Burrow, J.J. and Aybnore, I.A.G. 1984. Specific activity of phosphorus in mycorrhizal and non-mycorrhizal plants in relation to the availability of phosphate to all plants. *Soil Biol. Biochem.* **16**: 299-304.

5. Bolan, N.S. 1991. A critical review of the role of mycorrhizal fungi in the uptake of phosphorus by plants. *Plant Soil* **134**: 189-207.

6. Bowen, G.D. 1969. Nutrient status effects on loss of amides and loss of amino acids from pine roots. *Plant Soil* **30**: 139-142.

7. Caldwell, M.M., Eissenstat, D.M., Richards, J.H. and Allen, M.F. 1985. Composition for phosphorus, differential uptake from dual-isotope-labelled soil interspaces between shrub and grass. *Science* **229**: 284-286.

8. Camel, S.B.B., Reyes-Solis, G., Ferrera-Cerrato, R., Franson, R.L., Brown, M.S. and Bethlentalvay, G.J. 1991. Growth of VAM mycelium through bulk soil. *Soil Soc. Am. J.*, **55**: 385-393.

9. Chiarello, N., Hickman, J.C. and Mooney, H.A. 1982. Endomycorrhizal role of interspecific transfer of phosphorus in a community of animal plants. *Science* **217**: 941-943.

10. Christy, E., Sollins, P. and Trappe, J.M. 1982. First year sumival of *Tsuga heterophylla* without mycorrhizae and subsequent ectomycorrhizal development on decaying logs and mineral soil. *Can. J. Bot.* **60**: 1600-1605.

11. Cox, G. and Sanders, F.E. 1974. Ultrastructure of the host fungus interface in VAM. *New Phytol.* **73**: 901-912.

12. Dodd, J.C., Burton, C.C., Burns, R.G. and Jaffries, P. 1987. Phosphatase activity associated with the roots and the rhizosphere of plants infected with VA mycorrhizal fungi. *New Phytol.* **107** : 163-172.

13. Duce, D.H. 1987. Effects of VAM on *Agropyran smithi* growth under dronght strees and their influence on organic phosphorous minerlization. MS thesis, Utah State. University, Logan, UT.

14. Finlay, R.D. and Read, D.J. 1986. The structure and function of the vegetative ectomycorrhizal plants-I. Traslocation of ^{14}C-labelled carbon between plants interconnected by a common mycelium. *New Phytol.* **103**: 143-156.

15. Fogel, R. and Hunt, G. 1993. Contribution of mycorrhizae and soil fungi to nutrient cycling in a Dauglas fir ecosystems. *Can. J. For. Res.* **13**: 219-239.

16. Gadgil, R.L. and Gadgil, P.D. 1975. Suppression of litter decomposition by mycorrhizal roots of *Pinus radiata.Newzeal. J. For. Res.* **5**: 35-41.

17. Gianinazzi-Pearson, V. and Gianinazzi, S. 1986. The physiology of improved phosphate nutrition in mycorrhizal plants. In: *Physiological and Genetical Aspects of Mycorrhizae*, (Gianinazzi-Pearson, and S. Gianinazzi). INRA. Paris pp. 101-109.

18. Graham, J.H., Lindermann, R.C. and Menge, J.A. 1982. Development of external hyphae by different isolates of mycorrhizal *Glomus* spp. in relation to root colonization and growth of *Troyer citrange*. *New Phytol.* **91**: 183-189.

19. Gupta, R.K. 1991. Drought response in fungi and mycorrhizal plants. In : *Hand book of Applied Mycology*, Vol. 1, (eds. D.K. Arora, B. Rai, K.G. Mukerji and G.R. Knudsen). *Soil and Plants* Dekker, N.Y. pp. 55-75.

20. Harley, J.L. and Smith, S.E. 1983. *Mycorrhizal Symbiosis*. Academic Press. London, p. 483.

21. Harley, J.L. 1991. Introduction: The state of art. In: *Methods in Microbiology*. Vol. 23, (eds. J.R. Norris, D.J. Read, Varma, A.K.), Academic Press, London, pp. 1-24.

22. Harvey, A.E., Jurgenson, M.F. and Larsen, M.J. 1980. Clearcut harvesting and ectomycorrhizae: Survival of activity on residual roots and influence on a hardening forest stand in W. Montana. *Can. J. For. Res.* **10**: 300-303.

23. Herrera, R., Merida, T., Stark, N. and Jordon, C.E 1978. Direct phosphorus transfer from leaf litter to roots. *Naturwissendschaften* **65**: 208-209.

24. Heyman, D. S. 1987. VA mycorrhizae in field crop systems. In : *Ecophysiology of VA mycorrhizal Plants,* (ed. G.R. Safir), CRC. Press. Boca Raton FL. pp. 171-192.

25. Ho, J. and Trappe, J.M. 1975. Nitrate reducing capacity of two VA Mycorrhizal fungi. *Mycologia,* **67**: 886-888.

26. Isaac, R.A. and Kerber, J.D. 1971. Atomic absorption and flame photometry techniques and uses in soil, plant and water analysis. In: *Instrumental Methods for Analysis of Soil and Plant tissue.* Soil Soc. An. Inc. Madison, Wisconsiia, pp. 17-37.

27. Jackson, M.L. 1973. *Soil Chemical Analysis.* Prentice Hall, New Delhi p. 498.

28. Jalali, B.L. and Jalali, I. 1991. Mycorrhizae in plant disease control. In :*Hand Book of Applied Mycology,* Vol. 1, (eds. D.T. Arora, B. Rai, K.G. Mukerji and G.R. Knudsen). Marcel Dckker. N.Y. pp. 131-154.

29. Knight, W.G., Allen, M.F., Jurinak, J.J. and Dudley, L.M. 1989. Elevated $C0_2$ or solution phosphrus in soil with VAM western wheat grass. *Soil Sci. Soc. Ame. Jour.* **53**: 1075-1082.

30. Martin, F., Stewart, G.R., Genetet, J. and Le Tacon, F. 1986. Assimilation of NH_4 by beech ectomycorrhizas. *New Phytol.* **102**: 85-94.

31. Mosse, B. 1975. Specificity in VA mycorrhizas. In : *Endomycorrhizas,* (eds. F.E. Sanders, B. Mosse and P.B. Tinker), Academic Press, New York. pp. 469-484.

32. Mukerji, S., Mukerji, K.G. and Arora, D.K. 1991. Ectomycorrhizae. In : *Hand Book of Applied Mycology, Vol.*1. *Soil and Plants,* (eds. D.K. Arora, B. Rai, K.G. Mukeji and G.R. Knudsen), Marcel, Dekker, N.Y. pp. 187-215.

33. Oliver, A.J., Smith, S.E., Nicholas, D.J.D., Wallace, W. and Smith, R.S. 1983. Activity of nitrate reductase in *Trifolium subterraneum* : Effects of mycorrhizal infection and Phosphate nutrition. *New Phytol.* **94**: 63-79.

34. Piper, C.S. 1944. *Soil and Plant Analysis.* Adelaide, University of Adelaide.

35. Puppi, G. 1983. VA mycorrhizae and biological nitrogen fixation: A review. *Micol. Ital.* **12**: 3-10.

36. Ramarao, P. and Manoharachary, C. 1992. Report on studies on VAM association in *Tectona* or *Terminalia.* DST Project SP/GO/AO5/87 dt. 30-6-1989-92.

37. Read, D.J., Francis, R. and Finlay, R.D. 1989. Mycorrhizal mycelia and nutrient cycling in plant communities. In : *Ecological Interactions in Soil Plants, Microbe and Animals,* (ed. A.E. Fitter). Blackwell Sci. Publ. Boston. pp. 193-217.

38. Read, D.J. and Bajwa, R. 1985. Some nutritional aspects of the biology of ericacious mycorrhizas. *Proceedings of the Royal Society of Edinburgh,* **85B**: 317-332.

39. Rhodes, L.H. and Gerdemannm, J.W 1975. Phosphate uptake zones of mycorrhizal and non-mycorrhizal onions. *New Phytol.* **75**: 555-561.

40. Safir, G.R. 1987. *Ecophysiology of VA Mycorrhizal Plants.* CRC press, Inc., Boca Raton, Florida, p. 224.

41. Salick, J., Herra, R. and Jordan, C.F. 1983. Termitaria : Nutrient patchiness in nutrient deficient rain forests. *Biotropica* **15**: 1-7.

42. Smith, S.E. and Smith, F.A. 1990. Structure and function of the interfaces in biotrophic symbioses as they relate to nutrient transport. *New Phytol.* **114**: 1-38.

43. Tate, R.L. 1995. *Soil Microbiology.* John Wiley and Sons. Inc., p. 398

44. Vankessel, C., Singleton, P.W. and Hoben, H.J. 1985. Enhanced nitrogen-transfer from a soybean to maize by VAM fungi. *Plant Physiol.* **79**: 562-563.

45. Verma, A. and Hock, B. 1995. *Mycorrhiza (Structure, Function, Molecular Biology and Biotechnology).* Springer-Verlong, Berlin, N.Y. 747 pp.

46. Vogt, KA., Grier, C.C., Meier, C.E. and Edmonds, R.L. 1982. Mycorrhizal role in net production and nutrient cycling in *Abies amabilis* ecosystems in Washington. *Ecology* **63**: 370-380.

47. Vogt, KA., Grier, C.C. and Vogt, D.J. 1986. Production, turnover, and nutrient dynamics of above and below ground detritus of world forests. *Adv. Ecol. Res.* **15**: 303-377.

48. Went, F.W. and Stark, N. 1968. The biological and mechanical role of soil fungi. *Pro. Nati. Acad. of Scie.* USA. **60**: 4479-4504.

27 Mycorrhiza Under Stress Conditions

C.S. Singh

CONTENTS

ABSTRACT 446
1. INTRODUCTION 446
2. INFLUENCE OF STRESS CONDITIONS 447
   2.1 Drought 447
   2.2 Moisture and Flooding 448
   2.3 Soil Nutrients 448
      2.3.1 Nitrogen 448
      2.3.2 Phosphorus 449
   2.4 Hydrogen Ion Concentration (pH) 452
   2.5 Pesticides 452
   2.6 Heavy Metal 453
   2.7 Soil Salinity 454
   2.8 Light 454
   2.9 Temperature 456
3. OZONE EXPOSURE 457
   3.1 Soil Structure 457
      3.1.1 Compaction 457
      3.1.2 Erosion 457
   3.2 Carbon Dioxide and Oxygen 457
   3.3 Glucose, Activated Charcoal, Vermiculite Peat Moss 459
   3.4 Organic Matter 459
   3.5 Seasons 460
4. REFERENCES 460

Advances in Microbial Biotechnology
J.P. Tewari, T.N. Lakhanpal, Jagjit Singh, Rajni Gupta & B.P. Chamola (eds.)
APH Publishing Corporation, New Delhi - 110 002, India.

ABSTRACT

Mycorrhizal fungi can improve plant resistance to drought by increasing hydraulic conductivity, controlling stomatal opening and transpiration rate. It also contributes in improving phosphorus nutrition of host by altering the hormonal balance of the plant. Many VAM fungi can adopt high level of phosphorus. Germination, production and root colonization were best at optimum soil moisture condition congineal for the plant growth in increasing the floral tolerance of plant. Excessive moisture level or flooding interfere with the level of O_2 and CO_2, therefore, resulted in the reduction of VAM colonization. Decrease in VAM root colonization formation was also observed due to reduction photoperiod, clipping, defoliation, grazing, application of pesticide and concentration, due to their effect on phosphatase enzyme engaged in monitoring P uptake.

Occurrence and survival of various species of mycorrhizae depend on the concentration of heavy metals present in the environment. If present at lower concentration, heavy metals may be utilized as nutrient mediated by the fungi, but higher concentration has been reported to be detrimental to VAM and host also. In contrary, certain mycorrhizal fungi were found to eliveate tolerance of the host to heavy metals.

Hydrogen ion concentration has both direct and indirect effect on the performance of VAM. Directly it affect growth and multiplication of VAM because VAM germinate best at neutral pH of 7.0. Availability of various nutrients including phosphorus is dependent on the pH of medium, which play major role in the root colonization etc. Although pH 7.0 was noted as optimum but few VAM fungi (*Glomus clarum, Acaulospora laevis*) can germinate even at low pH of 4-5. Soil salinity (Chlorides) interfere with the uptake of NO_3 and P. However, certain species of VAM fungi improve plant tolerance to salinity due to increase in phosphorus nutrition, presence of heavy metals application of pesticide, high level of hydrogen-ion concentration, salinity, etc.

Temperature, ozone, light, CO_2 level also affect the spore germination, survival, growth and multiplication of VAM and are being discussed.

1. INTRODUCTION

In a particular soil, rhizosphere environment quantitatively and qualitatively is influenced by root-derived nutrients and microbial activity. Amongst, the predominant microorganisms in the vicinity of plant root are the vesicular-arbuscular mycorrhizal (VAM) fungi. These fungi are capable of forming functional mycorrhizal symbiotic association with a broad range of host plant taxa (34). As a rule mycorrhizal infection enhances plant growth. Prevalence of under these these fungi have been reported in all soils where plants growing under these conditions.

2. INFLUENCE OF STRESS CONDITIONS

2.1 Droughts

Mycorrhizal fungi are believed to improve water relation of host plant by increasing root hydraulic conductivity (44) increasing transpiration rate and lowering stomatal resistance (68, 97); or by altering the balance of plant hormones (3). Such physiological changes has been attributed to improved phosphorus nutrition of mycorrhizal plant and extension of root systems by extra-radical mycelium (12) exposed to drought stress. In all above experiments non-mycorrhizal plants were also fertilized to equal size with mycorrhizal plant. When this fertilizer level was applied to plants exposed to cyclic drought stress, mycorrhizal growth of onions (*Allium cepa* L.) was significantly greater than the growth of fertilized, non-mycorrhizal onions, implying that drought tolerance was improved by mycorrhizal colonization (83).

On contrary Hetrick *et al.* (51) were unable to demonstrate such drought tolerance in Corn (*Zea mays* L.), by using similar methods. In fact, mycorrhizal growth response was eliminated under drought stress and such contradictory results attributed to the more severe drought stress, was imposed on corn and difficult to regulate water stress with such large plant in the greenhouse. In later studies on the responses of mycorrhizal corn (*Zea mays* L.), Sudan grass [*Sorghum vulgare* (Piper) Hitch.], big blue stem (*Andropogon gerardii* Vitman) under drought stress, have indicated the benefit derived by each plant species under adequate water condition. The growth of mycorrhizal big bluestem was significantly greater than-mycorrhizal big bluestem even under severe drought stress (52). He also noted that ability of VAM to benefit plant growth under drought stress was apparently plant mediated and possibly related to the dependency of the plant on the mycorrhizal fungus. Under adequately watered condition, inoculated corn and Sudan grass were respectively 1.23 and 1.13 times larger than non-inoculated plant, while inoculated big bluestem was 6.56 fold larger than non-inoculated control plants (52).

In greenhouse studies, mycorrhizae have been shown to improve the drought resistance of cultivated crops such as wheat (27), soybean (14), onion (83), pepper (131) and red clover (31).

During periods of soil water deficit nutritional status of crops is reportedly improved following the enhancement of drought resistance (31). In addition, VAM colonization changes in leaf elasticity (5), improves leaf water and turgor potentials and also maintain stomatal openings and transpiration (5) and increases rooting length to depth (27, 61), probably affect the drought resistance.

Most of the information on the subject have been gathered from the result of experiments conducted in greenhouse or growth chamber environments and there is a paucity of information on the function of VAM in field environments. However, field studies suggested that VAM improve the drought resistance to plant (30, 120). White *et al.* (132) showed that irrigation schedule was more important than irrigation rate for enhancing establishment of functional VAM biomass.

Further more, spore germination (20), spore production (92) and VAM root colonization are reportedly maximum at optimal soil moisture supporting plant growth. Yocom *et al.* (133) observed increase in VAM infection of soybean (*Glycine max*) with *G. fasciculatum, G. mosseae, G. macrocarpum, G. deserticola* at high water level (25.00 cm rainfall). They also noted the differences in intensity of VAM infection between the above four species of *Glomus*. However, *G. deserticola* registered for higher root infection at both moisture level.

2.2 Moisture and Flooding

Soil water and aeration have considerable impact on the distribution and effectiveness of VAM due to their aerobic nature. Levels of oxygen and CO_2 were found to influence VAM response on plants, although root colonization with VAM fungi have been observed on acquatic plants.

Hartmond *et al.* (46) could not observe differences in the response mycorrhizal and non- mycorrhizal plants of citrus, planted after exposure to a 3 week flooding stress. In contrast, Keeley (59) reported the growth response of black gum (*Nyssa sylvatica*, March) to VAM colonization. He observed VAM colonization even after one year of waterlogging. It seems that VAM have no major role in enhancing flood tolerance of plant.

Singh and Tyagi (110) reported the root colonization of *Aeschynomene indica* by indigenous VAM fungi grown under normal moisture condition. On the other hand, no VAM fungal root infection was observed in *A. indica* plant uprooted from water logged conditions. However, when such plants were transplanted to potted soil conditions, after scissoring lateral roots and provided with normal moisture conditions, the new roots, thereafter developed, were highly infected with VAM fungi. In contrary, the roots of *A. indica* plants, which were provided with artificial water logging conditions, were devoid of VAM fungal infection (Table 1). These results have indicated the requirement of normal moisture conditions for better VAM fungal root infection of plants which is reduced due to higher moisture level (flooding). It has also been observed that roots of rice cultivars remained colonized with VAM fungi during their growth period under high moisture level and water logging condition (109). Subsequently, Gupta (40) worked on interaction of VAM and phosphate solubilizing bacteria for effective utilization of phosphorus in rice, found that *Glomus etunicatum* registered for best survival as compared to various species of *Glomus* and *Gigaspora*. He noted VAM fungal species can survive under condition and can colonize rice root to increase the uptake of relatively less mobile nutrients. Infection persisted through out the growth of flooded rice indicating that flooding did not affect colonization.

Table 1. Effect of water logging on VAM fungal root infection of *Aeschynomene indica* plant (110)

Moisture conditions	Per cent infection			Sm	C.D. 5%
	Initial	After 45 days of water-logging	After 45 days of under normal moisture conditions		
Water logged	Nil	Nil	80	1.54	3.64
Normal water holding capacity	73	Nil	83	1.29	3.05

2.3 Soil Nutrients

2.3.1 Nitrogen

High mineral availability, particularly of nitrogen is also known to reduce and finally inhibit mycorrhizal development (45). It has been concluded that nitrate has a stronger inhibitory effect than ammonium (95). The inhibitory effect of nitrogen has been attributed to a reduction of sugar concentration of the roots following the complete utilization of photosynthate for growth at higher N availability. However, Wallander and Nylund (128, 129) reported that high nitrogen availability did not significantly affect internal sugar concentrations rather concentration is elevated in the presence of mycorrhiza. They also recorded increase in mycelial biomass at low N level but inhibited at high N concentration.

The above cited work are related to ectomycorrhizae. However, high levels of fertilizer, especially P and N can decrease the proportion of root length colonized by VAM fungi and production of external hyphae (4, 76). Insufficient level of phosphorus restricts mycorrhizal development (11) therefore the balance of nitrogen and phosphorus in soil is essential for effective colonization (118). In addition, using ^{15}N dilution technique [$^{15}(NH_4)SO_4$], increase in total N yield of chickpea (115) and pigeonpea (*Cajanus cajan*) (Fig. 1 a, b) utilization of soil nitrogen and biological N_2 fixation with VAM inoculation, both in presence and absence of *Rhizobium* sp. at 20 Kg N ha^{-1} and 50 Kg P_2O_5 N ha^{-1} was also observed (109). Application of 29 Kg N ha^{-1} did not affect not colonization did not colonization by *G. macrocarpum* (107, 108) and *G. fasciculatum* (109).

2.3.3 Phosphorus

Colonization of plant roots by a VAM fungus is associated with enhanced mineral uptake and growth when the supply of relatively immobile nutrient ions is limiting (Ca). Most plant growth responses to VAM are due to improved phosphate uptake from soil. In general, these responses are greatest in P deficient soil and decrease with increasing addition of phosphates (45). Phosphate fertilizer can suppress the indigenous VA endophyte population more than selected endophyte introduced as inoculants (81), although long term application of phosphate can build up a phosphate tolerant endophyte population (91) and species native to high-P soils some time show higher P-tolerance than those in soils of lower P status (24, 118). Further more, Schubert and Hayman (102) noted reduction in VAM fungal colonization by increasing of phosphorus level in soil. Plant growth responses to mycorrhiza were more at low and medium phosphorus with *Glomus mosseae* and *G. epigaeum* and medium phosphorus with *Glomus mosseae* and *G. epigaeum* and medium phosphorus with *G. macrocarpum* and *Gigaspora margarita*. *G. caledonium* and *Glomus* sp "E_3" were generally effective at all low phosphorus level where as *G. clarum* was ineffective at low phosphorus. In addition, Singh and Tyagi (110) observed that root colonization by indigenous VAM fungi and dry weight of plant *Aeschynomene indica* was maximum at 50 kg P_2O_5 and 100 kg P_2O_5 ha^{-1} application, in the form of super phosphate, respectively. Amendment of 100 kg P_2O_5 ha^{-1} reduced VAM fungal root colonization (Fig. 2).

The reduction in VAM fungal root colonization due to amendment of high level of phosphorus(63, 110) caused reduction in activity and population of particular VAM fungal species (104). Sporulation of *Gigaspora margarita*, *G. clarum*, *G. mosseae* and *G. heterogama* was reportedly increased under pot cultures of bahiagrass drenched with superphosphate, where as sporulation of *G. etunicatum*, *G. macrocarpum* and *G. gigantea* was reduced by the superphosphate drenching (118)). Improved sporulation after application of superphosphate was associated with phosphorus (P) tolerance of the mycorrhizal fungi. *Gigaspora margarita, G. mosseae, G. clarum* were able to colonize the root in soil that had 199 mg kg-1 of acid extractable (P) (107, 108, 110, 118).

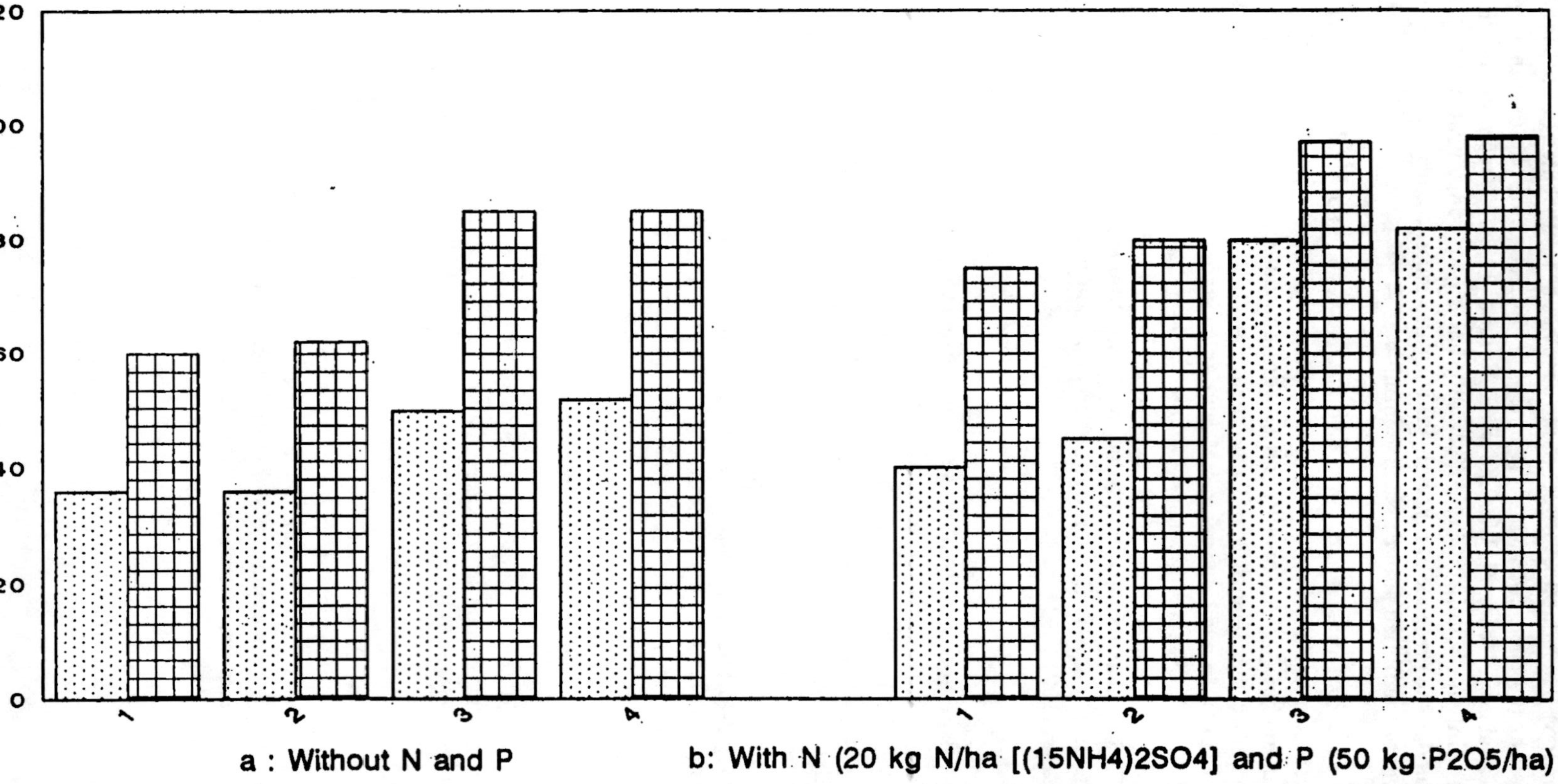

. **1a :** Effect of inoculation with *Rhizobium* sp. and *G. fasciculatum* on VAM fungi colonization of *C. cajan* var. *Prabhat* on 60th day

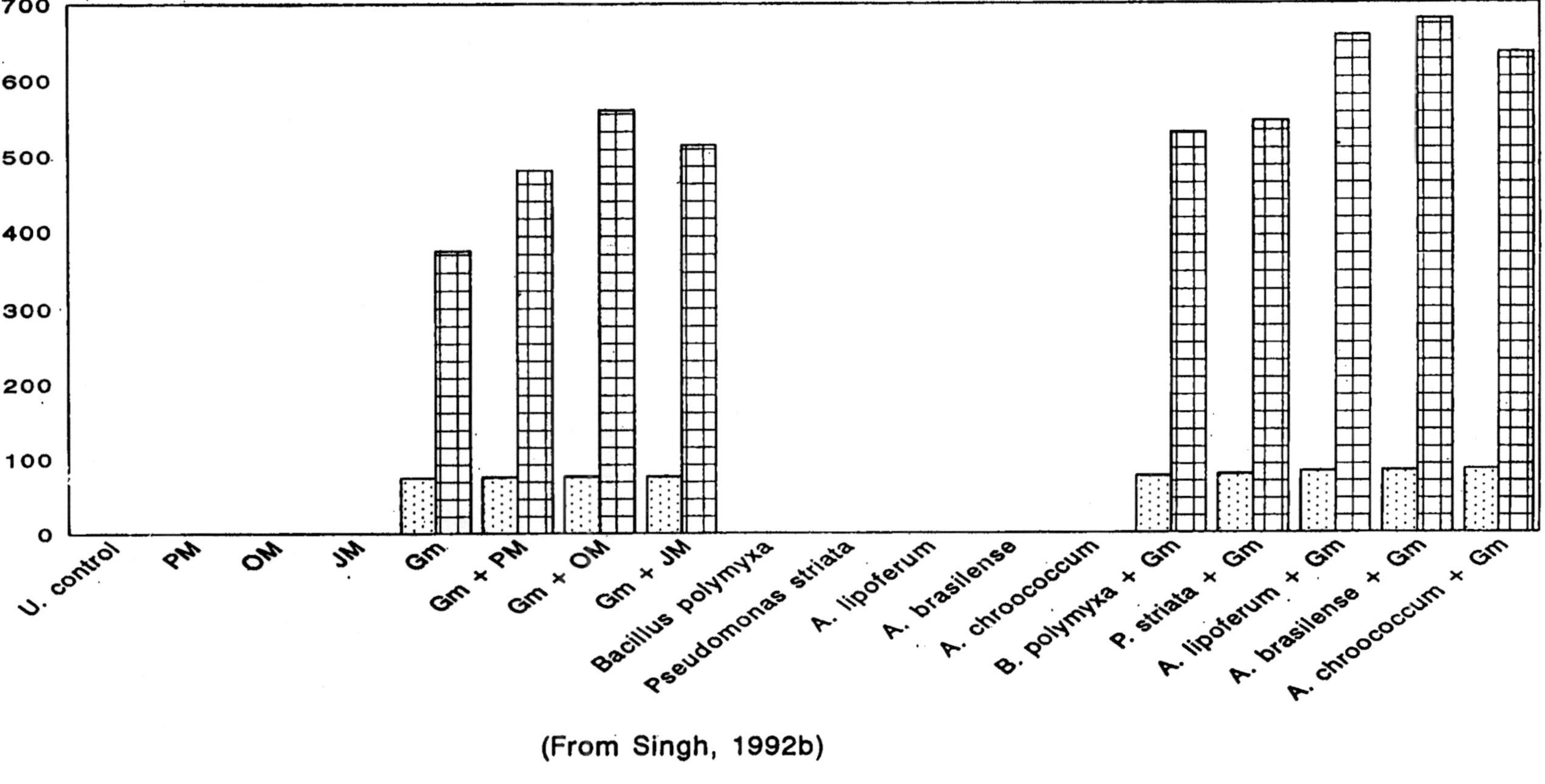

1b : Effect of N_2-fixing and P-solubilizing bacterial inoculation on the VA-mycorrhiza root infection and spore number, at 60 days of plant growht in the presence of 50kg P_2O_5 and 2kg N/ha (108).

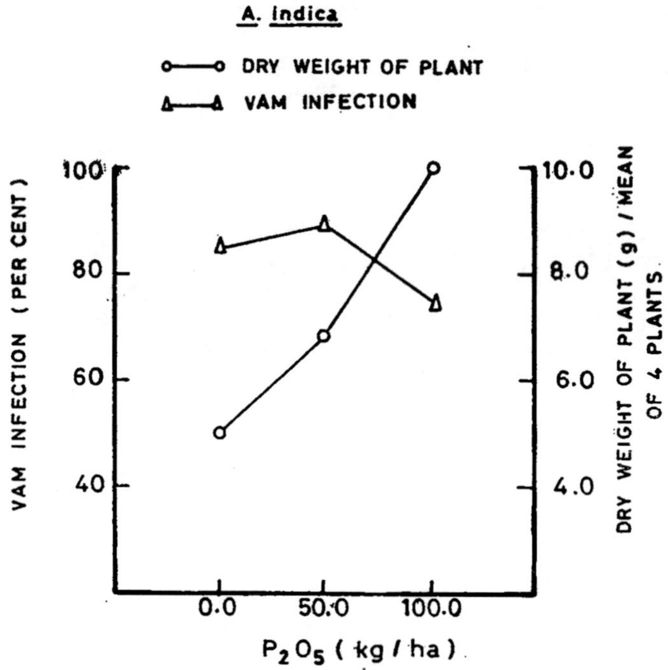

Fig. 2 : Effects of various levels of phosphorus (P_2O_5) on per cent VAM fungal infections and dry weight of plant at the age of 45d of plant growth (110).

2.4 Hydrogen Ion Concentration

Hydrogen ion concentration (pH) affects many soils characteristics and, therefore, availability of various vital nutrients such as phosphorus which has direct bearing on mycorrhizal formation (1, 23, 80, 130). Availability of other soil compounds is also pH dependent and affect the overall mycorrhizal performance. Most endophytes colonised roots at all pHs, even if they did not increase plant growth (22, 44, 55).

2.5 Pesticides

In general mycorrhizal fungi, under *in vitro*, condition, decline or die at high concentration of a particular pesticide or fungicide (19).

Their is decrease in the growth of onion, in soil treated with benomyl, due to suppression of the mycorrhizal fungus by this fungicide. Similarly decrease in growth, treated with captan, was found to be associated with the reduction in mycorrhizal population, rather than inhibition of the mycorrhizal fungus. Krough *et al.* (62) observed reduction in the growth of VA-mycorrhizal onion, but not of non-VA mycorrhizal plants due to application of benomyl and captan. Benomyl inhibit the formation of cytoskeleton and nuclear division by inhibiting the formation of microtubules (88). It also affects nuclear division, hyphal growth and delay or prevent effective symbiosis (43, 73, 74, 126). This fungicide is also known to inhibit activities of VAM fungal enzyme such as succinate dehydrogenase (62, 116), malate dehydrogenase and alkaline phosphatase (122). Since alkaline phosphatase monitor the uptake of mycorrhizal fungi, therefore, any kind of damage to this enzyme, affect the growth response of host (15, 30, 67).

Tommerup and Briggs (123) reported that germination or hyphal growth of VAM fungi (*Acaulospara laevis, G. caledonicum* and *G. monosporum*) was unaffected in sandy soil containing Bavistin (100 μg[1]). In contrary, Carr and Hinkley (16) observed inhibitory effect on *G. caledonicum*, when experimented on water agar, containing benomyl more than 10 mg ml^{-1}.

Effect of three fungicides (Benlate, Alette, Ridomil) on efficiency of P uptake from soil and transfer across the living plant-fungal interface of onion plants (*Allium cepa* L.) was determined with *Glomus* sp. (117) They observed that Benlate reduce P inflow and transfer across the interface. The rate of P uptake per m living external hyphae was not affected but, the development of living external hyphae in the soil was reduced. The contribution of the fungus to P uptake was small. Allette reduced growth of both shoots and roots, but apparently increased the accumulation of P in the tissues compared to control. However, Ridomil reduce P inflow per m of root and P uptake per m living external hypae, but had no effect on the rate of P transfer across the interface.

Certain fungicides (Simazine) initiated non-mycorrhizal plant (*Chenopodium quinona*) to mycorrhizal, perhaps due to enhanced secretion of sugar and amino acid from the roots (82, 103). It has been stated also that soil fumigants may not persist for long periods but delay in the reintroduction of mycorrhizal fungi may bring disastrous stunting of plant over years (73). However, if the pesticides do not persist, mycorrhizal fungi are reintroduced promptly, the deleterious effects may be short term, resulting in rapid recovery of host plant (125). Certain pesticides, under axenic condition have deleterious effect on spore germination at particular concentration there is no effect on the spore germination at low concentration (125).

2.6 Heavy Metal

Metal toxicity increase with decrease in pH. It has been found that Ericaceous plants, which inhabitates in acidic environment and colonized by Ericoid mycorrhizae, are resistant to both copper and zinc toxicity (13). Ectomycorrhizal fungi have lower resistance to copper and aluminium than those of ericoid fungi. Metabolically mediated absorption of Zn was 1.4 to 4.5 times more in ectomycorrhizal plants than non-mycorrhizal roots of *Pinus radiata*. It was also observed that at higher concentration of Zn extramatrical mycelium of some species was inhibited (18). Abundance of the extramatrical mycelium was shown to be important for heavy metal (HM) binding by the fungus. The heavy metals bind cell wall components like chitin, cellulose, cellulose derivates and melanins (33). However, few reports have also indicated interspecific differences within ectomycorrhizal fungi in relation to degree of resistance to heavy metals-Cu and Al e.g. *Suillus bovinus* has low resistance than *Amanita muscaria*. Sporocarps of ectomycorrhizae, have been found to accumulate high levels of metals (47, 134).

Gilden and Tinker (35) found that strains of *G. mosseae*, were more sensitive to heavy metals than were the roots of homologous hosts. It may be because VAM fungi are exposed to relatively low levels of metals. The degree of infection of onion with *G. mosseae* as found to be reduced by incorporation of Zn, Cu, Ni or Cd to soil and could be completely eliminated by heavy rates of application. In contrary, clover plants growing in the same area were relatively heavily with mycorrhizal fungi (28, 36). Uptake of heavy metals (Cu, Zn, Co) in very low concentration was enhanced by mycorrhizal roots (35, 64, 65, 66, 85, 93). Gilden and Tinker (36) observed increased yield due to enhanced supply of copper by mycorrhizal roots.

They have reported that increase uptake of Cu, Ni, Pb and Zn (at pH-3) by VAM can lead to reduction in the growth of roots and shoots in mycorrhizal *Erhart calyana* plants compared to non mycorrhizal plants at some concentration (60).

2.7 Soil Salinity

Soil salinity causes imbalance in availability of several nutrient required for the plant growth. Plaut and Grieve (89) reported that excess of Cl^- can interfere with NO_3 and PO_4 uptake, where as high Na^+ concentration can affect Ca^{++} and Mg^{++} acquisition. These ions have some significance in nutrient uptake and cause physiological changes in the plant (111). VAM fungi have been found to increase plants tolerance to salinity, there by increasing the yield (75, 90). Such effect was due to an improvement in phosphate nutrition and increase in K^+/Na^+ concentration in mycorrhizal plant.

Singh and Tyagi (110) observed root colonization of *Aschynomene indica* by indigenous VAM fungi in a soil amended with varing amounts of Sodium chloride (upto 2.0% weight basis in soil (Fig. 3). However, this host plant was unable to tolerate 2.0% level of sodium chloride, after 45 days (112).

2.8 Light

A reduction in supply of the light (photon irradiance or photoperiod) to the shoot of host plant has been correlated with reduction in colonization of plant root by VAM fungi (9), on contrary non-significant effects were also noted. Extent of colonization is dependent on the rate of growth of the fungus and rate of growth of the root. The fraction of the root length infected has also been shown to be slightly reduced at lower irradiance (121). The reduction in infection rate and/or extent of infection and growth of plant results in reduced photosynthesis and uptake of phosphorus (6, 29, 57).

It has been also found that a prolonged period of cloudy condition in the field or insufficient light in the green house culture may reduce the expected beneficial effects of inoculated VAM culture. Removal of aerial portion of the plant may also proved to be detrimental to VAM formation and mycorrhizal growth response. Bethlenfalvay and Dakessian (9) demonstrated that grazing reduced the extent of VAM fungal colonization in range grasses. The inhibition of VAM fungal colonization was due to reduction in foliage, increase in root/shoot ratio (Table 2).

Table 2. Effects of various light exposure on the VAM fungal root colonization, dry weight (g) of roots and shoots of *Aeschynomene indica*, under normal moisture holding capacity of soil (110)

Duration of light/dark (h)	Dry weight (g)		PercentVAM fungal root infection
	Shoot	Root	
0/24[1]	-	-	
6/18	0.50	0.30	32
12/12	2.84	1.95	88
24/0	3.0	2.04	77
S.Em.	0.51	0.49	4.91
C.D. at 5%	1.25	1.2	12.0

[1] Plant could survive only for 7d after transplanting under these conditions. (–) significant over corresponding control.

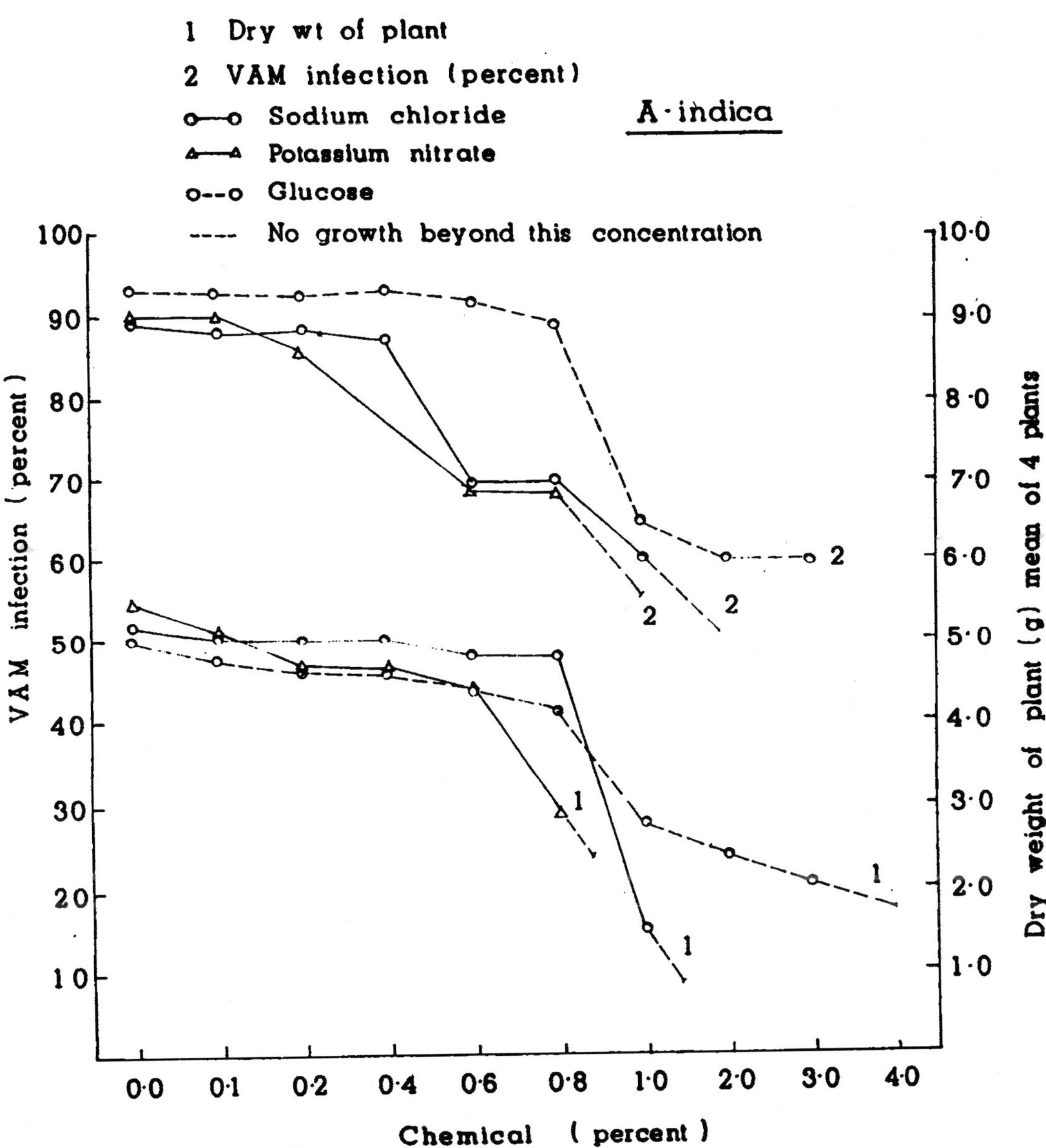

Fig. 3 : Effects of sodium chloride, potassium nitrate, glucose on per cent VAM fungi root infection and dry weight of plant at the age of 45 d of plant growth (110).

2.9 Temperature

Trappe (124) suggested that monitoring temperature requirements for the growth of ectomycorrhizal fungi might promote the matching of inoculum with the conditions at the amended site. Cline *et al.* (17) were able to monitor the growth of ectomycorrhizal fungi (*Pisolithus tinctorius, Cenococcum geophilum, Thedophora terrestris* and *Suillus granulatus*) over a wide temperature range. They found high degree of intraspecies variation with little correlation to geographic origin. However *P. tinctorious* was found to survive at high soil temperature. Hacskaylo *et al.* (42) studied the effect of temperature (35°C) on growth and respiration of six ectomycorrhizae (*Amanita rubescens, Suillus luteus, S. punctipes, Russula emetica, Cenococcum graniforme, Rhizopogon roseolus*). Most of the isolates were found to grow best at 24°C upto 12 days. Dry weight of *Rhizopogon roseolus*, was greatest at 18°C between 12 to 24 days, where as *Amanita rubescens* and *Suillus luteus* grew well at 24°C-29°C. Growth of *S. punctipes* was suppressed at 29°C. All isolates have shown very little growth at 2°C. Differences in respiration rates was observed in the six ectomycorrhizal fungi. *Cenococcum graniforme* and *Rhizopogon roseolus* respired at much greater rates at all temperature. Studies on effect of root zone temperature on ectomycorrhiza and VA-mycorrhiza formation have revealed that maximum formation of both ectomycorrhizae and VA mycorrhizae occurred at 18.5-24°C soil temperature. Mycorrhiza formation was moderate at 7.5°C (lowest) but reduced/ inhibited at or above 29.5°C (87). They also observed that treatment of soil at 35°C for week did not adversely affect viability of ectomycorrhizal fungal propagules. They suggested that the ability of native mycorrhizal fungi to grow at over a wide range is a greatest attribute because growth of the fungi even at low rates at low soil temperature could be highly significant in seeding established in areas where the early spring is cool. Evaluation study on the development of VA-mycorrhizae winter cereals has indicated substantial infection early in seedling growth under cooler soil temperature during months of November and December. Dodd and Jeffries (25) found different types of subfractions of VA-mycorrhizal fungal flora which were able to form VAM during winter months. Allen (2) observed that *Atriplex gardneri* was mycorrhizal at snow melt and became a non mycorrhizal form in summer, even though spores were present in the soil throughout seasons.

Furlan and Fortin (32) noted maximum plant response and sporulation of *Gi. calospora* at 21°C night and 26°C day temperature but same mycorrhizal plants grew less at 10°C night and 21°C day temperature less than the non-mycorrhizal controls. Schenck and Schroder (110) reported maximum sporulation of *Gi. margarita* on soybean (*Glycine max* L. Merr] at a soil temperature of 35°C. Maximum plant response was noted at 30 to 32°C but reduction in plant growth occurred at 17 to 18°C when compared to the non-mycorrhizal plant. Hayman (48) obtained reduction in growth response of onion to *Glomus mosseae* at 14°C as compared to 23°C. Menge *et al.* (72) observed that temperature cover 51.5°C for 10 min were proved lethal to *Glomus fasciculatum* in the roots of sudan grass (*Sorghum vulgare* Pers.). In another study *Gi. heterogama* and *Gi. gregaria* germinated best at 34°C, where as for *Glomus mosseae*, 20°C was found optimum.

Further more, Schenck and Smith (101) reported that *G. claroideum, G. clarum, Gi. pellucida* and *Gi. gregaria* produced greatest number of spores g^{-1} of colonized root at 24°C while *G. mosseae* and *Acaulospara laevis* reached their maximum at 30°C. *Gi. gregaria* had maximum root colonization at 36°C.

Nemec (84) found, that expcsure of *G. mosseae, G. interaradices* and *G. deserticola* to a temperature from 43 to 66°C reduced the activity. He found freezing and thawing of *Glomus* species inoculum before inoculation had no bad effect on the development in the root, because freezing reduced the spore dormancy factors. That spores were still infective after 12 weeks storage at 35°C and 45% r.h. but not at 14 week and 75% r.h. (20). Spores were found infective even after 12 weeks storage at 5°C at each r.h.

However survival of VAM fungi was noted at all storage temperature upto 14 weeks (Fig. 4).

3. OZONE EXPOSURE

Ozone is known as a photochemical air pollutant which inhibits process of photosynthesis. It induces stomatal closure, reduces photosynthetic capacity of chloroplast and causes leaf chlorosis and necrosis.

Besides, reduced mycorrhizal formation, in tomato and citrus, was observed due to plant damage caused by ozone exposure (71). Further more, Ho and Trappe (53) also observed reduced root weight and intensity of mycorrhizae formation due to exposure of mycorrhizal host plant, *Festuca arundinacea* to 0.1 ppm ozone, for 3 months.

3.1 Soil Structure

3.1.1 Compaction

Soil compaction resulted in reduced plant tillering, crown expansion and mycorrhizal colonization of bluestem (*Schizachyrium scoparium* Michx) plant (127). Simmons and Pope (105) worked on effect of soil compaction on mycorrhizal formation and growth of sweetgum (*Liquidambar styraciflua* L.) and yellow-poplar (*Lirioderdron tulipifera* L.) duly inoculated with *Glomus fasciculatum*) in soil with bulk densities ranging from 1.25 to 1.55 Mg m-3. They found variation in plant growth in compacted soil due to fungal species and plant.

It has been also reported that the biomass of non-VAM plants and fungal colonization, respectively, have positive and negative correlation with the increased coarse texture of soil, i.e. VAM will be more in silty than sandy soil. Dakisson *et al.* (21) observed absence of VAM infection with *Aschynomene indica* grown under water logged condition in clayey soil. But high density of VAM infection, was observed with *A. india* grown in rice held and moist condition.

3.1.2 Erosion

It is well established fact, that VAM fungi enhances nutrient uptake and root proliferation and plant cover, which in turn contribute to soil stability. There are few report, indicating large amount of hyphae bind soil particles together (77).

The chemical, physical and microbial properties of soil are badly affected due to erosion. It is known that soil erosion result in reduction of mycorrhizal propagules (41). Since the eroded soil have low fertility levels, therefore only limited potential benefits of VAM inoculation, used for rehabilitation, can be expected. However, results of the interaction effect mycorrhizal inoculation and fertility of soil on the establishment of plant in eroded area, have revealed beneficial effect of VAM on plant establishment (7, 8).

3.2 Carbon Dioxide and Oxygen

Atmospheric mole fraction of CO_2 average C 350 μ mol mol^{-1} at present and predicted to increase to over 800 μ mol mol^{-1} 'by the end of next century due to deforestation especially in

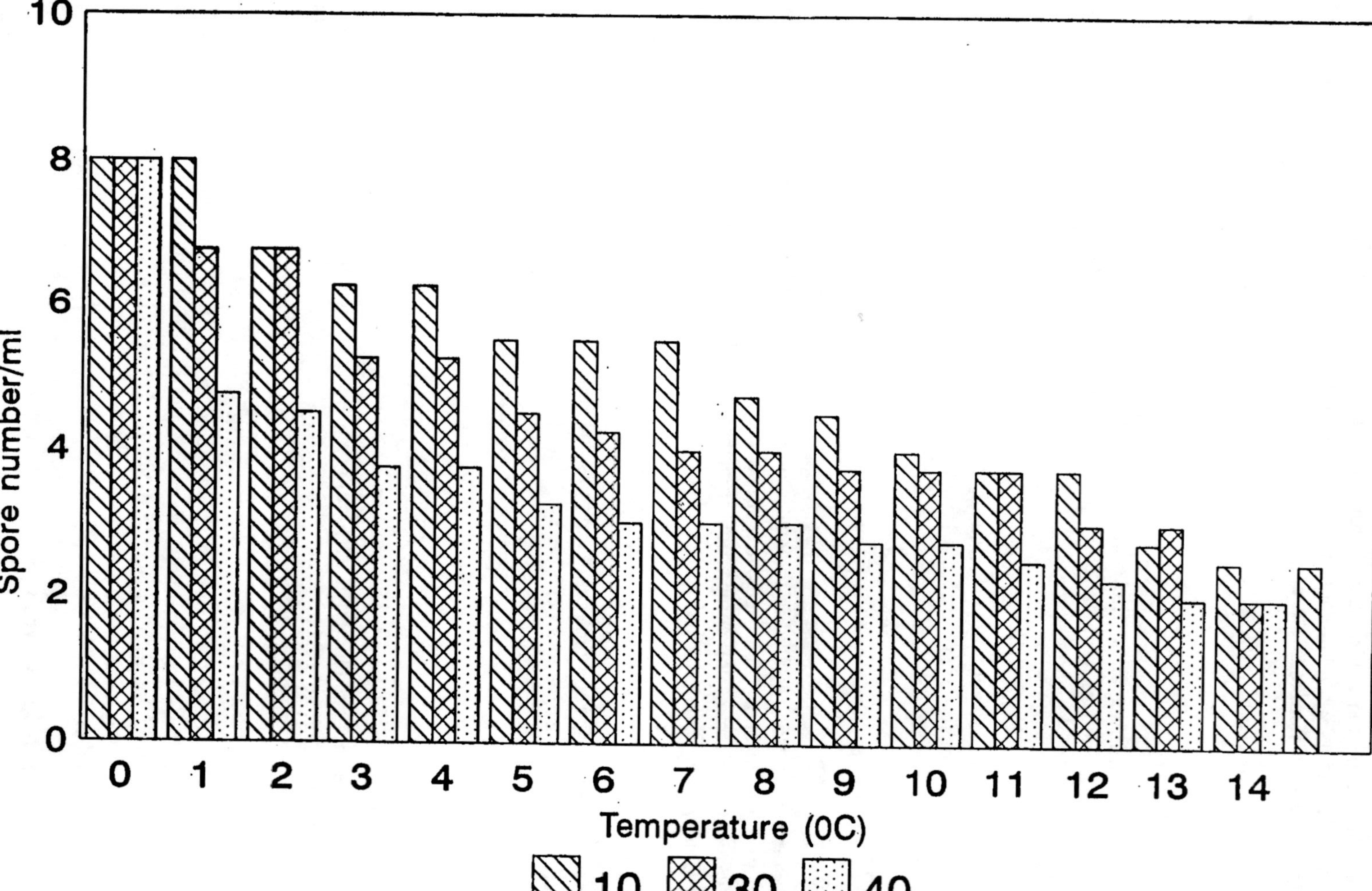

Fig. 4 : Effect of storage temperature on the survival of *Glomus macrocarpum* (unpublished)

tropics and to the combustion of fossil fuels (54). There are numerous reports on the effects of an elevated concentration of atmospheric CO_2 on morphology and physiology of various plant species (56). He also noted that scarcity of nutrients such as N, P can limit the capacity of plant to respond to elevated CO_2 by an average of 30% (N limitation) or 50% (P limitation). N limitation on the response of plant to CO_2 enrichment was also observed by Gouridaan and de Ruitter (39). The growth of *Trifolium subterraneum* at elevated CO_2 was noted with enhanced supply of P.

It was hypothesized that elevated CO_2 could favour plant growth by stimulating the activities of beneficial microorganisms e.g. mycorrhizal fungi. Increase in ectomycorrhiza formation in *Pinus echinata* and *Quercus alba* was noticed under elevated CO_2 (86). Elevated CO_2 has been reported to increase VAM colonization in C_4 grass-*Bauteloua gracilis* (78, 79) but it has no effect on the VAM colonization of the C_3 grasses-*Pascopyrum smithii* (78) and *Gossypium hirustum* (96) or a few members of a serpentine grassland community.

It was also speculated that elevated CO_2 might influence VAM fungi by increasing internal carbohydrate concentration in plant and increase in non-structural carbohydrate concentration in root tissue (58). Elevated CO_2 might affect mycorrhiza formation by modifying activity of rhizosphere due to changes in root exudation altering carbon nitrogen ratio (69). In addition, *T. repens* L. was grown for 38 days at ambient (350) μmol mol^{-1}) and elevated (700 μmol mol^{-1}) atmospheric CO_2, in absence and presence of *G. mosseae*. It was noted that elevated CO_2, significantly increased above and below-ground biomass and non-structural carbohydrates but no effect on root length colonized by *G. mosseae*. However, the percentage mycorrhizal infection and total mycorrhizal root length were increased with the exposure of plant to elevated CO_2 (58). Growth of mycorrhiza at various soil temperatures was influenced strongly by soil and CO_2. Efficiency of VA-mycorrhiza was found to be increased maximum at 30°C due to enhanced supply of oxygen (98).

3.3 Glucose, Activated Charcoal, Vermiculite Peat Moss

Duclos and Fortin (26) added a small quantity of glucose (0.5 g l^{-1}) to the culture medium and noted enhanced mycorrhiza formation on *Vaccinium angustifolium* Ait and *V. corymbosum* L. with *Pezizella ericae*. In contrary, increasing levels of glucose incorporation in soil progressively decreased the per cent root colonization by indigenous VAM and growth of the plant. Adverse effect was more pronounced at 4.0% glucose addition (Fig. 3). Moreover, incorporation of activated charcoal especially in presence of glucose or pectin stimulated ericoid mycorrhiza formation, by *P. ericae* on *V. angustifolium*. Active charcoal might absorb toxic compound from the rhizosphere.

Inoculum of *Pisolithus tinctorius*, prepared vermculite peat moss medium can be stored upto 19 weeks at 3°C, but *P. involutus* inoculum can not be stored for more than two weeks. However at 24°C neither inoculum retained viability for long.

3.4 Organic Matter

Organic matter amendment influences, pH nutrient and water holding capacity. All these factors influence directly and/or indirectly VAM development and efficiency. It has been reported that VAM hyphae were longer in the organic matter amendment treatment as compared to minus organic matter control (114). They also observed that inoculation with a mixed soil microflora inhibited mycelial growth in the treatment containing organic matter. Hepper and Warmer (50) also found out the importance of organic matter in enhancing hyphae formation and development. In contrary, Loree and Williams (70) found no correlation between-organic matter content and mycorrhizal rooting density.

Besides, water-soluble extract of the litter of shrubs and conifer have shown variable effect on the growth of ectomycorrhizal fungi (*Rhizopogon vinicolor, Pisolithus tinctorious, Laccaria laccata, Cenococcum geophilum*). However, low concentration of the extract stimulated fungal growth (94). Their green house studies has also shown that applied litter to the surface of a sand+soil mixture reduced the formation of *Rhizopogon* sp. on Dauglas-fir seedlings.

3.5 Seasons

Season of the year have impact on VAM infection and sporulation. Lower VAM infection was generally observed at the beginning of rainy season in all tropical forage crops. In the sandy loam soils increased,mycorrhizal infection was found in May and remained constant till July, but in clay loam soil mycorrhizal colonization increased steadily to a maximum in June and July (98). He also observed increase endogonacious spores from April to July which continued to be constant till November.

4. REFERENCES

1. Al-Agely, A.K. and Reeves, F.B. 1995. Inland sand dune mycorrhizae: Effect of soil depth, moisture and pH on colonization of *Oryzopsis hymenoides. Mycologia* **87** : 54-60.

2. Allen, M.F. 1981. Formation of vesicular-arbuscular mycorrhizae in *Atriplex gardneri* (Chenopodiaceae) : A seasonal response in a cold desert. *Mycologia* **75** : 773-776.

3. Allen, M.E., Moore, T.S. Jr., Chhristensen, M. 1982. Phytohormone changes in *Bouteloua gracilis* infected by vesicular-arbuscular mycorrhizae. 1. Cytokinin increases in the host plant. *Can. J. Bot.* **58** : 371-374.

4. Amigee, F., Tnker, P.B. and Stribley, D.D. 1989. The development of endomycorrhizal systems VII. A detailed study of the effect of soil phosphorus on colonization. *New Phytologist* **111** : 435-446.

5. Auge, R.M., Schekel, K.A. and Wample, R.L. 1987. Leaf water and carbohydrate status of V.A. Mycorrhizal rose exposed to drought stress. *Plant Soil* **99** : 291-302.

6. Azcon, R., Gomoz, M. and Tobar, R. 1992. Effects of nitrogen source on growth, nutrition, photosynthetic rate and nitrogen metabolism of mycorrhizal and phosphate fertilized plant of Lactuca sativa L. *New Phytologist* **121** : 227-234.

7. Aziz To. and Habte, M. 1990a. Stimulation of mycorrhizal activity *Vigna ungiculata* through low level fertilization of an oxisol subjected to imposed erosion. *Communication in Soil Science* and *Plant Analysis* **21** : 493-505.

8. Aziz, T. and Habte, M. 1990b. Enhancement of endomycorrhizal activity through nitrogen fertilization in cowpea grown in an oxisol subjected to stimulated erosion. *Arid soil Research and Rehabilitation* **4** : 131-139.

9. Bethlenfalvay, G.J. and Pacovsky, R.S. 1983. Light effects on mycorrhizal soybean. *Plant and Physiology* **73** : 969-972.

10. Bethlenfalvay, G.J. and Dakissian, S. 1984. Grazingeffects on mycorrhizal colonization and floristic composition of the vegetation on a semi arid range in northern Nevada. *Journal of Range Management* **37** : 312-316.

11. Bolan, N.S., Robson, A.D. and Barrow, N.J. 1984. Increasing phosphoru supply can increase the colonization of plant roots by vesicular-arbuscular mycorrhizal fungi. *Soil Biology and Biochemistry* **16** : 419-420.

12. Boyed, R., Furbank, R.T. and Read, D.J. 1986. Ectomycorrhiza and the water relation of trees. In: *Physiological and Genetical Aspect of Mycorrhizae* (eds V. Gianiazzi-Peerson and S. Gianinazzi) INRA Paris, pp 689-694.

13. Bradly, R., Burt, A.J. and Read D.J. 1982. Mycorrhizal infection and resistance to heavy metal toxicity in *Calluna vulgaris. Nature* **292** : 335-337.

14. Busse, M.D. and Ellis, J.R. 1985. Vesicular arbuscular mycorrhizal (Glomus fasciculatum) influence on soybean drought tolerance in highphosphorus soil. *Can. J. Bot.* **63** : 2290-2294.

15. Buttery, B.R., Park S.J., Findlay'W.I. and Dhanvantari, B.N. 1988. Effect of fumigation and fertilizer on growth yield, chemical composition and mycorrhizae in white bean and soybean. *Can. J. Plant Sci.* **68** : 677-686.

16. Carr, G.R. and Hinkley, M.A. 1985. Germination and hyphal growth of *Glomus caledonicum* on water agar containing benomyle. *Soil Biology and Bio Chemistry* **17** : 313-316.

17. Cline, M. L., Franc, R.C. and Reid, C.P.P. 1987. Intraspecific and interspecific growth variation of ectomycorrhizal fungi at different temperature. *Can. J. Bot.* **65** : 869-875.

18. Colpaert, J.V., Assche and Van, J.A. 1992. Zinc toxicity in ectomycorrhizal *Pinus sylvestris. Plant Soil* **143** : 201-211.

19. Cudlin, P., Mejstrik, V. and Skoupy, J. 1983. Effect of pesticides on ectomycorrhizae of *Pinus sylvestris* seedlings. *Plant Soil* **71** : 353-356.

20. Daft, M.J., Spencer, D. and Thomas, G.E. 1987. Infectivity of vesicular-arbuscular mycorrhizal inocula after storage under various environmental conditions. *Trans. Bri. Mycol. Soc.* **88** : 21-27.

21. Dakissian, S., Brown M.S. and Bethlenfalvay, G.J. 1986. Relation of mycorrhizal growth enhancement and plant growth with soil water and texture. *Plant Soil* **94** : 439-443.

22. Daniels, B.A. and Trappe, J.M. 1980. Factor affecting spore germination of the vesicular-arbuscular mycorrhizal fungus. *Glomus epigaeus. Mycologia* **72** : 457-471.

23. Davis, E.A., Young, J.L. and Linderman, R.G. 1983. Soil lime level (pH) and VA Mycorrhiza effect on growth responses of sweetgum seedlings. *Soil Science Society of America Journal* **47** : 251-256.

24. Davis, E.A., Young, J.L. and Rose, S.L. 1984. Detection of high phosphorus tolerant VAM-fungi colonizing hops and pepper mint. *Plant Soil* **81** : 29-36.

25. Dodd, J.C. and Jeffries, P. 1986. Early development of vesicular arbuscular mycorrhizas in autumn-sown cereals. *Soil Biology and Biochemistry* **18** : 149-154.

26. Duclos, J.L. and Fortin, J.A. 1983. Effect of glucose and active charcoal on *in vitro* synthesis of ericoid mycorrhiza with *Vaccinium* spp. *New Phytologist* **94** : 95-102.

27. Ellis, J.R., Larsen, H.J. and Boosalis, M.G. Drought resistance of wheat plants inoculated with vesicular-arbuscular mycorrhizae. *Plant Soil* **86** : 369-378.

28. Faber, B.A., Zasoski, R.J., Burau, R.G. and Uriru, K. 1990. Zinc uptake by corn as affected by vesicular arbuscular mycorrhizae. *Plant and Soil,* **129** : 121-130.

29. Fay, P., Mitchell, D.T. and Osborne, B.A. 1996. Photosynthesis and nutrient-use efficiency of barley in response to low arbuscular mycorrhizal colonization and addition of phosphorus. *New Phytologist* **132** : 425-433.

30. Fitter, A.H. 1986. Effect of benomyl on leaf phosphorus concentration in alpine grasslands : A test of mycorrhizal benefit. *New Phytologist* **103** : 767-776.

31. Fitter, A.H. 1988 relations of red clover *Trifolium pratense* L. as affected by VA mycorrhizal colonisation of phosphorus supply before and during drought. *Journal of Experimental Botany* **39** : 593-603.

32. Furlan, V. and Fortin, J.A. 1973. Formation of endomycorrhizae by *Endogone calospora* an Allium cepa under three temperature regimes. *Naturaliste Canadian* **100** : 467-477.

33. Galli, U., Schuep, H. and Brunold, C. 1994. Heavy binding by mycorrhizal fungi. *Physiologia Plantarum* **92** : 364-368.

34. Gerdemann, J.W. and Nicolson, T.H. 1963. Spores of mycorrhizal *Endogone* species extracted from soil by wet sieving and decanting. *Transactions of the British Mycological Society* **46** : 235-244.

35. Gildon, A. and Tinker, P.B. 1983a. Interaction of vesicular-arbuscular mycorrhizal infection and heavy metals in plants. I. The effects of heavy metals on the development of vesicular arbuscular-mycorrhizas. *New Phytologist* **95** : 247-261.

36. Gildon, A., and Tinker P.B. 1983b. Interaction of vesicular arbuscular mycorrhizal infection and heavy metals in plants II. The effect ofinfection uptake of copper. *New Phytologist* **95** : 263-268.

37. Gildon, A. and Tinker P.B. 1981. A heavy metal tolerant strain of mycorrhizal fungus. *Transaction of the British Mycological Society* **77** : 648-649.

38. Gilmore, A.E. 1971. The influence of endotrophic mycorrhizae on the growth of peach seedlings. *Journal of the American Society for Horticultural Science* **96** : 35-37.

39. Goudriaan, J. and De Ruiter, H.E. 1983. Plant growth in response to CO2 enrichment, at two levels of nitrogen and phophorus supply. Dry matter, leaf area and development. *Netherlands Journal of Agricultural Science* **31** : 157-169.

40. Gupta, S.B. 1996. Effective utilization of phosphorus in rice-wheat cropping system in vertisol through VA mycorrhizae and phosphorus solubilizer. PhD. Thesis I.A.R.I., New Delhi- 110 012, India.

41. Habte, M. 1989. Impact of stimulated erosion on the abundance and activity of indigenous vesicular-arbuscular mycorrhizal endophytes in an oxisol. *Biology and Fertility of Soil*, **7** : 164-167.

42. Hacskaylo, E., Palmer, J.G. and Vozzo, J.A. 1965. Effect of temperature on growth and respiration of ectotrophic mycorrhizal fungi. *Mycologia* **57** : 745-756.

43. Hale, K.A. and Sanders, F.F. 1982. Effect of benomyle on vesicular-arbuscular mycorrhical infection of red clover (*Trifolium paratense* L.) and consequences of phosphorus inflow. *Journal of Plant Nutrition* **5** : 1355-1367.

44. Hardie, K. and Leyton, L. 1981. The influence of vesicular-arbuscular mycorrhiza on growth and water relation of red clover. *New Phytologist* **89** : 544-608.

45. Harley, J.L. and Smith, S.E. 1983. Mycorrhizal symbiosis Academic Press, London.

46. Hartmond, U., Schaesberg, N.V., Graham, J.H. and Syvertsen, J.P. 1987. Salinity and flooding stress effect on mycorrhizal and non-mycorrhizal citrus rootstock seedlings. *Plant and Soil* **104** : 37-43.

47. Haselwanter, K., Berreck, M. and Brunner, P. 1988. Fungi as bio indicators of radio cesium contamination. Pre and Post chernobyl activities. *Transactions of British Mycological Society* **90** : 171-174.

48. Hayman, D.S. 1974. Plant growth responses to vesciular-arbuscular mycorrhiza. IV. Effect of light and temperature. *New Phytologist* **73** : 71-80.

49. Hayman D.S. and Tavares. 1985. Plant growth responses to vesicular arbuscular mycorrhiza XV influence of soil pH on the symbiotic efficiency of different endophytes. *New Phytologist* **100** : 367-377.

50. Hepper, C.M. and Warmer, A. 1983. Role of organic matter in growth of a vesicular-abuscular mycorrhizal fungus in soil. *Transaction of the British Mycological Society* **81** : 155-156.

51. Hetrick, B.A.D., Hetrick, J.A. and Bloom, J. 1984. Interaction of mycorrhizal infection, Phosphorus level and moisture stress in growth of field corn. Can. *J. Bot.* **62** : 2267-2271.

52. Hetrick, B.A.D., Kitt, D.G. and Wilson, G.T. 1987. Effect of drought stress on growth response in Corn, Sudan gras and big blue stem to *Glomus etunicatum. New Phytologist* **105** : 403-410.

53. Ho, I. and Trappe, J.M. 1984. Effect of Ozone exposure on mycorrhiza formation and growth of *Festuca arundinacea. Environmental and Experimental Botany* **24** : 71-74.

54. Houghton, J.T., Jenkin, G.J. and Epharaums, J.J. 1990. Climate change. The IPCC scientific assesment. Intergovernmental panel on climates change. Cambridge university Pressk, Cambridge.

55. Huang, Ling-Ling and Trappe, J.M. 1983. Growth variation between and within species of ectinomycorrhizal fungi in response to pH *in vitro. Mycologia* **72** : 234-241.

56. Idso, K.K. and Idso, S.B. 1994. Plant responses to atmospheric CO_2 enrichment in the face of environmental contraints. A review of the past 10 years research. *Agricultural and Forest Meterology* **69** : 153-203.

57. Jacob, J. and Hawlor, D.W. 1991. Stomatal and mesophyll limitations of photosynthesis in phosphate deficient sunflower, maize and wheat plants. *Journal of Experimental Botany* **42** : 1003-1011.

58. Jongen, M., Fay, P. and Jones, M.B. 1995. Effect of elevated carbon dioxide and arbuscular mycorrhizal infection on *Trifolium repens. New Phytologist* **132** : 412-423.

59. Keeley, J.E. 1980. Endomycorrhizae influence growthof Blackgum seedlings in flooded soils. *American Journal of Botany* **67** : 6-9.

60. Killham,K. and Firestone, M.K. 1983. Vesicular arbuscular mycorrhizal mediation of grass response to acid and heavy metal deposition. *Plant Soil* **72** : 39-48.

61. Kothari, S.K., Marschner, H. and George, E. 1990. Effect of VAMycorrhizal fungi and rhizosphere micoorganisms on root and shoot morphology, growth and water relations in maize. *New Phytologist* **116** : 303-311.

62. Krough, J.L. Gianinazzi-Pearson, V. and Gianinazzi, S. 1987. Depressed matabolic activity of vesicular-arbuscular mycorhizal fungi after fungicide applications. *New Phytologist* **106** : 707-715.

63. Krishna, K.R. and Daft, P.J. 1984. Effect of mycorrhizal inoculation and soluble phosphorus fertilizer on growth and phosphorus uptake of pearl millet. *Plant Soil* **81** : 247-256.

64. Lambert, D.H., Baker, D.E., Bake, D.E. and Cole Jr. H. 1979. The role of mycorrhizae in the interactions of phosphorus with zinc copper and other elements. *Soil Science Society of Amrica Journal* **43**: 976-980.

65. Lambert, D.H. and Wieidensaul, T.C. 1980a. Variation on the response of alfalfa clones and cultivars to mycorrhizae and phosphorus. *Crop Science* **20** : 615-618.

66. Lambert, D.H. and Weidensaul, T.C. 1991. Element uptake of mycorrhizal soybean from Sewage-sludge-treated soil. *Soil Science Society of America Journal* **55** : 393-398.

67. Larsen, J., Thingstrup, I. and Jakobsen, I. 1996. Benomyle inhibits phosphorus transport but not fungal alkaline phosphatase activity in a *Glomus*-cucumber symbiosis. *New Phytologist* **132** : 127-133.

68. Lery, Y. and Krikum, J. 1980 Effect of vesicular-arbuscular mycorrhiza on *Citrus jambhiri* water relations. *New Phytologist* **85** : 25-31.

69. Lincoln, D.E. and Couvet, D. 1989. The effect of carbon supply on allocation to allelochemicals and caterpillar consumption of pepper mint. *Oecologia* **78** : 112-114.

70. Loree, M.A.J. and Williams, S.E. 1987. Colonization of western wheatgrass (*Agropyron smithii* Rydb.) by vesicular-arbuscular mycorrhizal fungi during the revegetation of a surface mine. *New Phytologist* **106** : 735-744.

71. Mc Cool., P.M. and Menge, J.A. 1983. Influence of Ozone and carbon partitioning in tomato : Potential role of carbon flow in regulation of the mycorrhizal symbiosis under conditions of stress. *New Phytologist* **94** : 241-247.

72. Menge, J.A., Johnson, E.L. and Minassian, V. 1979. Effect of heat treatment and three pesticides upon the growth and reproduction of the mycorrhizal *Glomus fasciculatum. New Phytologist* **82** : 473-480.

73. Menge, J.A. 1982. Effect of soil fumigants and fungicides on vesicular-arbuscular fungi. *Phytopathology* **72** : 1125-1132.

74. Merry Wheather, J. and Fitter, A. 1996. Phosphorus nutrition of an obligately mycorrhizal plant treated with the fungicide benomyl in the field. *New Phytologist* **132** : 307-311.

75. Mertz, S.M., Heithaus, J.J. and Bush, R.L. 1979. Mass production of axenic spores of the endomycorrhizal fungus *Gigaspora margarita. Transactions of the British Mycological Society* **72** : 167-169.

76. Miller, D.E., Domato, P.A. and Walker, C. 1985. Colonization and efficacy of different endomycorrhizal fungi with apple seedlings at two phosphorus levels. *New Phytologist* **100** : 393-402.

77. Miller, R.M. and Jastrow, J.D. 1990. Hierarchy of root and mycorhizal fungal interactions with soil aggregation. *Soil Biology and Biochemistry* **22** : 579-584.

78. Monz, C.A., Hunt, H.W., Reeves, F.B. and Elliott, E.T. 1994. The response of mycorrhizal colonization to elevated CO_2 and climate change in *Pascopyrum smithu* and *Bouteloua gracilis. Plant Soil* **165** : 75-80.

79. Morgan, J.A., Knight, W.G., Dudley, L.M. and Hunt H.W. 1994. Enhanced root system C-sink activity, water relations and aspects of nutrient acquisition in mycotrophic *Bouteloua gracilis* subjected to CO_2 enrichment. *Plant Soil* **165** : 139-146.

80. Mosse, B. 1972a. The influence of soil type and *Endogone strain* on the growth of mycorrhizal plants in phosphate deficient soils. *Review of the Ecological Biological Society* **9** : 52a.

81. Mosse, B., Stribley, D.P. and Le Tacon, F. 1981. Ecology of mycorrhizae and mycorrhizal fungi. *Advances in Microbial Ecology* **5** : 137-210.

82. Nemec, S. 1980. Effect of 11 fungicides on endomycorrhizal development in sour orange. *Can. J. Bot.* **58** : 522-526.

83. Nelson, C.E. and Safir, G.R. 1982. The water relations of well-watered, mycorrhizal and non-mycorrhizal onion plants. *Journal of the American Society for Horticultural Science* **107** : 271.

84. Nemec, S. 1987. Effect of storage temperature and moisture on *Glomus* species and their subsequent effect on citrus root stock seedling growth and mycorrhiza development. *Transaction of the British Mycological Society* **89** : 205-212.

85. Nielsen, J.D. 1983. Influence of vesicular arbuscular mycorrhizal fungi on growth and and uptake of various nutrients as well as uptake ratio of fertilizer P for lucerne (*Medicago sativa*). *Plant Soil* **70** : 165-172.

86. O'Neill, E.G., Luxmoore, R.J. and Norby, R.J. 1987. Increases in mycorrhizal colonization and seedling growth in *Pinus echinata* and *Quercus alba* in an enriched CO_2 atmosphere. *Canadian Journal of Forestry Research* **17** : 878-883.

87. Parke, J.L., Linderman, R.G. and Trappe, J.M. 1983. Effect of root zone temperature on ectomycorrhiza and vesicular-arbuscular mycorrhiza formation in distirbed and undisturbed forests soils of south west Oregon. *Canadian Journal of Forest Research* **13** : 657-665.

88. Pedregosa, A.M., Rio, S., Monistrol, I.F. and Laborda, F. 1995. Effect of the microtubule inhibitor methyle benzimidazol-2-ylcarbomate (MBC) on protein secretion and microtubule distribution in *Cladosporium cucumerinum. Mycological Research* **99** : 43-48.

89. Plaut, Z. and Grieve, C.M. 1988. Photosynthesis of salt stressed maize as influenced by Ca:Na ratio in the nutrient solution. *Plant Soil* **105** : 283-286.

90. Pond, E.C., Menge, J.A. and Jarrell, W.M. 1984. Improved growth of tomato in salinized soil by vesicular-arbuscular mycorrhizal fungi collected from saline soil. *Mycologia* **79** : 74-84.

91. Porter, W.M., Abott, L.K. and Robson, A.D. 1978. Effect of applications of superphosphate on population of vesicular-arbuscular endophytes. *Australian Journal of Experimental Agriculture and Animal Husbandry* **18** : 573-578.

92. Red head, J.E. 1975. In : *Endomycorrhizas*. Academic Press, New York. pp. 447.

93. Rhodes, L.H. and Gerdemann, J.W. 1978. Translocation of calcium and phosphate by external hyphae of VA mycorrhizae. *Soil Science* **126** : 1125-126.

94. Rose, S.L., Perry, D.A., Pitz, D. and Schoeneberger, M.M. 1983. Allelopathic effect of litter on the growth and colonization of mycorrhizal fungi. *Journal of Chemical Ecology* **9** : 1153-1162.

95. Rudawska, M. 1986. Sugar metabolism of ectomycorrhizal scots pine seelings as influenced by different nitrogen froms ansd levels. In : *Mycorrhizae-Physiology and Genetics*, 1st ESM, Dijon, INRA Paris, 1-5 July 1985.

96. Runion, G.B., Curl, E.A., Rogers, H.H., Backman, P.A., Rodriguez Kabana, R. and Helms, B.E. 1994. Effect of free air CO_2 enrichment on microbial populations in the rhizosphere and phytosphere of cotton. *Agricultural and Forest Meteorology* **70** : 117-130.

97. Sanders, F.E. and Tinker, P.B. 1973. Phosphate flow into mycorrhizal roots. *Pesticide Science* **4** : 385-395.

98. Saif, S.R. 1986. Vesicular-arbuscular mycorrhizae in tropical forage species as influenced by season. Soil texture, fertilizer host species and ecotypes. *Angew. Botanik.* **60** : 125-139.

99. Saif, S.R. 1983. The influence of soil aeration on the efficiency of vesicular-arbuscular mycorrhizae II. Effect of soil oxygen on growth and mineral uptake in *Eupatorium odoratum* L., *Sorghum bicolor* (L) Moench and *Geuzotia abyssinica* (L.f.) Cass. inoculated with vesicular arbuscular mycorrhizal fungi. *New Phytologist* **94** : 405-417.

100. Schenck, N.S. and Schroder, V.N. 1974. Temperature response of *Endogone* mycorrhiza on soybean roots. *Mycologia* **66** : 600-605.

101. Schenck, N.C. and Smith G.S. 1982. Responses of six species of vesicular-arbuscular mycorrhizal fungi and their effects on soybean at four soil temperatures. *New Phytologist* **92** : 193-201.

102. Schubet, A. and Hayman, D.S. 1986. Plant growth responses to vesicular-arbuscular mycorrhizae. XVI. Effectiveness of different endophytes at different levels of soil phosphate. *New Phytologist*, **103** : 79-90.

103. Schwab, S.M., Johnson, E.L.V. and Menge, J.A. 1982. Influence of simazine on formation of vesicular-arbuscular mycorrhizae in *Chenopodium quiona* wild. *Plant Soil*, **64** : 283-287.

104. Sieverding, E. and Howaler, R.H. 1985. Influence of species of VAMycorrhizal fungi on cassava yield response to the phosphorus fertilization. *Plant Soil* **88** : 213-221.

105. Simmons, G.L. and Pope, P.E. 1988. Influence of soil water potential and mycorrhizal colonization on root growth of yellow-poplar and sweet gum seedings grown in compacted soil. *Can. J. For. Resea.* **18** : 1392-1396.

106. Singh, C.S. 1990. In : *Synergistic Effect of VAMycorrhiza with Rhizobia to Improve Nodulation and Nitrogen Fixation o Legumes*. A Internal Review Report, Indo-US S.S.P. Programme in Agriculture.

107. Singh, C.S. 1992 a. Mass inoculum, production of vesicular-arbuscular (VA) mycorrhizae. 1 Selection of host in the presence of *Azospirillum brasilense*. *Zentralblatt fur Mikrobiology* **147** : 447-453.

108. Singh, C.S. 1992b. Mass inoculum production of vesicular-arbuscular (VA) mycorrhizae. II. Influence of N_2 fixing and P solubilizing bacterial inoculation on VA-mycorrhiza. *Zntralblatt fur Mikrobiology* **147** : 503-508.

109. Singh, C.S. 1996. Arbuscular mycorrhiza (AM) in association with *Rhizobium* sp. improves nodulation, N_2 fixation and N utilization of pigeon pea (*Cajanus cajan*) as assessed with a ^{15}N. technique in pots. *Microbiological Research* **151** : 87-92.

110. Singh, C.S. and Tyagi, S.P. 1989. Study on the occurence of vesicular-arbuscular mycorrhizal (VAM) fungi in the root of *Aeschynomene indica* under the influence of various ecological factors. *Zentrablatt fur Mikrobiolgy* **144** : 241-248.

111. Smith, S.E. and Gianinazzi-Pearson, V. 1988. Physiological interactions between symbionts in vesicular-arbuscular mycorrhizal plants. *Annual Review of Plant Physiology and Plant Molecular Biology* **39** : 229-244.

112. Stahl, P.D. and Williams, S.E. 1986. Oil shale process water affects activity of vesicular-arbuscular fungi and *Rhizobium* 4 years after application to soil. *Soil Biology and Biochemistry* **18** : 451-455.

113. Stahl, P.D. and Christensen, M. 1991. Population variation in the mycorrhizal fungus *Glomus mosseae*. Breadth of environmental tolerance. *Mycological Research* **95** : 300-307.

114. St John, T.V., Coleman, D.C. and Reid, C.P.P. 1983. Association of vesicular-arbuscular mycorrhizal hyphae with soil organic particles. *Ecology* **64** : 957-959.

115. Subba Rao, N.S., Tilak, K.V.B.R. and Singh, C.S. 1986. Dual inoculation with *Rhizobium* sp. and Glomus *fasciculatum* enhances nodulation yield and nitrogen fixation in chickpea (*Cicer arietinum* Linn.). *Plant Soil* **95** : 351-359.

116. Sukarno, N., Smith, S.E. and Scott, F.S. 1993. The effect of fungicides on vesicular arbuscular mycorrhizal symbiosis. *New Phytologist* **25** : 139-147.

117. Sukarno, N., Smith, F.A., Smith, S.E. and Scott, E.S. 1996. The effect of fungicides on vesicular arbuscular mycorrhizal symbiosis. II. The effects on area of interface and efficiency of P uptake and transfer to plant. *New Phytologist* **132** : 583-592.

118. Sylvia, D.M. and Schenck, N.C. 1983. Application of superphosphate to mycorrhizal plant, stimulates sporulation of phosphorus tolerant vesicular-arbuscular mycorrhizal fungi. *NewPhytologist* **95** : 655-661.

119. Sylvia, D.M. and Neal, L.H. 1990. Nitrogen affects the phosphorus of VA mycorrhiza. *New Phytologist* **115** :303-310.

120. Sylvia, D.M., Hammond, L.C., Bennet, J.M., Haas, J.H. and Linda, S.B. 1993. Field response of maize to a VAM fungus and water management. *Agronomy Journal* **85** (in press).

121. Testerm, M. and Smith, F.A. 1985. Phosphate inflow into *Trifolium subterranean* L. Effect of photon iradiance and mycorrhizal infection. *Soil Biology and Biochemistry* **17** : 807-810.

122. Thingstrup, I. and Rosendahl, S. 1994. Quantification of fungal activity in arbuscular mycorrhizal symbiosis by polyacrylamide gel electrophoresis and densitometry of malate dehydrogenase. *Soil Biology and Biochemistry* **26** : 1483-1489.

123. Tommerup, I.C. and Briggs, G.G. 1981. Influence of agricultural chemicals on germination of vesicular-arbuscular endophyte spores. *Transactions of the British Mycological Society* **76** : 326-328.

124. Trappe,J.M., Molina, R. and Castellano, M. 1984. Reaction of mycorrhizal fungi and mycorrhiza formation to pesticides. *Annual Review of Phytopathology* **22** : 331-359.

125. Trappe, J.M. 1977. Selection of fungi for ectomycorrhizal inoculation in nurseries. *Annual Review of Pytopathology* **15** : 203-222.

126. Verkade, S.D. and Hamilton, D.F. 1983. Effects of benomyl on growth of *Liriodendron tulipifera* L. seedling inoculated with the vesicular-arbuscular fungus *Glomus fasiculatum. Scientia Horticulture* **21** : 252-260.

127. Wallace, L.L. 1987b. Effect of clipping and soil compaction on growth, morphology and mycorrhizal colonization of *Schizachyrium scoparium*, a C-4, bunchgrass. *Oecologia* **72** : 423-428.

128. Wallander, H. and Nylund, J.E. 1991. Effect of excess nitrogen on carbohydrate concentration and mycorrhizal development of *Pinus sylvestris* L. seedlings. *New Phytologist* **119** : 405-411.

129. Wallender, H. and Nylund, J.E. 1992. Effects of excess nitrogen and phosphorus starvation on the extramatrical mycelium of ectomycorrhizas of *Pinus sylvestris* L. *New Phytologist* **120** : 495-503.

130. Wang, G.M., Stribley, D.P., Tinker, B.P. and Walker, C. 1993. Effect of pH on an arbuscular mycorrhiza. 1. Field observation on long-term liming experiments at Rothamstead and woburn. *New Phytologist* **124** : 465-472.

131. Waterer, D.R. and Coltman, R.R. 1989. Response of mycorrhizal bell peppers to incoulation timing, phosphorus and water stress. *Horticultural Science* **24** : 688-690.

132. White, J.A., Pe Puit, E.J., Smith, J.L. and Williams, S.E. 1992. Vesicular-arbuscular mycorrhizal fungi and irrigated mined land reclamation in south western Wyoming. *Soil Science Society of America Journal* **56** : 1466-1471.

133. Yocom, D.H., Boosalis, M.G. and Flessner, T.R. 1987, In : *Mycorrhizae in the Next Decade* (eds. D.H. Sylvia, L.L. Hung and J.H. Graham), (NACOM), pp. 41.

134. Zabowski, D., Zasoski, R.J., Littke, W. and Ammirati,J. 1990. Metal content of fungal sporocarps from urban, rural and sludge-treated sites. *Journal Environmental Quality* **19** : 372-377.

28 Mycorrhiza Aided Biological Hardening of *in vitro* Raised Plantlets

Sudha, H. Kaur, A. Narula, Sanjeev Kumar, P.S. Srivastava and A. Varma

CONTENTS

ABSTRACT 470

1. INTRODUCTION 470
2. MICROPROPAGATION 471
3. PROBLEMS IN THE MICROPROPAGATION 472
4. MYCORRHIZATION AND HOST PLANT 473
5. IMPORTANCE OF BIOLOGICAL HARDENING 474
6. PROBLEM WITH MYCORRHIZA 476
7. CONCLUSION 478
8. REFERENCES 478

Advances in Microbial Biotechnology
J.P. Tewari, T.N. Lakhanpal, Jagjit Singh, Rajni Gupta & B.P. Chamola (eds.)
APH Publishing Corporation, New Delhi - 110 002, India.

ABSTRACT

Vegetative multiplication is the easiest way for clonal propagation of desirable plants of specific genotypes. In certain cases however vegetative propagation suffers because the micro-cuttings are recalcitrant to rooting. Therefore, another alternative method, micropropagation is being widely employed for mass production of plants of economic importance especially medicinal plants. Commercialization of micropropagated plants has therefore become an inescapable route in both developed and developing countries. Such *in-vitro* raised plants, however, require acclimatization phase that supports hardening and improves the initial transplant shock for survival after transfer to field conditions. Hardening is essential because the weak root system of *in-vitro* plants hampers the uptake of essential nutrients to the maximum from the soil. The problem can be partially overcome by providing physical and chemical hardening of plantlets in a mist chamber provided with variable light and temperature conditions. These conditions, although proved useful, but to a limited extent, because the plantlets lacked completely the exposure and interaction with microorganisms that the roots generally encounter in natural soil. It is, therefore, important that for transplantation and establishment of micropropagated plants the existence of mutualistic symbiosis of mycorrhizae and other associated rhizobacteria normally present in the soil are given consideration. It is being increasingly advocated now that the young plantlets be initially exposed to useful microorganisms such as growth promoting rhizobacteria or mycorrhizal fungi and other symbiotic and asymbiotic organisms. This technique of biological hardening is gaining acceptance by plant tissue culturist interested in commercial application of micropropagation. This article advocates an early inoculation of *in vitro* plantlets with selected mycorrhizae which would promise improved plant survival and performance and lesser use of synthetic fertilizers.

1. INTRODUCTION

The important global issue for current research is focussed on the protection and maintenance of sustainable environment. The idea is to minimize the use of synthetic chemicals, fertilizers and plant protecting materials such as herbicides and weedicides which are proving disastrous for the environment. The usage of enormous quantities of chemical fertilizers has resulted in exorbitant production cost as well as has affected environmental quality. Also, constant use of chemicals has lead to changes in the physical and chemical properties of the soil, which in turn, has some influence on the ecological equilibrium.

Vegetative propagation offers the easiest way to cultivate the plants. But, at the time of employing this technique, rooting of the micro-cutting is the most crucial step, therefore this does not help much in increasing the yield. In this direction, micropropagation is another widely used technique for mass production of important plants of economic importance, especially medicinal plants (1, 22, 41, 42, 62, 63, 80). Commercialization of micropropagated plants is the theme of the day in developed and developing countries. Unfortunately, the acclimatization phase (hardening) raises problems concerning the survival rate and initial "transplant shock" after transfer of the tissue culture-raised plantlets to the field. Their weak root system does not help them to uptake the maximum essential nutrients from the soil which are generally required for their better growth and development. At present this problem is partially overcome by providing physical and chemical hardening of the plantlets in a mist chamber under variable light and temperature conditions. This, however, does not significantly improve the inherent problems. The reason being their complete lack of exposure and interaction with microorganisms, e.g., bacteria, fungi and actinomycetes, normally found in the natural soil. The technology used for raising the micropropagated plants does not take into consideration the existence of mutualistic symbiosis of mycorrhizae and other associative rhizobacteria, which are normally present in soil. In the last few years, a new technique has been introduced where the young plantlets are also exposed to useful microorganisms, e.g., plant growth promoting rhizobacteria (PGPRs), or genetically-modified plant growth promoting rhizobacteria (gmPGPRs), mycorrhizal fungi and other symbiotic and asymbiotic organisms. In this procedure, the young plantlets are trained at an early stage of development to interact with mircoorganisms, which thereby learn to live in association and get familiar with the microorganisms normally expected to be encountered in nature. This technology has been termed as "biological hardening". Thus the modern technology envisages, physical, chemical and biological hardening of micropropagated plantlets. The plants obtained from these systems are normally non-mycorrhized. Such non-mycorrhized plants obtained from the micropropagation process eventually become mycorrhized when they are planted in field soil. However, an early inoculation of these plants with selected mycorrhizae promises to improve plant survival and performance and allow the lower chemical inputs.

With the current state of technology, inoculation of mycorrhiza is the best for transplanted crops. However, inoculum production of arbuscular mycorrhizal fungi (AMF) presents a very different problem. These fungi will not grow like any other fungi while separated from their hosts (obligate symbionts). Although genetic engineering of these fungi, may in future, provide the possibility of their production in the absence of the photobionts, the classical technique will still be in use for a considerable time. Obligate symbiotic mode of growth, non-availability of pure culture and expensive means of production and their unreliability for the beneficial effects have greatly undermined the mycorrhizal science, appropriately called a "Biological Enigma". Because of the absence of an authentic pure culture, the commercial production is the greatest bottleneck in the use and application of mycorrhizal biotechnology. The lack of authentic culture also slows down molecular analysis to understand the basis for the fungal interaction with plant-"cross talk" and their biotechnological applications.

Now there seems to be a silver lining for scientists dealing with mycorrhizal fungi. A new root endophyte designated as *Piriformospora indica*, gen. nov. sp. nov., discovered by scientists in Microbiology at the School of Life Sciences, Jawaharlal Nehru University, New Delhi, holds great promise in this regard. Ultra-structural and molecular analysis has indicated that it belongs to the Hymenomycetes (Basidiomycota) with relatively close relationship to *Rhizoctonia. P. indica* improves plant growth and overall biomass. The fungus can be easily cultivated on synthetic media. Hyphae colonizes the root and shows inter- and intracellular structures (vesicles and arbuscules-like). Chlamydospores are formed both inside the root tissues and externally into the environment.

2. MICROPROPAGATION

Commercialization of micropropagated plants has become a thrust area both in developed and developing countries (99). In plant biotechnology, the emphasis is on manpower rather than on expensive equipment. Among plant biotechnologies, micropropagation is often cited as the most successful example of a laboratory curiosity which has become an important commercial industry. In the vegetative propagation of many crops, rooting of (micro) cuttings is the most crucial step (49). Application of auxin is still the main tool to achieve rooting. Cuttings of many crops, however, do not readily undergo root formation after application of auxin. Rooting can be improved: i) by adapting the conditions of the (micro) cutting, and ii) optimizing the rooting treatment. Despite this, the progress has been steady and no other major factor has been shown to achieve rooting of recalcitrant crops. The propagation of plant species by cuttings or tissue culture is often limited by poor rooting. This limits mass scale production of desirable genotypes.

The structural and spatial morphology and dynamics of an individual plant root system have a significant effect on its ability to obtain water and acquire the nutrients. Root diameter and length are important in determining the basic uptake potential of a root system while structural and spatial morphology have a major impact upon actual uptake. The dynamics of the root system e.g., the length of time the root lives and the time of the year when they grow, are also important factors for maximum potential uptake when the growth and activity of the root system is synchronized with nutrient and water availability in the soil (40).

The prospects of micropropagation techniques have been amply elaborated earlier by many workers (2, 24, 46, 61). Researches highlighting achievements, problems and prospects related to micropropagation have been emphasized by Berlin (8).

The various methods of vegetative propagation through tissue culture that can be attempted are:

(i) direct formation of plants, shoot buds or embryos from the excised explants,

(ii) indirect regeneration of plants via callus culture,

(iii) androgenesis through anther/pollen culture, and

(iv) polyembryony.

It is well evident that the origin of majority of the material source of plants is the natural production system. Plant tissue culture techniques are being used globally for the *ex-situ* conservation and mass propagation of valuable germplasm. It would be worthwhile to adopt the methods of *in vitro* culture to multiply the species and monitor in the growth performance. Owing to the importance of plants as a source of several of the economically valuable products, alternative methods, such as *in vitro* culture, need to be explored to make quantitative and qualitative improvements and reclamation of land through plantation of such *in vitro* regenerants.

3. PROBLEMS WITH MICROPROPAGATION

The acclimatization phase raises the problems concerning survival and development of the plantlets. The survival rate is low and initial "transplant shock" on transfer of the plantlets to the field are very high. Stunted growth often leads to non-recovery of the plants. Often, they are attacked by soil fungi. At the weaning stage, about 10-40 per cent of the plantlets either die or do not attain market standards, therefore causing significant losses at the commercial level (12, 55, 63). Inoculation of micropropagated plantlets with plant growth promoting fungi (PGPF) appears to be critical for their survival and growth. This avoids "transient transplant shock" (95).

4. MYCORRHIZATION AND HOST PLANT

Mycorrhiza is a mutualistic symbiosis formed by most species in the plant kingdom, including almost all the plants currently being micropropagated. Mycorrhiza has been shown to positively affect plantlet growth by improving nutrient uptake to increase the rhizosphere soil volume and to alleviate biotic and abiotic stress. Mycorrhiza is also ecologically significant and is particularly important in new plantations in soil with problems (salinity, heavy metal contamination, low nutrient availability) or in sustainable agricultural systems where the non-biological inputs are minimized.

Plant responses to colonization by mycorrhizal fungi can range from dramatic growth promotion to growth depression. Factors affecting this response include the mycorrhizal dependency of host crop, the nutrient status of the soil, and the inoculum potential of the mycorrhizal fungi. Management practices such as tillage, crop rotation, and fallowing may adversely affect populations of mycorrhizal fungi in the field. Where native inoculum potential is low or ineffective, inoculation strategies may be helpful. With the current state of technology, the inoculation is most feasible for transplanted crops and in areas where soil disturbance has greatly reduced the native inoculum potential.

A method was developed by George *et al.* (29) to measure the nutrient uptake by mycorrhizal hyphae in conventional type pot experiments. Soil zones accessible only to hyphae, roots and hyphae, or neither to roots nor hyphae, were obtained by placing tubes with different pore sizes into the pot soil. Hyphae of the mycorrhizal fungus, *G. mosseae*, transported large amounts of phosphate, potassium and zinc to the host plant (rye grass). The possible mechanisms responsible for the hyphal translocation and measured rate of translocation were reviewed by Smith (79). The evolutionary success of arbuscular mycorrhizae (AMF) reflects the unique combination of a superior biotrophic mode of fungal carbon acquisition and the ability of the living plants to absorb nutrients, especially phosphorus from the fungus (47). Improved plant growth in response to mycorrhization is closely linked to this bi-directional exchange of carbon (C) and phosphorus (P) between the symbionts, an exchange which is strongly influenced by plant and fungal genotypes and by the environment. The primary advantage of mycorrhizal hyphae in P uptake as compared to root system is the ability of the hyphae to extend beyond the P depletion zone of the root (87, 88). It is also possible, however, that the hyphae may utilize P source of low plant availability. This aspect has been studied by means of isotopic dilution principle, where a tracer isotope (e.g. P^{32}) is allowed to exchange with the native soil P (34). A gene encoding a high affinity P transporter was cloned from a library prepared from mycorrhizae of *Medicago truncatula* and *Glomus versiforme*. The gene was expressed in the external hyphae of the fungus (43). Lei *et al.* (51) used the root-inducing (Ri) T-DNA transformed roots of carrot as the plant partner in a study of P^{32} absorption and plasmalemma ATPase activity in the hyphae of the vesicular arbuscular mycorrhizal fungus, *Gigaspora margarita* Becker & Hall. Hyphae from germinating spores were grown in the presence or absence of root exudates and volatile chemicals. In the presence of these root factors, P^{32} isotope labelling occurred in the hyphae, auxiliary cells and spores, while in the absence of these factors, the labelling occurred only in the fungal spores. This showed that root factors stimulate P^{32} uptake and plasmalemma ATPase activity in *Gigaspora margarita*. Chabot *et al.* (20) demonstrated the entire vegetative life cycle of the vesicular arbuscular mycorrhizal fungus, *Glomus intraradices* in a simple monoxenic culture system using RiTDNA-transformed carrot root and non-transformed tomato roots as the plant partners. The extent of internal root colonization by *G. intraradices* was relatively limited, but it was sufficient to permit the *in vitro* formation of hundreds of new spores free of contaminants, which were viable and capable of colonizing host roots. Molecular genetic studies of these fungi and their interaction with plants have been limited owing to the obligate symbiotic nature of the VAM fungi. The molecular mechanisms underlying fungal mediated uptake and translocation of phosphate from the soil to the plant, therefore, remain unknown. Maria and Buuren (57) investigated this process by identifying a complementary DNA that encodes a trans-membrane phosphate transporter (GvPT) from *Glomus versiforme*, a VAMycorrhizal fungus. The function of the protein encoded by GvPT was confirmed by complementation of a yeast phosphate transport mutant. Expression of GvPT was localized to the external hyphae of *G. versiforme* during mycorrhizal associations, these being the initial sites of phosphate uptake from the soil.

Although endomycorrhizal symbiosis represents the most widespread of plant-microbe interactions, the mechanisms leading to reciprocal functional compatibility are still unknown (66, 68). In order to better understand why plant defense reactions are only weakly and locally induced in endomycorrhizas, Graham and Graham (40), Gollotte *et al.* (37) and Xie *et al.* (100) determined whether endomycorrhizal fungi possess elicitors of plant defense reactions. They found that both ericoid and arbuscular mycorrhizal fungi contain chitin, ß-1, 3 glucans and/or chitosan and glycoproteins which could potentially induce plant defense reactions (89). However, important wall modifications occur in these endomycorrhizal fungi during the root colonization process and therefore, they could have strategies of self-camouflage so that they would not be recognized by the plant as non-self. Alternatively, plant defense reactions could be induced but then repressed. Evidence for the existence of such phenomena comes from studies of myc-1 nod$^-$ pea mutants which do not form arbuscular mycorrhizas nor nodules with rhizobial bacteria (39). Myc^{-1} mutants are characterized by the elicitation of strong plant defense reactions in epidermal and hypodermal cells in contact with fungal appresoria (31, 32, 76). Moreover, the fact that these mutants are resistant to both symbioses means that some plant genes are specifically involved in symbiosis establishment. Therefore, one role of symbiosis-related (SR) genes could be to repress a strong expression of plant defense reactions during endomycorrhization. Expression of these SR genes could somehow interact with fungal compatibility factors underlying these mechanisms (38, 71). The genetic factors that control the complex events of recognition, compatibility and/or resistance have only been examined in the last few years (17, 33, 36).

Hyphal wall structure and composition does not show a uniform pattern, as it differs between genera (35, 79) as well as between the different developmental stages of the fungus (15, 16, 17). Now there is evidence that the penetration of roots by AM fungi involves cell wall degrading hydrolytic enzymes (90). These enzymes may lead to an organized colonization of root (14). Cell wall degrading enzymes that have been detected in the mycorrhizal fungal wall are cellulases (14, 48), hemicellulases such as xylanases (65), pectinases (27), chitinases (50, 67, 70), and ß-1, 3 glucanases (40, 53). Significant quantities of xylanases have been detected from a new root-colonizer fungus *Piriformospora indica* gen. nov. sp. nov. while growing on oat and wheat xylan as a sole carbon source (98). An acid phosphatase, a mannoprotein, has also been revealed in both the fungal wall and the extracellular fibrils by immunocytochemistry. Though its presence is indicated by immunochemistry in the fungal wall of intracellular hyphae, it is not active. In contrast, in non-host plants, acid phosphatase activity remains in intracellular hyphae. This indicates that the host cells somehow inhibit acid phosphatase activity of intraradical hyphae (52). Moreover, an invertase, another mannoprotein, comigrates with one acid phosphatase isoform (81).

5. IMPORTANCE OF BIOLOGICAL HARDENING

The technology used for micropropagating plants takes into consideration the existence of mutualistic symbiosis of mycorrhizae and other associative "Plant Growth Promoting

Rhizobacteria" (PGPRs). The media used are devoid of mycorrhizal propagules and therefore, the plants obtained from these systems are non-mycorrhized. Such non-mycorrhized plants obtained from the micropropagation process eventually become mycorrhized when they are planted in the field. However, an early inoculation of these plants with selected mycorrhizal and associative bacterial inocula promises to improve plant survival, performance and allow lower chemical inputs. Micropropagated plants inoculated in ex vitro experiments are generally, high-value plants like grape vine (64, 73, 74), oil palm (12), apple (18), plum (25), pineapple (55), avacado (5), kiwi-fruit (75), strawberry (91), raspberry (93), hortensia (92, 93) and woody legumes (69). Therefore, there is high potential for introducing AMF into the micropropagation system of these plants and other high value plants.

There are several reports to suggest that treatment of micropropagated shrub legumes with mycorrhizal inoculum shortened their acclimatization process (56). It was found that mycorrhized apple plants were more uniform in size (99). From industrial and economic point of view the uniformity of the plants is a desirable characteristic in any nursery; AMF inoculated apple was 1/3 to 1/4 shorter than those with uninoculated plants. Lin *et al.* (54), based on a large scale survey of micropropagated potato, strawberry, blackberry, apple, rose, ginger, and pineapple with AMF strains, found blackberry and apple to be the most promising plants from which it would be feasible and advantageous to produce mycorrhizal nursery stocks.

The importance of soil is now recognized as an agricultural "resource base", but as a complex, living, and fragile system that must be protected and managed for its own sake to guarantee its long term stability and productivity. Bethlenfalvay and Schuepp (11) have described sustainability as "maximum plant production with minimum of soil loss". Studies have shown that AMF may alter the soil microflora by stimulating as well as inhibiting total bacterial counts or selected bacterial groups (21, 92). Soil microbes, in turn, may promote or antagonize mycorrhizae development. How do these complex interactions affect plant production and soil stability? A stimulation of plant growth may be achieved by manipulating specific groups of organisms, such as phosphate solubilizing (7, 28) or diazotrophic bacteria (10, 77, 78, 86) or rhizobacteria that may promote plant growth by various mechanisms (26).

Elmeskaoui *et al.* (23) developed a reliable method to obtain vesicular arbuscular mycorrhizae (VAM), *Glomus intraradices* in micropropagated plantlets of strawberry and determined their influence on growth. They found that colonized plantlets had more extensive root systems and better shoot growth than control plants. The VAM symbiosis reduced the osmotic potential of plantlets. They also assumed that this response may be a useful pre-adaptation for plantlets during transfer to the acclimatization stage.

Mycorrhization has been demonstrated in a commercial autotrophic micropropagation system where it has been shown to improve the performance of *Fragaria vesca* microplants at acclimatization (weaning) and in subsequent development (19). It was shown that photosynthesis *in vitro* is not impaired by mycorrhization. Microplants, mycorrhized *in vitro* or *in vivo* outperformed non-mycorrhized plants in fresh and dry weight gain after acclimatization. Varma and Schuepp (92) inoculated the micropropagated plantlets of hortensia (*Hydrangea* var. *leuchtfeuer*) with arbuscular mycorrhizal fungus *Glomus intraradices* Schenk and Smith. Roots were heavily mycorrhized at acclimatization stage. There was hundred per cent survival, shoot apices were active and no apparent "transient transfer shock" was noticed. Mycorrhized plants showed better growth and large leaf area. In contrast, the non-mycorrhized plants were severely wilted and produced reduced number of apical buds. It is postulated that there were benefits of early rejuvenation of natural microflora on the sterile substratum which helped the plantlets in early acclimatization, fast establishment, enhanced growth and possibly improved root mycorrhization (93).

Many of the reactions and interactions of the microflora and fauna that occur in the soil around roots are mediated by AMF that function to deliver mineral nutrients to the host plant in turn for a sustained carbon supply. They also impart other benefits to plants including stimulation of growth regulating substances and alteration of other chemical constituents in the plant, increased rate of photosynthesis, osmotic adjustment under drought stress, increased resistance to pest and tolerance to environmental stresses, and improved soil aggregation and thus, improved soil physical properties and stability. Membrane permeability is altered, and this change can greatly alter the quality and quantity of root exudates, resulting in a new microbial equilibrium in the soil surrounding that root now appropriately called "mycorrhizosphere". The influence of VAM fungi on rhizosphere mycoflora was studied together with the possible mechanisms involved in this process (6). Mukerji *et al.* (60) have described some techniques for isolating the microorganisms from mycorrhizosphere of the root surface (rhizoplane) as well as the soil surrounding the root system. Azaizeh *et al.* (4) found that the maize plants when grown in the presence of VAM fungus (*Glomus mosseae*) and other rhizosphere microorganisms together, the shoot/root ratios were tremendously increased in comparison to the non-mycorrhized plants. The general bacterial population was several folds higher from the rhizospheres of the mycorrhized roots over the control (91). The dominating bacterium was *Pseudomonas fluorescens* biovar I.

A large number of these mycorrhizosphere microbes act as Plant-Growth-Promoting Rhizobacteria (PGPRs) and/or genetically modified Plant-Growth-Promoting Rhizobacteria (gmPGPRs). These bacteria produce antibiotics or siderophores that suppress deleterious microbes, produce hormones or other growth-enhancing compounds that directly affect plant growth or enhance the nutritional status of the plants. It is still not understood whether growth enhancement is the result of the PGPRs-AMF interaction, or because of some other secondary metabolites. One could question whether fungi alone should be credited for all the benefits to plants; or more likely, a combination of mycorrhizal fungi and their microbial associates. Although the utility of such combinations has not been demonstrated, the logic of their use as a crop management strategy in sustainable agriculture is apparent.

Beneficial effects of mycorrhizal colonization of plant growth are both direct and indirect (30). Direct effects include nutrient uptake and protection against certain pathogens (45). The beneficial effect of mycorrhiza on soil structure, water holding capacity and biological activity (9) is equally important but are indirect factors. Thus, certain benefits of mycorrhiza for plants are inextricably bound with interactions between soil, plant and mycorrhiza. Several authors have reported that the phosphorus concentation of the host plant controls the colonization (58, 72, 84). However, in other experiments it was demonstrated that soil phosphorus exerted a direct effect on mycorrhizal colonization (59, 85).

6. PROBLEMS WITH MYCORRHIZA

The mycorrhiza-plant partnership is a basic, essential and integral part of plant survival and growth. In situations where native mycorrhizal fungal inoculum potential is low or ineffective, providing the appropriate fungi for plant production system is worth considering. With the current state of technology, inoculation is the best for transplanted crops and in areas where soil disturbance has reduced native inoculum potential. Inoculum production of Arbuscular Mycorrhiza Fungi (AMF) presents a very different problem. These fungi will not grow like any other fungi apart from with their hosts (obligate symbionts). Although genetic engineering of these fungi may in future provide the possibility of their production in the absence of the plants, the classical techniques will still be in use for a considerable time. Obligate symbiotic mode of growth, non-availability of pure cultures and expensive means of production and their unreliability for the beneficial effects have greatly jeopardized/undermined the mycorrhizal science, that may be appropriately called a "Biological Enigma". Because of the absence of an authentic pure culture, the commercial production is the greatest bottleneck in use and application of mycorrhizal biotechnology. The lack of authentic culture also slow down molecular analysis to understand the basics of the interaction of fungus with the plants- "cross talk" and their biotechnological applications.

A root endophyte was found initially in a sand desert in the interiors of Rajasthan and further investigations found this fungus in Marburg (Germany) and Philippines. In order to get a more precise idea about the closest relatives of this fungus, a part of the 18S rRNA was amplified, sequenced and compared with all corresponding data of different Basidiomycetes available in the gene bank sequence library, as well as of some Ascomycota and Basidiomycota (98). Based on these results the classification within Basidiomycota was made. It was named as *Piriformospora indica* for its characteristic spore morphology (Fig. 1). This new wonder fungus improves the growth and overall biomass production of different grasses, trees, and herbaceous species and can easily be cultivated on a number of complex and synthetic media. Significant quantitative and morphological changes have been detected when the fungus was grown on different media with no negative effect during cultivation. The new fungus produced chlamydospores at the apex of undifferentiated hyphae. These chlamydospores can be easily germinated on various synthetic media as reported by Verma *et al.* (98). There were only a few features to characterize the fungus morphologically and to group it according to the classical species concept. It was found that when the growing hyphae come in contact with the root surface of the maize plant, it penetrated into the root system.

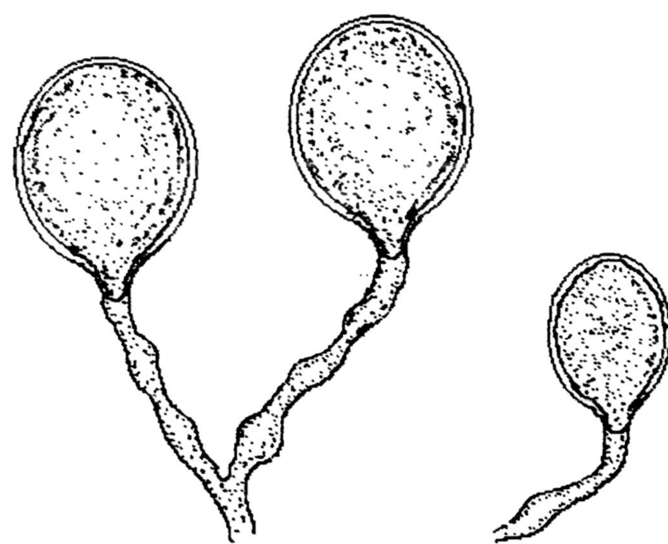

Fig. 1 : Pear shaped spores of *Piriformospora indica*

Fungal spores are multinucleated and hyphae are dimorphic and interwined when grown on synthetic medium. It has dolipore septum with a continuous and straight paranthesome on either side. It has also been found that the growth of *P. indica* was not affected by a virulent root and seed pathogen *Gaumannomyces graminis*. When they were inoculated together on the same synthetic medium, they formed a distinct demarcation zone. There is no existing fungus which covers all the characters of the new fungus. Other groups of Basidiomycota which may be related according to the molecular or ultra-structural data are not known to produce chlamydospores or have a different way of life. Therefore, a new genus was erected and the fungus was called *Piriformospora indica* (44, 68, 95, 96, 97, 98).

P. indica helps in nutrient uptake particularly phosphorus (82, 83) and improves the growth and overall production of different grasses, trees and herbaceous species. More recently it was found that in a monocot *Setaria italica*, the smallest genome next to rice, the fungal colonized plants were highly promoted by interacting with *P. indica* (95, 96, 97). The new fungus also promotes several tropical legumes. Interestingly, like AMF the new fungus did not invade the roots of *Brassica* (Brassicaceae) and the myco-minus mutants of Soybean (*Glycine max*) seeds. Blechert *et al.* (13) found the enhancement in seed germination of *Dactyorhiza purpurella* and *D. incarnata* (terrestrical orchids) in comparison to the control, on the oat medium when *P. indica* was inoculated. Control plants had well-developed leaves but very poor or almost lacking in roots. In contrast, plants inoculated with *P. indica* produced well-developed leaves and root system. In cross section of the roots, cortical cells with intracellular hyphal coils could be observed. The coiled hyphae of *P. indica* in the protocorm cells and the root cortex cells resemble the intracellular hyphae known from other basidiomycetes living in symbiosis with orchids.

7. CONCLUSION

Application of AMF to roots of micropropagated plantlets could, therefore, produce modification in root morphology and dynamics which aid the establishment and growth of plantlets. The extent of modifications in the plant root system morphology by AM fungi has only recently been appreciated and will have a major impact on our understanding of the mechanisms by which AM fungi influence plant uptake of water and the acquisition and cycling of nutrient in both natural and managed systems. This opens up a new thrust area of research as there appears to be great potential to manipulate root system morphology to benefit plant establishment and growth in the *ex vitro* weaning and growth stages of micropropagation and other systems where AM fungi are not normally present. The image-analysis system described by Atkinson *et al.* (3) provide a powerful tool to be used in the studies required to improve our understanding of nature and extent of these modifications and the mechanisms involved.

8. REFERENCES

1. Ali, G., Purohit, M., Saba, Iqbal, M. and Srivastava, P.S. 1997. Morphogenic response and isozymes of *Bacopa monniera* (L.) Wettst cultures grown under salt stress. *Phytomorphology* **47**: 97-106.

2. Ammirato, P.V. 1983. Embryogenesis. In: *Handbook of Plant Cell Culture I*, (eds. D.A. Evans, W.R. Sharp, P.V. Ammirato and Y. Yadmada). New York, pp. 82-123.

3. Atkinson, D., Berta, G. and Hooker, J.E. 1994. Impact of mycorrhizal colonization of growth regulators. In: *Impact of Arbuscular Mycorrhizas on Sustainable Agriculture and Natural Ecosystems*, (eds. S. Gianinazzi and H.Schuepp) Birkhauser Verlag, Basel, pp. 89-100.

4. Azaizeh, H.A., Marschner, H., Romheld, V. and Wittenmayer, L. 1995. Effects of a vesicular arbuscular mycorrhizal fungus and other soil microorganisms on growth, mineral nutrient acquisition and root exudation of soil grown maize plants. *Mycorrhiza* **5**: 321-327.

5. Azcon-Aguilar, C., Barcelo, A., Vidal, M.T. and De La Vina, G. 1992. Further studies on the growth and development of micropropagated Avocado plants. *Agronomie* **12**: 837-840.

6. Bansal, M. and Mukerji, K.G. 1994. Positive correlation between VAM-induced changes in root exudation and mycorrhizosphere mycoflora. *Mycorrhiza* **5**: 39-44.

7. Barea, J.M. and Jeffries, P. 1999. Arbuscular Mycorrhizas in Sustainable Soil-Plant Systems. In: Mycorrhiza, (second edition), (eds. A. Varma and B. Hock), Springer-Verlag, pp. 521-560.

8. Berlin, J. 1984. Plant cell cultures - a feature source of natural products. *Endeavour* **58**: 5-8.

9. Bethlenfalvay, G.J. 1992. Mycorrhizae in the agricultural plant-soil system. *Symbiosis* **14**: 413-425.

10. Bethlenfalvay, G.J. 1994. Vesicular arbuscular mycorrhizal fungi in nitrogen fixing legumes: Problems and Prospects. In: *Techniques for Mycorrhizal Research*, (eds. J.R. Norris, D.J. Read and A. Varma), Academic Press, London, pp. 835-850.

11. Bethlenfalvay, G.J. and Schuepp, H. 1994. Arbuscular mycorrhizas and agrosystem stability. In: *Impact of Arbuscular Mycorrhizas on Sustainable Agriculture and Natural Ecosystems*, (eds. S. Gianinazzi and H. Schuepp,) Birkhauser Verlag, Basel/Switzerland, pp. 117-132.

12. Blal, B., Morel, C., Gianinazzi-Pearson, V., Fardeau, J.C. and Gianinazzi, S. 1990. Influence of vesicular arbuscular mycorrhizae on phosphate fertilizer efficiency in two tropical acid soils planted with micropropagated oil palm (*Elaeis guinensis* Jacq.). *Biol. Fertl. Soils* **9**: 43-48.

13. Blechert, O., Kost, G., Hassel, A. and Rexer, K.H. 1999. First remarks on the interaction between *Piriformospora indica* and terrestrial orchids. In: *Mycorrhizae*, (second edition), (eds. A. Varma, B. and Hock), Springer-Verlag, Germany, pp. 683-688.

14. Blilou, I., Martin, J. and Ocampo, J.A. 1996. Influence of cellulase on the susceptibility of non-host cabbage to colonization by *Glomus intraradices*. In: *Mycorrhiza in Ontegrated Systems from Genes to Plant Development*, (eds.Azcon-Aguilar and J.M., Barea), ECSC-EC-EAEC, Brussels, pp. 215-217.

15. Bonfante-Fasolo, P. 1997. Gene expression in arbuscular mycorrhizae. COST-Action 8.21, Torino, Italy, pp. 16-17.

16. Bonfante-Fasolo, P., Faccio, A. and Schubert, A. 1990. Correlation between chitin distribution and cell wall morphology in the mycorrhizal fungus *Glomus versiforme*. *Mycol. Res.* **94**: 157-165.

17. Bonfante-Fasolo, P. and Perotto, S. 1995. Strategies of arbuscular mycorrhizal fungi when infecting host plants. *New Phytol.* **130**: 3-21.

18. Branzanti, B., Gianinazzi-Pearson, V. and Gianinazzi, S. 1992. Influence of phosphate fertilization on the growth and nutrient status of micropropagated apple infected with endomycorrhizal fungi during the weaning stage. *Agronomie* **12**: 841-846.

19. Cassells, A.C., Mark, G.L. and Periappuram, C. 1996. Establishment of arbuscular mycorrhizal fungi in autotrophic strawberry cultures *in vitro*. Comparison with inoculation of microplants *in vivo*. *Agronomie* **16**: 625-632.

20. Chabot, S., Becard, G. and Piche, Y. 1992. Life cycle of *Glomus intraradices* in root organ culture. *Mycologia* **84**: 315-321.

21. Christensen, H. and Jakobsen, I. 1993. Reduction of bacterial growth by a vesicular arbuscular mycorrhizal fungus in the rhizosphere of cucumber (*Cucumis sativus* L.). *Biol. Fert. Soils* **15**: 253-258.

22. Datta, A. and Srivastava, P.S. 1997. Variation in vinblastin production by *Catharanthus roseus* during *in vivo* and *in vitro* differentiation. *Phytochemistry* **46**: 135-137.

23. Elmeskaoui, A., Damont, J.P., Poulin, M.J., Piche, Y. and Desjardins, Y. 1995. A tripartite culture system for endomycorrhizal inoculation of micropropagated strawberry plantlets *in vitro*. *Mycorrhiza* **5**: 313-319.

24. Evans, D.A., Sharp, W.R. and Flick, C.E. 1981. Growth and behaviour of cell structure: Embryogenesis and organogenesis. In: *Plant Tissue Culture, Method and Application in Agriculture*, (ed. T.A. Thorpe). Academic Press, London, pp. 45-113.

25. Fortuna, P., Citernesi, S., Morini, S., Giovannetti, M. and Loreti, F. 1992. Infectivity and effectiveness of different species of arbuscular mycorrhizal fungi in micropropagated plants of Mr S 2/5 plum rootstock. *Agronomie* **12**: 825-830.

26. Garbaye, J. 1992. Biological interactions in the mycorrhizosphere. *Experimentia* **57**: 370-375.

27. Garcia-Romera, J.M., Garcia-Romera, I. and Ocampo, J.A. 1996. Purification of an arbuscular mycorrhizal endoglucanase from colonized roots. In: *Mycorrhiza in Integrated Systems from Genes to Plant Development*, (eds. Azcon-Aguilar and J.M. Barea), ECSC-EC-EAEC, Brussels, pp. 231-233.

28. Gaur, A.C. 1988. Phosphate solubilizing biofertilizers in crop productivity and their interaction with VA mycorrhizae. In: *Mycorrhizae Round Table*, (eds. A.K. Varma, A.K., Oka, K.G. Mukerji, K.V.B.R. Tilak and J. Raj), IDRC, Canada, pp. 505-529.

29. George, E., Gorgus, E., Schmeisser, A. and Marschner, H. 1996. A method to measure nutrient uptake from soil by mycorrhizal hyphae. In: *Mycorrhizas in Integrated Systems from Genes to Plant Development*, (eds. Azcon-Aguilar and J.M. Barea), EUR publication of European community, Luxembourg, pp. 535-538.

30. George, E., Haussler, K., Kothari, S.K., Li, X.L. and Marschner, H. 1992. Contribution of mycorrhizal hyphae to nutrient and water uptake of plants. In: *Mycorrhizas in Ecosystems*, (eds. D.L. Read, D.H. Lewis, A.H. Fitter and J.J. Alexander), CAB International, Wallingford, Oxon, UK, pp. 42-47.

31. Gianinazzi-Pearson, V., Gianinazzi, S., Guillemin, J.P., Trouvelot, A. and Duc, G. 1991. Genetics and cellular analysis of the resistance to vesicular arbuscular (VA) mycorrhizal fungi in pea mutants. In: *Advances in Molecular Genetics of Plant Microbe Interactions*, (eds. H. Hennecke and D.P.S. Verma, D.P.S.), Kluwer Academic Publishers, Boston, London, pp. 336-342.

32. Gianinazzi-Pearson, V., Gollotte, A., Franken, P. and Gianinazzi, S. 1994. Gene expression and molecular modifications associated with plant responses to infection by arbuscular mycorrhizal fungi. In: *Advances in Molecular Genetics of Plant Microbe Interactions*, (ed. M. Daniels), Kluwer Academic Publishers, Boston, London, pp. 179-186.

33. Gianinazzi-Pearson, V., Gollotte, A., Lherm, J., Tisserant, T.B., Franken, P., Dumas-Gaudot, E., Lemoine, M.C., Van Tuinen, D. and Gianinazzi, S. 1995. Cellular and molecular approaches in the characterization of symbiotic events in functional arbuscular mycorrhizal associations. *Can. J. Bot.* **73**: 526-532.

34. Gianinazzi-Pearson, V., Morandi, D., Dexheimer, J. and Gianinazzi, S. 1981. Ultrastrutural and ultracytochemical features of a Glomus tenuis mycorrhiza. *New Phytol.* **88**: 633-639.

35. Giovannetti, M. and Gianinazzi-Pearson, V. 1994. Biodiversity in arbuscular mycorrhizal fungi. *Mycol. Res.* **98**: 705-715.

36. Giovannetti, M., Sbrana, C., Citernesi, S.A. and Avio, L. 1996. Analysis factors involved in fungal recognition to host derived signals by arbuscular mycorrhizal fungi. *New Phytol.* **133**: 65-72.

37. Gollotte, A., Cordier, C., Lemoine, M.C. and Gianinazzi-Pearson, V. 1999. Role of fungal wall components in interactions between endomycorrhizal symbionts. In: *Mycorrhizae*, (eds. A. Varma and B. Hock), second edition, Springer-Verlag, Germany, pp. 412-428.

38. Gollotte, A., Gianinazzi-Pearson, V., Giovannetti, M., Sbrana, C., Avio, L. and Gianinazzi, S. 1993. Cellular localization and cytochemical probing of resistance reaction to arbuscular mycorrhizal fungi in a 'locus a' mutant of *Pisum sativum* (L.). *Planta* **191**: 112-122.

39. Gollotte, A., Gianinazzi-Pearson, V., Dumas-Gaudot, E., Giovannetti, M., Lherminier, J., Berta, G. and Gianinazzi, S. 1996. Plant mutants as model for studying the cellular and molecular bases of plant-fungal interactions in arbuscular mycorrhiza. In: *Mycorrhizas in Integrated Systems for Genes to Plant Development*, (eds. C. Azcon-Aguilar and J.M. Barea), European Commission, EUR 16728, Luxembourg, pp. 189-194.

40. Graham T.L. and Graham, M.Y. 1991. Cellular coordination of molecular responses in plant defense. *Mol. Plant-Microbe Interact.* **4**: 415-422.

41. Gulati, A., Bharel, S., Abdin, M.Z., Jain, S.K. and Srivastava, P.S. 1996. *In vitro* micropropagation and flowering in Artemisia annua. *J. Biochem. Biotech.* **5**: 31-35.

42. Gupta, N. and Srivastava, P.S. 1996. *In vitro* regeneration and isozyme patterns in *Zizyphus mauritiana. J. Plant Biochem. Biotech.* **5**: 87-90.

43. Harrison, M.J. and Buuren, M.L.V 1995. A phosphate transporter from the mycorrhizal fungus *Glomus versiforme. Nature* **378**: 626-629.

44. Hindav, R., Kumari, M., Mondal, N., Paul, J., Sahay, N.S., Singh, A., Sudha and Varma, A. 1998. One kilo tropical soil = 1 kilo gold : this is microbial science In: *Microbes for Health, Wealth and Sustainable Environment*, (ed. A. Varma), MPH, pp. 1-23.

45. Hooker, J.E. and Black, K.E. 1995. Arbuscular mycorrhizal fungi as components of sustainable soil-plant systems. *Crit. Rev. Biotech.* **15**: 201-212.

46. Hussey, G. (1986) Problems and prospects in the *in vitro* propagation of herbaceous plants. In: *Plant Tissue Culture and its Agricultural Applications*, (eds. L.A. Withers and P.G. Anderson), Butterworths, London, pp. 69-84.

47. Jakobsen, I. 1999. Transport of phosphorus and carbon in arbuscular mycorrhizas. In: *Mycorrhizae: Structure, Function, Molecular Biology and Biotechnology*, (eds. A. Varma and B. Hock), second edition, Springer-Verlag, Germany, pp. 305-322.

48. Kirche, J., Varma, A., Miller, D. and Mayer, F. (1994) Carboxymethyl cellulase from *Bacillus thermoalkalophilus*. Isolation, macromolecular organization and cellular localization. *J. Gen. Microbiol.* **40**: 52-62.

49. Klerk, G. J. De and Brugge, J.T. 1992. Factors affecting adventitious root formation in microcuttings of *Mallus. Agronomie* **12**: 747-755.

50. Lanfranco, L., Garnero, L., Dalpero, M. and Bonfante, P. 1995. Chitin synthase homologs in three ectomycorrhizal truffles. *FEMS Microbiol. Lett.* **134**: 109-114.

51. Lei, J., Becard, G., Catford, J.G. and Piche, Y. 1991. Root factors stimulate P^{32} uptake and plasmalemma ATPase activity in vesicular arbuscular mycorrhizal fungus, *Gigaspora margarita. New Phytol.* **118**: 289-294.

52. Lemoine, M.C., Gianinazzi-Pearson, V., Gianinazzi, S.A., and Straker, C.J. 1992. Occurrence and expression of acid phosphatase of Hymenoscyphus ericae (Read) Korf and Kernan, in isolation or associated with plant roots. *Mycorrhiza* **1**: 137-146.

53. Lemoine, M.C., Gollotte, A., Gianinazzi-Pearson, V. and Gianinazzi, S. 1995. ß(1-3) glucan localization in walls of the endomycorrhizal fungi *Glomus mosseae* and *Acaulospora laevis* during colonization of host roots. *New Phytol.* **129**: 97-105.

54. Lin, M.T., Lucena, F.B., Mattos, M.A.M., Paiva, M., Assis, M. and Caidas, L.S. 1987. Greenhouse production of mycorrhizal plants of nine transplanted crops. In: *Mycorrhizae for the Next Decade*, (eds. D.M. Sylvia, L.L. Hung and J.H. Graham), Practical application and research priorities, 7th NACOM, USA, pp. 281.

55. Lovato, A., Guillemin, J.P. and Gianinazzi, S. 1992. Application of commercial arbuscular endomycorrhizal fungal inoculants to the establishment of micropropagated grapevine rootstock and pineapple plants. *Agronomie* **12**: 873-880.

56. Lovato, P.E., Schuepp, H., Trouvelot, A. and Gianinazzi, S. (1999) Applications of arbuscular mycorrhizal fungi (AMF) in orchids and ornamental plants. In: *Mycorrhiza*, (eds. A. Varma and B. Hock), Springer, Berlin, Heidelberg, New York, pp. 443-468.

57. Maria, J.H. and Buuren, M. V. 1995. A phosphate transporter from the mycorrhizal fungus *Glomus versiformae*. *Nature* **378**: 626-629.

58. Menge, J.A., Steirle, D., Bagyaraj, D.J., Johnson, E.L.V., and Leonard, R.T. 1978. Phosphorus concentration in plants responsible for inhibition of mycorrhizal infection. *New Phytol.* **106**: 707-715.

59. Miranda, J.C.C. and Harris, P.J. 1994. The effect of soil phosphorus on the external mycelium growth of arbuscular mycorrhizal fungi during the early stages of mycorrhiza formation. *Plant Soil* **166**: 271-280.

60. Mukerji, K.G., Mandeep, Varma, A. 1998. Mycorrhizosphere microorganisms: screening and evaluation. In: *Mycorrhiza Manual*, (ed. A. Varma), Springer, Germany, pp. 85-98 .

61. Murashige, T. 1984. Parameters in regenerating plants in vitro. In: *Plant Tissue and Cell Culture Applications to Crop Improvement*, (eds. F.J. Novak, L. Havel and J. Dolezel), Czeck. Acad. Sci. Prague, pp. 23-31.

62. Purohit, M., Pande, D., Datta, A. and Srivastava, P.S. 1995a. Enhanced xanthotoxin content in regenerating cultures of *Ammi majus* and micropropagation. *Planta Med.* **61**: 481-482.

63. Purohit, M., Pande, D., Datta, A. and Srivastava, P.S. 1995b. *In vitro* flowering and high xanthotoxin in *Ammi majus* L. *J. Biochem. Biotech.* **4**: 73-76.

64. Ravolanirina, F., Gianinazzi, S., Trouvelot, A. and Carre, M. 1989. Mycorrhizal inoculation of micropropagated grapevine rootstocks. *Agric. Ecosystems Environ.* **29**: 323-327.

65. Rejon-Palmores, A., Garcia-Garrido, J.M., Ocampo. J.A. and Garcia-Romera, I. 1996. Presence of xyloglucan-hydrolyzing glucanases (xyloglucanases) in arbuscular mycorrhizal symbiosis. *Symbiosis* **21**: 249-261.

66. Roy, A., Singh, A., Sudha, Sahay, N.S. and Varma, A. 1998. Concepts in myco- and photobionts interactions - a review. *The Botonica* (in press).

67. Sahai, A.S., Balasubramanian, R. and Manocha, M.S. 1993. Immunofluorescence study of zygomyceteous fungi with two chitin binding probes. *Expt. Mycol.* **17**: 55-69.

68. Sahay, N.S., Sudha, Singh, A. and Varma, A. 1998. Trends in endomycorrhizal research. *Indian Journal of Experimental Biology* **36**: 1069-1086.

69. Salamanca, C. P., Herrera, M.A. and Barea, J.M. 1992. Mycorrhizal inoculation of micropropagated woody legumes used in revegetation programmes for desertified mediterranean ecosystems. *Agronomie* **12**: 869-872.

70. Salzer, P., Hubner, B., Sirrenberg, A. and Hager, A. 1997. Differential effect of purified spruce chitinase and beta-1, 3-glucanases on the activity of elicitors from ectomycorrhizal fungi. *Plant Physiol.* **114**: 957-968.

71. Sanchez, L.M., Doke, N. and Kawakita, K. 1993. Elicitor-induced chemiluminescence in cell suspension cultures of tomato, sweet pepper and tobacco plants and its inhibition by suppression from *Phytophthora* spp. *Plant Sci.* **88**: 141-148.

72. Sanders, F.E. 1975. The effect of foliar-applied phosphate on the mycorrhizal infection of onion roots in endomycorrhizas, B.M.& P.B.T.F.E.Sanders, Academic Press, London, pp. 261-276.

73. Scellenbaum, L., Berta, G., Ravolanirina, F., Tisserant, B., Gianinazzi, S. and Fitter, A.H. 1991. Influence of endomycorrhizal infection on root morphology in micropropagated woody plant species *Vitis vinifera* L. *Ann. Bot.* **68**: 15-41.

74. Schubert, A., Mazzitelli, M. and Ariusso, O. 1990. Effect of vesicular-arbuscular mycorrhizal fungi on micropropagated grapevines: influence of endophyte strain, P fertilization and growth medium. *Vitis* **29**: 5-13.

75. Schubert, A., Bodrino, C. and Gribaudo, I. 1992. Vesicular-arbuscular mycorrhizal inoculation of kiwifruit (*Actinidia deliciosa*) micropropagated plants. *Agronomie* **12**: 847-850.

76. Singh, A. and Varma, A. 1998. Arbuscular mycorrhizae: the promising biocontrol potentiators. In: *Biocontrol Potential and their Exploitation in Crop Disease Management*, (eds. R.L. Razak and R.K. Upadhayay), (in press).

77. Singh, K. and Varma, A.K. 1988. Mycorrhizal fungi stimulate legume growth and root nodulation in dry arid soils. I. Effect of dual infection of *Rhizobium* and VAM endomycorrhizal spores on tropical legume-bengal gram (*Cicer arietinum*). In: *Proc. Mycorrhizal Workshop*, (eds. A.K. Varma, K.G. Mukerji, K.V.B.R. Tilak and J. Raj), IDRC, Canada, pp. 356-371.

78. Singh, K., Kumar, D., Subbarao, N.S. and Varma, A. 1988. Mycorrhizal fungi stimulate legume growth and root nodulation in dry arid soils. II. Effect of dual infection of *Rhizobium* and VA endomycorrhizal spores on soybean (*Glycine max* Merril). In: *Proc. Mycorrhizal Workshop*, (eds. A. Varma, A.K. Oka, K.G. Mukerji, K.V.B.R. Tilak and J. Raj), IDRC, Canada, pp. 372-392.

79. Smith, S.E. 1995. Discoveries, discussions and directions in mycorrhizal research. In: *Mycorrhiza*, (eds. A. Varma and B. Hock), Springer-Verlag, pp. 3-24.

80. Srivastava, P.S., Purohit, M., Pande, D. and Datta, A. 1993. Phenotypic variation and alkaloid content in the androgenic plantlets of *Datura innoxia*. *Phytomorphology* **43**: 209-216.

81. Straker, C.J., Schnippenkoetter, W.H. and Lemoine, M.C. 1992. Analysis of acid invertase and comparison with acid phosphatase in the ericoid mycorrhizal fungus *Hymenoscyphus ericae* (Read) Korf and Kernan. *Mycorrhiza* **2**: 63-67.

82. Sudha, Hurek, T. and Varma, A. 1998a. Active translocation of phosphate (P^{32}) to rice and carrot by the deuteromycete, *Piriformospora indica*. In: *Second International Conference on Mycorrhiza*, Sweden, July 5-9, pp 162.

83. Sudha, Hurek, T. and Varma, A. 1998b. Translocation of phosphorus in a cultivable root colonising fungus, *Piriformospora indica*. ISME - 8 Halifax, Canada, pp. 316.

84. Tawaraya, K., Saito, M., Morioka, M. and Wagatsuma, T. 1994. Effect of phosphate application to arbuscular mycorrhizal onion on the development and succinate dehydrogenase activity of internal hyphae. *Soil Sci. Plant Nutr.* **40**: 667-673.

85. Thomson, B.D., Robson, A.D. and Abbott, L.K. 1991. Soil mediated effect of phosphorus supply on the formation of mycorrhizas by *Scutellospora calospora* on subterranean clover. *New Phytol.* **118**: 463-469.

86. Varma, A. 1979. Vesicular-arbuscular mycorrhizae and nodulation in soybean. *Folia Microbiol.* **24**: 501-502.

87. Varma, A. 1995a. Arbuscular mycorrhizal fungi: state of the art. *Critical Reviews in Biotechnology* **15**: 179-200.

88. Varma, A. 1995b. Ecophysiology and application of arbuscular mycorrhizal fungi in arid soil. In: *Mycorrhizae: Structure, Function, Molecular Biology and Biotechnology*, (eds. A. Varma and B. Hock), Springer-Verlag, Heidelberg, Germany, pp. 561-591.

89. Varma, A. 1999a. Hydrolytic enzymes from arbuscular mycorrhizae: the current status. In: *Mycorrhiza*, (eds. A. Varma and B. Hock), second edition, Springer-Verlag, Germany, pp. 373-390.

90. Varma, A. 1999b. Function and application of arbuscular mycorrhizal fungi in arid and semi-arid soils. In: *Mycorrhiza*, (eds. A. Varma and B. Hock), second edition, Springer-Verlag, Germany, pp. 521-556.

91. Varma, A. and Schüepp, H. 1994a. Positive influence of arbuscular mycorrhizal fungus on *in vitro* raised hortensia plantlets. *Angew. Bot.* **68**: 108-115 .

92. Varma, A. and Schüepp, H. 1994b. Infectivity and effectiveness of *Glomus intraradices. Mycorrhiza* **5**: 29-37.

93. Varma, A. and Schüepp, H. 1995. Mycorrhization of the commercially important micropropagated plants. In: *Critical Reviews in Biotechnology*, (eds. G.G. Stewart and I. Russell), CRC Press, Canada, pp. 313-328.

94. Varma, A. and Schuepp, H. 1996. Influence of mycorrhization on the growth of micropropagated plants. In: *Concepts in Mycorrhizal Research*, (ed. K.G. Mukerji), Handbook of Vegetation Science Series, Kluwer Academic Publisher, The Netherlands, pp. 113-132.

95. Varma, A., Sudha, Sahay, N.S., Singh, A., Kumari, M., Bharti, K., Sarbhoy, A.K., Maier, W., Walter, M.H., Strack, D. and Franken, P. 1998a. *Piriformospora indica* - A plant stimulator and pathogen inhibitor, arbuscular mycorrhiza - like fungus. In: *Central Pollution Control Board Symposium on Pollution Abatement Through Biological Treatment of Industrial Effluents.* New Delhi (in press).

96. Varma, A., Singh, A., Sudha, Sahay, N.S., Sharma, J., Roy, A., Kumari, M., Rana, D., Thakran, S. and Bharti, K. 1998b. *Piriformospora indica*, gen. et sp. nov. In: *Mycota IX*, (eds. K. Esser and P.A. Lemke), Springer-Verlag, Germany (in press).

97. Varma, A., Sudha, Sahay, N., Butehorn, B. and Franken, P. 1998c. *Piriformospora indica*, a cultivable plant growth-promoting root endophyte. *Applied and Environmental Microbiology* (in press).

98. Verma, S., Varma, A., Rexer, K.H., Sarbhoy, A., Bisen, P., Butehorn, B. and Franken, P. 1998. *Piriformospora indica*, gen. nov. sp. nov., a new root-colonizing fungus. *Mycologia* **90**: 896-903.

99. Vestberg, M. and Estaun V. 1994. Micropropagated plants, an opportunity to positively manage mycorrhizal activities. In: *Impact of Arbuscular Mycorrhizas on Sustainable Agriculture and Natural Ecosystems*, (eds. S. Gianinazzi and H. Schuepp), Birkhauser, Verlag, Basel, Boston, Berlin, pp. 217-226.

100. Xie, Z.P., Staehelin, C., Vierheilig, H., Wiemken, A., Jabbouri, S., Broughton, W.J., Vogeli-Lange, R. and Boller, T. 1995. Rhizobial nodulation factors stimulate mycorrhizal colonization of nodulating and nonnodulating soybeans. *Plant Physiol.* **108**: 1519-1525.

29 Nematophagous Fungi with Special Reference to Interaction with Nematodes

Geeta Saxena and Neelima Mittal

CONTENTS

ABSTRACT ... 486

1. INTRODUCTION ... 486
2. MORPHOLOGICAL ADAPTATIONS ... 487
   - 2.1 Predators ... 487
   - 2.2 Endoparasites ... 490
   - 2.3 Parasites of Cyst and Root-knot Nematodes ... 490
3. NUTRITIONAL REQUIREMENTS ... 490
   - 3.1 Carbon Requirements ... 490
   - 3.2 Nitrogen Requirements ... 491
   - 3.3 Vitamin Requirements ... 492
4. INDUCTION OF TRAP FORMATION ... 493
5. NEMATODE-FUNGAL INTERACTION ... 495
   - 5.1 Recognition (Host specificity) ... 495
   - 5.2 Chemoattraction of Nematodes ... 496
   - 5.3 Adhesion of Nematodes ... 496
   - 5.4 Penetration and Colonization of the Nematode ... 498
6. SURVIVAL STRATEGIES ... 499
   - 6.1 Conidial Traps ... 499
   - 6.2 Chlamydospores ... 499
   - 6.3 Mycoparasitism ... 499
   - 6.4 Antibiotic Production ... 500
7. CONCLUSIONS ... 500
8. REFERENCES ... 501

Advances in Microbial Biotechnology
J.P. Tewari, T.N. Lakhanpal, Jagjit Singh, Rajni Gupta & B.P. Chamola (eds.)
APH Publishing Corporation, New Delhi - 110 002, India.

ABSTRACT

Nematophagous fungi are unique organisms to set an example of host/parasite relationship by utilizing nematodes as source of nutrients. They are of three types : predators capturing nematodes by various trapping devices, endoparasites infecting by adhesive or palatable conidia or zoospores and thirdly parasitize cysts and eggs of nematodes. After recognition, nematodes are attracted towards these fungi, then adhesion and subsequent penetration and colonization occurs. This involves a series of events, which are discussed in this paper. Apart from this their nutritional requirements, trap formation and survival strategies have been discussed.

1. INTRODUCTION

Nematophagous (nematode-destroying) fungi which capture and consume nematodes leading to their complete-destruction, are one of the most interesting group of organisms to be found in soil. The nematophagous fungi occur in various kinds of substrates depending upon the nature and necessities of the fungi. These habitats may be dung, Bryophytes or decaying leaf litter. Since dung and Bryophytes have large fauna of nematodes and other microscopic animals they serve as good substrates for the occurrence of these nematophagous fungi, few species have also been recorded from aquatic habitats. The species of nematophagous fungi are widely distributed in soils from all over the world.

Nematodes constitute one of largest and most ubiquitous group of the animal kingdom. They are commonly referred to as 'eelworms' or 'round-worms' and cause diseases of several economically important crop plants. Most nematodes live either free (saprophytes) in soil or as parasites in plants. As seen under the microscope, they look like tiny eels, moving slowly/ rapidly with vigorous thrashing movements of their body, rather like a heraldic flash of lightening. Their bodies are elongate, non-segmented, cylindrical, usually tapering at both ends.

Of all the known predatory or parasitic associations, the relationship between fungi and nematodes is one of the most fascinating as predatism in the world of fungi is tremendously astonishing phenomenon to be ever believed.

Plant-parasitic nematodes are limiting factor in agricultural production in many parts of the world. Their world wide distribution, extensive host range, association with fungi, bacteria and viruses in disease complex presents a very challenging problems damaging world's food supply. Although the use of modern nematicidal chemicals provides effective control of the parasites, their use is confined to industrialized nations, or to high value crops in other countries. In many developing countries, the use of nematicides is impractical because of unavailability or the relatively high cost of the chemicals. Root-knot nematode, *Meloidogyne* spp. is regarded as one of the obstacles in the agricultural productivity. The nematophagous fungi have been an interesting subject for biological control of plant-parasitic nematodes.

2. MORPHOLOGICAL ADAPTATIONS

The nematophagous fungi have been placed into three categories, predatory, endoparasitic and parasites of cyst and root-knot nematodes depending upon their mode of attack. The predatory fungi have an extensive hyphal development on the substratum and capture nematodes by producing trapping organs. Hyphae are usually sparse and radiate out from the original point of primary capture with each hypha producing trapping devices at intervals along its length. Victims are captured either by adhesive or non-adhesive devices. Adhesive organs of capture have been categorized as hyphae, branches, knobs and nets (Fig. 1-5).

In case of endoparasites, there is no extensive hyphal development outside the body of the host and only evacuation tubes or conidiophores and conidia are produced externally. In the environment they exist principally in the form of spores. Some spores are persistent, while others are disseminative and following dispersal function as infection agents (Fig. 6-7)

2.1 Predators

2.1.1 Adhesive trapping devices

2.1.1.1 Adhesive hyphae

Adhesive hyphae are characteristic of lower fungi like Zygomycetes. Nematodes are captured by means of an adhesive produced directly on the hyphae. There is no region of a hypha that can be assigned secretory in function or in appearance. Hyphae are either coated with adhesive along their entire length or produce adhesive at any point in response to nematodes. A common example of such type of trapping device is *Stylopage hadra*. The adhesive hyphae are relatively stout and a thick yellowish adhesive material is formed at the point of capture. The captured nematode after a prolonged struggle becomes exhausted and is penetrated by the fungus. Following penetration, elongate unbranched absorptive hyphae grow in both directions along the length of the body. After complete exploitation of the contents, protoplasm is migrated back and a cross wall is formed. Thus, eventually the absorption hyphae become completely empty of contents.

2.1.1.2 Adhesive branches

Adhesive branches are characteristic of Deuteromycetes. They are the most primitive and morphologically the simplest form of capture organs produced in higher fungi. They are mostly a few cells in height. They arise as short laterals that grow erect from prostrate hyphae. A thin film of adhesive is secreted over the entire surface of the branch. The branches are arranged in a fairly close array such that a struggling nematode becomes attached at several points along its length. There are few nematophagous fungi such as *Monacrosporium cionopagum* and *M. gephyrophagum* in which the organs of capture are adhesive branches and even in these there is a tendency to form simple two dimensional nets. Adhesive branches are also reported in *Arthrobotrys ellipsospora* (111). At times nematodes break away from the adhesive contact by the violent thrashing action of their struggles and thus escape. This is partly influenced by environmental conditions.

2.1.1.3 Adhesive knobs

Adhesive knobs are morphologically distinct cells. They are found in Deuteromycetes and Basidiomycetes. They are either sessile or produced at the apex of a slender non-adhesive stalk composed of one to three cells. The knob is separated from the support stalk by a septum

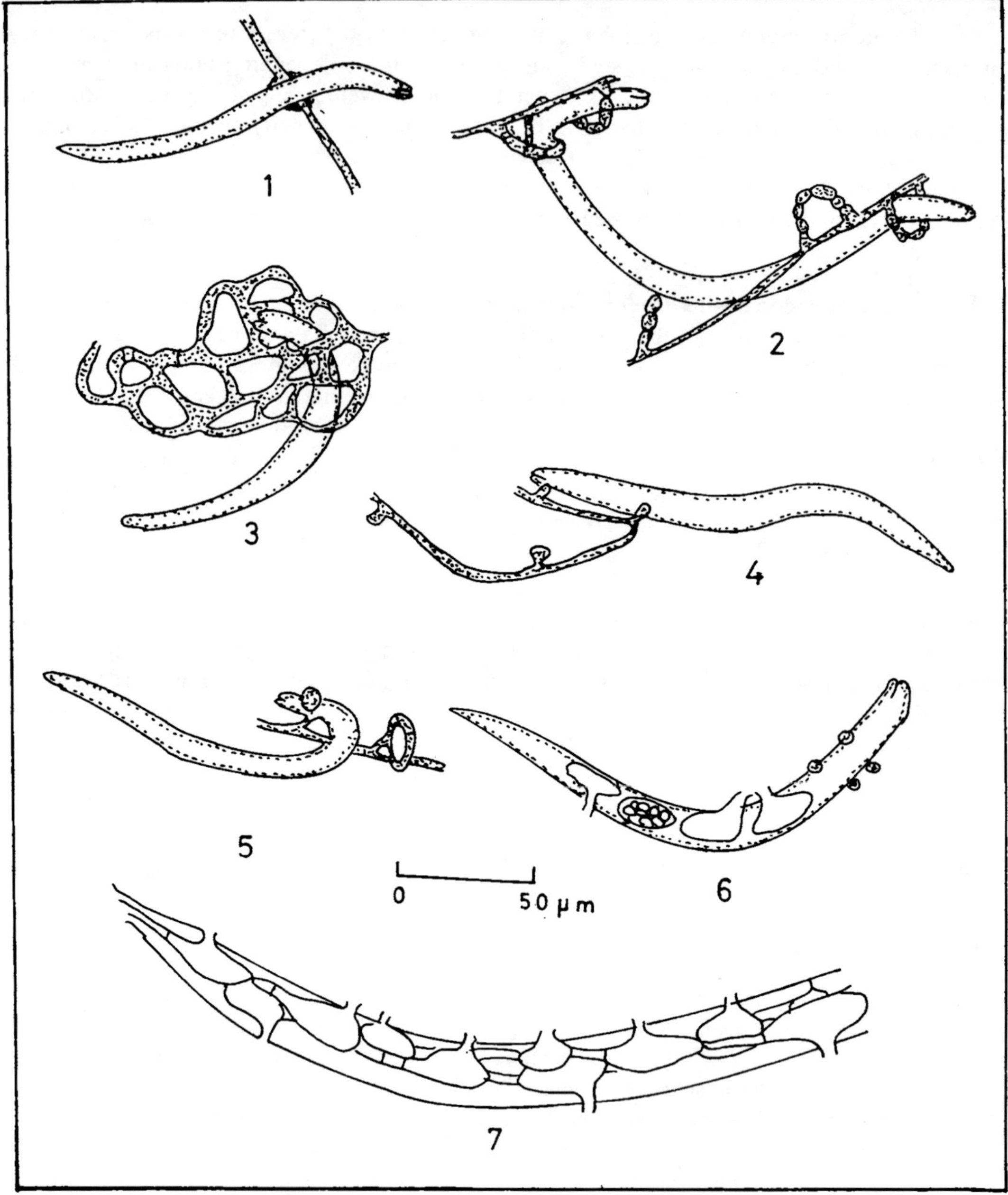

Fig.1-5 : Predatory fungi showing different trapping devices.

Fig. 6 - 7 : Endoparasitic fungi.

1. Adhesive hypha of *Stylophage leiohypha* which has secreted adhesive to capture the nematode.
2. Adhesive branches of *Monacrosporium gephyrophagum* which have united to form scalariform traps to capture nematodes.
3. Nematode captured in adhesive network of *Arthrobotrys oligospora.*
4. Adhesive knobs of *Monacrosporium parvicolle* capturing a nematode.
5. A nematode captured in constricting ring of *Arthrobotrys brochopaga.*
6. *Myzocytium papillatum* infected a nematode. Full as well as empty zoosporangia are seen.
7. A nematode full of zoosporangia of *Catenaria anguillulae.* Empty sporangia with evacuation tubes are seen after the release of zoospores.

at its point of origin. Nematodes adhere to a knob and eventually are caught by several knobs with subsequent penetration and destruction. Violent struggle for the escape usually leads to breaking off the knobs from the point of attachment to the stalk. The occurrence of detachable knobs is a distinct advantage to the parasite. The parasite is not only transported to a new site for further predation, but arrives with an immediately available food source and can quickly produce additional trapping devices.

2.1.1.4 Adhesive nets

Adhesive nets are the most commonly occurring trapping device (24). Net forming fungi are very aggressive in nature and ubiquitous in occurrence. They vary from simple loops as in *Arthrobotrys musiformis* to complex three-dimensional networks as in *Arthrobotrys oligospora*. Through anastomoses of lateral outgrowth with the parent hypha, a complex three dimensional network of hyphae is built up. Scanning electron microscopic studies by Nordbring-Hertz (66) have shown the presence of an adhesive substance, coating the surface of capture organs. The adhesive is highly effective and the prey is held fast. Its struggle results in contact with another net system and consequently the nematode is attached at several points. The violence of its efforts to escape, quickly exhausts the victim and it becomes moribund.

2.1.2 Non-adhesive trapping devices

The non-adhesive organs of capture are either constricting or non-constricting rings.

2.1.2.1 Non-constricting rings

Members of Deuteromycetes produce non-constricting rings on lateral branches arising from the prostrate hyphae. They are three-celled rings supported on a slender stalk. The juncture of the ring and its stalk is particularly weak and signs of collapse are often seen in mature rings. They are passive in their action. However, the nematode enters the ring at speed and its forward motion causes the ring to wedge around the body sufficiently tight to cause a marked constriction of the cuticle. During the struggle, the ring often breaks off and the host swims off with the ring around its body. Detached rings are viable, the prey is penetrated and contents consumed in the usual manner. Production of non constricting rings is often associated with adhesive knobs, as in *Dactylaria candida* and *D. lysipaga*.

2.1.2.2 Constricting rings

Constricting rings are most efficient mechanism for trapping nematodes. They are produced in the same manner as the non-constricting rings but the support stalk is usually shorter and much stouter. They capture nematodes by the garroting action of the ring cells, which by swelling grasp the nematode in a strangle hold from which there is no chance of escape. The ring cells are sensitive all along their inner edges. When a nematode enters a ring, the friction of its body stimulates the ring cells to swell rapidly inward. This almost obliterates the opening of the ring constricting the body of the nematode, events then follow their usual course. Several workers have investigated the mechanism of ring action. Couch (26) showed that the mechanism could be triggered by heat either dry or contact with water between 33^{0}C to 75°C. Muller (64) brought about the closure by means of hot scalpel in the vicinity of the ring. Commandon and deFonbrune (17,18) tried the rings to close by rubbing them with a micromanipulation needle. During swelling of the individual ring cell, a number of small vacuoles increase in size and finally condense to form a single large vacuole. A three fold increase of osmotically active substances takes place. Muller (64) has suggested that the stimulation of the inner wall of the ring cell induces an instantaneous decrease in wall pressure and an increase in membrane permeability. The resultant increase in the suction potential draws water from the surrounding medium. As this uptake proceeds, the level of osmotically active substances in the cell rises due to hydrolysis of large molecules present, which maintains the inflow of water. Couch (26) suggested that the additional water necessary was imbibed through the stalk or hyphal cells, a conclusion supported by Rudek (94). Zachariah (121) caused inflation of ring traps detached from axenic cultures of *Dactylella doedycoides* with chlorobutanol vapour on cornmeal agar. Isolated uninflated rings produced a single, slowly growing hypha from the stalk scar, it later produced branches. Isolated inflated rings produced several rapidly growing hyphae from the inner curvature of their three cells, mimicking the response elicited by trapping prey. Even after several hours of hyphal proliferation, isolated uninflated rings could be caused to inflate and inflation increased the growth rate of primary hypha.

2.2 Endoparasites

In the soil, the endozoic parasites of nematodes exist mainly as conidia. In the higher fungi conidia are passive and are either ingested, lodging in the digestive tract where they subsequently germinate or adhesive, becoming attached to the cuticle of the nematode and penetrating its skin. Some species of lower fungi are more aggressive as parasites and produce flagellated spores which are attached to hosts and encyst directly on the cuticle.

In case of Chytridiomycetes and Oomycetes flagellate spores are produced which either directly adhere or produce adhesive buds to get attached to the cuticle. In some cases, evacuation tubes are produced from the sporangia for the release of zoospores. In case of Deuteromycetes, conidia are either ingested by the nematode or they infect by adhering to the cuticle. The attached spore germinates, penetrates directly through the cuticle and forms an infection hyphae in the body cavity. The ingested spores may lodge in the muscle tissue of the oesophagous as in case of *Harposporium anguillulae* or in the buccal cavity as in the case of *H. bysmatosporum*. Whatever the case may be, hyphae grow and proliferate at the expense of the body contents and the infected animal gradually becomes lethargic and moribund. Within a few days the body of the nematode is filled with the hyphae of the parasite. External development is restricted to the development of conidiophores and conidia of the asexual reproductive stage.

There is one exceptional genus *Nematoctonus*, a member of Basidiomycetes, species of which have been reported both as predatory as well as endoparasitic. In predatory species, adhesive knobs are produced on external hyphae to capture nematodes. In endoparasitic species, nematodes are attacked by means of adhesive spores attaching to their cuticle.

2.3 Parasites of Cyst and Root-knot Nematodes

The third group is of parasites of cyst and root-knot nematodes which attack eggs or females of these nematodes by in-growth of vegetative hyphae. During parasitization, an appresorium like swelling may be involved. The cyst-nematode parasites infect mainly larvae, eggs or females of cyst nematodes. Fungi in this group have attracted much attention during recent years because of their use as biocontrol agents.

3. NUTRITIONAL REQUIREMENTS

3.1 Carbon Requirements

The nematode-destroying fungi are of two types : the predatory and the endozoic. An extraordinary feature of the first group is that despite their specialization, they do not have unusual nutritional requirements (34). All of them grow well in the absence of prey on very simple media. The endozoic types are more demanding, they produce little or no mycelium beyond the confines of the nematode suggesting a requirement for complex nutrients. An understanding of the general physiology and nutrition of nematode-trapping fungi has been investigated by few workers because of their peculiar habit of capturing, killing and subsisting on nematodes.

Although the predaceous habit or nematophagous fungi has a roused wide interest, surprisingly little has yet been learnt concerning their physiology. The available information does not provide at present reliable data on the metabolic specificity of all nematophagous fungi. Little information is available on the composition of the media suitable for them. Hyphomycetes are characterized by intensive utilization of carbohydrates, which supply the energy requirements. However, at the same time, growth of these fungi on synthetic carbohydrate media is rather scanty, which is probably due to the absence of certain growth factors or inadequate composition of carbohydrates. Soprunov (106) did preliminary investigations on the effect of vitamins on the fermentation of sugars. Cascarelli and Pramer (25) studied the nutritional requirement of *Arthrobotrys conoides*. *Arthrobotrys oligospora, Monacrosporium doedycoides* and *Dactylaria candida* grow reasonably well in a maltose nitrate salt medium, though growth was less on a medium prepared from natural source (11). Blackburn and Hayes (12) studied the effect of utilization of various carbon sources on the growth of *A. oligospora* and *A. robusta* in their saprophytic phase. Hayes and Blackburn (45) examined requirements of above mentioned fungi for different source of carbon. Olthof and Estey (81) found evidence that media low in dextrose tended to suppress the ability of the fungus to catch nematodes. Later on, Satchuthananthavale and Cooke (101) studied the effect of various carbon sources on four ring forming and three network forming fungi. Bricklebank and Cooke (15) studied the requirement of *Nematoctonus* spp. for various carbohydrate sources. Zachariah and Victor (122) studied the nutritional requirements of an auxotrophic strain of *Arthrobotrys dactyloides* and showed that in common with many other fungi, it required glucose.

3.2 Nitrogen Requirements

Blackburn and Hayes (12) studied the growth of *Arthrobotrys oligospora* and *A. robusta* on the medium supplemented with different sources of organic and inorganic nitrogen in their saprophytic phase. Organic sources of nitrogen included asparagine, urea and 20 amino acids. They made interesting observations that the two species differed in their ability to use varying nitrogen sources.

Hayes and Blackburn (45) examined their requirements for various nitrogen sources when feeding upon nematodes. The fungi were grown in sterile sand to which measured quantity of a minimal mineral medium was added. Nitrogen sources were added and inoculated with spores of fungi. *Panagrellus redivivus* was used to test the predaceous activity of fungi by comparing the nematode population on the culture medium without fungi. The percentage of non-survivors from the culture inoculated with the fungi compared with the control population was termed by them "predacity number." Nitrogen as ammonium considerably reduced activity compared with nitrate and nitrite. Predacity was reduced by a mixture of amino acids in the substrate, but organic nitrogen as urea and asparagine had little effect. In contrast to this, predacity of *A. robusta* was independent of chemical form of nitrogen except sodium nitrite which induced a marked reduction in nematode population. Olthof and Estey (81) investigated the effects of nitrogen levels in the medium on the predaceous activity of *A. oligospora*. They found that medium low in ammonium nitrate reduced the fungal ability of catching nematodes.

Satchuthananthavale and Cooke (100) studied the requirements of 7 nematode-trapping Hyphomycetes for various nitrogen sources. Four fungi with constricting rings and three with adhesive network as trapping organs were studied by them. A striking difference was found between net-forming and ring-forming species with respect to their ability to grow on inorganic nitrogen sources. The 3 network-forming species were able to use nitrate, nitrite, ammonium ion and organic nitrogen whereas the 4 ring-forming species could not use nitrite and showed reduced growth on nitrate. Glycine was poor nitrogen source for them. These findings correspond with that of Cooke (19,20,21) who divided these fungi into two ecological groups. The first group of species with adhesive network as good competitive saprophytes and species of the second group with constricting rings, adhesive knobs and branches as poor competitive saprophytes.

Zachariah and Victor (122) studied the nutritional requirements of a strain of *Arthrobotrys dactyloides*. It had an absolute requirement for histidine and methionine with a partial requirement for several other amino acids. Leucine was inhibitory for growth when supplemented in a mixture of amino acids. These fungi have certain specific metabolic features in particular, the property of breakdown and assimilation of animal protein. The protein breakdown in these fungi is less intensive than in saprophytic species. General observations from radioisotopic techniques show, if nematophagous fungi are given a choice between a compound dissolved in the medium and the same incorporated in nematode tissues, the compound from the medium is preferred.

3.3 Vitamin Requirements

Present knowledge of vitamin requirements of fungi indicates that they generally need only water-soluble vitamins of B-complex series including thiamine (B_1), riboflavin (B_2), pyridoxine (B_6), niacin (nicotinic acid), pantothenic acid, biotin, folic acid group, inositol, p-aminobenzoic acid and cyanocobalmin (B_{12}). None of the fat-soluble vitamins like A, D, E and K have so far been found to be synthesized by fungi and it appears that they do not require these growth factors. However, a number of growth factor requirements of fungi are still poor understood and therefore it is difficult to make a general conclusion.

Scanty observations are made about various growth factors required by nematophagous fungi. Different species of the same genus or the various isolates of the same species sometimes exhibit different physiological behaviour in general and nematophagous fungi in particular. Soprunov (106) did preliminary investigation on the effect of vitamins on the intensity of sporulation and fermentation of sugars. Despite the ability to capture, kill and subsist on nematodes, *Arthrobotrys conoides* does not show a unique nutritional pattern. Yeast extract, required for growth in a glucose-inorganic salts medium, was replaced by biotin, thiamine and zinc. Biotin biosynthesis by the fungus is blocked at the desthiobiotin to biotin conversion. *A. conoides* is unable to synthesize either of the two moieties of thiamine but, when supplied with pyrimidine and thiazole, completes biosynthesis of the vitamin via coupling reaction (25,41). Faust and Pramer (36) studied the nutritional requirements of *Monacrosporium ellipsosporum*. Its need were similar to that of *A. conoides*. Blackburn and Hayes (12) studied the utilization of 11 sources of vitamins in the saprophytic phase of two fungi *Arthrobotrys oligospora* and *A. robusta*. Both responded to thiamine, biotin and p-aminobenzoic acid, but neither species showed absolute dependence on an external source of vitamin. Hayes and Blackburn (45) studied the vitamin requirements of the above mentioned two species of *Arthrobotrys* in their predaceous phase, i.e., in the presence of nematodes. Sequential addition of biotin, thiamine and p-aminobenzoic acid to the minimal medium resulted in a marked growth response of both isolates. The same vitamins also markedly affected the predacity of *A. robusta*, but had little effect on this character in *A. oligospora*. The addition of thiamine and p-aminobenzoic acid suppressed the predaceous activity of *A. robusta* to the extent that nematode populations increased to levels higher than those in control cultures. The addition of other vitamins did not markedly affect this suppression. Satchuthananthavale and Cooke (99) studied the vitamin requirements of 7 nematode-trapping fungi. All of them and except one were deficient for thiamine and biotin respectively. Bricklebank and Cooke (15) studied carbohydrate and vitamin requirements of *Nematoctonus* spp. Zachariah and Victor (122) studied the vitamin requirement of an auxotrophic strain of *Arthrobotrys dactyloides*. It had an absolute requirement for one or more of B vitamins, in common with many other fungi.

Saxena *et al.* (98) studied utilization of 18 carbon and nitrogen sources and 9 vitamins by three network forming fungi *Arthrobotrys oligospora, A. conoides* and *Monascrosporium cystosporum*. They were able to utilize almost all the carbon and nitrogen sources except few and showed variation in their efficiency of utilization. They found that all the three fungi were found to utilize a wide range of carbon and nitrogen sources. It is evident from their result that amino acids substantially differ in their nutritive value for these fungi. These fungi showed requirement for vitamins like thiamine and biotin for their good growth and a partial requirement for 4-amino-benzoic acid.

All these work on nutritional requirement of nematode trapping fungi show that network forming species may be categorized as a group that is capable of saprophytic existence in soils being able to utilize a wide range of carbon and nitrogen compound when subject to nutritional stress, they form traps and colonize nematodes. Cooke (19,20) concluded that network forming fungi have a greater saprophytic ability than those with other types of trapping devices. Thus these network forming fungi are at an advantage as they may not have to depend so much on their predatory habit for survival in soil. Some of the network forming fungi such as *Arthrobotys oligospora* apparently remain strong saprotrophic competitors and potent cellulase producers but still catch nematodes as a nitrogen supplement when necessary and appropriate. But in the natural habitat nitrogen might be obtained in relatively small amounts i.e. total captures relative to the total hyphal system will be small. This will result in an infection of N that can be rapidly and efficiently absorbed. In petri-dish conditions, there will be large number of captures and a rapid injection of massive nitrogen supplements into the system which will result in a dramatic shift in the C/N ratio. There is a possibility that under these conditions the predatory mechanism is temporarily switched off until a more appropriate nutrient balance is achieved. This type of situation could be experienced in nature when large populations of nematodes develop rapidly as in manure or soils amended with carbohydrate (8).

Cooke (23) drew attention to our lack of understanding of the nematophagous habit. He suggested three approaches to solve the ecological problems of this group. First, establish the significance of the trapping habit by studying cultures of the fungus in association with suitable nematodes. Second, directly observe their behaviour under varied experimental conditions. Third, study them in culture with a view to define those characteristics which might have ecological importance.

4. INDUCTION OF TRAP FORMATION

To survive and compete in the demanding environment of the soil, nematophagous fungi have developed attributes, which give them a competitive advantage. By choosing nematodes as a food source, they have assured themselves of continuous and abundant source of available energy. For this, they have developed specialized capture organs and produce chemical substances to attract nematodes to the site of capture of predatory species and to the site of spore production in endoparasitic species. Biology of capture organ formation and attraction of nematodes towards these fungi are of main consideration here.

Predatory nematophagous fungi form capture organs in response to nematodes in their environment. These trapping devices are either adhesive or mechanical in function. Adhesive network type is of wide occurrence. Nematode-trapping fungi usually behave as saprophytes in pure culture and probably also in nature. The change from saprophytic to predaceous habit is characterized by the development of capture organs of different morphological appearance. The process of trap induction in nematode-trapping fungi has been a phenomenon that has interested many investigators for sometime. It has been known for many years that the fungi would form traps in the presence of nematodes or spontaneously. However, the circumstances under which this spontaneous capture organ formation occurs, is little understood. Spontaneous capture organ formation has been initiated by additions of animal sera and extracts (57,58,93).

Studies on morphogenesis induced by the presence of nematodes started with the discovery by Couch (26). Commandon and de Fonbrune (17) showed that sterile culture filtrates of nematodes were able to induce ring formation. It could be concluded from their work that nematodes secreted compounds which were responsible for the formation of trapping devices. Subsequent, studies on morphogenesis of nematophagous fungi revealed that there are a wide range of materials of animal origin such as human blood serum, extracts of earthworms, which effectively induced the formation of traps. Soprunov (106) tried a number of materials such as snow melted at room temperature, rain water preferably collected at the onset of storm, 1-2% solution of ethyl alcohol, few drops of N/10 NaOH (stored for 2-3 months) added to distilled water to a pH of 7.5 - 7.8 and many others. He suggested that ammonium compounds and carbon dioxide in rain water combined to form ammonium carbonate which induced the trap formation. Pramer and Stoll (87) and Pramer and Kuyama (86) suggested that a metabolic product of the nematode was responsible for the trap formation. They proposed the term 'nemin' for the substance or substances produced by the nematode which induce trap formation in the fungi. Pramer and Kuyama (86) suggested that nemin was a peptide of relatively low molecular weight or possibly even a single amino acid of rather common occurrence. Wootton and Pramer (119) reported the isolation of morphogenic agents from yeast extract, which was claimed to be valine, leucine and isoleucine. The uptake, transport and metabolism of valine in *A. conoides* have been studied extensively (42,43).

A limited supply of nutrients has been shown to govern the shift from saprophytic to predaceous phase, especially those where nematodes were present as an inducer of morphogenesis (35,65,120). The development of mycelium and predacity are not correlated i.e., a poor development of hyphae does not prevent high predaceous activity (45). Balan and Lechevalier (3) suggested that abundant trap formation in *A. dactyloides* was induced by adverse conditions of culture resulting in lack of nutrients and/or water. Physical factors such as water supply and adverse cultural conditions certainly play a role in this type of morphogenesis (26,58,59) but some sort of additional inducer seems to be necessary. For predaceous activity in nature, organic matter in a particular phase of decomposition is necessary (22) and readily available carbon sources should not have been completely exhausted (81). Proteins and amino acids are mentioned as the main sources of energy. Amino acids (especially valine) and peptides especially valyl peptides such as phenylalanyl-valine play a role in trap induction (50,65,67,68,89,119). In addition to having a peptide present, a low nutrient medium was also required for trap induction.

Feder *et al.* (37) proved that a single freshly killed nematode (*Panagrellus redivivus*) stimulated trap formation over a 1 cm diameter column containing *Dactylella doedycoides*. This indicated that a morphogenetic substance was present in the body and released in very small amounts. Feder *et al.* (38) suggested that the bulk of the nemin was in the nematode body as endogenous nemin (EN). They studied the effect of EN derived from 100,000 nematodes on several species of *Dactylella* and found that the fungi differed markedly in their responses. Monoson *et al.* (63) showed that endogenous nemin extracted from five different species of nematodes induced trap formation, but there were quantitative differences in the ability of nematodes of trap induction. They suggested the action of nemin at transcription level. The RNA-synthesis inhibited 6-methyl purine suppressed nemin, induced trap formation. In the presence of nemin, Bartnicki-Garcia *et al.* (10) showed that morphogenesis for two strains of *A. conoides* was CO_2 dependent. No traps were produced in its absence. These differences in response to CO_2 could be influential in governing strain distribution in the natural environment. Schenck and Pramer (103) studied, the effect of volatile compounds produced by the nematodes on trap formation by *Monacrosporium rutgeriensis*. They assumed that nematodes produce an active volatile when in direct contact with agar or a fungal metabolite.

Nordbring-Hertz and Odham (76) studied the effect of CO_2, NH_3 and two volatile organic compounds on the trap formation by *A. oligospora*. In their experiment, a strong inhibitory effect of high concentrations of CO_2 (5-10%) on the morphogenesis of this fungus in the presence of a trap-inducing peptide was found, which indicates that CO_2 directly interferes with trap formation. Ammonia (NH^{+3}, NH^{+4}) seemed to have a stimulatory effect on trap formation. Rosenzweig and Pramer (92) studied the effect of heavy metals Cd, Zn and Pb on the growth and the ability of trap formation by 5 species of *Arthrobotrys* and 2 species of *Monacrosporium*. Lysek and Nordbring-Hertz (59) studied the dependency of trap formation in *A. oligospora* on the endogenous rhythm after the addition of living nematodes to the colonies. The periodicity was independent of light-dark (LD) cycles of 24 h (10:14). Temperature influenced the hyphal elongation. Cayrol *et al.* (16) studied the trap formation in *A. oviformis* in the presence of nematodes associated with bacteria.

Saxena *et al.* (97) studied induction of trap formation in *Arthrobotrys oligospora*, *A. conoides* and *Monacrosporium cystosporum* in the presence of *Panagrellus redivivus* on various nutrient media. They also found that low nutrient mineral salt medium had the most pronounced effect on trap formation. Friman *et al.* (39) grew *A. oligospora* in liquid cultures supplemented with amino acids valine and phenylalanine to induce trap formation. Inclusion of phosphate in the medium inhibited trap formation completely (113). Formation of traps in the presence of the chitin-synthetase inhibitor polyoxin D (84) or after a shift of vegetative mycelium to a liquid medium containing D-alanine as the sole carbon source. Dijksterhius *et al.* (30) suggests that adverse environmental conditions may be an important factor in triggering trap induction. Thus we find that trap formation is dependent on a range of parameters like presence of peptides, living nematodes, adverse environmental conditions, volatile compounds, nutrient limitation, endogenous rhythm, etc.

5. NEMATODE-FUNGAL INTERACTION

5.1 Recognition (Host specificity)

There is no simple host specificity in any of the nematophagous species that would indicate a recognition on the molecular level between a fungus and its prey. In laboratory studies the predatory fungi do not seem to discriminate among different species of nematodes, as far as simply attachment is concerned. Although several laboratory studies point to a wide host range for most predatory and endoparasitic fungi, on the whole the endoparasites seem to possess a somewhat greater host specificity than the predatory fungi. This is especially true for those fungi that attack cyst-nematode eggs and females. The reason for this difference is not known. Recognition in nematode-nematophagous fungus system makes it clear that initial events would include both chemoattraction of nematodes to the vicinity of the predatory and adhesion of the nematodes to trapping or infective structures of these fungi.

5.2 Chemoattraction of Nematodes

Nematophagous fungi produce chemical substances that lure nematodes to the site of trap formation in predatory fungi or to the site of spores in endoparasitic species. It has been shown that following the formation of trapping organs, some predaceous fungi attract nematodes. Barron (5) in an experiment with *Harposporium helicoides*, found that *Rhabditis* nematodes readily consumed spores of the parasite. The presence of a chemical attractant on a spore surface makes it more palatable than others and greatly enhances the efficiency of endoparasites with ingested spores. If an egg is laid in the vicinity of *Rhopalomyces elegans*, hyphal branches quickly develop and grow more or less directly towards it. It is presumed that branches grown are in response to exudates from the egg involving one or more chemicals. Sayre and Keelay (102) noted that adult nematodes are readily parasitized by *Catenaria anguillulae* as compared to larval stages, may be due to the fact that large orifices allow substantial leakage of chemical attractants. Monoson and Ranieri (62) claimed the existence of a substance attractive to the nematode *Aphelenchus avenae* associated with the formation of traps in *Arthrobotrys musiformis*. Balan *et al.* (4) showed that *A. oligospora, A. conoides* and *M. rutgeriensis* produce a nematode attracting substance distinct from nematicidal compounds.

Jansson (46) studied the attraction of *P. redivivus* towards the 3 endoparasitic fungi. The predacity of few nematophagous fungi and attraction of *P. redivivus* towards them was also studied. There was a good correlation between attraction and predacity supporting the fact that the ability of the fungi to attract nematodes reflects the dependence of the fungi on nematodes for nutrients (74). *A. conoides* and *Monacrosporium cystosporum* were found to attract *P. redivivus* (97). We find that presence of infective structures and morphological adaptations influence the attractiveness of the fungi. The consistent pattern of strong attractiveness, in the more predaceous forms especially, should have some significance in the natural environment (70).

5.3 Adhesion of Nematodes

5.3.1 Lectin carbohydrate interaction

5.3.1.1 Predatory fungi

A lectin carbohydrate interaction has been suggested to operate in initial stages of the capture of nematodes by *Arthrobotrys oligospora* (75) and other fungi. These studies have been reviewed in detail (32,53,69,70,71,73,74,123).

This firmness of the attachment of the nematode to the adhering structure, despite the vigorous movements of the nematodes results from a series of events beginning with an interaction between complementary molecular configurations on the nematode and fungal surfaces. Adhesion starts with a physical contact between the living nematode and the trap. This contact is initiated by a lectin-carbohydrate binding between complementary molecules on the trap and nematode surfaces. Lectins are carbohydrate binding proteins or glycoproteins of non-immune origin that agglutinate cells and/or precipitate glycoconjugates (40). As with plant lectins, microbial lectins have received an increasing interest in recent years (61).

The first example of a lectin-carbohydrate interaction was detected in *A. oligospora*. The study was facilitated by the development of a suitable assay system using a dialysis membrane technique (75,78). Trap containing mycelium was flooded with hapten sugar solutions and adhesion was examined after adding the nematode *Panagrellus redivivus*. Capture of nematodes was inhibited mainly by N acetyl-D galactosamine (Gal NAC) out of 20 mono and disaccharides tested. This indicated the presence of a carbohydrate binding protein, a lectin on the traps of *A. oligospora* and N-acetylgalactosamine on the nematode surface. In a similar study with another strain of *A. oligospora*, the presence of a Gal NAC specific protein was confirmed (88). Gal NAC-specific and Ca^{2+} dependent lectin was isolated from trap bearing cultures of *A. oligospora* by affinity chromatography (13, 14).

Efforts have been made to detect trap lectins in other fungi with different types of traps. Rosenzweig and Ackroyd (90,91) found that glucose/mannose inhibited capture of the nematode by *A. conoides* and fucose prevented capture by *Monacrosporium eudermatum* and 2-deoxyglucose inhibited capture by *M. rutgeriensis*. Nordbring-Hertz *et al.* (77) found that 2-deoxyglucose inhibited capture by *Dactylaria candida*. It seems that different fungi carry lactins with different carbohydrate binding specificities. This has led to investigations into surface structures of both fungi and nematodes. Further, investigations were carried out to isolate the possible receptor of this lectin on the nematode surface (14) and to localize the lectin on the trap. These results showed that the lectin was located in cell walls of the traps, not in adhesive outside the cell wall.

5.3.1.2 Endoparasitic fungi

Adhesive conidia of endoparasites strongly attract nematodes, whereas non-adhesive conidia do not (46). Adhesive conidia of *Drechmeria coniospora* attach specifically to the cephalic region of nematodes with an adhesive bud which stick and subsequently penetrate the nematode cuticle. Some zoosporic fungi like *Catenaria anguillulae* and *Myzocytium* spp. infect nematodes by chemotaxis, adhesion and encystment on the nematode surface. Deacon and Saxena (29) studied in detail attraction and encystment of zoospores of *C. anguillulae* near the moth, excretory pore and anus of nematodes by video microscopy. They also found that *C. anguillulae* show consistent-orientation (Polarity) of zoospore encystment and cyst germination.

The involvement of a lectin-mediated adhesion has also been investigated in the endoparasite *D. coniospora*. Adhesion was partly inhibited by flooding the conidia with sialic acid (N-acetylneuraminic acid) or by treating *P. redivivus* with the enzyme sialidase (51). Adhesion of conidia to nematodes also decreased by treating the nematodes with limulin, a sialic acid specific lectin or by destroying surface proteins of the conidia with trypsin or glutaraldehyde (52). Later in 1993 Jansson (48) contradicted his previous hypothesis and said that there is no involvement of carbohydrates in the adhesion of *D. coniospora* and the nematode *P. redivivus*. His results suggested that the adhesion is mediated by protein(s) in the adhesive part of the conidium of *D. coniospora* binding to protein(s) excreted from the sensory organs of the nematodes. Jansson (49) studied adhesion of *D. coniospora* conidia to the cuticles of the wild type and mutants indicating that proteinaceous material emanating from the sensary. structures was rapidly replaced. Several studies have indicated that adhesion starts with a physical contact between the nematode and trap. This contact leads to several events including lectin-carbohydrate interaction, a reorganization of surface polymers and secretion of specific enzymes. Adhesion between fungus and nematode is a complex process involving protein and carbohydrate containing polymers on the fungi (113,115).

5.3.2 Enzymatic activity

Modification of surface polymers during adhesion might be alternatively, due to the activity of specific enzymes. These enzymes might be present in the extracellular polymer layer before the contact with the nematode or be secreted during adhesion. Acid phosphate activity (117) and peroxidase activity in the adhesive layer of *A. oligospora* has been detected. Persson and Friman (82) found intracellular proteolytic activity in mycelia of *A. oligospora* bearing nematode-trapping structures. Tunlid *et al.* (114) isolated extracellular serine protease from traps of *A. oligospora* growing in liquid cultures. Out of two fractions of protease FI and FII, FII immobilized *P. redivivus* indicating that the enzyme(s) might be involved in the infection of nematodes.

The penetration of nematode cuticles and eggshells may be either mechanical or enzymatic or both. Fungal collagenase has been suggested to work on collagen, a component of the nematode cuticle. They have been detected in cultures of several nematode trapping fungi (92,104). Egg shell consists of three layers made up of protein, chitin and lipid. Chitinase activity seems to be important for successful infection (60). Segers *et al.* (105) found the *Verticillium chlamydosporium* secreted several proteases in submerged culture, out of which one protease VCP1 was purified. It hydrolysed proteins in still from the outer layer of the egg shell of the host nematode *Meloidogyne incognita* and exposed its chitin layer. These observations suggest that proteases and other enzymes may play an important role in invasion of nematodes by these fungi.

5.3.3 Reorganization of surface polymers

Nematodes generally contain proteins and small amounts of carbohydrate and lipid in their thick cuticle. The surface coat has a strong negative charge and contains a number of sugar residues (47). Extracellular polymers supposed to be involved in the adhesion have been isolated from *Arthrobotrys oligospora*. In *Drechmeria coniospora*, extracellular polymer involved in adhesion appears to be composed of radiating fibrils present on the adhesive knobs of the conidia even before the attachment (31,95). In *Catenaria anguillulae* zoospores attach to the nematode cuticle with polymers. After encystment, the spores are surrounded by a fibrillar layer of extracellular polymer (115).

The structure of extracellular surface polymers, present in the intervening space between *A. oligospora* and *P. redivivus*, changes during adhesion (113). The mechanisms of reorganization of these polymers is not known, but these structural changes certainly lead to strong binding between the cells. Various factors like addition of Ca^{2+}, inclusion of other carbohydrates and proteins modify side chains and substitution of polymers.

5.4 Penetration and Colonization of the Nematode

After cuticle penetration by enzymatic activity, a large infection bulb is formed. Later trophic infection hyphae develop from this bulb. Dense bodies play crucial role in supplying building stones to facilitate the formation of trophic hyphae. In case of endoparasites, an appresorium is a prerequisite to facilitate effective infection (31). The trophic hyphae develop at the expense of the body contents of the host. The mechanisms involved in nematode colonization and digestion are similar among nematophagous fungi. Initial stages of infection are solely dedicated towards formation of invading hyphae. The main function of trophic hyphae is probably to digest nematode contents and convert its carbon-containing compounds into storage polymers and supply the vegetative mycelium and conidiophores with appropriate nutrients.

6. SURVIVAL STRATEGIES

The nematophagous fungi are at an additional advantage as far as their survival is concerned. A shift from a saprophytic mode of life to a predaceous one is an aid in their survival. The traps have high contents of dense bodies that may be considered as storage organelle to supply carbon and energy or nitrogen required for hyphal development (117, 118).

6.1 Conidial Traps

Conidial traps in nematode trapping fungi are adhesive or mechanical structures formed directly on germination of conidia, without an intermediate hyphal phase. Conidial traps have been reported in the constricting ring forming fungi *Dactylaria brochopaga* and *Arthrobotrys dactyloides* (6,7,33). Conidial traps of the hour glass type were reported on three species of *Nematoctonus*, *N. concurrens*, *N. leiasporus* and *N. robustus* (85). In *Arthrobotrys oligospora*, conidial traps were detected in response to diffusing substances from cow faeces (27). Saxena and Mittal (96) observed several species of *Monacrosporium* to form traps like adhesive branches, knobs and nets on their conidia. Hay (44) found that spore germination of net producing nematophagous fungi *A. conoides* and *A. cladodes* can lead to either immediate formation of the trap or produce single hypha upon which nets are eventually produced. Nordbring-Hertz *et al.* (79) isolated a mutant of *A. oligospora* which produced strict conidial traps in abundance on conidia. The properties of these fungi of forming traps on their conidia would increase their value as antagonists to nematodes. Trap formation on conidia seems to be an additional advantage to prey upon nematodes. These trap adhere to the surface of the passing nematode, and thus facilitate the dissemination of the fungus.

6.2 Chlamydospores

Typical fungal survival structures are found in nematode-trapping (e.g. *Dinddingtonia flagrans*), endoparasitic (e.g. *Harposporium anguillulae*) and parasites of eggs and cysts (e.g. *Cylindrocarpon destructans*). The thick-walled oospores of the female cyst nematode parasite *Nematophthora gynophila* can survive more than 5 years in soil (54). The importance of chlamydospores of nematophagous fungi for survival in the field has not been well understood.

6.3 Mycoparasitism

In addition to their ability to capture nematodes by special trapping devices, mycoparasitism has been encountered in this group of fungi. Several of the network-forming predatory fungi including *Arthrobotrys oligospora* were found mycoparasitic on several fungi including *Rhizoctonia solani* (83,116). Saxena *et al.* (98) found *A. superba* to behave as mycoparasite on *Rhizoctonia solani*. Persson and Friman (82) suggested that proteases produced by *A. oligospora* may be involved in the killing of *R. solani* either by hydrolysis of its cell wall proteins or by activation of cell degrading enzymes. It has been proposed that mycoparasites activate endogenous lytic enzymes of the host, resulting in localized host cell lysis (56). Several intracellular proteases specific to the mycoparasitic interaction between *A. oligospora* and *R. solani* have been detected.

6.4 Antibiotic Production

The involvement of toxins in the killing of the nematodes was postulated earlier but was not proved (4,80). The use of a toxin in the wood-decaying basidiomycetes *Pleurotus* has been detected. It was found that species of *Pleurotus* had developed a unique method of inactivating and eventually colonizing their nematode victims (9). On a water agar base, the hyphae of *Pleurotus* produced tiny droplets of toxin from minute spathulate secretory cells. Nematodes touching such droplets became more or less immobilized within a few minutes. Stimulated by leakage products from the immobilized nematode, hyphae in the vicinity produced directional hyphae which converged on the body orifices of the victims. On reaching the host these hyphae penetrated through the orifices, colonized and digested the victim. The *Pleurotus* method of capturing and consuming nematodes is more common than currently known. This behaviour is interpreted as a strategy of these fungi to supplement the low level of nitrogen available in wood by capturing nematodes in this way (110).

From cultures of *Pleurotus ostreatus*, a wood-rotting Basidiomycete, trans-2 decene-dioic acid was isolated as a weakly nematicidal compound, which in high concentration (300 mM/ml) immobilized *Panagrellus redivivus* within an hr. (55). The nematicidal compounds formed by *Pleurotus pulmonarius* are fatty acids namely linoleic acid and S-coriolic acid (109). Both acids exhibited high activity towards *Caenorhabditis elegans*.

Stadler *et al.* (107) carried out an investigation on the production of nematicidal metabolites in submerged cultures of predaceous fungi: Linoleic acid was shown to be the only detectable nematicidal agent in mycelial extracts of several predaceous fungi of the genus *Arthrobotrys*. In submerged cultures, the number of traps formed by *A. conoides* and *A. oligospora* was directly correlated to the increase of the concentration of linoleic acid. The immotility of the nematodes was accompanied by the apparent destruction of the intestine and the lack of response to light. Stadler *et al.* (108) isolated three antimicrobial metabolites Oligosporon, Oligosporol A, Oligosporol B from the culture filtrates of *Arthrobotrys oligospora*.

In parallel to the German work, Anderson *et al.* (1) from Australia independently isolated oligosporon and oligosporol B from an Australian isolate of *A. oligospora* and established their structures. They established the structures of three of related antibiotics namely 4' 5' dehydro-oligosporon, hydroxyoligosporon and 10', 11'-epoxyoligosporon. Anke *et al.* (2) screened 12 nematode-trapping fungi for antimicrobial and nematicidal activities and found three new antimicrobial metabolites from cultures of five *Arthrobotrys* strains. They found that *Nematoctonus robustus* and *Nematoctonus concurrens* produced pleurotin, dihydro-pleyrotinic acid and leucopleurotin metabolites previously isolated from cultures of *Hohenbuehelia* species suggesting that the same biosynthetic pathways function in both telomorph and anamorph.

7. CONCLUSIONS

Our understanding of interaction mechanisms between nematode-nematophagous systems has made significant advance during the past decade. Initial stages of interaction has been understood, but still we have insufficient information regarding biochemical and molecular mechanisms . Attempts should be made to exploit methods of biotechnology in order to understand various events occurring during nematophagous fungi and nematode interaction. For example, detailed study of the cuticle, and receptors of the nematode would be helpful to understand the interaction in a better way.

8. REFERENCES

1. Anderson, M.G., Jarman, T.B. and Rickards, R.W. 1995. Structures and absolute configurations of antibiotics of the oligosporon group from the nematode trapping fungus *Arthrobotrys oligospora. J. Antibiotics* **48(5)**: 391-398.

2. Anke, H., Stadler, M., Mayer, A. and Sterner, O. 1995. Secondary metabolites with nematicidal and antimicrobial activity from nematophagous fungi and ascomycetes. *Can. J. Bot.* **73(1)**: 932-939.

3. Balan, J. and Lechevalier, H.A. 1972. The predaceous fungus *Arthrobotrys dactyloides* : Induction of trap formation. *Mycologia* **64**: 919-922.

4. Balan, J., Krizkova, L., Nemec, P. and Vollek, V. 1974. Production of nematode attracting and nematicidal substances by predaceous fungi. *Folia Microbiol.* **19**: 512-519.

5. Barron, G.L. 1970. Nematophagous hyphomycetes : observations on *Harposporium helicoides. Can. J. Bot.* **48**: 329-331.

6. Barron, G.L. 1977. *The Nematode Destroying Fungi.* Canadian Biological Publications Guelph. Ontario. pp.140.

7. Barron, G.L. 1981. Predators and parasites of microscopic animals. In : *Biology of Conidial Fungi.* Vol.2. (eds. G. Cole and B. Kendrick), Academic Press, New York pp.167-200.

8. Barron, G.L. 1992. Lignolytic and cellulytic fungi as predators and parasites. In : *The Fungal Community, Its Organization and Role in the Ecosystem,* (eds. G.C. Carrol and D.T. Wicklow) Marcel Dekker, New York pp.311-326.

9. Barron, G.L. and Thorn, R.G. 1987. Destruction of nematodes by species of *Pleurotus. Can. J. Bot.* **65**: 774-778.

10. Bartnicki-Garcia, S., Eren, J. and Pramer, D. 1964. Carbon-dioxide dependent morphogenesis in *Arthrobotrys conoides. Nature,* **204**:804.

11. Blackburn, F. and Hayes, W.A. 1963. A chemically defined medium for the cultivation of nematophagous Hyphomycetes. *Trans. Br. Mycol. Soc.* **46**: 449-452.

12. Blackburn, F. and Hayes, W.A. 1966. Studies on the nutrition of *Arthrobotrys oligospora.* Fres. and *A. robusta* Dudd. I. The saprophytic phase. *Ann. Appl. Biol.* **58**: 43-50.

13. Borrebaeck, C.A.K., Mattiasson, B. and Nordbring-Hertz, B. 1984. Isolation and partial characterization of a carbohydrate-binding protein from a nematode-trapping fungus. *J. Bacteriol.* **159**: 53-56.

14. Borrebaeck, C.A.K., Mattiasson, B. and Nordbring-Hertz, B. 1985. A fungal lectin and its apparent receptors on a nematode surface. *FEMS Microbiol. Letters* **27**: 35-39.

15. Bricklebank, J. and Cooke, R.C. 1969. Utilization of polysaccharides by two nematode parasitic fungi. *Trans. Br. Mycol. Soc.* **52**: 347-349.

16. Cayrol, J.C., Combettes S. and Quiles, C. 1984. Influence of the association between nematodes and bacteria upon trap formation in the hyphomycete *Arthrobotrys oviformis. Cryptogamie Mycologie* **5**: 21.

17. Commandon, J. and de Fonbrune, P. 1938. Recherches experimentales sur les champignons predateurs de nematodes du sol. *C.R. Soc. Biol.*, Paris **129**: 619-625.

18. Commandon, J. and de Fonbrune, P. 1939. de la formation et du fonctionnement des pieges des champignons predateurs des nematodes. Recherches effectuees a l'aide de la micromanipulation et de la cinematographie. *C.R. Acad. Sci.* Paris, **207** : 304-305.

19. Cooke, R.C. 1962a. Ecological characteristics of nematode-trapping Hyphomycetes. I. Preliminary studies. *Ann. Appl. Biol.* **52**: 431-437.

20. Cooke, R.C. 1962b. The behaviour of nematode-trapping fungi during the decomposition of organic matter in the soil. *Trans. Br. Mycol. Soc.* **45**: 314-319.

21. Cooke, R.C. 1964. Ecological characteristics of nematode-trapping Hyphomycetes. II. Germination of conidia in soil. *Ann. Appl. Biol.* **54**: 375-379.

22. Cooke, R.C. 1968. Relationship between nematode-destroying fungi and soil-borne phytonematodes. *Phytopath.* **58**: 909-913.

23. Cooke, R.C. 1977. *The Biology of Symbiotic Fungi.* John Wiley and Sons, London.

24. Cooke, R.C. and Godfrey, B.E.S. 1964. A key to the nematode destroying fungi. *Trans. Br. Mycol. Soc.* **47**: 61-74.

25. Coscarelli, W. and Pramer, D. 1962. Nutrition and growth of *Arthrobotrys conoides. J. Bacteriol.* **84**: 60-64.

26. Couch, J.N. 1937. The formation and operation of the traps in the nematode catching fungus *Dactylella bembicodes* Drechsler. *J. Elisha Mitchell Sci. Soc.* **53**: 301-309.

27. Dackman, C. and Nordbring-Hertz, B. 1992. Conidial traps-a new survival structure of the nematode-trapping fungus *Arthrobotrys oligospora. Mycol. Res.* **96**: 194-198.

28. Dackman, C., Jansson, H.B. and Nordbring-Hertz, B. 1992. Nematophagous fungi and their activities in soil. In : *Soil Biochemistry* Vol.7 (eds. G. Stotzky and J.M. Bollag) Marcel Dekker, Inc. New York, pp.95-130.

29. Deacon, J.W. and Saxena, G. 1997. Orientated zoospore attachment and cyst germination in *Catenaria anguillulae*, a facultative endoparasite of nematodes. *Mycol. Res.* **101(5)**: 513-522.

30. Dijksterhuis, J., Harder, W. and Veenhuis, M. 1993. Proliferation and formation of microbodies in the nematophagous fungus *Arthrobotrys oligospora* during growth on oleic acid or D-Alanine as the sole carbon source. *FEMS Microb. Letters* **112(2)**: 125-130.

31. Dijksterhuis, J., Veenhuis, M. and Harder, W. 1990. Ultrastructural study of adhesion and initial stages of infection of nematodes by conidia of *DrechMeria coniospora. Mycol. Res.* **94**: 1-8.

32. Dijksterhuis, J., Veenhuis, M., Harder, W. and Nordbring-Hertz, B. 1994. Nematophagous fungi: Physiological aspects and structure-function relationship. In : *Advances in Microbial Physiology* Vol. 36. (eds. A.H. Rose and D.W. Tempest) Academic Press, London 111-143.

33. Dowe, A. 1987. Räuberische Pilze. Die Neue Brehm-Bücherei, *A.* Ziemsen Verlag, Wittenberg Lutherstadt, Germany.

34. Duddington, C.L. and Wyborn, C.H.E. 1972. Recent research on the nematophagous hyphomycetes. *Bot. Rev.* **38**: 545-565.

35. Eren, J. and Pramer, D. 1965. The most probable number of nematode-trapping fungi in soil. *Soil Sci.* **99**: 285.

36. Faust, M.A. and Pramer, D. 1964. Nutrition and growth of *Dactylella ellipsospora. Life Sc.* **3**: 141-143.

37. Feder, W.A., Everard, C.O.R. and Duddington, C.L. 1960. Heterocaryotic nature of ring formation in the predaceous fungus *Dactylella doedycoides. Science* **131**: 922-924.

38. Feder, W.A., Everad, C.O.R. and Wootton, L.M.O. 1963. Sensitivity of several species of the nematophagous fungus *Dactylella* to a morphogenic substance derived from free living nematodes. *Nematologica* **9**: 49-54.

39. Friman, E., Olsson, S. and Nordbring-Hertz, B. 1985. Heavy trap formation by *Arthrobotrys oligospora* in liquid culture. *FEMS Microb. Ecol.* **31**: 17-21.

40. Goldstein, I.J., Hughes, R.C., Monsigny, M., Osawa, T. and Sharon, N. 1980. What should be called a lectin? *Nature* **285**: 66.

41. Grant, C.L., Coscarelli, W. and Pramer, D. 1962. Statistical measurement of biotin, thiamine and zinc concentration required for maximal growth of *Arthrobotrys conoides. Appl. Microbiol.* **10**: 413-417.

42. Gupta, R.K. and Pramer, D. 1970a. Amino acid transport by the filamentous fungus *Arthrobotrys conoides. J. Bacteriol.* **103**: 120-130.

43. Gupta, R.K. and Pramer, D. 1970b. Metabolism of value by the filamentous fungus *Arthrobotrys conoides. J. Bacteriol.* **103**: 131-139.

44. Hay, F.S. 1995. Unusual germination of spores of *Arthrobotrys conoides* and *A. cladodes. Mycol. Res.* **99**: 981-981.

45. Hayes, W.A. and Blackburn, F. 1966. Studies on the nutrition of *Arthrobotrys oligospora* Fres. and *A. robusta* Dudd. II. The predaceous phase. *Ann. Appl. Biol.* **58**: 51-60.

46. Jansson, H.B. 1982. Attraction of nematodes to endoparasitic nematophagous fungi. *Trans. Br. Mycol. Soc.* **79**: 25-29.

47. Jansson, H.B. 1988. Receptors and recognition in nematodes. In : *Vistas on Nematology* (eds. J.A. Veech and D.W. Dickson). Society of Nematologists, Hyattsville, MD, USA, pp.153-158.

48. Jansson, H.B. 1993. Adhesion to nematodes of conidia from the nematophagous fungus *Drechmeria coniospora. J. Gen. Microbiol.* **139**: 1899-1906.

49. Jansson, H.B. 1994. Adhesion of conidia of *Drechmeria coniospora* to *Caenorhabditis elegans* wild type and mutants. *J. Nematol.* **26**: 430-435.

50. Jansson, H.B. and Nordbring-Hertz, B. 1981. Trap and conidiophore formation in *Arthrobotrys superba. Trans. Br. Mycol. Soc.* **77**: 205-207.

51. Jansson, H.B. and Nordbring-Hertz, B. 1983. The endoparasitic nematophagous fungus *Meria coniospora* infects nematodes specifically at the chemosensory organs. *J. Gen. Microbiol.* **129**: 1121-1126.

52. Jansson, H.B. and Nordbring-Hertz, B. 1984. Involvement of sialic acid in nematode chemotaxis and infection by an endoparasitic nematophagous fungus. *J. Gen. Microbiol.* **130**: 39-43.

53. Jansson, H.B. and Nordbring-Hertz, B. 1988. Infection events in the fungus-nematode system. In : *Diseases of Nematodes* Vol.II (eds. G.O. Poinar, Jr. and H.B. Jansson), CRC Press, Boca Raton, Florida pp.59-72.

54. Kerry, B.R. 1984. Nematophagous fungi and the regulation of nematode population in soil. *Helminth. Abstr. Ser.* B. **53**: 1-14.

55. Kwok, O.C.H., Plattner, R., Weisleder, D. and Wicklow, D.T. 1992. A nematicidal toxin from *Pleurotus ostreatus* NRRL 3526. *J. Chem. Ecol.*, **18**: 129-136.

56. Laing, S.A.K. and Deacon, J.W. 1991. Video microscopical comparison of mycoparasitism by *Pythium oligandrum*. Mycol. *Res.* **95**: 469-479.

57. Lamy, L. 1943. Intensite et Vitesse relatives de la formation des dispositifs capteurs chez les Hyphomycetes predateurs de Nématodes. *C.R. Soc. Biol.* (Paris) **137**: 337-339.

58. Lawton, J.R. 1957. The formation of constricting rings in nematode-catching Hyphomycetes grown in pure culture. *J. Exp. Bot.* **8**: 50-54.

59. Lawton, J.R. 1967. The formation and closure of constricting rings in two nematode catching Hyphomycetes. *Trans. Br. Mycol. Soc.* **50**: 195-205.

60. Lysek, G. and Nordbring-Hertz, B. 1981. An endogenous rhythm of trap formation in the nematophagous fungus *Arthrobotrys oligospora. Planta* **152**: 50-53.

61. Lysek, H. and Krajci, D. 1987. Penetration of ovicidal fungus *Verticillium chlamydosporium* through the *Ascaris lumbricoides* egg shells. *Folia Parasitol* **34**: 57-60.

62. Mirelman, D. 1986. *Microbial Lectins and Agglutinins : Properties and Biological Activity*. John Wiley and Sons, New York.

63. Monoson, H.L. and Ranieri, G.M. 1972. Nematode attraction by an extract of a predaceous fungus. *Mycologia* **64**: 628-631.

64. Monoson, H.L., Galsky, A.G. and Stephano, R.S. 1974. Studies on the ability of various nematodes to induce trap formation in a nematode-trapping fungus *Monacrosporium doedycoides. Nematologica* **20**: 96-102.

65. Muller, H.G. 1958. The constricting ring mechanisms of two predaceous Hyphomycetes. *Trans. Br. Mycol. Soc.* **41**: 341-364.

66. Nordbring-Hertz, B. 1968. The influence of medium composition and additions of animal origin on the formation of capture organs in *Arthrobotrys oligospora. Physiol. Plant.* **21**: 52-65.

67. Nordbring-Hertz, B. 1972. Scanning electron microscopy of the nematode trapping organs in *Arthrobotrys oligospora. Physiol. Plant.* **26**: 279-284.

68. Nordbring-Hertz, B. 1973. Peptide-induced morphogenesis in the nematode trapping fungus *Arthrobotrys oligospora. Physiol. Plant.* **29**: 223-233.

69. Nordbring-Hertz, B. 1977. Nematode-induced morphogenesis in the predaceous fungus *Arthrobotrys oligospora. Nematologica* **23**: 443-451.

70. Nordbring-Hertz, B. 1984. Mycelial development and lectin-carbohydrate interactions in nematode-trapping fungi. In : *The Ecology and Physiology of the Fungal Mycelium*, (eds. D.H. Jennings and A.D.M. Rayner) Cambridge University Press, Cambridge, pp.419-432.

71. Nordbring-Hertz, B. 1988. Ecology and recognition in the nematode-nematophagous fungus system. In : *Advances in Microbial Ecology*, (ed. K.C. Marshall) Plenum Publishing Corporation, New York, pp.81-114.

72. Nordbring-Hertz, B. 1988. Nematophagous fungi : Strategies for nematode exploitation and for survival. *Microbiol. Sci.* **5**: 108-116.

73. Nordbring-Hertz, B. and Chet, I. 1986. Fungal lectins and agglutinins. In : *Microbial Lectins and Agglutinins : Properties and Biological Activity* (ed. D. Mirelman) John Wiley, & Sons New York, pp.393-408.

74. Nordbring-Hertz, B. and Jansson, H.B. 1984. Fungal development, predacity and recognition of prey in nematode destroying fungi. In : *Current Perspectives in Microbial Ecology* (eds. D.H. Jennings and A.D.M. Rayner) Cambridge University Press, Cambridge, pp.419-432.

75. Nordbring-Hertz, B. and Mattiasson, B. 1979. Action of a nematode-trapping fungus shows lectin-mediated host-microorganism interaction. *Nature* **281**: 477-479.

76. Nordbring-Hertz, B. and Odham, G. 1980. Determination of volatile nematode exudates and their effects on a nematode-trapping fungus. *Microb. Ecol.* **6**: 241-251.

77. Nordbring-Hertz, B., Friman, E. and Mattiasson, B. 1982. A recognition mechanism in the adhesion of nematodes to nematode-trapping fungi. In : *Lectins, Biology, Biochemistry and Clinical Biochemistry* Vol.II (ed. T.C. Bog-Hansen) Gruyter, Berlin, pp. 83-90.

78. Nordbring-Hertz, B., Veenhuis, M. and Harder, W. 1984. Dialysis membrane technique for ultrastructural studies of microbial interactions. *Appl. Environ. Microbiol.* **47**: 195-197.

79. Nordbring-Hertz, B., Neumeister, H., Sjollema, K. and Veenhuis, M. 1995. A conidial trap-forming mutant of *Arthrobotrys oligospora. Mycol. Res.* **99(11)**: 1395-1398.

80. Olthof, H.A. and Estey, R.H. 1963. A nematoxin produced the nematophagous fungus *Arthrobotrys oligospora* Fresenius. *Nature* (London), **197**: 514-515.

81. Olthof, Th. H.A. and Estey, R.H. 1966. Carbon and nitrogen levels of a medium in relation to growth and nematophagous activity of *Arthrobotrys oligospora* Fresenius. *Nature* **209**: 1158.

82. Persson, Y. and Friman, E. 1993. Intracellular proteolytic activity in mycelia of *Arthrobotrys oligospora* bearing mycoparasitic or nematode-trapping structures. *Exp. Mycol.* **17(3)**: 182-190.

83. Persson, Y., Veenhuis, M. and Nordbring-Hertz, B. 1985. Morphogenesis and significance of hyphal coiling by nematode-trapping fungi in mycoparasitic relationships. *FEMS Microb. Ecol.* **31**: 283-291.

84. Persson, Y., Nordbring-Hertz, B. and Chet, I. 1990. The effect of polyoxin D on morphogenesis of the nematode-trapping fungus *Arthrobotrys oligospora. Mycol. Res.* **94**: 196-200.

85. Poloczek, E. and Webster, J. 1994. Conidial traps in *Nematoctonus* (nematophagous basidiomyces) *Nova Hedwigia* **59**: 201-205.

86. Pramer, D. and Kuyama, S. 1963. Symposium on biochemical bases of morphogenesis in fungi. II. Nemin and nematode-trapping fungi. *Bact. Rev.* **27**: 282-292.

87. Pramer, D. and Stoll, N.R. 1959. Nemin : a morphogenic substance causing trap formation by predaceous fungi. *Science* **129**: 966-967.

88. Premchandran, D. and Pramer, D. 1984. Role of N-acetylgalactosamine-specific protein in trapping of nematodes by *Arthrobotrys oligospora. Appl. Environ. Microbiol.* **47**: 1358-1359.

89. Rosenzweig, W.D. 1984. Role of amino acids, peptides and medium composition in trap formation by nematode-trapping fungi. *Can. J. Microbiol.* **30**: 265-267.

90. Rosenzweig, W.D. and Ackroyd, D. 1983. Binding characteristics of lectins involved in the trapping of nematodes by fungi. *Appl. Environ. Microbiol* **46**: 1093-1096.

91. Rosenzweig, W.D. and Ackroyd, D. 1984. Influence of soil microorganisms in the trapping of nematodes by nematophagous fungi. *Can. J. Microbiol.* **30**: 1437-1439.

92. Rosenzweig, W.D. and Pramer, D. 1980. Influence of cadmium, zinc and lead on growth trap formation and collagenase activity. *Appl. Environ. Microbiol.* **40**: 694-696.

93. Roubaud, M.E. and Deschiens, R. 1939. Capture des larves infectieuses de nematodes pathogenes par des champignons predateurs du sol. *C.R. Acad. Sci.*, Paris **208**: 245-247.

94. Rudek, W.T. 1975. The constriction of the trapping rings in *Dactylaria brochopaga. Mycopathologia* **55**: 193-197.

95. Saikawa, M. 1982. An electron microscope study of *Meria coniospora*, an endozoic nematophagous fungus. *Can. J. Bot.* **60**: 2019-2023.

96. Saxena, G. and Mittal, N. 1995. Trap formation by conidia of nematode-trapping *Monacrosporium* spp. *Mycol. Res.* **99(7)**: 839-840.

97. Saxena, G., Dayal, R. and Mukerji, K.G. 1987. Interaction of nematodes with nematophagous fungi : induction of trap formation, attraction and detection of attractants. *FEMS Microb. Ecol.* **45**: 319-327.

98. Saxena, G., Dayal, R. and Mukerji, K.G. 1989. Nutritional studies on nematode-trapping fungi. *Folia Microbiologica* **34(1)**: 42-48.

99. Satchuthananthavale, V. and Cooke, R.C. 1967a. Vitamin requirements of some of the nematode-trapping fungi. *Trans. Br. Mycol. Soc.* **50**: 221-228.

100. Satchuthananthavale, V. and Cooke, R.C. 1967b. Nitrogen nutrition of some nematode-trapping fungi. *Trans. Br. Mycol. Soc.* **50**: 423-428.

101. Satchuthananthavale, V. and Cooke, R.C. 1967c. Carbohydrate nutrition of some nematode trapping fungi. *Nature* **214**: 321-322.

102. Sayre, R.M. and Keelay, L.S. 1969. Factors influencing *Catenaria anguillulae* infections in a free living and a plant-parasitic nematode. *Nematologica* **15**: 492-502.

103. Schenck, S. and Pramer, D. 1975. The effects of volatile compounds from nematodes on trap formation by a nematode-trapping fungus. *App. Microbiol* **30**: 496-497.

104. Schenck, S., Chase, T. Jr., Rosenzweig, W.D. and Pramer, D. 1980. Collagenase production by nematode-trapping fungi. *Appl. Environ. Microbiol.* **40**: 567-570.

105. Segers, R. Butt, T.M., Keen, J.N., Kerry, B.R. and Peberdy, J.F. 1995. The subtilisins of the invertebrate mycopathogens *Verticillium chlamydosporium* and *Metarhizium anisopliae* are serologically and functionally related. *FEMS Microbiol. Letters.* **126(3)**: 227-231.

106. Soprunov, F.F. 1958. Predaceous hyphomycetes and their application in the control of pathogenic nematodes. Acad. Sci. Turkmen, SSR. Ashabed. (English Translation). 1966. Israel Program for Scientific Translations.

107. Stadler, M., Anke, H. and Sterner, O. 1993a. Linoleic acid - the nematicidal principle of several nematophagous fungi and its production in trap-forming submerged cultures. *Arch. Microbiol.* **160**: 401-405.

108. Stadler, M., Sterner, O. and Anke, H. 1993b. New biologically active compounds from nematode-trapping fungus *Arthrobotrys oligospora*. Fresn. *Z. Naturforsch.* **48**: 843-850.

109. Stadler, M., Mayer, A., Anke, H. and Sterner, O. 1994. Fatty acids and other compounds with nematicidal activity from cultures of Basidiomycetes. *Planta Med.* **60**: 128-132.

110. Thorn, R.G. and Barrow, G.L. 1984. Carnivorous mushrooms. *Science* **224**: 76-78.

111. Tubaki, K. and Yamanaka, K. 1984. An undescribed nematode-trapping species of *Arthrobotrys. Trans. Mycol. Soc. Japan* **25**: 349-353.

112. Tunlid, A., Jansson, H.B. and Nordbring-Hertz, B. 1992. Fungal attachment to nematodes. *Mycol. Res.* **96(6)**: 401-412.

113. Tunlid, A., Johnsson, T. and Nordbring-Hertz, B. 1991a. Surface polymers of the nematode-trapping fungus *Arthrobotrys oligospora. J. Gen. Microbiol.* **137**: 1231-1240.

114. Tunlid, A., Rosen, S.E.K.B. and Rask, L. 1994. Purification and characterization of an extracellular serine-protease from the nematode-trapping fungus *Arthrobotrys oligospora. Microbiology*-UK **140**: 1687-1695.

115. Tunlid, A., Nivens, D.E., Jansson, H.B. and White, D.C. 1991. Infrared monitoring of the adhesion of *Catenaria anguillulae* zoospores to solid surfaces. *Exp. Mycol.* **15**: 206-214.

116. Tzean, S.S. and Estey, R.H. 1978. Nematode-trapping fungi as mycopathogens. *Phytopathology* **68**: 1266-1270.

117. Veenhuis, M., Nordbring-Hertz, B. and Harder, W. 1985. An electron-microscopical analysis of capture and initial stages of penetration of nematodes by *Arthrobotrys oligospora. Antonie Van Leewenhoek* **51**: 385-398.

118. Veenhuis, M., Van Wijk, C. Wyss, U., Nordbring-Hertz, B. and Harder, W. 1989. Significance of electron dense microbodies in trap cells of the nematophagous fungus *Arthrobotrys oligospora. Antonie van Leeuwenhoek* **56**: 251-261.

119. Wootton, L.M.O. and Pramer, D. 1966. Valine-induced morphogensis in *Arthrobotrys conoides. Bacteriol. Proc.* p.75.

120. Wyborn, C.H.E., Priest, D. and Duddington, C.L. 1969. Selective techniques for the determination of nematophagous fungi in soils. *Soil. Biol. Biochem.* **1**: 101-102.

121. Zachariah, K. 1982. Growth responses of isolated ring traps of *Dactylella doedycoides. Can. J. Bot.* **60**: 580-585.

122. Zachariah, K. and Victor, J.R. 1983. A natural auxotroph of a nematode-trapping fungus. *Can. J. Bot.* **61**: 3255-3261.

123. Zuckerman, B.M. and Jansson, H.B. 1984. Nematode chemotaxis and possible mechanisms of host/prey recognition. *Ann. Rev. Phytopathol.* **22**: 95-113.

30 Head Smut of *Paspalum scrobiculatum* L. : Transmission and Management

A.K. Jain and M.N. Khare

CONTENTS

ABSTRACT 510

1. INTRODUCTION 510
2. OCCURRENCE 510
3. SYMPTOMS 510
4. PATHOGEN 511
5. NUTRITIONAL REQUIREMENT 512
6. DISEASE CYCLE 512
7. INOCULATION TECHNIQUE 512
8. HOST RANGE 512
9. MANAGEMENT OF THE DISEASE 513
   - 9.1 Resistance 513
   - 9.2 Cultural Practices 513
   - 9.3 Chemical Control 513
10. REFERENCES 514

Advances in Microbial Biotechnology
J.P. Tewari, T.N. Lakhanpal, Jagjit Singh, Rajni Gupta & B.P. Chamola (eds.)
APH Publishing Corporation, New Delhi - 110 002, India.

ABSTRACT

Paspalum scrobiculatum L. is attacked by *Sorosporium paspali thunbergii* McAlp. causing head smut resulting into heavy yield losses, as all the grains formed on an infected plant are smutted. The Pathogen is seed and soil borne and can be controlled by using resistant variety or seed treatment with specific fungicides. Screening method has been evolved for varietal screening.

1. INTRODUCTION

Paspalum scrobiculatum L. (common names Kodra, Kodo, Varagu) is an important component of dryland agriculture cultivated in many parts of the world in soils with poor fertility. It is consumed as food in Asia, Africa and some parts or Europe. In Australia and America it is cultivated for fodder. The species is widely distributed in damp habitates across the tropics and subtropics of the world. In India it is grown in Rajasthan, Madhya Pradesh, Orissa, Bihar, Maharastra, Tamil Nadu and Andhra Pradesh (5, 23). *P. scrobiculatum* has a long storage life (15), wider adaptability and assured harvest under extreme soil and climatic conditions. The crop suffers due to many diseases like rust (*Puccinia substriata* Ellis & Barth), sugary disease or ergot (*Claviceps paspali* Stev. & Hall), head smut (*Sorosporium paspali thunbergii* McAlp.) and foliar leaf spot and blight due to various fungi. Besides, the crop is also attacked by bacteria and viruses.

Among these diseases head smut is very important as it is responsible for heavy losses in grain yield (2). Viswanath (31) reported 30 to 40 percent loss due to the disease.

2. OCCURRENCE

Head smut was first reported from Queensland, Australia and was named *Sorosporium paspali* (17). The smut recorded in East Asia was first reported to be due to *Ustilago paspali thunbergii* Henn. Ito (7) renamed it as *Sorosporium paspali thunbergii* (P. Henn.) S.Ito. The same name was adopted by Fisher (6) and was universally accepted. Later, the disease was reported from India by Butler (2) and by Teng (28) in China.

In India, Butler (2) recorded it from Chamta Ghat and Monghyer, Thirumalachar and Mishra (29) from Bihar, Mundkur and Thirumalachar (22) from Bellur near Mysore. Mishra *et al.* (20) from Jabalpur and Jain and Verma (11) from Rewa. The disease was also reported from Karnataka, Andhra Pradesh and Tamil Nadu (23, 32). At present the disease is endemic in all the states of the country growing *P. scrobiculatum*. The severity of the disease has been reported upto 56.4 percent depending upon the soil, climate and host variety (10, 11, 13, 16).

3. SYMPTOMS

Typical symptoms are visible as the crop approaches flowering. The entire affected panicles are transformer into a long sorus, which are 5 to 8 cm long and about 0.6 cm broad. The entire sorus remains surrounded with a creamy membrane in early stage. Sometimes the sorus remains enclosed in the boot leaf and does not emerge fully. The sori destroy the whole inflorescence except the fibrovascular boundles, the rachis of which persists in the form of a bundle of fibres. At maturity the membrane of the sorus bursts and exposes the black mass of spores (2, 10, 22, 23, 24, 27, 32) (Fig. 1). Necrotic streaks on the boot leaf covering infected panicle have been reported by Ahmed (1).

Fig. 1 : Head smut of *Paspalum scrobiculatum*

4. PATHOGEN

The disease is caused by *Sorosporium paspali thunbergii.* Teliospores of the fungus are borne in loose spore ball like masses of 30 to 50μ in diameter (2, 22). Ahmed (1) also reported teliospores to remain in groups in the form of intact spore balls which get disintegrated into individual spores on getting little pressure. Individually the spores are globose, angular to roughly pear shaped, dark to yellowish brown with a thick smooth wall measuring 11-18 x 8-12μ in diamater (33). Ramakrishnan (24) noted that spores germinate by forming a promycelium bearing lateral and terminal sporidia. Ahmed (1) observed that the teleospores germinate by producing septate, single or branched hypha, constricted at the septum. The hyphae width was more but did not produce sporidia. According to him the teleospores were viable for seven months. The spores germinated on different media and maximum germination was recorded in sterile water with 2 percent sucrose solution, while the minimum germination was in distilled water. The spores did not germinate in sterile water.

Variation in colour, consistency, margin, lusture in monosporidial culture was observed by Ahmed (1) on eight culture media. Maximum radial growth of the fungus was on potato dextrose agar medium followed by Czapek's agar and *Paspalum* grain meal agar medium. Colour of the culture ranged from buff or vinaceous buff to whitish or hazel. The consistency of the culture was leathery on different media. The margin was smooth or fimbriate or wavy. Dull lusture was recorded in all the tested media except in *Echinochloa* grain meal agar on which it was shiny. Large variation in surface and topography of the culture was recorded. On potato dextrose agar, potato sucrose agar, *Paspalaum* plant extract agar, potato dextrose malt extract agar, potato sucrose malt extract agar, Czapek's agar and Richard's agar the mycelial growth was maximum on the 18th day (1).

Maximum temperature for germination of teliospores was between 24.5 to 37.5°C, while minimum was below 19°C. The optimum temperature for germination was 30°C (18, 19). Ahmed also recorded maximum growth of the fungus at 30°C and reported pH 4. 1 to be the best for the growth, however the fungus can grow on alkaline medium at 7.1 to 8.1 pH also.

5. NUTRITIONAL REQUIREMENT

The nutritional requirement of *S. paspali thunbergii* was studied by Ahmed (1). It utilized monosaccharides efficiently followed by oligosaccharides. The fungus preferred organic form of nitrogen. Urea was the poorest source.

6. DISEASE CYCLE

The disease is seed borne. The smut spore balls remain adhered on the seed surface as concomitant contaminant (19, 21, 26). The seed contamination results into systematic infection. Sattar (26) recorded infection from the crop residues of the previous crop persisting in soil. However, the infection was subjected to rains before sowing of the crop. Later Ramakrishnan (24) described it as soil borne disease also, which was confirmed by Ahmed (1). Jain and Gupta (10) also found infection through soil borne inoculum also leads to systemic infection. It could be therefore concluded that head smut of *P. scrobiculatum* is both seed and soil borne.

Seedling infection occurs by penetration of germtube through the cell wall. The resulting mycelium develops both inter and intra cellularly in the host tissues (1, 26). The mycelium becomes systemic and enters the meristematic tissues which finally infect the earheads (24). Teliospore formation progressed from centre to epidermis. No sign of hypertrophy or hyperplasia of infected cells was observed. Hyphae get established in all the infected tissues except xylem vessels (1).

7. INOCULATION TECHNIQUE

Since uniform infection does not occur under natural conditions, a suitable method was needed to screen varieties against head smut. Seedling inoculation with viable sporidial suspension resulted in maximum incidence of the disease (1). Jain and Gupta (9) reported mixing of 2g viable teliospores of the fungus with one Kg seed before sowing to be the best method as it gave desired artificial epiphytotic of the disease. The soil infestation resulted in comparatively poor disease incidence. The method is easy, simple, less time consuming and reproducible.

8. HOST RANGE

The knowledge about the host range of a pathogen is important to know about its perpetuation. Ahmed (1) tested other small millets as collecteral host of the pathogen, but the fungus failed to survive on *Eleusine coracana, Setaria italica, Panicum miliaceum, Panicum, miliare, Echinochloa frumentacea* and *E. colonum.*

9. MANAGEMENT OF THE DISEASE

The disease can be managed by using resistant sources, optimum cultural practices and cheaper chemicals.

9.1 Resistance

Jain *et al.* (12, 14) observed very few lines exhibiting resistance to *S. paspali thunbergii* on screening all the available cultivars and germplasm. The cultivars and lines KMV 8, KMV 20, JK 41, JNK 117, IGBKK 1, DPS 34 and DPS 77 have been identified as promising doner for resistance against head smut. Various workers have also identified sources of resistance (25). Some such sources are KMV 8, KMV 20, RPS 136-1 (31); IGBKK 1, 3, 5, 6, 7, 8, KMV 8, KMV 20, D 73 (4); JK 41, JNK 117, KMV 8 (9, 10); GPLM 717, 779, 786 (12, 13); DPS 34, 77, 84, 90, 100, 158, 159, JNK 117, RPS 136-1, KMV 20 (8).

9.2 Cultural Practices

Field sanitation by destroying infected plants and burning their straw reduced the primary inoculum. Verma (30) reported that the intensity of the disease was more in black soil as compared to red gravel and red soils. Early sowing though suffers with slight higher incidence but gives higher yields (8). Delayed sowing reduces the incidence with low yield (8, 31). The deeper sowing (5-10 cm) and delayed emergence of coleoptile from the soil provide more chance for infection by the fungus and shallow sowing is helpful in early emergence resulting in less incidence of head smut (8). The increase in nitrogen level increases the incidence of head smut. The application of balanced fertilizers is desirable to minimise the head smut incidence in *P. scrobiculatum* (8).

9.3 Chemical Control

McRae (19) and Sattar (26) observed that steeping of *P. scrobiculatum* seeds in 1.5 percent copper solution or treatment with copper carbonate 1Oz per 10 lb seed reduced the incidence of the smut by nearly 50 percent. Ahmed (1) reported complete control of head smut by treatment with Carboxin, Carbendazim, Emisan and Thiram. Chalam *et al.* (3) tested six fungicides as seed treatment of which Bevistin (1g/kg seed) followed by Dithane M-45 (Mancozeb) (3g/Kg seed) gave better control of head smut. Jain and Gupta (9) tested five fungicides as seed treatment in field for two years and found that Bevistin 25-SD, Dithane M-45 and Parasan 2g/Kg seed were most effective in reducing the disease incidence with higher yields.

Small millets form an important component for rainfed areas with marginal land, they are all infected by smut fungi with different disease cycles, hence specific control measures are needed. Grain smut of *Eleusine coracana* and *Echinochloa frumentacea* are seed and soil borne. The sporidia formed in soil are transmitted through air to individual flowers causing infection. In smut of *Setaria italica, Echinochloa frumentacea, Paspalum scrobiculatum* the smut spores get associated with seeds as concomitant contaminant and result into systemic infection of the plant. It is essential to find susceptibility of different small millets by one or the other smut fungi by cross inoculation. Studies on variability in these smut fungi are needed. Predisposing factors leading to epidemiology need to be further worked out. Small millets did not get proper attention of pathologists but as this group of crops is mostly concerned with poor tribals of the country, special attention for detailed investigations is needed.

10. REFERENCES

1. Ahmed, N.N. 1991. Biology of Smut fungi of the minor millets *Paspalum scrobiculatum* Linn. and *Echinochloa frumentacea* (Robx) Link in Karnataka. M.Sc. (Ag) Thesis. Univ. of Agriculture Sciences, Bangalore, India, pp. 1-91.

2. Butler, E.J. 1918. *Fungi and Diseases of Plants*. Thacker Spink and Co., Calcutta.

3. Chalam, T.V., Reddy, K.S. and Rao, G.N. 1981. Seed treatment for the control of head smut of *Paspalum scrobiculatum* (Kodo millet). *Millets News Letter* **8** : 56.

4. Dantre, R.K. and Rao, S.S. 1993. Evaluation of kodo millet genotypes for smut resistance. *Indian Journal of Mycology and Plant Pathology* **23** **(2)** : 206.

5. deWet, J.M.J. 1989. Origin evaluation and systematics of minor cereals. In : *Small Millets in Global Agriculture,* (eds. A. Seetharam, K.W. Rilew and G. Harinarayana). Oxford & IBH Pub. Co. Pvt. ltd.

6. Fisher, G.W. 1953. *A Manual of the North American Smut Fungi*, Ronald Press Co., New York.

7. Ito, S. 1935. Notae Mycologicae Asiae Orientalies (Mycological notes from Eastern Asia II), Trans. Sapporonat Hist. Soc. pp. 87-96.

8. Jain, A.K. 1995. Management of head smut kodo millet. (*Paspalum scrobiculatum* L.) *Annals of Agricultureal Research* **16(2)**: 172–178.

9. Jain, A.K. and Gupta, J.C. 1993. Chemical control of *Sorosporium paspali thunbergii* McAlp causing head smut of Kodo millet. *Advances in Plant Sciences* **6** **(1)** Suppl. : 164–167.

10. Jain, A.K. and Gupta, J.C. 1993, Comprative efficecy of techniques to evaluate kodo millet varieties against head smut. *Advances in Plant Sciences* **6** **(1)** Suppl.: 18-23.

11. Jain, A.K. and Verma, S.N.P. 1989. Field Survey of head smut of Kodo millet. *Millets News Letter* **8** : 57-58.

12. Jain, A.K., Gupta, J.C. and Yadava, H.S. 1993. Diversity in the genotypes of small millets for reaction to major diseases. Proceeding National Seminar on "Biodiversity: Strategies for conservation and future Chalanges", pp. 25-30.

13. Jain, A.K. Gupta, J.C. and Yadav, H.S. 1993 Sources of resistance to head smut in Kodo millet. *Agriculture Science Digest* **13(1)**: 9-12.

14. Jain, A.K., Yadava, H.S. and Jain, S.K. 1996. Genetic resistance against microbes in small millets. Abstract Proceedings and Souvenir 55 APSI Scientists Meet & National Seminar, Hardware, pp. 17-18.

15. Linge Gowda, B.K., Ashok, E.G. and Chandrappa 1986. Agronomic investigations on small millets in Karnataka. International Workshop on Small Millet held at Bangalore, Sep. 29 to Oct. 3 1986, pp. 25.

16. Mantur, S.G., Lucy Channamma, K.A. and Viswanath, S. 1988. Reactions of Kodo millet (*Paspalum scrobiculatum* L.) varieties to smut (*Sorosporium paspali thunbergii* McAlp). *Millets News Letter* **7** : 55.

17. McAlpine, D. 1910. *The Smuts of Australia*, Melborne, p. 285.

18. McRae, W. 1928. Report of the Imperial Mycologists. Sci. Repts. Agric. Res. Inst. Pusa, 1926-27. pp. 25-55.

19. McRae, W. 1930. Report of the Imperial Mycologists. Sci. Repts. Agric. Res. Inst. Pusa, 1928-29. pp. 51-66.

20. Mishra, R.P., Pall, B.S. and Nem, K.G. 1976. Ustilaginales of Jabalpur, Madhya Pradesh II. *JNKVV Research Journal* **10 (2)** : 189.

21. Mundkar, B.B. 1945. Studies in Indian careal smut VIII. Nomenclature of Indian smut fungi and probable mode of their transmission. *Indian Journal of Agricultural Science* **15**: 108–110.

22. Mundkar, B.B. and Thirumalachar, M.J. 1952. Ustilaginales of India. *CMI Kew, Survey* pp. 34.

23. Pall, B.S., Jain, A.C. and Singh, S.P. 1980. Diseases of lesser millets, JNKVV, Jabalpur, M.P., India pp. 61-64.

24. Ramakrishnan, T.S. 1963. *Diseases of Millets.* ICAR, New Delhi, pp. 152.

25. Rao, A.S., Reddy, P.N., Reddy, R.R., Reddy, P.N., Reddy, T.D. and Narayana, D. 1988. Evaluation of kodo millet genotypes for smut resistance. *Millets News Letter* **7** : 55-56.

26. Sattar, A. 1930. *Sorosporium paspali* McAlp on *Paspalum scrobiculatum* L. Kodra Smut. *Bull. Imp. Agrc. Res. Pusa* **201** : 16

27. Singh, Onkar, Jain, A.K., Gupta, J.C. and Yadava, H.S. 1993. Status report on small millets JNKVV, Campus Rewa, M.P. pp. 19-24.

28. Teng, S.C. 1947. Addition to the myxomycetes and the carpomycetes of China. *Bot. Bull. Acad. Sinica* pp. 25-44.

29. Thirumalachar, M.J. and Mishra, J.W. 1953. Contribution to the study of fungi of Bihar, India-I. *Sydowia* **7**: 29-83.

30. Verma, S.N.P. 1989. Researches on small millets at JNKVV. College of Agriculture, Rewa, M.P., p. 12.

31. Viswanath, S. 1992. Management of biotic factors (Diseases) in small millets. 6th Annual Workshop on Small Millets, BAU, Kanke. from April 30 to May, 2, 1992.

32. Viswanath, S. and Seetharam, A. 1989. Diseases of small millets and their management in India. PP. 250. In : *Small millets in Global Agriculture*, (eds. A. Seetharam, K.W. Rilew and G. Harinarayana). Oxford & IBH Pub. Co. Pvt. Ltd., USA, p. 250.

33. Zundel, G.L. 1953. Ustilaginales of the world. School of Agriculture, State college. Pennsylvania pp. 69.

31 A Re-appraisal of the Genus *Morchella* in India

T.N. Lakhanpal and O.S. Shad

CONTENTS

ABSTRACT 518

1. INTRODUCTION 518
2. MORPHOLOGICAL ASPECTS 518
3. GENETICAL ASPECTS 521
4. CONCLUSION 521
5. REFERENCES 522

Advances in Microbial Biotechnology
J.P. Tewari, T.N. Lakhanpal, Jagjit Singh, Rajni Gupta & B.P. Chamola (eds.)
APH Publishing Corporation, New Delhi - 110 002, India.

ABSTRACT

Morchella has six species and all of them are reported from India. During the last, ten years, the authors have extensively collected *Morchella* specimens from Himachal Himalaya. All the six species have been collected. Based on these observations, a reappraisal of the genus *Morchella* is presented.

1. INTRODUCTION

Morels (*Morchella* spp.) are highly prized and most sought after wild edible fungi. A good deal of work and efforts have gone physiology of morels (2, 3, 4, 7, 9, 10), but only recently some success has been achieved in studying life cycle in the laboratory (20) or understanding natural propagation in the wild (23). Inspite of the fact that *Morchella* is one of the oldest known genus of Pezizales, established as early as 1719, there is little consensus on the systematic status of the species composing the genus; some workers recognising 3-6 species while others favouring as many as 50–or more (8, 11, 14, 15, 21, 25, 26). This problem has stemmed from the fact that there is lack of detailed study of different developmental stages of the various species and also lack of data on mating and reproduction (6).

In India various attempts have been made periodically by researchers (1, 12, 13, 27, 28, 29), to collect and describe the representative species in the genus but never have all the species been collected simultaneously except by us (16, 17). Among these reports, it is only Wakode's (l.c.), report which pertains to outside Himalayan region. In all these studies [except Batra and Batra (1) and Lakhanpal and Shad (17)] no mention has been made if developmental stages have ever been collected to see stability of morphological characters used in distinguishing the species. Therefore, a need was felt for systematic collection and investigation of the species.

We systematically collected morels from Himachal Himalayas (1800 m to 2700 m), from 1980 onwards and there are about 500 specimens having representatives of all the six species viz. *Morchella angusticeps* Peck, *M. conica* Pers., *M. crassipes* (Vent.) Pers., *M. deliciosa* Fries, *M. esculenta* (L.) Pers, and *M. semilibera* DC ex Fr. Attempt was made to collect sporophores of all these species at various developmental stages to study intermediates, if any. We succeed in collecting such specimens of all the species except *M. semilibera*, which was collected only in mature stage. All other species in our collections are represented by small primordia to medium and mature sporocarps. A reappraisal of the genus is presented on the observations made on these populations.

2. MORPHOLOGICAL ASPECTS

Presently, six species are recognised in the genus *Morchella* (24, 25, 26). Overholts (19) recognised *M. deliciosa, M. esculenta* and *M. crassipes* as distinct species. He differentiated *M. deliciosa* from others by its markedly smaller size. He considered *M. conica* to be a form of *M. esculenta*. Groves and Hoare (8), regarded *M. deliciosa* and *M. esculenta* as different developmental states of *M. crassipes*, the former being the earliest stage. They considered

M. conica as an immature stage of *M. angusticeps*. They based their conclusions on long time field observations of fruiting body development. Batra and Batra (1) merged *M. conica, M. deliciosa* and *M. crassipes* with *M. esculenta* and pointed out that there is no difference in the tissue structure of these fungi and that when a large collection is available from a single locality, various stages of development at stages can be seen merging into one another.

In the six species concept, which seems to be primarily, based on mature frutifications, the species have been keyed by the following criteria by Seavers (25):

Pileus adnate to the stem of the base

Ribs of the pileus similar in colour to the interior of the pits or lighter.

Pileus large, reaching a length of 4-8 cm at maturity.

Pits large, usually shallow; ribs thin; stem strongly enlarged at the base 1. *M. crassipes*

Pits small, deep; ribs thick; stem only slightly enlarged at the base.

Pileus subglobose or only slightly enongated 2. *M. esculenta*

Pileus elongated or strongly attenuated upwards: 3. *M. conica*

Pileus, small, not exceeding 2-3 cm in length 4. *M. deliciosa*

Ribs of the pileus much darker than the interior

of the pits, becoming smoky brown: 5. *M. angusticeps*

Pileus free from the stem at the base. 6. *M. hybrida* (*M. semilibra*)

On the basis of these criteria, two species, i.e. *M. semilibera* and *M. angusticeps* stand out distinctly whereas others frequently pose problems as their characters overlap each other. Inspite of this difficulty, these bases of demarcation are the best till date for identifying different species.

A study of mature sporocarps of all these species from our collections revealed that some specimens were typical and easily identifiable whereas others, especially in the species, viz, *M. esculenta, M. conica, M. deliciosa,* and *M. crassipes* were represented by intermediate forms supporting Batra and Batra's (1) contention that when a large collection is available from a single locality, specimens representing each of these species can be found. In collections from Himachal Pradesh, each species is represented by ample samples so that inter–and intraspecific variations in morphology could be easily studied. In populations from Himachal *M. semilibera* and *M. angusticeps* are clearly distinct, the former by virtue of its half free pileus and the latter by its dark coloured ribs. *Morchella conica* is also fairly distinct in some specimens, only if its shape is given prime consideration but on the basis of size, it does not separate out without some overlaps. We have specimens which fit into *M. deliciosa* and/or *M. esculenta* fairly well, but they otherwise have typically conical pileus of *M. conica*. Further in these species, some specimens have smaller stipes, resembling those of *M. esculenta* and *M. crassipes*. In other words, a great variation has been observed in the distinguishing features of *M. conica, M. deliciosa, M. esculenta* and *M. crassipes*. These may be taxonomically good species when described from a few isolated specimens, but certainly pose problems of appropriate placement when represented by larger populations. In such instances they all seem to be just one complex morphological species.

Most of the species in *Morchella* have been described from mature specimens, giving no consideration to immature and smaller fruitifications. This has been mainly the reason for the prevailing confusion in the species concept of *Morchella* and for the recognition of variable number of species by various workers. A study of the younger stages of the five species in *Morchella* by us, viz. *M. conica, M. angusticeps, M. deliciosa, M. crassipes* and *M. esculenta*, revealed that it was not only difficult but impossible to distinguish just emerging fruiting body initials by themselves, if they were not accompanied by mature sporophores, because they all looked alike. Therefore, distinction between small fructifications of different species was just not possible. It was difficult to place them in different species as the characters which are used to distinguish species, had not yet appeared in them. The distinction in them became apparent only when they became medium sized. Such initials, however, usually accompanied mature fructifications of specific species, and hence this indicated of which *Morchella* species developmental stages these were. Otherwise, the initial stages of all the five species were similarly shaped and structured and not at all distinguishable. So these observations clearly point out that not only *M. conica, M. deliciosa, M. crassipes* and *M. esculenta*, but *M. angusticeps*, also is similar to them in the initial or primordial stages of development, further complicating the species concept in *Morchella.*

Lakhanpal and Shad (16, 17) recorded how distinctly colour in different species of *Morchella* varied with different forest types and altitudes. No wonder then that such colour variants have been given different names from time to time, by various workers. Such morphological and geographic types have also been accorded specific ranks by some, but all these now need to be investigated more critically in the light of Gessner *et al.* (6) recent genetical segregation of *M. deliciosa* and *M. esculenta* before discarding them altogether as mere colour variants. Hence some of the species now reduced to synonymy or described as new, deserve some discussion.

Recently, Zang (30) described *M. tibelica* Zang, a species in the genus *Morchella* from China. This species is stated to be a close relative of *M. angusticeps*, differing from it in having a stipe with rough surface. Our specimens of *M. angusticeps* from Himachal Pradesh are comprised of two types of specimens: in one type, the frutifications are comparatively smaller and stipe smooth; in the second type, the fruitbodies are taller and stipe possesses a rough surface. The latter populations, therefore, represent what has been recently described l.c. (sensu Seaver) by Zang as *M. tibelica* and the former is typical *M. angusticeps*. It has been observed that rough stipe is characteristic of taller fructifications and smooth of smaller ones always, to be merely coincidental and therefore, these certainly represent two distinct groups of specimens in the same or in different species, as has been done by Zang (30). But whether the rough stiped, taller populations really need a separate name remains to be proven by studies on genetical behaviour of these two groups. Presently it seems best to treat them separately from *M. angusticeps* till such studies prove them otherwise, taking clue from the studies of Gessner *et al.* (6).

If these populations are treated as two distinct species, some specimens which we have and which are characterised by clear white ridges/ribs, not recorded in any of the other species, also deserve some place in this discussion. These specimens, represented by three good populations, possess clear-white ridges. Miller (18) included such specimens in *M. deliciosa*. But on the corollary of *M. tibelica*, and observations of Gessner *et al.* (6) these also need to be treated separately till some genetical basis proves them identical. Similarly it is also felt that *M. rotunda*, merged presently into *M. esculenta*, also needs reinvestigation and should be regarded a distinct species because of its typical shape just as is regarded *M. conica*. In our collections we have enough specimens of *Morchella* which have been at present placed in *M. deliciosa* and/or *M. esculenta,* but which possess a typically rounded pileus apex, characteristic of *M. rotunda*. Therefore, such specimens have been treated as *M. rotunda* on morphological basis.

3. GENETICAL ASPECTS

It emerges from the above discussion that inspite of all overlaps and integrates, not only the six species, but three more seem to be distinct so long as they are described from mature fruiting bodies. The data presented on intermediate stages by Groves and Hoare (8) and Batra and Batra (1) suggest only three species to be clearly distinct morphologically viz. *M. semilibera*, *M. angusticeps* and *M. esculenta*. However, as pointed out recently by Gessner *et al.* (6) problem preventing the satisfactory resolution of this issue is the lack of data regarding mating and reproduction in all the species of *Morchella*. Cultural studies of Hervey *et al.* (10) of single ascospore isolates of *M. esculenta* suggested that mating types do exist and sexual reproduction is involved in the genus *Morchella*. They studied characteristics of cultures derived from single ascospores of several isolates of *M. esculenta*. They observed that some cultural isolates exhibited the phenomenon of finite growth, whereas others did not and these also did not produce ascospores. So they wanted to determine whether the infertility of the single spore cultures was due to self incompatiblity of a heterothallic fungus. They combined such cultures derived from single spores of the same fruiting body, interfruit and intra stock combinations. In their 'cross' if the cultures were from the same single spore culture, the line of contact between two colonies was indistinguishable. However, in all non-self combinations the line of contact where the mycelium grew together had a build up of aerial hyphae. They suggested that these results indicate, either, (i) the ascus fusion nuclei are heterozygous for certain genetic factors, or (ii) the asci of a singe ascocarp do not arise from the same pair of nuclei.

Gessner *et al.* (1) presented electrophoretic data that serves to confirm that allelic variations exist and segregate in natural populations of individuals of *Morchella*, indicating the possibility of sexual reproduction. They obtained data for single ascospore isolates of *M. deliciosa* from individual ascocarps. The loci studied exhibited allelic differences between ascospore isolates indicating that genes from different parental genomes exist in individual offspring from a single ascocarp. *M. esculenta* did not demonstrate as much genetic variations as *M. deliciosa*, but ascospore isolates from individual ascocarps exhibited allelic differences. They suggest that presence of polymorphisms, the possibility of mating types, and production of ascospores indicate that *M. deliciosa* and *M. esculenta* exist as Mendelian populations. Their data further suggests that these two species are reproductively isolated, and exist as separate gene pools or distinct species. On the basis of these findings Gessner *et al.* (6) do not subscribe to Groves and Hoare's (8) assumption of treating *M. deliciosa* as an immature form of *M. esculenta* but maintain them as two separate taxa.

4. CONCLUSION

As given in previous pages, on the basis of morphological features, out of the four species in the *M. esculenta* complex, (*M. esculenta, M. deliciosa, M. conica* and *M. crassipes*), *M. conica* and *M. crassipes* have at least one distinguishing characteristic each (conical shape in the former and swollen stipe in the latter) by which they can be easily identified and placed in the taxonomic hierarchy. But there is hardly any such definitive feature that separates *M. esculenta* and *M. deliciosa*, so that these have been more often regarded as a single species. However, the data of Gessner *et al.* (6) on allelic variation and segregation in these two species clearly point that these two are distinct taxa. This data plus data of Hervey *et al.* (10) suggests that the potential for gene exchange may be used as a means of discriminating taxa within the genus *Morchella*. It emerges therefore, that similar types of studies are needed on other species of *Morchella*; not only on the six distinct species, but also on other relatively less known ones, e.g. *M. rotunda, M. tibelica* and *Morchella* sp., before they are merged into the synonymy of other related species. Because of these observations it is pertinent to treat the white ribbed specimens as a new species; relegate the rough stiped-longer fruiting bodies in *M. angusticeps,* to *M. tibelica* and to place the round-apexed fructifications in *M. rotunda*, as these may prove to be different if investigated. Till then it is better not to merge them on morphological considerations alone. As pointed out by Gessner *et al.* (6), "definite taxonomic conclusions will require further sampling from geographically distant populations for all these species".

ACKNOWLEDGEMENTS

Thanks are due to DST, New Delhi for grant which facilitated the collection work.

5. REFERENCES

1. Batra, L.R, and Batra, S. W. T. 1963. Indian Discomycetes. *Univ. Kansas Sci. Bull.* **44**: 109-256.
2. Brock, T.D. 1951. Studies on the nutrition of *Morchella esculenta* Fries, *Mycologia* **43**: 402-422.
3. Cailleux, R. 1968. Pent on cultiver les morilles, *Rev. Mycol.* **33**: 304-308
4. Delmas, J. and Poitou, N. 1974. Introduction Al'etude des morilles on France, *Mushroom Sci.* **9**: 847-857.
5. Eckblad, F. E. 1968 The genera of operculate Discomycetes. A re-evaluation of their taxonomy, phylogeny and nomenclature. *Nytt. Mag. Bot.* 1-192.
6. Gessner, R.V., Remano, M.A. and Schutz R.W. 1987. Allelic variation and segregation in *Morchella deliciosa* and *M. esculenta. Mycologia* **79**: 683-687.
7. Gilbert, F.A. 1960. The submerged culture of *Morchella. Mycologia* **52**: 201-205.
8. Groves, J.W. and Hoare, S.C. 1953. The Heluellaceae of ottawa district. *Canad Fld. Nat.* **67**: 95-102.
9. Heim, R. 1936. La culture des morilles. *Rev Mycol* I Suppl. **I 10-11 Suppl 2**: 19-25.
10. Hervey, A. Bistis, G. and Leong, I. 1978. Cultural studies of single ascospore isolates of *Morchella esculenta. Mycologia* **70**: 1269-1274.
11. Jacquetant, E. 1984. Les Morilles La Biblietheque des Arts, (Paris) pp. 114.
12. Kaul, T.N., Kachroo, J.L. and Raina, A. 1976. Common edible mushrooms of Jammu and Kashmir Proc First Symposium survey and cultivation of edible musrooms in India, **11**: 135-150.
13. Kaul, T.N. 1981. Common edible mushrooms of Jammu and Kashmir. Proc. 11th Intern. Sci. Cong. on the cultivation of edible fungi, Australia. *Mushroom Science* **11**: 79-82.
14. Kimbroug, J.W. 1970. Current trends in the classification of Discomycetes. *Bot. Rev.* **36** : 91-161.
15. Korf, R.P. 1973. Discomycetes and Tuberales. In : *The Fungi,* Vol. IV A (eds. G.C. Ainsworth, F.K. Sparrow and A.S. Sussman), Academic Press, New York, pp. 249-319.
16. Lakhanpal, T.N. and Shad, O. 1986a. Studies on wild edible mushrooms of Himachal Pradesh (N W Himalayas) I Ethnomycology, Production and Trade of *Morchella* species, (Guchi) *Ind. J. Mush.* **12** : 5-14.

17. Lakhanpal, T.N. and Shad, O. 1986b. Studies of wild edible mushrooms of Himachal Pradesh (N W Himalayas) II Ecological relationship of *Morchella* spp. *Ind. J. Mush.* **12**: 15-20.

18. Miller, Orson, 1981. *Mushrooms of North America.* Dutton & Co. New York.

19. Overholts, L.O. 1934. The morels of Pennsylvania. *Proc. Penn. Acad. Sci.* **8**: 108-114.

20. Ower, R. 1982. Notes on the development of the morel ascoarp *Morchella esculenta. Mycologia* **74**: 142-168.

21. Rifai, M.A. 1968. The Australiasian Pezizales in the hervarium of the Royal Botanical gardens Kew. *Verh. K. Ned. Akad. Wet. Afd. Natur. Kd. Reeks* **57** : 1-295.

22. Robbins, W.J. and Hervey, A. 1965. Manganese calcium and filtrate factor for *Morchella crassipes, Mycologia* **57**: 262-274.

23. Schmidt, E.L. 1983. Spore germination and carbohydrate colonization by *Morchella esculenta* at different soil temperatures. *Mycologia* **75**: 870-875.

24. Seaver, F. J. 1928. *The North American Cupfungi.* (operaculates) New York.

25. Seaver, F.J. 1942. *The North American Cupfungi* (operaculates). Hafner Publishing Co. New York 377p.

26. Smith, A.H., Smith, H.V. and Weber, N.S. 1980. *The Mushroom Hunter's Field Guide* (Ann Arbor). Univ. of Michigan Press.

27. Sohi, H.S, Kumar, S. and Seth, P.K. 1965. Some interesting fleshy fungi from Himachal Pradesh. *J.I.B.S.* **54**: 69-73.

28. Wakode, D.D. 1983. Distribution of *Morchella* in time and space with special reference to Meighat forests. Silver Jubilae Symposium on science and cultivation of edible fungi Regional Research Lab Srinagar.

29. Waraitch, K.S. 1976. The genus *Morchella* in India, *Kavaka*, 4:69-76.

30. Zang, M. 1987 Some new and noteworthy higher fungi from eastern Himalayan China. *Acta Bot Yunnanica* **9**: 81-88.

32 The Laboulbeniomycetes : Developmental Patterns

Anupama Pathak

CONTENTS

ABSTRACT 526

1. INTRODUCTION 526
2. MORPHOLOGY 526
3. DEVELOPMENT 529
4. HOSTS 529
5. SPECIFICITY 530
6. TRANSMISSION 530
7. TAXONOMIC POSITION 531
8. CLASSIFICATION 531
9. DEVELOPMENT 532
10. LIFE CYCLE OF *LABOULEBENIA INDICA* AND *CORETHROMYCES SUGIYAMAE* 546
11. REFERENCES 553

Advances in Microbial Biotechnology
J.P. Tewari, T.N. Lakhanpal, Jagjit Singh, Rajni Gupta & B.P. Chamola (eds.)
APH Publishing Corporation, New Delhi - 110 002, India.

ABSTRACT

The order Laboulbeniales belongs to the class Ascomycetes and is characterized by the formation of a specialised type of perithecium. At the same time, the order is totally unlike any other in the class. It is unique to itself as it lacks a mycelium through out its life cycle. It forms a thallus, which may be monoecious or dioecious. The carpogonium is the female sexual organ and the antheridia are the male sexual organ. After fertilization the perithecia are formed on the monoecious thallus or the female thallus in dioecious species. They produce the ascospores which germinate to form a new thallus. Thus, mycelium is not formed at any stage.

1. INTRODUCTION

The members of Laboulbeniales are unique in occurring as obligate exoparasites on the cuticle of various kinds of insects, mites, termites and mallophaga. Although it's an obligate exoparasite, it has not been seen to cause any harm to the insect hosts. Only in a few cases there has been report of dissolution of insect tissue where the fungal haustorium has penetrated the insect cuticle eg. *Trenomyces histophthorus* (16). A haustorium has been reported in the species of *Herpomyces* and also in *Microsomyces telephanus* where it is so large that it could not be extracted completely without breakage but no effect on tissue was seen (8).

Although a haustorium has not been reported in all species under the order. Scheloske (22) reported the transfer of nile blue pigment from the insect tissue to the foot of the fungus. This shows that the fungus is deriving nutrition from the insect. Their exact relationship though remains to be studied.

These fungi are small in size and generally not visible to the naked eye. It ranges from 0. 1 mm to 1.0 mm in total size. It is multilayered with each layer being uni- or multi-cellular. The thalli vary from being expanded in one direction only (i.e. filamentous eg. *Filariomyces*) or in both directions thereby becoming massive in size eg. *Zodiomyces* sp.

2. MORPHOLOGY

Each thallus consists of a receptacle, primary and secondary appendages, and, the sex organs consisting of the antheridia and the carpogonial apparatus.

2.1 Receptacle

The receptacle is composed of a basal foot which is the organ for anchoring the fungus to the host. The foot is conical, black and leads to the basal cell of the receptacle. The receptacle is generally composed of three to five layers of cells which may be multicellular eg. *Peyritschiella* sp. or unicellular eg. *Laboulbenia* sp. Depending on the shape there can be six major types of receptacles -

2.1.1 Cylindrical receptacles

These have one longitudinal series of layers. Most layers are of one cell. The distal portion may be branched or simple and usually narrower towards the tip. eg. *Misgomyces, Autoicomyces, Laboulbenia,* etc.

2.1.2 Simple filamentous receptacles

These have one longitudinal series of numerous layers of cells. They taper towards the base as well as distally eg. *Enathromyces, Rhachomyces,* etc.

2.1.3 Branched filamentous receptacles

These consist of numerous layers of cells. The receptacle is dichotomously branched, each branch consisting of layers of two laterally arranged cells eg. *Rickia dichotoma.*

2.1.4 Fan shaped receptacles

The number of cells in each layer increasing from the sub-basal layer to the distal layer. This gives the fungus a fan-like shape eg. *Peyritschiella* sp.

2.1.5 Simple leaf like receptacles

These consist of a few to many layers of cells. Each layer consists of a few to many cells arranged transversely eg. *Kainomyces*, some species of *Rickia.*

2.1.6 Branched leaf like receptacles

These have many layers of cells, there is a main leaf like body consisting of many cells arranged transversely with few to many branches produced at the subbasal portion of the receptacle. Each branch is similar to the main body eg. *Rickia kawasakii.*

2.2 Appendages

The appendages are generally simple or branched, cylindrical or filamentous. They may originate from various parts of the receptacle.

Thaxter classified them into two types - Primary and Secondary. The primary appendages are formed from the apical/distal cell of the ascospore while the secondary appendages are formed from any cell of the receptacle. What he described as the primary appendages probably occur in all species of the Laboulbeniales. In many genera, the primary appendages are distinctly separated from the receptacle by special cells or constriction at the base. In others they are not distinct from the receptacle (3).

The location of the appendages seems to be a characteristic of the genus and may thus be used as one of the taxonomic criteria.

Depending on the location, there can be four types of appendages. These are -

(i) Single to many simple appendages formed at the terminal portion of the receptacle eg. *Hydraeomyces, Peyritschiella.*

(ii) Numerous simple or branched appendages formed laterally on one side of the receptacle eg. *Euzodiomyces, Kainomyces, Rhachomyces,* etc.

(iv) Numerous, simple appendages found in a row on both sides of the flat, leaf-like receptacle eg. *Rickia*.

(v) Few to many, simple or branched appendages formed at the side of the cylindrical receptacle eg. *Autoicomyces*, *Enathromyces*, *Laboulbenia*, etc.

2.3 Antheridium

The male reproductive organ is the antheridium. It produces the male gametes or the spermatia. According to Thaxter order Laboulbeniales can be divided into two groups depending on the difference in formation of spermatia. These two groups are - Ceratomycetineae and the Laboulbenineae (33).

(i) Ceratomycetineae - Here the spermatia are produced exogenously at the tips of antheridial branches.

(ii) Laboulbenineae - Here the spermatia are produced endoaenously within the antheridia. This group was further divided by him into two families - Laboulbeniaceae and Peyritschiellaceae.

(a) Family Laboulbeniaceae - Here the antheridia are simple, unicellular, flask shaped structures with terminal ostioles. They are produced at the distal end of a basal cell on the receptacle or the appendage. They may be produced singularly or in small clusters, eg. *Laboulbenia*, *Dioicomyces*, *Corethromyces*, etc.

(b) Family Peyritschiehaceae - Here compound antheridia are present. These are multicellular, with several antheridiferous cells and a common chamber with an ostiole. Each cell releases the spermatia into the chamber and they are released from there eg. *Peyritschiella*, *Dimeromyces*, etc (33).

2.4 Perithecium

The perithecium develops from the female sexual organ - carpogonium, formed on the receptacle from a single initial cell. The initial develops into the perithecium on a stalk with the wall layers enclosing the ascogonial apparatus. The stalk may be short or long and consists of around five to six cells. The cells of the stalk may be united to the cells of the receptacle, or may be partially or fully free. The number of cells of the stalk united to the receptacle as well as the place where the stalk cell is found on the receptacle, are definite in a particular species and can be used as a taxonomic character for identification of taxa.

The perithecia are generally elliptical or ovate in shape with mostly a terminal ostiole through which the ascospores are discharged. The perithecial wall is composed of two layers of cells - an outer layer and an inner layer. Each layer consists of four to five rows of cells arranged in three tiers. The number of tiers though varies from three in *Laboulbenia* to nine in *Herpomyces*.

Some perithecia show projections on their walls which may be multicellular and are generally derived from an outer wall cell eg. *Distolomyces euborelliae*. Some species show twist in the arrangement of their outer wall cells eg. *Laboulbenia orientalis*, *L. rupae*, etc.

Perithecia differ in their number on the thallus. The number produced is constant in many genera. Sugiyama distinguished four types of perithecia depending on the position and the number of perithecia, which are characteristic of each genus. These four types are -

(i) One or more perithecia formed laterally on the receptcle eg. *Enathromyces*, *Euzodiomyces*, etc.

(ii) Only one perithecium formed laterally on the receptacle eg. *Laboulbenia*; *Corethromyces*, *Cantharomyces*, etc.

(iii) Only one perithecium formed terminally on the receptacle eg. *Rickia*, etc.

(iv) Two to many perithecia found terminally and aggregated on a disc eg. *Peyritschiella* sp. (24).

3. DEVELOPMENT

The fungi of the order start their development from the ascospore. The ascospore is a two celled structure. The two cells are unequal; they may be subequal in some genera. Each spore is surrounded by a gelatinous sheath. The ascospore is released from the perithecium and when it contacts a suitable site, it attaches itself and germinates. The smaller or the apical cell forms only the primary appendages while the bigger, basal cell forms the attaching foot and the receptacle. Divisions of the basal cell form the multicellular, multilayered receptacle. The primary appendage forms the simple antheridia singly or in bunches, while, the compound antheridia are formed on cells of the receptacle. On the receptacle a cell starts to function as the perithecial initial cel*L*. It is cut off by an oblique division. The initial cell divides and forms two cells. The basal cell after further divisions forms two stalk cells and three to four basal cells of the perithecium. The upper cell also divides transversely and forms the carpogonial apparatus consisting of the carpogenic cell, trichophoric cell and trichogyne.

One exception in the development of perithecia was seen in the genus *Herpomyces*. In all other genera, the carpogonial cell is formed first and the wall cells grow up around it. While in *Herpomyces* two tiers of wall cells are formed, then, one of the cells in the lower tier moves inwards and then extends upwards and forms the carpogonial filaments.

After fertilization the perithecial wall cells enclose the developing ascogenous cells. The carpogenic cell forms the ascogonial filaments where the ascogonial cells are cut off. These form the asci. The number of ascogenous cells vary from one to two to four within the species of Laboulbeniales.

Each ascus has eight ascospores. The asci are generally known to be deliquescing while Benjamin (7) observed empty asci which showed that the spores may be released too. After release, the ascospores begin fresh cycle of development.

4. HOSTS

The fungi of the order Laboulbeniales, develop on the superficial integuments of a very wide range of hosts. The host range lies mostly in the class Insecta, with the exception of a few Myriapods (Millipedes) and Acarina (mites). Among the Class Insecta, the Order Coleoptera was most common as the host to these fungi.

Other than Coleoptera, the hosts belonged to -

Diptera or flies - hosts belong to nine families
Dermaptera or ear wigs - hosts from two families
Blattodea or cockroaches - hosts from twelve genera
Hymenoptera or Ants - hosts from seven genera

Acarina or mites - hosts from four genera
Isoptera or termites - hosts from three genera
Hemiptera or aquatic insects - hosts from six families
Mallophaga or bird lice - hosts from three genera.

From the order Coleoptera, the most commonly infested families are Carabidae, Staphylinidae, Hydrophilidae, Chrysomelidae, Cucujidae, Dytiscidae, etc.

5. SPECIFICITY

The order shows specificity to a varying degree. Some genera (like *Laboulbenia*) have numerous species on numerous hosts with most species showing no specificity and occurring on a large number of unrelated hosts. Many species of *Rickia* also grow on beetles and on mites infesting the beetles.

One species of *Laboulbenia* occurs on ants of the genus *Eciton* in Brazil, and also on a Coleopteran beetle inhabiting the ant nest, and also on the mites that occur on the beetles. But, in some cases, a high degree of specificity is exhibited.

Specificity of the fungus may be for the host, sex-of host and/or position-on-host. Some fungi are found restricted to a single species of host insect or a certain position on the host insect and thus appear to be as limited in their host range as the obligate fungi parasitizing angiosperms. Some other Laboulbeniales occur on only closely related species within a genus or on closely related genera (4).

Specificity for a particular position (site) on the host has been reported for a number of species. eg *Herpomyces periplanetae* occurs only on the antennae of the cockroach *Periplaneta americana*. As such also, the entire *genus* of *Herpomyces* occurs only on the twelve genera of cockroaches.

An extraordinary example of specificity was recorded by and on the aquatic gyrinid beetle *Orectogyrus specularis*. The beetle bears sixteen species of the fungus *Chitonomyces*, with each of the species occurring only on a definite and restricted area. The position is so specific that the species could be identified by its point of occurrence. Along with position specificity, sex of host specificity was also observed. Only one species occurred on both male and female insects,while six species are found only on males and nine species on females (4, 36).

6. TRANSMISSION

It was believed that the localization of the fungus was related to some recurring movements of the host, probably the mating habits (36).

It was also found after experimenting with *Stigmatomyces baeri* maintained on houseflies, found that transfer from one infested fly to a new, uninfested one occurred through direct contact. The fungus occurred on the upper surface of females and lower surface of males.

Arwidsson and Lindroth (5) studied transmission of *Laboulbenia* sp. on *Harpalus* (Carabidae, Coleoptera). They found transmission through spore infested soil to be more effective than direct contact. They also found that the fungus could be transmitted from its original host species to only the closely related species and not to all its species.

Whisler studied infection of *Stigmatomyces ceratophorus* on *Fannia canicularis*. Infection was seen to occur on almost any external portion of the host. In primary infections though, the male is infested ventrally and the female dorsally. Movements of the flies spread the infection to secondary locations on the flies (7).

Blackwell (12) after studying host specificity of *Arthrorhynchus* sp. concluded that mating behaviour could account for the transmission of fungi in the genital region or between the ventral abdomen of male and dorsal abdomen of female, but other kinds of contact would be required for transmission to other body regions.

Thus, it can be summed up from the various experiments done or observations made that the transmission of fungus from one infected host to another is mostly due to direct contact. In direct contact, both the mating behaviour and other movements are jointly responsible for bringing about transmission of the fungus. Though, transmission can occur via spore infested soil and vegetative debris through which insects move.

7. TAXONOMIC POSITION

Thaxter (31) the pioneer worker on the order Laboulbeniales maintained that they were true Ascomycetes. But, since the order is totally unlike other orders of Ascomycetes, many authors did not agree with him. Chadefaud (Ainsworth *et al.*) included them in a separate class Laboulbeniomycetes.

Von Arx disposed them as an "Anhang" to the Laboulbeniales pointing out that Laboulbeniales hardly have a close relationship to the Ascomycetes.

Gaumann assigned them with Aspergillales and other orders to the sub class Prototunicate, Ascomycetes that are characterized by deliquescing asci.

Muller and Loeffer also listed Laboulbeniales under Prototunicatae.

Benjamin (24) suggested a relationship with Pyrenomycetous forms because of the presence of ostiolate perithecia.

Kohlmeyer (15) discovered fungal parasites of marine algae and formed a new order Spathulosporales. They showed many similarities with Laboulbeniales for instance both are host specific, weak parasites; both have reduced thalli, with true hyphae absent; both have spermatia and trichogynes and ostiolate perithecia. The chitosan reaction tests positively for both the orders. Spathulosporales showed similarities to Sphaeriales also in having typical pyrenomycetous ascocarp. Kohlmeyer thus proposed that in view of its debatable position Laboulbeniales should be placed under Unitunicatae (= Ascohymeniales) with Spathulosporales linking Laboulbeniales with Sphaeriales. Thus, the taxonomic position of these fungi is still debated (15).

8. CLASSIFICATION

Three systems of classification have been proposed for the order Laboulbeniales. They vary mostly because of a couple of characters, that is the type of antheridium found, and, the type of perithecium.

Thaxter proposed the first system of classification. He defined three families in the order. These are -

8.1 Family–Laboulbeniaceae

It includes genera in which spermatia are formed endogenously and the antheridia are of the simple type.

8.2 Family–Peyritschiellaceae

It includes genera where spermatia are formed endogenously in compound antheridia.

8.3 Family–Ceratomycetaceae

It includes those genera where the spermatia are formed exogenously.

This system proposed by Thaxter is the one followed by most mycologists. Colla considered homothallism or heterothallism as a taxonomic character of these fungi. She thus divided the order into four families. She retained the families Peyritschiellaceae and Ceratomycetaceae but divided the family Laboulbeniaceae into two families on the basis of monoecism and dioecism. Her system though is not widely accepted as many genera are known to have both monoecious and dioecious species. This is even more significant, considering that the genus *Laboulbenia* has a dioecious species along with other monoecious species. It is thus obvious that monoecism vs dioecism is not a sufficiently significant character with which to separate genera into families.

Bessey recognized four families in the order. He retained the two families Laboulbeniaceae and Peyritschiellaceae but elevated the two subfamilies under Ceratomycetaceae of Thaxter's system to family level - Ceratomycetaceae and Zodiomycetaceae. Family Ceratomycetaceae includes genera forming spermatia exogenously. Family Zodiomycetaceae includes genera forming spermatia exogenously and having massive receptacles. According to Shanor this system is also not right as illustrated in the case of *Euzodiomyces*. Thaxter diagnosed it and assigned it to family Ceratomycetaceae because of its massive receptacle. Benjamin and Shanor (4) carried out detailed study and found the antheridia to be of simple endogenous type. It was then transferred to Laboulbeniaceae. Thus, the nature of antheridum remains the most significant character for classification.

According to Sugiyama (24) though, both Bessey's and Thaxter's classification has a major problem which is the fact that antheridial nature is considered significant while, one third of the genera do not even show the occurrence of antheridia. There thus seems to be a drawback with all systems of classification.

9. DEVELOPMENT

The fungi are of two types - monoecious and dioecious.

The monoecious thalli show the presence of the male antheridia along with the female perithecia.

The dioecious species on the other hand, have separate thalli for the different sexes. The male thalli may differ a lot from the female thalli (eg. *Herpomyces* Thaxt.) or may differ only very indistinctly in the final divisions (eg. *Laboulbenia formicarum* Thaxt.).

There has been a lot of taxonomic work on the order since it has been discovered in the 1850s. But, there have been only a handful of reports where the life cycles of the species under the order have been described.

The pioneer worker on this order, Thaxter described a few life cycles and established a general pattern for the development of two genera of the family Laboulbeniaceae - *Laboulbenia* and *Stigmatomyces* and two genera of the family Peyritschiellaceae - *Peyritschiella* and *Enathromyces*.

Thaxter actually gave the complete account of development of the species *Stigmatomyces baeri* (Knoch) Peyr. He also gave similarities and differences in certain species of *Laboulbenia, Peyritschiella* and *Enathromyces*. Cytology of two species of the genus *Laboulbenia* namely *L. chaetophora* Thax. and *L. gyrinidarum* Thax., were studied by Faull.

With these studies, the basic pattern of development for the family Laboulbeniaceae was established.

Benjamin and Shanor (2) described the stages in the development of the dioecious species *Laboulbenia formicarum*. The development of this species has not been described earlier and this is also the only dioecious species of the genus of *Laboulbenia* having about 450 species.

The fungus, *L. formicarum* is found only on ants. The hosts are nineteen species or forms of the ants from the genera - *Lasius, Formica, Prenolepis, Polyergus*. Their collection was mostly made from winged male ants, with the wings being heavily infested with all stages of the fungus. This enabled them to study the development in great detail. Whole mounts of the fungus or the wings of the ants with the fungus, were more useful in their study than the whole ants.

They studied the life cycle beginning from the ascospore stage. Four ascospores are formed per ascus. The spores are present with their basal cells placed upper in orientation in the perithecium. The spores are discharged in pairs. Each ascospore has a sticky, gelatinous sheath, surrounding it. When the spore reaches the host, this helps in attaching it to the host cuticle. Thereafter, the sheath contracts due to drying but persists as a thin, tough membrane and envelops the fungus throughout.

The foot is formed from the basal end of the basal-cell and it begins with the formation of a brown-black area which contacts the host. With further growth this growth of surrounding tissue (probably) from the area becomes surrounded by out gelatinous sheath. The mature foot has a central dark area and an expanded transparent portion surrounding it. Thaxter believed this region to be the principal absorbing region of the foot. As the fungus grows older, the membrane around the foot, darkens so much as to hide the internal structure.

After firm attachment of the spore to the host cuticle, there is a slight overall enlargement. After this, the apical, smaller cell of the spore divides a number of times to form four superposed cells. These form the primary appendage. The apical cell of the primary appendage, elongates a lot during early growth. The elongation is more in the female thallus than in the male.

Later, the cell may divide further and forms upto six cells in the primary appendage. Benjamin and Shanor (3) reported that the formation of one or more lateral branches is seen very rarely.

The septum between the basal cell of the spore and the basal cell of the primary appendage is called the primary septum. The three septa of the appendage, above the primary septum, become black and the basal cells swell up, becoming characteristically barrel shaped.

The receptacle develops from the basal cell of the spore. Initially, there is no difference morphologically between male and female thalli, except that the female thalli are slightly bigger and have elongated more rapidly. The first two division follow each other rapidly. As a result, three cells are formed, cell I at the basal end, functions as the single cell of the first layer. Cell I is separated from cell 'c' by a horizontal septum and the cell 'c' is separated from cell 'b' by man oblique septum. According to the authors the exact sequence is not known but the oblique division perhaps takes place first.

After this stage, the thallus again undergoes enlargement. Cell 'c'then divides, obliquely to form the cells 'd' and I. The cell II is towards cell I and remains as such as the subbasal cell of the receptacle i.e. the cell of the second layer of the receptacle. Cell 'd' develops into the sexual apparatus and the perithecium in the female thalli only. In the male thalli, the cell remains as such.

The basal cell of the primary appendage, in both male and female thalli, undergoes further divisions and forms four cells cell 'e' at the base, cell 'f' at its posterior end and two cells 'g' at its anterior end. The cell 'e' forms the flat, black insertion cell of the appendage. The cell'f remains unchanged. The cells 'g' form the secondary appendages. In the female thalli, these secondary appendages remain as one to two celled, finger like outgrowths and are sterile throughout the development. In the male thalli , the cells 'g' form two antheridia with their basal cells. The antheridia are simple, flask-shaped structures, with a terminal ostiole through which the spermatia are discharged, at maturity.

In the female thalli, the cell 'd' begins to grow upwards and outwards before any division. It divides once, horizontally to form the cell 'i' at the distal end and the cell'h'at the base. The cell'i'functions as primordial cell of the procarp, and cell'h' as the primordial cell of the perithecium proper including the stalk and basal cells.

The cell 'h' first shows further development. It divides by an oblique, longitudinal division and forms the cells 'j' and 'k'. Cell 'k' elongates and forms the cell VI at the base and cell 'm' at its upper end.

The cell VI remains as such and functions as the primary stalk cell of the perithecium while cell 'm' is the basal cell. Cell 'j' also divides and forms cell VII at its lower end and cells'n' and 'n"at its upper end. In a focus only one of the 'n' cells is visible. The cell VII is the secondary stalk cell and the 'n' and 'n" are the basal cells.

In the mean time, the cell 'b' has already divided to form the remaining cells of the receptade - cellsIII, IV and V. The cell 'b' first cuts off the cell IV and 'b' . 'b" then divides and gives rise to the cells III and IV. The cells IV and V lie side by side. At maturity the cell V is smaller than cell IV and triangular in shape. These five cells i.e. I, II, III, IV and V are the same in both male and female thalli.

The basal cells 'm' , 'n' and 'n" cut off four cells, these are called the 'o' cells and are the primary wall cells. Around the same time, the procarp initial 'i' divides horizontally and forms the basal carpogenic cell, subbasal trichophoric cell and a several celled trichogyne at the apex. The cells 'o' envelope the carpogenic cell and the trichophoiic cell. At this time, probably the fertilization takes place. After this stage, the cells 'o' cut off four cells the 'o" cells which are the additional wall cells and form the second tier of the outer wall cells. The cells 'm', 'n' and 'n" also cut off four parietal cells (cells 'p'). These are placed internal to the 'o' cells and form the inner wall cells. The cells 'p' also divide horizontally and form four primary canal cells 'q'. At a later stage each of the 'o" cells divide horizontally giving rise to two cells 'w' and 'w". These two cells from all the four 'o" cells constitute the two upper tiers of outer wall cells. Following this division the four 'q' cells also divide and form eight cells four 'x' and four 'x" these are the canal cells and form the inner wall cells. Apart from the receptacular cells, there are thus, two stalk cells VI and VII, three basal cells 'm', 'n' and 'n", twelve outer wall cells in three tiers of four cells each 'o', 'w' and 'w"; and twelve inner wall cells arranged similarly 'P' 'x' and 'x' (Plate I).

According to Benjamin and Shanor (3) the maturation and discharge of the ascospores results in the destruction of the parietal and canal cells of the mature perithecium.

The mature female thallus thus has five cells of the receptacle, the perithecium with its complement of cells, the primary appendage and the secondary appendages. The male thalli on the other hand have antheridia on the secondary appendage and no perithecium while the rest of the thallus is similar to the female thallus.

Benjamin and Shanor concluded that, the developmental pattern of this fungus was close to that described by Thaxter for the fungus *Stigmatomyces baeri* though it is a monoecious fungus while *L. formicarum* is a dioecious species. The male and female thalli do not differ much, the male thallus being slightly smaller in size, and the carpogonial initial cell not developing into the perithecium. The female thallus on the other hand does not show the presence of antheridia on the secondary appendages.

In other dioecious genera of the subfamilies Amorphomyceteae and Herpomyceteae of Laboulbeniaceae; of Dimorphomyceteae; and of family Peyritschiellaceae, the male thallus is very conspicuously reduced in size and complexity. This along with the fact that *L. formicarum* is the only dioecious species in the genus *Laboulbenia* of more than 450 species, led Benjamin and Shanor to surmise that the species represents a recent establishment of dioecism. They studied several thousand pairs of the fungus and found that in no case do the antheridia and the perithecia occur on the same individual. This consistent development of unisexual individuals in the species showed that the dioecism is genetically controlled.

Shanor (23) described the characteristics and morphology of a new genus on an earwig *Filariomyces forficulae*. Already twenty six species of Laboulbeniales belonging to four genera have been reported from the pycnidia of male earwigs and on cerci of the female earwig of the species *Prolabia pulchella*. Shanor also described the developmental stages of the perithecium and the mature thallus. He isolated the fungus from the host integument and mounted them on slides with cotton blue and lactophenol which gave better results for studying cellular details, than acid fuchsin in Hoyer's medium. He concluded from his studies that in this genus also, there was no marked deviation from the general pattern of development as elucidated by the previous workers.

Filarlomyces forficulae has simple antheridia on the basis of which it has been kept in the family Laboulbeniaceae. It also has a filamentous receptacle which separates it from other genera. The mature thalli consist of the filamentous receptacle made up of a large and indefinite number of cells, which terminates in a primary appendage. It produces secondary appendages which are sterile and those bearing antheridia. The receptacle also produces perithecia singularly or two to three in number. The development of the simple antheridia, the perithecia, as well as the receptacle, conforms to the pattern elucidated previously . There is a peculiar development of the posterior basal cell of the perithecium and one of the anterior basal cells, into a long

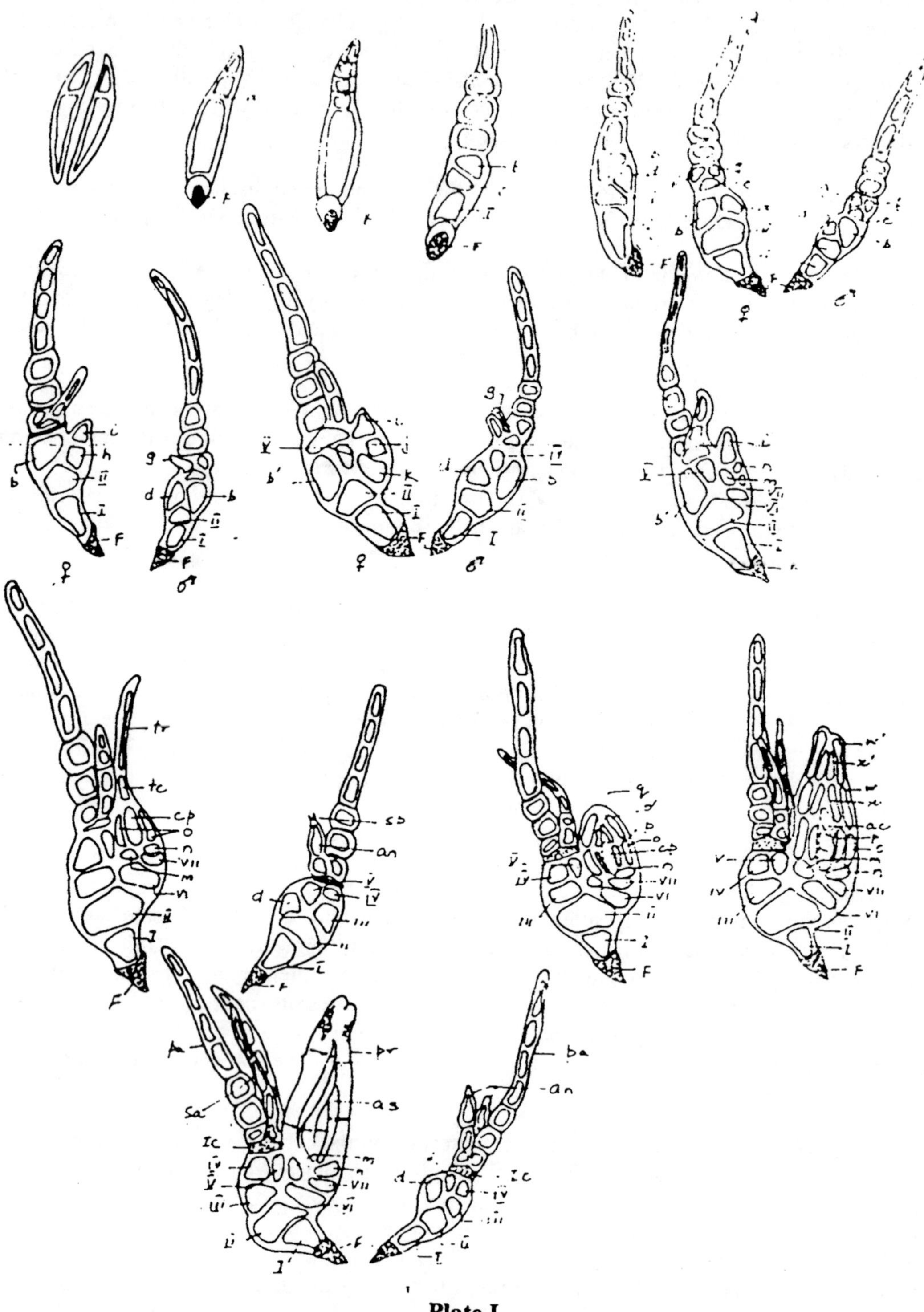

Plate I

Figs.1-8. Developmental stages of *L. indica*. (Camera lucida drawings, X 400).

extension, which moves backwards along the stalk cell and establishes connection with a cell of the receptacle just above the cell which forms the perithecium. Although this development is not seen in other genera, it is not of significance in the overall developmental pattern (30).

Richards and Smith (21) at the University of Minnesota studied the morphology of the fungus. They undertook an extensive study on the infection of cockroaches with *Herpomyces*. In this study they dealt with a number of aspects like the development, life history, pathogenicity, histology, Histopathology and experimental studies with host specificity. They described the development of *Herpomyces stylopygae* Spegazzini, occurring on the oriental cockroach *Blatta orientalis* L. They gave a pictorial presentation of the development of both the male and female thalli, with a special reference to the differences between growth on setae and on the general antennal surface.

The fungus is very different from the other genera in lacking the usual black foot, in having multiple perithecia and in the peculiar multicellular, basal shield formed from the secondary receptacle. They found that there were major fundamental differences between male and female development. The male thallus is very different from the female unlike in the previously described dioecious species of *Laboulbenia, L. formicarum*. The male thallus, here is extremely simple, and, consists of only the primary receptacle and the antheridial branches which bear simple antheridia. This was in contrast to the female thalli which had the secondary receptacle forming the shield the perithecia, as well as the primary receptacle.

For the male thallus, the two celled ascospore divides to become four celled, then the apical cell and sometimes subapical cell divides a number of times and form a small bunch of simple antheridia.

In the female thallus formed on the seta, in the four celled structure the third cell divides and forms two to three cells. All these cells together constitute the six to seven celled primary receptacle. The fourth cell or the basal cell divides and forms an initial cell for the perithecia, another initial cell for the secondary receptacle and the haustorium. The shield is not formed in these thalli.

For the female thalli developing on the surface of the antennae, the third cell produces a cell which gives rise to two cells which are the initial cells for the secondary receptacle. Simultaneously, other cells basal to the third cell also produce initial cells for the secondary receptacle and the perithecium in the same manner. The first secondary receptacle cell then produces the haustorium and later hypertrophies to form the shield at the base of the thallus.

Thus, in the male thalli the apical cell of ascospores contributes to the formation of antheridia, whereas in the female thalli only the basal cell contributes to the formation of perithecia and secondary receptacle. The cell lineage is fundamentally different in the two sexes (19).

Batra (1) described the developmental stages of a new species that she described from an ant from India. She found the black ant *Camponotus* to be infested by a new monoecious species of the genus *Laboulbenia - L. camponoti*. She compared the developmental stages with those of the allied species on ants, *L. formicarum*. She concluded that the new species differed from the previous species in being monoecious, having a longer and a branched appendage, longer trichogyne, cell II was smaller, cells III and IV did not bulge outward and the appendage does not show constricted septa between the three basal cells. She also noticed abortion of perithecia in about 0.5% of the fungus, while there were no aborted antheridia. She thus concluded that the two species are morphologically similar and appear to be closely related and that the monoecious *L. camponoti* may have been ancestral to the dioecious *L. formicarum*.

Tavares (25) investigated the development of the fungus *Herpomyces paranensis* Thaxter, parasitizing cockroaches. The characteristic feature of this species is the slender, flexuous prolongation of the terminal cell of the female primary axis. The mature female thallus consists of two series of perithecia. Each series is joined by a short branch to the primary axis. In each series, there is a broad, straight shield of the initial perithecium which covers the shields of all subsequent perithecia. The perithecia are three spined.

Earlier studies by Thaxter did not describe the exact growth pattern of the thallus, the manner in which the primary axes and the perithecia develop (32, 33).

Tavares described the development of *H. paranensis* in minute detail, she also explained the extent of similarity of the perithecial and secondary receptacular development between *H. paranensis* and *H. stylopygae*. The germination of the two celled ascospore takes place on the minute setae of the antennae. In both sexes, the two celled spore divides to form a hyaline four celled primary axis. The foot is undifferentiated. The apical cell forms the antheridial branch in male thalli and extends upwards into a long filament. The subapical cell may also contribute and form an antheridial branch in the male thalli.

In the female thalli, the third cell or the suprabasal cell, produces two extensions at the base, each of these gives rise to two series of perithecia. The extension forms an initial cell, the base of this produces a perithecial upgrowth. The basal cell of each upgrowth produces three secondary receptacle cells, one of these produces the shield and another upgrowth. The apical cell of each upgrowth forms a perithecium. There thus, occurs a continuously branching system of shield and secondary receptacle.

For the formation of the perithecium, the apical cell undergoes two transverse divisions and forms two basal cells and an apical cell. Then three vertical divisions occur in the apical cell to form the initial four wall cells of the perithecium. The second basal cell cuts off a third basal cell. The four outer wall cells divide transversely and form a second tier of four wall cells. At the two tier stage, a first tier wall cell grows inwards and upwards and then forms the carpogonial upgrowth later (at the four tier stage). The inner wall cells are formed from the third tier of outer wall cells. Eventually six tiers of inner wall cells and nine tiers of outer wall cells are formed. The three spines of the perithecia develop from the fifth tier cells in two rows.

Tavares (25) also noted that the abortion of perithecia occurs frequently. When it occurs at a late stage, the base of the perithecium forms antheridial branches. Shields of such perithecia produce new primordia which also undergo abortion at various stages.

Tavares (26) described the structure and development of *Herpomyces stylopygae* Speg. A series of papers were already published by Richards and Smith (20) on the same fungus. Tavares in her paper extended their work done on development and explained the sequences of internal cell divisions which form the ascospores and the inner cells of perithecial neck. Also, explanation was provided for the development of the trichogyne and the exact manner of germination on the cuticle. She gave different observations on some points.

She stated that the earlier authors Thaxter, Richards and Smith did not notice that the shield forming thalli were attached to the minute setae on the antennae rather than on the surface of the antennae. Although Richards and Smith emphasized that the basal cell in the female forms a downward projection, according to Tavares it was possible that the projection was from one of the cells produced by the suprabasal cell while the basal cell becomes inconspicuous. Also, the lower end of the primary axis of the male thallus forms no elongation, contrary to what the earlier authors suggested. Richards and Smith also reported that the apical cell alone contributes to the antheridial branches, whereas according to Tavares, the upper three cells of the primary axis contribute to the antheridia.

According to Richards and Smith the primary axis downgrowth divides first horizontally and forms the secondary receptacle primordium and the perithecial primordium. The perithecial primordium, then forms a vertical row of three cells, the middle one of which produces the basal cells while the apical cell undergoes a transverse division and then a longitudinal division to form the perithecium. According to Tavares the middle cell does not divide. Richards and Smith supposed the apical cell cut off from the initial central carpogonial upgrowth to be the base of initial trichogyne. While, Tavares found the nucleus in this cell to be a sister nucleus to the one in the carpogonium. Richards and Smith reported the presence of trichogynes covered with spermatia which were not observed by Tavares who did not find any evidence of the migration downwards of spermatia nuclei. She found that the trichogynes that were sometimes seen, did not possess any nuclei and appeared to be deteriorating.

Tavares also enumerated the characteristic features by which *H. stylopygae* could be distinguished from *H. periplanetae*. *H. stylopygae* has a conical apical cell of the female primary axis, lacks the conspicuous large rounded cells in the basal cell region of the perithecium and base of perithecial neck, and also differs in its ability to grow to maturity on large setae (19, 20, 26).

Whisler attempted experimental studies on a new species of *Stigmatomyces* - *S. ceratophorus* Whisler on the lesser housefly *Fannia conicularis*. Along with the experimental studies, he also described briefly the developmental stages. He found that many times, the whole group of thalli were at the same stage of development. He found the development to be similar to that illustrated by Thaxter for *Stigmatomyces baeri*. The apical (smaller) cell of the ascospore gives rise to the primary appendage which branches sympodially with each branch producing a simple antheridium at its tip. A thallus produces approximately eleven antheridia. The terminal cell of the appendage often produces a blunt adhesive tip instead of an antheridium. The basal (bigger) cell of the spore produces the receptacle and the perithecium.

Tavares described the development in three species namely *Amorphomyces* stenusae Thax.; *A. schistogeniae* Thax. and *A. falagriae*. The genus is dioecious and peculiar since the two celled appendage produced by the apical spore segment in the female is embedded in the outer wall of the perithecium. She also found that there are no separate cells of the second and third layers. The perithecium which was proposed to be terminal on the two celled receptacle (Thaxter) is said to be lateral in relation to the primary axis of the spore. According to Tavares development of the perithecium is more or less similar to that found in *Laboulbenia*. The initial is cut off obliquely from the cell II - III. The appendage cells are above the cell II - III and remain small in size. They become pressed against the outer wall of the enlarged perithecial wall cells. They appear as two minute cell lumina in the perithecial wall by maturity. The species of *Amorphomyces* are very similar and their development follows the same pattern (27).

Tavares and Balazuc (28), described the development of the antheridium and of the outgrowths of perithecium in the fungus species *Misgomyces mastacis* Balazuc. According to

Thaxter simple phialides of the antheridia were present in the genus. Picard on the contrary, believed, the inflated multicellular region above the receptacle to be a compound antheridium. He also considered the multicellular region between the two dark septa in *M. dyschirius* to be part of the receptacle. Although Tavares and Balazuc considered this part to be the lower part of the apendage.

According to them in *Misgomyces*, the spore septum becomes blackened. The two lower cells of the appendage enlarge. By the time the lower most cell divides, the upper dark septum becomes displaced to the edge of the dome-like structure and the narrow upper appendage looks to be lateral in position. At the trichogyne stage, when there is one tier of perithecial wall cells, the lower most cell of the appendage divides. It forms an antheridial cell which moves to the base of the neck of the compound antheridium; before producing spermatia. There may be three to four antheridial cells. The aperture of the neck is close to the base of the trichogyne. It was thus found that some of the cells of the multicellular structure that Picard observed do produce spermatia which emerge through an aperture close to the trichogyne.

They found the perithecium of *M. mastacis* to have a subterminal spine and a lateral protuberance. The protuberance develops from the apex of a long lower outer wall cell. This cell develops from one of the basal cells. The spine develops from the third tier of two outer wall cell rows (28).

A new genus with a single species *Osoriomyces rhizophorus* was reported from Taiwan, which was found growing on the elytra, leg and other body parts of *Osorius formosae*, Beinhauer belonging to Staphylinidae, Coleoptera Terada, also described the developmental stages of the fungus, briefly.

According to Terada (30) after spore germination, a series of horizontal divisions occur and form a uniseriate thallus with a dark, constricted septum. Vertical divisions are laid down in the lower cells resulting in biseriate lower receptacle. Rhizoid like appendages extend outwards just above the base of the thallus while gelatinous appendages arise erect, and lateral at a higher level. The primary axis terminates in a cluster of gelatinous appendages. By maturity, five or six simple, long, rhizoid like appendages were seen that extend downward from the suprabasal group of cells in the primary axis. Triangular, subsessile cells are formed on one or both upper corners of the upper cells of the receptacle. Each of these cells produced a short gelatinous branch. Secondary axes that arise from the upper part of the primary axis could degenerate later.

In the material examined, Terada observed one specimen to bear gelatinous, unicellular, elongate structures directly on triangular subsessile cells or in clusters on a stalk cell produced by a subsessile cell. Terada regarded these structures to be simple antheridia since they are very narrow apically. They are ephemeral and are soon converted into sterile appendages.

The perithecia develop on variable positions, with the lowermost one usually occurring on the suprabasal portion of the primary axis. During perithecial ontogeny, Terada observed the following sequence of events. A single cell arises laterally from the receptacle which divides obliquely into an upper cell and a lower cell. The lower cell divides further into two cells. These undergo further divisions and finally form the perithecial walls. The upper cell produces the female sexual apparatus. It forms initially three superposed cell, a lower carpogenic cell a median trichophoric cell and a terminal trichogyne. Only one or two celled trichogynes were observed. By the time, there are three tiers of outer wall cells, an ascogenous cell with three supporting cells is found below the trichophoric cell within the perithecium.

Terada related this pattern of development to that described in detail by Thaxter for *Stigmatomyces baeri* (Knoch) Peyritsch, and, other taxa.

By the time perithecial wall consists of four tiers of cells, asci have been formed. The second tier of cells in the fully mature perithecia is three to four times taller than the lowest tier. Sometimes, the apical outer wall cell on the inner side of the perithecium also divides. The stalk cells and the basal cells are strongly appressed to the receptacle so that the perithecia appeared to be sessile (30).

Benjamin in his paper on comparative morphology of *Idiomyces* and its possible allies *Diplomyces*, *Sandersoniomyces*, *Symplectromyces* and *Teratomyces* (Ascomycetes, Laboulbeniales), described the pertinent stages of development of the receptacle, perithecium, and primary and secondary appendages of *Idiomyces peyritischii* and compared it with the type species of the genera *Diplomyces*, *Sandersoniomyces*, *Symplectromyces* and *Teratomyces*. He also studied the relationship between the genera due to their similarities.

He found the ascospores of both *Idiomyces peyritschii* and *Teratomyces mirificus* to have modified tip of the larger cell. *I. peyritschii* has a hyaline end piece terminating the truncate foot-forming cell which is unique in the Laboulbeniales. In *T. mirificus* there is a conspicuous subterminal constriction of the basal cell.

The primary receptacle of *I. peyritschii* consists of five cells with three cells being derived from cell II. The cell III proliferates and forms a cellular complex that gives rise to secondary appendages. Perithecia sometimes arise from these cells though generally they arise from a group of cells derived from cell II. The receptacles of the genera *Diplomyces*, *Sandersoniomyces*, *Teratomyces* and *Symplectromyces* are found to be similar by Benjamin in the three to four celled young stages of development and later they become different. Their receptacles are of the indeterminate type (i.e. of a variable and large number of cells). The primary receptacle though is determinate, consisting of three cells,cells I, II, and III; a fourth one is added by division of the cell II. The cell III gives rise to a complex of cells in the secondary development. These cells give rise to the secondary appendages and the perithecia.

The development is similar in *Diplomyces* and *Sandersoniomyces*. The cell III cuts of a pair of anterior cells. In *Diplomyces* these give rise to perithecia, secondary appendages and secondary cellular axes bearing secondary appendages and perithecia. In *Sandersoniomyces divaricatus* these cells form only secondary cellular axes which bear perithecia and secondary appendages (Benjamin). In all the genera, the primary appendages are more or less branched, and multicellular. Some branches give rise to seriate antheridia in *Symplectromyces vulgarism* This fungus also shows well developed branched, fertile secondary appendages, as seen by Benjamin.

In *D. actobianus*, *D. clavifer* and *S. divaricatus*, secondary appendage consists of a single basal cell bearing externally a three celled, sterile, beaked branched and internally several branchlets forming a series of intercalary antheridia. In *S. vulgaris*, the elongated, simple or branched appendage never bears the beaked sterile branchlets. In *T. mirificus*, the secondary appendages become sympodially branched, and, bear free antheridia and sterile branchlets with beaked terminal cells. The secondary appendages of *I. peyritschii* may be branched or unbranched. They often have one or more elongate, cellular branchlets arising from basal and distal cells with each cell in between delimiting two or three simple, sessile adnate antheridia, separated from the mother cell by broad septa.

According to Benjamin perithecial development is like in other members of the suborder Laboulbenineae. The walls of the perithecium grow up around the carpogonium. In all these five genera, the stalk and basal cells persist and are well defined in age. The perithecia in all are pedicellate with the basal cell 'm' and VII along with cell VI contributing to the stalk. The wall cells are in four tiers of cells. The terminal tier always has very small cells and surrounds the terminal ostiole. The number of ascogenic cells varied from four to eight in *Symplectromyces vulgaris* for the small and large perithecia respectively. Perithecia of *T. mirificus*, *D. actobianus* and *Sandersoniomyces divaricatus* have two ascogenic cells.

Idioomyces peyritschii has comparatively short and stout perithecia. The length of the entire perithecial body including the basal cells 'n" and 'n" is often less than that of cell 'm' upto half of the length of the perithecial stalk consists of the combination of cells 'm' and VII. Wall cells of lower most tier are large and 30-40% longer than sub-basal tier. The lower two tiers constitute 3/4th of the total length of perithecium. There are four ascogenic cells (7).

Benjamin also described the development of the trichogyne in *Synandromyces telephani.* The genus is found on cucujiform beetles. The receptacle consists of three elongate cells, cell I is centrally located, surrounded anteriorly by cell II and posteriorly by cell III. Appendage is subtended by cell III and has three to four superposed cells. The basal cell separates and on the inner side forms a supernumerary cell bearing a single simple antheridium. The other cells above the basal cells bear single antheridia distally. The immature perithecia are with the large bicellular trichogynes with the upper cell simple or furcate bearing conspicuous receptive, globoid prominences. There is a well defined stalk and basal cells and five tiers of the outer wall cells. There is a single ascogenic cell present.

Benjamin gave a key to aid in the identification of nine species of the genus. He described the development of perithecia and the trichogyne in the species *S. telephani.* He noted that right from the beginning, the trichophoric-trichogynic initial develops a bud, towards the antheridia. This bud keeps enlarging. At the time when the basal cell cuts off an outer wall cell, the trichogyne cell is cut off from the initial. The large cell begins to show signs of divariation. By the time the cell has cut off the first wall cell, the trichogyne becomes bi-cellular by a septum near the base of the bud. The bifurcation keeps proceeding in the upper cell of the trichogyne. At the stage of development where the lowermost tier of all four wall cells are present, in the upper bifurcate cell of the trichogyne, the arms keep elongating and begin to form the characteristic prominences at the tips. As the upper cell of the trichogyne matures, the two arms elongate and form numerous globoid or ovoid outgrowths apically. Around this time, the wall cells cut off a second tier of wall cells.

After fertilization, the free part of the trichogyne falls away or degenerates, while the lower part of the basal cell persists as the trichogynic scar on the posterior wall of the perithecium. His studies were based on the specimens of *S. telephani* collected from Illinois. Thaxter and Benjamin both observed on specimens of the same species collected from Argentina, and Guatemala, the trichogynes which bore the globoid prominences all along the surface of the long, robust upper cell. Thus, Benjamin concluded that trichogyne morphology indicated genetic differences between *S. telephani* populations occurring in Illinois and in Central and South America (8).

Benjamin described the morphology of *Microsomyces telephanus* Benjamin. The species was found on Coleoptera beetle. The receptacle has three cell layers each of which is single celled in young stage. The primary appendage is formed from the upper cell of the spore and has two cells terminating in an antheridium. The typical feature of this species is that in early stage only, the cell II starts cutting off the first cell of a succession of primary corticating cells. By maturity, the divisions of the corticating cells have formed a large number of small cells. Upgrowths of some of the corticating cells produce secondary appendages of variable number. The secondary appendage is generally one or two celled and is terminated by an antheridium. The corticating cells also produce the perithecial initial. Generally one perithecium is formed on each thallus, but Benjamin reported upto five perithecia on a thallus.

The perithecial development is similar to the pattern established previously. It has the usual stalk cell, secondary stalk cell, basal cells and the outer and inner wall layers. There are the usual four cells in each tier of the two wall layers. The lower two tiers constitute nearly 80% of the total length of the wall layers. The perithecium is broadest below (starting at the ascigenous part) and tapers to a narrow apical region. The ascogenic cells proliferate and form an extended layer at the base of the venter and form a number of asci.

Another important characteristic of the genus is the deep haustorium that it sends into tissue under the integument of the host. Benjamin found it impossible to remove the haustorium in its entirely (9).

Benjamin described the features of development for six species of the genus *Triceromyces* Majewski. This genus includes both monoecious and dioecious forms. The receptacle is three celled. Cell I, cells II and cell III which are laterally adnate and parallel to each other. A unique feature is that the cell III gets divided into an upper nucleate cell and a lower empty segment. Primary appendage arising from cell III has two to three sterile cells or more subtending it which form one simple antheridium internally. The perithecium arises from the cell II and has the usual stalk cells and basal cells with a single ascogenic cell. The perithecium may or may not have one to three prominent outgrowths below the apex. In the dioecious species, the female form is as described above with the appendage consisting only of two to three superposed sterile cells. The male thallus consists of three superposed cells terminated by a single simple antheridium.

A special feature of the genus is in the form of the antheridiiferous part of the appendage in the monoecious form. After the appendage has become four celled, the apical cell shows a diagonal cross wall forming two cells. They may remain two celled or the upper cell may form another cell. Later a succession of four to seven cells is formed. In *T. hydrometrae* each of the four to seven cells grow upwards and outwards before cutting off a single antheridium which thus, become free. Each of these cells in all the species produces a single antheridium distally. The antheridia are thus united at the base with each other, with their efferent tubes being free.

The appendage with its antheridia was always found to be maturing before the trichogyne matured.

The development of the perithecium is similar to the previous pattern. The trichophoric cell persists at the tip of the perithecium disintegrating only when the ascospores are mature enough to be released. In four of the species that Benjamin studied, he found there to be five cells in one row of outer wallcells, in one species he saw a fifth cell in two rows and in the sixth species, the fifth cell could be made out in three of the four rows of cells (10).

Benjamin investigated the development in eight species of another genus-*Aporomyces*. The fungus is dioecious with a female thallus and a much smaller male thallus. The male thallus consists only of a few superposed cells terminated by a single, simple antheridium. The female thallus has a receptacle consisting of a uniseriate row of a number of cells. These cells are derived basically from cell II and some from Cell III. Cell III subtends the simple or branched primary appendage, which may arise from cell III or receptacular cells above the perithecium. The perithecium develops from a single cell which may be terminal or intercalary. The perithecium is unique in having five basal cells and five rows of inner and outer wall cells.

The ascospores are dimorphic being of unequal size. The small ones forming the male and large ones forming the female thalli. The two cells of the spore are subequal, the cross-wall being nearly median.

Five basal cells are formed instead of three. Cell 'k' cuts off two cells instead of one. One of the two cells then cuts off another small cell.

All the cells of the perithecium are formed by the protoplast of the perithecium initial and are thus bound by the expanding outer wall of the initial. As development proceeds, the area occupied by the cell VI enlarges and eventually forms the major portion of the venter enclosing the ascogenous cells and the asci. The cell VII, basal cells, and wall cells are all displaced upward. The wall cells are formed in three tiers. One upper outer wall cell in a few species forms a slender, cellular extension that grows upwards through the ruptured tip of the perithecial initial. This opening functions as the ostiole while the true ostiole formed by the upper true wall cells is at some distance below. In a few species, the true ostiole is emergent or almost emergent (11).

Benjamin also described the taxonomy and morphology of *Acompsomyces* Thaxter., after his study of five species. The genus shows a receptacle consisting of three cells-basal cell I, subbasal cell II, and terminal cell III which is united laterally with cell II. Primary appendage is made up of four to five superposed cells, subtended by cell III. The basal cell cuts off, a small cell distally and internally, which gives rise to three simple adnate antheridia having free necks. The second and, third cells are sterile, the distal cells form single antheridia directly. In some species the terminal cell separates a small cell externally. The perithecium is formed on the cell II, it has four rows of wall cells of five cells each. There is a single ascogenic cell. The trichogyne is three celled with a basal cell and two divergent, elongate cells which have several distal globoid papillae.

Prior to perithecial maturity, the third outer wall cell tier which is the upper tier, divides and forms the fourth and fifth tiers. The lower part of each upper cell projects upwards and outwards. This forms the typical, conspicuous, broad, transverse or inclined perithecial tip characteristic of the genus. The tips of the distal wall cells form the ostiole. A division of the distal parietal cells takes place and the resulting inner wall cells form the central passage through which the ascospores pass on their way to the ostiole.

The ascospore is two-celled, the cross wall is sub median. The tip of the upper segment hardens and forms a spinose structure which persists at the tip of the mature appendage (12).

Till recently, there had been very little work done on this group of fungi in India.

The first and most extensive work done was by Thaxter who in his monographs described species from South and South East Asia. From the Indian peninsula he reported fifteen species of the genus *Laboulbenia* and one species of the genus *Stigmatomyces* i.e. S. *hylemyiae* (31,33, 35, 36, 37).

His work was extended by Batra who reported three new species from India. She also found new host and locality records for two species of the fungi from among those previously described. These two species were *L. vulgaris* Peyr. and *S. hylemyiae* Thaxter. She collected *L. vulgaris* from Kashmir on four species of the carabid beetle *Bembidion* - on both elytra of *B. lysander* Andr.; on both elytra of *B. babaulti* Andr.; on elytra, prothorax and legs of *B. acquilum* Andr.; on the metathoracic coxae of *B. gagates* Andr. She noticed that the mature and immature fungal specimens on the insect *B. gagates* had peculiar proliferations of the appendages. The mature specimens showed the replacement of antheridia by short appendages.

She collected the specimens of *S. hylemyiae* Thaxter on the anthomyiid fly *Fannia* sp. from Pahalgam, Kashmir. These specimens were almost twice as long as the type because of the elongated subbasal cell of the receptacle. The venter of the perithecium was seen to be pale yellow rather than brown as in the type species.

Batra also described a new species of *Stigmatomyces*. - *S. scatellae* from the third and fourth abdominal tergite of the Ephyd-fid fly *Scatella* sp. This fly was collected from Srinagar, Kashmir. The receptacle of the fungus was hyaline, with the basal cell being longer than the sub-basal cell. Appendages were long and wide, with basal sterile cell giving rise to three cells bearing single antheridia and ending in terminal cell. The perithecium was pale, golden-yellow in colour. Venter is inflated at the base, and bears two broad flat ridges, which extend for one-fourth of a spiral turn. The ridges become more distinct apically and end abruptly at the neck. The neck of the perithecium is relatively slender, with a subterminal elevation. Apex of the perithecium bears two projections. The ascospores are also hyaline, fusiform with a knob at the end of the longer cell.

The other two new species that she described belonged to the genus *Laboulbenia*. *L. olitskyi* Batra was collected from the carabid *Chlaeminus biguttatus* from Hauzkhas, Delhi. She described the thallus as having a hyaline receptacle with smoky patches in the septa, and at the lower end of the basal cell. The cell III and cell VI are subequal in size and shape. Cell IV is smaller than cell III. The insertion cell is black in colour. The appendages are two to five in number, long, hyaline and slender. They are borne on one or rarely two basal cells. The inner basal cell and sub-basal cell bear numerous antheridia in groups of three on branch-cells. The perithecium was observed to be pale yellow-brown, it is three-fourths free with the inner margin being evenly convex and the outer margin straight. The apex has a nipple-like shape with a black spot on each side of its base. The lips are hyaline, turned outward. The ascospores are hyaline, fusiform, slightly sigmoid in shape.

The second new species that she described was *L. camponoti* which, she collected from all parts of the male ants *Camponotus* from Hauz Khas, Delhi. She observed the individuals to be occurring in pairs. Receptacles were hyaline with the cells III and IV being sub equal and bulging, the insertion cell was black. There was a single appendage, long, hyaline, usually branched. One or two antheridia are borne on a basal cell. The perithecium was seen to be faintly olivaceous, two-thirds free; inner margin is straight, outer margin is convex. Apex is blunt, with blackened sub-apical region and hyaline tip.

Batra also illustrated the developmental stages of the fungus. She compared it with the developmental stages of *L. formicarum* the other species of the genus occurring on ants, as described by Benjamin and Shanor (3). She concluded from the comparison that *L. formicarum* differed from *L. camponoti* in being dioecious; having smaller; unbranched appendage and trichogyne; slight difference in the sequence of cell division and maturation; cells III and IV not bulging outward; cell II enlarged; appendage with constrictions at the septa between three lowest cells. She also observed abortion of perithecia in about 0.5% of the sample of *L. came.* She did not see any aborted antheridia or antheridia replaced by secondary appendages.

She found these two species to be closely related as well as morphologically similar and thought that the monoecious species *L. camponoti* may have been ancestral to the dioecious *L. formicarum* (1).

After the publication of Batras work there followed a period of almost three decades where this group of fungi was totally ignored. It was in 1989 that work was started in this group of fungi in the Applied Mycology Laboratory, Department of Botany, University of Delhi. Work was started simultaneously and while Surinder Kaur worked on the taxonomic and ecological aspects of the Laboulbeniales, I worked on the taxonomic and developmental studies of the forms.

Kaur (14) reported nineteen species of these fungi belonging to nine genera. These were *Laboulbenia* (eight species), *Peyritschiella* (three species) *Herpomyces* (two species) and one species each of *Rickia*, *Misgomyces*, *Cantharomyces*, *Dimeromyces*, *Dioicomyces* and *Aporomyces.*

10. LIFE CYCLE OF *LABOULEBENIA INDICA* AND *CORETHROMYCES SUGIYAMAE*

Life cycles of the fungi isolated are given. For describing the life cycles of the fungi, the terminology used by Thaxter has been followed. Small case letters are used to designate the immature cells of the receptacle that have yet to undergo divisions. When the cells of the receptacle have reached the condition found in the mature fungus, Roman numerals are used to designate the cells. When the layer of the receptacle is single celled, the cell of that layer is referred to with the number of the layer in the text eg. the cell of the single celled first layer is referred to as cell I.

The diagrams have been labelled using following abbreviations :

F – foot; I – first layer; II – second layer; III – third layer; IV– fourth layer; Ic insertion cell; pa – primary appendages; sa – secondary appendage; d – perithecial primordium; h – basal cell; i – carpogonial initial; k – stalk cell primordium; j secondary stalk cell primordiurn; 'm', 'n', and 'n'' – basal cells; VI – stalk cell; VII–secondary stalk cell; 'o'– basal ties of outer wall cells; 'w' and 'w'' – upper tiers of outer wall cells; p – basal tier of inner wall cell; 'x' and 'x'' – upper tiers of inner wall cells; pr – perithecium; as – ascospore; ac – ascus; an – antheridium; bc– basal cell; dt – distal part of receptacle.

10.1 *Laboulbenia indica* Pathak and Mukerji (Plate: IIa, b)

This species of *Laboulbenia* was isolated from all over the body of a Coleopteran beetle. The beetle was collected from under a flower pot. It was found in a single collection. The fungus was found on the legs, the elytra, the abdomen, and on the thorax. The fungus is yellowish brown in colour. Found basally on the fungus is the conical, black foot subtending the receptacle. The receptacle consists of four layers of cells. Each layer is single celled except the fourth. Primary appendages are two to three in number and are stout structures. The perithecium develops on the cell II, and is swollen symmetrically. It has a blackened and constricted region present sub apically while the apex remains hyaline.

10.1.1 Specimens examined

The fungus was isolated in Delhi, on 18.7.94 and deposited in the Delhi University Herbarium under DU/MAP/998. Portion of the type has been deposited with Y. Nakamura, Shizuoka City, Japan.

10.1.2 Development

The life cycle starts with the ascospores. Each ascospore is hyaline, two celled, the cross wall being subterminal forming thereby a bigger cell (the basal cell) and a smaller cell (the apical cell). The bigger cell undergoes a division at its basal end to cut off the foot. The rest of the spore remains filiform (Plate - II). Later, the basal cell enlarges and then divides to form cell I at the base and a cell 'ii' at its distal end. The upper cell of the spore is undivided as yet and later gives rise to the primary appendage and the simple antheridia (Plate - II).

The next division is transverse in the upper cell of the spore to form two cells of the appendage.

The cell 'ii' and the upper cell of the spore undergo another horizontal division each. Cell 'ii' forms the cell 'iii' (Plate - IIa, b) Cell 'ii' then divides obliquely and forms the cell 'd', and cell II. Cell 'iii' divides horizontally and forms the cell 'iv' and cell III.

The cell 'd' gives rise to the perithecium after further divisions. First it divides and forms the cells 'h' and 'i' . Cell 'h' is found at the base of cell 'i' . Cell 'h' gives rise to the stalk cell basal cells and wall cells , while cell 'i' forms the carpogonial apparatus (Plate - IIa, b).

At the same time the basal cell of the appendage divides obliquely and cuts off a small cell at the level of the second cell of the appendage (Plate - II). The basal cell then flattens and forms the insertion cell 'IC'. Gradually it turns black, to take the typical form (Plate - IIa, b). The cell 'iv' divides obliquely at its distal end and forms a small triangular cell. These two cells then constitute the two celled fourth layer of the receptacle (Plate - IIa, b).

The cells of the receptacle are thus fully formed, they only undergo enlargement from this point onward. The cells I and II are rectangular in shape, longer than broad. The cell I is narrow at the base and gradually broadens to its distal end where it leads to the cell II. The cell II is attached to cell III and the cell of the perithecium at its distal end.

The cells III and IV are almost as long as broad and have a bulging outline. Of the two cells at the base of the primary appendage, one divides horizontally to form another cell at its distal end which functions as the basal cell to the antheridium. The basal cell may divide vertically to form one or two more cells, each of which then functions as a basal cell and cuts off an antheridium each at their distal end. The antheridia are simple, flask shaped, with short efferent tubes (Plate - IIa, b).

The subsequent divisions take place in both the cells 'h' and 'i'. The cell 'i' divides once to form cell 'cp' at the base and cell 'e' at the distal end. Cell 'cp' is carpogenic cell. Cell 'e' divides again and forms the terminal trichogyne cell and the basal trichophoric cell. By this time cell 'h' divides horizontally and forms cells 'j' and 'k'. The cell 'k' is towards the base and elongates upwards and cuts off a cell 'm' at its upper end and becomes the stalk cell, cell VI. The cell 'j' divides and forms the cell VII at the lower end and two cells 'n' and 'n'' (Plate

- IIa, b). The cell VII is the secondary stalk cell and the cells 'n' and 'n'' along with 'm' function as the basal cells. The basal cells cut off four cells ('m' and 'n', cut off one cell each and 'n'' cuts off two cells), which function as the outer wall cells 'O'. The cells 'O' further divide and form another set of four cells in an upper tier. The trichogyne elongates and may persist till the second tier of wall cells is formed (Plate - IIa, b). The cells 'm', 'n' and 'n'' cut off another set of four cells which are placed internally to the 'O' cells and function as the inner wall cells 'p'.

After fertilization has been effected, the trichogyne regresses and the wall cells close up around the carpogonial apparatus.

The cells of the primary appendage also divide a few times and form a stout three to four celled appendage. Generally, it branches at the base to form two appendages of similar dimensions (Plate - IIa, b).

The antheridia are ephemeral structures and are not seen in the maturing thallus.

With the formation of the ascogonial filaments and then the asci, the perithecium increases in size, the wall cells increase in number and in size to keep up with the perithecium. In total three tiers of four cells each are formed of the outer and the inner wall cells. As the perithecium approaches maturity, a slight constriction appears in the subapical region which blackens. The tip remains concolourous with the thallus. The upper tier of outer wall cells form the ostiole while the inner wall cells form the passageway, through which the mature ascospores are released out of the perithecium.

The mature thallus consists of the receptacle subtended by the black foot. The receptacle has four layers of cells. The basal cell is cell I which is rectangular, longer than broad and narrows to the foot at the base. Distally, it gives rise to cell II which is also rectangular, longer than broad. This subtends the cell III and the stalk cell of the perithecium, cell VI. Cell III and layer IV are rectangular too but shorter than cells I and II. They are also bulging outwards and have slightly constricted septa. The bigger cell of the fourth layer shows a small triangular cell present anteroposteriorly, at its distal end. The stalk cell is almost square in shape, the secondary stalk cell - cell VII is smaller and almost round in shape as are the basal cells. 'm', 'n' and 'n''. Only one of the 'n' cells is visible in one plane. The wall cells are flattened, elongated cells and enclose the ascigenous cavity. The perithecium is obovate in shape and is symmetrically inflated. The apex is tapered, but, blunt, having a constriction in the subapical region. This region is blackened, while, the apex remains hyaline. An ostiole is present terminally and the lip cells are slightly unequal in size. The ascospores produced are long, fusiform consisting of two unequal cells, and, surrounded by a gelatinous sheath (Plate - IIa, b).

10.2 *Corethromyces sugiyamae*, Pathak and Mukerji (Plate - III)

This fungus was found on the abdomen, the elytra and the legs of the beetle *Scydmaenus* (Scydmaenidae, Coleoptera). The beetle is very small and was found under leaf litter. It was collected thrice during the month of October, 1993.

The fungus resembles the genus *Laboulbenia* but is differnt in having a single celled fourth layer and a completely free perithecium arising from the 2nd layer of the receptacle. The perithecium is symmetrically inflated and has a blunt apex.

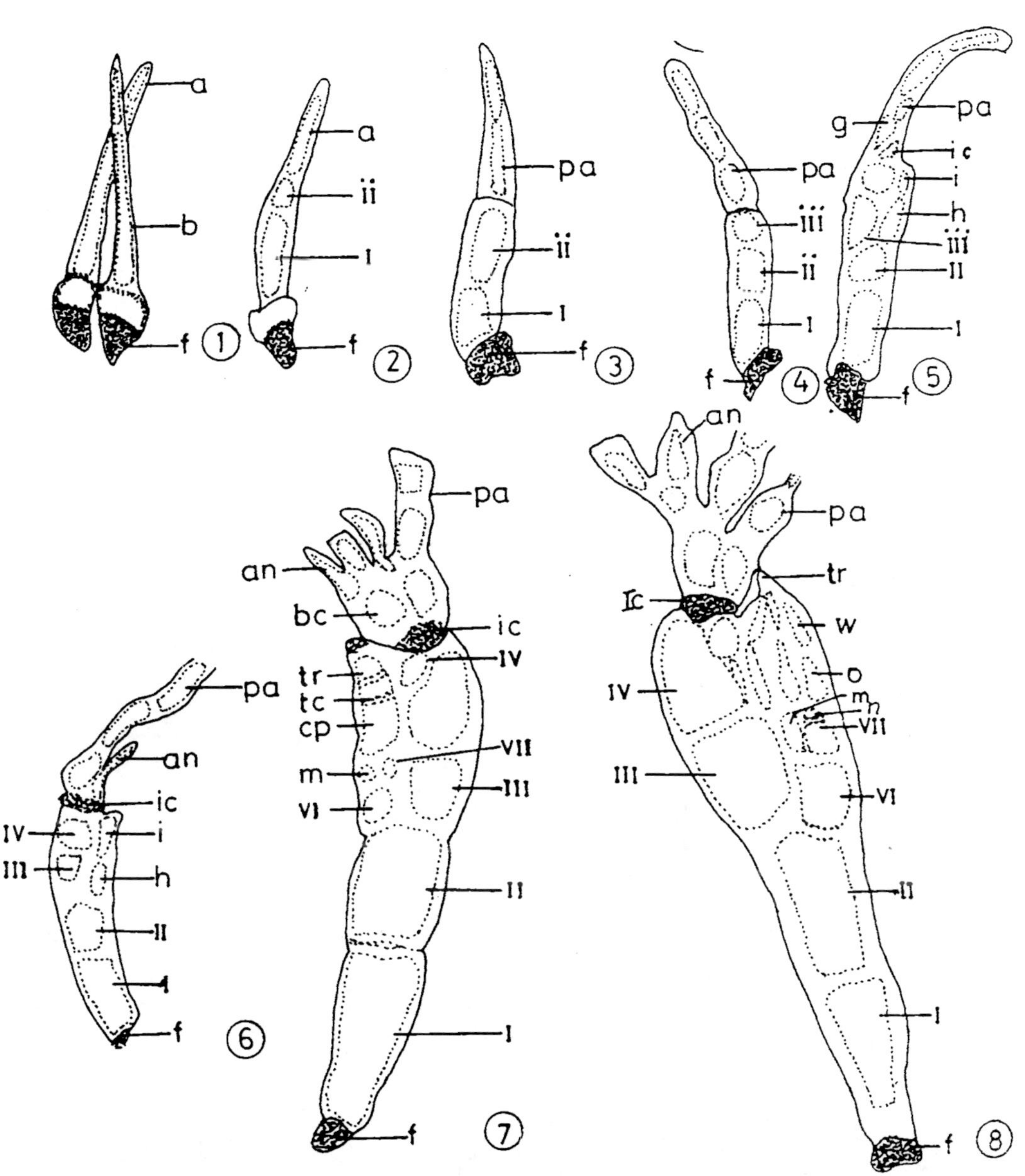
a
b
f
1
a
ii
i
f
2
pa
ii
I
f
3
pa
iii
ii
I
f
4
pa
g
ic
i
h
iii
II
I
f
5
pa
an
ic
IV
i
III
h
II
I
f
6
pa
an
bc
ic
IV
tr
tc
cp
VII
m
III
VI
II
I
f
7
an
pa
tr
Ic
w
IV
o
m
n
VII
III
VI
II
I
f
8

PLATE - IIa

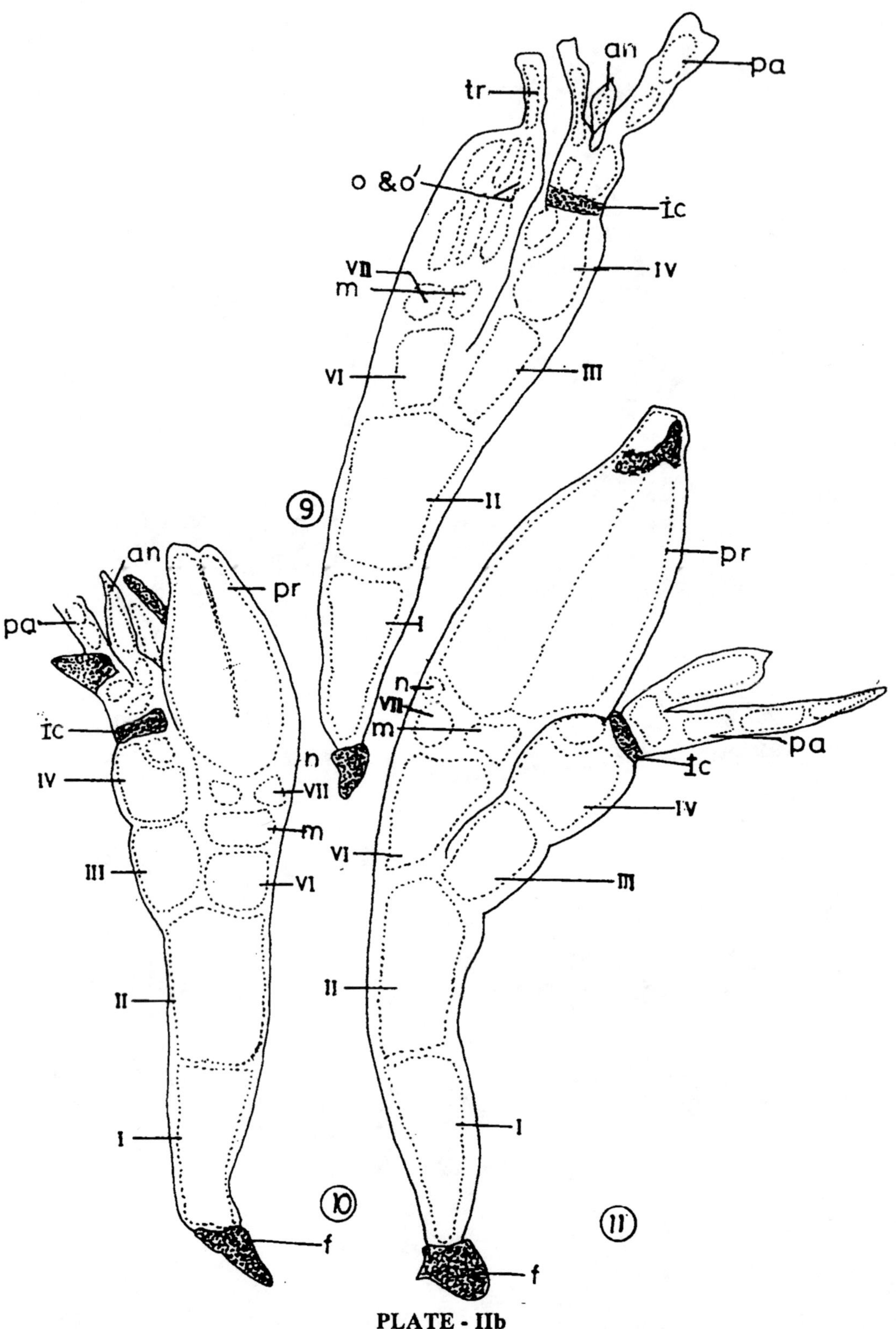

PLATE - IIb

Figs. 9-11. Developmental stages of *L. indica*. (Camera lucida drawings, X 400).

It is characteristic in having a completely hyaline thalli. The receptacle is in two parts, basal and distal. The basal part has three single celled layers and the distal part has four to five superposed cells which may branch distally.

10.2.1 Specimens examined

Delhi, 7.10.93 DU/MAP/1011; 15.10.93 DU/MAP/1012; 16.10.93 DU/MAP/1013.

10.2.2. Development

Each perithecium has four asci. Each ascus produces eight ascospores. The ascospores are oblong to spindle shaped and consist of two unequal cells. The smaller cell is termed as the upper cell and the larger cell is called the basal cell. With the beginning of development, the ascospore divides at its basal terminal end to cut off a small black foot. This serves to attach the fungus to the host cuticle. The basal cell enlarges slightly to become broader than the apical cell. While the apical cell remains elongated and rectangular in shape (Fig. 1 Plate - III)

The basal cell divides by a horizontal division to form two cells of equal size. These cells I and III are both rectangular hyaline,and form two single celled layers of the receptacle.

The cell I enlarges and cuts off a cell by an oblique division. The cell is designated as cell 'ii'. The apical cell, at the same time, also divides horizontally to form two elongated, cells in a row (Fig.2, Plate - III), these constitute the distal part of the receptacle.

The cell 'ii' is barrel shaped slightly swollen and divides by a horizontal division to form a cell 'd' apically and a cell II at the base. The cell 'd' functions as the perithecial primordium (Fig.4, Plate - III).

The basal cell of the distal part of the receptacle is the cell IV. The terminal cell of the distal part of the receptacle divides longitudinally and forms two cells at the same level. The cell on the anterior side, forms a slender hyaline cell. This cell divides horizontally, and forms upto three cells in a row. The cells are slender, cylindrical cells. This constitutes the single appendage of the thallus (Fig.6, Plate - III).

The other cell formed by the longitudinal division elongates and later divides to form two to three long, rectangular cells. These cells, continue as the distal part of the receptacle.

The cell 'd' divides horizontally and forms a cell 'i' at its distal end, and the stalk cell - cell 'h' at the base (Fig.5, Plate - III). The cell 'h' is slightly swollen and this cell also looks barrel shaped, like the other cells of the receptacle, but has a slight constriction towards its upper half. The cell 'h' now divides by oblique septa and forms two cells. These two cells function as the stalk cell VI and a smaller cell at the distal end, the secondary stalk cell VII. The cell VI is the bigger cell and divides at its distal end to form a basal cell 'm' which is present at the same level as cell VII. The cell VII also divides to form two basal cells 'n' and 'n" at its distal end, although only one cell is visible in one focus. The cells 'm' and VII are triangular shaped, small cells in contact with each other at one angle, with their bases in contact with cell VI (Fig.8, Plate - III).

The cell 'i' functions as the carpogonial initial. It divides by two horizontal divisions and forms three cells in a vertical row. The basal cell is the carpogonial cell, the sub basal cell is the trichophoric cell and the apical cell is the trichogyne (Fig.7, Plate - III). This whole apparatus along with the basal cells, and the secondary stalk cell forms a distinct rhomboid

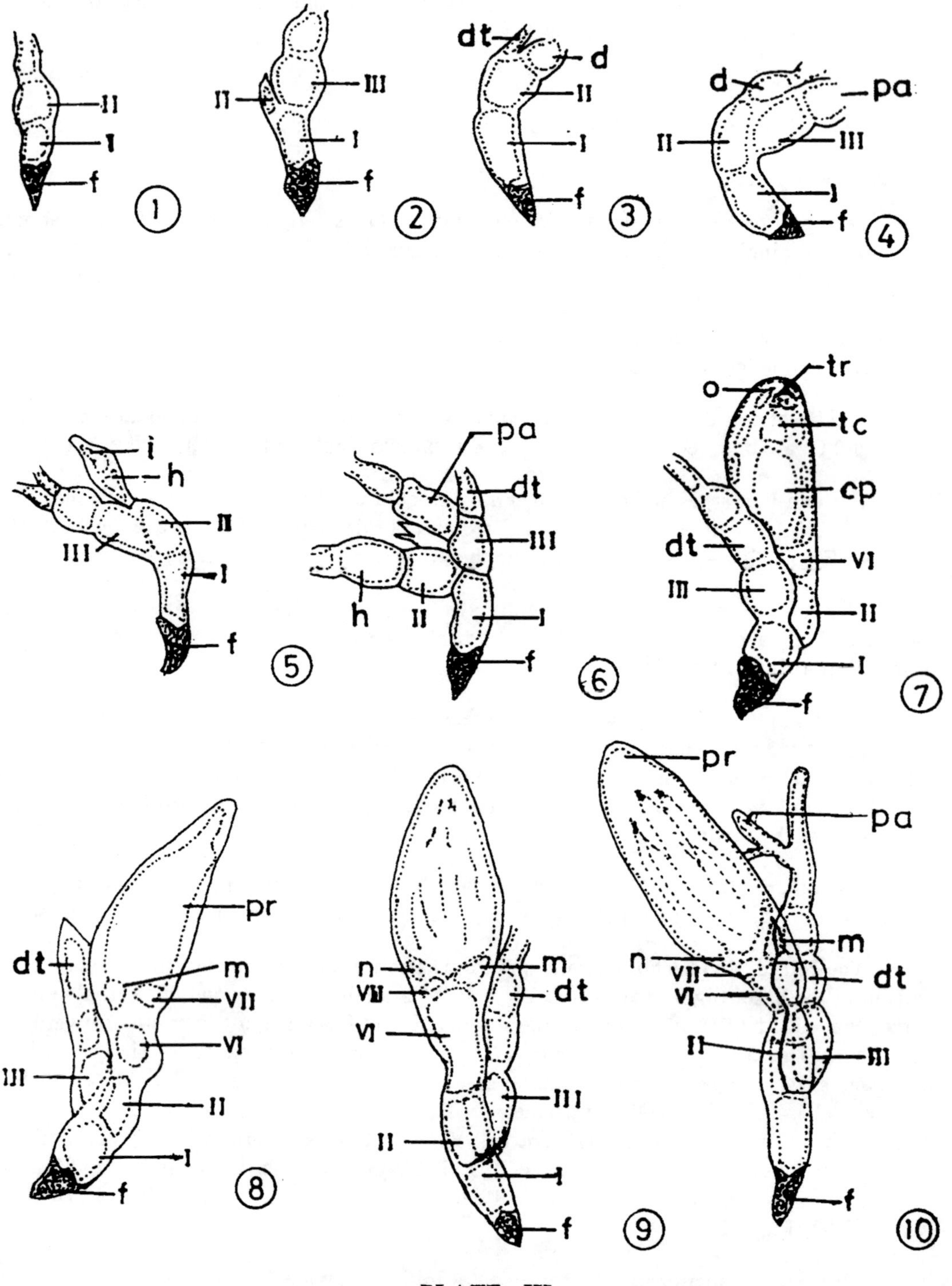

PLATE - III

Figs.1-10. Developmental stages of *Corethromyces sugiyamae* sp. nov. (Camera lucida drawings, X 400).

shaped structure with a broad base where it is attached to the stalk cell (Fig.9, Plate - III). The stalk cell shows a slight constriction at its upper 1/3rd point.

The basal cells 'm', 'n' and 'n", divide horizontally and form four cells. These four cells labelled as 'o' cells are long and slender, they envelope the carpogonial apparatus and function as the outer wall cells. All the four 'o' cells divide horizontally and form four more cells. The cells are arranged in a tier. There forms thus, two tiers of four cells each. The basal cells divide again and form four cells, which are arranged interior to the 'o' cells. These cells are also thin and elongated and function as the inner wall cells 'p'.

With the growth of the carpogonial apparatus, the wall cells also elongate and divide.

The trichogyne, projects out beyond the enveloping wall cells in the form of a long rectangular process. It may even be branched at its base to form a fork (Fig. 7, Plate - III). In the distal part of the receptacle, more horizontal divisions result in the formation of three to four rectangular cells arranged in a row. This part may curve slightly. After fertilization, trichogyne regresses back. The carpogonium produces the carpogonial filaments which form the ascogonial cells. These form the four asci. In each ascus, eight ascospores are formed. Along with the growth in the carpogonial apparatus, the wall cells also keep up. They divide and form upto three tiers each (i.e. in both outer and inner wall cells). The perithecium inflates laterally and then tapers to form a narrow blunt apex (Fig.8, Plate - III).

By maturity, the whole perithecium becomes symmetrically inflated. The stalk cell is narrow at its base but swells up towards its distal end. It's septum with the basal cells is slightly constricted. The basal part of the perithecium is swollen then constricts very slightly towards the middle before swelling again towards the apex. The apex is blunt.

The simple thallus, (Fig.10, Plate - III) thus, consists of a basal black conical foot. The receptacle consists of a basal part and a distal part. The basal part consists of the three cells of the lst, 2nd and 3rd layer. The distal part consists of two barrel shaped cells and then tapers to form two branches. Both the branches, consist of rectangular cells arranged in a cylindrical row. Arising from the layer I is the perithecium. It is almost completely free of the recpetacle, only the stalk cell arising from the cell II.

The perithecium consists of a stalk cell, a secondary stalk cell, three basal cells and the two wall layers. Each wall layer consists of three tiers of four cells each. The inner wall cells form the ostiole through which the ascospores are released. Most of the cells of the receptacle like the cells I, II, III and IV are barrel shaped with distinct, slightly constricted septa. The distal branches consist of rectangular cells. The secondary stalk cell and the basal cells are small and triangular in shape. The wall cells are thin and elongated. The perithecium proper is swollen more at the base with a very slight constriction at the upper middle region. It has a blunt apex.

11. REFERENCES

1. Batra Suzanne W. Tubby 1963. Some *Laboulbenia*ceae (Ascomycetes) on insects from India and Indonesia. *Am. J. Bot.* **50(10)** : 986-992.
2. Benjamin, R.K. and Leland Shanor. 1950. The development of male and female individuals in the dioecious species *Laboulbenia formicarum* Thaxter. Amer.
3. Benjamin, R.K. and Leland Shanor. 1950. Discovery of dioecism in *Laboulbenia formicarum*. Science **111** : 33-34.

4. Benjamin, R.K. and Leland Shanor. 1952. Sex of host specificity and position specificity of certain species of *Laboulbenia* on *Bembidion picipes*. *Amer. J. Bot.* **39** : 125-131.

5. Benjamin, Richard K. 1965. Study in Specificity. *Natural History* **74** : 42-49.

6. Benjamin, Richard, K. 1974. *Sympodomyces*, a new genus of *Laboulbenia*les from New Guinea. *ALISO* **8(l)** : 1-6.

7. Benjamin, Richard, K. 1984. Comparative morphology of Idiomyces and its possible allies *Diplomyces*, *Sandersoniomyces*, *Symplectromyces* and *Teratomyces* (Ascomycetes, Laboulbeniales). *ALISO* **10(3)** : 345-382.

8. Benjamin, Richard, K. 1985. *Synandromyces telephani* from Illinois and development of its trichogyne. *ALISO* **10(4)** : 489-504.

9. Benjamin, Richard, K. 1986. Morphology and relationships, of a new species of *Microsomyces* from Illinois, USA. *ALISO* **11(2)** : 127-137.

10. Benjamin, Richard, K. 1989. Taxonomy and Morphology of Aporomyces (Laboulbeniales). *ALISO* **12(2)** : 335-367.

11. Benjamin, Richard, K. 1989. Taxonomy and Morphology of Acomposmyces (Laboulbeniales), with notes on two excluded species, *Acompsomyces stenchini* and *Acompsomyces lasiochili*. *ALISO* **11(2)** : 201-232.

12. Blackwell, Meredith. 1980. Incidence, host specificity, distribution and morphological variation in *Arthrorhynchus nycteribiae* and *A. eucampsipodae*. *Mycologia* **72(l)** : 143-158.

13. Blackwell, Meredith. 1980. Developmental morphology and taxonomic characters of *Arthrorhynchus nycteribiae* and *A. eucampsipodae*. *Mycologia* **72(l)** : 159-168.

14. Kaur, S., Pathak, A. and Mukerji, K.G. 1993. Studies on Indian Laboulberiiomycetes I. On three unrecorded species of the genus *Laboulbenia*. *Crypt. Bot.* **3** : 357-360.

15. Kohlmeyer, J. 1974. Spathulosporales, a new order and possible missing link between Laboulbeniales and Pyrenomycetes. *Mycologia* **65(3)** : 614-647.

16. Meola, Shirlee and Joyce Devaney. 1977. Parasitism of *Mallophaga* by *Trenomyces histophtorus*. *J. Invertebr. Pathol.* **28(2)** : 195-201.

17. Richards, A.G. 1954. Similarity in histochemical differentiation of insect cuticle to walls of parasitic fungus. *Science* **120** : 761-762.

18. Richards, A.G. and Smith, M.W. 1954. Infection of cockroaches with *Herpomyces* (Laboulbeniales) : III. Experimental studies on host specificity. *Botanical Gazette* pp. 195-198.

19. Richards, A.G., and Smith, M.W. 1955a. Infection of cockroaches with *Herpomyces* (Laboulbeniales) I. Life History Studies. *Biol. Bull.* **108** : 206-218.

20. Richards, A.G., and Smith, M.W. 1955b. Infection of cockroaches with *Herpomyces* (Laboulbeniales) IV. Development of *H. stylopygae* Spegazzini. *Biol. Bull.* **109** : 306-315.

21. Richards, A.G., and Smith, M.W. 1956. Infection of cockroaches with *Herpomyces* (*Laboulbenia*les).*II*. Histology and histopathology. *Ann. Entomol. Soc. Amer.* **49** : 85-93.

22. Scheloske, Hans-Werner. 1970. Contribution to the biology, ecology, and systematics of Laboulbeniales with particular reference to the parasite-host relationship. *Parasitological Series* **No.19** : 176.

23. Shanor, Leland. 1952. The characteristics and morphology of a new genus of the Laboulbeniales on an earwig. *Am. J. Bot.* **39** : 498-504.

24. Sugiyama, Keiichi, 1973. Species and genera of Laboulbeniales (Ascomycetes) in Japan. *Ginkgoana* **2** : 1 - 97.

25. Taveres, Isabelle. I. 1965. Thallus Development in *Herpomyces paranensis* (Laboulbeniales). *Mycologia* Vol. VII **No.5** : 704-721.

26. Tavares, Isabelle. I. 1966. Structure and development, of *Herpomyces stylopyge* (Laboulbeniales). *Am. J. Bot.* **53(4)** : 311-318.

27. Tavares, Isabelle. 1970. The appendage of *Amorphomyces. Mycologia* **62(4)** : 741-749.

28. Tavares, Isabelle and Balazuc, J. 1978. *Misgomyces mastacis* (Laboulbeniales) : Its structure and host relationship. *Mycologia* **69(5)** : 1069-1073.

29. Tavares, Isabelle, I. 1980. Notes on Perithecial development in the Euceratomycetaceae Fam. Nov. (Laboulbeniales, Laboulbeniineae) and *Herpomyces* (Herpomycetinae). *Mycotaxon* **11(2)** : 485-492.

30. Terada, Katsuyuki. 1981. New or interesting species of the Laboulbeniales found on some Coleopterous insects of Japan. *Trans. Mycol. Soc. Jpn.* **21(2)** : 193-203.

31. Thaxter, R. 1896. Contribution towards a monograph of the Laboulbeniaceae, Part I. *Mem. Amer. Acad. Arts and Sci.* **12** : 187-429.

32. Thaxter, R. 1902. Preliminary diagnoses of new species of Laboulbeniaceae - V. *Proc. Amer. Acad. Arts and Sci.* **38** : 39.

33. Thaxter, R. 1908. Contribution towards a monograph of the Laboulbeniaceae, Part II. *Mem. Amer. Acad. Arts and Sci.* **13** : 217-469.

34. Thaxter, R. 1914. On certain peculiar fungus parasites of living insects. *Bot. Gaz.* **58** : 235-252.

35. Thaxter, R. 1924. Contribution towards a monograph of the Laboulbeniaceae, Part III. *Mem. Amer. Acad. Arts and Sci.* **14** : 309-426.

36. Thaxter, R. 1926. Contribution towards a monograph of the Laboulbeniaceae, Part IV. *Mem. Amer. Acad. Arts. and Sci.* **15** : 427-580.

37. Thaxter, R. 1931. Contribution towards a monograph of the Laboulbeniaceae, Part V. *Mem. Amer. Acad. Arts and Sci.* **16** : 1-435.

38. Whisler, Howard, C. 1969. Experimental studies with a new species of *Stigmatomyces* (Laboulbeniales). *Mycologia* **60(l)** : 65-75.

Index

A

Abelmoschus esculentus, 396
Ables balsamea, 57
Absidia repens, 172
Abutilon theophrasti, 219,220,261, 263,266
Acacia arabica, 383
Acacia saligna, 218
Acaulospora laevis, 446,457
Acaulospora trappei, 424
Acaulospora sp., 410
Acerpseudoplatanus sp., 366
Acetobacter aceti, 159
Acetobacter sp., 112
Achromobacter sp., 49
Acinetobacter sp., 49
Acompsomyces sp., 544
Acremonium falciformi, 356
Actinomucor elegans, 100
Aegle mermelos, 383
Aerobacter aerogens, 99
Aeromonas sp., 133,134
Aeschynomene indica, 236,448,453, 460
Aeschynomene virginica, 218,260,263, 264
Aflatoxin, 141,146,149,152,154,156, 165, 163
Ageratina reparia, 218
Agrobacterium sp., 133
Albugo tragopognis, 263
Alcaligenes eutrophus, 28,32,33,36
Alcaligenes sp., 32
Allium cepa, 383,425,447,457
Alnus rubra, 60
Alternaria alternata, 355,370,372
Alternaria brassicae, 222
Alternaria carthami, 233
Alternaria cassiae, 220,225,226,228, 229,230, 231, 262,263,266
Alternaria eichhorniae, 219
Alternaria euphobiicola, 263
Alternaria helianthi, 233
Alternaria macrospora, 230,233,384
Alternaria stemphylium, 371
Alternaria tennuis, 370
Alternaria solani, 383
Alternaria sp., 263
AM fungi, 57,409,410,411,412, 413,433,474,478
AM, 57,397,410
Amanita muscaria, 458
Amanita rubescens, 456
Amaranthus retroflexus, 260
Ambrosia artemisiifolia, 236,263
Ambrosia trifida, 219,220
Amorphomyces stenusae, 539
Amorphomyces sp., 539
Ampelomyces quisqualis, 316
Amphobotrys ricini, 220,263
Andropogon gerardii, 447
Anoda cristana, 220,230
Antagonist, 313
Aphanomyces cochliobolus, 355
Aphanomyces euteiches, 385
Aphelenchus avenae, 496
Aporomyces sp., 543,546
Arachis hypogea, 425
Arthrobacter viscousus, 339
Arthrobacter sp., 46,112

Arthrobotrys brochopaga, 488

Arthrobotrys conoides, 492

Arthrobotrys dactyloide, 492,493,499

Arthrobotrys ellipsospora, 487

Arthrobotrys musiformis, 488,496

Arthrobotrys oligospora, 488,493,495, 496,498,499, 500

Arthrobotrys sp., 488,492,495,500

Arthrorhynchus sp., 531

Aschynomene indica, 457, 458

Ascochyta hyalospoara, 263

Ascochyta sp., 382

Aspergillus awamorii, 99,194

Aspergillus carbonarius, 99,100,101

Aspergillus ficuum, 100,105

Aspergillus flavicepes, 194

Aspergillus flavus, 99,142,143,144, 145,146,149,154,155,159, 160,161, 162,167,172,173,

Aspergillus flavus-oryzae, 155

Aspergillus fumigatus, 50,210,397

Aspergillus nidulans, 193,194,195

Aspergillus niger, 46,50,58,100,101, 102, 103,104,105,161,194,172,331, 340,356,397

Aspergillus niger (syn) *A. ficuum*, 99

Aspergillus niger var. *awamori*, 102

Aspergillus oryzae, 89,100,101,155

Aspergillus parasiticus, 142,143,151, 154,155,156,157,158,159,173

Aspergillus terreus, 99,233

Aspergillus versicolor, 99

Aspergillus wentii, 99

Aspergillus sp., 46,99,100,143, 154,161,169,194,208,371,374,394,433

Atriplex gardneri, 456

Auercus alba, 459

Aureobasidium sp., 316

Aureobasidium pullulans, 340,366, 371,374,371

Avena fatua, 219

Azadirachta indica, 383

Azospirillum sp., 112

Azotobacter sp., 112

B

Bacillus cercus, 331

Bacillus lactobacillus, 208

Bacillus licheniformis, 81,84

Bacillus megaterium, 154

Bacillus polymyxa, 134

Bacillus sphaericus, 7,8

Bacillus subtilis, 56,98,99,319,335

Bacillus thermophilus, 134

Bacillus thuringiensis, 7,8,51,324

Bacillus sp., 8,28,46,47,112,133, 134,135,136,318, 331,335

Bautelous gracilis, 459

Beauveria barssiana, 51,195

Bebaryomyces hansenii, 321

Benzyl Penicillin, 70

Betula pendiula, 56

Bipolaris halepense, 219

Bipolaris oryzae, 195

Bipolaris setariae, 219

Bipolaris sorghicola, 263

Botrytis cinerea, 195,314,315,318

Boutelova gracilis, 425

Brassica campestris, 116

Brassica oleracae, 222

Brevibacterium ammoniagenes, 47

Brevibacterium sp., 49

Bromus tectorum, 236,263

Bt, 9, 11,12,13,15,16,17,18,19
Bt var. *kurstaki*, 9
Bt var. *berliner*, 16
Bt var. *israeliensis*, 9
Bt var. *tenebrionis*, 9
Burkholderia cepacia, 122

C

Caenorhabditis elegans, 500
Cajanus cajan, 453
Caldocellum saccharolyticum, 134,438
Callistemon lanceolatu, 384
Callophyllum inophyllum, 384
Calluna vulgaris, 56,57
Calotropis procera, 384
Candida albicans, 193,194
Candida apicoda, 191
Candida glabrata, 338
Candida tropicalis, 47,194
Candida utilis, 44
Cantharomyces sp., 529,546
Caperonia palustrsi, 263
Capsicum annum, 425,426
Carduus tenuiflorus, 263
Carduus thoermeri, 218
Carissa sp., 222
Cassia alata, 226
Cassia obtusifolia, 219,220,222,225, 261,263,255
Cassia occidentalis, 220,261,263
Casurina equisetifolia, 384
Catenaria anguillulae, 488,497,498
Catenaria sp., 496
Catharanthus roseus, 221
Cenchrus sp., 439
Cenococcum geophilum, 460
Cenococcum graniforme, 456
Cenococcum sp.,456
Centaurea diffusa, 218
Centaurea maculosa, 269
Cephalosporium diospyn, 261
Cephalosporium sp. 134,370
Ceratosystis fimbricata, 355
Cercospora caricis, 219
Cercospora rodmanii, 229,262,263
Chaetomium globosum, 372
Chainia sp. 134
Chenopodium album, 219
Chenopodium quinona, 457
Chenopodium sp., 263
Chlaeminus biguttatus, 545
Chlorella emersonii, 337
Chlorella pyrenoidosa, 340
Chlorella vulgaris, 331,336,340
Chlorella sp., 48
Chondrilla juncea, 216,217,218
Chondrostereum purpureum, 261,266
Chorochoa saye, 160
Ciliaris sp., 439
Cianopagum sp., 487
Cirsium arvense, 236,270
Citrobacter sp., 27,335,337
Cladosporium cladosporioides, 370
Cladosporium cucumerium, 226
Cladosporium, *epicoccum*, 371
Cladosporium herbarum, 318,370,371, 372,374
Cladosporium sp., 210
Claviceps paspali, 510
Claviceps purpurea, 46,142
Clostridium acetobutylicum, 46
Clostridium stercorarium, 134

Clostridium sp., 208

Cochliobolus lunatus, 219,229

Colletotrichum coccoides, 219,261, 263,266

Colletotrichum gloeosporioides, 217, 218,220,229,230, 260,261,263,270,318

Colletotrichum graminicola, 219

Colletotrichum lindemuthanum, 60,225

Colletotrichum malvacum, 262,263

Colletotrichum orbiculare, 261,263

Colletortichum truncatum, 231,263

Colletotrichum sp., 60,263,270

Conidiobolus coronatus, 84

Coniophora puteana, 302

Convolvulus arvensis, 219,262

Corethromyces sugiyamae, 525,548, 552

Corethromyces sp., 528

Corynbacterium dismutans, 46,47

Corynbacterium rubrum, 172

Crotalaria spectabilis, 220,261,263

Cryphonectria parasitica, 193

Cryptococcus sp., 319

Cucumis melo, 226

Cucumis sativus, 223

Cucurbita texana, 229,262

Curvularia lunata, 47,195,219,433

Cuscuta sp., 263

Cylinderocarpon destructans, 367,368, 499

Cymbopogon citratus, 55

Cyperus esculentus, 219,260,263

Cyperus rotundus, 263

Cystosporum sp., 493

D

Dactilospheara viticola, 60

Dactylaria brochopaga, 499

Dactylaria candida, 488,497

Dactylaria candida, 497

Dactylella doedycoides, 490,495

Dactylium denrodides,172

Datura stramonium, 226,263,384,386

Dactylaria candida, 419

de novo, 69

Debaryomyces hansenii, 319,320,322

Debaryomyces sp., 316,317,319, 320,328

Desmodium tortuosum, 230,263

Detergent, 79

Dextran, 51

Dichotomophthora indica, 219

Dictyoglomus sp., 133,135,137

Digitaria sanguinalis, 219,220

Dimeromyces sp., 528,541,546

Diospyros cordifolia, 384

Diospyros virginiana, 261

Diplomyces sp., 541

Distolomyces euborelliae, 528

Drechslera sp., 263

Drechmeria coniospora, 497,498

E

Echinochloa crus-galli, 219,263

Echinochloa frumentaces, 512

Echinochloa frumentaces, 519

Echinochloa sp., 263

Eichhorinia crassipes, 219,229,262, 263, 266

Elasmopalus lignosellus, 161

Eleocharis kuroganwai, 263

Elesuine indica, 219,220

Elesuine coracana, 512
Enterobacter cloacae, 337
Enterobacter sp., 98,
Enterococcus hirae, 27,30,31
Entyloma ageratinae, 218
Enzyme, 78,79,132,474
Epicoccum pupurascens, 371
Erharts calyana, 458
Erica cinerea, 57
Erwinia sp., 236
Erysiphe cichoracearum, 369
Erysiphe graminis, 225,226
Eschericia coli, 28,46,47,70,154,338
Eucalyptus globulus, 384
Eucalyptus sp., 384
Euphorbia sp., 263
Euzodiomyces sp., 525,529,532
ex vitro, 478
Exserohilum monoceras, 220
ex-situ, 472

F

Fagopyrum esculentum, 372
Fannia canicularis, 531,539
Fannia sp., 545
Festuca arundinaces, 457
Filariomyces forficulae., 535
Filariomyces sp., 525
Flavobacter sp., 49,112
Flavobacterium aurantiacum,172
Fomes annosus, 60
Fragaria ananssa, 319
Fragaria vesca, 475
Fusarium culmorum, 367
Fusarium graminearum, 50
Fusarium laterititium, 219,220,230, 231,263
Fusarium lini, 384
Fusarium moniliforme, 195,232,383, 384,385,397
Fusarium orbanches, 263
Fusarium oxysporum, 231,263,355, 367,385,397
Fusarium oxysporum f. sp. *ciceri*, 384
Fusarium oxysporum f. sp. *cubense*, 385
Fusarium oxysporum f. sp. *lycopersici*, 193,195,385
Fusarium oxysporum f. sp. *lentis*, 384
Fusarium oxysporum f. sp., *lyscopersici*, 193,195
Fusarium oxysporum f. sp. *udum*, 384
Fusarium roseum, 263
Fusarium solani, 58,299,355,384,397
Fusarium solani f. sp. *cucurbitae*, 262
Fusarium tricinctum, 263
Fusarium udum, 384,386
Fusarium sp., 142,143,193,226,263, 266, 392,393,384

G

Gaumannomyces graminis, 478
Gaeumannomyces graminis var. *tritici*, 123
Ganoderma capsisci, 385
Geotrichum candidum, 99
Gibberella fujikuroi, 46
Gibberella zeae, 193
Gigaspora calospora, 456
Gigaspora decipiens, 424
Gigaspora gigantea, 423
Gigaspora gregaria, 456
Gigaspora heterogama, 456

Gigaspora margarita, 453,456,473
Gigaspora sp., 410,448,453
Gliocladium roseum, 318,367
Gliocladium virens, 233,260
Gliocladium sp., 397
Gloeocercospora sorghi, 219
Gloeosporium papayee, 385
Glomerella cingulata, 236
Glomus caleadonium, 439,440,453,457
Glomus clarum, 446,453,457
Glomus deserticola, 447,456
Glomus epigaeum, 423,453
Glomus etunicatum, 448,453
Glomus fasciculatum, 58,425,438,439, 440,447,450,453, 456, 457
Glomus heterogama, 453
Glomus interaradices, 425,456,473, 475
Glomus macrocarpum, 58,392,436, 447,453
Glomus monosporum, 457
Glomus mossae, 58,59,60,437,447,453, 456,457,458, 459,473,476
Glomus versiforme, 473,474
Glomus sp., 410,423,438,447,448, 453,456,457
Glycine max, 225,447,456,478
Gossypium hirsutum, 370,459
Gram-negative bacteria, 73

H

Hansemaspora uvarum, 319
Harpalus sp., 530
Harposporium anguillulae, 490,499
Harposporium helicoides, 496
Helianthus annuus, 228
Helicobacter pylori, 30
Heliothis virescens, 19
Helix pomatia, 188
Helminthosporium avenae, 227
Helminthosporium maydis, 168
Helminthosporium oryzae, 233
Herpomyces paranensis, 538
Herpomyces periplanetae, 530
Herpomyces stylopgae, 537,538
Herpomyces sp., 525,528,529, 530,532,546
Histoplasma capsulatum, 208
Hordeum vulgare, 222
Humicola sp., 89
Hydrilla verticillata, 263
Hyphomycetes sp., 263

I

Idiomyces peyritischii, 541,542
in situ, 122,142,
in vitro, 9,13,15,17,18,78,142,154,165, 167,222,226, 316,371,383,384,397, 469,472,475
in vivo, 154,167,222,226,383
Ipomoea batats, 355

J

Jussiaea decurrens, 262

K

Kalmia sp., 56,57
Klebsiella aerogenes, 98,339
Klebsiella terrigena, 101
Kluveromyces lactis, 134

L

Laboulbenia corethromyces, 529

Laboulbenia formicarum, 532,533
Laboulbenia indica, 537, 546
Laboulbenia orientalis, 528
Laboulbenia sp., 525,528,530,533,535, 537,539,544, 545,546,548
Laccaria laccata, 460
Lactic acid bacteria, 99
Lactobacillus acidophilus, 172
Lactobacillus amylovorus, 98,105
Lactobacillus delbrueckii, 47
Lactobacillus sp., 44,98
Leucaena leucocephala, 357,384
Linum usitatissimum, 117,425
Lippia alba, 384
Liquidambar styraciflua, 457
Liriodendron tulipifera, 457
Lolium perenne, 367
Lupinus albus, 228
Lweptinotarsa decemlineate, 195
Lycopersicon esculentum, 117,223,425
Lygodesmia juncea, 216,217
Lygus herperus, 160

M

Macrophomina phaseolina, 383,397
Macrospora roridum, 384
Malbranchea sulfurea, 195
Malus pamila, 319
Malva parviflora, 228
Malva pusilla, 218,261,263
Manduca sexta, 16,17
Melilotus officinalis, 385
Meloidogyne incognita, 498
Meloidogyne sp., 489
Merulius lacrymans, 290
Metarhizium anisopliae, 51
Methanobacterium sp., 46
Methylococcus sp., 49
Methylophilus methylotrophus, 49
Microbacterium ammoniaphila, 47
Micrococcus sp., 47
Microorganism, 51,54,55,58,311,330, 333,334,395,459
Microsomyces telephanus, 525,542
Monacrosporium cystosporum, 495, 496
Monacrosporium eudermatum, 497
Monacrosporium gephyrophagum, 488
Monacrosporium parvicolle, 488
Monacrosporium rutgeriensis, 495
Monacrosporium sp., 419,487,492, 493,495,499
Morchella angusticeps, 518
Morchella conica, 519
Morchella sp., 518,520,521,522
Morrenia odorata, 217,218,261, 262,263
Mortierella substitissima, 56
Mucor ambigus, 172
Mucor dispersus, 100
Mucor piriformis, 99
Mucor racemosus, 372
Mucor sp., 46,99
Mycobacterium leprae, 34
Mycobacterium tuberculosis, 74
Mycorrhizoshere, 394,367
Myrothecium verrucaria, 220
Myzocytium papillatum, 488
Myzocytium sp., 497

N

Neisseria gonorrhoea, 70
Nematoctonus concurrens, 500
Neurospora crassa, 188,195

Nicotiana tobacum, 223
Nyssa sylvatica, 448

O

Oidium lini, 397
Oryza sativa, 218

P

Pachysolen tannophilus, 195
Panagrellus redivivus, 419,495,497,500
Panicum miliaceum, 512
Panicum, miliare, 512
Papulaspora sp., 56
Parthenium argentasum, 425
Pascopyrum smithii, 459
Paspalum scrobiculatum, 509,510, 511,519
Penicillium chrysogenum,47,194
Penicillium digitatum, 316
Penicillium purpurogenum, 189
Penicillium rubrum, 56,
Penicillium sp., 46,99,100,208,315, 331,373,433
Peniophora gigantea, 60
Peyritschiella sp., 525,529,533,546
Pezizlla eicae, 459
Phalaris arundinacea, 227
Phaseolus vulgaris, 60,105,223
Phoma tracheiphila, 193
Phoma sp., 331
Phomopsis convolvulus, 219,262
Photobacterium, 120
Photosynthetic microorganisms, 46
Phragmidium violaceum, 218
Phyllachora cyperi, 219
Phymatootrichum ominvorum, 385
Phytophthor cryptogea, 226
Phytophthora capsici, 385
Phytophthora cinnamimi, 60, 383
Phytophthora drechsleri f.sp. *cajani*, 384
Phytophthora infestans, 223,225
Phytophthora megasperma f.sp. *glycinea*, 229
Phytophthora nicotiana var. *parasitica*, 61
Phytophthora palmivora, 217,218,230, 261,262,266
Phytophthora parasitica var. *nicotianae*, 226
Picea abies, 56,58
Pichia guilliermondii, 319,320
Pichia stipitis, 195
Pinus echinata, 459
Pinus radiata, 56,357,458
Pinus resinosa, 57
Piriformospora indica, 471,474, 476,478
Pisolithus tinctorious, 58,356,459,460
Pisum sativium, 222
Plasmodiophora brasscae, 386
Pleurotus ostreatus, 193
Pleurotus pulmonarius, 500
Pleurotus ostreatus, 193
Pleurotus sp., 195,500
Plodia interpunctella, 19
Plutella xylostella, 18,19
Poa annua, 263
Pongamia glabra, 384
Populus tremula, 56
Poria weirii, 60
Portulaca oleracea, 219
Potamogeton sp., 220
Prolabia pulchella, 535
Prosopis juliflora, 384

Protea cyanarodies, 383
Protoplast fusion, 188,192,195
Prunus armentane, 319
Prunus serotina, 261
Pseudomonas aeruginosa, 47,118,331, 334
Pseudomonas cepacia, 319
Pseudomonas fluorescens, 476
Pseudomonas putida, 27,36,37,355
Pseudomonas syringae, 36,233
Pseudomonas syringae pv. *tagetis*, 236
Pseudomonas sp., 36,47,49,89,97,100, 112,113,115, 117,119,121,133,236, 263,334,339,433
Puccinia canliculata, 260,263
Puccinia carduorum, 218
Puccinia chondrilla, 216,217,218
Puccinia coronata, 226
Puccinia jaceae, 218
Puccinia substriata, 510
Pyricularia grisea, 219,220
Pythium aphanidermatum, 384
Pythium ultimum, 123,124,195
Pythium sp., 122,392,393

R

Ranuculus scleratus, 383,384
Rhiozopus arrhizus, 335
Rhizobium sp., 56,59,112,355,450
Rhizoctonia solani, 123,124,159,195, 266, 384,394,397,499
Rhizoctonia sp., 392,393
Rhizoplane, 367,396
Rhizopogon roseolus, 456
Rhizopogon vinicolor, 460
Rhizopogon sp., 460
Rhizopus arrhizus, 27, 172,331,340
Rhizopus nigricans, 159,397
Rhizopus oryzae, 99
Rhizopus stolonier, 320
Rhizopus sp., 46
Rhizoshere, 367,396
Rhodotorula mucilaginosa, 320
Rhodotorula rubra, 190
Rhopalomyces elegans, 496
Rickia dichotoma, 525
Rickia kawasakii., 525
Rickia, misgomyces, 546
Rickia sp., 525,528,529
Rosea odorata, 386
Rottboellia cochinchinensis, 219
Rubes sp., 218
Russula sp., 456

S

Saccharmomyces cerevisiae, 44,46,47, 99,159,188, 192,195,331,335
Sandersoniomyces divaricatus, 542
Scatella sp., 545
Scelerotium rolfsili, 195,383
Scelerotium sclerotium, 269,270, 220,384
Schizachyrium scoparium, 457
Schizosaccharomyces pombe, 338
Sclerotinia homoeocarpa, 195
Schwanniomyces casteillii, 101,103
Schwanniomyces occidentalis, 103
Schwanniomyces sp., 99
Sclerotium cepivorum, 59
Scopulariopsis brevicaulis, 159
Scutellospora calospora, 424
Septoria cirsii, 263
Septoria tritici f.sp. *avenae*, 219

Sequicillium condelabrum, 368
Serbania exaltata, 263
Serpula himantioides, 300
Serpula lacrymans, 284,287,288,290, 291,294,295,297,299,300,301,304
Serratia marcescens, 195
Sesbania aegyptiaca, 425
Sesbania exaltata, 222
Sesbania grandiflora, 425
Setaria italica, 478,512,519
Shizosaccharomyces pombe, 190
Shorea robusta, 56,
Sida spinosa, 220,263
Sigmatomyces baeri, 541
Sitophilus zeamais, 174,
Solanum carolinensc, 262,263
Solanum ptycanthum, 263
Solanum tuberosum, 217
Sorghum bicolor, 222,355
Sorghum halepense, 219,263
Sorghum vulgare, 395,456
Sorghum sp., 447
Sphacelotheca holci, 219
Spodoptera frugiperda, 17
Spongospora subterranea, 386
Sporothrix flocculosa, 316
Spriulina platensis, 336
Stachybotrys atra, 210
Stachybotrys sp., 208
Staphylococcus aureus, 27,56,69,70
Stigmatomyces baeri, 530,535,539
Stigmatomyces ceratophorus, 531
Stigmatomyces sp., 533,539,544,545
Streptococcus pyogenes, 70
Streptococcus sp., 44,98
Streptomyces clavuligerus, 47
Streptomyces fradiae, 47
Streptomyces pilosus, 340
Streptomyces viridochromogenes, 233
Streptomyces sp. 46,133,134,232,233, 330,433
Stylopage hadra., 487
Stylopage leiohypha, 487
Suillus bovinus, 458
Suillus luteus, 456
Synandromyces telephan, 542
Synnmukerjiomyces thermophile, 1,2,3, 4,5

T

Tagetes minuta, 55,386
Tagetes sp., 236
Tectona grandis, 432,439,440
Tectona sp., 439,441
Teratomyces mirificus, 541
Terminalia arjuna, 432,438,439,440
Tetrahymena pyriformis, 172
Thedophora terrestris, 456
Thermomonospora fusca, 134
Thevatia peruviana, 384
Thielaviopsis basicola, 123,124,385
Thiobacillus ferrooxidans, 331
Tilletia caries, 372
Tilletia indica, 382
Tinospora conicocordia, 385
Tobacco Mosaic Virus, 263
Trenomyces histophthorus, 525
Triceromyces sp., 543
Trichoderma cladosporium, 374
Trichoderma harzianum, 189,190,191, 193,195, 295,318,
Trichoderma koningi, 58
Trichoderma longibrachiatum, 295

Trichoderma sp., 46,189,190,195, 294,295,296,318, 319,373,397,433
Trichoderma viride, 56,58,159,172, 295,368,373,394, 397
Trifolium repens, 367
Trifolium subterraneum, 355,459
Trigonella fornum graceum, 58,397
Triticum aestivum, 227
Triticum spelta, 411
Tritirachium album, 84

U

Ulocladium chyartarum, 318
Uromycladium tepperianum, 218
Ustilago hordei, 370
Ustilago paspali, 510

V

Vaccinium angustifolium, 459
VAM fungi, 57,58,356,357, 395,397,399,410,411,412,421,423,424, 425, 432,433,438,439,446,448,450, 453,457,458,459,476
VAM fungus, 410
VAM plants, 57
VAM, 57,58,356,357,392,395,396,410
Venturia inaequalis, 195,372
Verticillium albo-atrum, 58,193,355,386
Verticillium chlamydosporium, 498
Verticillium dhahliae, 355
Verticillium sp., 193,373,385
Vibrio sp. 120
Vigna radiata, 385
Vigna umbellata, 391,396,398,399,400
Vigna ungiculata, 223
Vinca rosea, 384
Vitis vinifera, 319

Xanthium spinosum, 261,263
Xanthium sp., 263
Xanthomonas compestris, 263
Xanthomonas campestris pv. *campestris*, 236
Xanthomonas campestris pv. *malvacearum*, 384
Xanthomonas campestris pv. *versicatoria*, 385
Xanthomonas sp., 46,115,133,236,

Y

Yeast, 331
Yersinia sp., 34

Z

Zea mays, 226,424,447
Zodiomyces sp., 525
Zogloea sp., 331
Zooloea ramigera, 339